Advances in Intelligent Systems and Computing

Volume 928

The series "Advances in Intelligent Systems and Computing" contains publications on theory, applications, and design methods of Intelligent Systems and Intelligent Computing. Virtually all disciplines such as engineering, natural sciences, computer and information science, ICT, economics, business, e-commerce, environment, healthcare, life science are covered. The list of topics spans all the areas of modern intelligent systems and computing such as: computational intelligence, soft computing including neural networks, fuzzy systems, evolutionary computing and the fusion of these paradigms, social intelligence, ambient intelligence, computational neuroscience, artificial life, virtual worlds and society, cognitive science and systems, Perception and Vision, DNA and immune based systems, self-organizing and adaptive systems, e-Learning and teaching, human-centered and human-centric computing, recommender systems, intelligent control, robotics and mechatronics including human-machine teaming, knowledge-based paradigms, learning paradigms, machine ethics, intelligent data analysis, knowledge management, intelligent agents, intelligent decision making and support, intelligent network security, trust management, interactive entertainment, Web intelligence and multimedia.

The publications within "Advances in Intelligent Systems and Computing" are primarily proceedings of important conferences, symposia and congresses. They cover significant recent developments in the field, both of a foundational and applicable character. An important characteristic feature of the series is the short publication time and world-wide distribution. This permits a rapid and broad dissemination of research results.

**** Indexing: The books of this series are submitted to ISI Proceedings, EI-Compendex, DBLP, SCOPUS, Google Scholar and Springerlink ****

More information about this series at http://www.springer.com/series/11156

Zheng Xu · Kim-Kwang Raymond Choo ·
Ali Dehghantanha · Reza Parizi ·
Mohammad Hammoudeh
Editors

Cyber Security Intelligence and Analytics

Set 1

Springer

Editors
Zheng Xu
Shanghai University
Shanghai, China

Kim-Kwang Raymond Choo
University of Texas at San Antonio
San Antonio, TX, USA

Ali Dehghantanha
University of Guelph
Guelph, ON, Canada

Reza Parizi
Kennesaw State University
Marietta, GA, USA

Mohammad Hammoudeh
Manchester Metropolitan University
Stockport, UK

ISSN 2194-5357 ISSN 2194-5365 (electronic)
Advances in Intelligent Systems and Computing
ISBN 978-3-030-15234-5 ISBN 978-3-030-15235-2 (eBook)
https://doi.org/10.1007/978-3-030-15235-2

Library of Congress Control Number: 2019933893

This Springer imprint is published by the registered company Springer Nature Switzerland AG
The registered company address is: Gewerbestrasse 11, 6330 Cham, Switzerland

Preface

The 2019 International Conference on Cyber Security Intelligence and Analytics (CSIA 2019) is an international conference dedicated to promoting novel theoretical and applied research advances in the interdisciplinary agenda of cyber security, particularly focusing on threat intelligence and analytics and countering cybercrime. Cyber security experts, including those in data analytics, incident response, and digital forensics, need to be able to rapidly detect, analyze, and defend against a diverse range of cyber threats in near real-time conditions. For example, when a significant amount of data is collected from or generated by different security monitoring solutions, intelligent and next-generation big data analytical techniques are necessary to mine, interpret, and extract knowledge of this (big) data. Cyber threat intelligence and analytics are among the fastest-growing interdisciplinary fields of research bringing together researchers from different fields such as digital forensics, political and security studies, criminology, cyber security, big data analytics, and machine learning to detect, contain, and mitigate advanced persistent threats and fight against organized cybercrimes.

Each paper was reviewed by at least three independent experts, and the acceptance rate was 45%. The conference would not have been possible without the contributions of the authors. We sincerely thank all the authors for their valuable contributions. We would like to express our appreciation to all members of the Program Committee for their valuable efforts in the review process that helped us to guarantee the highest quality of the selected papers for the conference.

We would like to express our thanks to Prof. Shunxiang Zhang, Anhui University of Science and Technology, and Prof. Xiangfeng Luo, Shanghai University, for being the keynote speakers at the conference. We would also thank the Steering Committee, General Chairs, Program Committee Chairs, Organizing Chairs, and Workshop Chairs. The local organizers' and the students' help is highly appreciated.

Our special thanks are due also to the editors for publishing the proceedings in Advances in Intelligent Systems and Computing of Springer.

Zheng Xu
Kim-Kwang Raymond Choo
Ali Dehghantanha
Reza Parizi
Mohammad Hammoudeh

Organization

General Chair

Zheng Xu	TRIMS, China

Program Committee Chairs

Ali Dehghantanha	University of Guelph, Canada
Mohammad Hammoudeh	Manchester Metropolitan University, UK
Reza M. Parizi	Kennesaw State University, USA

Publication Chairs

Shuai Ding	Hefei University of Science and Technology, China
Kun Gao	Zhejiang Wanli University, China

Publicity Chairs

Xianchao Wang	Fuyang Normal University, China
Jemal Abawajy	Deakin University, Australia

Local Organizing Chairs

Xueping Liu	Anhui Polytechnic University, China
Qingjun Wang	Anhui Polytechnic University, China

Program Committee Members

William Bradley Glisson	Sam Houston State University, USA
George Grispos	University of Nebraska Omaha, USA
V. Vijayakumar	VIT Chennai, India
Aniello Castiglione	Universit di Salerno, Italy
Neil Yen	University of Aizu, Japan
Shunxiang Zhang	Anhui University of Science and Technology, China
Guangli Zhu	Anhui University of Science and Technology, China
Tao Liao	Anhui University of Science and Technology, China
Xiaobo Yin	Anhui University of Science and Technology, China
Kelei Sun	Anhui University of Science and Technology, China
Xiangfeng Luo	Shanghai University, China
Xiao Wei	Shanghai University, China
Huan Du	Shanghai University, China
Zhiguo Yan	Fudan University, China
Abdulbasit Darem	Northern Border University, Saudi Arabia
Hairulnizam Mahdin	Universiti Tun Hussein Onn Malaysia
Anil Kumar K. M.	JSS Science & Technology University, Mysore, Karnataka, India
Haruna Chiroma	Abubakar Tafawa Balewa University, Bauchi, Nigeria
Yong Ge	University of North Carolina at Charlotte, USA
Yi Liu	Tsinghua University, China
Foluso Ladeinde	SUNY Korea
Kuien Liu	Pivotal Inc., USA
Feng Lu	Institute of Geographic Science and Natural Resources Research, Chinese Academy of Sciences
Ricardo J. Soares Magalhaes	University of Queensland, Australia
Alan Murray	Drexel University, USA
Yasuhide Okuyama	University of Kitakyushu, Japan
Wei Xu	Renmin University of China
Chaowei Phil Yang	George Mason University, USA
Hengshu Zhu	Baidu Inc., China
Morshed Chowdhury	Deakin University, Australia
Elfizar	University of Riau, Indonesia
Rohaya Latip	Universiti Putra Malaysia

Contents

Regular Papers

Short Papers

Regular Papers

Application of the CT Image Processing in the Urology Surgery

Xuan Huang[1], Juan Peng[1(✉)], and Laiqing Sen[2]

[1] Hubei 672 Chinese and Western Medicine Combined Orthopaedic Hospital, Wuhan 430000, Hubei, China
huyp0213@126.com

[2] Hubei University of Chinese Medicine, Wuhan 430000, Hubei, China

Abstract. The spiral CT can obtain the three-dimensional or the two-dimensional images of different levels and angles, which can provide the more intuitive and comprehensive imaging data, and clarify the size and location of the lesions and their relationship with the surrounding tissues. The spiral CT scanning time is short and the image quality is good. Compared with IVU and B-mode ultrasonography, the spiral CT scanning has more guiding value in the diagnosis and treatment of the urinary diseases.

Keywords: CT image · Technical processing · Urology surgery · Application strategy

With the improvement of the CT performances and the development of the functional software, the post-processing functions are becoming more and more diverse, including the image processing technology and the image measurement and juicing technology [1]. The measurement and calculation contents mainly include: the CT value, length, distance, perimeter, area, volume and other data, which are not described here one by one. This paper only talks about the image post processing technology.

1 CT Image Processing

Computed Tomography (CT), which uses the precisely collimated X-ray beams, gamma rays, ultrasound and other detectors to scan one section after another around a certain part of the human body, has the characteristics of the fast scanning time and the clear images, and can be used for the examination of various diseases. According to the different X-rays used, it can be divided into the X-ray CT (X-CT), the ultrasound CT (UCT), and the γ-ray CT (γ-CT).

The working procedure of CT is as follows. According to the difference of the X-ray absorption and transmittance of different tissues of the human body, it uses the highly sensitive instruments to measure the human body, and then inputs the measured data into the computer. After the data are processed by the computer, the cross-section or the three-dimensional images of the examined part of the human body can be taken, and the minor diseases of any part of the body can be found [2]. The adjustment of the window width and the window position is a routine content in the digital image

Z. Xu et al. (Eds.): CSIA 2019, AISC 928, pp. 3–8, 2020.
https://doi.org/10.1007/978-3-030-15235-2_1

post-processing, and is also the most important function in the image display technology. The correct selection and application of the window technology is an important means to obtain the high-quality images and improve the diagnostic rate [3].

In the image display, the magnification technique can be used to observe the micro-lesions and the fine anatomical structures. There are two forms of the image enlargement. One is to enlarge the scanning field, that is, to narrow the scanning field, and the other is to enlarge the image electronically [4]. The image enlargement in the post-processing is different from that in the scanning. It is an electronic enlargement. With the increase of the enlargement ratio, the image sharpness decreases. In addition, the enlarged image also needs to adjust the window width and the window position properly in order to better observe the image. There are two commonly used methods for enlarging the region of the interest. One is to move the cursor (+) to the center of the image to be enlarged and input the enlargement factor to get the corresponding enlarged image. The other is to enlarge the image directly with the box. The smaller the box is, the larger the image enlargement will be.

The subtraction usually takes place between two images. Usually, one image is selected as the subtracting image and the other as the subtracted image [5]. That is to say after subtracting the two images, the subtracted image can be obtained. The filtration can be processed in a single frame. According to the effect of the filtering, there is the smoothing, average, edge enhancement and shadow display. The filtering method is that the computer uses different image algorithms to reprocess the images in order to achieve some effects. Among the three methods mentioned above, the image enlargement is the most commonly used in the clinic, usually for the diagnostic purposes, in order to make up for some deficiencies in the scanning.

The multidirectional and the three-dimensional reconstruction are also used as the image post-processing. In fact, they are all the functions of displaying the images in different ways after the image post-processing on the basis of the cross-sectional scanning. Generally, the cross-sectional images can be composed of the coronal, sagittal, inclined or arbitrary surfaces, which are called the multi-directional reorganization. The advantages of the multi-directional reorganization are as follows. Firstly, it can observe the specific anatomical structures, and secondly, it can help to determine the size of the lesions or the fractures, which is helpful for the diagnosis. The biggest disadvantage of this method is that it is reconstructed on the basis of the cross-sectional scanning, and its image quality is affected by the cross-sectional scanning image quality. In the three-dimensional reconstruction method, the realistic and the stereoscopic display can be obtained through the reconstruction of the cross-sectional images. Like the multi-position reconstruction, this method needs to be based on the thin-layer scanning to obtain the satisfactory images. Usually, the thinner the scanning layer, the better the effect of the reconstruction will be.

At present, there are many advantages of the multi-orientation reconstruction using the spiral CT scanning. The volume scanning in a short time is difficult to move and the volume data acquisition is complete due to the short scanning time. The spiral CT can use a thicker scanning layer, while the reconstruction can use the thinnest reconstruction interval, with the retrospective reconstruction at any number of times, but the patient's radiation does not increase. Usually the post-processing reorganization is a method used to better observe the location, the extent of lesions and their relationship with the surrounding tissues.

It should be pointed out that the image reorganization and the image reconstruction are completely different concepts. The reconstruction is to get the display data matrix by calculating the original data with a computer. The reorganization is to take out the relevant display data from the existing layers and combine them into the new layers of the images. Normally, the quality of the reconstructed image is not as good as that of the directly reconstructed image. When voxels are isotropic, this situation will be significantly improved.

The CT image is composed of a certain number of the pixels with the different gray levels from black to white arranged in matrix. These pixels reflect the X-ray absorption coefficients of the corresponding voxels. The size and the number of pixels of the images obtained by different CT devices are different. The size can be 1.0×1.0 mm and 0.5×0.5 mm, while the number can be 256×256, that is 65536, or 512×512, that is 262144. Obviously, the smaller the number of the pixels, the more detailed the image is, that is, the higher the spatial resolution will be. The spatial resolution of the CT images is not as high as that of the X-ray images.

2 Analysis of the Nursing Requirements in the Urology Surgery

We should find out the distribution of the pathogens and the risk factors of the nosocomial infection in the urological patients, as well as the solutions, to improve the therapeutic effect of the patients, and guarantee the safety of the patients' lives and the quality of their hospitalization, and promote the rapid development of the medical field, which are of great value. Results are analyzed through the corresponding experiments, and the hospital obtained some important information. There was no significant difference in the distribution of the departments between the two groups, but the number of the urological patients was more. There was no significant difference in the proportion of the pseudomonas aeruginosa infection patients and the escherichia coli infection patients. The drug resistance analysis showed that there was no significant difference between the two strains. But apart from the piperacillin/tazobactam, this antibiotic has a very good therapeutic effect and needs to be paid attention by hospitals. Conclusion: Through the detailed research and analysis of the experiment, it is found that the corresponding analysis of the urinary surgical patients with the infections can find drugs for treatment, which has very good analytical value and can provide very important guidance for the clinical treatment, so that the disease can be effectively controlled and the therapeutic effect can be improved.

The health education is mainly divided into the pre-operative education and the post-operative health education. The preoperative health education is mainly to strengthen the patients' self-understanding, to tell the patients the causes of the urological diseases, symptoms, health hazards, ways of infection, medication notes, dietary structures and nursing knowledge mainly through the nurses to explain their own anatomical characteristics and physiological knowledge of the urinary system and the reproductive system, in order to prepare for the later treatment. Suitable dietary plans should be formulated according to the specific conditions of the patients.

The preoperative nursing mainly includes several aspects. The appeasing the patients' emotions: Patients will have anxiety before the operation, and the serious emotions may affect the effect of the surgery, so the nursing staff should understand the causes of the patients' anxiety, and adopt the appropriate appeasement measures, inform the patients of the main contents of the surgery and the safety of the surgery, explain the successful cases to the patients, and encourage the patients to face it calmly. Carry out the physical examinations and the health assessment of the patients. Some operations may require the use of the antibiotics to control the infection. In the application of the antibiotics, we should strictly comply with the doctor's advice. For some patients with the serious anxiety and the emotional excitement, give the appropriate oral or intramuscular injection of diazepam, and communicate with the encouraging language, and assist the anesthesiologists to position them, and closely observe the electrocardiogram. Notify the surgeon of any abnormalities.

From the beginning of the hospitalization, the receptionists should be friendly, eliminate the strangeness of the patients and increase the relationship between the doctors and the patients. Carry out the patient inquiry about the patient's conditions and symptoms. For some privacy issues, we should use the euphemism, and the attending physicians should be confident of the hospital and their own strength and enhance the patient safety. For the patients who are more nervous before the operation, nurses should do a good job of their psychological counseling and reassurance, introduce the operation related situations in details, and tell the safety of the operations, so that the patients can have a general understanding of the operation, and then eliminate the fear caused by the unknown. After the operation, nurses greet the patients in time and congratulate them on the success of the operation. This psychological encouragement and comfort is conducive to the recovery of the patients.

3 Application of the CT Image Processing in the Urology Surgery

The CT images are expressed by different gray levels, reflecting the degrees of the X-ray absorption of the organs and the tissues. Therefore, like the black-and-white images shown in the X-ray images, the black shadows represent the low absorption areas, i.e. the low density areas, such as the lung with more gas, and the white shadows represent the high absorption areas, i.e. the high density areas, such as bones. But compared with the X-ray images, CT has a high density resolution. Therefore, although the density difference of the human soft tissue is small and the absorption coefficient is close to water, it can also form the contrast and the image. This is the outstanding advantage of CT. Therefore, CT can better display the organs composed of the soft tissue, such as brain, spinal cord, mediastinum, lung, liver, gallbladder, pancreas and pelvic organs, and display lesions in a good anatomical image background.

The CT urography (CTU) is a non-invasive examination, which is easy to operate and can display the three-dimensional structure of the urinary system. While displaying the hydronephrosis, it can help to clarify the cause of the obstruction and observe the secretion and excretion function of the kidney at the same time. The 64-slice CT enhanced delayed urography scanning can understand the physiological functional changes of the urinary system lesions, and help to grasp the location and morphology of the lesions. It has great significance in the diagnosis and differentiation of the congenital malformations of the urinary system, the pathological changes of the urinary tract obstruction and the traumatic diseases. The z-axis resolution of the 64-slice spiral CT images is high, which brings convenience for the two-dimensional and the three-dimensional image reconstruction. The image post-processing techniques commonly used in the CTU include the volume rendering, the maximum density projection, the multi-plane reconstruction and the surface reconstruction.

VR can show the three-dimensional spatial relationship of the urinary system, and combined with the multi-angle rotation, it can better show the adjacent lesions, the size and the number of the stones. MIP can segment the urinary system and the lesions, show the slight density changes, and accurately measure the lesion ranges. MPR and CPR images, with the good density resolution, can also display the whole urinary tract when the contrast agent is not well filled, which is helpful to the qualitative diagnosis of the lesion. After the CT imaging, the size, shape, number and three-dimensional image of the calculi can be clearly understood by the volume rendering, and the relationship between the calculi and the pelvis and the calyx can be understood. Before the operation, the surgeon can know that the calculi can be removed to the greatest extent by puncture from the posterior calyx of the kidney or by incision from the dorsal calyx of the kidney. It is best to take out and remove the calculi.

The ureteroscopic surgery for the ureteral calculi, and especially after the difficult catheterization under the ureteroscope, can be clearly seen by the volume reappearance after the CT examination whether there are stones around the double J tubes, how many stones there are, and the size of the stones compared with the double J tubes, so as to determine whether the extracorporeal shock wave lithotripsy is needed, or whether it is necessary to retain a period of the self-lithotripsy, or can be directly extubated. The CT fast thin-slice volume scanning and the three-dimensional reconstruction have the high sensitivity and specificity in the diagnosis of the ureteral diseases. At the same time, they can be multi-directional and multi-plane stereo imaging, which makes the location and the qualitative of the lesions more accurate. It is one of the best imaging methods for the diagnosis of the ureteral diseases at present.

4 Conclusion

The volume reconstruction imaging after the multi-slice spiral CT scanning can clearly see the three-dimensional shape of the stones in many directions, which can improve the success rate of the stone surgery or the effect of the stone clearance. The volume reproduction can clearly show the shape of the double J tube, with or without stones around it, which is very clear. Especially in the middle ureter, other examination methods are defective. The CT enhanced scanning can show the shape of the urinary tract. It can show the injury or the obstruction of the ureter very clearly, and the image is intuitive, so it is convenient to communicate with the patients. Therefore, we believe that in the diagnosis of the urinary calculi and other obstructive diseases, the CT volume rendering in the application of the urological diseases has the clear, intuitive and accurate characteristics, which is worth promoting.

References

1. Cong Y, Lu H (2017) Continuous nursing practice after the permanent urostomy. J Shanxi Med Coll Contin Educ 01:51–53
2. Li J (2017) Distribution of the pathogenic bacteria and the analysis of the infection factors in patients with the clinical infection in urology. J Clin Med Lit 12:155–156
3. Wang J, Guo S (2018) The influence of the seamless nursing management on the quality of the urological nursing. Clin Study 10:132–133
4. He H, Li Q, Zhang J (2018) Observation of the clinical application effect of the rehabilitation surgery concept combined with the psychological nursing for urological patients. Health Vocat Educ 11:119–120
5. Yu L (2018) Distribution and risk factors of the nosocomial infection in urological patients. World Latest Med Inf 11:138–139

Application of the Data Mining Technology in the Economic Management in the Age of Big Data

Wenbiao Luo(✉)

Guangzhou Vocational College of Science and Technology, Guangzhou 510550, Guangdong, China
objuu7@yeah.net

Abstract. Under the background of the continuous acceleration of the reform process of the domestic market economy, relying on the continuous improvement of the information technology level, the Internet era has come in an all-round way. Under this background, the traditional economic management model obviously cannot meet the needs of the current economic market. In view of the problems existing in the traditional economic management, this paper discusses the informatization of the economic management under the background of the big data era.

Keywords: Big data era · Data mining technology · Economic management · Application mechanism

For the rapid growth of the economic data, the data mining technology, on the one hand, expands its own database to meet the needs of the large-scale search, and on the other hand, improves the accuracy of the data retrieval, so that users can quickly find the most needed data from a large number of the data.

1 Analysis of the Characteristics of the Age of the Big Data

McKinsey, a world-renowned consulting firm, was the first to propose the era of the "big data". McKinsey said, "Data has penetrated into every industry and business function areas and become an important factor in our production [1]. The mining and application of the massive data heralds a new wave of the productivity growth and consumer surplus". "Big data" has existed for some time in the fields of physics, biology, environmental ecology, military, financial, communications and other industries, but in recent years, due to the development of the Internet and the information industry, it has attracted people's attention.

The big data technology is widely used in all walks of life, bringing about changes in the management mode, management concepts and management methods of all walks of life. In the field of the economic management, we should conform to the needs of the development of the science and technology, renew the management concepts, actively use the big data to implement the economic management, share the technological

Z. Xu et al. (Eds.): CSIA 2019, AISC 928, pp. 9–14, 2020.
https://doi.org/10.1007/978-3-030-15235-2_2

dividends brought by the scientific and technological progress to our economic management, and effectively enhance the implementation effect of our economic management [2].

2 The Impact of the Big Data Technology on the Economic Management

Our country has a vast territory and many industries. All walks of life achieve common development in the process of the mutual integration. A large amount of the data and information will be generated in every industry and every field at every moment. These data and information are also the original basis for the relevant decision-making departments to correctly judge the macroeconomic situations, and then implement the branches in the social and economic fields [3]. The study of the economic management provides the decision-making guarantee for the healthy development of the social economy. The emergence and application of the big data technology provides the technical support for the collection, generation, analysis, collation and processing of the economic management data, and helps to make the economic management decisions.

2.1 Data Sets Can Be Provided for the Economic Management

In order to carry out the economic management, the relevant decision-making departments and the management departments of our country need statistics from all walks of life. At the same time, the economic development of our country involves a large number of the industrial fields, such as industry, agriculture and tertiary industry, and so on. Every industry, every field, and even every enterprise will produce a large number of the statistical data [4]. The accumulation and disposal of the data is a big project. When the big data technology is used to analyze and collate these statistical data, we can make use of the technical advantages to ensure that the total amount of the data is doubled, and classify all kinds of the complex data information in an orderly manner [5]. The practice of the big data in the field of the economic management also shows that the economic management data provided by the big data technology has the characteristics of the continuous increase in the total amount, the convenient data processing and the clearer statistical results, and can form the relevant data sets according to the needs of the management agencies, which is convenient for the query and use.

2.2 Data Information Processing Efficiency Is More Efficient

Our country has a vast territory, and a large number of the data transmission needs a period of time. Moreover, the most critical factor in making the economic decisions in the economic management is the timeliness of the data acquisition [6]. That is to say, the economic management departments and the decision-making bodies must collect the latest data and information in the first time, and the value of the old and outdated data will drop dramatically. The big data technology can send the relevant information to the decision-making organs in a timely manner, improve the speed of the data dissemination and processing efficiency, and provide the data information for the economic management organs and the decision-making departments in a timely manner.

2.3 Big Data Can Provide a Variety of the Data Information

Compared with the traditional data processing technologies, the data information provided by the big data technology is more diversified. In addition to the basic digital information, the big data can also use the related technologies to process some structured and unstructured data. It can also provide texts, pictures, clicks, web pages, videos, links and geographical locations. In this way, the decision-making of the economic management will be more scientific, more specific, more targeted, and more grounded, demonstrating the working ability of our economic management personnel. The big data technology can also make random analysis and random sampling of these data, taking into account the whole, as well as the local and the individual. The volume makes the coverage of the information wider, and it is convenient for the sampling and application of the management organizations.

With the arrival of the era of the big data, the society has become more stringent on the requirement of "mining" the data. Every precise result has its own "value". At this time, the new attribute of the era of the big data – "value" has been interpreted vividly. Data mining (DM) is a new interdisciplinary subject, which can converge many disciplines. It is an extraordinary process, that is, extracting the unknown, implicit and potentially valuable information from the huge data. In August 1989, at the 11th Joint Conference on Artificial Intelligence held in Detroit, the United States, the knowledge discovery in database (KDD) was first proposed by scientists. At the same time, some people call the knowledge discovery the data mining, but the two are not exactly the same. In 1995, the term KDD was accepted at the first International Conference on Knowledge Discovery and Data Mining held in Montreal, Canada, which analyzed the entire process of the data mining. In essence, the data mining is a sub-process of the knowledge discovery.

After about 20 years of development, the data mining research has made considerable achievements, and gradually formed a set of the basic theoretical basis, including classification, clustering, pattern mining and rule extraction. The data mining is a technology to "mine" the potential and unprecedented knowledge from the massive data in life. Processing the big data requires a comprehensive, complex and multi-directional system. There are many processing modules in the system. The data mining technology exists in the entire system of processing the big data as an independent identity, and complements and coordinates with other modules. In the era of the big data, the status of the data mining technology is incomparable.

3 Research Status of the Data Mining

The data mining integrates the theories and technologies of the high-performance computing, machine learning, artificial intelligence, pattern recognition, statistics, data visualization, database technology and expert system. The era of the big data is both an opportunity and a challenge for the data mining. Analyzing the big data, establishing an appropriate system, and constantly optimizing and improving the accuracy of the decision-making are more conducive to grasping and adapting to the multi-terminal changes of the market. In the era of the big data, the data mining as the most commonly used means of the data analysis has been recognized in various fields. At present, scholars at home and abroad mainly study the application of the classification, optimization, identification, prediction and other technologies in the data mining in many fields.

The data mining technology refers to the technology of quickly finding the effective information that can be used by oneself from a large number of the economic statistics. This technology is a deep processing and development of the original data. After finding the useful information from the huge database, we should analyze and integrate it, and finally arrange the required data information clearly in front of the readers. The information data in the original economic statistics information database has many characteristics, such as the large amount of information, the high degree of the information complexity, the poor relevance, the random occurrence of various kinds of the information, and the frequent errors in the data information and so on. It is very difficult to find the useful information from such a large and complex database. If the information in the database cannot be reasonably selected and used, the database itself will lose its original role.

The data mining technology is a complex technology form involving many disciplines, including mathematical statistics, database, pattern recognition, rough set, fuzzy mathematics and other disciplines, but its main application is the data statistics. The data mining technology can automatically extract those raw data which are closely related to the required data by analyzing and summarizing the huge and disorderly raw data. After the extraction, they will be arranged in a certain order, and finally the data needed by the data extractor will be formed. The data mining technology is a combination of the database knowledge, the pattern recognition system, the fuzzy concept technology and other knowledge systems. These knowledge systems are the industry's cutting-edge knowledge systems. Only with powerful computing, screening and integration capabilities can the information in the database be effectively analyzed, extracted and collated.

The characteristics of the data mining technology have the following aspects. The data information contained in the data mining technology is huge, and the data information it can process is also very large, and the information processing can reach the level of GB and TB. For those users who are not sure what information they want to query, as long as the users input the simple keywords, the data mining technology can find the most closely related resources from the huge database and rank them according to the degrees of the contact. The data mining can analyze a large amount of the information in the database and find out the intrinsic correlation of various kinds of the information. It can also use the data mining to show the development of the economic situation in the previous stage and predict the future trend of the economic situation.

4 Application of the Data Mining Technology in the Economic Management in the Age of the Big Data

As a new technology, the big data is widely used in the process of the modern economic management, which promotes the changes of the management concepts and the management modes. The decision-making organs of the modern economic management in China should actively use the big data technology to meet the challenges of the modern science and technology and share the achievements of the modern science and technology for the social and economic life.

4.1 Use the Big Data Technology to Strengthen the Process Control of the Data Information Analysis

At present and in the future, China's economic management decision-making bodies should actively use the big data technology. The core of the big data technology is the process of analyzing and processing the big data. The process of analyzing and processing the big data is closely related to the decision-making process of the economic management. The decision-making of the economic management implementation in China should first collect various data based on the problems and the data, and use the big data technology to sort out these data, and transmit the data to the data platform. After the analysis, the decision-making process of the economic management should be formulated. The implementation plan of the economic management decision-making is constantly improved in the process of the implementation. At the same time, the data prediction and monitoring must be carried out in the process of the data analysis, and the problems should be found in time, so as to provide basis for the new decision-making.

4.2 Use the Big Data Technology to Improve the Scientificality of the Economic Management Decision-Making

The use of the big data technology can change the decision-making body of the economic management in our country. The decision-making body will change from the top management to the grass-roots staff and popularization. Because the grass-roots staffs are familiar with the front line, they can discover problems in time and strengthen the collective effectiveness by analyzing the front-line data. In the decision-making mode, the big data technology can also analyze the correlation between various factors, get rid of the past empirical and intuitive decision-making methods, and adopt the ways of collecting and analyzing the data to make decisions, so as to reduce the risks of the decision-making.

4.3 Use the Big Data Technology to Establish the Classified Economic Management Database

At present, all walks of life in our country are flourishing. The rising-sun industry emerges in endlessly in the new period, which has become a new growth point in the process of the economic management in our country. Therefore, it is imperative to speed up the construction of the classified economic management database at present, which will enhance the economic benefits of the emerging industries and realize the new production. The efficient development of the industries is of great significance. With the help of the classified databases of the big data, we can realize the convergence of all factors and resources, scientifically grasp the laws of China's economic operation, promote the transformation and upgrading of the emerging industries, improve the qualities and efficiency, and ultimately realize the sound and rapid development of China's national economy. At present, China's economic management and decision-making institutions can build the large databases of the economic management, the financial supervision, the energy and economic management, the agricultural economic management and other databases, providing the most accurate raw data for the formulation of the economic management and the decision-making in the related industries.

5 Conclusion

With the rapid development of the information technologies, it has brought the new big data technology, and the development of the new technology has also brought new opportunities and challenges. Nowadays, the big data technology is widely used in all walks of life and all aspects of our society. The existence of the big data technology also brings the management concepts and the management models to our society. Management has brought about innovation. Of course, China's economic management should also be thoroughly innovated with the arrival of the big data, to enjoy the technology dividend brought to us by the science and technology, and the practical economic management effect.

References

1. Feng R (2017) Exploring the impact of statistics on the contemporary economic management. Econ Res Guide (05):111–112
2. Sun Y (2018) Brief analysis of the application of the big data in the modern economic management. Mod Bus (06):121–122
3. Xu Y (2018) Application of the big data in the modern economic management. Telecommunications (08):132
4. Yang Q (2018) Research on the informatization of the economic management in the internet era. Inner Mongolia Coal Econ (09):133–134
5. Zhou Q (2018) Multi-layer affective computing model based on emotional psychology. Electron Commer Res 18(1):109–124
6. Xu J (2018) Challenges and changes of the enterprise financial management in the context of the big data. China Commer Forum (11):105–106

Comprehensive Evaluation System Which Adapt to the Layout of Power Facilities and the Utilization of Channel Resource Based on the New-Type Urbanization

Bin Ye[1](✉), Xuli Wang[1], Bin Hu[1], Junxi Ren[1], Xuemei Shi[1], and Lei Wang[2]

[1] State Grid Anhui Economic Research Institute, Hefei, China
kingguogang@sohu.com
[2] School of Electrical and Automation Engineering, Hefei University of Technology, Hefei, China

Abstract. In view of the lack of corresponding evaluation methods for distribution network planning under the background of new urbanization, this paper studies the evaluation methods for the coordinated layout of power facilities and comprehensive utilization of channel resources under the background of new urbanization, and puts forward a comprehensive evaluation system for the layout of power grid facilities and utilization of channel resources suitable for new towns. The evaluation system includes the characteristics of saving and intensification. In addition, from the four aspects of technology, economy, adaptability and comprehensive utilization of social resources, 34 basic indexes for comprehensive evaluation of distribution network planning are designed. Applying the comprehensive evaluation system proposed in this paper, two sets of distribution network planning schemes of a town in Anhui Province are evaluated respectively. The practical examples show that the evaluation system can adapt to the characteristics of distribution network planning under the new town, and guide the direction of power network planning under the new urbanization.

Keywords: New-type urbanization · Distribution network · Power facilities · Channel resources · Comprehensive evaluation

1 Introduction

In recent years, the process of new-type urbanization has developed rapidly. In particular, the "National New-type Urbanization Plan (2014–2020)" has put forward new requirements on the demand for the resources saving and the intensive use of the model of the grid facilities layout and channel. However, there is a lack of understanding of the collaborative layout of power facilities and comprehensive utilization of channel resources based on the new-type urbanization. First of all problems is the power grid is poorly savings, especially the efficiency of facilities utilization and the occupancy rate of land resources; furthermore, the coordination and intensiveness of grid facilities and urban planning, land planning and municipal pipeline resources are poor; last but not

Z. Xu et al. (Eds.): CSIA 2019, AISC 928, pp. 15–23, 2020.
https://doi.org/10.1007/978-3-030-15235-2_3

least, there are differences in the types of towns and stages of urban development. Therefore, it is urgent to study the scientific comprehensive evaluation system to optimize the construction mode, channel construction mode, comprehensive pipeline construction mode and construction strategy of urban and rural distribution facilities in the form of differentiated development.

Many studies have carried out on the traditional comprehensive evaluation of distribution network [1–3]. One study proposes to use combinatorial weighting method to determine the weight of technical indicators of power quality, and make the weight more objective and comprehensive, which improves the evaluation effect [4]. A macro-control indicator system that reflects the scale of power grid construction has been established which is basic on AHP [5]. One study proposes a comprehensive evaluation of power grid planning methods [6]. A comprehensive urban distribution network comprehensive evaluation system establishes by systematic quantitative comprehensive evaluation theory [7]. One study based on the planning of power grid studies the evaluation of power grid planning and proposes the evaluation indicator system and evaluation method [8]. A hybrid evaluation method combines the fuzzy consistent matrix method and the fuzzy optimization method to deal with all kinds of qualitative and quantitative evaluation [9]. However, these studies

However, these researches have carried out a detailed evaluation of power grid planning, but few evaluate the comprehensive utilization of channels. Moreover, they cannot adapt to the construction of new urban power grids with obvious differences in different urbanization stages.

Therefore, this paper gives a comprehensive evaluation system, which adapt to the layout of power facilities and the utilization of channel resource based on the new-type urbanization.

2 The Construction Principle of Comprehensive Evaluation System

Comprehensive evaluation system is composed of a number of individual evaluation indicators, which reflects the requirements of the indicator, and it needs to be scientific, reasonable and practical. Therefore, the construction of evaluation system should follow some principles:

(1) Consistency: The indicators are the target of the specific, behavioral and operational. It should be consistent with the development goals and management objectives of object. In addition, two conflicting indicators could not put together.
(2) Testability: In order to get a clear conclusion, the indicator should measure directly by the actual observation.
(3) Comparability: The indicators should reflect the comparability of the evaluated objects. In addition, the operability of the evaluation process should ensure.
(4) Independence: The evaluation system is a combination of indicators. However, the indicators are independent from each other. There is no inclusion, no overlap and no causal relationship among the indicators at the same level. Otherwise, the feasibility of the assessment will reduced.

(5) Completeness: The system cannot miss any important indicators. It should comprehensively reproduce and reflect the development and management objectives of the object.

According to these principles, two points should take in the selection of indicators when establishing the evaluation indictor system of power grid. On the one hand, the actual situation should reflected, as fully as possible and important indicator should not be missed; on the other hand, the validity of the information, the difficulty of data acquisition, and the amount of calculation should not be repeated neither missed.

The comprehensive evaluation system established in this paper should include more principles:

(1) The setting of the comprehensive evaluation system should be compatible with the differentiated development pattern of new urbanization and the differentiated power characteristics of different urbanization stages;
(2) The establishment of comprehensive evaluation system should base on the core concept of urban-rural infrastructure integration, including the saving of resource requirements and the intensive arrangement of forms;
(3) Comparing with the traditional grid evaluation system, the comprehensive evaluation system includes the power grid planning, and the comprehensive utilization of channel resources.
(4) The comprehensive evaluation system should take into account the synergy of grid planning and urban planning.

3 A Comprehensive Evaluation System

3.1 The Influencing Factors of the System

The influencing factors of demand saving: poor distribution network efficiency mainly reflected in the low demand for power grid equipment and the occupancy rate of equipment. However, optimization of power grid equipment affected by time coincidence with customer needs economy of operation and satisfaction degree of power facilities. The occupancy rate of equipment is affected by the type of equipment, different levels of voltage lines corridor layout and different types of equipment coordinated arrangement.

The influencing factors of intensive arrangement: intensive grid cooperation mainly reflected in the coordination of urban planning, plan of land utilization, urban types, municipal facilities pipeline, land for municipal construction, Greenfield, roads, etc. Specifically, including technical support for construction, synergism of urban types, the level of urban development; the will of land department, reservation of land, costs of demolition compensation, technical requirements of site, requirements of operation; contradiction of land demand; inaccurate power pipe gallery planning.

3.2 Evaluation System

According to comprehensive evaluation system and AHP (analytic hierarchy process) [10, 11], combining with the characteristics and purposes of power grid planning, a clear hierarchy is set up for the evaluation objectives, and indicators of evaluation system are proposed. The weight of the same level indicators set by Delphi method, and then calculates the weight of the indicator according to AHP. At last, the indicator score standard is determined by using fuzzy membership method.

(1) System indicators

AHP method used to decompose the optimal evaluation indicators of urban and rural electric power facilities layout and channel resources comprehensive utilization planning, and the model of three hierarchical relations consists of target layer, criterion layer and indicator layer.

(1) The target layer (level 1): the layout of the optimal power facilities and channel planning;
(2) The standard layer (level 2): technical, economic, adaptability and comprehensive utilization of social resources;
(3) The indicator layer (level 3): indicator layer is the division of the standard layer.

The three-level indicator system adopted by the comprehensive evaluation system plan shown in Table 1.

Table 1. The table of comprehensive evaluation system indicator

Level 1 indicators	Level 2 indicators	Level 3 indicators
Demand savings	Optimization of power grid equipment demand	Corridor and grid construction time collaboration rate; corridor space satisfaction rate; low transmission loss; power grid loss reduction
	Area savings of power grid equipment	The total construction investment indicator; corridor utilization rate; unit channel cost; the annual value of unit channel; substation facilities annual operating costs; annual operating cost of unit channel
Intensive arrangement	Synergy of grid equipment and urban planning	Matching degree of urbanization; matching degree of corridor resources and urban development level; level of matching between substation facilities and urban development; satisfaction degree of environmental; environmental impact indicator
	Synergy of grid equipment and land planning	The conformity of land's nature; land price; cost of demolition; government attitudes; public attitudes; land expropriation impact indicator; transport facilitation; convenience of access road; convenience of water supply; convenience of station power; convenience of line access; distance from load center; comfort of power cabin
	Synergy of grid equipment and other municipal planning	Satisfaction with other municipal pipeline safety distance; the proportion of integrated corridor construction
	Synergy of grid equipment and road construction	The time cooperation rate of integrated pipe network and road network construction; the consistency between demand width of power cabins and section of road; length ratio of corridor; location of corridor

(2) The weight of indicators

The weight of indicators is set by multi-indicator comprehensive score analysis method, and the specific process is shown in Fig. 1.

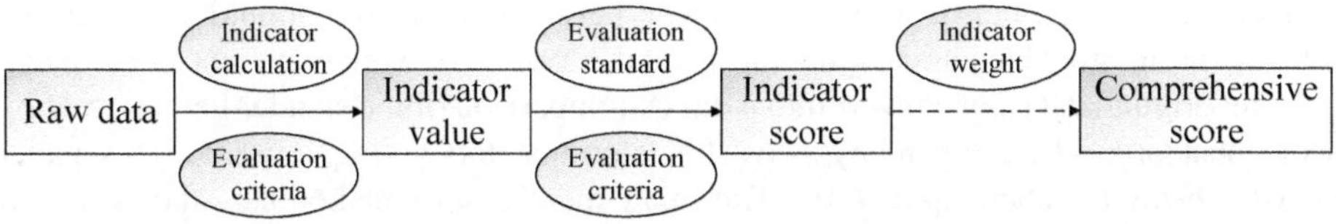

Fig. 1. Evaluation process

The solid arrow in Fig. 1 shows the operation on the single indicator and the dotted arrow shows the operation on the system. As can be seen from the figure: first, the evaluation criteria for the indicator is the threshold value to determine the value of the indicator; and then the indicator calculate the score according to the indicator evaluation criteria; finally, calculate the evaluation object's comprehensive score according to the weight of indicators.

Calculation steps of indicator weight value:

(1) According to the reciprocity scale theory, the judgment matrix derived from the comparison of same level indicators;
(2) Checking the consistency of judgment matrix;
(3) If the judgment matrix consistency is qualified, then calculate the weight of judgment matrix; otherwise re-determine the judgment matrix and calculate the weight;
(4) Calculating the comprehensive weight.

Comprehensive score is decided by in (1):

$$B^n = \sum_{i=1}^{k} a_i^{(n+1)} A_i^{(n+1)} \tag{1}$$

In which, B^n is the score of the level n indicator in the comprehensive evaluation system; k is the number of the level n + 1 indicator of B; $A_i^{(n+1)}$ is the score of the ith sub-indicator at the level n + 1; $a_i^{(n+1)}$ is the weight of the ith sub-indicator.

The higher the score gets, the better the grid plan is. The manifestations of scoring system: hundred-point system, ten-point system and five-point system. The comprehensive evaluation results are divide into to five levels: 0–60 means Level 5; 60–70 means Level 4; 70–80 means Level 3; 80–90 means Level 2; 90–100 means Level 1. Level 1 is the best.

(3) The criterion of score

In the process of evaluation, the level 3 indicators need to define in a reasonable range. The range means the criteria for the indicator, which has two manifestations: point, interval. For the former, the indicator value is satisfactory if it exceeds or does not exceed this point; for the latter, the indicator value is only within or outside this interval to meet the requirements. The evaluation criteria of indicators should consider the relevant professional technical guidelines and the long-term work experience of experts.

The original data can convert into a direct comparison of the standardized format by score standard. There are many ways to determine the scoring criteria. This paper chooses fuzzy membership method. The fuzzy membership method uses the degree of membership to describe different degrees of ambiguity. It is a description of fuzziness in an accurate mathematical language. The membership function creates a mapping on [0, 1] to reflect the object that has some ambiguity or belongs to fuzzy concept.

In this paper, five-level of assessment used to deal with the fuzzy membership degree of indicators. Three-level of assessment is used for some indicators that cannot be defined.

4 The Process of Comprehensive Evaluation System

The process of the comprehensive evaluation system shown in Fig. 2, which includes the setting of evaluation indicator and the evaluation of planning programs.

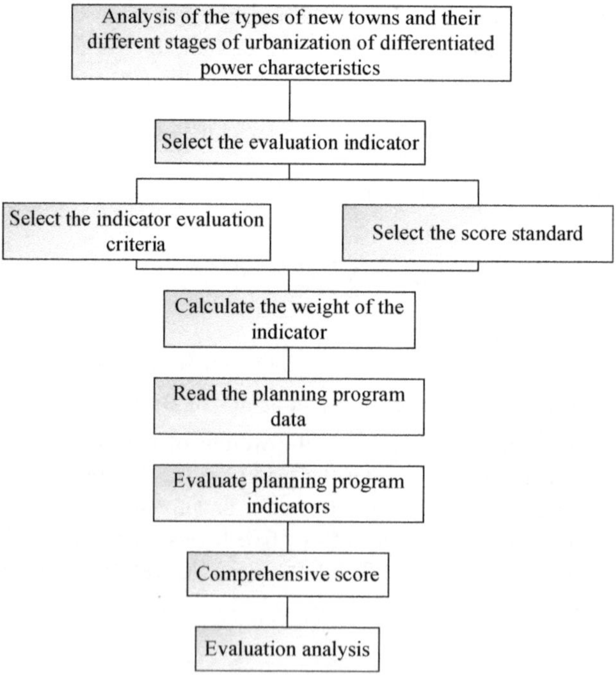

Fig. 2. The flow chart of comprehensive evaluation system

In the stage of setting, we should analyze the types of towns and the differentiated power characteristics in different development stages of the town. Then the evaluation indicator, the evaluation criteria and the evaluation criteria are selected to calculate the weight of the indicator.

In the stage of the evaluation of planning program, the original data of the distribution network planning evaluation is read in. After that, the evaluation is conducted for different planning programs. Each indicator is evaluated, and then the score is calculated based on the weight. Finally, comparing different planning programs to measure the pros and cons according to the score.

5 Sample

In this paper, the distribution network planning of a town in Anhui Province is selected as a sample. The two urban planning programs are evaluated separately. The weight of the indicators is shown in Table 2. The values in parentheses are the weight values for the corresponding indicator.

Evaluating indicators of two different planning program respectively. Then using formula (1) to calculate the comprehensive score shown in Table 3 according to the weight.

Table 2. The indicator weight of comprehensive evaluation system

Level 1 indicators	Level 2 indicators	Level 3 indicators
Demand savings (0.5)	Optimization of power grid equipment demand (0.6)	Corridor and grid construction time collaboration rate (0.559)
		Corridor space satisfaction rate (0.261)
		Low transmission loss (0.048)
		Power grid loss reduction (0.133)
	Area savings of power grid equipment (0.4)	The total construction investment indicator (0.143)
		Corridor utilization rate (0.431)
		Unit channel cost (0.275)
		The annual value of unit channel (0.036)
		Substation facilities annual operating costs (0.077)
		Annual operating cost of unit channel (0.037)
Intensive arrangement (0.5)	Synergy of grid equipment and urban planning (0.558)	Matching degree of urbanization (0.463)
		Matching degree of corridor resources and urban development level (0.195)
		Level of matching between substation facilities and urban development (0.195)
		Satisfaction degree of environmental (0.073)
		Environmental impact indicator (0.073)

(*continued*)

Table 2. (*continued*)

Level 1 indicators	Level 2 indicators	Level 3 indicators
	Synergy of grid equipment and land planning (0.25)	The conformity of land's nature (0.232)
		Land price (0.105)
		Cost of demolition (0.075)
		Government attitude (0.181)
		Public attitude (0.075)
		Land expropriation impact indicator (0.144)
		Transport facilitation (0.018)
		Convenience of access road (0.018)
		Convenience of water supply (0.018)
		Convenience of station power (0.018)
		Convenience of line access (0.048)
		Distance from load center (0.030)
		Comfort of power cabin (0.039)
	Synergy of grid equipment and other municipal planning (0.096)	Satisfaction with other municipal pipeline safety distance (0.070)
		The proportion of integrated corridor construction (0.030)
	Synergy of grid equipment and road construction (0.096)	The time cooperation rate of integrated pipe network and road network construction (0.558)
		The consistency between demand width of power cabins and section of road (0.249)
		Length ratio of corridor (0.096)
		Location of corridor (0.096)

Table 3. Comprehensive score

Planning program number	Demand savings	Intensive layout	Comprehensive score
1	83.52	79.11	81.32
2	82.40	88.67	85.54

The final composite score shows that the evaluation results of planning schemes 1 and 2 are Level 2, scheme 1 is better in demand savings, and scheme 2 is better in intensive layout, and the comprehensive score of scheme 2 is better than scheme 1. Program can chose according to the actual situation.

6 Conclusion

Comprehensive evaluation system is based on AHP to establish a clear hierarchy of evaluation objectives, and put forward the indicators of the standard layer and the indicator layer; the weight of indicators is given by Delphi method, and the value of the weight is calculated according to AHP; the standard of score is determined by the fuzzy membership method. The system is based on three levels indicators, including two level 1 indicators, six level 2 indicators, thirty-four level 3 indicators.

The comprehensive evaluation system of the comprehensive utilization of urban and rural power facilities coordination layout and channel resources is the quantification of the evaluation indicator in the influencing factors. Through quantitative analysis, it is possible to find the correlation of the importance of each indicator and to indicate the direction of subsequent technical optimization methods.

The comprehensive evaluation system for the comprehensive utilization of urban and rural power facilities coordination layout and channel resources takes into account the characteristics of saving in demand and intensive arrangement, and provides a quantitative basis for the planning of the project from a systematic perspective.

References

1. Luo G, Mei M, Lin Z (2011) Evaluation for distribution network planning program and completed project. Shaanxi Electric Power 39(6):16–20
2. Yang L, Wang S, Lu Z (2012) Indices of distribution network intelligent planning evaluation. Power Syst Technol 36(12):83–87
3. Tang H, Peng J (2003) Research on synthetic and quantification appraisal index of power quality based on fuzzy theory. Power Syst Technol 27(12):85–88
4. Li L, Yao J, Long L et al (2007) Application of combination weighing method in fuzzy synthetic evaluation of power quality. Autom Electric Power Syst 31(4):56–60
5. Li X, Li P, Zhu X et al (2007) Evaluation construction size of district high voltage distribution networks. Autom Electric Power Syst 31(4):46–50
6. Xiao J, Luo F, Wang C et al (2005) Design and application of a multi-attribute decision-making system for power system planning. Power Syst Technol 29(2):9–13
7. Xiao J, Gao H, Ge S et al (2005) Evaluation method and case study of urban medium voltage distribution network. Power Syst Technol 29(20):77–81
8. Xiao J, Cui Y, Wang J et al (2008) A hierarchical performance assessment method on the distribution network planning. Autom Electric Power Syst 32(15):36–40
9. Zhou Q, Lou J, Xie F, Liu Q (2011) The method research on optimum selection of agricultural machinery. Int J Digit Content Technol Appl 5(6):334–337
10. Wang R, Zhao W (2013) A fuzzy decision-based method to evaluate planning scheme for urban high voltage transmission network. Power Syst Technol 37(2):488–492
11. Yang X, Li H, Yin Z et al (2013) Energy efficiency index system for distribution network based on analytic hierarchy process. Autom Electric Power Syst 37(21):146–159+195

Computer Desktop Image Compression Technology Based on the Clustering Algorithm

Liu Fei(✉)

Lianyungang Technical College, Lianyungang 222006, Jiangsu, China
dengpanqiao@126.com

Abstract. The clustering algorithm has a very important application in the data mining technology, and can achieve good results in the data classification operation. With the rapid development of the network communication technology and the personal computers and other digital devices, the real-time computer desktop image transmission technology has been widely used. The computer desktop image compression algorithm based on the block classification can effectively realize the compression and storage of the computer desktop images, and significantly improve the speed and quality of the computer desktop image transmission.

Keywords: Clustering algorithm · Computer · Desktop image · Compression technology

Aiming at the problem of the computer desktop image compression, an image compression algorithm based on the color clustering is proposed. The desktop images are divided into 16×16 non-overlapping blocks, which are classified into the text/graphics blocks, the natural image blocks and the mixed blocks [1]. The color clustering is used to reduce the number of the color categories of the color-rich text/graphics blocks, and the lossless compression is performed. H.264 intra prediction coding method is used for the natural image blocks and the hybrid coding method is used for the mixed blocks.

1 An Overview of the Computer Desktop Image Compression Algorithms

The computer desktop image is a resource that contains texts, graphics and the image information. It is often called the mixed image together with the scanned document images [2]. The following will explain the computer desktop images from two aspects: the basic attributes of the computer desktop images and the classification methods.

1.1 Basic Properties of the Computer Desktop Images

The first is the pixels. When scanning the computer desktop image, the small image read is called the image element, which is called the pixel for short [3]. The second is the spatial resolution. It is the smallest detail to distinguish the computer desktop

Z. Xu et al. (Eds.): CSIA 2019, AISC 928, pp. 24–29, 2020.
https://doi.org/10.1007/978-3-030-15235-2_4

image. The third is the gray level resolution, which refers to the series of the brightness changes contained in the unit magnitude, which is the smallest change in the gray level to distinguish the computer desktop image. The fourth is the gray histogram. As a function of gray level, it can accurately describe the number of the gray level pixels in a computer desktop image. That is, it can roughly describe the outline of a computer desktop image. The fifth is the texture feature. The texture feature of the computer desktop image reflects the spatial distribution of the gray value of the pixels.

1.2 Classification Methods of the Computer Desktop Image

The classification methods of the computer desktop images mainly include the image space and the feature space. The following will be explained separately. The first is the classification method based on the image space. This is a method to classify the computer desktop images by using the gray levels, textures, colors, shapes, locations, underlying features and other attributes of the images. The second is the classification method based on the feature space. This is a classification method which is closely related to the method and effect of the feature extraction [4]. The classification method based on the feature space can effectively reduce the dimension of the image data and the complexity of the image data calculation.

1.3 Explanation of the Existing Computer Desktop Image Compression Algorithms

Nowadays, most of the existing desktop image compression algorithms are based on the blocks. The common desktop image compression algorithms include the virtual network computing algorithm, the hybrid desktop image compression algorithm and the hybrid desktop image compression algorithm based on the fine classification [5]. The reduction algorithm will be explained separately below. The virtual network computing algorithm is a lossless compression algorithm based on the sub-rectangular classification, but when applied to the natural image compression, the compression efficiency is too low. The hybrid desktop image compression algorithm is an improvement of the virtual network computing algorithm. The compression algorithm divides 8×8 blocks into two types of the text/graphics blocks and the image blocks. JPEG-LS and the dynamic JPEG compression algorithms are used to compress the desktop images.

1.4 Hybrid Desktop Image Compression Algorithm Based on the Fine Classification

As an improvement of the hybrid desktop image compression algorithm, Lin et al. proposed a hybrid desktop image compression algorithm based on the fine classification. This algorithm can effectively extract the text/graphics elements and classify and encode them in the image blocks, thus greatly improving the coding quality of the computer desktop images. However, the algorithm has the characteristics of the slow encoding speed, and the number of the frames processed at one time is relatively small.

2 Procedure of the Compression Algorithms for the Computer Desktop Images

The computer desktop image compression algorithm firstly divides the desktop image into 16 × 16 non-overlapping blocks with the JPEG lossy coding, and then divides these non-overlapping blocks into the text/graphics blocks, the image blocks and the mixed blocks according to the different attributes of each block. For the text/graphics blocks, the lossless compression coding is used, and the JPEG compression coding is used for the image blocks, while the hybrid compression algorithm based on the fine classification is used for the desktop image compression.

2.1 Principles of the Image Compression

From the point of view of the information theory, the image is a source, and the data describing the source is the sum of the information quantity and the information redundancy. So there are a lot of the redundancies in the representation of the image data, such as the time redundancy, the spatial redundancy, the knowledge redundancy, and the visual redundancy and so on. We can use some characteristics of the image itself and the visual characteristics of the human eyes to remove these redundant data, which can greatly reduce the amount of the original image data, thus solving the problem of the huge amount of the image data, and realize the image data compression.

2.2 Classical Image Coding

Many image coding principles are to achieve the purpose of the compression by eliminating the redundancy of the images, while the transform coding is to change the expression of the redundancy. The original data is expressed in another more compact way, and sometimes the higher data compression can be achieved. The discrete cosine transform (DCT) is a kind of the coding of the fractal transform. The advantage of the DTC is that it can divide most of the images into the image blocks, so that the energy of the image blocks can be concentrated on a few low-frequency DTC coefficients, so that the DCT can concentrate the energy of the images to a large extent and lay a foundation for the compression. The principle of the entropy encoding is to find the optimal match between the probability and the length of the codeword according to the distribution characteristics of the occurrence probability of the messages or the message sequence.

3 Computer Desktop Image Compression Technology Based on the Clustering Algorithms

With the in-depth study and application of the clustering algorithms, many clustering algorithms have been proposed based on different ideas, and they have different advantages and disadvantages. In view of the current research status and construction ideas of various clustering algorithms, we can roughly classify the current clustering algorithms into the following categories: the partition-based method, the level-based method, the density-based method, the grid-based method and the model-based method.

3.1 The Partition-Based Method

The basic idea of the clustering algorithm method based on the partitioning is to give the data set to be clustered (assuming n data) and the number of classes to be generated (assuming K, $k \leq n$). Firstly, the data set is constructed into k initial partitions according to certain rules. One partition represents a cluster, and each class should contain at least one. There is one data object, and the same data object can only belong to one class. Then, using the iterative relocation technology, the data objects in the initial class are moved continuously to change the partition contents. The relocation each time will improve the similarity of the data objects in the same class. This kind of the iterative relocation operation will stop until the data in all kinds of the partitions meet certain criteria.

3.2 The Level-Based Method

The level-based clustering method decomposes a given set of the data objects layer by layer. During the iteration each time, the number of the classes and the data members in the class will change. The hierarchical clustering algorithm can also be divided into the agglomerative clustering method and the split clustering method. The basic idea of the agglomerative clustering method is as follows. First, each data is separately grouped, and then two nearest clusters are merged and the process is repeated, so that the larger and larger classes will be formed gradually until all data objects belong to the same class or a set termination condition.

3.3 The Density-Based Method

The main idea of the density-based clustering is to cluster according to the distribution density of the data objects. Firstly, the algorithm connects the regions containing more than a certain number of the data objects into a group, and then clusters the regions without classification, as long as the density of the data objects in a region is higher than the pre-set threshold value, so that the data objects of this region and its latest clustering are merged into the same class. The density-based clustering algorithm overcomes the disadvantage that the distance-based clustering algorithm cannot find the complex shape clustering, and the density-based clustering algorithm does not need to input the number of the clusters in advance, and can be used to filter the "noise" data.

3.4 The Grid-Based Method

The grid-based clustering algorithm uses a multi-resolution grid data structure. First, the data space is divided into a finite number of cells, and then the cells on the grid are clustered. The advantage of the clustering algorithm based on the network is the fast processing speed, because the processing time of this algorithm does not depend on the number of the data objects, but only on the number of the partitioned units. In addition, the clustering results of this method are independent of the order of the data input, and have good compatibility. It can deal with many types of the data, but this operation will bring a certain cost, which will reduce the quality and accuracy of the clustering. The disadvantage of the grid-based clustering algorithm is that only the vertical or the horizontal edges can be found, but the oblique edges cannot be found.

3.5 The Model-Based Method

The basic idea of the model-based clustering method is to set up the multiple data models for the data set to be clustered, and then try to find the best match between the data set and these mathematical models. The operation of this method is based on the assumption that the data set has an intrinsic mixed probability distribution. At present, most model-based algorithms are mainly used in bionics. There are two typical model-based clustering methods: the statistical method and the neural network method. The image signal always has a very large amount of the data, but because the image is sparse in the wavelet domain or the discrete cosine domain, it can be compressed. No matter whether the image signal is transformed by wavelet transform or the discrete cosine transform, the large coefficients are concentrated in the lower-frequency domain, while the coefficients in other frequency domains are smaller. Compared with large coefficients, the coefficients in the domains are smaller and can be considered to be approximately zero. It can be considered that the image signal is sparse in the wavelet domain and the discrete cosine domain.

Therefore, the compression sensing theory can be used to process the traditional image compression process. The orthogonal basis transformation can be used to transform the images. Only a small part of the data with the large absolute values is obtained. Most of the other data are very small and approximate zero. After these small data are zeroed, the data with the large residual values will be processed. The encoding reduces the amount of the image data and achieves the purpose of the image compression. The image decoding process of the image compression technology based on the compression sensing theory is that the compressed data is nullified and then inversely transformed to the original image data at the coding end. In this way, only a few low-frequency components with the larger values are preserved, while the high-frequency components representing the details of the image are discarded. Thus, the compressed details of the images will be distorted to a certain extent, but the human eye is only sensitive to the low-frequency components of the images, so this series of the compression processes are acceptable.

At the decoding end of the images, the observed data are reconstructed by the compressive sensing signal, and the original image data is reconstructed. Because in the process of the image processing, the process of the orthogonal basis transformation and zero of the small coefficients in the traditional image compressive methods is replaced by the random observation process of the compressive sensing, this improvement reduces the computational complexity and reduces the complexity of the coding operation in the image compression. In the process of the image compression, three key problems of the compressive sensing are solved smoothly. Because the natural image signal is sparse in the discrete cosine domain and the wavelet domain, the sparse transform of the original signal can adopt the wavelet basis or the discrete cosine basis, and the random observation matrix can adopt the random generated Gauss random matrix pair. The original signal is observed. Finally, the compressed data is reconstructed to the image data by using the common methods of the compressive sensing signal reconstruction such as OMP, BP and GP.

4 Conclusion

The compressive sensing is widely used in the field of the image compression. Its compression process includes the following steps. The original image is observed randomly. After the observation, the data is the compressed data of the images. The compressive sensing image coding system has the orthogonal transform and the orthogonal inverse transform, the linear measurement and the signal reconstruction, and the quantization and inverse quantization. At the coding end, the orthogonal module transforms the image data to concentrate the energy, and the linear measurement module compresses the input coefficients and outputs the reduced dimension data. The quantization module and the coding module complete the compression and the entropy coding of the transform coefficients. In this way, better image restoration quality can still be obtained with only a few sampling points reserved.

References

1. Ma W, Wang X (2016) Robot vision image processing based on the smooth denoising block K-means algorithm. J SSSRI 12:51–53
2. Zhang H, Cao B, Liu W (2017) Hyperspectral image compression with the multi-band and look-up table prediction correction. Laser Infrared 9.101–102
3. Zhu Y, Zheng W, Pan S (2017) Hyperspectral image compression based on the block sparse expression mode. J Huazhong Univ Sci Technol (Nat Sci Ed) 12:246–247
4. Gu H, Zhang J (2018) Application of the outlier detection algorithm based on the deviation degree in the cluster analysis. J Jiamusi Univ (Nat Sci Ed) 7:132–133
5. Zhao Y, Zhu Q, Hu R, Qu X (2018) Hybrid recommendation algorithms based on the self-coding machine and clustering. Microelectron & Comput 11:93–94

Construction of the English Teaching Model Based on the Interactive Network Streaming Media

Mao Zhi[1(✉)] and Wang Ping[2]

[1] Sichuan Vocational and Technical College of Communications, Chengdu 610000, Sichuan, China
lingjun09419531@163.com
[2] Chengdu Aeronautic Polytechnic, Chengdu 610000, Sichuan, China

Abstract. There are limitations in the time, regions and contents in the traditional English teaching, and there are some limitations in the traditional multimedia network teaching. The design of the English teaching mode based on the interactive network streaming media has brought a profound change to the English teaching. As the interactive network streaming media provides a good platform for people's communication, the role of teachers in the English teaching based on the interactive network streaming media is no longer the "leader" of the teaching, but the "guide" of the teaching. Teachers provide the hardware conditions, the learning resources and the communication objects. Students can choose their own way of learning. When they encounter problems, they can consult their teachers or their friends through the network streaming media platform.

Keywords: Interactive · Online streaming media · English teaching mode · Construction mechanism

The distance learning will provide more people with access to education. The instructor notifies the audience to visit a URL before the lecture begins. When the lecture begins, the audience can see the pictures of the speaker's speech and hear his voice [1]. The entire lecture can also be recorded in the form of the media files for their later on-demand broadcasting.

1 Streaming Media Technology

The streaming media technology refers to a transmission technology that transmits the audio and video information at a given rate independently of the network load between the sender and the receiver. The streaming media has the characteristics of the implicit time dimension, the real-time transmission, the isochronism and the high throughput [2]. The Internet is difficult to meet the real-time requirement of the streaming media due to the limitations of the insufficient bandwidth and the weak quality control mechanism. In the streaming transmission, voices and images are transmitted continuously and in the real time to the users through the network [3]. The users do not have

Z. Xu et al. (Eds.): CSIA 2019, AISC 928, pp. 30–36, 2020.
https://doi.org/10.1007/978-3-030-15235-2_5

to wait until the entire file is downloaded, but can watch them only after several seconds or more start-up delay.

The streaming media technology is a technology that uses the streaming to transmit the continuous time-based media. The streaming transmission is to compress the videos, audios and other media into one compression package, which is transmitted continuously and in the real time from the video server to the user's computers. It only needs to cache enough playable video capacity at the user end to start playing [4].

The streaming media system consists of the coding tools, the streaming media data, the servers, the networks and the players. The coding tools: That is to say, it is used to create, capture and edit the multimedia data to form the streaming media format. Using the media acquisition equipment, produce the streaming media. It includes a series of tools, ranging from the independent videos, sounds, pictures, and text combinations to the production of the rich streaming media. The streaming media files generated by these tools can be stored in a fixed format for the use by the publisher [5].

The streaming media data is the carrier of the media information. The commonly used streaming media data formats are .ASF and .RM and so on. The server: That is to store the media data. In order to store the large-capacity film and television data, the system must be equipped with the large-capacity disk arrays, with the high-performance data reading and writing capabilities, the high-speed transmission of the external data requests, the high scalability, and the compatibility, and can support the standard interfaces [6]. This system configuration can satisfy thousands of hours of the video data storage and realize the mass storage of the chip source. The network: That is a network suitable for the multimedia transmission protocol or even the real-time transmission protocol. The streaming media technology develops with the development of the Internet technology. It adds the multimedia service platform on the basis of the existing Internet.

The player: That is the software for users to appreciate the online media. The streaming media supports the live audio and video broadcasting and on demand. It can be embedded in the popular browsers, play a variety of the popular media formats, and support a variety of the media forms in the streaming media, such as texts, pictures, Web pages, audios and video integrated forms. When the bandwidth is abundant, the streaming media player can automatically detect the connection status of the video server and select more suitable videos to achieve the better results.

2 Transfer Modes of the Streaming Media

2.1 Progressive Streaming Transport

That is, at a given time, users can only watch the downloaded part, but cannot skip the part that has not been downloaded. It cannot be adjusted according to the speed of the user connection during the transmission, as in the real-time streaming transmission. Because the viewing part of the files before playback is downloaded without prejudice, while the sequential streaming transmission is suitable for the high-quality short clips, such as the film head, the film tail and the advertisements, which is not suitable for the long clips and videos, lectures, speeches and demonstrations with the random access requirements, and does not support the live broadcasting. Strictly speaking, it is a VOD technology.

2.2 Real-Time Streaming Transport

The information bandwidth can automatically and dynamically adapt to the network bandwidth when the media is transmitted to ensure that the media signal bandwidth matches the network connection so that the media can be viewed in the real time. Unlike the sequential streaming, it requires the dedicated streaming media servers and the transport protocols. The real-time streaming transmission is suitable for the on-site events and also supports the random access. Users can fast forward or backward to view the front or the back contents. In the theory, the real-time streams can be continuously watched once they are played, but in the practice, the periodic pauses may occur.

2.3 A Comparison Between the Two

In terms of the video quality, the real-time streaming transmission must match the connection the bandwidth. Because the missing information is neglected, the video quality will decline when the network is congested or problems arise. If the video quality is to be guaranteed, the sequential streaming transmission is better. The real-time streaming requires the specific servers, which allow more levels of the control over the media delivery. The system settings and the management are more complex than the standard HTTP servers. The real-time streaming transmission also requires the special network protocols, which sometimes cause problems when there are firewalls, so that users cannot see the real-time contents of some locations, while the sequential streaming transmission has nothing to do with the firewalls.

3 The Background of the Construction of the English Teaching Model Based on the Interactive Network Streaming Media

The so-called "interactive" teaching mode is to regard the educational activities as a kind of the life-life interaction and communication between the teachers and the students, and the teaching process as a dynamic and developing process of the interaction and interaction between the teaching and the learning. In this process, by optimizing the ways of the "teaching interaction", that is, by adjusting the relationship between the teachers and the students and their interaction, a harmonious relationship can be formed. The teacher-student interaction, the student-student interaction, the interaction between the learning individuals and the teaching intermediaries, strengthen the interaction between people and the environment, to produce the teaching resonance, and to improve the teaching effect.

3.1 Construct the Facilities and Technical Foundation of the Interactive Teaching Between Teachers and Students Under the Network Environment

As the cost of the computer manufacturing and the production has been greatly reduced, especially in the urban life, most households have purchased the home computers. The home Internet can be used in many ways, such as the telephone dialing, ADSL, the optical fiber, and the cable TV and so on. According to the survey, in recent years, more than 90% of households have computers and access to the Internet. And most students can easily and skillfully use the computers to browse the information on the Internet. In the parents of the students, there are also quite a number of people with the high educational levels, often using computers to work and deal with some affairs. It is the popularity of the computer network in families and schools, and the wide application of the computer operation and the application technologies in students and teachers that lay a solid material and technical foundation for the further development of the interactive teaching under the network conditions. The teacher-student interaction has a very clear educational characteristic. Its purpose is to promote the learning, their cognition and the social development of both the teachers and the students, and especially the students.

3.2 The Characteristics of the Teacher-Student Interactive Teaching Under the Network Environment

In the interactive teaching under the network environment, the teaching media has evolved into a unique and rich network environment, which has an important influence and restriction on the operation of the interactive teaching system. The interactive teaching between the teachers and the students under the network environment provides the favorable conditions for the reform of the teaching mode. In the user interface, teachers explain the course videos and the students learn the course videos. The teachers and the students communicate and evaluate through the online interaction, including the students' mutual evaluation and other teaching links. Through the combination of the design and the teacher's curriculum design, we can make full use of the streaming media technology to integrate the teaching and the learning into a unified teaching platform.

Firstly, the teaching management module is established, and the online video-on-demand based on the streaming media and the teacher-student interaction network learning is adopted. The main purpose is to construct the virtual classroom and establish the network teaching close to the real situation at different times or in different places, so as to realize the users' safe access, on-demand, whiteboard demonstration and other functions. Secondly, it establishes the information management module, including the information management of teachers, students and curriculum contents, and establishes the corresponding login and operation authority to ensure the normal operation of the system according to the relevant data information. Finally, the curriculum result module is established to record the specific information about the courses that have been taken, and the teaching evaluation results are obtained through the teaching evaluation.

4 The Construction Mechanism of the English Teaching Model Based on the Interactive Network Streaming Media

The database of the network learning platform based on the streaming media technology follows the third paradigm of the database design. In the design of the system, the redundancy and the identification columns of the database are moderately increased, mainly including the following registration forms: the teacher registration form, the student registration form, the main network course selection registration form, and the class situation record form. In order to make the resources on the network learning platform be used properly and reasonably, the security and reliability of the network learning platform should be considered. In the network layer of the information system, it realizes the security control of the remote teaching information system, manages the login rights of the end users, and prevents the wrong operation of the end users. The middle layer of the remote teaching information system should establish an independent security design with the third-party unauthorized system to ensure that the security information of the system will not leak in the non-secure system when configuring the system.

From the point of view of the existing teaching resources and the teaching modes, there is still a lack of the interactive platform for the online teaching in schools. Most schools still adopt the traditional Internet learning mode, such as the multimedia teaching mode. Therefore, after defining the principles of the interactive teaching mode, it is necessary to build a network interaction platform suitable for their students' foundation and the school resources, which require the teachers to fully understand the campus network environment and the resource advantages, as well as some cooperation resources in and outside the schools. At the same time, the teachers need to focus on understanding the learning characteristics of the students in their respective schools, and build a scientific and reasonable network teaching platform under the preceding premise, so as to improve the students' learning enthusiasms.

The main contents of the platform construction can be divided into four modules: the courseware, cases, discussion and homework. The four modules can cover the main contents of all the lectures and the PPT original. The homework platform can meet the needs of different students for the difficulty of their homework. The construction of the discussion platform can collect and accept the suggestions of the students for making the case courseware of the teachers. Teachers can improve and perfect the teaching courseware and methods by collecting the information regularly or collecting suggestions that can be adopted, so as to design the teaching courseware more in line with the students' personality characteristics and their learning characteristics.

The multimedia courseware into the English classrooms undoubtedly injects the new vitality into the English teaching. It creates scenes by means of pictures and texts, sounds and images, and dynamic and static expressions. It can not only arouse the students' interests and enrich the teaching contents, but can also strengthen the teaching effect, thus bringing the network education into a new realm.

Using the modern network technology to serve the English teaching has incomparable advantages over the traditional teaching mode. It not only conforms to the trend of the educational informatization, but also conforms to the requirements of the college

English teaching and the natural psychological process of the students' language learning. The English learning itself is a hard work. In the vast sea of the words and the complicated grammar phenomena, the traditional printed texts and the teacher-centered teaching methods are difficult and seldom take into account the enthusiasms and interests of the students' participation, and the students' participation and interests are the key to the language learning. Under the condition of the multimedia and the network technology, students have better participation because of the use of the graphical interface and the window interactive operation to realize the human-computer interaction. The ways students receive the English knowledge are no longer confined to the form of words, but in a three-dimensional world of cartoons, movies, music, pictures, words and sounds, which fully mobilizes all kinds of the senses and enables the students to participate in their learning more actively.

The difference between the "network streaming media" and the "multimedia network" is not only in the name, but also in the traditional form of the "multimedia + network". There is no organic combination between them. The main mode is that students access the multimedia materials on the server through LAN, such as movies and music and so on. The "network streaming media" implements the data transmission of the multimedia (mainly in the form of the streaming media) supported by the broadband network. It emphasizes the bidirectionality and the real-time of the communication. Of course, it also has the non-regional characteristics of the traditional network.

5 Conclusion

With the development of the science and technology, the modern information technology has entered the field of our education and teaching, gradually affecting and changing the ways people used to accept knowledge. As an important part of the assistant teaching, the multimedia courseware of the network teaching has been favored by more and more people. The classrooms will no longer be confined to schools, and we can communicate with the network anytime and anywhere to learn. The rational use of the multimedia technology in the network teaching can stimulate the students' interests, promote the students' development, and improve our teaching efficiency.

Acknowledgment. Foundation Project: This paper is a phased achievement of Sichuan Applied Foreign Language Research Association's scientific research project "The Impact of Background Music on Chinese and English Reading Comprehension" (Project No. SW2018G015).

References

1. Liu J (2017) Research on improving the college english teaching by using the flipped classroom. Engl Square 12:131–132
2. Zhu B (2018) Innovation of the "Internet+" college English teaching mode. Overseas Engl 11:128–129

3. Jin N (2018) Construction of the "student-centered" pre-class and post-class integrative teaching mode – taking "comprehensive business English" course of Dalian Neusoft University of Information as an example. Mod Commun 12:146–147
4. Liu X, Han X, Xia H (2018) Survey of the effectiveness of "on-line + off-line" hybrid teaching model – taking the college English teaching as an example. Mod Commun 12:115–116
5. Bao W, Zhou H, Xie F (2017) An optimal design of web based online English teaching by using struts framework. Boletin Tecnico/Tech Bull 55(10):58–63
6. Wang A (2018) Exploration of the integrative teaching model of reading and writing in college english guided by thinking. J Shangqiu Vocat Techn Coll 12:133–134

Design and Implementation of the Image Processing Software Based on the Infrared Image Feature Matching

Fei Tian(✉) and Shilin Cui

Nanyang Institute of Technology, Nanyang 473000, Henan, China
cupjianyang@yeah.net

Abstract. The infrared image has some disadvantages, such as the narrow dynamic range, the low contrast and the susceptibility to the noise pollution. The traditional infrared image denoising algorithm filters out the image details while removing the noise. A new method of the infrared image denoising based on the sparse representation is proposed. Firstly, the original infrared image is clustered, and then each clustered sub-image is decomposed into a dictionary, and the denoised infrared image is reconstructed from the sparse coefficient matrix.

Keywords: Infrared image · Feature matching · Image processing · Software design · Implementation results

The infrared imaging system shows that the infrared radiation of the observed scenery can be converted into the infrared video image only through the atmospheric transmission, the optical imaging, the photoelectric conversion and the preprocessing [1]. Therefore, the infrared images are generally characterized by the low signal-to-noise ratio, the low background and target contrast, the blurred edges and the poor details. With the improvement of the requirement of the infrared image enhancement effect, keeping the details of the original infrared image can better adapt to the computer analysis and processing.

1 Characteristic Connotation Analysis of the Infrared Images

At present, most scene matching algorithms extract the global and invariant features, which can solve the consistency decision problem of the same object very well, but it is difficult to eliminate the image distortion [2]. When the imaging distortion between the images is very complex, it is very difficult to use the global information for matching. Especially when there is the local occlusion, the whole image features will change accordingly. Based on the above characteristics, the infrared image matching, recognition and tracking are generally feature-based methods. Generally, the local invariant features are used to process and recognize the infrared images.

The local feature extraction is to segment the whole image into several parts and extract the global features for each part. The segmentation here is not what we think intuitively [3]. Ideally, people always want the local features to correspond to part of

Z. Xu et al. (Eds.): CSIA 2019, AISC 928, pp. 37–42, 2020.
https://doi.org/10.1007/978-3-030-15235-2_6

the object in the objective world, but this is unrealistic, and often needs to understand the high-level scenes with the help of the image processing technology. The infrared image reflects the spatial distribution of the infrared radiation of the target and the background. The distribution of the radiation brightness is mainly determined by the temperature and the emissivity of the observed scenery [4]. The high-temperature part of the observed scenery is shown in the large gray value part of the infrared image histogram, and the low-temperature part of the observed scenery is shown in the small gray value part of the infrared image histogram. Therefore, the infrared video image approximately reflects the infrared radiation spatial distribution, the temperature difference or the radiation difference of the observed scenery.

The histogram equalization not only enhances the contrast, but also enlarges the noise and results in the loss of the image information. The wavelet transform can separate the high-frequency and the low-frequency signals of the infrared images, while the high-frequency signals often represent the details and the noise of infrared images [5]. The Retinex theory is the method of the feature-based contrast enhancement. The target part of the infrared image can be enhanced and the noise can be effectively suppressed. The infrared image is a gray image which shows the alternating light and shade of the object's heat distribution. It has the characteristics of the low signal-to-noise ratio, the low background, the low target contrast, the blurred edge and the poor detail performance. According to these characteristics of the infrared images and the analysis of the histogram equalization, the wavelet transform and the Retinex algorithm, the details-preserving Retinex infrared image enhancement algorithm in this paper synthesizes the advantages of the three algorithms.

Because the wavelet transform can decompose the infrared image into the low-frequency subband image and the high-frequency subband image, the details of the infrared image can be preserved in the high-frequency subband image, and the noise can be effectively filtered in the Retinex algorithm, so the high-frequency part can be enhanced by the Retinex algorithm, and the original details of the infrared image can be preserved. The high-frequency part of the information enhances the image and can effectively filter the noise. For the low-frequency images, the histogram equalization algorithm will be used to enhance the contrast, so that the contrast of the infrared image is higher than that of the infrared image enhanced by the Retinex algorithm or the wavelet transform algorithm.

2 Functions and Value Requirements of the Image Processing Software Design

2.1 Implementation of the Document Management Functions

This software has the document management functions of other commercial software. It can realize the document management functions by selecting the relevant setting items when building a new project with the help of VC.

2.2 Implementing the Image Preprocessing

The software has the strong pertinence to realize various image processing. Its principle is based on the selection of the thresholds in the image binarization processing. In the adjustment of the image saturation, contrast and brightness, it can be achieved in the form of the mode dialog box. In addition, there are many other image pre-processing functions, such as the multi-medium band color synthesis function, the image histogram equalization function and the image median smoothing function.

2.3 Achieving the Target Feature Extraction

The software effectively achieves the function of the target feature extraction in the related aspects. The line-surface feature is one of the extraction contents of the target feature. It can not only realize the effective description of the related aspects such as the target mid-level vision, but also is one of its main components. The existing carrier of the target line-surface is mainly the remote sensing image and so on. In the remote sensing image, there are a large number of the target faces, so for the image processing, extracting the relevant features of the target lines has always been the main research focus in this field. At the same time, the software has a large number of the algorithms and modes in the field of the on-line surface correlation feature extraction, so it effectively solves the difficult problems of extracting the line features effectively because of the complex imaging conditions and the related space conditions. It uses the edge operators to carry out the effective acquisition of the target correlation edge characteristics in turn, the connection processing between targets and the integration of the target features after completing the first two steps. On the basis of the human-machine coordination, the seed points are selected manually, and then the related skeleton of the line and surface targets is acquired, and the external contours of the face areas are acquired. After tracking and extraction, the line and the surface targets can be implemented in order to effectively reduce the storage and facilitate the efficient processing of other aspects. Freeman chain code is the main tool used to characterize the line and surface targets.

2.4 Realizing the Online Help

In order to establish the help function of the software effectively and realize its function efficiently, the most important task is to establish a reasonable topic of the help-seeking. The main body of the help-seeking can be composed of graphics and the text content in the structure and the content, that is, a combination of the graphics and texts. Its display form is that we are in the help-seeking box. Often you can see a window page with texts and images. It can have the multiple texts and graphics at the same time. At the same time, the help topic also has different features from its appearance, namely, the jumping and identification features. As for the contents presented by the help-seeking topic, it can be edited according to the actual needs. The principle is the same as that of using the WORD software to edit relevant texts. It can add the required text contents by inputting. In addition, if the relevant image is needed to cooperate with the text contents, it can also be inserted by the file. Insert the way into the main page.

For the RTF files, it can be compatible with the following types of the graphics files. First, the extension of the Windows metafile is the wmf file, and second, the extension of the hypergraph file is the shg file, and third, the extension of the Windows bitmap file is the dib and bmp file.

3 System Design

The system is designed to collect the analog image signals from the front-end wave infrared detectors, convert them to the analog and digital signals, process the real-time images, and display all the analog TV signals with 768 * 576 CCIR-656 single frame pixels. In order to meet the requirements of the high speed and the real-time performance of the entire system, the system adopts the design idea of combining DSP with FPGA. The SRAM A, B, C buffer memory is mapped to the CE space of the DSP by the FPGA, and the seamless interface between the co-processing systems is realized. FPGA is an image data acquisition unit, and the collected image data is pre-processed and transmitted to the DSP. The DSP part is the core unit of the system image processing system, which determines the performances of the entire system. After the DSP completes the processing task, the results will be returned to the FPGA, which will be responsible for displaying the information and providing the human-computer interaction as the frequency display unit of the system's videos. The overall process of the system is shown in Fig. 1.

Because the analog image signal output by the detector is converted into the digital image signal through the four A/D converters, the FPGA will be responsible for the acquisition of the digital signals. The main and subordinate mode of the work is adopted between the FPGA and the DSP. The DSP controls the work of the FPGA by writing the state to the FPGA, and executes the corresponding functions. When an effective detector signal is collected, it is first corrected by the non-uniformity, brightness/contrast control and the blind element replacement. The processed image data is cached in a piece of SRAM A, triggering a read interrupt of the DSP when a full field is stored. The data is written into the L2 buffer of the chip of the DSP through the EDMA transmission mode of the DSP. The image display is transformed by DSP, and the 14 bit image data is transformed into 8 bit image data. Then real-time image enhancement and other algorithms are processed to improve the image quality. Because the two image data in each frame are the same, only one image data need to be processed in the system. When an image data is processed by an image processing algorithm, a write interrupt of the DSP is triggered, and the data is written into two SRAM B and C external to the FPGA by the EDMA transmission mode of the DSP. According to the synchronous signal logic, the switching control is carried out by the FPGA. The processed image data is fed into the digital-to-analog video conversion of the video encoder, and finally the CCIR-656 compatible fully analog TV signal is output.

The entire hardware system is divided into the DSP module and the FPGA module, in which the DSP and the FPGA are the core devices. The DSP module has the functions of the system management and control, the image display transformation, the non-uniformity correction calibration, the blind element detection, and the digital image enhancement processing and so on. The FPGA module has the functions of the real-time pixel acquisition, the image preprocessing, the image data input/output field caching, the video encoder control, and the asynchronous serial communication and so on. The master-slave mode is used between the DSP module and the FPGA module. The DSP module is the main module, which controls the task scheduling of the entire system and the transmission of the image data between the DSP module and the FPGA module. After the image processing, the system outputs the image data to the video encoder according to the timing requirement of the CCIR-656 full-TV signal. It includes the video synthesis and the D/A digital-to-analog converter, and then outputs the analog full-TV signal to the TV monitor. The circuit part is divided into two big modules: the FPGA and the DSP. Each module has peripheral circuits and peripheral devices. Bus is used to connect the two modules.

The DSP module includes the digital signal processor, the large-capacity dynamic memory, the FLASH memory, the JTAG debugging interface, the clock and reset circuit and the corresponding power supply circuit. The FPGA module includes the Stratix programmable logic device, the input and output frame buffer memory, the video encoder, the configuration chip of the programmable logic device and its configuration interface, the asynchronous serial communication, and the corresponding power supply and clock circuit. The Sratix series programmable logic devices have abundant pin resources, but the external circuit is very concise, so only the corresponding configuration chip can work normally. The programmable logic chips are mainly used to deal with the real-time video streams, including the input and the output. The corresponding image data is accumulated after a full session and then interrupted by the DSP to read. Similarly, the interrupted refresh data is generated immediately after the DSP processes the data. There are three frame buffers, one input buffer and two output buffers, which are assumed by three SRAMs. The video D/A part uses the video coding chip to control the synchronization and the blanking signals of all the TV signals generated by FPGA. The control interface of the video encoding chip is the I2C interface, which is connected to the FPGA. The chip sends configuration commands directly for the configuration. The standard asynchronous serial communication control module generated by the programming of the FPGA cooperates with the level conversion chip to realize the asynchronous serial communication, which improves the modular design of the system and the ability of the multi-device communication.

4 Conclusion

Because of the high real-time requirement and the large amount of data for the infrared image acquisition, the traditional infrared image acquisition system is more complex. The proposal of the programmable on-chip system (SOPC) provides an effective solution for the construction of the infrared image acquisition system with the good real-time performance and the simple structure. The experimental results show that the system can not only complete the function of the infrared image acquisition under different light conditions, but also has the characteristics of the simple structure, flexibility and high efficiency.

Acknowledgment. Henan key task project in scientific and technological research, Project Number: 172102210414.

References

1. Xu J, Fu T, Yang J, Su F (2016) Improving the infrared and visible image registration using the saliency analysis and the edge direction histogram features. Opt Precis Eng 11:111–112
2. Zeng W, Liao S, Wang W, Wei H (2017) An improved infrared image enhancement algorithm and its implementation on FPGA. Electron Des Eng 12:181–182
3. (2018) Extraction method of the key technical features of volleyball based on the infrared image sequence. Nat Sci J Xiangtan Univ (04):162–163
4. Huang S, Xu C (2018) Design of the night driver fatigue detection system based on the infrared image. Automot Pract Technol (07):135–136
5. Shen W, Zhang L, Huang B, Wei C, Liu C (2018) Improved K-means infrared image transformer segmentation method. Transducer Microsyst Technol (10):103–104

Design and Implementation of the Image Processing Software Based on the Wavelet Transform

Shilin Cui(✉) and Zhanzheng Wan

Nanyang Institute of Technology, Nanyang 473000, Henan, China
dingda444660@163.com

Abstract. The wavelet transform technology is widely used in many fields, such as the graphics processing, the image processing, the video-voice processing, and the digital signal processing, and has achieved good application results. Starting from the principles and types of the image processing technology based on the wavelet transform, two processing methods of the continuous wavelet transform and the discrete wavelet transform are described in detail, and their specific applications and advantages in the image compression and the image denoising are enumerated.

Keywords: Wavelet transform · Image processing · Software design · Implementation mechanism

1 Design of the Image Processing Software

With the rapid development of the wavelet theories, the wavelet transform has been successfully applied to the image processing. Firstly, the definition of the wavelet transform and its application in the image processing are given [1]. The applications of the wavelet transform in the image compression, the image denoising, the image smoothing, the image enhancement and the image fusion are listed, and the simulation results of MATLAB are given. The simulation results show that the application of the wavelet transform technology in the image processing has good effect and the high engineering values, which is worthy of our further study [2].

The main theoretical basis for the design of the software is the relevant theoretical knowledge of software engineering. The specific design tool is the VC high-level language programming software. The main modules of the software include the following.

1.1 Document Management

The document management also has the following functions. Firstly, the opening function of the documents can display all the data images related to the integrated system, including the original remote sensing images and so on in the form of windows, and different forms of images can be read according to the independent documents. The second is the document storage function, which can store various types of

Z. Xu et al. (Eds.): CSIA 2019, AISC 928, pp. 43–48, 2020.
https://doi.org/10.1007/978-3-030-15235-2_7

images through the conversion format [3]. The third is the document closure function, which can close multiple images simultaneously, and the fourth is the recent file function, which can open the required images by using the shortcut keys.

1.2 Image Preprocessing

The function of this system is to realize the visualization of images and facilitate the follow-up operations. The main types of images it processes are various remote sensing images. Its functions mainly include the following aspects [4]. The first is the binary processing function. The software first gives users a threshold initial value by statistics, and then obtains the accurate threshold according to the user experience. The threshold function mainly includes two kinds of the thresholds of the double thresholds and the fixed thresholds. The second is the visual adjustment function, which is mainly based on the tool of VC. It can effectively adjust the contrast and the brightness of the images visually, so as to meet various needs. The gray standard processing function of this module can help improve the image quality very well. The third is the smoothing processing function. In order to reduce the interference of the smoothing noise to other parts, usually through setting the relevant parameters, we can achieve the control [5]. The fourth is the sharpening processing function, mainly for the primary and the secondary differential and various operators. The fifth is the black-area processing function, which is based on the binary processing. Its main functions are the small-area removal and the feature-point extraction. The sixth is the multi-band pseudo-color synthesis function, which uses the framework function of VC to realize the pseudo-color synthesis and present it in the form of the color bitmap.

1.3 Line Target Extraction

Its main functions are as follows. The first is the line processing function, which belongs to the global linear extraction target except the seed point extraction, and mainly refers to the operator target, including Deutch/Rosenfield and Hildreth operator skeleton extraction, in addition to the vector chain code storage of the simple line processing and so on. The second is the edge extraction function. The processing function should be based on the actual situations of the images and combined with their own experience, but there is no universal method. The visualization software extraction algorithm lard has the operator automatic extraction and so on. The extraction results can be optimized by the deburring processing method. The third is the area extraction function, whose main principle is based on the gray structures, so that the continuous regions can be tracked and acquired. Its main functions are the image segmentation and so on.

In addition, there are also the windows management and the online modules and so on. The main functions of the window management module include the construction of the new windows, the smooth roaming, the image size adjustment, the directed mirror and the window arrangement and the closing order and so on. The online helping module can help users master the correct software operation methods and guide users to carry out the visualization of the image processing.

2 Wavelet Transform

2.1 Wavelet and Its Related Concepts

The wavelet is a kind of the function that decays rapidly to zero in a finite interval. The wavelet analysis is to decompose the signals into the wavelets with different displacements and expansions of the original wavelet function (also called wavelet basis). The wavelet transform uses the theory of the wavelets to process the original signals, so that it has some more suitable time-frequency characteristics for the subsequent processing. The wavelet transform is especially suitable for processing the non-stationary signals because of its good spatial and frequency localization, multi-resolution and low time complexity. The digital image is a typical non-stationary two-dimensional signal.

2.2 Wavelet Transform of Images

The image wavelet transform adopts the fast algorithm of the two-dimensional wavelet transform, which is to decompose the previous images into the four sub-bands. With the original images as the initial signal, the original signal is decomposed into the four sub-bands through a set of the high-pass and the low-pass filters, namely a low-frequency sub-band (LL) and three high-frequency sub-bands (HL, LH, HH). Among them, LL is an approximate image, while HL is a horizontal detail image, and LH is a vertical detail image, and HH is a diagonal detail image. This is called the first-order wavelet decomposition, which can be iterated, but only for the lower-frequency sub-images of the previous level. It can be decomposed infinitely in theory. But in the image compression, the quality of the reconstructed images needs to be considered, so it is better not to exceed five levels and we generally use the three-level wavelet decomposition.

After the wavelet transform of the images, the energy of the images is not compressed, but the signal energy of the whole image is redistributed. The low-frequency sub-images contain most of the image information, and the high-frequency sub-images have the values of most points close to 0. The higher the frequency, the more obvious this phenomenon is. For an image, the most important part of representing an image is the low-frequency part. So we can make full use of the characteristics of the transform and organize the transformed wavelet coefficients by appropriate methods. The most common method is to keep the low-frequency coefficients and minimize them, and to set the high-frequency coefficients as large as possible to 0, so as to realize the effective compression of the image information. The process of the image compression based on the wavelet transform is to transform the input original images into the wavelet transform, quantify and encode the coefficient matrix of the wavelet transform according to the processing needs, and reconstruct the images through the inverse wavelet transform.

The wavelet transform is a time-scale analysis method of the signals. It has the characteristics of the multi-resolution analysis, and has the ability to characterize the local characteristics of the signals in both the time and the frequency domains. It is a time-frequency localization analysis method with the fixed window size but variable shapes and variable time and frequency windows. That is to say, the lower-frequency part has the higher frequency resolution and time resolution, and the higher frequency part has the higher time resolution and lower frequency resolution. It is very suitable for detecting the transient anomalies entrapped in the normal signals and displaying their components, so it is known as the analysis signal microscope.

The wavelet analysis decomposes the signals into two parts: the low-frequency al and the high-frequency dl. In the decomposition, the lost information in the low-frequency al is captured by the high-frequency dl. In the next decomposition, al is decomposed into two parts: the low frequency a2 and the high frequency d2. The information lost in the low frequency a2 is captured by the high frequency d2. By analogy, further decomposition can be carried out.

The two-dimensional wavelet function is obtained by the tensor product transformation of the one-dimensional wavelet function. The two-dimensional wavelet function decomposition is to decompose the low-frequency part of the scale j into four parts: the low-frequency part of scale j + 1 and the high-frequency part of three directions (horizontal, vertical and oblique). Discarding has little effect on the image quality, so the characteristics of the wavelet transform give a good tool for the image compression. After decomposing the original images with the wavelet transform, a threshold is set for the high-frequency information. If the value of the point is less than a, then zero is set so that the low amplitude and the high frequency information with little influence in the images is discarded and restored. The quality of the images did not decrease significantly, but the space occupied became smaller.

3 Design and Implementation of the Image Processing Software Based on the Wavelet Transform

3.1 RGB Images

The three components R, G and B represent the colors of a pixel. If we want to read the pixel values in the image (100, 50), we can view the ternary data (100, 50, 1:3). The RGB images can be stored in the double precision with the brightness range of [0, 1]. A more customary storage method is to use the unsigned integer storage with the brightness range of [0, 255]. There are two structures. One is the palette and the other is the image data matrix. The palette is a color mapping matrix with three columns and several rows. Each row of the matrix represents a color, and three columns represent the double-precision numbers of red, green and blue intensity respectively. Note: In the palette color intensity [0, 1] of MATLAB, 0 represents the darkest, while 1 represents the brightest.

3.2 Image Denoising

The purpose of the image denoising is to reduce the image noise on the premise of retaining as many image features as possible. The wavelet transform has been applied in the image denoising for its multi-resolution, decorrelation and base selection flexibility. The useful signal and noise in the images show different characteristics after the wavelet transform. The useful signal corresponds to the coefficients with the larger amplitude and the noise corresponds to the coefficients with the smaller amplitude. So the noise can be filtered by setting threshold. The specific method of the wavelet denoising is first to decompose the wavelet and then to quantify the threshold of the high-frequency coefficients, which is the key of the entire denoising process. The selection and processing of the threshold is related to the quality of the image signal denoising. When dealing with the threshold, we can choose the ways of the hard threshold or the soft threshold, as is shown in the hard threshold function (1) and the soft threshold function (2). In the equate (1) and (2), w denotes the wavelet coefficients, and T denotes a given threshold, while sign (*) denotes a symbolic function. After the threshold processing, the wavelet is reconstructed. That is to say, the entire process of the image denoising is completed.

3.3 Image Retrieval

The wavelet transform has the good performance in analyzing the point singular signals, which makes it play a great advantage in the image feature extraction, and the feature extraction is the key technology in the image retrieval, so the theory of the wavelet is becoming more and more prominent in the retrieval. In the process of the retrieval, the input image is firstly transformed by the wavelet transform, and the coefficient matrix of the wavelet transform is obtained. The mean and variance of the matrix are calculated to form the texture feature vectors of the image. Then the feature matching is carried out to retrieve the image feature database and calculate the distance between the feature vectors of the input image and the feature database. The image most similar to the sample image is produced.

3.4 Image Compression

With the increasing amount of the data in today's information system, how to store the information quickly and accurately becomes more and more important, so the compression technology can solve this problem very well. The wavelet transform is actually equivalent to a low-pass filter. After a single wavelet transform, the image signals are divided into four frequency bands, i.e. the high-frequency part and the low-frequency part in the horizontal, vertical and diagonal directions. The energy of the image is mainly concentrated in the low-frequency part, while the energy of the high-frequency part is less. The wavelet transform is used to realize the image compression. That is, the high-frequency part of the image is removed by the wavelet decomposition, while the low-frequency part is retained. The low-frequency part continues to decompose, so the image signal is decomposed into many sub-image signals with different spatial resolutions, different frequency characteristics and direction characteristics. Different wavelet decomposition times realize the different degrees of the image compression to meet the requirements of the image data storage.

3.5 Image Enhancement

The image enhancement refers to the process of highlighting some information in an image according to the specific needs, while weakening or removing some unnecessary information. The image enhancement can be achieved by two different methods: the time domain and the frequency domain. The wavelet transform combines the advantages of these two methods with its multi-scale frequency characteristics. After the image is decomposed by the wavelet transform, the low-frequency part of the energy concentration area is enhanced, and the high-frequency part is attenuated to achieve the image enhancement. The experimental results show that compared with the traditional wavelet threshold denoising algorithm, this method can obtain the higher signal-to-noise ratio (SNR) and the peak signal-to-noise ratio (PSNR). The results show that this method can not only remove noise more effectively, but also achieve better visual effect.

4 Conclusion

When using the traditional system to process the blurred images, there are some problems such as the long time-consuming and the poor processing effect. Therefore, the design of the fuzzy image processing system based on the wavelet transform is proposed. According to the interface layer and the logic layer of the system architecture, the hardware and the software of the system are designed. The core part of the hardware architecture is composed of the chips, the field programmable gate array and the integrated circuit. According to the characteristics of the wavelet, the coefficients are analyzed, and the image processing technology is combined to reverse the transform, so the software function is designed. The blurred image is decomposed and reconstructed by the wavelet transform, and the wavelet algorithm is implemented in the integrated circuits.

Acknowledgment. Henan key task project in scientific and technological research, Project Number: 172102210414.

References

1. Li H, Zheng J (2016) Blind detection method of the forged images based on the noise variance estimation. Appl Res Comput (04):111–112
2. Wu Y, Jiang Z, Wang J, Xu X (2017) Application of the improved wavelet denoising algorithm in the bearing defect images. Comb Mach Tool Autom Mach Technol (11):121–122
3. Gulimige M, Turhongjiang A (2017) Remote sensing image fusion method based on the dyadic wavelet transform. J Xinjiang Normal Univ (Nat Sci Ed) (12):132
4. Tan X, Li P (2018) Multiwavelet construction and implementation in the image processing. Electron Technol (10):133–134
5. Zhai J, Dai J, Wang J, Ying J (2018) Improving the threshold of the wavelet image denoising. Sci Technol Innov (11):105–106

Design and Implementation of the School Sports Management System Based on WEB

Guanghui Sun(✉)

Zhengzhou University of Science and Technology,
Zhengzhou 450064, Henan, China
sunrenjin@126.com

Abstract. With the continuous improvement of the levels of the sports informationization, the sports information management system is more and more widely used in school sports, competitive sports and social sports. From the single-machine system to the file/server model-based system, to the client/server (C/S) model-based system to the browser/server-based system, various forms of the management information systems serve the sports management work at the same time. With the development of the network technology, the Browser/Server-based system has become the mainstream network system.

Keywords: WEB · School physical education · Management system · Design idea · Implementation mechanism

The school physical education management system based on the WEB fully combines the massive resources of the Internet web and the data mining analysis technology, which can carry out the real-time analysis of the students' learning stages and results, provide a basis for the students to master their own learning progress and for the teachers to develop their teaching promotion programs, and promote the improvement of the sports teaching levels [1].

1 Web and the Database Interconnection Technology

1.1 Web Technology

The Web technology, invented by the Geneva Particle Physics Laboratory, belongs to the hypertext technology. It can link any point of a file to any point of another file, so as to achieve the rapid information browsing. The Web technology has two standards. One is the Hypertext Transfer Protocol (HTTP) and the other is the Hypertext Markup Language (HTML). With these two presentation layer grammars, the clients of the Web are traditionally called the browser, and the text written in the HTML is the home page [2]. The next home page is transmitted through the HTTP protocol. In this case, the use of the CGI technology can make the application of the Web broader.

Z. Xu et al. (Eds.): CSIA 2019, AISC 928, pp. 49–54, 2020.
https://doi.org/10.1007/978-3-030-15235-2_8

1.2 Database Technology

The database processing is an important part of the computer applications. Data can be processed through the data collection, storage, processing and dissemination. The data management refers to the classification, organization, coding and storage of the data. The structure of the database system is constantly developing, and various unused structures meet different needs. First of all, we must determine the system architecture mode, and the most basic are the C/S structure and the B/S structure [3]. The advantage of the B/S is to simplify the system maintenance and upgrade, reduce the costs and expand the choice opportunities, but it may make the server run with the heavy data load. The C/S structure should go through at least three layers. The first layer is the client, and the second layer is the business logic server, while the third layer is the database [4]. The advantage of this structure is that it can store the data centrally, and the business logic and the security rules can be defined once on the server, while the backup and the recovery are easier, and the network traffic is reduced.

1.3 Interconnection Technology

There are many kinds of the Web and the database interconnection technologies, including the general gateway interface, the PHP, the dynamic server pages, and the ASP.NET and other technologies. The CGI is the interface between the Web and the database. The main function of CGI is to transmit the customer requirements to the database form the homepage, and then send it to the customers by the Web server [5]. This kind of the intervention makes the Web more powerful and can build the dynamic Web pages. Although it is inefficient, it is the first one to use this technology. The PHP can support multiple platforms. The extended PHP can communicate with other network teaching. The dynamic server pages are abbreviated as ASP, which can support a variety of the scripting languages. Therefore, they can quickly complete the application of the website. They are compatible with many languages. They use the simple text editors and contain six built-in objects. The ASP.NET provides the powerful functions, which can be used to develop the programs to support the exception control and the type safety and so on.

2 Design Background of the School Sports Management System

The WEB teaching plays an important role in the physical education. According to the different levels of students, it can make the appropriate WEB teaching mechanisms to meet the needs of the students at different levels. You can learn without occupying the class time, and as long as you can surf the Internet, you can learn the knowledge and skills of the relevant parties. The technical, tactical and physical training, the practice methods, the competition methods and the refereeing issues can be more comprehensively learnt and understood. Therefore, we can greatly increase the enthusiasms and initiatives of the sports enthusiasts, so that we can consciously and voluntarily use their knowledge to guide them to be engaged in more extracurricular physical exercises, improve their own sports levels and play the purpose of fitness.

In traditional physical education, teachers disseminate the knowledge and technologies to the students. Because teachers always play a leading role and students are in a passive role, the teaching form between the teachers and the students is a relationship of "you tell and I listen, and you do and I practice", so that the students' learning efficiency is relatively low. Compared with the traditional teaching mode, the WEB teaching mechanism of the physical education takes the form of the words and graphics, combines audios, videos and animations, and displays the teaching contents in the three-dimensional way. Its manifestations and means are more abundant and flexible, which can fully reflect its unique advantages. The modern WEB teaching technology, which integrates texts, images, sounds, animations and the interactive network, can make the teaching process vivid and full of pictures and texts. At the same time, the use of the WEB teaching technology can establish the campus network, connect with the Internet, and enter the world teaching sharing system.

The traditional teaching mode divides the educational groups into the educators and the educatees. Educators use their knowledge carriers to convey the knowledge and develop the skills of the educatees by means of the media. Educators themselves participate in the teaching activities as the knowledge carriers. As no one can fully inherit the scientific and cultural heritage of the human beings, the carriers used by educators are limited to books, educational tools and teachers themselves. The way for the educators to acquire the knowledge is relatively simple. Teachers emerge as the main subjects in the teaching process. The control of the knowledge transmission, the selection of the teaching materials and the teaching methods are all decided by the teachers. Students can only passively receive the education under this mode.

With the continuous development of the society and the progress of the science and technology, the management efficiency of the higher education has been greatly improved. Especially with the expansion of the enrollment leading to the expansion of the scale of schools, the improvement of the management efficiency has important and practical significance for our daily operation. As one of the most important compulsory items in our education, and especially the changes of the current national sports system, the requirements of the sports management are also obviously strengthened. Because of the comparative advantages of the disciplines, the trend in the cross-disciplinary research has brought the potential huge room for the improvement of the school sports management. Therefore, how to design and develop an effective management system has become an urgent practical issue for the school sports management departments. The development of the sports management system can make the sports teaching and management informationized, highly efficient and standardized, which can further improve the efficiency of the sports management.

This paper studies how to improve the efficiency of the school physical education management by using the WEB. Through the design and development of the school physical education management system, it can provide an efficient and standardized management operation process for the school physical education teachers, thus helping them to complete the functions and objectives of the sports management. Through the in-depth analysis of the school physical education management, the JAVA technology and the SQL Server2008 database technology are used as the basic tools for the system development and application, and the system's expansibility and maintainability are improved. In the field of sports, the research of the large-scale management information system based on the B/S mode is less, which hinders the improvement of the overall level of the sports information in China to a certain extent. Therefore, it is necessary to study the management information system based on the B/S mode.

3 Design and Implementation of the School Sports Management System Based on the WEB

The main goal of the system development is to design and build an online course selection system which integrates the information, the network and the automation according to the characteristics of the school physical education. The system is designed in the B/S mode. With the support of the TCP/IP, the client accesses the technology and the structure of the Web server and the background database through the browser. This system mainly includes the design and implementation of the administrator-side functional module, the student-side functional module, the teacher-side functional module and the database design, the three-tier B/S system architecture, the system platform building, the user interface design, and the system security design and so on. The system realizes the office automation of the course selection management and can effectively promote the scientific level of the sports teaching management.

In the process of developing the school sports management system, firstly, we need to enrich the database of the teaching resources, arrange the sports-related resources through the network, books, and expert talks and so on, and store them in the database in a unified format. After confirming the database resources, we should design the functional modules of the school sports management system. Each functional module is an important part of the system, and we can draw the block diagram of the system structure. We design and develop the functions from the aspects of the system management, the course selection, the strengthening exercises, and the user login and so on. The design follows the modular design concept, which can facilitate the expansion and maintenance of the system in the future.

From the point of view of the functional module, the school sports management system uses the multi-level menu options. After the students and the teachers use their respective roles to carry out the system, they can complete the input and opening of the information by clicking on different options, and obtain the corresponding resources from the system. The school physical education management system adopts the web to design, and uses the MVC design concept to separate the display, the control and the model. At the same time, it combines the ASP web page technology to design the human-computer interaction pages, and integrates the physical education and the network organically to form a perfect sports teaching management system. The data mining technology is introduced into the database. By analyzing the data of the students' operation in the system and the results of the daily exercises, the direction and the weak links of the students' interests in sports are excavated, and the sports resources that the students are interested in are precisely pushed forward, and the differentiated counseling strategies can be formulated for the teachers to make a reference.

The users of the school sports management system are mainly divided into three rights, namely administrators, teachers and students. Users with different permissions will have different permissions to operate when they enter the system. For example, administrators are the maintainers of the entire school sports management system. They can operate on the curriculum management, the personnel permission setting, the management of the site facilities, the data addition, deletion and modification, and the group announcement information publication and so on. They have the highest permissions. After entering the system, the students can inquire about the courses they choose, modify their personal basic information, browse the physical education resources, and query the results, which mainly involve the operation of the students themselves. Teachers can grasp the students' learning situations in the real time in the system, and can formulate the different teaching promotion programs to promote the all-round development of the sports knowledge of the students.

The Web teaching assistant platform provides the abundant teaching resources, facilitates the classroom teaching, and also provides the teachers with the sufficient time to pay attention to the students' learning behaviors, and more time to adjust their teaching behaviors according to their learning conditions. Under the Web-assisted platform, "learning to teach" is the main characteristic of the teachers' teaching behaviors. The evaluation of the learning effect mainly uses the form of the self-evaluation or the mutual evaluation to evaluate the completion of the works, to explore the solutions to the typical problems, to exchange and summarize the experience, and to share the excellent works.

The database of the system fully considers the division of the data tables and their relations in the design process. It contains the following database tables: the teacher information table, the student information table, the teacher status, the student status, the curriculum information table, the student selection table, the class curriculum timetable, the grading question bank, the questionnaire question bank, various grading states, the teacher evaluation records, the teacher mutual evaluation records, the teacher self-evaluation record, the student evaluation records, the teacher score statistics, the questionnaire survey records, and the record of the suggestions and opinions. In the trial run stage of the system, the system functional test and the system performance test are mainly carried out. The purpose of the functional testing is to ensure that the functions of the software meet the requirements of the software. Based on different testing purposes, testers need a series of the tests, such as designing the language testing, the database testing, the form testing, the Cookies testing, and the link testing and so on. The performance testing is one of the most difficult steps in the entire testing process, which is aimed at the testing of the entire system, such as the connection performance test, the load test, the concurrent test, and the large data measurement test and so on.

4 Conclusion

With the progress of our society, people have gradually entered the information age. The Web technology not only plays an important role in people's daily life, but also plays an irreplaceable role in our sports teaching. At present, in terms of the teaching and management, the application level of the WEB technology is not very high, and the advantages of this technology in the field of the physical education have not been brought into full play. In order to gradually change this situation and let the Web technology play a greater role in our physical education, this paper focuses on the discussion of the development of the sports management system based on the Web technology.

Acknowledgement. The study is supported by the following projects: (1) "The Chinese disabled person sports culture in the construction of the rule of law practice and promoting research" (2016&ZX062). (2) The topic "independent college physical test study of students in Henan province" from the project of Henan Science and Technology department (2017A004). (3) The topic "Internet+" movement, under the background of mobile phone APP to join the university sports teaching evaluation effectiveness research (SKL-2018-2299).

References

1. Li L, Ma Q (2015) Development and application of the computer comprehensive management system for the school physical education based on network. Electron Test (01):115–116
2. Qiu B, Ding W, Ma Q (2015) Analysis of the development and application the school sports computer integrated management system. New Technol New Process (08):100–101
3. Lu B, Yin J, Gao S, Jin K, Liao Y, Ren Y (2017) Computer science education: opportunities for all – characteristics and enlightenment of K-12 computer science framework in the United States. E-educ Res (03):133–134
4. Zhong L, Luo Z (2016) Design and implementation of the web-based management system for the student attendance – taking the sports schools of Shenzhen City as an example. Inf Technol Educ (08):123–124
5. Kong Y (2017) Development and research of the school physical education computer teaching and the automation management system. Autom Instrum (09):134–135

Design and Research on the Chinese Medicine Health Management System Based on the Wireless Sensor Network

Xiang Gao[1], Liuyun Lin[2], Tianying Lan[1], and Xinyan Gan[1(✉)]

[1] Guangxi University of Chinese Medicine, Guangxi 530200, Nanning, China
thpanjsu@yeah.net
[2] The First Affiliated Hospital of Guangxi University of Chinese Medicine, Guangxi 530023, Nanning, China

Abstract. In view of the rapid growth of China's aging population and the rising demand for the medical services, it is more necessary to build a health management system of the traditional Chinese medicine based on the wireless sensor networks. This design combines the ZigBee technology, the city grid management and the virtual reality technology, to establish a real-time, accurate and fine health monitoring system, so as to comprehensively improve the efficiency, quality and level of people's health management.

Keywords: Wireless sensor network · ZigBee technology · Real-time monitoring · Health management

1 Introduction

The wireless sensor network is introduced into the health management of traditional Chinese medicine. According to the architecture and the performance standards of the wireless sensor networks, some special requirements for the health management in traditional Chinese medicine are proposed. A set of the experimental system is designed. On this basis, several network protocols and applications are developed, to realize the multi-hop routing, the multi-parameter acquisition and the network invulnerability [1]. The load location and the loosening diagnosis of the fastening connectors are accomplished by using these programs. Experiments show that the wireless sensor network has a broad application prospect in the health management of the traditional Chinese medicine [2].

2 The Wireless Sensor Network and the Health Management System

The health management testing system comes from bionics and is the common product of the interaction of various disciplines. It uses an embedded or surface-mounted sensor system as the nervous system, which can sense and predict the internal defects and damages of the structure [3]. With the development of sensors, micro-electro-

Z. Xu et al. (Eds.): CSIA 2019, AISC 928, pp. 55–61, 2020.
https://doi.org/10.1007/978-3-030-15235-2_9

mechanical systems and networks, the wireless sensor network, a new technology of the information acquisition and processing, has emerged. Many researchers predict that with the development of the semiconductor processors, the sensor network units can be reduced to the extent that they can be seen by a microscope [4]. The communication, the sensors using the microelectronics and the low-power semiconductor logic technologies can be integrated into a very small chip, which can spread to all kinds of the physical environment like dust, and even the surrounding energy can be used as the electricity.

The architecture of the wireless sensor networks consists of three main parts: the sensor nodes, the terminal nodes (Sink) and the observation objects. The sensor nodes scatter in the observation areas to collect the data related to the observation objects, and transmit the data after the cooperative processing to Sink. Sink can communicate with the task management node through the Internet or the communication satellite.

There are the following factors affecting the design of the routing protocols in the architecture of the wireless sensor networks:

2.1 Network Dynamics

Most of the network architectures assume that the sensor nodes are static while Sink is mobile. The moving or the stationary objects depend on the requirements of different applications.

2.2 Network Topology

There are two kinds of the topology configurations: the fixed and the self-organized types [5]. In the fixed cases, the nodes are manually configured, and the data is transmitted through a predetermined path. In the self-organizing cases, the nodes are randomly distributed in the adhoc mode.

2.3 Data Transmission Mode

According to different applications, the data transmission mode can be divided into the continuous mode, the event-driven mode, the query-driven mode and the hybrid mode.

2.4 Node Types

Usually all the sensor nodes are isomorphic. The heterogeneous sensor nodes also exist in the sensor applications requiring different functions. Recently, it has been proposed to use the special energy-limited nodes as the sensor nodes with three functions: forwarding, sensing and aggregating data.

2.5 Path Selection

There are two choices of the multi-hop and the single-hop. The transmission energy of the radio frequency is proportional to the square of the distance, and the energy consumption of the multi-hop path is less than that of the single-hop path, so the multi-hop path is often used. However, due to the overhead of the topology management and the link connection of the multi-hop paths, the single-hop paths are more effective when the distance between the sensor nodes and the Sink nodes is short.

To evaluate the success of the routing design in a wireless sensor network, the following performance criteria are often used:

Energy Efficiency/Life Cycle
The energy efficiency is an important factor to be considered in the design of the sensor networks. Reducing energy consumption as much as possible and prolonging the network life cycle are the primary goals of the network design.

Reliability/Fault Tolerance
The sensor nodes are vulnerable to the failure due to the energy depletion or the environmental interference. The failure of some sensor nodes should not affect the tasks of the entire network.

Scalability
In some applications, hundreds of the sensor nodes may be required, and the routing design should be able to meet the collaboration of a large number of the nodes.

Delay
The delay time in the sensor networks refers to the time it takes for an observer to send a request to receive the response information, and the delay must be minimized as much as possible.

3 Special Requirements of the Wireless Sensor Networks in Traditional Chinese Medicine Health Management

The ideal Chinese medicine health management method should be able to accurately detect the damages in the early stage, locate and determine the extent of the damages, and then provide the safety assessment of the structure, and predict the residual life of the damaged structures. In this particular field, the application of the wireless sensor networks has the following characteristics:

The health monitoring is to distribute multiple sensors in the key parts of the objects under the test, so once the sensor nodes are arranged, they will not move. At this time, without considering the mobility of the nodes, the routing nodes can be manually configured, and the data can be sent according to the predetermined path, without the self-organization to form the network topology and routing.

The real-time and accurate acquisition of the sensing signals and the effective analysis of the signals are the keys to the whole work. Therefore, the network needs the fast and real-time response and the reliable communication, and we should complete the data acquisition and the preliminary processing as far as possible in the possibly shortest time, and sent them to the PCs.

In some cases, in order to accomplish the task of the health monitoring, the multi-parameter acquisition is even needed in some key parts (different parameters may be collected in different regions, such as the Lamb wave, lights, strains, displacement, pressure, acceleration and temperature and so on, or a node may need to complete the acquisition of the multiple parameters). Therefore, the wireless sensor networks should have the function of the selective multi-parameter acquisition according to needs.

4 Design of the Chinese Medicine Health Management System Based on the Wireless Sensor Network

4.1 Technical Study

ZigBee Technology

The ZigBee technology is a kind of the wireless communication technology developed in recent years, which is suitable to connect various electronic devices with the short distance, the simple structure, the low transmission rate and the low cost. ZigBee's basic protocol is based on IEEE 802.15.4. The whole protocol architecture is divided into four layers: the physical layer, the MAC layer, the network layer and the application layer. It supports three basic network structures: the star, the network and the tree. The good scalability and security make it widely used in the industry, commerce, agriculture, health care, family, video game industry and many other fields.

The wireless sensor network contains tens of millions of the sensor nodes. The high-density distribution makes the network have the highly reliable and stable data transmission capability. The combination of the ZigBee technology and the wireless sensor network will expand the application fields of both sides infinitely. The node devices in the ZigBee network can be divided into two categories: the full-function devices and the semi-functional devices. The full-function devices have the functions of all types of the devices in the network and can be converted at will. Generally, they are used as routers, coordinators or terminal devices. The function of the ZigBee coordinator is to build the wireless network, select the idle channel and a panid to build the basic network, wait for the request of the peripheral sensor nodes to join, and establish the security management in the network layer to respond to the various binding requests of the corresponding layer. After the completion of the network construction, the function of the coordinator is changed into the router, which collects the data from the nodes and forwards it to the gateway nodes through the routing algorithm calculation, and uploads it to the management system. The router expands the coverage of the network, allows the multiple nodes to join the network, and can continuously expand the transmission range of the signal in the process of the work. The terminal device is in a dormant state when it is not working. When the detection target is found, it is activated. Compared with the minimal power consumption, it can work for half a year to two years.

Design of the Wireless Sensor Health Management System

The system design combines the advanced computer technologies and the scientific management means, and injects the humanistic care factors to mobilize all the human and material resources around it. Set up a self-checking network within a certain range, establish a separate management model for each person, and monitor their daily behaviors, medical care, and basic health parameters (such as the blood pressure, the body temperature, and the pulse and so on). And compare them with the basic values set by the system, and analyze, predict and warn the potential risk factors, so as to improve the quality of the life of the population in an all-round way and better safeguard the healthy development of people's physics and psychologies.

The wireless sensor health management system is an important part of the medical security of people's lives. It is mainly based on the ZigBee technology and integrates the embedded technology, the communication technology and the computer technology. It comprehensively monitors people's daily behaviors, health indicators, psychological qualities, disease development and other parameters. It also analyses and compares the results of the generated data and presents the early warning reports, to achieve the goal of improving the quality of our life and ensuring our physical health and longevity. The focus of the study is to monitor the vital signs and the movement status of the human beings, and to put forward the corresponding healthy lifestyle plans or send out the alarm signals in time when abnormal situations occur.

The whole system mainly includes the wireless sensor network, the grid management platform, the intelligent health monitoring system and the visual service interface. The wireless sensor networks include the acquisition nodes and the sensor nodes. The coordinator controls the processing of the data collected by each node to establish the binding. Other routers or sensor device nodes automatically discover the network and join the network. People's vital sign data collected by the sensor device nodes are transmitted through the GPRS module or the external network to the PC ends, and will be parsed by the upper computer software. The parsed data are displayed graphically on the PCs for the managers of the monitoring center to check people's health status and put forward a targeted early warning program.

4.2 Design and Implementation of the System Hardware

The hardware of the system mainly includes the ZigBee wireless sensor nodes, PCs, the GPRS modules, the network cards and the embedded server. The establishment of the ZigBee wireless network mainly includes two types: the acquisition nodes and the sensor nodes. There are two main functions of the acquisition nodes. First, as a coordinator, the channel is chosen to establish a new ZigBee wireless network, and second, as a router, it establishes a connection with the PCs, receives the instructions from the PC software, and sends the node information back to the PCs for analysis and processing. As the terminal device, the sensor nodes are designed to be portable. The network detects that the terminal device will automatically bind the nodes, collect the pulse, the blood pressure, the body temperature and other data in real time, send the data through the GPRS module or the gateway, and display them on the PC software for the professional analysis and processing.

The formation of the network is mainly through ZigBee coordinator, to write hardware drivers in the physical layer and the control layer to complete two levels of the control operations, and prepare the coding of the network layer and the application layer. When the user presses the xm1 sensor node to be set as a coordinator, the network is initialized and the data is exchanged with the gateway using the universal interface. When not set, the node completes the routing function as a routing device. When the keyboard presses xm2, the network is initialized first, and each acquisition node is bound to the coordinator self-built network. When the data is collected, it is used as a terminal device. When the data needs to be jumped, the node is configured as a router to realize the function of addressing and forwarding the messages.

Each hardware node device has its own parameters. The configuration parameters of the devices in the same coordinator self-built network are the same. In the networking, the terminal device sends a request to wait for a response, allowing the joining system to automatically assign the address to the terminal node (16-bit short address or 64-bit long address). When all the terminal nodes join the network, the upper position machine can send out instructions to collect the data information.

4.3 Design and Implementation of the System Software

Based on the grid technology, the information management platform in the system divides the space information, the time information and the management information into the standard cells, establishes the relatively independent supervision, the early warning and the assistance mode, and actively uses the vital sign tester and sensor to track and detect the human vital signs and behaviors. The establishment of the platform makes full use of the simulation technologies to build an integrated unit grid data, the three-dimensional space environment data, the component management data, the location coding data and other data resources to simulate the real environment. At the same time, it combines the wireless sensor network, GPS, the remote warning, the virtual reality and other technologies to establish a scientific and efficient and integrated information management platform with the friendly interface.

In the grid management and the health surveillance of the science and technology community, the visual management of the system has attracted more and more attention. This system mainly uses the latest development results of the virtual reality technology integrating the computer graphics, the computer simulation, the artificial intelligence, the sensor technology, the display technology, and the network parallel processing technology and so on. It designs and develops a visual health monitoring management system, which runs on the PCs of the management side.

5 Conclusion

In this paper, we build a wireless sensor health management system to ensure that people in the community can get the timely help when they encounter unexpected situations in their daily life and sports, and send out the disease warning signals through the statistical analysis of the daily health sign data of people, and ensure the healthy development of the human body combined with effective lifestyle intervention programs, which has a long-term significance to the development of the community medical and health undertakings.

Acknowledgment. Foundation Project: Guangxi Key Research and Development Plan (Guilin Science AB18126099), Development and Demonstration Application of Guangxi "Internet+" TCM Disease Prevention and Health Management Service Cloud Platform.

References

1. Zhou M, Xu K (2017) Conception of the construction of "Internet + Health Management" platform. Electron J Clin Med Lit 4(65):12861–12862
2. Wang B (2008) Intelligent Health Management for Attention. China Computer World, 27 August 2012
3. Chang Q, Yuan S, Qiu L, Dong C, Liang D, Hu B (2012) Structural health management system based on the knowledge and wireless sensor networks. Vib Testing Diagn 32(01):105–110+165
4. Chang Q (2011) Research on the key technologies of the knowledge-based structural health management system. Nanjing University of Aeronautics and Astronautics
5. Fan C, Yu G (2011) The remote health management based on the internet of things. China Digit Med 6(01):58–60

Design of the In-depth Intelligence Education System Based on the Intercultural Communication

Minzi Song(✉)

School of Foreign Languages, Chongqing Technology and Business University, Chongqing 400067, China
taojiangzj@126.com

Abstract. With the globalization of the economy and the informationization of the society, more and more attention has been paid to the cultivation of the foreign language talents' intercultural communicative competence. The foreign language teaching model based on the intelligent education system has also become an important research topic in the reform of the foreign language education in colleges and universities. Especially the popularization of the information technology in the field of our education provides more diversified ways and technical support for the design of the in-depth intelligent education system. From the perspective of the intercultural communication, this paper studies and analyzes the design of the in-depth intelligence education system.

Keywords: Intercultural communication · In-depth intelligence education system · Design

1 Introduction

With the rapid development of the network technology nowadays, more and more students begin to use the Internet to carry out their learning. With the help of the mobility, interaction and dynamics of the Internet, an in-depth intelligent education system of the cross-cultural communication is constructed, which can not only create a vivid language learning situation for students, but can also enhance their flexibility, individuality and interests in learning. The in-depth intelligence education system can make up for the shortage of the students' communicative activities in the traditional education, promote the perfection of the current foreign language teaching contents and forms, and integrate the teaching resources, thus improving the cross-cultural communicative competence of the foreign language talents [1].

2 Connotation and Importance of the Intercultural Competence

In such an era of the economic globalization and the cultural diversity, the cultural exchanges between countries become more frequent. With the implementation of the strategy of the economic development along the Belt and Road Initiative, our country

Z. Xu et al. (Eds.): CSIA 2019, AISC 928, pp. 62–67, 2020.
https://doi.org/10.1007/978-3-030-15235-2_10

pays more attention to the cultivation of the intercultural communication talents [2]. Major universities in China are also constructing the education platforms of the cross-cultural communication.

2.1 Overview of the Cross-Cultural Competence

Students' cross-cultural training is defined as: on the basis of combining the individual differences in their real life and learning, students can combine the language and the cultural expertise through teachers' constant encouragement and guidance, and then gradually form their own learning styles [3]. This process is not only a cross-cultural knowledge accumulation, but also a process of building the cross-cultural attitudes, and ultimately students can form an effective thinking system, which can accurately analyze and solve the problems in the cross-cultural situations.

2.2 Importance of the Intercultural Competence Training

The cultivation of the students' intercultural communicative competence can help students to form the correct cultural values, form the ideological concept of the humanistic care in the trend of the multi-cultural integration, and have a certain sense of social mission. The cultivation of the cross-cultural competence cannot be achieved overnight, but requires a process. This process is a combination of the languages, cultures and other contents, and is closely related to the situations. A person with the cross-cultural competence can not only respect other people's cultural identity, but can also understand the cultural feelings and values of different cultures from the right perspectives, to achieve the cross-cultural communication in the true sense.

3 Design of the In-depth Intelligence Education System for the Intercultural Communication

When carrying out the cross-cultural teaching in colleges and universities, we should not simply rely on some fixed courses, but should achieve it relying on the advanced teaching methods and information technology. Through the construction of the in-depth intelligence education system and the cross-cultural communication projects, we can achieve the cross-cultural ability of the students and the accumulation of knowledge.

3.1 Framework Design of the Intelligent Education

Based on the in-depth intelligent education system from the perspective of the cross-cultural communication, the educational framework of the intelligent education system is the overall direction of the system. Its design is particularly important. The intelligent education system relying on the information technology is of great help to the cultivation of the students' cross-cultural communication ability.

System Implementation
In the design of the intelligent education system based on the cross-cultural communication, the convenience of the operation between the students and the teachers, that is, the operability of the system, should be considered first.

First of all, it is necessary for students to log on to the cross-cultural communication education platform and get the relevant foreign language learning materials and data from it. The system is designed to run through the web pages and servers, and has access to the service ports, so that students can directly click on the corresponding pages and have access to the web pages at any time to carry out the autonomous learning, or to complete the teacher's assignment tasks. Through the web pages, we can query, play or download all the videos and information in the system, to implement the language communication practice.

Secondly, it is convenient for the teachers to maintain and update later, and upload the relevant educational videos and data. Teachers can implement related operation of the video corpus through the web pages.

Finally, it is the timeliness of the system upgrade and the corpus updating. Administrators can achieve the updating of the corpus in two ways. After the administrators upload the video data, the system JV-Finder can automatically analyze the relevant videos and store the data with the context of the system and the related search, so that the video corpus can be well and completely stored in the database.

Teacher Client

Video Uploading
In this system, all the video corpus sources are from the teachers. Teachers collect some useful video corpus and upload the video corpus to the system. When uploading files, teachers first need to ensure that the file format is MP4, and its subtitle format is SRT or ASS. Only such files can be uploaded to the system, which is also convenient for the teachers to upload. Teachers can upload the video corpus according to their own convenience, as long as there are two ways to upload: one is to upload through the client, and the other is to upload by choosing the local application client.

When the teachers upload the video corpus, in the uploading interface, there are several names, which are Path; Video, Duration, and URL. These are the upload paths of the caption files, the names, duration and paths of videos. After the teachers have input the information, click the corresponding button to upload the videos.

Video Corpus Subtitle Processing
In the teaching process of the teachers, through the video upload function, through the system function, the teaching contents are sorted, which are stored in an effective retrieval. In the intelligent education system, the effective words, parts of speech and pronunciation labels are stored, which are arranged through the effective sequence, to achieve the convenience of the retrieval.

The design process of the intelligent education system is based on the JAVA language, which is a foreign language modal parser developed in the C language. Through the effective processing, it can provide a more word-based function to classify the foreign sentences in accordance with the learning usage, so as to construct a cross-cultural communication language learning system. In this process, through the effective design arrangements, we can obtain more effective teaching contents, including the language pronunciation, continuation, and the important meanings and attributes of the relevant foreign language vocabularies.

Through the systematic processing, the video corpus and the subtitles are processed more conveniently. With the use of the effective design, we can realize the subtitle word segmentation, and can extract the word segmentation results, including the study-related information, which can provide more convenience for the students to retrieve, to construct the richer retrieval conditions, thus realizing the effectiveness of the teaching and the study.

For example, in the process of learning the word types by the students, including verbs, adjectives, and adjectival nouns, through a more detailed search, we can achieve a scientific system design. At the same time, according to the corresponding relationship between the texts and time in the subtitle files, the retrieval information of the words can be better divided and their effective association can be realized. In addition, in the subtitle files, the Chinese vocabulary and the foreign words correspond to each other, realizing the convenience of translation. Through this deep design, the teaching task is effectively carried out. Students can learn more easily through the flexible use of words, pronunciation and morphology, and at the same time improve their learning efficiency. Their learning contents can be more easily grasped. In this process, the deep intelligent education system under the cross-cultural communication also realizes the convenience of the fuzzy condition query and the multiple retrieval methods.

Subtitle and Video Matching

The videos exist in the system mainly in the form of the multimedia files, which cannot be combined with the videos directly by the keywords to a certain extent, but the matching of the subtitles and the videos forms an effective corresponding relationship. Through the keywords, we can find the location of the subtitles, and combined with the location of the subtitles, we can better intercept the relevant video clips. In this way, the learning process becomes more convenient, and it can be better directed learning. It can be said that after the in-depth development of the system design, the perfect matching between the subtitles and the videos can be achieved, and the scientific positioning can be achieved, which is very conducive to the students' better learning of the effective foreign language knowledge.

Video Online Play

JW Player web media player can play the related types of videos on-line, and the media supported by Adobe Flash Player can be played on-line. The common playback forms include: MP4/MP3 file format, HTTP video type, flexible setting to play a variety of the real-time video streams, and broadcasting a variety of the list formats, and supporting the Javascript API system.

In addition, it can also extend the player to meet the diverse appearance needs and expand the playback functionality. Then it provides data for analysis, reference and utilization, and provides convenience for the data sharing, searching and storage.

Video Label Management

After the video data is uploaded, it can be classified by the administrator operation after uploading the anticipation and processing the subtitles. The common classification criteria include dialog, function, scene, relation, and free classes. According to the standard classification, the administrator annotates the label, which is convenient for the users to extract and invoke.

3.2 Student Client

In order to give full play to the convenience and efficiency of the retrieval function, the system provides the students with a concise and comprehensive retrieval function interface in the process of the continuous improvement, so that students can search and consult the relevant information instantly through the browser, and have access to JV-Finder instantly. In this system, the student client provides the keyword search captions, the video playback retrieval, the download page display video sources, the adjustable video playback time and other functions for the learning users and the students can set through the related operations. The video playback length can be determined according to the video playback time, and the sentence length can be adjusted. In general, the default value of the system time is 26 s, while the longest video time is 60 s around the keyword. Search and test the keywords can get the relevant contents of each of the 10 sentences before and after the sentence, and the system default setting is the two sentences before and after the standard. When playing the video clips, students' page displays the scenes, includes both the foreign language and the bilingual subtitles. The key words are displayed in the red standard, and the download function of the video clips is provided for the students.

4 Reflective Assessment of the In-depth Intelligence Education System in Intercultural Communication

With the help of the information technologies, we can achieve the design of the cross-cultural communication depth intelligent education system, whose educational effect and teaching situation need to be tested in practice by educators. Therefore, combined with the above system design, the following is a reflective assessment of it:

Through guiding the students to participate in the cross-cultural communication activities, give full play to the values and create the values, so that a comprehensive evaluation of the students is formed. The evaluation contents include: the interdisciplinary knowledge accumulation, the cultural awareness formation, the self-learning attitude, the mastery of skills, the critical thinking, the communication and communication skills. Through the judgment of the multiple factors, this paper investigates and evaluates the completion of the students' cross-cultural communication training, and judges the students' potential and creativity.

When conducting the cross-cultural teaching and learning, teachers should select the commentary words to evaluate the students' gradual accomplishments, and evaluate the students' accomplishments of their cross-cultural communication learning tasks by comparing the results of the stages with the final report. In this process, students are the main participants in the cross-cultural communication activities, and designers are the organizers. Therefore, in the design of the in-depth intelligent education system, the teaching activities should be formulated according to the real levels of the knowledge accumulation and abilities. After continuous discussion and reflection, various operations should be carried out, which can be monitored and adjusted immediately and at any time. Adjust the teaching progress, to ensure the completion rate of the students' cross-cultural communicative competence.

Through the construction of the formative and reflective evaluation mechanism, students can deeply understand the connotation of the cross-cultural education, enhance the effectiveness of the theoretical teaching, and enhance their self-reflection, self-cognition, self-reconstruction and self-improvement abilities. Using the reflective method, stimulate the students' learning motivation and mobilize the students' learning consciousness. The common forms of the reflective assessment include the self-assessment, the student-to-student assessment, the teacher assessment, and the establishment of a comprehensive mechanism for the reflective assessment with reference to the research journals and writing the teaching activity documents. It can be seen that the use of the formative assessment and reflective assessment can greatly stimulate the students' internal motivation to learn, so that students can improve their ability on the basis of a full understanding of themselves, truly and accurately grasp their own learning state, have the targeted solutions to their learning problems, and enhance the comprehensive abilities of their cross-cultural learning.

5 Conclusion

Based on the concept of the cross-cultural education, the cultivation model of the in-depth intelligent talents is characterized by interdisciplinary, open and reflective characteristics, focusing on stimulating the students' intrinsic learning motivation, guiding the students to integrate languages, cultures and professional knowledge organically and creating individualized cross-culture through the construction of a comprehensive educational action plan for the comprehensive evaluation, so as to help the students understand themselves more deeply, and gradually improve their cross-cultural comprehensive abilities, to achieve the construction of their self-concept and their social and cultural identity.

Acknowledgment. Project foundation: 2017 Chongqing social science planning project, Project Grant No.: 2017WYZX01.

References

1. Zhang Y (2018) Design of the business foreign language teaching based on the intercultural communication competence. J Heilongjiang Inst Educ (09):145–147
2. Xu L (2018) Application of the reflective teaching model in the course of the intercultural communication in higher vocational colleges. J Mudanjiang Univ (09):141–143+148
3. Bao W, Zhou H, Lu W, Xie F (2016) The system of knowledge management using web based learning and education technology. Comput Syst Sci Eng 31(6):469–473

Design of the Network Security Intrusion Detection System Based on the Cloud Computing

Meng Di(✉)

University of Science and Technology Liaoning,
Anshan 114001, Liaoning, China
coolidl@126.com

Abstract. The intrusion detection system usually first carries out the real-time analysis and collection through the core data information of the computer host system and network, and then makes the correct judgment and the corresponding reflection on the abuse of the resources by some legitimate users and the intrusion behaviors of the illegal users. On the basis of the traditional network security technology, the intrusion detection system has completed the response and detection, played a full defensive function, dealt with the network security incidents, realized the transition from the post-discovery to the pre-warning and the automatic response, and also provided the more effective evidence to investigate the legal responsibility of the intruders. The emergence of this technology means that the research in the field of the network security has entered a new era.

Keywords: Cloud computing · Network security · Intrusion detection system · Design mechanism

The intrusion detection technologies used in the computer network security applications mainly include the host-based and the network-based intrusion detection systems. The main detection object of the host-based intrusion detection system is the computer. By setting up the host in advance, the host can be judged whether it has received the illegal intrusion or not according to its running state and the related parameters [1]. The host-based intrusion detection system can judge whether the current attack is successful or not, to provide a reliable basis for the host to take the appropriate measures. The intrusion detection systems based on the intrusion networks usually set up the multiple security points.

1 Network Security Problems Faced by the Cloud Computing

When it comes to the cloud computing, the security issues cannot be avoided. In fact, this is the biggest problem encountered in the process of the popularization of the cloud computing applications. Although the cloud computing service providers are trying to dilute or avoid this topic, as the end users of the cloud computing, this is exactly a

Z. Xu et al. (Eds.): CSIA 2019, AISC 928, pp. 68–73, 2020.
https://doi.org/10.1007/978-3-030-15235-2_11

major focus of their attention [2]. At present, the commercial value of the cloud computing has been confirmed, and at the same time, these "clouds" have begun to become the targets of hackers or malicious organizations. In a word, with the development and success of the cloud computing, the security problems of the cloud computing are becoming more and more worrying, which are embodied in the following aspects.

The cloud computing mode determines that a large amount of the data of users should be stored in the cloud, which can reduce their investment in the IT equipment and resources, but can also bring a variety of conveniences. However, the more the data stored in the "cloud", the greater the dependence on the cloud will be, and once the cloud data is damaged or lost, the loss to the users will be very huge.

In general, the enterprise IDC keeps a large number of the private data, which often represents the core competitiveness of the enterprises, such as the customer information, the financial information, and the key business processes and so on. In the cloud computing mode, when enterprises transmit the data to the cloud computing service providers through the network for processing, they face several problems: first, how to ensure that the data of the enterprises are strictly encrypted and not stolen in the process of the network transmission; second, how to ensure that the cloud computing service providers do not disclose the confidential data when they obtain the data; third, how to ensure that the access users are strictly authorized and legitimate the data access and ensure that enterprises can access their own data safely at any time when storing in the cloud computing service providers.

In the cloud computing environment, how can the cloud computing providers provide the necessary data support while ensuring that they do not bring risks and interference to the data calculation of other enterprises, so as to assist the third-party organizations to audit the security and accuracy of the data generation and achieve the compliance requirements of enterprises? In the process of the business sustainability development certification of the enterprises by the cloud computing services, there is also a potential security risk in how to ensure that the cloud computing service providers can not only provide the effective data, but can also not harm the interests of other existing customers, so that the enterprises can choose a long-term cloud computing service providers with the technological strength for their business delivery [3].

2 Analysis of the Common Intrusion Detection Technologies for the Network Security

The intrusion detection system mainly refers to the monitoring of the computer network system, collecting and analyzing the information by using its main key points, and then discovering some behaviors that are contrary to the security strategy, or finding some signs of the attack, and making the timely and automatic response [4]. Usually its functions are: monitoring and analyzing the behaviors of the systems and the users, timely auditing and checking the configuration and vulnerabilities of the system, evaluating the integrity of the data files and systems, identifying and statistical analysis of the known attacks or patterns, tracking and auditing the operating system, and managing and identifying some user behaviors that are contrary to the security policy and so on [5]. The intrusion detection system plays an important role in the computer network security: the anomaly detection and the misuse detection.

In the statistical anomaly detection, the main activities can be observed by the anomaly detector, and then the critical behavior contours of these activities can be depicted. The statistical analysis of the anomaly detection methods are as follows. The first is the operational model. This model assumes that the anomalies can be obtained by comparing the measured results with some fixed indicators. The fixed indicators can be obtained by the empirical values or the statistical averages over a period of time [6]. For example, the multiple failed logins in a short time are likely to be the password attempt attacks.

The second is the variance. Calculate the variance of the parameters and set the confidence interval. When the measured value exceeds the confidence interval, it indicates that it may be the abnormal behavior. The third is the multivariate model. The extension of the operation model, through the simultaneous analysis of the multiple parameters, achieves the detection. The fourth is the time series analysis. The resource consumption is sequenced according to the time and the event counts. If a low-probability event occurs in a certain period of time, it can be preliminarily determined as a possible invasion. The fifth is the Markov process model. The first line defines the state of the system as each type of the event, and then the change of the state is represented by the state transition matrix, which may be abnormal when the probability of the transition is small or the event of the defined type occurs.

The anomaly detection of the neural network mainly trains some continuous information units in the neural network. In the input layer, there are the users' existing input commands and the commands after the operations have been completed. The commands completed by the users can be used by the neural network, and then the users' next input commands are pre-processed. There are three stages in the training of the neural networks. (1) The network construction and training: In the training stage, the weights of the network are adjusted according to the errors of the actual output mode and the expected output mode of the neural network. (2) The network pruning: Delete the connection between the redundant hidden layer nodes and the redundant nodes. (3) Rule extraction: That is, extract the classification rules from the pruned networks.

The anomaly detection of the feature selection mainly selects a subset of the main measures of some intrusions from a set of the measures, and then classifies or predicts the intrusions that have been effectively detected. The key problem of this method is to make an effective judgment on the intrusion and the anomaly activities, but it is difficult to make the judgment consistent with the actual measurement. Because the proper selection of the measurement subset is mainly based on all the intrusion types that have been effectively detected, it is difficult for all the intrusion classes to have a certain measurement set. It is also difficult to detect some intrusive behaviors in the special environments by some measures that have been determined beforehand. The construction of the feature subset is as follows. The construction method of feature subset is as follows. Assume that there are n measures related to the intrusion potential, and the number of the subsets constituted by these n measures is 2n. Because of the exponential relationship between the metric and the search space, it is ineffective to find the best degree of the quantum set. Maccabe can propose a genetic method to search the whole quantum space and find the correct quantum set. The method is to use the learning classifier to generate the gene mutation operator and the crossover operator, to remove the lower predictive invasive quantum set, and to use the intensity quantum set generated by the genetic operator. Combining this method with a higher predictive quantum set, it is more effective than other heuristic search techniques in the allowable search space.

The anomaly detection of the pattern prediction requires certain assumptions. That is, the sequence of events must be the regular and discernible patterns rather than random. This method mainly considers the relationship and the sequence of the events. By observing the user's behaviors, a rule set is summarized, and then the main outline framework of the users is constructed. If the observed sequence of the events deviates from the subsequent events, it indicates that the user's operation is abnormal.

3 Design of the Network Security Intrusion Detection System Based on the Cloud Computing

Generally speaking, the network intrusion detection system includes the multi-level architecture, that is, the agent, the control and the management layer. The control layer assumes the functions of collecting the information from the agents and displaying the attacked items, so as to realize the management and configuration of the agent. The agent undertakes the monitoring function of the network data package, and sends the detected data information and the attack behaviors to the management.

3.1 Design of the Data Source Acquisition Module

In the design of the data source acquisition module, the author realizes the function of the data source acquisition module based on the packet acquisition mechanism of Libpcap. This design refers to the design structures of sniffer, Snort and other sniffers in the network. In the data source acquisition module involved by the author, it can realize the functions without affecting the normal protocol stack processing of the data package by the operating system.

3.2 Design of the Data Preprocessing Nodule

In the design of the data preprocessing module, the author refers to the KD99 data set, which comes from the Lincoln Laboratory of Massachusetts Institute of Technology and is specially developed for the intrusion detection evaluation project. This data set has denial of the service attacks, the detection attacks, the unauthorized access from the remote hosts and the unauthorized locality, the super user privileged access and other four categories of 22 kinds of the intrusion behaviors. Combined with this information, the author uses the process of calculating the average value attributes of the feature variables, calculating the mean variance of the sample features and normalizing to standardize the data processing, which is based on the python processing language.

3.3 Design of the Feature Extraction Module

In the design of the feature extraction module, the author applies the basic principle of the principal component analysis, which can reduce the correlation and the redundant information between the original indicators, and reduce the amount of the calculation in the follow-up work. The manager manages all kinds of the alarms and logs, displays the detected attack information and the security information, responds to the attack warning and the configuration information, executes all kinds of the commands issued by the console effectively, and transmits the warning attack information issued by the agent to the console, and finally completes the entire measuring process of the intrusion detection. According to this process, we formulate a scientific application strategy of the intrusion detection technology.

3.4 Design of the Neural Network Training and the Detection Module

As the core of the computer network intrusion detection system studied in this paper, in the design of the training and detection module of the neural network, the author refers to the design of the Kohonen neural network, and improves the BP algorithm neural network according to this design. This improvement can be achieved mainly through the application of the additional momentum method, the adaptive learning rate method, the conjugate gradient method and the quasi-Newton algorithm.

3.5 Design of the System Response Module

In the design of the system response module, the author designs two kinds of the corresponding modes, the active response mode and the passive response mode. Among them, the active response mode can counterattack after detecting the network attacks. The author also designs two kinds of the modes for this system: cutting off the network connection with the attacker and filtering the packet from the intruder's IP address, so that we can better ensure the practicability of the entire system. The intrusion management system is a unified platform for the integrated management, from the system level to solve the intrusion behaviors. The intrusion management technology is a process, which considers what vulnerabilities exist in the network or what kind of intrusion risks it faces before its behavior occurs. We should not only detect the intrusions, but also actively interrupt and terminate them. After the occurrence of the intrusions, we should analyze the intrusions in depth to determine whether there is another attack.

4 Conclusion

The intrusion detection technology of the computer network security is an important means to maintain the computer network security. It collects and analyses the files and data in the computer system, detects the intrusions that violate the computer network security, and alarms and blocks these intrusions. The intrusion detection technology mainly maintains the network security, analyzes and monitors the system and the user activities, audits the weaknesses and structures of the system, implements the response recognition and reporting of the known attack mode and activities, analyzes the statistical abnormal behaviors, evaluates the data files and the important system integrity, tracks the implementation of the audit operating system management, and identifies the activities that violate the security and other actions, to ensure the reliability and security of the computer network systems.

References

1. Shen Q, Wang J, Xu L, Peng H, Yang W, Hou YX (2017) Innovation design and system construction of the exhibition service supply chain – based on the "Internet+" and the "cloud" computing angles. J Commer Econ (12):108–109 (2017)
2. Wang N, Shi H, Wang Z, Ge S (2018) Information security threat response behavior – an empirical study based on the cloud computing context. Syst Eng-Theory Methodol Appl (08):122–123
3. Wang Y, Jing L, Zhao Z (2018) Effective combination and application of the sensor technology based on the big data, the cloud computing and the internet of things. China Comput Commun (Theor Ed) (11):123–124
4. Wu Z, Chen Y, Chen M (2018) Attribute-based outsourcing signature scheme with the fixed signature length in the cloud computing environment. J Nanjing Univ Posts Telecommun (Nat Sci Ed) (11):106–107
5. Zhou Q, Liu R (2016) Strategy optimization of resource scheduling based on cluster rendering. Clust Comput 19(4):2109–2117
6. Wang Y, Shi P, Liu L, Zhou Y (2018) Research on the resource allocation optimization algorithms of the equivalent golden section based on the green cloud computing platform. J Chifeng Univ (Nat Sci Ed) (10):131–132

Design of the Small Microstrip Antenna Based on the WLAN/UWB Wireless Communication System

Jiangyun Qiu(✉)

Hangzhou Dianzi University, Hangzhou 310000, Zhejiang, China
lingyuxu0215@yeah.net

Abstract. The antenna, as the component of receiving and sending signals in the front end of the wireless communication system, plays an important role in the communication quality, and is often the main obstacle to the development of the miniaturization of the wireless communication system. The microstrip antenna is composed of a metal radiation patch which can be pasted on the dielectric substrate. The choice of the shape of the metal patch can be arbitrary. The conventional shape is usually chosen as the metal patch to simplify the analysis and prediction of its performance. The dielectric constant of the microstrip antenna substrate is relatively low in order to enhance the field intensity of the edge radiation field. However, the dielectric constant of the substrate needs to be greater than 5 in order to meet the needs of other performance of the microstrip antenna, so that the radiation edge effect cannot be enhanced, resulting in the greater contradictions.

Keywords: WLAN technology · UWB technology · Wireless communication system · Small microstrip antenna design

As one of the indispensable components in the field of the wireless communication, the antenna will have the most direct impact on the performance of the wireless communication system. Some of the traditional antennas have the good transmission performance, but their size is too large to meet the needs of the contemporary miniaturization development. The microstrip antenna has many advantages, such as the light weight, the low cost, the small size, the easy manufacture and the integration with other RF microwave circuits. It has great application value and has become one of the hottest topics in the antenna research.

1 Technical Characteristics of the WLAN/UWB Wireless Communication System

With the progress of our mankind and the rapid development of the science and technology, the information and the knowledge have become the strategic resources and the basic elements of the social and economic development in people's daily life and work. Humans need broader bandwidth wireless communication technology to

Z. Xu et al. (Eds.): CSIA 2019, AISC 928, pp. 74–79, 2020.
https://doi.org/10.1007/978-3-030-15235-2_12

meet the needs of the contemporary wireless communication equipment [1]. The wireless communication equipment is also gradually developing towards the goal of the multimedia, popularization, diversification and globalization. Among them, the diversification of the communication network puts forward the stricter requirements for the wireless communication system, which will further promote the development of the radio communication technology [2].

The wireless local area network (WLAN) is a technology that combines the computer network with the wireless communication technology. Based on the IEEE802.11 standard, the WLAN can use 2.4 GHz and 5 GHz bands in the ISM band for the wireless connection in the local area network environment. Because of the development of this band without the authorization and the wireless communication technology, the WLAN has been applied more and more widely. As a key component in the WLAN system, the antenna must adapt to the development of the WLAN communication technology. Therefore, it is necessary to design the high-performance antenna that can meet the requirements of the WLAN communication, and it has the important practical significance and the application value. At present, there are three commonly used standards for WLAN, namely IEEE802.11a, IEEE802.11b and IEEE802.11g. Aiming at the three standards of WLAN, two new dual-band microstrip antennas are designed using the HFSS software [3]. Both the antennas use the common FR4 material as the dielectric substrate.

The ultra-wideband technology is a new type of the wireless communication technology. By directly modulating the impulse pulses with the steep rising and the falling edges, the signal has the GHz bandwidth. It has the advantages of the insensitivity to the channel fading, the low power spectral density, the low interception ability, the low system complexity and the centimeter-level positioning accuracy [4]. Therefore, the UWB technology has the great potential in the field of the short-distance wireless communication, and it is one of the hot issues in the field of the communication. The analysis of the system performances is very important to the research of the UWB wireless communication system, so we focus on the performance analysis of the UWB system. In the UWB wireless communication system, the inter-symbol interference and the multi-user interference become the main factors affecting the system performances [5]. The inter-symbol interference (ISI) is mainly caused by the dense multipath channels, while the multi-user interference (MUI) is caused by the multi-users in the same frequency band.

2 Design Concepts of the Small Microstrip Antenna Cased on the WLAN/UWB Wireless Communication System

Firstly, the basic theory of the microstrip antenna is briefly summarized. Through the analysis of the structures and principles of the microstrip antenna, the ways of the miniaturization, broadband and multi-frequency technology of the microstrip antenna are given. In the analysis of each technology, we start with the concept, and then we demonstrate the feasibility and the shortcomings of each technology. With the rapid development of the modern science and technology, various communication devices and interconnection schemes emerge as the times require. Compared with the traditional wired network, the wireless network has the mobility, the flexibility, the low cost and the simple installation.

Then, the antenna in 2.4 GHz/5 GHz band is explored. The design technology of the small dual-band microstrip antenna based on the wireless local area network (WLAN) is studied in this paper. Based on the research and improvement of the traditional microstrip antenna, a small dual-band planar monopole antenna for the wireless local area network (WLAN) is designed by improving the shapes of the radiation patches with different shapes. The microstrip patch of the antenna is a rectangular ring structure, which is compatible with the RF circuit of the general interface. It is very suitable for the integration with the RF front-end circuit. The simulation and the actual measurement results show that the antenna can completely cover the working band of the WLAN, and has the good radiation characteristics.

Finally, the design technology of the small UWB antenna is studied. Based on the research and improvement of the traditional circular patch antenna, a small coplanar waveguide feeding antenna structure which can cover all the frequency bands of the UWB wireless communication system is proposed. On this basis, the U-slot structure is introduced to make it have the notch characteristics in 4.9 GHz–6.0 GHz band, thus avoiding the interference with other communication systems in this band. Through the simulation and the experiment, it can be concluded that both antennas can fully work in the UWB wireless communication system, and have the excellent radiation characteristics.

3 Design Ideas of the Small Microstrip Antenna Based on the WLAN/UWB Wireless Communication System

The microstrip antenna is a new type of the antenna developed gradually in recent 30 years. The concept of the microstrip antenna was proposed as early as 1953, but it has not attracted the attention of the engineering circles. There were only the sporadic studies in the 1950s and the 1960s, and the real development and application was in the 1970s. A common type of the microstrip antenna is a thin dielectric base (such as the PTFE glass fiber laminate), with a thin metal layer attached as a grounding plate on one side, and a metal patch with a certain shape made by photoetching on the other side. The patch is fed by the microstrip lines and the axis probes, which constitutes a microstrip antenna. When the patch is an area unit, it is called a microstrip antenna. If the patch is a slender strip, it is called a microstrip dipole antenna. Another kind of the microstrip antenna is the microstrip slot antenna. The grounding plate is carved out of a window, i.e. a slot, and a microstrip line is printed on the other side of the dielectric substrate to feed the slot.

The microstrip antennas are classified into two categories according to their structural characteristics, namely the microstrip patch antennas and the microstrip slot antennas, and they can also be classified into the rectangular, the circular and the circular microstrip antennas according to their shapes. According to the principle of their operation, any kind of the antenna can be divided into the resonant (standing wave) and the non-pulling (traveling wave) microstrip antennas. The former type of the antenna has a specific resonant size, which can only work near the resonant frequency.

The latter type of the antenna has no resonant size limitation, and its end must be matched with a load to ensure the transmission of the traveling waves. The microstrip antenna can be analogous to a closed resonator, and its resonant characteristics can be equivalent to a parallel resonant circuit with a high Q value. Generally speaking, the microstrip antenna has the inherent shortcoming of the narrow impedance bandwidth, which limits its wide application. Therefore, it is necessary to expand the bandwidth of the microstrip antenna.

The antenna bandwidth is determined by the voltage standing wave coefficient at the input end of the microstrip antenna. The range is that the voltage standing wave coefficient is lower than the frequency band corresponding to a specific value. The antenna bandwidth is expressed by the voltage standing wave ratio of the microstrip antenna. It can be seen from Formula (1) that when the standing wave ratio (VSWR) of the microwave antenna is constant, the fundamental way to improve the antenna bandwidth is to reduce the antenna quality factor Q. We can start from the following aspects:

When a thick substrate is chosen, the radiation conductivity increases with the increase of its thickness, and the corresponding Qr and the total Q value of the radiation decrease accordingly. However, the effect of this method is limited. The thickness of the substrate leads to the excessive ratio of the thickness of the substrate to the wavelength, which generates the surface wave excitation. At the same time, the weight of the substrate will increase and the space occupied will increase. In this paper, the coaxial cable is used to feed the microstrip antenna, and the thickness of the antenna will increase the reactance of the probe, which will reduce the efficiency of the antenna. In addition, the reduction of er can reduce the binding of the medium to the field, make it easy to radiate, and consequently reduce the energy storage of the antenna, making the corresponding Qr of the radiation smaller and the bandwidth wider. But there are also limitations. The minimum value is 1, equivalent to the use of the air media. However, the width of the patch element and the microstrip feed is relatively wide, and the radiation loss needs to be suppressed, and the corresponding value is relatively large. The ferrite substrate material can improve the bandwidth by changing the magnetic field, but its loss is too high.

The shape of the patch is changed to maximize the radiation efficiency of the stored energy, such as using the rectangular or the circular patches. The rectangular ring patch is relatively easy to fabricate. In this paper, the rectangular patch is selected to increase the bandwidth. Use a stepped substrate. Because the thickness of the two resonator substrates at the two radiation ports is different, the double-loop phenomenon will be caused by the step capacitance coupling between them, so the bandwidth will be expanded by the changes of the substrate shapes.

Since a microstrip antenna patch can be equivalent to a parallel resonant circuit, the multiple patches can superimpose the multiple similar resonant frequencies to increase the bandwidth. Therefore, a planar multi-resonant structure can be adopted. That is, there are multiple patches on the same plane, one of which is fed directly, and the other patches are the parasitic patches. The multi-layer structure consists of two or more patches attached to different dielectric layers and stacked together. Its feeding modes are divided into the electromagnetic coupling type and the aperture coupling type. The position of the feed probe can be adjusted by adding the short-circuit pins to the microstrip antenna, thus stimulating a variety of the similar resonant frequencies.

The single feed point can realize the circular polarization radiation without any phase shifting network and the power divider. Based on the theory of the cavity model, he uses the degenerate modes of two orthogonal polarized radiation waves and introduces some asymmetry in the cavity to eliminate the degeneracy of these two modes. The geometric structure of the single-feed circularly polarized microstrip antenna has many forms, such as the quasi-square, the ellipse, the circle and the polygon. This kind of the back-fed single-feed circularly polarized microstrip antenna is fed by a point on the back of the antenna substrate. In Type A, the feed point F is placed on the x-axis, and in Type B, the feed point is set on the diagonal line. The attenuation of the degenerate mode is solved by adding a degenerate and separating unit ΔS.

The antenna designed in this paper is a double-layer patch antenna. The upper and the lower patches are machined on the dielectric materials with the thickness of 3 mm and the dielectric constant of 9.2. The probe is directly connected to the upper patch antenna through the hole of the lower patch antenna. The lower patch antenna is the parasitic unit of the upper patch antenna. It is fed by the electromagnetic coupling through the upper patch antenna without feeding alone. The resonant frequency of the dual-frequency antenna is determined by the size of the upper and the lower microstrip patches. Among them, c is the speed of the light in the free space, and L is the actual length of the microstrip patch antenna, and Δl is the available extension length of the admittance caused by the edge effect, and is the relative dielectric constant of the dielectric plate of the microstrip antenna. The circular polarization radiation of the microstrip antenna is realized by choosing the size of the square tangent angle.

The ANSOFT HFSS software based on the finite element method is used for calculation. The feed uses a microwave port, which is added to the coaxial port. The solution frequency is set to 1.57 GHz, and the fast sweep frequency is used between 1 and 2 GHz. The self-adaptive solution process is adopted, and the number of the iterations is set to 6. Each calculation increases 20% compared with the previous one. When the amplitude and the phase change of the S parameter is less than 0.2 or 6 times, the solution is completed. The calculated S parameter reflects the reflection of the input port energy and the utilization of the energy. The axis ratio reflects the performances of the circularly polarized wave radiated by the antenna.

4 Conclusion

Nowadays, the wireless local area network (WLAN), one of the most effective wireless access networks, has become a research hotspot in recent years. WLAN uses the wireless communication technology to transmit the data and the information in the air. Users can exchange the information anytime and anywhere. In recent years, the practical needs of the low cost, the miniaturization and the easy processing of the WLAN antennas have attracted extensive researches. Various miniaturized multi-band WLAN antennas have been proposed one after another. The bandwidth of the linearly polarized microstrip antenna is limited by the impedance bandwidth. The input reactance of the antenna can be reduced by using the feeder matching technology, and thus the bandwidth is expanded.

References

1. Wang Q, Liu J (2017) Design of the miniaturized broadband microstrip antenna based on the artificial electromagnetic structure. J Microw (12):99–100
2. Zhu H, Wang J, Zhang H (2018) Structure design of an airborne logarithmic periodic antenna. Radio Commun Technol (10):212–213
3. Wang P, Cheng B, Yang L (2018) Design and implementation of the wearable antenna applied to 5.8 GHz. J Anhui Univ (Nat Sci Ed) (11):102–103
4. Dong J, Jin F, Qin W, Li Y, Wang S (2018) Antenna rapid optimization platform design based on the intelligent algorithms and the agent models. Telecommun Eng (11):71–72
5. Yang Y, Li Y, He H (2018) Design and implementation of the large logarithmic periodic antenna array servo control system. Hebei J Ind Sci Technol (12):61–63

Development and Application of the Computer Integrated Management System for the Sports Teaching Platform

Shoucheng Zhang(✉)

Chongqing Real Estate College, Chongqing 401331, China
zhengpu371730@126.com

Abstract. The computer integrated management system of the physical education in colleges and universities can also promote the students to master the sports-related action essentials. Unlike other arts and science courses, the physical education courses are not based on the theoretical knowledge, but the most important thing is practice. However, due to the limitations of the venues and the teachers, sometimes students cannot grasp the correct essentials of a sports project, such as the movement standardization and the promotion of the sports levels and so on. The computer integrated management system for university sports can integrate the Internet resources, have access to the latest and most complete sports resources, and display them through the way of the computer integrated management system, which can provide the students with the most authoritative and the most standard sports display, so that the students can clearly grasp the main points of the items, so as to improve their levels of sports.

Keywords: Sports teaching platform · Computer · Integrated management system · Development mechanism · Application analysis

The computer integrated management system is the first type of the teaching technology that has been applied in the teaching practice since the reform of our teaching. The growing maturity and development of the computer integrated management system provide more opportunities for our teaching to share and optimize the teaching framework [1]. The flexibility brings many benefits to our teaching. However, the use of the computer integrated management system in our physical education is not very extensive, so we need to actively explore the application and development of the computer integrated management system in our physical education.

1 Development and Application of the Computer Integrated Management System for the Sports Teaching Platform

The physical education and especially the indoor physical education is the theoretical part of the sports teaching, and the application of the media technologies can give more play to the flexibility of the physical education [2]. But the application is not very

Z. Xu et al. (Eds.): CSIA 2019, AISC 928, pp. 80–85, 2020.
https://doi.org/10.1007/978-3-030-15235-2_13

extensive, so we need to do more research work in this area, and give full play to the advantages of the computer integrated management system. The advantage of the computer integrated management system is that it can make the rigid content vivid, so as to enhance students' interests in their learning. In addition, for some video teaching contents, students can have active interests in their learning, and enhance the learning motivation of their physical education [3]. Therefore, the use of the computer integrated management system teaching technology, to a large extent, can improve the efficiency of the students' learning, and for the sports disciplines, it is also a useful teaching method. The development of the new teaching mode is the focus of the discussion in the field of our education. Therefore, the application of the computer integrated management system is inevitable.

The most successful application of the computer integrated management system in our teaching lies in the reasonable integration of the traditional media with the modern media, and the comprehensive application of the computer integrated management system in our physical education, which fundamentally renews the means of the physical education [4]. First of all, in the teaching methods, various computer-aided teaching tools and slides, videos and other teaching methods can enable the students to experience more vivid teaching scenes in the teaching process, greatly stimulate students' creativity and enhance their initiatives in their sports learning. No matter whether in the overall implementation of the course, or in the extracurricular learning, students can benefit greatly [5]. For example, in the gymnastics teaching subjects, teachers make slides to show to the students, which can not only lighten the burden of the teachers, but can also enable the students to better grasp the essentials of gymnastics and improve their gymnastic skills by watching the demonstration films. The application of the computer integrated management system in the physical education is also manifested in the full sharing of resources. The network information sharing based on the computer integrated management system can greatly reflect the informatization of our teaching. The physical education in different regions and schools can make common progress by sharing the information of the computer integrated management system, so as to improve the teaching levels in an all-round way. At the same time, the physical education is a very practical subject. In many cases, the quality of our teaching directly depends on the overall quality of the physical education teachers.

2 Development Ideas of the Computer Integrated Management System for College Sports

In the process of developing the computer integrated management system of physical education in colleges and universities, firstly, the database of the teaching resources should be enriched, and the sports-related resources should be sorted out by means of the network, books and expert discussions, and should be stored in the database in a unified format. After the database resources are determined, the functional modules of the comprehensive management system of the school sports computer are designed. Each functional module is an important part of the system, and the system structure block diagram is drawn. From the system management, the course selection, strengthening exercises, the user login and other aspects, we can carry out the function design and development, and follow the concept of the modular design, to facilitate the expansion and maintenance of the system in the future. Finally, the network technology should be popularized in colleges and universities. After all, the comprehensive management system of sports computers in colleges and universities is a new teaching platform based on the Internet. The sports teaching work of students and teachers is based on the platform.

From the point of view of the functional module, the university sports computer integrated management system uses the multi-level menu options. After the students and the teachers use their own identity roles to carry out the system, they can complete the information input and opening by clicking on different options, and obtain the corresponding resources from the system. The university sports computer integrated management system adopts the B/S structure to design, and uses the MVC design concept to separate the display, control and model. At the same time, it combines the ASP web-page technology to design the human-computer interaction pages, which integrates the physical education and the network organically and forms a perfect physical education management system. The data mining technology is introduced into the database. By analyzing the data of the students' operation in the system and the results of their daily exercises, the direction and the weak links of the students' interests in sports are excavated, and the sports resources that the students are interested in are precisely pushed forward and the differentiated counseling strategies can be formulated for the teachers to make reference. The computer integrated management system of physical education in colleges and universities fully combines the massive resources of the Internet and the data mining analysis technology to analyze the students' learning stages and achievements in real time, which provides a basis for the students to master their own learning progress and for the teachers to formulate the teaching promotion programs, and promotes the improvement of the teaching levels.

The users of the computer integrated management system of physical education in colleges and universities are mainly divided into three rights, namely administrators, teachers and students. Users with different permissions will operate within different permissions when they enter the system. For example, the administrator is the maintainer of the comprehensive management system of the university sports computer. He can operate on the course management, the personnel permission setting, the management of the site facilities, the data addition, deletion and modification, and the information publication of the group announcement and so on, and has the highest authority. When the students enter the system, they can inquire about the courses they can choose, modify their basic information, browse the sports teaching resources, and inquire about their achievements. They can also ask and answer the sports questions to other students or the teachers in the discussion area. Teachers take the class as the unit to manage the students. After entering the system, teachers can check the progress of the physical education, and test the reports, the physical examinations and the health of students in each stage. They can also upload the sports teaching resources, courseware and videos collected by the lock to the system for the students to browse and learn. Teachers can grasp the learning situations of each student in real time in the system, and can formulate differentiated teaching promotion programs and promote the all-round development of the students' sports knowledge.

3 Several Problems that Should Be Paid Attention to in Implementing the Computer Application Management System in Colleges and Universities

3.1 The Leadership of the Computer Application in Colleges and Universities

This work should consist of the main leaders of the sports department (room) and the experienced intermediate and senior teachers. The main responsibility is to formulate the computer application programs, the implementation methods, the application funds, the staffing and the personnel training, and solve problems arising in the implementation and application process.

3.2 The Cultivation of the Multi-level Talents

A series of the characteristics of the sports computer application management require a multi-level and reasonable knowledge and intelligent group. There are the senior programmers (responsible for the design of the sports system, the implementation, optimization and application of the sports system, and the entire process of activities), the programmers (responsible for the implementation, optimization and application of the system), and the operators (responsible for the application of the sports system). From the perspective of the development and the long-term consideration, the application and management of the sports computer in colleges and universities should be well done, and the knowledge and intelligent group structure of all kinds of the personnel should be reasonable. Therefore, we must pay attention to:

In the computer software personnel, we must pay attention to the training of the senior and the intermediate personnel responsible for the development and the daily maintenance and optimization of the sports system. In the application of the sports system, we should pay attention to the training of the operators, provide the necessary assistants for the middle and the senior personnel, and meet the requirements of the input and the application of the daily sports information. Because of the marginality of the sports computer application, while training the above-mentioned personnel, we should also pay attention to training them with the corresponding levels of the hardware and software knowledge and all kinds of the sports knowledge, so that they can gradually develop into the high-level dual-disciplinary marginal talents at different levels.

Enhance the second development function of the software. As more and more computer projects are applied in the field of the college sports, the scope is expanding, and the functions are improving, and the links between the developed systems are getting closer and closer. The continuous improvement of the development and management of the school physical education will put forward higher requirements for the application of computers in our physical education, which requires the preparation of the second development system and the enhancement of the maintainability, flexibility and portability of the software.

The data interface is reserved to prepare for the opening of the information high speed channel. The sports computer application management system is closely related to other information systems. With the wider application of computers in various departments, the widespread use of the information networks and the opening of the information networks, and the wider interconnection between various departments and schools, the sports computer management information system cannot be isolated and closed. Besides collecting and using the data from within the system, it must also be from outside the system. Other sports management systems collect the information and data to achieve the sharing of the sports information and data.

4 Application Analysis of the University Sports Computer Integrated Management System

After the development of the computer integrated management system of physical education in colleges and universities, the real application of the system in our physical education in colleges and universities can improve the quality of our physical education, mainly reflected in the following aspects. The use of the computer integrated management system can change the display forms and the cognitive ways of each component in college physical education. For example, the teachers' explanations of the sports theoretical knowledge can be presented in the form of the computer integrated management system courseware, assisted by the vivid action decomposition animation, so that students can fully understand the physical movements and the key points needed by the sports projects, master the technical points of the project, and at the same time make use of the massive teaching funds in the Internet. Source can better expand students' understanding and research of the sports knowledge, and form a perfect knowledge system of the sports theory, so that students can achieve the all-round improvement of their sports levels.

With the emergence of the computer integrated management system, teachers can also achieve the goal of the distance teaching. Only by loading the required teaching courseware into the system in advance, students can learn at anytime and anywhere by landing on the system Web page, and they can practice and exercise at home and outdoors at any time, which greatly makes up for the lack of the teaching time of the physical education curriculums, and the limitation of the venues and facilities, which can improve the efficiency of the physical education in colleges and universities.

Teachers can grasp the learning situations of each student at different stages in real time by using the comprehensive management system of physical education computer in colleges and universities, and make scientific and reasonable teaching plans according to the different learning degree. They can arrange physical education teaching resources according to the students' interests and weak links, and take interests as guidance, to achieve the improvement of the students' sports levels. College sports computer integrated management system can also promote the students to master the sports-related action essentials. Unlike other courses of arts and sciences, the physical education curriculum is not based on the theoretical knowledge, but the most important thing is practice. However, due to the limitations of the venues and teachers, sometimes students cannot grasp the correct essentials of a sports project, such as the standardization of the actions and the improvement of their sports levels and so on. Through the computer integrated management system of physical education in colleges and universities, we can integrate the Internet resources, obtain the latest and most complete sports resources, and display them through the way of the computer integrated management system, which can provide the students with the most authoritative and standardized sports demonstration, so that the students can clearly grasp the essentials of the project, and we can really improve our sports levels.

5 Conclusion

Specifically, it can display more than two kinds of the information media at the same time. Our common information media include words, pictures, sounds and videos, and PPT is a familiar example of the teaching technology of the computer integrated management system, which is widely used in the teaching of various courses. It can explain the abstract contents with the media of words, images or sounds. In fact, the computer integrated management system is a technical form that integrates the media, the computer technologies and the network technologies.

References

1. Jian K, Liu Y, Wang L, Hu L (2017) Exploration and practice of the computer curriculum teaching reform in universities – taking city college of dongguan institute of technology as an example. J Dongguan Univ Technol (02):115–116
2. Li H, Wu C, Han G, Dong Y, Guo Q, Wang J, Wan S, Ma H, Wang Q (2017) Discussion of the key points of the quality assurance of computerized system in GLP laboratories. Drug Eval Res (04):100–101
3. Lu B, Yin J, Gao S, Jin K, Liao Y, Ren Y (2017) Computer science education: opportunities for all – characteristics and enlightenment of K-12 computer science framework in the United States. E-Educ Res (03):133–134
4. Li H, Chen H (2018) Dynamic science knowledge mapping analysis of the machine translation research in China (2007–2016) – based on the contrastive statistics of the word frequency between linguistic and computer science periodicals. J Xi'an Int Stud Univ (06):123–124
5. Feng X, Chen D (2018) Research on the nature of the advertising behavior of browser shielding video website – reflections on the case of the unfair competition dispute between Shenzhen Computer System Co., Ltd. and Beijing Science and Technology Co., Ltd. Hebei Law Sci (04):134–135

Development and Design of the Virtual Design System Based on the Visual Communication

Bai Xue(✉)

School of Art and Design, Wuhan University of Science and Technology, Wuhan 430065, Hubei, China
xianghongxuel@yeah.net

Abstract. The virtual design is the design concept based on the three-dimensional space, which is different from our existing two-dimensional design concept. It no longer divides the expression of space into six or more aspects to carry out the design. The virtual space design starts the entire design process from the perspective of the space. Just like building the architecture and the landscape in reality, the virtual space design uses the virtual space design platform to fully display the ideas of the architects and planners in computers, so that designers can better control their own design, constantly adjust their own design, and make a comparison between different plans.

Keywords: Visual communication · Virtual design · System development · Design thinking · Artistic connotation

1 Definition and Connotation Analysis of the Visual Communication Design

In the contemporary information society, the visual communication design has gone far beyond the traditional meaning of the design art itself. It is not only the "language" in the traditional design sense, but is also a new expression of the social civilization and progress, which is a comprehensive cultural phenomenon with the omnidirectional description existing in the contemporary society [1]. The virtual design is an innovative design method based on the computer aided design.

The visual communication design refers to the design that uses the visual symbols to transmit various kinds of the information. The designer is the sender of the information and the object of the communication is the recipient of the information. It is abbreviated as the visual design [2]. The visual communication is the communication between people in the form of "seeing", which is the way of expression and communication through the visual language. Different regions, skin colors, ages, genders and people who speak different languages can communicate the information, emotions and cultures through the vision and media [3]. The visual observation and experience can overcome the barriers of different languages and eliminate the barriers of different words, and obtain the understandings and interaction with the help of the visual consensus of "pictures" – images, graphics, patterns, pictures, graph methods and schemas.

Z. Xu et al. (Eds.): CSIA 2019, AISC 928, pp. 86–91, 2020.
https://doi.org/10.1007/978-3-030-15235-2_14

The visual communication includes two basic concepts: the visual symbol and the communication. The so-called "visual symbols", as its name implies, refer to the symbols that the human visual organs – the eyes can see and can express certain properties of things, such as photography, television, film, plastic arts, buildings, various designs, urban architecture and various sciences, and words, as well as the stage design, music, heraldry, ancient money and so on, which can be seen with the eyes. They are all the visual symbols [4]. The so-called "communication" refers to the process in which the sender uses symbols to convey the information to the recipient. It can be an internal communication or an individual communication, such as the information communication between all living things, between the human beings and nature, between the human beings and the environment, and within the human body.

The visual communication design is the design that is expressed and conveyed to the audience through the visual media, which embodies the characteristics of the design era, the graphic design and the rich connotation [5]. The emergence of the digital multimedia constantly challenges and enriches the traditional ways of the visual communication, and expands the extension of the contemporary visual communication design. The visual communication has gradually changed from the planarization and the static form in the past to the dynamic and comprehensive direction, from the single media to the multimedia, from the two-dimensional plane to the three-dimensional space, and from the traditional printing design products into the transmission of the virtual information images. Advertising, as an important form of the visual communication, has also undergone a qualitative leap under the impetus of the digital multimedia technology. The generation of the network advertisement, the digital film and television advertisements, the multimedia electronic display screens, the multimedia interactive advertisements and other new advertising visual communication methods penetrated into all aspects of our social life with the trend of the rapid development.

2 Analysis of the Connotation and Characteristics of the Virtual Design

The virtual design refers to the designer's design in the virtual environment, which mainly manifests that the designers can modify the parametric models in the virtual environment by different interactive means. The virtual reality technology has a strong practicality. In the virtual design of a variety of the human-computer interaction, a number of the self-developed typical application systems are introduced, and the contents are novel, and the author explained the profound theories in simple languages. And the practicability is strong. The main contents include: Overview, Common tools and software of the virtual design system, Multi-channel human-computer interaction and the gesture recognition based on the data gloves, Virtual design and assembly based on the three-dimensional mouse, The real-time interactive virtual customization of products under the Web environment, Active and passive three-dimensional display system, Speech recognition technology and the virtual assembly, and Multi-parameter coupling sliding bearing virtual experiment based on the stereo display technology and so on. The application of the virtual design technology has many advantages, such as the visualization of the simulation technology, which is convenient for improving and revising the original design.

As far as the "design" is concerned, all the design work of the traditional design is aimed at the physical prototype (or the conceptual model), while all the design work of the virtual design is centered on the virtual prototype. As long as the virtual prototype can meet the design requirements, the actual products will certainly meet the design requirements. As a high-tech and practical technology, the virtual reality technology integrates the computer graphics, the computer simulation, the man-machine interface, the multimedia, the sensors and the measurement technology, and has been widely used in engineering, military, aerospace, architecture, medical treatment, training, tourism and advertising. Among them, the virtual design and the virtual manufacturing technology play an important role in the field of engineering.

The virtual reality technology has the strong practicality. The virtual design technology is a comprehensive system technology formed by the multi-disciplinary advanced knowledge. Its essence is based on the computer-supported simulation technology. In the stage of the product design, the entire process of the product development and its impact on the product design are simulated in the real time and parallel, and the product performance is predicted. The manufacturing costs, the manufacturability, the maintainability and the disassembly of products can improve the one-time success rate of the product design. It is also conducive to organizing the manufacturing more effectively, economically and flexibly, making the design and the layout of factories and workshops more reasonable and effective, so as to minimize the product development cycles and costs, optimize the product design qualities and maximize the production efficiency.

3 Development and Design of the Virtual Design System Based on the Visual Communication

As far as the "virtual" is concerned, the designer of the traditional design outlines the conceptual design with lines and wireframes on the drawings, while the virtual designer improves the prototype interactively, real-time and visually at any time in the immersed or non-immersed environment, and can see the results of the modification immediately.

3.1 Development Requirement of the Virtual Design System Based on the Visual Communication

A virtual design system has three functions: the 3D user interface, the parameter selection and the data transmission mechanism. The 3D user interface designers no longer use the 2D mouse or keyboard as the interactive means, but use the gestures, voices, 3D virtual menu, ball cursors, game joystick, touch screen and other ways to interact. The selection parameter designer uses various interactive ways to select or activate a data in the virtual environment to modify the original data. After the parameter modification, the model in the virtual environment becomes a new model. The data generated by the model modification of the data transmission mechanism is transmitted to the CAD/CAM system which works with the virtual environment, and sometimes the data is returned to the virtual environment from the CAD/CAM system. This virtual design system includes an independent CAD/CAM system, which provides the function of building the model for the virtual environment. The modified models in the virtual environments are sometimes returned to CAD/CAM systems for the precise processing and the graphics output. Therefore, this bidirectional data transmission mechanism is necessary in a virtual design system.

The traditional virtual design methods have some problems, such as the inaccurate data collection and the unreal virtual images. People are no longer satisfied with browsing the virtual environment. Therefore, a virtual system design based on the three-dimensional landscape visualization is proposed. With the help of the big data analysis and the computer aided design, the contents of the landscape such as the terrain, planning and planting are analyzed, and the functions of different modules are developed to complete the hardware design of the system. The terrain is designed by the regular grid method, and the design principle is analyzed.

3.2 Design Strategies of the Virtual Design System Based on the Visual Communication

To construct a reasonable logical structure, first of all, we should divide the source data into levels, clarify the relationship between levels and levels, between the information and the information, and between the information and the levels, and complete the preliminary construction of the structure by grouping and classifying the information. The information of different dimensions, natures, magnitudes and orientations is mapped through the graphic design. According to the user's information needs and the logical thinking relationship, the objective laws of the problems are highlighted by sorting the structures of all kinds and dimensions, so that people can think more clearly when they do the visual thinking activities. The semantic comparison tables can help to complete the visual communication process quickly and reduce the cognitive costs of the information.

The design of the hardware part of the system emphasizes the design methods, and divides all the designers into several working groups to work together and integrate their ideas into them. Finally, one person reviews and finalizes the design results in order to meet the needs of customers. The virtual design service center can provide the virtual product software platform and the related demonstration tools for the system, and promote the design of the system in the integrated and collaborative environment. It is necessary to aggregate various tools to form the information transmission and the process integration between tools. According to the use of the relevant tools, the system design can transform the products from the physical state to the virtual state in the three-dimensional virtual environment, from the material form to the intangible form, and from the tangible to the intangible, thus completing various virtual control of the system design.

After determining the design schemes, it is necessary to construct a virtual model for the design. This process can make full use of the virtual technology to provide the technical support for the model construction. The virtual technology can provide the three-dimensional environment for the visual communication effect. Through the direct observation of the visual images, the data information of each part of the model can be obtained and designed in the virtual environment, and the model can be modified, which can improve the accuracy of the model building. Using the virtual technology, we can make the model fully adapt to the virtual environment, and give full play to the imagination and creativity, and be consistent with the actual situations of the product model and highly close to the real products, without the need for a physical prototype, so that the product results can be analyzed from all directions and multi-angles. Every stage of the product virtual model construction plays an important role. The virtual technology can provide the advanced training means, and the scale of the network utilization is expanding day by day. All links in the products can be carried out efficiently, and the information feedback can be carried out at different stages to optimize the product performances under certain conditions.

The difference of the visual design determines its “pop-up effect”. The better the pop-up effect, the easier the user’s attention screening will be. Therefore, by changing the shapes, scales, and locations of the icons and other visual effects, the target information can be highlighted from the background or the non-target information. The visual clarity is a trade-off between the individuals and the collectives. If multiple visual effects are changed at the same time, the target information can be simulated, visualized and realistic. At the same time, it should be noted that the target information should be highlighted, noticeable and not too striking. Therefore, we should also maintain the unity of the whole, that is, the consistency of the overall interface style, so that the user’s attention will not be too distracted, which will hinder the memory and understanding.

With the help of the experience knowledge, users can always learn and adapt to the new information visualization products by relying on the existing knowledge of the cognitive experience. Therefore, the information visualization products should not only balance the new display forms, functions and carriers, but also consider the user’s experience knowledge, and also improve the readability and comprehension of the information through the new visual effects. Through the continuous training of users’ visual thinking, enrich the users’ experience and knowledge, and make the visual activities more effective. The digital creative means are represented by the virtual reality technology. According to the different objects applied by the virtual reality technology, it can be manifested in different forms. For example, implement the visualization and operability of some conceptual designs or conceptions to achieve the realistic on-site effect. It can be a real reflection of a particular real world or a purely conceived world. The sensing devices include the three-dimensional helmet, the data gloves, the data clothes and other devices worn on the body and sensors in the real environment.

4 Conclusion

In the process of the system design and the effect presentation, we face the customers in the form of the picture showing. It is difficult to specify the characteristics and the image expression of the virtual structure. If we want to change the model, we must redesign it. It takes a lot of time to make the stereoscopic rendering through the network, and more time to make the physical objects. To change the shapes and the outsourcing design, we need to complete them according to the above design steps. In order to ensure the efficiency and security of the system, the hierarchical structure of the system is designed by the virtual technology, and the visual image is analyzed based on the visual communication in the three-dimensional environment. The data information of each part can be obtained, and thus the virtual model can be constructed. According to the system software flow, the virtual organizational structures and the high-quality virtual rendering can be achieved. We can carry on the analysis of the beautiful appearance from the visual transmission angle, and thus obtain the virtual impression drawing. The experimental results show that the virtual design effect of the system is good.

References

1. Fu D, Liao P, Zheng X (2016) Research on the virtual design method. West Leather (01):150–151
2. Lu G, Gan H, Zheng H, Shi X (2016) Virtual design of the ship compressed air system based on unity 3D. Comput Meas Control (09):194–195
3. Li M, Wang M (2017) Philosophical evolution of the aesthetic value orientation in the contemporary digital virtual design. Hubei Soc Sci (05):188–189
4. Wang C, Hao X (2018) Research on the virtual design method of the interior decoration based on the APICloud. Mod Electron Techn (01):165–166
5. Zhao Q, He H (2018) Design of the virtual design system based on the visual communication. Mod Electron Techn (09):142–143

Innovation of the School Computer Teaching Model Based on the Information Network and the Intelligent Campus

Zhang Qian(✉)

School Attached to Huazhong Agricultural University, Wuhan 430070, Hubei, China
zhongxiao081758@126.com

Abstract. With the deepening of the teaching reform, the computer has become a compulsory course in colleges and universities. The teaching of the computer course has the distinct practicality and practical ability, and the teaching characteristics requires that we should carry out the analysis in terms of the shortcomings of the computer course in our country, and put forward the relevant measures to solve the problems, so as to promote the sound development of the computer curriculum teaching and increase the students' interests in the computer courses. With the rapid development of the network information technologies and the deepening of our educational reform, more and more people pay attention to the computer teaching. Therefore, the whole teaching activities need to use the information network and the intelligent campus to constantly reform and develop.

Keywords: Information network · Intelligent campus · School computer · Teaching mode · Innovative ideas

The school management organizations should actively set up a team with the sound core talents, carry out the publicity and education on the basic knowledge such as the information and the "Internet+" for the service managers, the teaching staffs and the students, so that the computer information and the network knowledge can be popularized, so as to solve the most basic problems. At the same time, the campus managers should spend some funds introducing the high-quality IT talents to serve the construction of the smart campus [1]. Only in this way can the smart platform continue to serve the students.

1 Innovative Background of the School Computer Teaching Model Based on the Information Network and the Intelligent Campus

As a product of the era of the knowledge economy, the computer plays an important role in all fields of our society in the new era. However, in the practical education and teaching, due to the rapid technological updating, the teaching mode of the computer

Z. Xu et al. (Eds.): CSIA 2019, AISC 928, pp. 92–98, 2020.
https://doi.org/10.1007/978-3-030-15235-2_15

education cannot keep pace with the times to a large extent, and the teaching method seems relatively backward, which leads to some students' lack of interests in the information technology [2].

The computer application technology is a subject closely related to the theory and practice, which develops rapidly and is widely used. It is also highly operable and involves many media, such as graphics, sounds, animations and images [3]. Therefore, we are required to explore the new computer teaching mode. In the actual teaching process, I found that some students are fresh and curious when they first come into contact with the computer courses, but once they really get the textbooks and sit in the classrooms, they feel boring. Some feel that it is very abstruse, thus slowly losing interests in their learning.

Through the train of thought and the practice of the teaching reform of the "computer" course, the corresponding experimental teaching objectives and methods are clarified, and the cultivation of the students' innovative thinking abilities and engineering practice abilities is strengthened, and the three-dimensional teaching structure model of the "computer" course is established, which effectively improves the comprehensive application abilities of the livelihood computer technologies and enables the students to have the combination of the software and the hardware [4]. The way of thinking of the rational problems, using the computer as the main technology to analyze and solve the practical problems in the field of this specialty, achieves the goal of training the excellent talents with the innovative and applied computer abilities for each specialty.

2 Application Advantages of the School Computer Teaching Model Based on the Information Network and the Intelligent Campus

The intelligent campus refers to the integrative environment of the campus work, the learning and the life based on the Internet of things [5]. This integrative environment takes various application service systems as the carriers, and fully integrates the teaching, the scientific research, the management and the campus life. In 2010, in the "12th Five-Year Plan" of the informationization, it was proposed to build an "exciting" smart campus. This blueprint depicts the ubiquitous network learning, the integration of the innovative network scientific research, the transparent and efficient school administration, the rich and colorful campus culture, and the convenient and thoughtful campus life. In short, we should establish "a safe, stable, environmentally friendly, and energy-saving campus".

The three core features of the smart campus are as follows. Firstly, it provides a comprehensive intelligent perception environment and the comprehensive information service platform for the teachers and the students, and provides the personalized and customized services based on their roles. Secondly, it integrates the information services based on the computer network into various applications and service fields of schools to realize the interconnection and cooperation. And thirdly, through the intelligent perception environment and the comprehensive information service platform, it provides an interface for the schools and the outside world to communicate and perceive each other.

The application of the computer network technologies and the Internet technologies has provided the important technical prerequisites for the construction of the smart campus. In recent years, the concept of the "Internet+" has further promoted the development and application of the smart campus mode. Not only China but almost the entire world has entered the "smart campus" and "scientific and technological campus". The realization of the intelligent campus is an inevitable requirement for the sustainable development of colleges and universities. Firstly, the information development situation based on the cloud data provides an important resource guarantee for the development of the smart campus. Secondly, the new generation of the students and the young people has their own qualities of keeping pace with the times and synchronizing with the science and technological information teaching.

3 Functional Analysis of the School Computer Teaching Based on the Information Network and the Intelligent Campus

The technological system of the smart campus includes three parts: the digital resources, the application services and the infrastructure. The construction of the digital resources should highlight the characteristics of the specialty and the school-enterprise cooperation according to their own conditions. In the process of its construction, we should follow the principles of giving consideration to the social services, giving priority to the introduction, being cautious and self-construction, application in the construction, the combination of the construction with the application, and being open and sharing. The construction of the application services requires colleges and universities to select and implement the information management requirements at the ministry, provincial and municipal levels according to their own conditions, to do their best and gradually improve it. The infrastructure construction should make full use of the advantages of the network interconnection and the cloud computing technology, and actively use the social forces to coordinate the construction of the university central computer rooms, servers, networks, simulation training system environment and other infrastructure, combining the digital services from the outside school with the school-based services, providing the digital services economically and efficiently for students, teachers and managers.

Comprehensive environmental awareness: The comprehensive perception in the smart campus includes two aspects. One is that the sensors can perceive, capture and transmit the information about people, equipment and resources anytime and anywhere. The other is the perception, capture and transmission of the learners' individual characteristics (their learning preferences, cognitive characteristics, attention status, and learning styles and so on) and their learning situations (learning time, learning space, learning partners, and learning activities and so on). The network should be of the seamlessly interoperability. Based on the network and the communication technologies, and especially the mobile Internet technologies, the smart campus supports the connection of all the software systems and the hardware devices. The information can be transmitted quickly and in the real time after being sensed, which is the basis for all the users to collaborate in their learning and working in a new way.

There is the support of the massive data. According to the data mining and modeling technology, the smart campus can construct the models based on the "massive" campus data, establish the prediction methods, and make the trend analysis, prospect and prediction of the new information. At the same time, the smart campus can integrate all aspects of the data, the information, rules and other contents. Through the intelligent reasoning, we should make rapid response and the active response, and more reflect the intelligence and the smart characteristics.

Open learning environment: The core idea of our education is the cultivation of the innovative ability. The campus is facing the demand of moving from "closed" to "open". The wisdom campus supports the expansion of the resources and the environment so that the students can break through the limitations of the textbooks. It supports the expansion of the time environment, so that the learning can expand from the class to after class. It supports the expansion of the space environment, so that the effective learning can occur in the real and the virtual situations.

Personality services for teachers and students: The intelligent campus environment and its functions are based on the concept of the personality service. The application of various key technologies is aimed at effectively solving many practical needs of the teachers and the students in their campus life and their learning and work, and has become an indispensable part of the reality.

4 Innovation Mechanisms of the School Computer Teaching Model Based on the Information Network and the Intelligent Campus

The intelligent campus refers to an open educational teaching environment and a comfortable living environment, which is based on the idea of the personalized services for teachers and students, which can comprehensively perceive the physical environment, identify the learners' individual characteristics and their learning situations, provide the seamless network communication, and effectively support the analysis, evaluation and the intelligent decision-making of the teaching process. In order to improve the students' interests, enthusiasms and consciousness in their learning and achieve the satisfactory teaching results, the following modes can be used for our teaching:

4.1 Improve the Students' Self-study Abilities by the Interlacing Method

Design the teaching principle of taking the students as the main body, and select the teaching contents and strengthen the cultivation of the students' self-study abilities. Students can't and shouldn't follow their teachers all their lives, especially in the era of advocating the lifelong education. Their ultimate goal of learning should be to learn their own ability to acquire the knowledge. Therefore, teachers are required to grasp the textbooks in a general way, to have a thorough understanding of the contents, and then to find out the key points, and select the teaching contents carefully, so as to be detailed, appropriate and clear in priority and the secondary. In the process of our teaching, teachers cannot explain in the order of the textbooks' catalogues and chapters, but can adjust the teaching order accordingly. This can not only arouse the students' thirst for knowledge, but also prevent them from forgetting the former while learning the later, so as to consolidate the knowledge they have learned.

For example, when teaching "Basic Operations of Windows XP" in Section 2 of Chapter 2 in "Computer Application Foundation and Training", this chapter just introduces the composition of the taskbar in the front, and does not mention that it can be moved and hidden. Before I explained this chapter, I explained to my classmates that the task bar can be hidden and moved, and told them that there are two ways of the operation to attract the students' attention, and the classroom atmosphere was also immediately activated, so as to mobilize the students' learning enthusiasms and stimulate their thirst for knowledge. This not only improves the students' enthusiasms for learning, but also points out their shortcomings in time.

Through this way of teaching, we can not only "force" those students who are less conscious to study by themselves, but also urge those students who are more conscious to study and explore themselves more actively, so as to improve the self-study abilities and the practical abilities of all the students. Teachers can also shorten the lecture hours of some theoretical courses and increase the students' computer operation hours. This kind of the scientific, interactive and timely explanation is very popular with the students, and has achieved twice the results with half the efforts.

4.2 Implement the Targeted Requirements for Students to Repeatedly Operate, and Deepen Their Understanding and Mastery

The computer is a highly operational subject, but it cannot be separated from a solid theoretical basis. Some students do not attach importance to the learning of the basic theories, so that when they do not know where to start the tasks assigned by the teachers while going to the computer, and they are always unclear about the concepts. Therefore, in our teaching, teachers must carefully coordinate the relationship between the "practice" and the "presentation". In view of the characteristics of the weak foundation, we should carry out the repeated operation and the intensive practice, to deepen their understanding and mastery of the basic concepts, so that they can enhance their self-confidence, stimulate their interests in the success of the operation, and eliminate their psychological fear of the learning pressure.

For example, the composition, characteristics and differences between the windows and the dialog boxes have been explained in detail in class. But as soon as I got on the computer, I opened a window and asked the students whether it was a window or a dialog box. Few students could answer it. In terms of this kind of situation, we can set some specific operation tasks for the students in the computer room, so that the students can operate by themselves. For example, first minimize the operation of the window, and then see whether the dialog box can minimize the operation. Then move the window, and then move the dialog box and so on. Through these purposeful and targeted operations, they have the opportunity to think independently, analyze and compare them, and find out the differences. In this way, students not only digest the contents of the textbooks, but also deepen the understandings and memory of the knowledge they have learned.

4.3 Use the "Reverse" Thinking Method to Stimulate the Students' Learning Potentials

The experimental steps of the computer can be either positive. That is, each step is prescribed, and then the expected results can be obtained step by step according to the prescribed steps. It can also be reverse. That is, give an experimental result first, and then ask the students to operate by themselves without the help of other students, so as to improve the students' practical operation abilities.

For example, after learning the Word, students are required to make a simple Word form - the curriculum form, so that students can use the comprehensive knowledge they have learned to conceive the best and fastest implementation steps, so that students can talk about their own design and production process, and evaluate a best feasible scheme. Through this way of talking and discussing, students can be guided not to stick to an attempt, a method, an example and a conclusion. In this process, students will fully mobilize their potentials. This not only cultivates the independence and tenacity of the students' wills and qualities in the process of the operation, but also helps to improve their practical abilities, overcome the dependence of the students in the process of the operation, and cultivate their spirits of being willing to think, being good at thinking and being diligent in thinking.

5 Conclusion

To sum up, the popularity of the smart campus platform under the background of the "Internet+" is an inevitable choice for schools to remain their competitiveness with their competitors. In order to ensure the construction of the information technology, enhance the openness of the campus network information, and introduce a strong team of the building and maintaining talents, only by starting from these basic problems can we make full use of the resources in implementing the construction services of the smart campus. The information integration can provide the better services for the teachers and the students, so as to promote the process of the education in China.

References

1. Li Y (2017) Reform and design of the computer classroom teaching in the higher vocational colleges under the background of the "smart campus". Educ Mod (02):99–100
2. Hou J, Zhang W (2018) An empirical study of the hybrid teaching model based on the We-Chat public platform – taking "university computer" course as an example. China Health Ind (11):212–213
3. Chen X (2018) Innovation of the computer teaching model based on the computational thinking. China Comput Commun (Theor Ed) (11):102–103

4. Zhou K, Li M, Wang L, Yang W, Yang Y (2018) Research and practice of the mixed teaching mode in the computer courses in universities under the background of the "Internet+". Microcomput Appl (11):111–112
5. Li G (2018) Design of the intelligent campus model architecture of vocational schools based on the "index system" – taking the construction of the intelligent campus of Jiangsu vocational schools as an example. Jiangsu Educ Res (03):22–23

Integration Theory and Optimal Application of the Traffic Accident Management System

Junhui Zhang(✉)

Beijing Jiaotong University, Beijing 100044, China
cs2017eijournals@163.com

Abstract. With the development of the social economy, the traditional transportation mode will be replaced by the new mode which emphasizes the integration and cooperation between the transportation modes. The advent of automobilization in China has further intensified the pressure of the road traffic. The traffic jam alarm caused by the traffic accidents is increasing year by year. The minor accidents that can be dealt with quickly account for more than 70% of the traffic accidents. Traditionally, when dealing with the minor accidents, car owners often choose to stay on the spot and wait for the traffic police to deal with them. It is easy to cause "minor accidents and big traffic jams", which lead to the congestion of a road, a bridge and even a block.

Keywords: Traffic accident · Management system · Integration theory · Optimization mechanism · Application system

With the development of the cities and the improvement of people's living standards, the number of vehicles in the cities is increasing, and the traffic accidents are becoming increasingly prominent. The urban traffic accident handling is an important work of the traffic management department [1]. Its policy and legal characteristics make it impossible to be careless in the process of the accident handling. Therefore, how to effectively manage the information records in the process of dealing with the daily traffic accidents, so that the information can be quickly and conveniently queried for record and the efficient statistical analysis, is an important task facing the current traffic management departments.

1 Integrated Theory and Optimized Application Background of the Traffic Accident Management System

Based on the investigation and analysis of many information management systems and the business processes of the traffic accidents at the present stage, this paper designs and implements a set of the road traffic accident information management system for the traffic management department by using the popular ASP.NET technology and the SQL database management technology based on the three-tier architecture design idea of the B/S mode [2]. The system includes five functional modules: the accident information management, the accident assistant management, the inquiry statistics,

Z. Xu et al. (Eds.): CSIA 2019, AISC 928, pp. 99–104, 2020.
https://doi.org/10.1007/978-3-030-15235-2_16

the basic information management and the system management. This paper mainly studies the following three aspects:

Because the data formats of the document materials and the form templates involved in the system businesses are various, in order to ensure that the information processing in the system meets the business requirements, the system writes the commonly used formats into the user controls. When the operation needs, the user controls of the corresponding formats can be invoked only by calling the corresponding categories [3].

Because of the frequent interaction of some business information and the large amount of the information in the system, in order to make the system achieve the good operability and the user experience, the system introduces the client-based AJAX asynchronous interaction technology, which achieves the effect of the asynchronous processing of many complex businesses and no refresh operation of the user interface in the system [4]. In some basic modules, when the data format is relatively standardized and the amount of the data is relatively large, the system realizes the function of the data batch import, which makes the input operation of these data more convenient and fast, and avoids the manual input by the staff.

After the coding of the system is completed, the unit testing, the integration testing and the system testing are carried out respectively. The problems in the testing process are the timely feedback and modification, and finally all the testing work of the system is completed. At present, the system has passed the user acceptance, and has been put into operation [5]. From the operation situation, the system indicators have reached the expected requirements. Through the construction and implementation of the system, the road traffic management department not only standardizes the management department's fair handling of the traffic accidents and handling procedures according to the law, but also realizes the supervision and control of all aspects of the accident handling, making the accident handling process more transparent and open.

2 Integrated Theory and Optimized Application Function of the Traffic Accident Management System

2.1 Accident Coordinating Input

This module mainly provides the normative and necessary data for the data analysis and the decision analysis. Determine the location of the accidents. The electronic map is used to standardize the location of the incidents. The electronic map is made by Mapinfo and the program is called by ActiveX control. According to the attributes of the electronic map itself, we use two coordinates to represent the location of the incident location simultaneously: the plane coordinate description and the intersection section description. At the same time, we can meet the different needs of the incident location query. Input the incident location coordinate. A data is drawn from the traffic accident database, and the corresponding location is found on the electronic map according to the incident location described. The irregular location description in the original database is transformed into coordinates that can be recognized by the system and the map, and other relevant attributes of the traffic accident are stored in the new database for further classification. Considering the need of computing the next three modules of the project, some related attributes in the original database, such as the time, the number, the weather, the road, the environment, and the reason and so on, need to be saved.

2.2 Investigation of the Accident-Prone Points

The module of the accident-prone point section adopts the method of the accident plane coordinate clustering. By analyzing the traffic accident data stored in the database, the module obtains the basic situation of the accident scene in each area or section, and finds out the accident-prone point section, so as to play a fast, accurate, intuitive, efficient, concise, comprehensive and reliable role in the future road traffic accident analysis and decision-making. In addition, the data of the road traffic accidents have the orientation. That is, the data of the road traffic accidents have quite obvious geographical characteristics. In the process of speeding up, accurately and visualizing the decision-making information of the road traffic accidents, electronic maps are introduced. By binding with the electronic maps, the results of the investigation are displayed intuitively. It is displayed on the electronic map and provides a friendly interface for the users. This module introduces the road database, and compares the road conditions of each intersection or section with those of the accident database when the accident occurs, and finds out the potential accident-prone points.

2.3 Accident Alarm Reception and Response

Accident alarm reception: The monitoring center receives the alarm through various ways, including the section service telephone, the group monitoring center transfer alarm, the 110 center transfer alarm, the road administration report, the toll collector internal telephone report, and the monitoring video discovery and so on. After receiving the police, the supervisor will simply enter the accident situation (the pile number, direction, casualties, congestion, and dangerous chemicals leakage and so on). The system will preliminarily judge the accident grade according to the set conditions.

Accident confirmation: Once the monitor enters the accident pile number system, it will automatically send the video switching signal to the video monitoring system, and switch several camera guns near the accident point to the large screen for the monitor to confirm the accident. And after the completion of the accident information input, send a short message to the road administration team on-duty leader, and request to send the personnel to the scene to further check the facts.

Accident information display: The system displays all kinds of the information of the accident points on the GIS map, including the video images (the fixed camera, the mobile video, and the mobile phone photographs), the traffic flow, the traffic flow near the toll stations, the surrounding rescue, the traffic police, the government, the medical, the fire and other institutional information, the meteorological information, the national and provincial road interaction information, the emergency material storage point information, the variable intelligence board information, the tunnel monitoring equipment information, and the working vehicle GPS positioning and so on.

2.4 Statistical Analysis of the Accident Data

Based on the characteristics of the traffic accidents, such as the stage, causality, contingency, latency and hazards, it is very meaningful to analyze the time distribution, the space distribution and the personnel distribution of the traffic accidents. This module mainly uses the method of the mathematical statistics to analyze the traffic accident data.

The traffic accident statistics, in fact, is a method to quantitatively analyze the various forms of the accidents and the related factors within a certain range and to find out the laws of the accidents. In this paper, the causes and forms of the accidents are taken as the objects of the analysis, and the effects of other factors on the causes and forms of the accidents are calculated. The accident pattern refers to the form of the traffic accident, that is, the specific state of affairs shown by the conflict between the traffic accident participants or the uncontrolled accident itself. It can be divided into seven categories: collision, rolling, scraping, rollover, car crash, explosion and fire.

The statistical items in this module are divided into the statistical analysis table of the accident causes and the statistical analysis table of the accident forms. In the process of the analysis, users can determine the data objects to be analyzed by specifying the time and the region ranges, and select the items to be analyzed for the statistics. The statistical results are presented in the form of charts.

2.5 Decision Analysis

The decision analysis module is mainly to provide support for the decision makers to make the correct decisions. The module is based on the rough set theory and the association rule data mining idea, and builds the corresponding traffic accident analysis model based on the data mining theory. The collected data are analyzed, and the concise and effective decision rules are obtained to improve the efficiency of the decision makers. The decision rules judge the main factors of the traffic accidents by comparing the probability of the occurrence between the cause and the effect. This form of expression is described by the causal words.

In this module, the decision-maker can select the data to be analyzed according to the specified time and space ranges, set the corresponding parameters according to the actual situation, make the preliminary statistics according to the accident causes or accident forms, and then conduct the in-depth analysis according to the specific reasons, and get the rules with the support and confidence. Through these rules, policymakers can understand and grasp the comprehensive factors that form the various conditions of the accidents. According to the support and the trust of the rules, we can fully understand the various conditions and factors affecting the severity of the accidents.

3 Integration Theory and Optimized Application Mechanism for the Traffic Accident Management System

In the process modeling methods, the IDEF3 method, the role activity diagram method and the Gantt diagram method and so on are commonly used. In the aspect of the process integration, the concurrent engineering technology, the workflow technology and the business process reengineering technology are mainly used to design, optimize and control the process to achieve the goal of the process integration. The research on the functional integration mainly focuses on the methods of the functional modeling, such as the IDEF0 method and the functional decomposition graph method and so on. The domestic research on the knowledge integration is mainly in the civil field, which is embodied in improving the competitiveness of the enterprises, and integrating various types of the knowledge within the enterprises, and carrying out a series of the researches around the virtual enterprises and the dynamic enterprise process reengineering (BPR) and achieving certain results, including various knowledge integration models, methods and technologies and so on.

The map operation function: The layer control of the electronic maps can realize the functions of enlarging, reducing, roaming, measuring, annotating, editing and querying graphics, as well as overlapping display of different layers. The data input and management: The system directly imports all kinds of the data files related to the accidents or adds the accident records through the input interface. Through the special database structure, the system can associate the added data with the basic geographic information, the road network information and the traffic facilities information in the GIS database. The maintenance of the database can be operated on the map.

The accident statistics and reports: According to the statistical norms prescribed by the traffic regulations, the accident data in a specific period of time are subjected to the item statistics, and the statistical graphs such as the report forms and the histograms, the pie charts, the scatter charts and the line charts are generated to visually and intuitively reflect the temporal and spatial variation of the traffic accidents. Accident information inquiry: It is divided into the detailed query and the conditional query. Implement the detailed inquiry for each accident. According to the administrative area code and the accident number, inquire the accident details, and the inquiry results include the accident information table and the personnel information table on the accident collection table. The condition inquiry is the user-defined inquiry conditions, which can support the joint condition inquiry under three conditions. The location of the accident is shown. Accidents can be classified and displayed on the electronic maps in different colors. Accident black spots can be selected according to certain criteria (such as the accident number and the accident rate) according to the accident situations at various points, intersections, sections and the whole area. Accident black spots can be automatically identified on the maps to visually display the statistical analysis results and the time and space distribution of the accidents.

Spatial data analysis function: On the basis of the special database, an analysis model based on the spatial information is established. With the powerful analysis functions of GIS (such as the buffer analysis and the overlay analysis and so on), the causes of the road traffic accidents are analyzed and counted from the aspects of the time and the space distribution, the accident characteristics, the causes, the relationship with the road spatial structures and the traffic management facilities. User management: In order to ensure the normal operation of the system and the security of the system data, the system authorizes different users to login through passwords with different rights. Compared with the ordinary users, the advanced users have the function of maintaining and managing the system data. Output printing: Data analysis results are output by texts, charts and thematic maps, such as the point mark of the accident location, the coloring of the accident impact buffer, and the expression of the accident-prone period and so on. Various statistical reports and thematic map windows can be printed directly through the layout.

4 Conclusion

The traffic accident rapid processing system passes through the construction concepts of photographing first and post-processing, and quick withdrawal and direct compensation, to achieve the purpose of the evacuation in 3 min for the minor traffic accidents. Through the extensive use, the road congestion can be reduced by nearly 30%. Through the Internet + Quick compensation platform, we can effectively provide the convenient services to the public. Through the high-precision collection of the accident data, it lays a good foundation for the analysis and judgment of the large-scale accident data, and the construction of a one-stop service of the long-distance accident handling responsibility, damage deciding, settlement of claims, and car repair.

References

1. Zhou Y, He L (2014) Functional module analysis and implementation of the traffic accident management system based on MapX. J Shunde Polytech (01):52–53
2. Yang C (2016) Cause analysis and management of the urban road traffic accidents. Bus Forum (06):124–125
3. Xu L, Chi J (2017) On the development of the traffic accident information management system. J Hubei Corresp Univ (06):110–111
4. Chen B (2018) Discussion of improving the emergency management mechanism of the emergency environmental events caused by the traffic accidents. Pollut Control Technol (10):14–15
5. Wang C, Wang Y, Wang L, Sheng J, Yu Y (2018) A comparative study of the road traffic accident management systems at home and abroad. Traffic Eng (08):122–123

Paleomagnetic and Sedimentary Evidence for the Multi-stage Paleoenvironmental Evolution of Prydz Bay Based on Big Data Analytics

Zhongyang Lin[1,4], Xibin Han[2], Xiangl ong Jin[1,2(✉)], Liang Yi[3], and Zhenggang Li[2]

[1] College of Marine Science and Technology, China University of Geosciences, Wuhan 430074, China
zhihuo705253@163.com

[2] The Second Institute of Oceanography, State Oceanic Administration, Hangzhou 310012, China

[3] State Key Laboratory of Marine Geology, Tongji University, Shanghai 200092, China

[4] Zhejiang Institute of Geological Survey, Hangzhou 311203, China

Abstract. A 324-cm-long sediment core was obtained from Prydz Bay, Antarctica, during the 29th Chinese National Antarctic Research Expedition. Based upon sediment grain size, ^{14}C, TOC, $\delta^{13}C$, and magnetism data, we show that the major magnetic minerals are ferrimagnetic pseudo single domain (PSD)-multi domain (MD) magnetite. Variations in the paleoenvironmental proxy records allow us to define four zones in the core. These zones broadly delineate climatic variations in the region since the late Early Pleistocene, with a warm period, a transitional period, and a cold period. The magnetic particle assemblage varies with glacial-interglacial cycles. Abrupt changes in particle size, TOC content, and magnetism occur at 102–90 cm depth in the core, indicating a sudden warming in the Antarctic region, signaling the onset of the Holocene. We identified three additional climatic signals in the middle part of the core (232–162 cm) that show unexpected cooling events during the warm period in Prydz Bay.

Keywords: Paleomagnetism · Paleoenvironment · Climate change · Late quaternary · Prydz Bay · Antarctica

1 Introduction

The parameters of magnetic minerals, which are controlled by the concentration, grain size, and type of magnetic minerals present, have been widely utilized in sedimentary stratigraphy and paleoenvironmental reconstructions [1–4]. Analyses of sediment grain-size distribution provide vital information for reconstructing past and present sediment transport dynamics and depositional conditions, and are widely used in sedimentological studies [1–5]. In recent years, the study of environmental magnetism

Z. Xu et al. (Eds.): CSIA 2019, AISC 928, pp. 105–113, 2020.
https://doi.org/10.1007/978-3-030-15235-2_17

in combination with paleomagnetism in marine sediments has gained interest [3, 5]. Due to its unique geographic location and sensitivity to global climate change, the Antarctic region plays an important role in paleoclimate research [6–9]. However, paleoenvironmental reconstructions from sedimentary records in the Antarctic region are rare because of extremely low sedimentation rates and difficulties in establishing suitable chronologies.

Prydz Bay is adjacent to the Lambert Graben, which extends for more than 700 km along East Antarctica (Fig. 1). The Lambert Glacier drains 20% of the ice sheet in East Antarctica, flows onto the Lambert Graben, and feeds the Amery Ice Shelf. Prydz Bay has a deep inner shelf that defines its southwestern margin. The bay also features a large sedimentary fan on the continental slope and the Prydz Channel, which runs longitudinally across the western part of the outer shelf. In this paper we examine the paleomagnetic, environmental magnetic, and sedimentary characteristics of a recently collected gravity core from Prydz Bay [6, 7]. The aim is to reconstruct environmental change during the Holocene and Late Pleistocene, in order to provide critical information for regional paleoclimatological reconstructions.

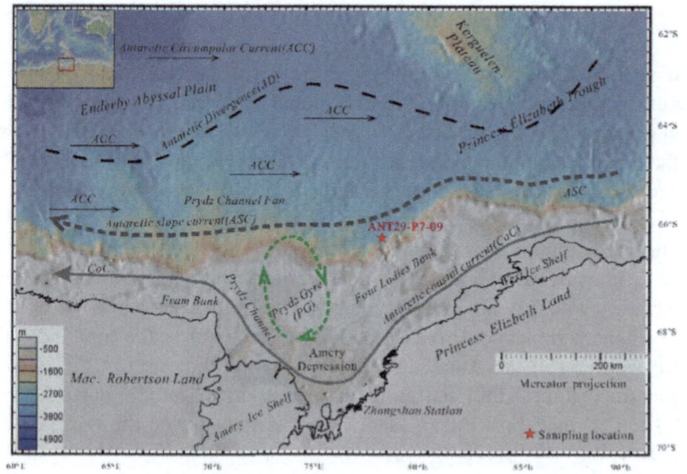

Fig. 1. Map of Prydz Bay and location of the core.

2 Materials and Methods

Core ANT-P7-09 was collected at a water depth of 1118.50 m on the outer continental slope of Prydz Bay (66°31.758′S, 78°03.883′E) during the 29th Chinese National Antarctic Research Expedition (Fig. 1). The core has a length of 324 cm and was split in two longitudinally, with one half sub-sampled at 2 cm depth intervals [8–11]. The samples were freeze-dried prior to analysis.

After pretreating the samples with 5% H2O2 and 0.2 M HCl to remove organic matter and biogenic carbonate, respectively, sediment grain size was measured using a laser size analyzer (Coulter LQ-100Q). Before analysis, 0.5 M sodium hexametaphosphate

((NaPO3)6) was added and the samples were ultrasonically disaggregated [10]. Grain size parameters were calculating following Poizot et al. [11].

Approximately 0.5 g of each sample was acidified with 10% HCl to remove carbonates. Then, TOC and $\delta^{13}C$ were measured using a continuous flow isotope ratio mass spectrometer with an accuracy of $\leq 0.2\%$ in weight percent. $\delta^{13}C$ data are reported here with standard errors, and the analytical precision was $\leq 0.2‰$.

Bulk sediment dating often uses AMS^{14}C dating methods. Here the bulk sediment ages were determined using an ultralow-background liquid scintillation analyzer (1220 Quantulus) at the Guangzhou Institute of Geochemistry, Chinese Academy of Sciences, China. We obtained a total of six ^{14}C ages and used the CALIB software to correct the age for each depth that is estimated according to the equal interval interpolation method and what are except for calendar ages and dating mensuration point. The calibrated ages are discussed in this paper [12–16].

For the paleomagnetic measurements, a total of 162 specimens were collected using 2 × 2 cm non-magnetic plastic cubes. We measured the natural remanent magnetization (NRM) and then subjected the samples to stepwise alternating field demagnetization up to 80 mT at 5–10 mT intervals.

For environmental magnetic analyses, a Bartington Instruments MS2 magnetic susceptibility meter was used to measure low-field magnetic susceptibility (χ). Anhysteretic remanent magnetization (ARM) was imparted with a Schonstedt GSD-1 alternating field demagnetizer. Isothermal remanent magnetization (IRM) was measured with a 2-G Enterprises model 660 pulse magnetizer, and IRM obtained in a 1T pulsed field is hereafter referred to as SIRM [17–20]. All magnetic measurements were conducted in the Paleomagnetism and Geochronology Laboratory, Institute of Geology and Geophysics, Chinese Academy of Sciences, Beijing, China [21, 22].

3 Results

3.1 Grain Size, TOC, $\delta^{13}C$, and AMS^{14}C

The samples have a clay (<4 μm grain size) content of 2.40%–51.07%, and silt (4–63 μm) content varies between 12.34% and 71.56%. The sand (>63 μm) fraction ranges from 2.89% to 84.32%, and the average concentrations of clay, silt, and sand are 19.53%, 40.51%, and 39.96%, respectively. Sand content is less than 18% from 0 to 136 cm core depth, and the average percentage of clay and silt is 82%. In contrast, the 136–300 cm interval is dominated by silt and sand, with an average clay content of 9%. In the 300–323 cm interval, the average clay and sand contents are all greater than 20%. The core ANT-P7-09 sediments are mainly silty clay and clayey sand (Fig. 2).

TOC concentrations range from 0.06% to 1.28%, with an average value of 0.42%, and $\delta^{13}C$ values range from −29.19‰ to −24.56‰, with an average value of −26.32‰. Variations in TOC and $\delta^{13}C$ are similar to those in grain size.

Sediment samples with high carbon content were dated from depths of 1, 7, 21, 41, 91, and 123 cm. The results are listed in Table 1. Samples from depths of 1 and 7 cm yielded corrected ages of 3875–3995 and 11,615–11,960 Cal BP, respectively; samples from 21 and 41 cm gave respective corrected ages of 15,760–15,990 and 22,305–22,460 Cal BP; and at 91 cm and 123 cm gave respective corrected ages of 25,530–25,800 and 25,210–25,560 Cal BP. Therefore, we inferred a Late Pleistocene to early Holocene age for the core. Employing these data, interpolation and extrapolation methods with linear interpolation were used to obtain corresponding depths.

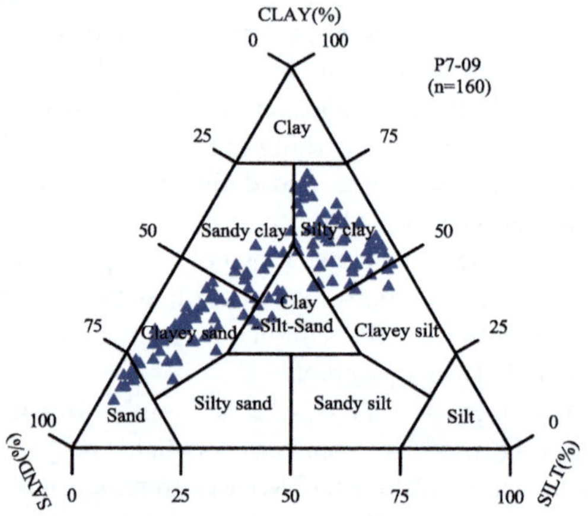

Fig. 2. Ternary diagram of particle size data for the core sediments.

Table 1. Radiocarbon dating results of Borehole ANT-P7-09

Beta number	Depth(cm)	Cal. BP	$\delta^{13}C$(‰)	AMS^{14}C age	Calibrated age range (2σ)
427282	1	3945	−25.1	3680 ± 30 BP	Cal BC 2045 to 1925 (Cal BP 3995 to 3875)
422902	7	10,180	−25.1	10,180 ± 40 BP	Cal BC 10,010 to 9665 (Cal BP 11,960 to 11,615)
422903	21	13,250	−24.4	13,250 ± 40 BP	Cal BC 14,040 to 13,810 (Cal BP 15,990 to 15,760)
422904	41	18,540	−25.4	18,540 ± 60 BP	Cal BC 20,510 to 20,355 (Cal BP 22,460 to 22,305)
422905	91	21,360	−25.1	21,360 ± 80 BP	Cal BC 23,850 to 23,580 (Cal BP 25,800 to 25,530)
427281	123	25,370	−24.5	21,070 ± 80 BP	Cal BC 23,610 to 23,260 (Cal BP 25,560 to 25,210)

3.2 Magnetic Mineralogy

The samples generally show a gentle increase in susceptibility on heating to ~300 °C followed by a decrease in susceptibility at ~500 °C. A sharp Hopkinson peak and a rapid decrease at ~580 °C indicate the presence of single-domain magnetite. In addition, the sample from a depth of 14–16 cm shows a continuing decrease in susceptibility until 680 °C, suggesting the presence of hematite. The sample from a depth of 108–110 cm also shows a continuing decrease in susceptibility above 580 °C, indicating the presence of oxidized magnetite [12, 13].

Hysteresis loops generally saturate below 300 mT, indicating the dominance of soft magnetic minerals such as magnetite. Remanent coercivity is relatively low (25–60 mT), indicating that the magnetic minerals in the samples are dominated by ferrimagnetic minerals.

3.3 Environmental Magnetic Properties

Stratigraphic variations in the concentration of related magnetic parameters (χ, ARM, and SIRM) are shown in Fig. 4a–c. χ values range from 1.56×10^{-8} to 12.02×10^{-8} m^3 kg^{-1}, with an average value of 4.77×10^{-8} m^3 kg^{-1}. χ values are lower and more stable below 1.5 m, and increase upward to 1.3 m. From 1.3 to 0.2 m, χ values are higher and stable except for fluctuations between 1.2 and 0.8 m. The uppermost 0.2 m gives slightly lower χ values. The concentration-related parameters ARM and SIRM show similar trends to that of χ, except that they show a declining trend from 1.3 to 0.2 m.

3.4 Paleomagnetism

Results for magnetic declination (Dec), inclination (Inc), maximum angular deviation (MAD), and virtual geomagnetic pole (VGP) are shown in Fig. 4d–g. The variations in declination shows a simple changing model, with samples at the top part (0–180 cm) showing Dec > 90°, and samples below 180 cm mainly showing Dec < 90°. Figure 4f shows that core MAD values are mostly <5°, part of layer >5°, a small amount of layer can reach more than 10°, the region of high MAD values generally corresponds to an interval during which the angle of inclination changes.

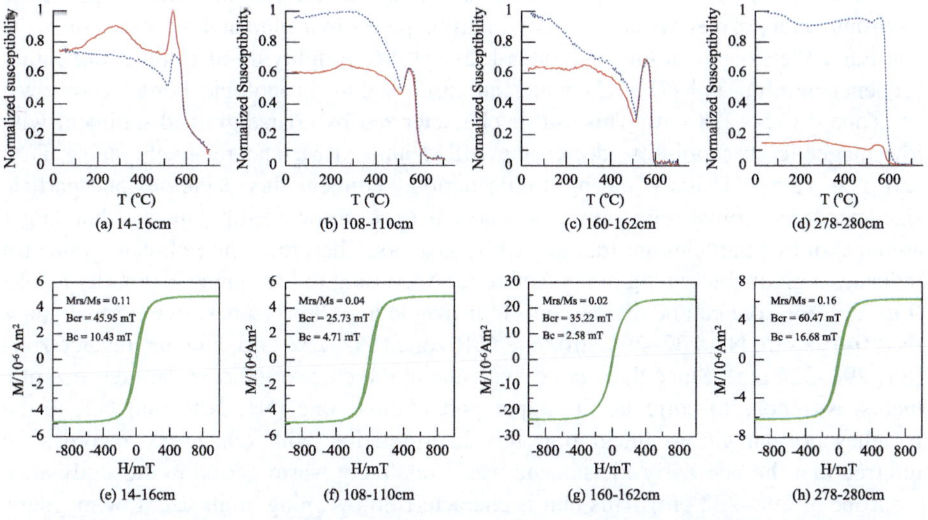

Fig. 3. Thermomagnetic curves for selected samples.

Red and blue curves represent the heating and cooling curves, respectively (a–d). Hysteresis loops after slope correction for paramagnetic contribution (e–h).

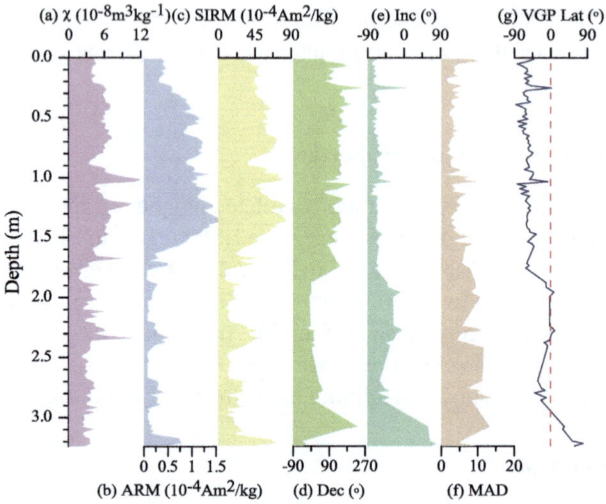

Fig. 4. Downcore changes in magnetic parameters.

4 Discussion

Previous studies have shown that sediment particle size is an important influence on magnetic characteristics [14]. TOC reflects the amount of initial productivity in the study area, and $\delta^{13}C$ values can be used to decipher the source of organic material in the sediment. The ARM/SIRM ratio is an index of magnetic particle size. Based upon these environmental proxies, we attempted to interpret paleoenvironmental processes since the late Early Pleistocene in the study area [15–17]. We roughly divided this sedimentary sequence into four units (Fig. 5), which are discussed in chronological order as follows:

Zone 4 (324–298 cm): This unit is characterized by coarse-grained sediment with low magnetic susceptibility, decreasing SIRM and ARM, and relatively stable TOC and $\delta^{13}C$ values. Usually, during a cold period, sediment flux decreases and particle size increases; conversely, during a warm period, snow melting means that larger amounts of fine particles are transported by glaciers. Therefore, the sediment grain size indicates a gradual warming in the Antarctic. According to the magnetic polarity results (Fig. 5p), the core can be divided into four magnetic polarity zones: two normal zones (N1, 0–192 cm; N2, 232–298 cm), one N/R zone (192–232 cm), and one reverse zone (R1, 298–324 cm). Since there is no evidence in the core to indicate the presence of a hiatus, we chose to correlate the upper part of this zone (N1, N/R, and N2) to the Brunhes normal chron. Integrating this interpretation with our proxy records, we inferred that the late Early Pleistocene was a relatively warm period in the study area.

Zone 3 (298–232 cm): This unit is characterized by fining grain size, low magnetic susceptibility, decreasing TOC and $\delta^{13}C$ values, and relatively stable SIRM and ARM, suggesting that depositional processes changed little during this time and that the climate was warm. In addition, the silt content in the depth interval of 250–232 cm is low, likely suggesting a climate transition from warm to cold.

Zone 2 (232–162 cm): This unit shows alternating appearing grain size with gravels, fluctuating magnetic susceptibility values, and decreasing SIRM, ARM, TOC, and $\delta^{13}C$ values, suggesting significant variations in the depositional environment and a warm–cold climate. Furthermore, the depth interval 232–192 cm records alternations between positive and negative inclination. Given the environmental magnetism parameters and the frequent changes in $\delta^{13}C$ values in this zone, this period might correspond to the climatic transition during the Mid-Brunhes event at $\sim$430,000 years ago in the Antarctic region.

Zone 1 (162–0 cm): This unit is characterized by fining grain size, and increasing SIRM, ARM, TOC, and $\delta^{13}C$ values, indicating gradual climatic cooling. From 102–96 cm, the unit is coarse-grained and shows large fluctuations in magnetic susceptibility, decreasing SIRM and ARM, and relatively stable TOC and $\delta^{13}C$ values, suggesting the frequent occurrence of short-lived snow-melt events [18, 19]. From 96–0 cm, the lithology is clayey silt with a gradual upwards decrease in clay content, χ, ARM, and SIRM, and increase in TOC content. It is thus inferred that the climate was stable in the context of global warming [20].

A lower ARM/SIRM ratio indicates a coarser sediment particle size and that the sediments contain more multidomain (MD) particles. Based on our ARM/SIRM ratio record (Fig. 5k), in conjunction with the smaller coercivity of remanence (Bcr) characteristics (Fig. 3e–h), it is inferred that the core sediments are mostly associated with pseudo single domain (PSD)-MD particles.

Reports on the Brunhes–Matuyama boundary in the Antarctica region are relatively scarce. Pandit et al. [21]. conducted paleomagnetism research on several cores in this area, including one 400-cm-long core that recorded the polarity reversal 780,000 years ago. In addition, Jovane et al. [22]. recorded the Brunhes–Matuyama polarity reversal at $\sim$500 cm depth in a core from the Ross Sea. Nevertheless, whether the reversed polarity intervals in the core sediments in this article represent the same intervals as the afore-mentioned studies needs to be confirmed by further empirical research.

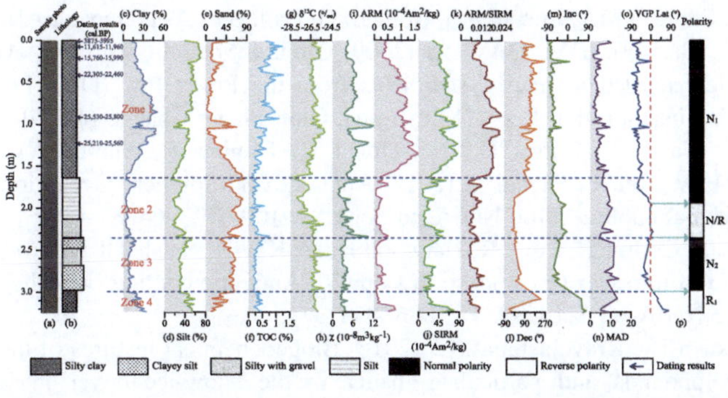

Fig. 5. Core photograph (a), lithology (b), grain size (c–e), magnetic parameters (h–k) and paleomagnetic data (l–o). Magnetic polarity of the core is shown on the right-hand side (p).

5 Conclusion

Rock magnetism results show that ferrimagnetic minerals with low coercivities dominate the core, and ARM/SIRM ratios are low, indicating the presence of magnetite minerals in PSD and MD particles. Variations in the paleoenvironmental proxy records allow us to distinguish four intervals, which appear to be associated with different depositional environments and climatic conditions. In summary, we infer that (a) the climate was relatively warm in the late Early Pleistocene, (b) this was followed by continuous cooling since the Middle Pleistocene, and (c) short-term snow-melt events are superimposed on a background warming trend that probably corresponds to the Holocene. These results confirm that paleoenvironmental processes were complex in the East Antarctic region and that more analyses are required to clarify their driving forces.

Acknowledgments. We thank Linggang Tang for assisting with magnetic hysteresis loops and χ–T curve experiments, as well as the scientific staff and the crew of R/V Xuelong during the 29th Chinese National Antarctic Research Expedition.

References

1. Maher BA (2008) Environmental magnetism and climate change. Contemp Phys 48:247–274
2. Liu Q, Roberts AP, Larrasoana JC, Banerjee SK, Guyodo Y, Tauxe L, Oldfield F (2012) Environmental magnetism: principles and applications. Rev Geophys
3. Raj R, Chamyal LS, Juyal N, Phartiyal B, Ali SN, Thakur B (2016) Late Quaternary fluvio-aeolian interaction: palaeoenvironment and palaeoclimatic conditions in the pediment zone of Vatrak River basin, western India. Z Geomorphol 60:151–169
4. Levitan MA, Balukhovskii AN, Antonova TA, Gelvi TN (2013) Quantitative parameters of pleistocene pelagic sedimentation in the Pacific Ocean. Geochem Int 51:345–352
5. Avramidis P, Iliopoulos G, Panagiotaras D, Papoulis D, Lambropoulou P, Kontopoulos N, Siavalas G, Christanis K (2014) Tracking Mid- to Late Holocene depositional environments by applying sedimentological, palaeontological and geochemical proxies, Amvrakikos coastal lagoon sediments, Western Greece, Mediterranean Sea. Quat Int 332:19–36
6. Escutia C, Bárcena MA, Lucchi RG, Romero O, Ballegeer AM, Gonzalez JJ, Harwood DM, Florindo F, Harwood DM, Levy RH (2009) Circum-Antarctic warming events between 4 and 3.5 Ma recorded in marine sediments from the Prydz Bay (ODP Leg 188) and the Antarctic Peninsula (ODP Leg 178) margins. Glob Planet Change 69:170–184
7. Mckay R, Naish T, Carter L, Riesselman C, Dunbar R, Sjunneskog C, Winter D, Sangiorgi F, Warren C, Pagani M (2012) Antarctic and Southern Ocean influences on Late Pliocene global cooling. Proc Nat Acad Sci USA 109:6423–6428
8. Crampton JS, Cody RD, Levy R, Harwood D, McKay R, Naish TR (2016) Southern Ocean phytoplankton turnover in response to stepwise Antarctic cooling over the past 15 million years. Proc Nat Acad Sci USA 113:6868–6873
9. Nemirovskaya IA, Kravchishina MD (2015) Biogeochemical features of the distribution of organic compounds and particulate matter in the snow-ice cover in East Antarctica. Geochem Int 53:430–440

10. Zhang W, Yu L (2003) Magnetic properties of tidal flat sediments of the Yangtze Estuary and its relationship with particle size. Sci China Ser D 46:954–966
11. Poizot E, Méar Y, Biscara L (2008) Sediment trend analysis through the variation of granulometric parameters: a review of theories and applications. Earth-Sci Rev 86:15–41
12. Duan Z, Liu Q, Yang X, Xing G, Su Y (2014) Magnetism of the Huguangyan Maar Lake sediments, Southeast China and its paleoenvironmental implications. Palaeogeog Palaeoclimatol Palaeoecol 395:158–167
13. Ouyang T, Tian C, Zhu Z, Qiu Y, Appel E, Fu S (2014) Magnetic characteristics and its environmental implications of core YSJD-86GC sediments from the southern South China Sea. Chin Sci Bull 59:3176–3187
14. Li P, Li P, Zhang X, Cao C, Xu X, Du J, Liu L, Li P, Li P, Zhang X (2005) Magnetic properties of different grain-sized particles of sediments from the Okinawa Trough and their relationships to sedimentary environment. Chin Sci Bull 50:696–703
15. Macrì P, Sagnotti L, Camerlenghi A, Rebesco M (2001) Environmental magnetic record of the SEDANO deep sea cores (Pacific margin of the AntarcticPeninsula) and interhemispheric correlation of late-Pleistocene climatic events. In: International ANTOSTRAT symposium
16. Hamdan MA, Ibrahim MIA, Shiha MA, Flower RJ, Hassan FA, Eltelet SAM (2016) An exploratory Early and Middle Holocene sedimentary record with palynoforms and diatoms from Faiyum lake, Egypt. Quatern Int 410:30–42
17. Scarciglia F, Mercatante G, Fondevilla V, Anadón P, Oms O, Donato P, Agnini C, Papini M, Rook L, Ghinassi M (2018) Pleistocene paleosol development and paleoenvironmental dynamics in East Africa: a multiproxy record from the Homo-bearing Aalat pedostratigraphic succession, Dandiero basin (Eritrea). Quatern Sci Rev 191:275–298
18. Fabio F, Steven MB, Patrick SE, Carl R, Andrew PR, Patricia AW, Jason MW (2003) Magnetobiostratigraphic chronology and palaeoenvironmental history of Cenozoic sequences from ODP sites 1165 and 1166, Prydz Bay, Antarctica. Palaeogeogr Palaeoclimatol Palaeoecol 198:69–100
19. Hou H, Luo Y, Zheng H, Wang B, Tang X (1998) Heinrich layer in Antarctic marine sediments and its significance to global changes. Chin Sci Bull 43:1830–1834
20. Sagnotti L, Macrí P, Camerlenghi A, Rebesco M (2001) Environmental magnetism of Antarctic Late Pleistocene sediments and interhemispheric correlation of climatic events. Earth Planet Sci Lett 192:65–80
21. Pandit MK, De Wall H (2014) Preliminary magnetic susceptibility characteristics of the Bharati promontory (Grovness Peninsula), Larsemann Hills, Prydz Bay region, East Antarctica. J Geol Soc India 84:163–173
22. Jovane L, Acton G, Florindo F, Verosub KL (2008) Geomagnetic field behavior at high latitudes from a paleomagnetic record from Eltanin core 27–21 in the Ross Sea sector, Antarctica. Earth Planet Sci Lett 267:435–443

Preliminary Exploration of the Distribution Characteristics and Influencing Factors of the Intangible Cultural Heritage in Sichuan Province Based on GIS

Yiming Xu[1], Siyang Gu[1], and Ying Liu[2](✉)

[1] Sichuan Agricultural University, Chengdu 611130, Sichuan, China
[2] Sichuan Agricultural University, Ya'an 625000, Sichuan, China
chinacloudy@yeah.net

Abstract. This paper analyzed the distribution of the intangible cultural heritage in Sichuan province by using the nuclear density method in Arcgis 10.2, and obtained the provincial and nuclear density distribution maps of the intangible cultural heritage in Sichuan province, as well as the structural and spatial distribution characteristics of the intangible cultural heritage in Sichuan province. The authors analyzed and summarized the influencing factors of these characteristics and put forward the suggestions on strengthening the infrastructure construction in Western Sichuan province and encouraging exchanges and cooperation among the ethnic groups.

Keywords: GIS · Landscape culture · Intangible cultural heritage

1 Introduction

Protecting the intangible cultural heritage is not only the need of the cultural inheritance in China, but also the historical mission that cannot be ignored. The intangible cultural heritage is also an important part of the landscape culture in the urban development. The landscape culture is the spiritual concept formed by people's long-term production practice with the surrounding environment and adapted to the landscape, which exactly echoes the intangible nature of the intangible cultural heritage and its unique spiritual connotation. At present, the research on the intangible cultural heritage in China pays more attention to the differences between the different intangible cultural heritages and the particularity of the project itself [1]. Many disciplines such as ecology, anthropology, folklore, aesthetics and communication have been applied to the research of the intangible cultural heritage. However, the research on the spatial distribution by using the method of GIS is rare. The purpose of this paper is to analyze the spatial structures, the spatial characteristics and the influencing factors of the intangible cultural heritage in Sichuan province by means of GIS, summarize their characteristics and put forward suggestions for improvement, so as to provide reference for the construction of the landscape culture in Sichuan province by using the intangible cultural heritage.

Z. Xu et al. (Eds.): CSIA 2019, AISC 928, pp. 114–122, 2020.
https://doi.org/10.1007/978-3-030-15235-2_18

2 Research Methods and Data Processing

2.1 Research Methods

The GIS spatial analysis method is usually used to study the residential spatial differentiation, because it can generate the new data by calculating the spatial entity data, and can analyze the vertical relationship between the data by overlapping and merging functions, so that the research results are not limited to the single-factor static analysis, but tend to the multi-factor dynamic analysis process [2]. This paper uses Arcgis10.2 to analyze the number, types and distribution characteristics of the intangible cultural heritage in Sichuan province, calculate their nuclear density, and study their spatial location, spatial distribution characteristics and influencing factors of the intangible cultural heritage in Sichuan province.

2.2 Data Sources and Processing

The data of this paper are derived from the national directory and successor list of China's intangible cultural heritage network. There are 9 categories, with 56 items.

In order to clarify the specific distribution points of the intangible cultural heritage, the list of the inheritors is introduced in the data processing, and the two lists are matched one by one, and thus the list of the intangible cultural heritage projects in Sichuan province is summarized (Table 1). In order to facilitate the use of the GIS for the data processing, 13 items of the intangible cultural heritage projects which have not found the specific locations or scopes are too broad and duplicate with other areas are removed. This table is divided into townships according to their administrative divisions.

Table 1. Sichuan intangible cultural heritage lists

Name	Type	Location
Ke Zhi of Yi Nationality	Folk Literature	Sichuan - Liangshan Yi Autonomous Prefecture - Meigu County
Song of Bashan Back-carriers	Traditional Music	Bazhong, Sichuan
Multi-voice Folk Songs (Qiang Multi-voice Folk Songs)	Traditional Music	Sichuan - Aba Tibetan and Qiang Autonomous Prefecture - Songpan County
Buddhist Music (Capsac Sanskrit)	Traditional Music	Sichuan Province - Aba Tibetan and Qiang Autonomous Prefecture - Rutang County
Weeding Gong and Drum in North Sichuan	Traditional Music	Qingchuan County, Sichuan Province
Performing and Making Skills of Qiang Flute	Traditional Music	Maoxian, Sichuan
Derong Xue Qiang	Traditional Dance	Sichuan - Ganzi Tibetan Autonomous Prefecture - Rongxian

(*continued*)

Table 1. (*continued*)

Name	Type	Location
Guo Zhuang Dance (Ganzi Guo Zhuang)	Traditional Dance	Sichuan - Ganzi Tibetan Autonomous Prefecture - Xinlong County
Qiang's sheepskin dance	Traditional Dance	Sichuan Province - Aba Tibetan and Qiang Autonomous Prefecture - Wenchuan County
Dragon Dance (Tongliang Dragon Dance, Zhanjiang Human Dragon Dance, Shanwei Tumbling Golden Dragon, Pujiang Bench Dragon, Changxing Louver Dragon, Fenghua Cloth Dragon, Luzhou Rain-altar Color Dragon)	Traditional Dance	Luxian County, Sichuan
Mountain-Crossing Hinge	Traditional Dance	Sichuan - Bazhong - Pingchang County
Castram Dance	Traditional Dance	Heishui county, Sichuan Province
Shadow Play (Sichuan Shadow Play)	Traditional Opera	Sichuan - Nanchong - Langzhong City
Lantern Opera (Liangshan Lantern Opera, North Sichuan Lantern Opera)	Traditional Opera	Nanchong, Sichuan
Bamboo Weaving (Quxian Liu's Bamboo Weaving)	Traditional Art	Sichuan - Dazhou - Quxian County
Clay Sculpture (Xu's Clay Painted Sculpture)	Traditional Art	Sichuan - Suining - Daying County
Mianzhu Woodblock New Year Pictures	Traditional Art	Deyang, Sichuan
Tibetan Knitting and Cross-stitch Embroidery	Traditional Art	Sichuan - Aba Tibetan and Qiang Autonomous Prefecture
Tibetan Thang-ga (Mian Tang School, Qinze School and Karma Gazi School)	Traditional Art	Sichuan - Ganzi Tibetan Autonomous Prefecture
Qiang Embroidery	Traditional Art	Sichuan - Aba Tibetan and Qiang Autonomous Prefecture - Wenchuan County
Sugar Sculpture (Chengdu Sugar Painting)	Traditional Art	Chengdu, Sichuan
Sichuan Embroidery	Traditional Art	Chengdu, Sichuan
Sichuan Figured Satin Weaving Skills	Traditional Skill	Chengdu, Sichuan
Chengdu Lacquer Art	Traditional Skill	Chengdu, Sichuan
Brewing Techniques of Luzhou Laojiao Liquor	Traditional Skill	Luzhou, Sichuan

(*continued*)

Table 1. (*continued*)

Name	Type	Location
Silver Jewelry Manufacturing Techniques (Yi Silver Jewelry Manufacturing Techniques)	Traditional Skill	Sichuan - Liangshan Yi Autonomous Prefecture - Butuo County
Deep Drilling and Drawing Techniques of Salt in Zigong Well	Traditional Skill	Daying County, Sichuan
Traditional Production Techniques of Bean Paste (Pixian Bean Paste Traditional Production Techniques)	Traditional Skill	Sichuan - Chengdu - Pixian
Lacquerware Decoration Techniques of Yi Nationality	Traditional Skill	Sichuan - Liangshan Yi Autonomous Prefecture - Xide County
Umbrella Making Skills (Oil-paper Umbrella Making Skills)	Traditional Skill	Sichuan - Luzhou - Jiangyang District
Traditional Brewing Techniques of Distilled Liquor (Traditional Brewing Techniques of Shuijingfang Liquor)	Traditional Skill	Chengdu, Sichuan
Traditional Brewing Techniques of Distilled Liquor (Traditional Brewing Techniques of Tuobai Qujiu Liquor)	Traditional Skill	Sichuan - Suining - Shehong County
Traditional Brewing Techniques of Distilled Liquor (Traditional Brewing Techniques of Wuliangye Liquor)	Traditional Skill	Yibin, Sichuan
Bamboo paper making techniques	Traditional Skill	Jiajiang County, Sichuan Province
Tibetan Metal Forging (Tibetan Copper Forging)	Traditional Skill	Baiyu County, Sichuan Ganzi Tibetan Autonomous Prefecture
Black Tea Making Skills (South Road Tea Making Skills)	Traditional Skill	Ya'an, Sichuan
Tibetan Engraving and Printing Techniques of the Sutra printing house of Dege	Traditional Skill	Dege County, Sichuan
Tibetan medicine (Mercury Washing Method of Lhasa Northern School Tibetan Medicine and Compatibility Skills of Renqing Changjue Tibetan Medicine, Ganzi Southern School Tibetan Medicine)	Traditional Skill	Ganzi Tibetan Autonomous Prefecture, Sichuan
New Year of Qiang Nationality	Folk Custom	Sichuan - Aba Tibetan and Qiang Autonomous Prefecture - Maoxian
Lifting Pavilion (Core, Iron Branch, Piaose) (High Dam Loading)	Folk Custom	Sichuan - Yibin - Xingwen County, Sichuan - Mianyang - Jiangyou City
Sanhui Choi Ting Hui	Folk Custom	Sichuan - Dazhou - Quxian County
Gold Finger Plate	Quyi	Chengdu, Sichuan
Sichaun Anklong	Quyi	Chengdu, Sichuan

2.3 Summary

According to the items listed in the table, Sichuan province is rich in the intangible cultural heritages. At present, there are nine categories of the intangible cultural heritages classified by the state, each of which is involved. The specific number and distribution of different categories of projects are different and will be analyzed in the following chapters. The data used in this paper cannot fully reflect the spatial distribution characteristics of the intangible cultural heritages in Sichuan province after some processing, which has relativity and limitations.

3 Analysis of the Distribution of the Intangible Cultural Heritages in Sichuan and the Results

3.1 The Types and Structures of the Intangible Cultural Heritages in Sichuan

Based on the data of China intangible cultural heritage network, Table 1 divides Sichuan intangible cultural heritage projects into nine categories: folk literature, traditional music, traditional dance, traditional drama, traditional art, traditional skills, traditional medicine, folk custom and folk art. The number of the items was 1, 5, 6, 2, 8, 15, 1, 3 and 2 respectively. The ratio of these nine types of projects to the total projects is 2.4%, 11.6%, 13.9%, 4.6%, 18.6%, 34.9%, 2.4%, 7% and 4.6%.

It can be seen that among the intangible cultural heritage projects in Sichuan province, the number of the items belonging to the traditional skills is the largest (15 items), accounting for 34.9% of the total, followed by the traditional arts (8 items), accounting for 18.6% of the total. The traditional dance (6 items), the traditional music (5 items), the folk customs (3 items), the traditional drama (2 items) and Quyi (2 items), the folk literature (1 item) and the traditional medicine (1 item) accounted for 13.9%, 11.6%, 7%, 4.6%, 4.6%, 2.4% and 2.4% of the total.

In summary, the types and structures of the intangible cultural heritages in Sichuan province are mainly the traditional arts, followed by the traditional arts and the traditional music, with fewer items of the folk custom, the traditional drama, the folk art, the folk literature and the traditional medicine.

3.2 Spatial Distribution Structures of the Intangible Cultural Heritages in Sichuan

3.2.1 Provincial Distribution Characteristics

In Arcgis10.2, the distribution of the intangible cultural heritages in Sichuan is visualized, and Fig. 1 is obtained.

From Fig. 1, we can see that the spatial distribution of the intangible cultural heritages in Sichuan province is uneven, with the characteristics of the regional differences and the ethnic differences, and the distribution of the intangible cultural heritages in groups. Among them, Chengdu is the most densely distributed area, with more and more densely distributed in the east, followed by the south, and more sparse and fewer in the west and the north.

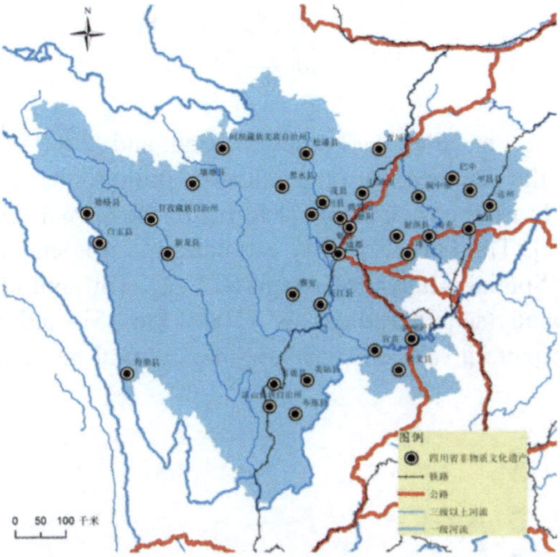

Fig. 1. Distribution map of the intangible cultural heritages in Sichuan (Source: Self drawn by the author)

It can be seen that the distribution of the intangible cultural heritages in Sichuan has the following characteristics:

(1) Groups are formed according to the distribution of the ethnic settlements. Because the geomorphology, climates and the number of resources of different ethnic settlements are different, these environmental components will make different settlements form different cultures, thus producing different intangible cultural heritages.
(2) The distribution density of the water system and the road network is relatively high. The river system in Sichuan province belongs to the Yangtze River and its tributaries, while the road network has developed in the eastern part of the railways and highways, and underdeveloped in the western part, which is related to the topography and the climate. The abundant water resources in the eastern part of China can guarantee the self-sufficiency of the human life and the production. On this basis, the construction of highways and railways will help to spread and disseminate the cultures of the inhabited areas. As can be seen from Fig. 1, the number of the intangible cultural heritage projects in the eastern part is larger than that in the western part.
(3) There are universal intangible cultural heritages in the minority autonomous prefectures and their vicinity.

3.2.2 Analysis of the Spatial Distribution and the Accumulation Areas

In this paper, the nuclear density is used to measure the spatial aggregation areas. Its formula is:

$$f_n(x) = \frac{1}{nh}\sum_{i=1}^{n} k(\frac{x - X_1}{h})$$

In this equation, $k()$ is called the kernel function, and $h > 0$ is the bandwidth, while $(x - X_1)$ represents the distance from the valuation point to the event. The main factors affecting K, D and E are the mathematical forms of the kernel function and the values of the bandwidth [3]. The kernel density estimation is implemented using the Kernel Density tool in the Spatial Analyst of Arcgis10.2 [4]. Through many experiments, we choose the broadband (search radius) as 1500.0 km [5], and generate the nuclear density distribution pattern of Sichuan intangible cultural heritages (Fig. 2).

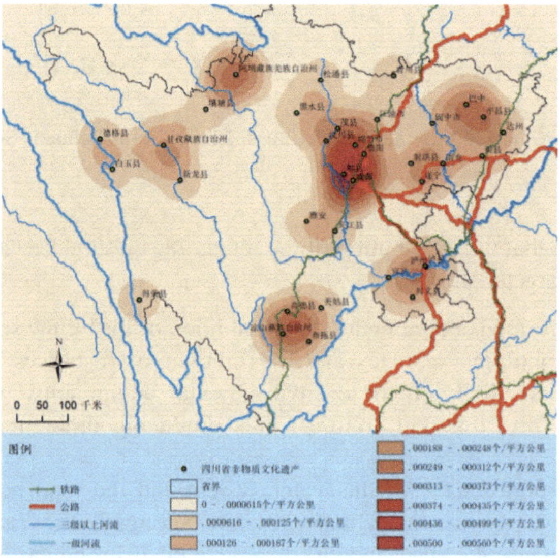

Fig. 2. Nuclear density distribution map of the intangible cultural heritages in Sichuan (Source: Self drawn by the author)

As can be seen from Fig. 2, the distribution of the intangible cultural heritages in Sichuan province has formed one high-density core circle, five sub-core circles and three sub-core circles. Among them, the high-density core circle is spread outward with Chengdu as the center, and its radiation range includes Songpan County, Heishui County, Ya'an, Jiajiang County and Qingchuan County. Taking Aba Tibetan Autonomous Prefecture, Ganzi Tibetan Autonomous Prefecture, Liangshan Yi Autonomous Prefecture, Luzhou and Bazhong as the core areas, the radiation scope includes Xinlong County, Xide County, Meigu County, Butuo County, Luxian County, Yibin, Xingwen County, Langzhong City, Pingchang County, Dazhou County, and Quxian County. The small core areas take Dege County, Derong County, and Sui Ning County as the core areas.

4 Summary of Factors Affecting the Distribution of the Intangible Cultural Heritages in Sichuan

4.1 Environmental Factors

In the analysis of Fig. 1, the distribution characteristics of the intangible cultural heritages in Sichuan province have been put forward. It can be seen that the geographical and environmental factors such as topography, river network, climate characteristics and the generation of the intangible cultural heritages are inextricably linked. Firstly, the topography and the physiognomy affect the distribution of settlements. The number of the settlements in flat areas is obviously more dense than that in the plateau or the mountainous areas, and the distribution points in the alpine and the cold climatic areas are more sparse. Secondly, the distribution of the water systems further refines the scopes of the settlement. In Fig. 2, the high-density core and the sub-core areas of the intangible cultural heritages are distributed around the water systems. Finally, the areas with more settlements tend to be more economically developed, which is conducive to the exchange and integration of the settlements' cultures, and is easier to produce new cultures and can ultimately impel the formation of the new intangible cultural heritages.

4.2 Policy Factors

In the process of declaration and approval of the intangible cultural heritages in China, most of the projects can successfully pass the declaration and approval, but there are still some projects that cannot pass the examination and approval. The reasons why these projects cannot pass the examination and approval can be many-sided, but at present, China has not yet decided a very detailed assessment method on the identification of the intangible cultural heritages, so there is no lack of the subjective evaluation methods. How to quantify and evaluate the importance, endangerment and values of each project is the key point for the relevant scholars to make the scientific research and policy as soon as possible.

5 Conclusion

The intangible cultural heritages are the indispensable part of our cultural heritages. The types and contents of the intangible cultural heritages in different regions are essentially different. In the distribution characteristics of Sichuan intangible cultural heritages, both the provincial distribution and the spatial aggregation and distribution are centered on Chengdu, which shows that the economic prosperity can promote the results of the cultural development. The traditional skills, the traditional arts and the traditional dances accounts for a large proportion of the intangible cultural heritages in Sichuan province, and other categories are relatively scarce, showing the uneven geographical distribution and the distribution of clusters. The intangible cultural heritages in Sichuan province have formed a high-density core circle, five sub-core circles and three small core circles. The factors affecting the spatial distribution of the intangible cultural heritages in Sichuan are the environmental factors and the policy factors.

According to the distribution of the intangible cultural heritages in Sichuan province, I put forward two suggestions. First, we should strengthen the infrastructure construction in the western part of Sichuan province to help the western region develop its economy. Second, we should encourage the exchanges and cooperation among the minority inhabited areas in the western part of Sichuan province, and at the same time, we should formulate a more detailed content and support the relevant policies to declare and examine the intangible cultural heritages.

Based on the intangible cultural heritage data of the intangible cultural heritage network of Sichuan province of China, this paper uses Arcgis10.2 to analyze the processed data, obtains the distribution characteristics of the intangible cultural heritages in Sichuan province and preliminarily summarizes the influencing factors, and does not classify different types of the intangible cultural heritages. Further research is needed to expand the follow-up work.

References

1. Wu K, Guo M (2018) Integration of the intangible cultural heritages and the national identity of the landscape cultures. Hunan Packag 33(01):35–38
2. Li X, Liu K (2007) GIS and spatial analysis – principles and methods. Science Press, Beijing, pp 6–7
3. Ouyang Z (2012) RMP analysis of the tourism development of the silk road intangible cultural heritage. J Arid Land Resour Environ 12:203–208
4. Zheng G, Chao Y (2010) Analysis and application of the geographic information system. Publishing House of Electronics Industry, Beijing, pp 36–37
5. Ouyang X (2010) A complete collection of the ArcGIS geographic information systems. Science Press, Beijing, pp 391–392

Real-Time Video Edge Enhancement IP Core Based on FPGA and Sobel Operator

Boyang Liu(✉)

Shandong University, Jinan, China
juyun1982@126.com

Abstract. In order to solve the problems of inefficient retrieval target, visual fatigue and complex algorithm implementation in the monitoring system, this paper proposes an innovative method of image edge enhancement by superimposing edge detection image and source image. Based on this idea, an IP module of edge detection and visual enhancement was designed. This module realized Sobel edge detection algorithm, video alignment and video superposition. I did algorithm verification and code generation on Matlab, and generated IP core in Vivado development environment of Xilinx. The experimental results showed that the system could achieve real-time video processing and enhance the effect of displaying video targets.

Keywords: Sobel · IP · Edge detection · Visual enhancement

1 An Overview of the Edge Detection

With the development of science and technology, video surveillance system has been widely used in banking, transportation, security facilities. However, most of the current video surveillance systems usually only have video recording and return visiting functions. It is still necessary for the surveillance persons to recognize targets by using eye recognition and brain analysis [1]. Faced with massive, long-term, high-intensity and low-recognition information sources, it is easy to cause visual fatigue, visual impairment of supervisors. Therefore, it is urgently needed to realize the technology of real-time visual enhancement to process the mass surveillance videos.

It is found that the contour of video image, i.e. edge information, is an important feature of the image and also the most informative part for the people who use it. Edge enhancement can not only reduce the amount of data in image processing, but also show the boundary or trajectory of the target more clearly [2]. The core of edge enhancement is to highlight the pixels whose gray scale changes step by step around the image through edge detection technology. The commonly used edge detection methods are gradient edge detection based on first-order differential operation and edge detection operators based on second-order differential operation, such as sobel, Roberts, Prewitt, Laplace, canny and so on [3]. Among them, Sobel algorithm is a very effective and commonly used edge detection method, which can not only obtain more accurate edge information, but also has a certain smoothing effect for noise, and is suitable for real-time video signal processing.

Z. Xu et al. (Eds.): CSIA 2019, AISC 928, pp. 123–129, 2020.
https://doi.org/10.1007/978-3-030-15235-2_19

Although edge detection operator can retain image edge information, the loss of image details is too much. In this paper, an innovative method of image edge enhancement is proposed by superimposing edge detection image and source image. It can not only highlight the edge information of the image, but also retain the details of the image, which facilitates the recognition of monitors [4].

Software for image edge enhancement has the characteristics of low cost, flexibility and convenience, but the processing speed of software is not ideal. The field programmable gate array (FPGA), which has the characteristics of parallel processing, pipeline structure, fast configuration and simple burning, can meet the real-time requirements of image edge detection and enhancement. In this paper, based on the edge detection operators of FPGA and Sobel, a real-time video edge enhancement IP core is designed, which realizes the functions of Sobel edge detection algorithm, video alignment and video overlay. The algorithm is validated on Matlab, and the code is written and simulated under the Vivado development environment of Xilinx.

2 Sobel Edge Detection Operator

The Sobel operator is an edge detection algorithm based on first-order differential operation. It includes two 3 * 3 matrices, which take 8 neighborhood pixels around the center point, the left one is vertical, and the right one is horizontal. The operator coefficients of the four pixels adjacent to the central pixel are 2, and the operator coefficients of the four diagonally oblique pixels are 1. This is because the closer the neighborhood points are to the central pixel, the greater the influence is, on the contrary, the farther the distance is, the smaller the influence is. By plane convolution of Sobel operator and image, the approximate values of vertical and horizontal brightness difference can be obtained respectively. The Sobel operator is as follows (Fig. 1):

-1	-2	-1
0	0	0
1	2	1

-1	0	1
-2	0	2
-1	0	1

Fig. 1. The Sobel operator

If I is used to represent the original image, G_x and G_y are used to represent the gray values detected by horizontal and vertical edges respectively. The formulas are as follows:

$$G_x = \begin{bmatrix} -1 & 0 & 1 \\ -2 & 0 & 2 \\ -1 & 0 & 1 \end{bmatrix} * I, G_y = \begin{bmatrix} -1 & -2 & -1 \\ 0 & 0 & 0 \\ 1 & 2 & 1 \end{bmatrix} * I \tag{1}$$

Then we get the approximate gray value of this point through the lower formula:

$$G = \sqrt{G_x^2 + G_y^2} \tag{2}$$

The following approximation formulas are usually used in real-time processing systems with high computational speed.

$$G = |G_x| + |G_y| \tag{3}$$

3 Model Design and Verification

The real-time video edge enhancement model designed in this paper includes three parts: edge detection module, video alignment module and video overlay module. The auxiliary module includes frame-to-pixel conversion module and image-to-frame conversion module. The algorithm is realized and simulated by using MATLAB and Simulink. The model structure is shown in the following Fig. 2.

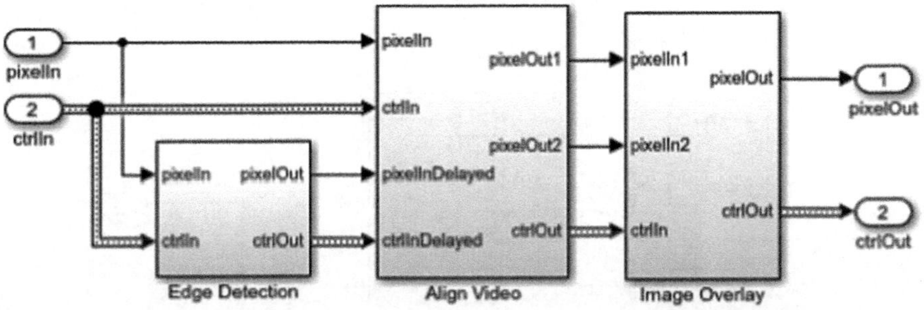

Fig. 2. Schematic diagram of real-time video edge enhancement model based on Simulink

3.1 Frame-to-Pixel Conversion Module

Because video image input is in full frame mode, and edge detection can only process a single pixel, it is necessary to design a frame-to-pixel conversion module before edge detection. In order to simplify the computational complexity and improve the system speed, this design uses gray image temporarily, and the video stream is 204 × 320 pixels.

3.2 Edge Detection Module

When using Sobel operator to detect the edge of an image, firstly, the horizontal gradient and the vertical gradient are calculated by Sobel operator, then the absolute gradient values of the two directions are added together. Finally, the appropriate threshold is selected, and the image edge is judged by the threshold processing module and the result of edge detection is output.

Because the absolute values of Sobel operators are 0, 1 and 2, the combinational logic circuit is used to realize the operation of 3 × 3 window data and horizontal and vertical operators. After detecting the valid frames represented by hStart and vStart, the edge detection module starts processing. In this module, a threshold value is set. When the calculated value is greater than the threshold value, the pixel is judged as an edge. The threshold value is between 0 and 256. The higher the threshold value, the less edge information detected. Because Sobel operator needs to process the data of the first row and the last row of the row of a single pixel, it is necessary to design a data buffer module in the edge detection module. In this design, a frame image has 320 pixels per row, and the shift register has 960 registers.

3.3 Image Alignment Module

Because of the characteristics of pixel stream processing, edge detection will delay, so the output image of edge detection can not be directly superimposed on the source image. To solve this problem, the video alignment module is used to synchronize the pixel stream before the video overlay. The usual way to synchronize pixel streams is to use FIFO structure to push the valid pixels that appeared before, and then pop up the corresponding pixels from the FIFO in time according to the valid signal of the delayed pixels. In the video alignment subsystem, FIFO detects active frames with control signals. Once the valid signals are true, it begins to buffer frames. The structure of the video alignment module is as follows (Fig. 3):

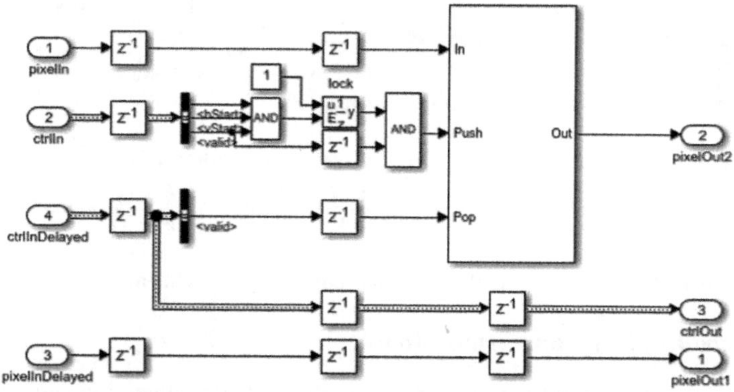

Fig. 3. The structure of the video alignment module

3.4 Image Overlay Module

The module is used to merge and overlay the edge-detected image and the time-aligned source image. In this module, alpha value is designed, and the effective value is between 0 and 1. It is the ratio of the output of edge detection and the original image. If Alpha is 0, the output image is the edge detection result, and if Alpha is 1, the output image is the source image.

$$\text{Superposition image} = \text{alpha} \times \text{source image} + (1 - \text{alpha}) \times \text{edge detection image}$$

3.5 Pixel to Frame Conversion Module

Because edge detection and video alignment process a single pixel, and video output is a two-dimensional matrix of a complete image, that is, full frame mode. Similar to the frame-to-pixel conversion module, a pixel-to-frame conversion module is designed to convert the pixel stream back to the full frame before outputting the image.

3.6 Model Simulation

The design model was simulated in Simulink, and 240 × 320 pixels gray video was input. First, edge detection was carried out, and then the detection results were aligned with the source image and superimposed. The following Figs. 4, 5 and 6 shows source video, edge detection video and overlay video, respectively.

Fig. 4. Source video

Fig. 5. Edge detection video

Fig. 6. Overlapped video

From the simulation results, it can be seen that the system can achieve real-time edge detection of video images. The overlapped images generated by the alignment of the edge detection image and the source image are completely coincident, and there is no image dislocation due to the delay. After the superimposed image, the edge of the target is enhanced, which is very easy to recognize and achieves the desired purpose.

4 IP Core Generation and Synthesis

Using HDL Coder in Matlab, the real-time video edge enhancement model was generated into VHDL code. A new project was built in the integrated design environment Vivado of Xilinx Company. The VHDL code generated by HDL Coder was imported to generate real-time video edge enhancement IP core. The RTL level schematic diagram of IP core is shown as follows (Fig. 7):

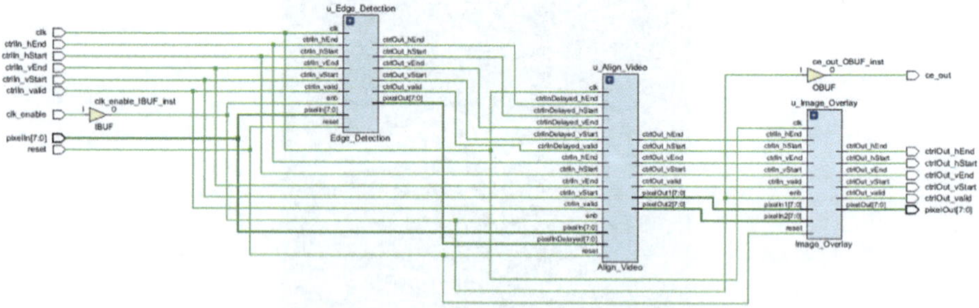

Fig. 7. The RTL level schematic diagram of IP core

Based on the Xilinx Zynq-7000 SoC ZC702 development suite, the IP core of real-time video edge enhancement was synthesized in Vivado, and the synthesis results were obtained. The amount of resources used is shown in the figure (Fig. 8):

Resource	Utilization	Available	Utilization %
LUT	28	303600	0.01
FF	100	607200	0.02
BRAM	0.50	1030	0.05
IO	48	600	8.00
BUFG	1	32	3.13

Fig. 8. The amount of resources

It can be seen from the comprehensive results that the IP core uses only a small part of the FPGA resources and has the characteristics of less resources and higher real-time performance, which is in line with the practical needs.

5 Conclusions

In this paper, the real-time video edge enhancement model was designed and simulated by Matlab and Simulink. The VHDL code of the model was generated by HDL Coder, and the IP core was generated by the Vivado development environment. Experimental results show that the IP core can achieve the edge enhancement function of video image and achieve the desired goal. The design is aimed at 240 × 320 pixel grayscale video, but considering that the monitoring video is mostly 1080P color image in reality, the next step is to improve the design, so that it can be applied to reality Environment.

References

1. Wang H, Zhang Y, Shen H, Zhang J (2017) An overview of image enhancement algorithms. Chin Opt 10(4)
2. Liu B, Gao L (2016) Sobel image edge detection method based on FPGA. Appl Sci Technol 43(6):59–61
3. Sun B, Leng J, Zhao J Research on Sobel edge detection algorithm based on FPGA. Chem Autom Instrum 45
4. Jiao S, Guan C (2014) FPGA implementation of Sobel edge detection for real-time video signals. Electron Sci Tech

Research and Implementation of the Intelligent Tourism System Based on VRGIS

Yongqi Liao(✉)

Sichuan Technology and Business College, Dujiangyan 611830, Sichuan, China
shuxiao4617683@163.com

Abstract. In order to further study the research and implementation of the VRGIS Intelligent Tourism system, this paper discusses the application of the VRGIS intelligent tourism system in the tourism industry, analyses the implementation technology of the virtual display system, gives the overall solution to the virtual display system, and summarizes the application of the virtual reality technology in the tourism. It is hoped that through the further research, the knowledge of the VRGIS's intelligent tourism system can be improved, which can also provide the effective reference for the relevant technical personnel.

Keywords: VRGIS · Intelligent tourism system · Research and implementation

1 Introduction

With the continuous innovation of the computer technologies and the vigorous development of the Internet industries, the virtual reality industry is developing rapidly all over the world. In the 12th Five-Year Plan, the State Council has listed the virtual reality technology as one of the leading technologies in the field of the information. As an important technology in Digital China, the virtual reality technology is widely used in education, industry, tourism, municipal planning, military simulation, exhibition and other fields [1]. The virtual reality technology can not only reduce the waste of resources, but also give users a sense of immersion. The application of the VRGIS intelligent tourism system in the tourism industry is mainly embodied in the production of the architectural effect map, building the animations and the virtual reality technology, and the related service items. With the development of our country's economy, the continuous warming of the tourism industry has led to the development of its related industries. In recent years, the VRGIS intelligent tourism system has been applied in some large-scale planning projects, which shows that the application of the VRGIS intelligent tourism system in the tourism industry has become more and more popular [2].

2 VRGIS Summary

VRGIS is a kind of the geographic information technology. Like other geographic information technology, the geographic information system can be closely integrated with the virtual reality technology. It is the frontier technology under the continuous

Z. Xu et al. (Eds.): CSIA 2019, AISC 928, pp. 130–135, 2020.
https://doi.org/10.1007/978-3-030-15235-2_20

development of the geographic information technology, and also the development direction of the geographic information technology. The geographic information technology has been applied in China as early as the end of the 20th century, and the experiments on the geographic information system have been done [3]. After the test results have been widely recognized, the VRGIS geographic information technology has been put forward. Specifically, the core technology of VRGIS is VR. Through VR, the interface interaction and communication can be realized. At the same time, the data storage, the data processing and the data analysis can be realized. It is the product of the integration of various geographic information technologies [4].

With the rapid development of the geographic information technology, people's requirements for the tourism resource management are gradually increasing. The application of the VRGIS technology in the tourism resource management is not limited to the two-dimensional spatial geometry. The real three-dimensional spatial coordinates of the VRGIS technology and the targeting data have become an important trend of the tourism resource management in the future [5]. The geographic information technology can restore the topography and geomorphology of any place by virtualization of its information system, which provides more intuitive data and spatial support for the better understanding of the tourism resources, and also provides more scientific reference for the resource allocation and the resource development. To this end, people at all levels can grasp the most accurate data, make the management of the command and the coordination become more conducive, and make the tourism management and services become more efficient, such as the virtual scenic spots and the tourism. The virtual reality (VR) is also known as the psychic technology or the artificial environment. It is a computer system that can create and experience the virtual world. This system generates various virtual environments, which act on the user's vision, hearing and touch, and make the user feel immersed in the virtual world. The virtual reality technology is widely used in tourism, entertainment, education, art and military industry. The virtual scenic spots and the virtual tourism are based on the realistic tourism landscape modeling. Through the simulation, we can build a virtual environment which can make the users feel immersed and interactive. There are three forms of the virtual tourist attractions in China, namely, the two-dimensional virtual tourist attractions, the three-dimensional virtual tourist attractions and the 360° panoramic roaming virtual tourist attractions. Compared with the developed countries, the virtual scenic spots in China started late. Although the development speed is faster, the overall level is not high, and the interaction is not strong, and the attraction is not enough. At present, due to the deficiencies in the forms of restoring the scenic spots in the domestic virtual tourism system, the interaction of the tourism system is not strong. Users can only passively accept the pictures of the scenic spots, but cannot actively conduct the in-depth interaction and exploration of the scenic spots, which leads to a great discount on the attractiveness of the tourism system to the users.

3 Virtual Tourism System Architecture

3.1 Client Architecture

At present, many virtual tourism services are mostly displayed in the form of the websites, but the response speed of the web pages and the servers will ultimately affect whether the virtual tourism websites can be widely used. The flash player is used to play the scenes of the virtual reality, and the flash is used to play the scenes on the web pages. The front-end mainly provides the services for the users, including two major parts: the virtual tourism and the tourism services.

3.2 Design and Production Arrangement of the System

The design and development process of the virtual display system is divided into three parts: the pre-preparation, the mid-preparation and the VRP post-production.

The preliminary preparation: It is necessary to investigate the environment of the scenic belts and their surrounding environment, to explore the manufacturing methods of the surrounding buildings, and to analyze the feasibility of the system. Finally, we need to take pictures on the spots, and then divide the tasks.

The mid-term preparation: After the analysis of the system requirements is completed, the next step is to start making the models, making them beautiful and less faceted, and PS the mapping. The maps must be made with high clarity, and the size of the maps should also be noted that the final saved size must be nth power of 2. Lastly, stick the map on the model and try the rendering effect. If it is not real enough, it still needs to be revised again. The baking needs to be done according to the setting of the baking parameters, so that the computer can drive, and there will be no problems, and the effect is better. After the baking, VRP is produced later.

The post-test: After the system is designed and manufactured, it is necessary to check whether there is any problem in the 3D model made by ourselves. If the problem is found, it should be modified immediately.

3.3 Server-Side Architecture

The background of this system will use the IIS server and the SQL Server database management system, which can avoid the congestion caused by the multi-users entering the system at the same time. The parallelism of the database can ensure the simultaneous operation of each module of the system. The back-end implements the management function, which mainly includes three modules:

The first is the design system of virtual reality. In this part, the virtual landscape interaction is implemented, and the image acquisition of the landscape is carried out. The Sphere panoramic technology is realized by the RIA technology, and then the shortest path algorithm is used to plan the most reasonable and achievable path.

The second is the tourism management system. It implements the whole landscape management, controls the amount of the landscape collection, and finds the balance between the distance of the adjacent image display sites and the user's visual perception.

The third is the user system. Users can log in at the front desk, upload the photos, upload the strategies, modify the passwords, view the members, add or delete the types, add the hotel information, delete the hotel information, and the message board functions and so on.

3.4 Virtual Tourism Includes

The first is the panoramic roaming. Using the virtual reality technology, the past, the present and the future contents of the scenic spots are digitized and virtually realized on computers.

The second is the path planning. Aiming at the problems that the roamers in the virtual system are easy to lose their directions, a two-dimensional navigation view and a compass are established in the project to provide the real-time geographic information of the roamers, to plan the reasonable paths for the donkey friends and enhance the reality of the roamers.

The third is the recommendation of the scenic spots. Customize a set of the strategies for different users. If new users are registered, we will recommend the scenic spots with the high evaluation of the tourists, and provide the scenic spots according to the users' personal preferences and their browsing records.

The fourth is the immersion experience. By using the sound simulation method, the guide commentaries are allocated for the donkey friends in the tourism routes, so as to improve the immersion of the scenic spots and provide the complete preparatory services for the personal experience in the future.

3.5 Tourism Services Include

The first is the scenic information. The system provides the tourists or the users with the information about the scenic spots, including the types of the scenic spots, and the historical and cultural backgrounds, and the combination with the local cultural customs. During the tour, both the scenic spots and the cultural edification can be enjoyed.

The second is the map search. Baidu Map provides the location information search function, the basic map display function, the search function and the location function, which facilitate the user route navigation, their search of the destination and the current location confirmation.

The third is the hotel transportation. By directly adding the hotel traffic services to the system, it can provide the integrated services for the users, so the system can provide the users with full services, eliminating the search and ordering trouble of other software.

The fourth is the tour guide service. According to the location of the visitors to the scenic spots, we can provide the users with the real-time scenic spot guide mode, and enjoy the scenic spot explanation and tour route planning.

3.6 Function Realization

The virtual tourism system includes the human-computer interaction, and the collision detection and other functions. The human-computer interaction is the focus of the roaming. In this paper, the first-person control role is used to traverse the virtual scene to achieve the human-computer interaction. The collision detection is to control the physical reaction between the roles and the models. In Unity3D, objects are added to the corresponding colliders and the obstacles are set. When roles encounter obstacles, there will be no bugs such as the interpolation and the collapse. Simulate the physical response of an object when it encounters obstacles. For example, when the character encounters a wall, it cannot continue to move forward.

After adding the colliders, we need to add the lightmapping to the model to enhance the stereo sense of the scenes and make the scenes look more realistic. In Unity3D, first select the model and then check the Static check box in the Inspector view so that the model is marked as Static. Unity3D will default that these objects are the static objects that do not move. Then these models marked as Static will be affected by the scene lights. Next, we modify the lighting parameters, such as the directions of the light source, the Bounces values, and the Resolution values and so on and finally generate the Lightmaps.

4 Key Technologies of the Virtual Tourism System

4.1 Navigation Based on the Path Planning

In the virtual tourism navigation technology, the selected path is pre-generated by the grid-based path algorithm. The virtual tour guides use the shortest path algorithm to automatically generate the optimal path after the path planning, which provides a reference for the users. The path planning and the navigation have realized many functions for us. Firstly, it can reach the destination safely on the premise of avoiding all the obstacles automatically. Secondly, it takes the shortest time and path for the users to reach their destination. The real-time geographic information of the roamers is provided to enhance the reality of the roamers aiming at the problem that the roamers are easy to lose their directions when roaming in the virtual system.

4.2 Panoramic Walkthrough of Virtual Reality

The GPS surveying technology is used to provide the precise three-dimensional coordinates of the point, line and surface elements and other related information quickly, efficiently and accurately. Because of its all-weather, high precision, automation and high efficiency, this technology is also suitable for the outdoor camping and exploration activities of tourists. The navigation system achieves different levels of functions, uses the precise positioning, surpasses the limitations of the traditional navigation, and achieves the real man-machine synchronization. In terms of the humanized services, the beautiful voice interpretation is adopted and recorded by the professional voice. There are also the super-beautiful pictures of the scenery and scenes on the map. The tour guide is equipped with the abundant information such as the ticket discount, the hotel discount, and the food shopping and so on, which are integrated into a complete urban strategy.

4.3 Roaming Architecture Design

The Web VR technology is used to design the virtual reality of the scenic tourism. The roaming platform will have the characteristics of the real-time online, the deep immersion and the high interaction. Therefore, when dealing with the scenic scenes, we should consider not only the perception effect of the virtual landscape, but also the fluency of the browsing. It requires the backstage database to store a large number of the panoramas, 3D models, and virtual scenes and so on. This technology solves the congestion caused by the parallel use of the system by a large number of users. Through the above technologies, the system surpasses other ones. Aiming at the problems existing in other projects, the system has three characteristics. First, experience the tourism in the virtual environment beforehand. Secondly, we should optimize the realistic tourism routes and plan an efficient tourism model. Third, attract the potential consumers to join in the tourism industry.

5 Conclusion

With the development of the science and technology in China, the tourism industry has begun to enter the road of the informatization. Under the background of the capital investment, more and more tourism industries begin to choose the application of virtual reality, namely the VR technology. Based on this, this paper takes the intelligent tourism system of VRGIS as the research object, and elaborates the application analysis of the related technologies in the virtual tourism from the above aspects through the research of the related technologies in the virtual tourism application, hoping to promote the development of the tourism through further researches on the intelligent tourism system of VRGIS.

References

1. Ma Q (2018) Development and design of the B/S-based intelligent tourism system. Autom Instrum (10):102–104
2. (2018) Global intelligent tourism integrated location service system based on BeiDou. Radio Commun Technol 44(06):528+645
3. Li B, Yuan Y (2018) Design of the BeiDou global intelligent tourism integrated location service system. Digit Commun World (10):11–12+23
4. Wang X (2018) Innovation research on the intelligent tourism system platform in Qinhuangdao under the internet of things. China Market (24):180–181
5. Shi H (2018) Design and implementation of the intelligent tourism system based on the three-tier architecture. J Shangluo Univ 32(04):14–18

Research on Robot Control Based on Reinforcement Learning

Gang Liu(✉)

Department of Mechanical Engineering,
Hefei University, Hefei 230601, Anhui, China
ncydyedu@yeah.net

Abstract. This paper reviews the rise and the development of the deep reinforcement learning (DRL). Then, the deep reinforcement learning algorithms for the high-dimensional continuous action space are divided into three categories of the algorithm based on the value function approximation, the algorithm based on the strategy approximation and the algorithm based on other structures. The latest representative algorithms and their characteristics of the deep reinforcement learning are explained in details, and their ideas, advantages and disadvantages are emphasized. Finally, combined with the development direction of the deep reinforcement learning algorithm, the control mechanism of using the deep reinforcement learning method to solve the control mechanism in the robotics problems is prospected.

Keywords: Reinforcement learning · Robot control · Environment restoration · Operation method

At present, several different theories and methods of the behavior learning for the intelligent robots have been proposed internationally. For example, Brooks, Reinforcement Learning and Evolutionary Algorithms based on the Behavior-based Robotics are proposed. To understand the simulation process of the reinforcement learning method so that the intelligent robots can produce the autonomous behaviors, we must first understand the concepts and principles of the reinforcement learning method.

1 Connotation and Application of the Reinforcement Learning

The reinforcement learning is a method of the autonomous learning for the intelligent robots [1]. It refers to the process in which the intelligent robots select their own actions according to the external environment in their actual environmental work, then perform the work, evaluate the actions according to the results of the actions, and use the results of the evaluation to learn their own rules of behaviors. The reinforcement learning is to regard the learning as a process of the trial and error evaluation and the self-correction [2]. This process is similar to the human trial and error. When people do not know how to operate the best in many cases, they choose an operation method to achieve the best

Z. Xu et al. (Eds.): CSIA 2019, AISC 928, pp. 136–141, 2020.
https://doi.org/10.1007/978-3-030-15235-2_21

effect by evaluating the self-modified behaviors after their own operation, to achieve the purpose of the self-study.

Agent sends the action A to the environment. After the environment accepts the action, it changes the state S and sends out the reinforcement signal r. The reinforcement signal r is the successful or the failed feedback information [3]. This feedback information is transmitted to Agent. When the reinforcement signal r tends to be the successful feedback information, Agent will choose to do the next action according to the successful feedback information and the current state. However, the selected action will immediately affect the reinforcement value, and will also affect the state of the next action and the final reinforcement value. In the reinforcement learning, the reinforcement signal r produced by Agent's actions to the environment is only a criterion for evaluating the success and the failure, but it does not directly tell the Agent the correct action [4]. Because the external environment gives the Agent a single signal, the Agent needs to react to the environment through the continuous actions, and the environment transmits the feedback reinforcement signal r to the Agent to determine the correctness of the action. Through the repeated operations, determine the success and failure, to constantly correct their own actions, and learn a series of the correct actions from this constant repair, so as to achieve the purpose of adapting to the environment [5].

The computational method of the reinforcement learning system is a dynamic parameter adjustment to maximize the reinforcement signal. If the gradient information of the action A and the reinforcement signal is known, then Agent can use the supervised learning algorithm [6, 7]. But because the action A and the reinforcement signal cannot get the expression of the correlation function, the gradient information of the action A and the reinforcement signal cannot be obtained [8]. If a random unit is adopted which can make Agent search in the action space and produce the correct action, then the reinforcement learning system can make Agent learn autonomously.

2 Research on the Robot Control

2.1 Motor Control

The engine, commonly known as "motor", refers to an electromagnetic device which realizes the conversion or transmission of the electric energy according to the laws of the electromagnetic induction. In the robot control system, the motor is used to drive the robot's joints, so that the robot can complete the formulation work. Specifically, the robot electric drive system uses the torque and the force generated by various motors to drive the body of the robot directly or indirectly to obtain the executing mechanism of the various movements of the robot. The robot has a high demand for the driving motor. Rapidity is the first. After the motor gets the command signal, it should enter the working state quickly. The shorter the response time of the command signal, the higher the working efficiency will be. The starting torque inertia ratio of the motor is larger. In the case of the driving load, it is required that the starting torque of the motor of the robot is larger and the rotational inertia is smaller. The motor should have a wide speed range. At present, the DC servo motors are widely used in the industrial robots. Their high starting torque, their large torque and their low inertia make the industrial production efficiency improved.

2.2 Access of the Sensor Signal

Robots are the computer-controlled machines and devices to help and serve the human beings. Robots have limbs similar to human beings and some sensory functions. They are flexible and intelligent, and can be independent of the human operations to a certain extent. Sensors play a very important role in the realization of the more intelligent robots. The robot sensors can be divided into the internal sensors and the external sensors. The internal sensors are the sensors to detect the state of the robot, which are mostly the sensors to detect the positions and the angles. The external sensors are the media of the communication between the robots and the external information. The external sensors mainly include the tactile sensors, the force sensors, the auditory sensors, and the visual sensors and so on. They are like the human five senses and the skin. When working, these sensors carry out the omni-directional detection of the external objects, the detection of brightness, the color concentration, the detection of the object shapes and the detection of object position and so on. The detected signals are integrated into the micro-computer of the robot, so that the robot can make more accurate judgments of the work. For example, the robot avoids the obstacles. The infrared detection is carried out by the sensors, and the distance between the obstacles is judged according to the specific frequency of the infrared radiation emitted by the reflection.

2.3 Output of the Control Signal

The output of the robot control signal is an important part of the control system. The signal output of the robot refers to a digital output signal to represent the current state of the operation. The output signal is generally divided into the analog signal and the digital signal. Taking the ABB robot as an example, the analog signal can output 0–10 V voltage and the digital signal can output 24 V voltage. The typical applications of the signal output are the execution error, the collision occurrence or the emergency stop state output of the robot. Here are some signal output functions, the Motion Supervision Triggered robot collision detection function being triggered, the power fail error, and the TCP Speed robot speed and so on. The output of the control signal is equivalent to the expression of the internal operation language of the robot, which makes the work of the robot more systematic.

2.4 Design of the Robot Control System

Throughout the world, as a strategic high-tech, the technology of robotics has attracted great attention from all countries in terms of promoting the development of the intelligent manufacturing equipment, the resource development and the future robotics industry. In this trend, the performances of robots are constantly improved, and are constantly developing towards the direction of intelligence, standardization and networking. The robot control technology, as the mainstream research direction, is very popular for the design of the new systems. The robot control system can be divided into the open type and the closed type. Since the birth of robots, most control systems have been the closed type. However, with the complexity of the tasks, the limitations of the closed system are increasingly apparent, so the design and development of the open control system is the current trend.

3 Research on the Robot Control Based on the Reinforcement Learning

The deep reinforcement learning technology combines the deep learning with the reinforcement learning technology. The deep reinforcement learning is a new research hotspot in the field of the artificial intelligence. It combines the perception ability of the in-depth learning with the decision-making ability of the reinforcement learning in a general form, and can realize the direct control from the original input to the output through the end-to-end learning. Since it was proposed, in many tasks requiring the perception of the high-dimensional raw input data and the decision-making control, and the method of the in-depth reinforcement learning has made substantial breakthroughs. Three main kinds of the deep reinforcement learning methods are described, including the deep reinforcement learning based on the value function, the deep reinforcement learning based on the strategy gradient and the deep reinforcement learning based on the search and supervision. Secondly, some frontier research directions in the field of the deep reinforcement learning are summarized, including the hierarchical deep reinforcement learning, the multi-task migration deep reinforcement learning and the multi-agent deep reinforcement. In the end, the successful application and the future development trend of the deep reinforcement learning in several fields are summarized.

In recent years, the application of robots has become more and more extensive. However, the actual environment is dynamic and full of uncertainty. In order to make the application of the robots penetrate into all aspects of our society, robots must have the high intelligence. The main work and the innovations include the detailed analysis of the current research status of the mobile robot control technology at home and abroad. Through the study of the mobile robots, a control strategy of the mobile robot based on the reinforcement learning is proposed, which combines the Q learning algorithm with the BP neural network. Use the artificial potential field method to determine the reinforcement function value, and then realize the obstacle avoidance of the mobile robots in the uncertain environment and ultimately control the mobile robots to reach the targets. The remote monitoring interface of the mobile robot is designed. Through the remote monitoring of the robots, we can monitor the movement of the mobile robots in the environment in the real time, so as to avoid the collision. The research status of the vision navigation technology for the mobile robots is analyzed in details. A dynamic vision device with pitching and swaying is applied to obtain the information in a wider field of the vision and to fuse it with the information obtained by the distance sensors to guide the robot actions. Through the image acquisition card, the collected images are processed, and finally the image matching is realized. Visual C++ is mainly used to realize various image processing algorithms.

In order to solve the cumulative error caused by the long-distance navigation, the principle and the algorithm of the predictive positioning by the encoder are used. On the premise that the distance between the ultrasonic sensor and the known landmark is accurately measured in the course of the robot moving, the cumulative positioning error is corrected by the Kalman filtering algorithm. A panoramic vision system has been developed in the laboratory. By using the advantages of the wide field of the vision and the natural characteristics of the indoor environment, a multi-landmark location method based on the panoramic vision has been realized, and a pose calculation algorithm based on the natural features of the panoramic images has been given.

A new visual positioning method is proposed to measure the relative distance between the mobile robots and the targeting objects. When the targeting object is in the range of the vision of the mobile robots, the mobile robots act accordingly according to the distance measured. The motion of the camera is analyzed in details, and the motion of the camera is planned. The camera is mainly controlled by the single chip computer through the serial port. The hardware structure of the RIRAII mobile robot is designed, and the hardware structure of the RIRAII mobile robot is analyzed and discussed. Its hardware structure is mainly composed of the wireless transmission module, the driver module, the ranging module, and the visual module and so on. The control cores of each module are the single chip computer and the PC.

Aiming at the problem of the obstacle avoidance in the intelligent agent movement, this paper combines the characteristics of the reinforcement learning, such as the trial-and-error, the environment interaction, the strategy of choosing actions in a certain state, and the online learning without tutors. The experimental results show that the algorithms such as the Q-learning and the reinforcement learning can effectively solve the obstacle avoidance problems in different situations. In order to adapt to the complex tasks and the environments, the multi-robot reinforcement learning is realized by the Markov game theory. Combining the advantages of the MAXQ-based multi-agent hierarchical reinforcement learning, a multi-robot coordination method based on the MAXQ-based multi-agent hierarchical reinforcement learning is proposed, which enhances the adaptability and the self-coordination ability of the robots in the complex environments. In addition, the multi-robot simulation platform and the mobile robot experimental system developed in the research process are briefly introduced. The simulation results proved the effectiveness of various algorithms.

4 Conclusion

The research of the multi-robot system and the multi-robot coordination has become a dynamic and promising research direction in robotics. It is of great theoretical and practical significance to study the basic theories of the multi-robot system architecture and the multi-robot coordination and cooperation. Finally, the simulation and the experiments are carried out, and the experimental results are analyzed. The results show that the research results can effectively realize the searching and the tracking of the targeting objects in the unknown environments of the mobile robots.

Acknowledgment. Foundation Project: Key Natural Science Research Projects of Anhui Universities (Project Grant No. KJ2018A0552).

References

1. Liu Q, Zhai J, Zhang Z, Zhong S, Zhou Q, Zhang P, Xu J (2017) Summary of the deep reinforcement learning. Chin J Comput (01):99–100
2. Wang R, Kong M, Zhang X (2017) Analysis and application of the robot system control method based on android. Mech Des Manuf Eng (10):112–113
3. Wang F, Qi H, Zhou X, Wang J (2017) Demonstration programming and optimization methods of the cooperative robots based on the multi-source information fusion. Robot (07):108–109
4. Xue T (2018) Principle of the deep reinforcement learning and its application in the robot motion control. Telecommunications (08):75–76
5. Duo N, Lv Q, Lin H, Wei H (2018) Stepping into the high-dimensional continuous space: application of the deep reinforcement learning in the robot field. Robot (09):61–62
6. Beom HR, Cho HS (1995) A sensor-based navigation for a mobile robot using fuzzy logic and reinforcement learning. IEEE Trans Syst Man Cybern 25(3):464–477
7. Mataric MJ (1997) Behaviour-based control: examples from navigation, learning, and group behaviour. J Exp Theor Artif Intell 9(2–3):323–336
8. Bianchi RAC, Santos PE, da Silva IJ et al (2017) Heuristically accelerated reinforcement learning by means of case-based reasoning and transfer learning. J Intell Rob Syst 91:1–12

Research on the Agricultural Remote Sensing Image Enhancement Technology Based on the Mixed Entropy Model

Youzhi Zhang(✉)

Remote Sensing Technology Center,
Heilongjiang Academy of Agricultural Sciences,
Harbin 150000, Heilongjiang, China
liwangirsa@yeah.net

Abstract. Uncertainty is the most important factor affecting the quality of the remote sensing image classification. Aiming at the characteristics of both the random and the fuzzy uncertainties in the process of the remote sensing image classification, a method based on the mixed entropy model is proposed to measure these two uncertainties comprehensively, and a multi-scale evaluation index is established. Based on the analysis of the basic principles of the mixed entropy model, a method of using the statistical data of the feature space and the fuzzy classifier to establish the information entropy, the fuzzy entropy and the mixed entropy is proposed. At the same time, on the scale of the pixel and the category, the index of the mixed entropy of the pixel and the mixed entropy of the category are established to evaluate the uncertainty of the classification.

Keywords: Mixed entropy model · Agricultural remote sensing · Image enhancement · Technology research

Because the quality of the enhanced remote sensing images is mainly evaluated by the human vision and the visual evaluation is a highly subjective process. Therefore, for a specific purpose, a specific processing method is adopted to obtain the specific images, and the evaluation methods and criteria for the quality are also specific. Therefore, at present, it is difficult to determine a unified evaluation criterion with the general adaptation, and the theory of the remote sensing image enhancement needs to be further improved [1]. Therefore, the exploration of the remote sensing image enhancement technology is experimental and diverse, and its enhancement methods are often targeted.

1 Analysis of the Application Background of the Agricultural Remote Sensing Image Enhancement Technology

The image processing includes the image denoising, the image enhancement, the image restoration and the image segmentation. The image enhancement is an important technology in the image processing. Through the image enhancement processing, useful information in the images is enhanced, and the original unclear images become

Z. Xu et al. (Eds.): CSIA 2019, AISC 928, pp. 142–147, 2020.
https://doi.org/10.1007/978-3-030-15235-2_22

clear, and the effect of the image interpretation and recognition is strengthened to meet the needs of some special analysis [2]. The satellite remote sensing images are affected by the image acquisition instruments, the atmospheric light, the electromagnetic interference radiation and the remote sensing transmission media in the imaging process, which results in the insufficient resolution and the poor emphasizing ability of the useful features in the satellite remote sensing image acquisition. It is necessary to enhance the satellite remote sensing images by the image enhancement processing [3]. The satellite remote sensing image enhancement technology will have the high application values in the geophysical prospecting, the meteorology, the remote detection, the laser infrared detection and other fields, and the related algorithm research has attracted people's attention.

For the problem of how to divide the gray scale intervals and set the endpoints in the piecewise linear transformation method, a remote sensing image enhancement algorithm based on the weighted fuzzy C-means clustering is presented. The gray scale partition of images is realized quickly, and the corresponding transformations are adopted to enhance each interval [4]. The experimental results show that the algorithm in this paper effectively improves the images. The signal-to-noise ratio (SNR) enhances the contrast and clarity, and is easy to implement. In view of the shortcomings of the traditional fuzzy enhancement methods, a remote sensing image fuzzy enhancement algorithm based on the maximum Tsallis entropy principle is proposed. First, the Tsallis entropy is extended to the multiple thresholds. Based on the maximum entropy principle, the genetic algorithm is used to calculate the Tsallis entropy [5]. The optimal classification threshold of the image gray level is obtained, and then the improved membership function and the enhancement operator are used to enhance the images [6]. Finally, the multi-threshold image enhancement method is extended. The experimental results show that the algorithm can automatically and effectively select the threshold, and make the brightness distribution of the images balanced and the contrast enhanced significantly [7–9]. It has the good visual effect.

2 Agricultural Remote Sensing Image Enhancement Technology Based on the Mixed Entropy Model

The agricultural remote sensing is an important branch of the remote sensing science. This paper reviews the development history of the agricultural remote sensing researches in the past 100 years. It is believed that the application of the remote sensing in agriculture is expanding in breadth and depth [10]. The agricultural remote sensing is deepening from acquiring the traditional factors such as the total output, areas and the unit yield to more monitoring factors, and expanding from the traditional resources and environment to the plant protection and agronomy. The agricultural remote sensing is gradually becoming the basic and key technology of the agricultural science. This paper summarizes the achievements and important results of the agricultural remote sensing researches in China in recent years from the fields of the agricultural remote sensing, the agricultural disaster remote sensing and the agricultural resources and the environment remote sensing. The international frontier of the agricultural remote sensing researches and development is pointed out from the aspects of the agricultural quantitative remote sensing, the UAV remote sensing and the crop phenotype remote sensing. The domestic agricultural remote sensing technology has great prospects in the fields of the big data acquisition of the agricultural remote sensing, and the intelligent extraction and mining of the information such as the sky-ground integration, the artificial intelligence and the big data.

Whether it is the land use types, the classification and recognition of the crop species, or the quantitative remote sensing of the crop growth status and environmental factors, it is a very complex cognitive process. Because of the correlation between the bands of the remote sensing data, the limitation of the design bands of the remote sensors, and the spectral complexity of the same object and the same spectrum of the same object, the remote sensing information extraction and the intelligent mining have ill-posed problems, and there are many uncertainties. The development of the artificial intelligence and the big data technology has provided a new technical way for the retrieval, extraction and application of the agricultural resources and the environment information.

The integration of the remote sensing observation with the navigation and positioning, the Internet, the Internet of Things, the big data and other technologies, combined with other disciplines in the field of the agriculture, can promote the development of its own disciplines from the methodological point of view. At the same time, the interdisciplinary applications will also expand the application fields. It is necessary to further establish an integrated agricultural management system to promote the application of the sky-ground cooperative remote sensing observation in precision, the intelligent agriculture, the crop breeding phenotype, the monitoring and evaluation of the agricultural insurance, the green development of the agriculture and the evaluation of the agricultural policy effects. The remote sensing technology is an advanced and practical detection technology. In China's agricultural application, from the early study of the land use and the land covering area estimation and the crop yield estimation by the remote sensing, it has been extended to the current 3S integration for the real-time diagnosis of the crop growth, and the application of the hyperspectral remote sensing data for the important organisms, the inversion of the agronomic parameters, the mechanism of the hyperspectral agronomic remote sensing, the research and application of the models, the grassland yield estimation, the forest dynamic monitoring and other multiple levels and multiple aspects.

In China's agricultural remote sensing, it is necessary to grasp the basic issues of the hyperspectral remote sensing and the precision agriculture research in the future, such as the study of the remote sensing mechanism and the remote sensing markers under the environmental stress, the diagnostic theory of the integration of the remote sensing and the GIS on the crop stress, the diagnostic theory of the crop stress by integration of the remote sensing and GIS, the spatial difference mechanism of the crop growth environment and the real distribution of the harvest yield and the quantitative relation between the environmental stress and the yield formation by the remote sensing. It is necessary to grasp the key technologies such as the hyperspectral, high-resolution, radar and the "3S" integration to analyze and estimate the biophysical parameters of the vegetation, such as the leaf area index biomass, the total nitrogen and the total phosphorus. It is necessary to design and establish the basic agricultural information system, to extract the crop symptom information and to diagnose the agriculture with the support of GIS. It is necessary to study the spectral characteristics of the crops and the agronomic remote sensing mechanism, and apply it directly to the agricultural production decision-making and management.

3 Basic Principles of the Mixed Entropy Model

The classification accuracy evaluation models based on the entropy theory include the entropy, the information entropy, the fuzzy entropy, the cross-entropy, the mixed entropy, the divergence, the correlation coefficient, the point product and the Euclidean distance. In order to consider the total uncertainties caused by the randomness and fuzziness in a unified way, a hybrid entropy model is introduced to measure the random and fuzzy uncertainties of the classification. The mixed entropy model is based on the information theory and the fuzzy set theory.

According to the basic principle of the mixed entropy model, the quantitative expression of the random uncertainty can be corresponded to the statistical entropy at the pixel scale, and the quantitative expression of the fuzzy uncertainty can be corresponded to the fuzzy entropy, and the comprehensive influence of the two kinds of the uncertainty can be corresponded to the mixed entropy. Therefore, in order to evaluate the comprehensive uncertainty of the pixel scale, the modeling method of the pixel mixing entropy is discussed.

In the remote sensing image processing, the information entropy is used to measure the dispersion and uniformity of the image brightness values of the same kind of the objects, i.e. to measure the random uncertainty of the classification, so as to evaluate the resulting uncertainty of the "homologous foreign object" and the "isomorphic foreign body". The determination of the statistical entropy mainly depends on the probability density distribution function of the feature space. Due to the influence of the random factors such as the atmospheric conditions, the background, the object orientation, and the sensor noise and so on, it is impossible for each sampling point of the same kind of objects to represent only the same point in the spectral feature space, but to form a relatively clustered point cluster.

The uncertainty in the fuzzy set is to measure the uncertainty of the classification and evaluate the uncertainty of the "mixed pixels" caused by it. It can be seen that the determination of the fuzzy entropy mainly depends on the membership function. In order to obtain the membership function, we use the fuzzy classifier to process the remote sensing images. The most common fuzzy classifier is the fuzzy maximum likelihood classification. Based on the fuzzy theory, this method treats the probability vectors generated in the maximum likelihood classification as a fuzzy set, which can well express the useful uncertain information in the classification process.

4 Agricultural Remote Sensing Image Enhancement Technology Based on the Mixed Entropy Model

In this paper, a hybrid entropy model is introduced to evaluate the uncertainty of the remote sensing image classification. Based on the analysis of the principle of the hybrid entropy model, a method of using the statistical information of the feature space and the fuzzy classifier to obtain the modeling parameters is proposed, which provides a unified theoretical framework for evaluating the uncertainty of the classification. To improve the ability of the feature detection and the recognition of the satellite remote sensing images, the image enhancement is needed. The satellite remote sensing images under the complex background interference are accurately fused and recognized, and the image feature extraction and the image acquisition are carried out. The affine invariance of the images is used to collect the feature parameters of the satellite remote sensing images. Under the interference of the complex backgrounds, the satellite remote sensing images are built in an irregular triangular mesh model.

According to the above analysis, the multi-scale Retinex color feature components of the satellite remote sensing images are extracted from the feature templates, and the edge contour feature points of the satellite remote sensing images are fused on the basis of the Retinex color features. The moving average modeling of the satellite remote sensing images is realized by the template adaptive matching method, to realize the enhancement of the feature points of the spatial domain information of the images. Compared with the traditional methods, this algorithm can improve the peak signal-to-noise ratio (PSNR) of the image output, reduce the execution time and the computational cost, and increase the number of the information feature points. The results show that the method can effectively enhance the image information and improve the ability of the image analysis and recognition. Through the image enhancement processing, the useful information in the images is enhanced, and the effect of the image interpretation and recognition is strengthened. The satellite remote sensing image enhancement processing can effectively improve the ability of extracting and recognizing the useful feature points of the satellite remote sensing images.

In the actual practice, to find an effective method, extensive experiments must be carried out. It is very difficult to predict the effectiveness of a specific method when there is no prior knowledge of how to reduce the quality of a given image. Considering the specific circumstances, we must consider the factors that need to be enhanced to select the transformation parameters and algorithms. And through the repeated experiments, observe the transformation results, and carry out the constant adjustment until the results are satisfactory. The commonly used method is to use a combination of the several gray-scale enhancement techniques or to use the method of adjusting the parameters. In order to achieve a better improvement effect on an image, sometimes we need to use a combination of various enhancement methods to give full play to the advantages of each method, which requires us to understand all kinds of the remote sensing. According to the characteristics of the remote sensing image structures and the requirements of the image processing, the corresponding enhancement methods should be selected.

For a specific image enhancement method, the selection of the remote sensing image enhancement method can be accelerated by viewing the effect of the enhanced image and analyzing the characteristics of the image with better effect. Adjusting parameters is a method often used in the remote sensing image enhancement. How to determine the optimal value of the parameters is the key factor to achieve the better image effect. Therefore, the biggest difficulty of the image enhancement is that it is difficult to quantify the results of the enhancement. It can only be evaluated by the experience and the subjective feeling. At the same time, to obtain a satisfactory enhancement effect, the human-computer interaction is often required.

5 Conclusion

A satellite remote sensing image enhancement technology based on the improved algorithm is proposed. The image acquisition model of the satellite remote sensing images is constructed. The denoising pretreatment of the satellite remote sensing images is carried out by using the wavelet multi-scale decomposition method. The multi-scale color feature components of the satellite remote sensing image are extracted from the feature template and fused on the basis of the color features. The edge contour feature points of the satellite remote sensing images are adaptively matched by the template to realize the moving average modeling of the satellite remote sensing images and the enhancement of the information feature points in the image spatial domain. The research shows that this algorithm can effectively reduce the noise interference, suppress the uninterested features, improve the peak signal-to-noise ratio of the output images, and then improve the recognition ability of the satellite remote sensing images. It has the high application value.

References

1. Li J, Zhou R, Wang D, Zhang Q (2017) Variance-mixed entropy portfolio optimization model with the yager entropy constraints. J Beijing Univ Chem Technol (Nat Sci Ed) (07):118–119
2. Wang H, Wang Y, Duan Y (2018) Visual analysis of the research hotspots of the remote sensing in the agricultural field based on the method of the scientific knowledge map. Chin Agric Sci Bull (03):132–133
3. Li T, Liu W, Zhou X (2018) Fuzzy comprehensive evaluation of the internal linkage efficiency of the airport land passenger transportation based on the mixed entropy weight. J Changsha Univ Sci Technol (Nat Sci Ed) (09):145–146
4. Mao X, Yao Y, Chen S, Liu J, Du Z, Wei J (2018) QuickBird remote sensing image forest classification recognition based on the yikang software – taking jiangle forest farm in fujian province as an example. J Nanjing Forest Univ (Nat Sci Ed) (10):114–115
5. Guo Z, Song P, Zhang Y, Yan M, Sun X, Sun H (2018) Aircraft target detection in the remote sensing images based on the deep convolution neural network. J Electr Inf Technol (11):106–107
6. Blaschke T (2010) Object based image analysis for remote sensing. ISPRS J Photogramm Remote Sens 65(1):2–16
7. Lu D, Chen Q, Wang G et al (2016) A survey of remote sensing-based aboveground biomass estimation methods in forest ecosystems. Int J Digit Earth 9(1):63–105
8. Wang L, Song W, Liu P (2016) Link the remote sensing big data to the image features via wavelet transformation. Cluster Comput 19(2):793–810
9. Niphadkar M, Nagendra H, Tarantino C et al (2017) Comparing pixel and object-based approaches to map an understorey invasive shrub in tropical mixed forests. Front Plant Sci 8:892
10. Xie L, Li G, Peng L et al (2017) Band selection algorithm based on information entropy for hyperspectral image classification. J. Appl. Remote Sens. 11(2):026018

Research on the Application of Computer Aided Translation CAT in the Field of Translation

Wang Yang(✉)

Computer Science Editorial, Shanghai University of Engineering Science, Shanghai 200437, China
huangmingbucea@126.com

Abstract. With the continuous development of the network information age, how to effectively use modern means to eliminate language barriers between people has become a common problem facing all mankind. Translation is a powerful means to overcome language barriers. However, due to its high cost and slow translation speed, traditional manual translation methods can not meet the needs of the translation market, and machine translation emerges as the times require. Machine Translation (MT) is the process of transforming one natural language into another by computer. Although machine translation is simple and convenient, its accuracy still needs to be improved. At present, it can only be used as an auxiliary means of manual translation. For this reason, computer-aided translation (CAT), that is, computer-aided translation, has attracted people's attention and favor.

Keywords: Computer aided translation system

1 Introduction

In 1954, the first public demonstration of machine translation in human history at Georgetown University opened the door for people to study machine translation. Subsequently, machine translation is on the rise. However, the ALPAC (Automatic Language Processing Advisory Committee) report issued by the American Academy of Sciences in 1966 severely damaged the booming machine translation at that time. After a relatively stable period of development, machine translation began to recover nationwide in the 1980s, and made great progress in the 1990s. Professor B. Vauquois, a famous French expert in machine translation, summarizes the method of machine translation based on language rules into the following figure (Fig. 1). We can make the figure MT_Pyramid:

Z. Xu et al. (Eds.): CSIA 2019, AISC 928, pp. 148–154, 2020.
https://doi.org/10.1007/978-3-030-15235-2_23

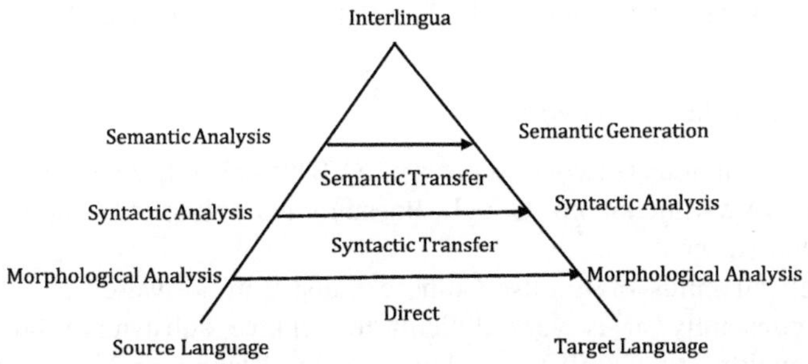

Fig. 1. Machine translation pyramid

The research of MT in China can be traced back to the mid-1950s. Due to the influence of the 1966 ALPAC report of the United States, the study of machine translation has been seriously affected and retrogressed. It was not until the mid-1970s that machine translation began to revive and flourish in China. In the past 20 years or so, the research methods of machine translation have been called "a hundred flowers blossom and a hundred schools of thought contend" [1]. They are rule-based, knowledge-based, statistical and corpus-based, example-based, dialogue-based, etc. From another perspective, there are direct method, conversion method, interlanguage method and so on.

At present, the developed machine translation software can be roughly divided into four categories:

(1) Dictionary Translation Software

Dictionary translation software, e-dictionary, is mainly designed to facilitate people to read foreign language materials. There are famous Jinshan Ciba in China and Babylon (BABYLON) abroad [2].

(2) Chinese Translation Software

Chinese translation software is mainly designed to reduce the difficulties encountered by users who do not understand English when using computers and surfing the Internet. This kind of software mainly focuses on domestic products, such as Oriental Express, Jinshan Express Translation, etc.

(3) Professional Translation System with Automatic Translation as the Core

Professional translation systems with automatic translation as the core, such as LogoMedia, Dr. eye Translator, IBM Translator, Jinshan Express Translator, Oriental

Express, Times Translator and Bible Translator, can automatically translate the full text [3].

(4) Online Translation Network

Online translation network such as: SYSTRAN http://babelfish.altavista.com; Huajian http://www.hjtek.com; Google Jinshan Ciba online translation; ICB http://fy.iciba.com and so on.

Although all kinds of English-Chinese translation software are selling well and upgrading constantly, it is very difficult to achieve substantive breakthroughs in translation quality. As we all know, human natural language is an open system. To enable computers to understand natural language, not only precise mathematical models are needed to describe and optimize the algorithm as a guide, but also a huge language knowledge base and grammar knowledge base to support. Now we are facing the dilemma that automatic translation has reached a "bottleneck", and in the short term, it is difficult to improve its quality greatly [4].

Due to the urgent need to improve the quality of machine translation, many machine translation researchers have turned to MAT (Machine Aided Translation) or CAT (Computer Aided Translation) mode. It can be said that computer aided translation is the product of the resistance of machine automatic translation. Computer-aided translation (CAT) is a kind of translation with the participation of language database and human beings. The quality of translation is the same or even better as that of manual translation, and it can greatly improve the efficiency of translation [5].

2 Computer Aided Translation Software and Its Advantages

Computer-aided translation (CAT) integrates machine memory translation, grammatical analysis translation and interpersonal interactive translation. It handles mechanical, repetitive and trivial tasks in the process of translation to the computer. In this way, the translator only needs to concentrate on the translation [6, 7]. It can not only guarantee the quality, but also improve the efficiency.

2.1 Computer Aided Translation Software

CAT technology has a long history in foreign countries. There are many manufacturers developing translation memory software. The most famous ones are Trados, DJV, TransStar, IBM Translation Manager, WordFisher, Wordfast, OmegaT, etc. Among them, the most famous one, the most representative one and widely used is TRADOS in Germany (Tado). In China, the concept of CAT is still in its infancy, and there are few institutions engaged in CAT technology research and product development, such as Athens CATS, Huajian of Chinese Academy of Sciences, Wenjie WinMAT, ITM of Beijing Yongbang Bodian and translation environment of translators, etc.

These translation software will undoubtedly greatly improve the efficiency of translation, but the premise is that the translator must have certain computer operation skills, skilled operation of commonly used office software; will install and use important programs and electronic resources such as the electronic version of

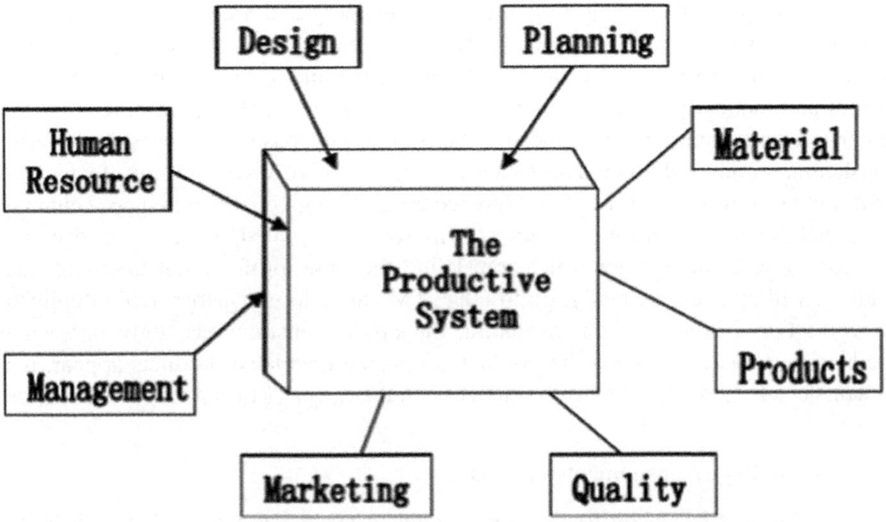

Fig. 2. An analysis of the current situation and development trend of computer-aided translation technology

Encyclopedia Britannica, and be familiar with the retrieval and acquisition of network resources and so on [8]. Only in this way can we use the translation software with various functions appropriately (Fig. 2).

2.2 Principles and Advantages of Computer-Aided Translation

Computer-aided translation software is recognized by most people because of its unique operation principle and advantages. Let's take DJV as an example to see the working principle and advantages of CAT.

(1) Project management

Project management is the basic working interface of DJV, which provides a bilingual translation platform. After the translator inputs the materials to be translated manually or electronically, he creates a new translation project, sets up the source language and the target language, imports the translation materials, and then translates them under the project page. This method is easy to operate and has a neat and orderly interface. It solves the problem of bilingual difficulty in comparison and is easy to translate.

In addition, the project has the function of automatic preservation, and can generate translation memory through relevant operations, so as to prepare for future translation. It can also generate a terminology library, which is the basis of DJV operation.

(2) Translation Memory (TM)

The core of CAT technology is translation memory technology. While translating, translators learn and store new translations in the background, and establish language database. In the process of translation, whenever the same or similar phrases appear, the system automatically searches for the same or similar translation resources (such as sentences, paragraphs, etc.) in the translation memory base, and gives the reference translation, so that users can avoid unnecessary repetitive work and only focus on the translation of new content. For the reference translation given, the translator can copy it completely or use it after modification. If he feels unsatisfied, he can also discard it.

According to the survey, when translating the same professional literature, many contents will appear repeatedly, up to 60%. TM can help translators avoid duplication of work. For example, when translating the product instructions, only one version needs to be translated. When the product is updated or related products appear, many contents can be directly taken over and used. In this way, TM is an extension of human memory.

(3) Terminology management function

It is always a headache for us to encounter or specialize in translation. It is time-consuming and laborious to input familiar words over and over again. Looking up dictionary terminology also affects speed. If the system automatically displays the meaning of the translated word, and you just press the button, the word will run to your mouse. The term DJV does this. There is a management tool in the translation memory software to regulate all the professional terms. Translators only need to establish one or more standard terms list at one time. When translating, the translation memory software will automatically identify those words or structures as defined terms, and give the corresponding terms translation to ensure the unification of terms. As long as you accumulate certain terminology in your usual translation, your translation will be "more, faster, better and less".

(4) Alignment of documents

Aligning documents is a very useful function of DJV. In addition to our own translation work, if valuable bilingual materials are collected in our daily reading, we can use the function of DJV to align documents. According to the operation hints, the materials are imported into the alignment document interface, and the bilingual materials are aligned according to the functions of the dialog boxes. At this time, the translation memory library is automatically generated by clicking the button on the page, and the terminology library is added manually. In the long run, the translator's memory base and terminology base will become larger and larger, which will greatly facilitate your translation work. In addition to the four advantages mentioned above, DJV can also import and export translations in a timely manner so that translators in non-DJV environments can complete revision and proofreading of translations. The

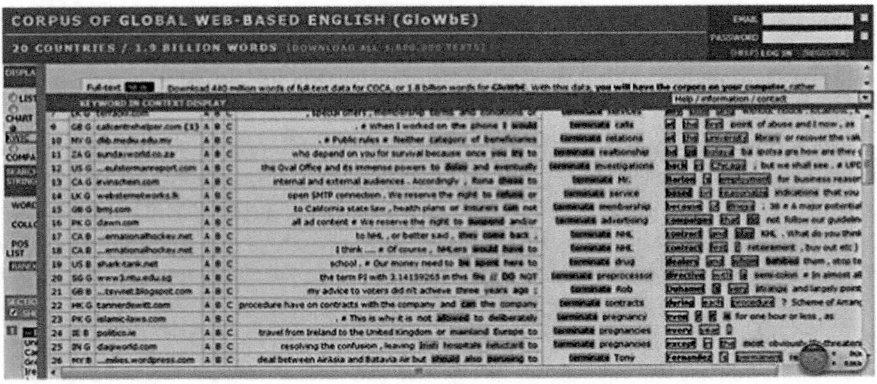

Fig. 3. Computer aided translation practice

more translation work is done, the more obvious the advantage of DJV is. Other translation software, such as Wordfast, Trado s and so on, also have their own unique advantages, which are not exemplified here (Fig. 3).

3 Limitations and Improvement Strategies of CAT

CAT provides a platform for translation, but it has its own shortcomings. Firstly, translation teaching in China only focuses on theory, does not provide students with developing practice, rarely involves CAT professors, and does not provide students with good bait to develop and utilize this new technology. Secondly, the promotion of CAT translation software is weak. Influenced by some subjective and objective factors, many people think of online translation, but they don't know much about CAT translation software. Finally, the development of translation software is not perfect. Since translation software is mainly developed by computer scholars, and translation is about human language activities, some functions are not very practical.

Signed here, the author believes that in the future development and utilization, we should improve from three aspects. Firstly, we should strengthen CAT teaching. Today, with the rapid development of computers, it shows students the wonderful translation mode of CAT, and improves their interest in translation. Secondly, to enhance the translator's computer operation level and ability to accept new things. Only by possessing a certain knowledge reserve and calmly accepting and learning new things can we benefit the tide of world development. Finally, the development of CAT software requires the cooperation of computer linguists, translators, mathematicians and linguists. CAT is not only a technology but also an art. It is difficult for computer experts to form practical translation techniques without the theoretical guidance of linguists. At the same time, without the practical guidance of translators, it is difficult for computer scientists to notice the problems often encountered in translation. Therefore, only with the cooperation of all aspects, can more excellent translation software come out. Advanced memory technology, convenient human-computer interaction mechanism and open lexicon, memory and terminology database of computer-aided translation provide professional translators with an efficient and high-quality translation platform, which makes them work like fish in water, greatly improves the translation environment and improves the efficiency of their translation and quality.

Due to the differences between languages and the differences in social background and culture, full automatic translation of meaning can not be regarded as the ultimate goal. We need to use machine-aided translation services. From machine translation to computer-aided translation, information technology such as Internet and cloud computing has played a very important role in promoting the development of computer-aided translation. A comprehensive understanding of the development trend of computer-aided translation technology is more conducive to the effective use of computers to carry out translation work.

References

1. Gong M (2009) Brief discussion on the development history and prospects of machine translation. J Jilin Inst Educ 25(223):135–136
2. Feng Z (2004) Machine Translation Research. China Foreign Translation Publishing Company, Beijing, pp 74–76
3. Luo R (2012) A brief talk on the present situation of machine translation. Jiannan Lit 12 (12):236–237
4. Du J, Zhang M, Zong C et al (2013) The basis and challenge of Chinese machine translation research - summary and prospect of the eighth national machine translation seminar. Chin J Inf 27(4):2–8
5. Zhang J, Peng Y (2009) Text context machine translation quality Shandong foreign language teaching. Chin Lang Learn 15(6):73–77
6. Liang S (2004) Comparative analysis of machine translation and computer aided translation foreign language telephone teaching. Stud Lang Linguist 20(100):42–45
7. Bundgaard K, Christensen TP, Schjoldager A (2016) Translator-computer interaction in action–an observational process study of computer-aided translation. J Spec Transl 25:106–130
8. Shadiev R, Huang YM (2016) Facilitating cross-cultural understanding with learning activities supported by speech-to-text recognition and computer-aided translation. Comput. Educ. 98:130–141

Research on the Application of the Computer Big Data Technology in the Urban Power Energy Conservation

Jieran Shen[1(✉)], Baocai Zhao[2], Rui Liu[2], and Feng Jiao[3]

[1] University of California San Diego, La Jolla, USA
sheng_xc@yeah.net
[2] State Grid Liaoning Electric Power Supply Co., Ltd., Shenyang 110000, Liaoning, China
[3] Beijing Bochao Times Software Co., Ltd., Beijing 100001, China

Abstract. With the arrival of the era of the big data, the power energy conservation has attracted much attention. The smart grid technology is a deep integration of the traditional power technology and the information, and the control and automation technology. By collecting a large amount of the data in the power generation, transmission, distribution, the power consumption and dispatching, marketing and other links, we can conduct the in-depth analysis and mining of the big data information to guide the decision-making and optimization in all links, in order to improve the efficiency of the power enterprises, enhance the stability of the power grid operation, better meet the needs of the power customers, and further ensure the efficiency of the urban power energy conservation.

Keywords: Computer big data · Urban power · Energy saving · Application

1 Introduction

With the continuous advancement of the smart grid construction, the scale of the smart grid is expanding day by day. Various kinds of the heterogeneous distributed data such as the smart meters, the sensors and the information systems continue to generate large amounts of the data, which is called the power big data. The power big data is regarded as an important support for the intellectualization of the power grid. Because of the basic function of the big data of the electric power, the collection, transmission, storage, processing and mining of the big data of the electric power have improved the levels of the energy-saving of the urban electric power [1].

2 Power Big Data and Its Characteristics

The power big data refers to a large number of the data generated by the smart grid in the power generation, distribution, transmission, marketing and management. A large number of the sensors deployed on various devices, the smart meters installed in households of various users, the customer feedback collected by the marketing system

Z. Xu et al. (Eds.): CSIA 2019, AISC 928, pp. 155–161, 2020.
https://doi.org/10.1007/978-3-030-15235-2_24

and other data sources are generated and assembled into a centralized data center for the unified storage and management [2].

The power big data is one of the supporting elements for building a stable, reliable, efficient and energy-saving smart grid. Through the analysis of the power big data, we can improve the lean management levels of the smart grid, formulate more scientific production plans, optimize the energy transmission and dispatch, and establish more accurate user behavior models [3].

2.1 Large Volume

With the further development of the smart grid construction, the terminal data collection devices such as the device sensors and the smart meters have been intensively deployed, and the scale of the data collected will increase exponentially, reaching the TB or even the PB level. Taking Zhejiang Province as an example, there are 22 million electricity users in the whole province. If all the smart meters are installed, according to the requirement of the State Grid Corporation to collect one electricity information data every 15 min, 2.1 billion new electricity records are added every day.

2.2 Many Types

In addition to the traditional structured data, the production management, the marketing and other systems produce a large number of the semi-structured and the unstructured data, such as the audio and the video data [4]. The diversity of the data types requires the diversity of the storage and processing technologies. This paper focuses on the data acquisition and processing system of the electricity information, and still focuses on the structured data, without discussing the processing of the semi-structured and the unstructured data.

2.3 Fast Speed

The big data acquisition and processing of the electric power have a very fast speed. The surge in the number of the terminals requires the storage systems to meet the demand for the high-throughput data access hundreds of thousands of times per second [5].

2.4 Data Is Energy

The big data of the electric power contains very important information such as the user's electricity consumption rule, and the optimal transmission dispatching strategy and so on. These kinds of the information play a unique and the tremendous role in the rational arrangement of the production and reduction of the energy consumption loss, which promotes the reduction of the energy consumption and the sustainable development of the power grid, thus embodying the characteristics of the data as the energy.

2.5 Data Is Interaction

The big data of the electric power industry is aggregated through the interaction with the big data of other industries, and is deeply excavated and analyzed. The information contained in the big data has very important reference value for the high-level decision-making and the judgment of the economic situation of the country.

2.6 Data Is Empathy

The big data of the power supply provides a new way for the State Grid Company to discover and meet the needs of the users in time and accurately. The empathic concern is the indebted feeling as if it were received in person. With the help of the big data of the electric power, the production and the marketing provide more high-quality, safe and reliable electric power services to the vast number of the electric power users, so as to achieve the goal of the common development.

3 Analysis of the Application Strategy of the Computer Big Data Technology in the Urban Power Energy Conservation

Improving the efficiency of the terminal power consumption is a technical transformation of the power-saving equipment. While satisfying the diversified services, it saves the electricity and reduces the consumption of electricity, including the technical improvement and the scientific management of lighting, air-conditioning, refrigerators, motors, transformers and distribution equipment, so as to achieve the goal of reducing the electricity consumption. Therefore, to evaluate the energy-saving potentials and energy-saving levels of the power users is to find out which electrical equipment of the power users has the potential of the technological transformation and the size of the potential.

At present, the smart watt-hour meters are becoming more and more popular in the power grid. The smart watt-hour meters are bounded by the total line of the power users. They play a certain role in the total load monitoring and the realization of the sequential power consumption. However, the monitoring scope of the smart watt-hour meters has not been able to penetrate into the internal of the power users, and develop the effective information interaction with the power equipment, let alone provide the energy efficiency analysis and the rational suggestions on the power consumption of the users, so as to restrict the promotion of the energy efficiency management. However, at present, some smart community users are equipped with the smart sockets in their homes. The intelligent socket can collect the electricity consumption of the specific electrical equipment and send it to the monitoring terminals. Then it can be displayed on the intelligent monitoring terminals. Users can clearly understand their own electricity structures, which is conducive to discovering which equipment has the potential of the technological transformation. The intelligent devices similar to the intelligent sockets are the important foundation of the energy efficiency management in the future. Therefore, it is imperative to vigorously develop and promote such equipment, so as to promote the development of the energy efficiency management.

This paper considers that the principle of the energy efficiency analysis under the background of the big data is very simple, which mainly includes two steps: collecting the energy consumption information of the power equipment and the comparative analysis of the energy consumption information.

Collect the energy consumption information of the power equipment. Usually, according to the different needs, flexible setting of the working point parameters can generate the energy consumption reports and the energy consumption curves of the power equipment in different periods, such as the equipment consumption, the production line consumption and the team consumption. At the same time, for the large-scale equipment, the energy consumption information can be collected in the form of "one device with one energy meter". For the ordinary residential electrical equipment, the intelligent sockets can be used to monitor the energy consumption information of the multiple devices at the same time.

Contrastive analysis of the energy consumption information: It is usually necessary to make statistics of the energy consumption information of the specific electrical equipment, then compare it with the international and the domestic standards at the same time, horizontally and vertically, and compare it with the industry energy efficiency benchmark value provided by the third-party evaluation agencies, so as to determine whether the equipment has the energy-saving potential and the size of its potential.

3.1 Analysis of the Fine and Orderly Power Consumption Strategy Based on the Big Data

China's power load characteristics are undergoing tremendous changes. When the provincial power grid companies carry out the power balance, there are frequent phenomena that the power can be balanced, and the power gap is difficult to make up. Therefore, the current power shortage problem is not the traditional power shortage, but the load shortage. However, the peak power consumption time in a year is generally short, and it may occur in some seasons of the year. Therefore, simply relying on expanding the investment scale and increasing the installed capacity to compensate for the peak power shortage will inevitably lead to raising the power supply costs due to the insufficient utilization of the related power equipment. Therefore, effectively organizing the power users to participate in the orderly power consumption work such as the peak staggering, the peak avoidance, the power limit and the switch-off is the key to balancing the load, absorbing the power supply and demand gap and improving the economy of the power grid operation in the peak time or the sudden situation. The traditional orderly management of the electricity consumption is extensive and subjective, which is unreasonable and impartial to the choice of the users. There is no scientific basis for the formulation of the peak avoidance mode and the peak avoidance quantity of the orderly electricity consumption. Under the background of the big data, we can make full use of the existing massive data, combined with the data mining algorithm, to deeply mine the user's electricity consumption behaviors, so that different orderly power consumption schemes can be proposed for different types of the power users.

This paper argues that the implementation of the orderly power consumption under the background of the big data should include three aspects, namely, the short-term load forecasting based on the big data, the error-avoiding peak period and the capacity based on the big data mining, and the evaluation of the comprehensive value of the power users.

The short-term load forecasting based on the big data is the basis for judging the grade of the power gap and the premise of the orderly power consumption, and plays an important guiding role. Based on the historical load data, the big data analysis technology is used to analyze the main factors affecting the power load. A short-term load forecasting model based on the artificial intelligence algorithm is constructed to improve the accuracy and stability of the short-term load forecasting.

The errorable peak avoidance time and capacity of the users based on the big data mining: The quadratic clustering is used to determine the participating time of the users in the category dimension. Firstly, the K-means algorithm is used to cluster roughly based on the average load, the load rate and the peak-valley ratio, and then the fuzzy mean algorithm is used to cluster quadratically based on the user load characteristic curve, so as to refine the power users. Finally, according to the results of the user segmentation, the time period of the user participation in the peak avoidance is identified. In the time dimension, based on the user load characteristic curve, the participating users are clustered separately, and then different power consumption patterns of the participating users within one year are identified. Finally, according to the different power consumption patterns of the participating users, the capacity of the users participating in the peak avoidance in a specific period of time is identified.

User value comprehensive assessment: The user value comprehensive assessment is to determine the priority list of the users who participate in the orderly use of electricity. It mainly considers two aspects of the customer power value and the customer applicability, and establishes a fuzzy comprehensive evaluation model. The customer electricity value includes the customer's social contribution, the customer's electricity credit and the customer's electricity potential. The customer applicability includes the customer's typical load shape similarity and the customer's willingness level.

Generally speaking, the refined and orderly power consumption management process based on the big data is as follows. Firstly, the short-term load forecasting model is used to forecast the load of the power grid and determine the actual power gap. Secondly, the data mining method is used to determine the wrong peak avoidance period and the capacity of all the participants. Thirdly, the user comprehensive value evaluation model is used to establish the priority list of the participants. Finally, based on the results of the first three steps, a more scientific and refined orderly power consumption scheme is worked out.

3.2 Design of the Power Communication Platform

Through the design of the power communication platform, the application value of the big data is guaranteed. In the smart grid, the power communication can be divided into the internal communication and the external communication. The former refers to the information communication and the interaction within the communication system, and the main contents of the communication are the information of the working parameters, the operation status and the attributes of the communication equipment. The external communication mainly refers to the internal communication of the power control system relying on the communication system, which is mainly used to detect the operation status of the power system and the users' application of the power equipment and so on, and the device sends the operation instructions through the control center to all the electric power devices. With the continuous development of the power grid construction, the types of the power generation equipment in the power grid are increasingly diversified, and the power communication protocols are updated rapidly. In the process of the internal and the external communication, it is necessary to simultaneously compatible with various communication modes and record and store the big data communication information, so as to provide the reliable reference for the power system management decisions. According to the above analysis, the overall design objectives of the power communication platform can be defined:

Data Classification and Storage Capacity
All kinds of the information transmitted by the power communication system include the operation status and the operation parameters of the power equipment. By classifying and storing the above data, the development and utilization level of the power system communication data can be improved, and the reliable reference basis for the power system control and decision-making can be provided.

Compatibility with the Different Communication Protocols
With the continuous development and expansion of the smart grid, the types and kinds of the internal equipment of the power communication systems are increasingly diversified, and the types of the communication protocols are gradually increasing. In order to ensure the normal development of the power communication, the power communication platform must have compatibility of various communication protocols.

Ability to Dock with the Data Centers
The power data center contains all the equipment, the personnel working parameters and the data information of the power system operation, and also configures a variety of the computer software systems to realize the analysis and utilization of the power data. In the era of the big data, the data center is a key supporting system. Therefore, the power communication system must maintain the good connection and communication with the power data center, which is an important guarantee to achieve the goal of the communication of the power system.

Data Separation and Storage Module
The main function of the data separation and storage module is to analyze the communication data, classify and store them, and prepare for the data mining in the data center. The data separation and storage module is embedded in the data receiving part of the power communication system, and is responsible for replicating the communication transmission data. Its main working flow is as follows: the data recognition, the data classification, the data de-noising and the data conversion.

The data recognition refers to the analysis and identification of the communication data, the selection of the appropriate data receivers according to data types and communication protocols, and the corresponding analysis results. In the process of the above operation, first of all, we need to obtain the current protocol information through the protocol adaptation module, and determine the types of the data receivers.

The data classification is to classify the operation parameters and the data of the power system into different types according to the data types. The classified data are stored in the corresponding data registers to improve the level of the subsequent data development and utilization.

4 Conclusion

In short, the computer big data technology plays an important role in the process of the urban power energy conservation. As a new technology and management concept, the power big data is in the initial stage of development, but it has shown the great energy. Therefore, it is necessary to study the specific application strategies of the power management in China under the background of the big data.

References

1. Zhou K, Ma Z, Xu Z (2018) Situation awareness analysis of the smart grid based on the big data technology. Electr Appl Energy Effi Manag Technol (21):70–76
2. Xue Z, Zhou Y, Gao Y (2018) Research on the key technologies of the power big data for the smart grid application. Sci Technol Innov (32):103–104
3. Song G (2018) Research on the marketing management innovation of the power supply enterprises in the background of the big data. New Technol New Prod China (20):137–138
4. Gao Y (2018) Application of the big data technology in the information security protection of the power system. Inf Commun (10):141–142
5. Ling D, Huang T, Guan X, Zhao Y (2018) Research and practice of the power customer behavior analysis system based on the big data. Power Big Data 21(10):13–17

Research on the Data Mining Technology in College Students' Attendance System Based on the Big Data Architecture

Zhong Jian(✉)

Tianhe College of Guangdong Polytechnical Normal University, Guangzhou 510540, Guangdong, China
zhongfei463460@126.com

Abstract. With the development and improvement of the information technologies, the increasing of the upper application systems and the rapid expansion of the data accumulated in the campus information environment, a typical campus big data environment has initially been formed. Because of the characteristics of the higher education, students' mobility is great and their learning environment is uncertain, so that the students' attendance mostly used the manual naming. The student attendance system based on the big data architecture is relying on the campus network, and adopting the appropriate sensors. Through the data mining technology, combined with the campus One Card solution, we can realize the management of the attendance without naming in class. It can not only strengthen the management of the students, but can also improve the management levels of the colleges and universities.

Keywords: Big data architecture system · College students · Attendance system · Data mining technology

Because the amount of the big data is very large and comes from different fields and industries, it can be obtained through various ways. For the unstructured data, when it is analyzed, for the data in the field of communication, it can be realized by the query and inductive statistics, and the valuable information can be obtained in a relatively short time. For this kind of the data processing, the query analysis mode is adopted.

1 Main Classification Algorithms of the Data Mining

In the context of the data mining, different classification algorithms are needed to deal with the specific situations of the data. Based on this, there are many existing data classification algorithms [1]. Their emergence can effectively promote the data classification and processing, and then promote the development of the computer technologies. About the types of the data classification algorithms, the author made a related summary, and the specific contents are as follows.

Z. Xu et al. (Eds.): CSIA 2019, AISC 928, pp. 162–167, 2020.
https://doi.org/10.1007/978-3-030-15235-2_25

1.1 Decision Tree Classification Algorithms

In fact, the traditional mode of the decision tree algorithm is the C4.5 algorithm, which has the advantages of simple rules and easy operation. The so-called decision tree classification is also called the greedy algorithm. This algorithm adopts the top-down divide-and-conquer method. Its greatest advantage is that it can deduce the classification rules in the form of the decision tree in the chaotic cases and data. In fact, this classification algorithm classifies and processes the related data on the basis of examples [2]. Because of its robustness to the noise data processing in the actual application process, it has gradually become the most common algorithm adopted in the process of the data classification processing in various fields. In the process of building the decision tree algorithm, each node represents the test of a certain attribute, while the decentralization represents the test output of the data. In the process of analyzing the unknown data samples, the method of comparing the attribute values of samples with the decision trees is often adopted [3].

1.2 CBA Clustered Data Algorithm

The CBA clustered data algorithm is a data algorithm based on the association rules. The implementation of this algorithm often relies on the data to construct the classifiers. At present, the CBA algorithm mainly depends on the use of the Apriori algorithm technology in the process of its operation [4]. The advantage of this technology is that it can surface the potential data association rules, so as to facilitate the induction and collation of the data. In fact, the CBA clustered data algorithm also has many shortcomings [5]. For example, because the data classification is vulnerable to the loopholes, the related optimization roles will be difficult to play, and ultimately the efficiency of this algorithm will be reduced.

2 Data Analysis Technologies Under the Big Data

2.1 Specific Analysis Process of the Big Data

In the era of the big data, the data analysis and processing needs to follow certain rules and procedures. In the real life, all kinds of the data have penetrated into every field and every industry, and these data are changing between different regions and different industries. The data analysis in the era of the big data is different from the previous data analysis requirements, so the big data analysis has gradually replaced the traditional data analysis technology [6]. In the era of the big data, the data analysis technology should collect all kinds of the information from various fields and regions of the society, then transform the information into the effective data, and then use the data analysis technology to classify, integrate and analyze the data [7, 8]. In the process of the data analysis, we should delete the useless information, deeply analyze and process the useful information, and apply it to the real life and work, so as to give full play to the advantages of the data.

2.2 Analysis Models of the Big Data

According to the sources of the data, we can divide the big data into the following categories: ① Various statistical data in different industries and fields in China; ② The click-through rate and the browsing times of the users to various information and resources on the Internet, including the changes of various information on the dynamic web pages and social websites. When searching this kind of the data, we can get all kinds of the data in the Internet through the real-time monitoring data and searching keywords. ③ The data in the field of communication, including the personal user information, the call recording time and other related data. ④ The log data such as the user's behavior and the user's operation on the systems. For the log data, the click-browse behaviors of users can be analyzed to understand the operation behaviors of their log and the system. If the data is generated in the Internet, it should be retrieved, processed and analyzed through the search engines. Because of the continuous progress of the science and technology in China, the analysis engine of the big data has become very convenient and accurate, and we can obtain the most valuable and useful data in the mass of the information. For the web page data and the log data, the actual processing process is relatively close. Through the careful analysis of the data, the valuable information can be obtained.

2.3 The Big Data Analysis Technology

In the era of the big data, the amount of the data is very large and its sources are very complex. These data resources are constantly changing. In order to analyze and obtain the useful information from the constantly changing and rapidly expanding data, advanced data analysis technologies should be adopted. Only in this way can we deal with a large amount of the data. Because there are many kinds and structures of the data and the main data types are the unstructured data types and the semi-structured data types, the MapReduce-based big data processing technology can be used to process and analyze the technology.

In the actual process of the data processing, the MapReduce-based join algorithm can effectively improve the efficiency of the querying related data. In the era of the big data, the amount of the data is very large. This join algorithm can improve the execution efficiency of the algorithm, and has a good effect on improving the query efficiency. By adding an index, the transmission amount of the network is reduced, and the efficiency and the quality of the query task are improved.

3 Research on the Data Mining Technology in College Student Attendance System Based on the Big Data Architecture

At present, the information of the students attending classes is registered and counted manually by the teachers. Counselors have a great lag in understanding the attendance of each class. This attendance system is managed by computers. Counselors can timely understand the attendance of the students and timely correct the students who are absent and late without reasons. The timely progress puts forward stricter requirements and challenges for the solution of the traditional student attendance management. In order to make the development of the system successful, the solution to the student attendance system should be in line with the actual teaching needs and the development trends. Before implementing the design of the attendance system, a series of problems must be scientifically demonstrated, such as the demand analysis of the system, the overall planning of the system, the functions and implementation plans of the student attendance system, the software and the hardware configuration of running the student attendance system, and the management methods of the student attendance system.

3.1 System Functional Modules

The student attendance management system mainly provides the fast and accurate statistics of the students' absenteeism and the information of their leave, and realizes the computerization of the student attendance work. Users of the system are divided into students, teachers, counselors (head teachers) and administrators of departments and colleges. Students: Through the student attendance management system, we can apply for a leave online and inquire the attendance information of the classes. Teachers: Through the student attendance management system, students' leave information can be inquired online, and the real-time input of their attendance information can be completed. Counselors (head teacher): Through the student attendance management system, students' leave information can be examined and approved online, and the students' attendance information can be inquired, and their class truancy and the information of their being late can be counted through the student attendance management system.

3.2 Database Design

The data in the database is the core of the entire system. The reasonable establishment of the database and the data table will directly affect the performances of the entire database system. According to the characteristics and requirements of the attendance management system, the following are the key table information of the database. Student: It is used to store the basic information of the students, such as their school numbers, names, genders, telephones, departments, classes and places of origin and so on, which are mainly input by the students themselves to improve the information or by the department counselors (head teachers). Course: It is used to store the students' curriculum information and their class situations, such as the numbers of a certain course, names, class locations, class teachers, and class time and so on. Absence: It is used to record students' attendance information, such as the student numbers, course numbers, being late, early leave, absence, and leave and so on. Leave: It is used for asking for a leave online, and there are the following fields, such as: the leave number, the student number, the affiliated department, the leave reason, and leave time and so on.

3.3 A Brief Introduction to the Basic Mining Algorithms of the Association Rules

The most basic algorithm for the mining association rules is the Apriori algorithm first proposed by Rakesh Agrawal. Two concepts of the support and the confidence are introduced to describe the user's interests in the mining rules. By traversing a large number of the transaction data, count the starts from a single item. After each traversal of all the transactions, the items whose support is less than the user's given support are cut down, and then gradually extended to the multiple transactions. Finally, the frequent item-sets are saved to generate the association rules by subset the generation method, and then the association rules whose confidence is lower than the minimum confidence specified by the users are removed. Finally, the association rules that satisfy the user's needs are left.

3.4 The Core of the Mining System Based on the Oracle Database System

The data logic processing of the conventional C/S or the B/S database application systems is usually placed on the application server layer, and the database is mainly managed as a data storage medium. The specific data in the database is extracted through a special programming language, and then processed and returned to the database. The Oracle database is a kind of the very complete data storage, data conversion and data application of the central system. The data mining systems often need to scan the entire data table and the massive data traversally, and put as much as possible the whole data conversion process and the processing process in the database system for the execution. The efficiency is generally higher than that placed in the application server layer.

This multi-tier architecture is the application server-centric service, trying to isolate the data layer. In addition to making the system more versatile and portable, this also can further reduce the burden of the database. However, the conventional J2EE database service architecture is not focused on the large-scale data processing, but on the data application. When traversing the massive data, the application server is usually used as the center, which leads to a lot of the unnecessary data exchange between the application server and the database. These large-scale data processing architectures are implemented in the database system, and many optimized processing functions and stored procedures within the database system can also be utilized. The Oracle data is a very complete data system. In addition to the data storage and query, the Oracle data can be processed and transformed by writing a series of the storage processes and triggers.

Because the mining algorithm of the association rule needs to traverse all the data information, if it is the massive data, the conventional application server-centric data processing architecture is not easy in the association mining algorithm. This paper introduces a data processing architecture based on the Oracle database, which can efficiently mine the association rules. At the same time, this implementation architecture relies on the processing environment of the Oracle data system, which has the strong scalability and can deal with the massive data. Student workers' management and education provide the important and valuable information or knowledge, which has an immeasurable effect. It is of practical significance to popularize and apply it in the management of the college students. What needs further study is to use the data mining technology to develop the intelligent student management and education system.

4 Conclusion

The student attendance is an indispensable task in every college classroom. In order to improve the quality of our teaching, teachers often regard the student attendance as part of the final examination results. The big data brings opportunities and challenges to the management of the students in colleges and universities, and also puts forward new topics. Only by reconstructing the existing student management information and improving the collection and analysis system of the semi-structured and the unstructured data on the basis of the existing student management data mining can we continuously optimize the student management strategy, more clearly describe the overall information of the students, and more scientifically divide the student groups. The goal of the individualization and refinement of the student management in colleges and universities will be realized.

Acknowledgment. Foundation Project: 2017 Youth Innovative Talents Project by Guangdong Provincial Department of Education (Natural Science), "Design and Application of the Intelligent Classroom Attendance and the Mobile Phone Use Restriction System Based on the WiFi Probe" (2017KQNCX269).

References

1. Lu Y, Du Q (2017) Application of the data mining in the university information management. Sci Technol Innov Herald 1:101–102
2. Li H (2017) Research on the informatization of the teaching management in colleges and universities based on the data mining technology. Comput Knowl Technol 10:222–224
3. Zhou W (2018) Research on the campus network security management technology based on the big data. Netw Secur Technol Appl 10:132–133
4. Guan X, Zeng Y (2018) Progress and trend of the parallel data processing, analysis and mining in the background of the spatiotemporal big data. Prog Geogr 10:151–153
5. Yin Z (2018) Modeling and analysis of the book lending flow in university libraries based on the data mining. Microelectron Comput 11:193–194
6. Romero C, Ventura S (2013) Data mining in education. Wiley Interdiscip Rev: Data Min Knowl Discov 3(1):12–27
7. Bharamagoudar SR, Geeta RB, Totad SG (2013) Web based student information management system. Int J Adv Res Comput Commun Eng 2(6):2342–2348
8. Sun Y, Yan H, Lu C et al (2014) Constructing the web of events from raw data in the web of things. Mob Inf Syst 10(1):105–125

Research on the Directional Control of Direct Magnetic Field of the Locomotive Asynchronous Traction Motor

Tiezhu Zhang(✉) and Qiongjie Zhang

Zhengzhou Railway Vocational and Technical College,
Zhengzhou 450000, Henan, China
yangyu5739629@163.com

Abstract. The directional control of the direct magnetic fields of the induction motor can improve the dynamic response speed of the motor, and its flux observation link is located in the feedback channel, so it has better parameter insensitivity than the indirect field-oriented control. In addition, the traditional method is to use the photoelectric encoder to obtain the speed signal needed for the flux observation, but in some bad conditions, the installation of the photoelectric encoder reduces the reliability of the motor, and improves the cost and the volume of the motor. In this paper, a new method is presented, which uses the S speed estimation algorithm of the model reference adaptive system (MRA) to observe the flux signal through the voltage and the current of the motor terminal and the calculated speed signal, so as to achieve the goal of the directional control of the direct magnetic field.

Keywords: Locomotive · Asynchronous traction motor · Magnetic field directional control · Flux observer

The directional control of the indirect magnetic field uses the dynamic slip frequency and the rotor speed to get the synchronous frequency, and the angle of the rotor flux is obtained after the integration. The directional control of the indirect magnetic field uses the flux observer to calculate the motor flux in the real-time manner, and then coordinate the conversion to implement the control. The flux observer includes the voltage, the current and the hybrid types [1]. It can be divided into the open-loop and the closed-loop forms according to its structures. Its performances directly affect the accuracy of the field orientation and the control of the motor.

1 Fundamentals of the Directional Control of the Direct Magnetic Field

The directional control of the magnetic field of the induction motor eliminates the cross-coupling between the torque current and the magnetic field current in the stator current of the motor, realizes the real trouble-solving operation, and adjusts the excitation current and the torque current of the motor respectively, so as to achieve a faster

Z. Xu et al. (Eds.): CSIA 2019, AISC 928, pp. 168–173, 2020.
https://doi.org/10.1007/978-3-030-15235-2_26

dynamic response [2]. The basic field-oriented control (FOC) technology can be divided into two methods. One is the indirect field-oriented control (IFO), i.e. the slip frequency estimation vector control. Although the algorithm is simple, because the calculation of the slip frequency is located in the feed-forward channel, the control results are very sensitive to the changes of the rotor parameters, so sometimes it is impossible to achieve the real lotus root. The other is the direct field-oriented control (DFO), which is a real field-oriented technology. The reliability of the system decreases because of the additional magnetic field detection elements [3]. Generally, the stator or the rotor magnetic field is estimated by detecting the terminal voltage and current of the stator, i.e. the flux observer. The accurate field orientation is achieved by calculating the flux vector. To achieve the correct coordinate transformation, the coordinate d axis after the transformation always coincides with the direction of the magnetic flux, so as to achieve the goal of the complete torque current and the magnetic field current [4].

In direct field-oriented control, because the flux observer is located in the feedback channel of the control system, its control results are relatively insensitive to the parameter changes, and by optimizing the observer design, the flux can be estimated relatively accurately when the motor parameters change. In addition, the bandwidth of the observer (motor speed range) is widened to meet the requirements of the wide range accurate speed regulation. The field-oriented control technology is also called the vector control [5]. Based on the coordinate transformation principle, by adjusting the phase, frequency and amplitude of the motor voltage, the current, excitation and torque levels of the motor are directly controlled, so that the excitation and torque components of the motor current are decoupled. The motor has good dynamic and static characteristics, which is conducive to the vehicle adhesion control and is suitable for the high speed [6]. The traction and electric braking control of the high-power motors is one of the mainstream technologies in the field of the traction drive [7].

Aiming at the complex problems of the multi-variable, non-linear and strong coupling of the bearingless induction motor, an inverse system decoupling control method based on the air gap magnetic field orientation is proposed. Firstly, the dynamic mathematical model of the bearingless induction motor is analyzed and established, and the reversibility of the system is analyzed, and the dynamic mathematical model of the inverse system is established [8]. The decoupling control strategy decouples the bearingless induction motor into four linearized subsystems: the motor speed, the air gap flux linkage and two radial displacement components. The structure of the decoupling control system is given. Finally, the system simulation is carried out. The simulation results show that the system has the good dynamic decoupling control performance, the fast response speed and the strong anti-load torque disturbance ability.

2 Analysis of the Characteristics of the Flux Observer

For the analysis of the flux observer, firstly, the mathematical model of the motor is established. The mathematical equation of the motor in the stationary coordinate system expressed by the complex phasor is simple, and it is easy to deduce and analyze the frequency response function. The variables of the motor are expressed in the form of the complex phasors. Different flux observer equations can be derived. The current flux observer is calculated according to the input current, the voltage flux observer and the hybrid flux observer are calculated according to the voltage.

Because the operation parameters of the motor will change, the excitation inductance of the motor has the non-linear characteristics. For example, the stator and the rotor resistance will change with the increase of the temperature of the motor in the operation, and the measurement error of the motor parameters itself will lead to the inaccurate flux observation in the controller and affect the control effect. Based on the frequency response function, the characteristics of all kinds of the flux observers are analyzed. The slip frequency estimation vector control is simple, but the control result is very sensitive to the change of the motor parameters, which affects the performances of the whole system. Because of the decrease of the system reliability caused by the additional magnetic field detection elements, the accurate field orientation can be achieved by detecting the stator voltage and current, estimating the flux linkage, and calculating the flux vector, so as to achieve the goal of the complete decoupling control of the torque current and the excitation current.

Combining the analysis of the response function (FRF), the effect of the motor parameters on the flux observation accuracy and the performance of the observer at different speeds can be obtained. The amplitude-frequency function can show the magnitude deviation between the observed flux and the actual flux, and the phase-frequency function can indicate the phase difference between the two, that is, the deviation of the observed flux angle. In different slip ranges, the observer is affected by different parameters. Because the fundamental frequency is the rated frequency of the motor, it is equivalent to the difference of the observer characteristics of different motor speeds. When the slip is large and the speed is low, the rotor resistance deviation has a greater impact on the error of observing the flux magnitude, but at a high speed, it is smaller. On the contrary, the error of the excitation inductance is greater when the slip is small and the speed is high, but smaller when the speed is low.

The current flux observer is not affected by the stator resistance, while the voltage flux observer is affected by the stator resistance at a low speed, and is less affected by other parameters. Therefore, the two observers are complementary to each other. A hybrid flux observer based on the combination of the two observers and the introduction of the closed-loop correction link can achieve good observation results. The direct flux orientation control can be achieved by observing the flux linkage based on the hybrid flux observer. We can carry out the simulation research in the Matlab/Simulink, and the motor model in Simulink can output the actual flux angle and magnitude, which is convenient for comparing the accuracy of the flux observer in calculating the flux linkage.

3 Direct Field Oriented Control of the Asynchronous Traction Motor for the Locomotive

The simulation results show that the flux estimation of the sensorless induction motor direct field-oriented control and the speed estimation using the estimated flux are the feasible methods, and the algorithm is stable. In the practical applications, the application of these two algorithms can be achieved by the DSP and we can achieve the high dynamic performances. The simulation results show that the curve of the flux estimation is basically unchanged, which indicates that the algorithm is insensitive to some parameters. The accuracy of the velocity estimation using the estimated flux can be fully satisfied. The open-loop speed estimation is based on the mathematical equation of the asynchronous motor in the static reference frame. Because of the exact parameters of the motor, the steady state velocity error may occur. However, the structure of the speed estimator is relatively simple.

The asynchronous motor is widely used in the traction drive system of the high-speed train because of its simple structure, durability, small maintenance and good mechanical torque characteristics. The control of the asynchronous motor belongs to the field of the AC drive technology. Since the 1970s, the AC drive technology has developed rapidly. The software and the hardware technology of the power electronic power converter and the control system have made great progress. Many advanced AC motor control strategies have emerged, such as the slip frequency control, the magnetic field orientation control and the direct torque control. These control strategies are based on the steady-state or the dynamic mathematical model of the motor. The closed-loop control is applied to the motor, which greatly improves the operation characteristics of the motor.

In order to reduce the switching loss of the high power traction drive system, after the traction motor enters the weak magnetic region, the inverters usually enter the square wave mode to make full use of the DC bus voltage and reduce the switching frequency. Under the condition of the square wave, the traditional vector control algorithm is limited by the single degree of the freedom of the voltage vector angle. At the same time, the accuracy of the magnetic field orientation directly determines the performances of the motor control in the whole speed range, and the accuracy of the magnetic field orientation is affected by such factors as the changes of the motor parameters. In view of the above problems, this paper studies the motor control under the square wave condition and the field orientation correction of the full-speed motor.

Firstly, the mathematical model of the asynchronous motor is established. Considering the limitation of the current and the voltage under the square wave, the current distribution strategy which maximizes the motor's torque output under the square wave is deduced. According to the different operating conditions, the current instructions of the dq axis are reasonably designed. Then the limitation of the traditional double current loop vector control under the square wave is analyzed. The motor voltage under the square wave is limited by the DC-side voltage, and the current controller loses its regulating function because of the zero adjustable voltage margins. Secondly, a single-current closed-loop control strategy is proposed for the special conditions of the square wave. Only the d-axis current loop is reserved, and the q-axis voltage is calculated by the d-axis voltage. This strategy solves the problem that the traditional vector control cannot operate in the square wave conditions. Based on this, a dual-mode motor control method suitable for the hybrid multi-mode pulse width modulation is proposed, which realizes the full speed control of the induction traction motor.

Thirdly, based on the proposed full-speed dual-mode control method, according to the different control strategies before and after the square wave, the influence of the change of the rotor time constant on the variables in the motor control is analyzed, and the corresponding field-oriented correction strategy is proposed. In the low and medium speed ranges, the flux orientation correction strategy based on the torque observation is adopted. Under the square wave condition, the single current closed-loop control of the square wave proposed in this paper can self-tune when the flux orientation is not on time, so as to ensure the accurate output of the motor torque. Finally, the proposed control strategy is validated on the CRH5 EMU hardware-in-the-loop platform provided by the Sifang Vehicle Research Institute of China Railway, and verified on the 5.5 kW asynchronous motor Trailer platform in the campus laboratory. The simulation and experimental results verify the validity and feasibility of the proposed closed-loop control strategy of the square wave single current and the field-oriented correction strategy based on the torque observation.

The relationship between the stator dq axis current and the stator flux is analyzed. The demagnetization component of the stator magnetic field caused by the q axis current is deduced. The component is compensated by increasing the d axis current when the load changes and the decomposition of the stator current dq axis component are realized. The vector control with the open-loop flux linkage is adopted to avoid the complexity and inaccuracy of the traditional field-oriented vector control in the magnetic field observation at a low speed. However, in order to ensure the accuracy of the field orientation, a field orientation correction method based on the stator current error is proposed. By analyzing the relationship between the given current and the feedback current of the dq axis in advance and a lag time of the field orientation, the q-axis current can be compensated, and then the slip can be changed to compensate for the errors caused by the inaccuracy of the field orientation. Finally, the simulation model and the experimental platform of the asynchronous motor based on the MATLAB are built. The results show that the magnetic field orientation has a good correction effect when the steady-state operation and the motor parameters change, and the static and the dynamic performances of the control system are guaranteed.

The frequency response function analysis of the voltage-type, the current-type and the hybrid closed-loop flux observer in the direct field-oriented control system shows that the current-type observer is not affected by the stator resistance, while the voltage-type observer is greatly affected by the stator resistance at a low speed, but is not affected by the observer rotor deviation. The hybrid flux observer combining the advantages of the two methods, and the flux angle can be calculated accurately through the feedback closed-loop, and the parameter robustness is strong. It is suitable for the direct field oriented control system. The simulation is verified. In the further engineering implementation, it is necessary to solve the problems of the accurate voltage and the current measurement and the processing methods, and overcome the influence of the low switching frequency and the high current harmonics on the motor control characteristics.

4 Conclusion

The rotor field oriented control technology is used in the locomotive traction system, such as CRH1, CRH2, CRH3 and some locomotives. In this paper, the flux observer in the direct rotor field-oriented control is studied thoroughly and comprehensively.

Taking a traction motor for a certain type of the EMU as an example, the accuracy of the flux observer and its sensitivity to the parameter changes are analyzed, and the flux observer for the traction control of the high-speed train is designed. The direct field-oriented control based on the hybrid flux observer is realized by simulation. The simulation results show that the designed flux observer is accurate and the decoupling control of the traction motor is realized.

References

1. Wang Y, Xie H, Qin M, Jiang Y (2016) Traction control strategy of the permanent magnet synchronous motor based on the rotor field orientation. Mod Electron Tech (07):133–134
2. Jia Y, Yu N, Yang C, Zhang X, Duan H, He T, Yu Y (2017) Impact of the model-based iterative reconstruction on the airway analysis of the computer-aided detection system under the low-dose lung CT condition. Sci Technol Innov (12):133–134
3. Hou H, Tang W, Tang H, Li H, Zhong J, Li Z, Li C, Huang Y (2017) The simulation analysis of the influence of the high thermal conductivity insulation structure on the temperature rise of the permanent magnet synchronous traction motor. Insul Mater (09):19–20
4. Liu Z, Luo Y, Rao S, Hu Y, Zhang F, Gong H, Liu J (2018) Directional solidification of the polycrystalline silicon under the axial magnetic field. Hot Working Technol (01):156–157
5. Zhang C, Liu H, Lan J, Yu J, Yang Z, Ren Z (2018) Effect of the longitudinal magnetic field on the directional solidification microstructure of Co-8.8Al-9.8W-2Ta cobalt-based superalloy. Shanghai Met (05):126–127
6. Zheng L, Fletcher JE, Williams BW et al (2011) A novel direct torque control scheme for a sensorless five-phase induction motor drive. IEEE Trans Ind Electron 58(2):503–513
7. Zhou Y, Lin X, Cheng M (2016) A fault-tolerant direct torque control for six-phase permanent magnet synchronous motor with arbitrary two opened phases based on modified variables. IEEE Trans Energy Convers 31(2):549–556
8. Diao L, Tang J, Loh PC et al (2018) An efficient DSP–FPGA-based implementation of hybrid PWM for electric rail traction induction motor control. IEEE Trans Power Electron 33 (4):3276–3288

Research on the English Teaching Resource Library Management System Based on the Computer Technology

Junmin Wu(✉)

Zhengzhou University of Science and Technology,
Zhengzhou 450064, Henan, China
xazrll@yeah.net

Abstract. With the continuous development of the computer network technology and the network education, colleges and universities have gradually carried out the construction of the college English multimedia network teaching resource library. The construction of the English teaching resource library management system based on the computer technology is the core of the network teaching. The management system of the English teaching resource library based on the computer technology should integrate the elements of the English teaching into a whole, so that it has the functions of the teaching content publishing and management, the online teaching interaction, the online teaching evaluation, the collaborative learning and the developmental teaching evaluation. According to the characteristics of the teaching objectives, the construction of the resource library should establish a multi-level, multi-functional and multi-module resource system.

Keywords: Computer technology · English teaching · Resource library · Management system

With the rapid development of the computer technologies and the deepening of the reform of the educational system, the traditional English teaching mode and the management methods are increasingly challenged by the network education from the computer technologies. The important influence of the computer technologies on our education is the resource-based learning [1]. The management system of the English teaching resource library based on the computer technologies is the premise and basis of the network education. Basically, the core of the computer technology based education is the construction of the English teaching resources. Therefore, it is the most important task to develop an English teaching resource management system that can meet the needs of the current education development and give full play to the role of various existing English teaching resources.

Z. Xu et al. (Eds.): CSIA 2019, AISC 928, pp. 174–180, 2020.
https://doi.org/10.1007/978-3-030-15235-2_27

1 Design Requirements of the English Teaching Resource Library Management System Based on the Computer Technologies

Based on the computer technologies, the teaching system of the English teaching resource library can realize the diversification of the teaching resources. Students can query various resources according to their own preferences, as well as their learning progress and achievements [2]. At the same time, it is also convenient for the teachers to supervise students' learning situations on the Internet and to inquire students' learning results quickly.

Firstly, we should pay attention to both the knowledge and the interest of the corpus. The basic purpose of the construction of the reading resource library is to help the users of the corpus in the resource library acquire certain knowledge or abilities through the use of the corpus in the resource library [3]. At the same time, we must realize whether the corpus can arouse the students' interests in their learning. Only when interest and knowledgeable are well combined can college students be motivated to learn the English knowledge and the reading materials actively, so that they can read with interest, and then make full use of the corpus in the resource bank to maintain and enhance their interests in reading and learning [4]. Therefore, when designing the resource library and choosing certain corpus, we should fully consider the knowledge and interest of the selected corpus so as to stimulate students' reading motivation, drive them to independently try various learning strategies in reading, conduct the deep reading, and constantly achieve the leaps and bounds in the knowledge acquisition, the development of their language skills and the improvement of the humanistic literacy.

Secondly, pay attention to the hierarchy and the gradualness of the corpus. The corpus selection in the construction of the college English reading resource bank should be based on college students' English language proficiency, reading abilities, academic levels, and professional characteristics and hobbies. Otherwise, it will easily fall into the pit of all things and fail to achieve the effect of teaching the students in accordance with their aptitude [5]. Of course, the hierarchy and gradualness here should be based on the requirements of the college English curriculum teaching, take their reading abilities as the starting point, take into account the individual differences of their language learning levels, and then improve and develop step by step from the "general requirements" to the "higher requirements" and to the "highest requirements" according to the hierarchical requirements of the foundation. In this way, not only can the idea of the hierarchical college English teaching be effectively implemented in the corpus grading, but college students can also choose the corpus suitable for themselves and obtain pleasure from reading [6].

Finally, the emphasis should be laid on both the professionalism and the universality of the corpus. The language education should take into account both the instrumentality and the humanism of the language [7]. However, in recent years, the idea of dividing the main contents of the college English teaching into three parts: the general English, the special purpose English and the cross-cultural communication has been more and more embodied in the concrete teaching, on the basis of which three kinds of the courses have been formed in order to cultivate their relevant language competence. Of course, such classification has its full practical needs. Because, after all, the acceptance of the English education should meet the language needs of the professional learning and the future occupational development. For example, the business majors have the obvious professional register characteristics in terms of the vocabulary composition, the grammar requirements and the format [8]. Therefore, in the construction of the reading resource library, it is necessary to determine the specialty and the universality of the corpus according to the learning purposes of the college students, so as to provide the rich and individualized corpus resources for the college students.

2 Design Requirements of the English Teaching Resource Library Management System Based on the Computer Technology

The network resource library is important for the students [9, 10]. The really effective learning resources are the well-designed educational resources that can meet the needs of the learners' autonomous learning. Making full use of the network resources and innovating the English teaching can create a more individualized English learning environment for the students, so that students can independently choose the English learning contents and control the English learning speed according to their own situations. Give full play to the enormous advantages of the English teaching resource library, and especially strengthen the links between the curriculum and the students' learning, life and society, so as to truly achieve the educational goal of letting students actively learn English, experience English, appreciate English, and achieve their own acquisition of knowledge, self-renewal and even create the new knowledge [11].

The network teaching in the process of the English teaching in the network resource library does not allow the students to browse on the Internet without definite purposes. For the English teaching resource library, we should closely focus on the goal of our English teaching. The construction of the resource library should first highlight the specialty and the practicability of the English subject. It should also screen and process the resources suitable for the English teaching and integrate them into the carrier of the campus network so as to make them have the functions of seeing, hearing, appreciating and analyzing. The contents of the resource library should fully stimulate the students to carry out their English learning, their English expression and their English aesthetic activities.

The construction of the English teaching resource library needs the management of the professional and technical personnel. As the construction of the resource library is a process of the continuous improvement of the database structures and the gradual increase of various resources, the sustainable development of the resource library depends on the continuous enrichment and the timely updating of the educational resources in the database, as well as the credibility and the authority of the information provided. It must be managed by the specialized technicians. Managers select a suitable server platform and the database system, and classify and manage the resource database according to certain rules, and carry out the electronic conversion of the data, and maintain and update it regularly. The appropriate resource management platform should have a complete user management system, which can assign different levels of the authority to different users and make different types of the educational resources relatively independent.

The construction of the English teaching resource library needs the participation of a large number of the English teachers. The construction of the network resource database includes the construction of the resource database platform and the teaching resources. English teachers are one of the most important forces in the construction of the English teaching resources. They understand the most urgent needs in our English teaching. It is necessary for them to participate in the construction of the English teaching resources. Therefore, we must mobilize all the positive factors and establish a development team composed of the front-line English teaching backbones and the computer technicians to discuss and build it together. Organize the English teachers to participate in the design, planning and construction of the network resource library, so as to make it more subject-specific and meet the needs of our English teaching, including the teachers' teaching preparation and the students' autonomous learning. To meet the needs of the teachers in preparing the lessons for teaching is that the database can provide the teachers with the resources needed for preparing the lessons for the English teaching related to the contents of the textbooks, including the teaching plans, the courseware, and the materials for preparing the lessons and so on.

3 Research on the English Teaching Resource Library Management System Based on the Computer Technologies

3.1 Architectures of the Resource Library System

The system adopts the typical three-tier architecture of B/S based on the Web: the presentation layer, the business layer and the data access layer, and realizes the communication between the sites by combining the technology of Web Services. The superior system performances of the B/S architecture are also adopted. In the past, the client-server system was composed of two layers: the client program + the database server. The client directly connects to the database system. They can be divided into two modes: the fat client and the fat server. The feature of the fat client is that the data computing and the data processing are centralized in the client side, which is prone to the access conflicts on the database side when the users connect more than one time. The feature of the fat server is that the data computing and the data processing are centralized in the database server. When the users connect more, the system will be blocked and the performance will decline sharply, which cannot meet the needs of the multi-users at all. In the three-tier client-server system, the data computing and the data processing are concentrated in the middle-tier components, and can achieve the distributed computing functions.

It has the excellent security performances. In the traditional C/S structure, the client directly connects with the database, and the database is completely exposed to the outside, so that there are great security risks. In the three-tier structure, the client is only responsible for sending the requests, not directly connected with the database, but connected with the database by the application server. There is a natural barrier between the external application and the database, and at the same time, the data security and the integrity are guaranteed to the greatest extent. In the traditional C/S structure, each user needs to establish a connection with the database server. In the three-tier structure, users can share the database connection, which reduces the burden of the database and improves the performances of the system.

It is easy to maintain and upgrade. In the C/S structure, upgrading and maintenance of the system must be carried out at both the client and the server. Many clients are difficult to maintain, which also increases the costs of the maintenance. If the program changes, each of the clients must be upgraded, which also increases the risks of the system upgrading. Under the B/S structure, the maintenance or the upgrading of the system is only on the server side, and there is no need to change the client side. This reduces the maintenance costs and risks of the system, while the upgrading does not affect the daily work and the business of the educational network.

3.2 The Design Flow of the Resource Library System Program

For a user, the specific flow process of using the resource description and the management system is as follows. After the user successfully logs in through the home page, the system displays the basic information of the user after confirming the consistency, and then the user's rights will be given, and the operation of the relevant rights consistent with the user can be carried out. The system will load different system files in the interface according to the different user identities. In these system files, the user's privileges are set. If an administrator user or a teacher user logs in, the system adds some special management functions to the user besides having the lowest functional privileges as the student user. Users can view, add, modify and delete the resources and the sub-resources contained in the resources to see if there are any required teaching resources. If the local resource library does not have the resource files needed by users, it can also interact with other resource libraries and transfer the resource files from other servers to the local resource server. The information processing and transmission process of the whole database is as follows. The information is generally searched by the retrieval tools, and the information needed is found in the educational resource database. After processing of the information processing tools, the finished products are generated on the production platform and displayed on the display platform. Finally, after the evaluation and identification, it is determined whether to save the finished products. This greatly improves the use values of the educational resource library.

3.3 Design Contents of the Resource Library System Program

The teaching resource database mainly includes four sections: the teaching guidance, the teaching courseware, the learning guidance, and the teaching resources. In each large section, it is divided into several small sections. The following is a specific description of each section. The first is the teaching guidance. The instructional guidance includes the syllabus, the teaching plans and the curriculum plans of the basic module and the vocational module of the English textbooks in vocational schools.

Secondly, the teaching courseware mainly includes the network teaching materials of these two modules, as well as the teaching courseware independently developed by the teachers. The last is the learning guidance. It mainly includes various strategies of learning the English knowledge, such as how to learn the listening, how to learn the pronunciation, how to learn the vocabulary, how to learn the reading, and how to learn the writing and so on. Of course, it also includes introducing and exchanging the learning experiences, such as the English lectures, and the experience of excellent students and so on. Among them, the English lectures mainly include the English academic reports conducted by our teachers and the English lectures brought by the English experts. Through the cooperation and communication between the teachers and the students, it not only helps to timely feedback the teaching results, but also fully stimulates the students' interest in their learning and motivates them to actively participate in the learning activities.

4 Conclusion

The construction of the English teaching resource library needs to consider many factors such as the teaching, the teachers, the students, and equipment and so on. The construction of the English teaching resource library should be based on the actual needs of the English teaching in schools, taking into account the differences in the school equipment, the computer levels of teachers and students, and the school characteristics in different areas. To measure the qualities of an English resource library, we should pay attention not only to the amount of the resources, but also to whether the resource library can effectively serve the students' learning and the teachers' teaching.

Acknowledgement. The study was supported by the teaching reform project from Zhengzhou Education Bureau: Zhengzhou high educational English teaching team construction project [2016344]; Zhengzhou University of Science and Technology teaching quality project [201667]: Exploration of "New Engineering" talent cultivation mode for Foreign Language Majors under the background of "Belt and Road" [ZZJG-B8021].

References

1. Liu F, Ma L, Cai Z, Zhang J (2016) Design and implementation of the multimedia teaching resource management system based on the PHP. Softw Eng (07):145–146
2. Li J, Chen K (2017) Construction and implementation of the network management system for primary school teaching resources. Course Educ Res (08):105–106
3. Zhang J (2018) Design and implementation of the teaching resource management system based on .NET. Electron Des Eng (05):141–142
4. Chen Y (2018) Research on the application of the multimedia teaching resource management system in the higher vocational colleges. Inf Rec Mater (04):122–123
5. Feng X, Song Y (2018) Design of the network teaching resource management system based on the MVC model. China Comput Commun (Theor Ed) (09):110–111
6. Bao W, Zhou H, Lu W, Xie F (2016) The system of knowledge management using web based learning and education technology. Comput Syst Sci Eng 31(6):469–473

7. Palomo-Duarte M, Berns A, Cejas A et al (2016) Assessing foreign language learning through mobile game-based learning environments. Int J Hum Cap Inf Technol Prof (IJHCITP) 7(2):53–67
8. Huang YM, Chiu PS (2015) The effectiveness of a meaningful learning-based evaluation model for context-aware mobile learning. Br J Edu Technol 46(2):437–447
9. Anshari M, Alas Y, Guan LS (2016) Developing online learning resources: big data, social networks, and cloud computing to support pervasive knowledge. Educ Inf Technol 21 (6):1663–1677
10. Sarrab M, Elbasir M, Alnaeli S (2016) Towards a quality model of technical aspects for mobile learning services: an empirical investigation. Comput Hum Behav 55:100–112
11. Zheng B, Warschauer M, Lin CH et al (2016) Learning in one-to-one laptop environments: a meta-analysis and research synthesis. Rev Educ Res 86(4):1052–1084

Research on the Ideological and Political Education Service Platform Based on the Campus Network Environment

Xinglong Du(✉)

Zhengzhou University of Science and Technology, Zhengzhou 450064, Henan, China
myjf75@yeah.net

Abstract. Nowadays, with the increasingly mature construction of the campus network, new changes have taken place in the way students participate in their learning and understanding of the knowledge. For the ideological and political educational activities, we need to build a brand new ideological and political education service platform based on the new changes in the campus network environment. Starting from the analysis of the characteristics, connotation and the teaching impact of the campus network environment, this paper analyses the new changes of the ideological and political education services under the campus network environment, and combines the values and significances of the ideological and political education services under the campus network environment, so as to analyze the implementation strategy of constructing the ideological and political education service platform based on the campus network environment.

Keywords: Campus network environment · Ideological and political education · Service platform · Innovation mechanism · Informatization

1 Analysis of the Characteristics, Connotation and Teaching Impact of the Campus Network Environment

The college campus network plays an important role in the operation of the entire school, and it is also an important channel and platform for the ideological and political education of college students. With the development of the modern science and technology, the Internet technology has gradually popularized. With its advantages, it has an important impact on the ideological and political education, and its contents have also played a wide role in our society [1]. The network has brought new changes to the ideological and political education of college students. Therefore, in order to further improve and promote the ideological and political education of college students and realize the virtuous circle of the ideological and political education on the network, colleges and universities have put forward the requirements of how to improve the ideological and political education of college students in the network information age [2].

Z. Xu et al. (Eds.): CSIA 2019, AISC 928, pp. 181–187, 2020.
https://doi.org/10.1007/978-3-030-15235-2_28

The campus network provides the information-based teaching platform for the teachers to carry out the teaching and scientific researches, and for the students to participate in their learning and even the information exchange activities. The campus network, as a broadband application, is the local area network with the interactive function and the professionalism, which includes the teaching platform, the multimedia demonstration, the lesson preparation system and the database containing various types of the resources and so on [3]. With the help of the platform advantages of the Internet, whether it is the contents of various teaching knowledge, or the extension of the extracurricular reading, all can be implemented by means of the network. Of course, for any subject, we should combine the advantages of the campus network environment to build the best teaching mechanism.

In today's information construction, with the help of the campus network built by the Internet and the intelligent equipment, the libraries, classrooms, teaching resources and other resources are integrated, so that the teaching resources can be fully and deeply shared. In the operation of the campus network, by using the Windows operating system, the Microsoft System Management Server system to manage the service software and other software, combined with the education network and the optical fiber broadband, we can meet the demands of the teachers and the students for the Internet applications, to achieve the effective integration of the data resources.

The campus network is not an application product, but a network operating system [4]. It is a technology platform like WINDOWS. Many web application-level products can be developed and extended on the campus network platform [5]. It is a new network operating system for the individuals and the small companies who do not have the server resources but have the powerful information creation capabilities and need a large number of the free network applications. The campus network is a SNS network platform based on the theme of the campus learning, life, entertainment, games and entrepreneurship. It is an open WEB operating system integrating blogs, space, community forums, alumni directories, groups, topics, albums, music, WEB games, online teaching, P2P, streaming media and many other Web2.0 applications, which is implemented with the most advanced WEB desktop operating system technology.

2 New Changes of the Ideological and Political Education Service Under the Campus Network Environment

Under the influence of the Internet, new and complex changes have taken place in the entire educational system. In order to achieve the best results in our teaching, we need to consider students' learning habits, the overall requirements of the times, and the new changes in the teaching contents. The ideological and political education activities are different from the general teaching. They have many unique characteristics. If we want to carry out this educational activity well, we need to stand on its unique basis, choose the appropriate ways, integrate the students' demands, enrich our educational resources, and achieve the best communication. Generally speaking, the new changes in the ideological and political education services are mainly embodied in:

First, from the contents of the ideological and political education, with the development and progress of the science and technology, the traditional ideological and political education cannot meet the needs of the contemporary college students because it cannot keep up with the development trend of the times. Therefore, the ideological and political education in contemporary colleges and universities is facing a more severe challenge. Under this situation, university leaders begin to pay attention to the adjustment and improvement of the ideological and political work in the process of our education, and further think about how to apply the new concept of "smart campus spring" to the ideological and political work of schools. This is the difficulty facing the popularization of higher learning institutions at the present stage.

Secondly, from the perspective of the ways of the ideological and political education activities, under this change, the ideological and political courses in colleges and universities urgently need to realize the renewal and transformation of the teaching means. The development of the modern information technology and the gradual popularization and application of the multimedia in the ideological and political teaching in colleges and universities provide the conditions for this transformation. The construction of the diversified network teaching platform will become an effective carrier of the ideological and political classroom teaching in colleges and universities. The teaching information publishing platform, the library information service platform, the virtual practice platform, the interactive teaching platform and the learning evaluation platform based on the network will provide the strong support for the subjectivity teaching of the ideological and political courses in colleges and universities.

Thirdly, from the perspective of the impact of the ideological and political education, the trend of the ideological and political education in universities towards the network teaching is the trend of the people's hearts and the general trend, and it is inevitable. However, the reform of the ideological and political education in universities is facing a series of difficulties, including the contradiction between the "decentralization" of the network information and the "centralization" of the ideological and political education in universities, and the contradictions between the "changeability" of the network contents and the "stability" of the ideological and political education in colleges and universities, and the contradiction between "low cost" of the network communication and the "high cost" of updating and maintaining the ideological and political network teaching, and the contradiction between the "voluntary" and "random" of the network learning mode and the "compulsory" and "seriousness" of the ideological and political education in colleges and universities. Breakthroughs in the predicaments require that we should adhere to the principle of adapting to the time and circumstances, and the basic principle of teaching students in accordance with their aptitude. In terms of the specific countermeasures and suggestions, we should construct a new mode of the interaction between the teachers and the students in the ideological and political education in colleges and universities, cultivate a professional and full-time team of teachers in the network teaching of the ideological and political education in colleges and universities, explore the shared development mode of the network teaching of the ideological and political education in colleges and universities, and enhance the novelty and attractiveness of the network teaching means and forms of the ideological and political education in colleges and universities.

3 The Values and Significances of the Ideological and Political Education Service Under the Campus Network Environment

With the rapid development of the network technology and the communication technology, the emergence of the new media provides a new way for the innovation of the higher education, and also promotes the deepening reform of the ideological and political education of college students to a certain extent. As a new campus media, the emergence and application of the platform in the ideological and political education can comprehensively innovate the development model of the ideological and political education, break the bottleneck of its development, and explore a new way to promote the sustainable development of the ideological and political education. In recent years, with the continuous development of the teaching technologies and the continuous advancement of the educational reform in colleges and universities, the construction of the network teaching platform has been gradually strengthened in the process of the ideological and political education in colleges and universities in China. In fact, the emergence of this new teaching method has greatly promoted the development of the ideological and political education in China's colleges and universities, and led to the completion of the teaching tasks and objectives. At present, due to the limitation of the realistic environment, there are some problems in the construction of the ideological and political education network platform in colleges and universities in China.

In the report of the Nineteenth National Congress of the CPC, Xi Jinping mentioned the Internet eight times, emphasizing the important roles of the Internet in the development of the new era of our country in the decision to build a well-off society in an all-round way, and in building a strong socialist modernization country. The "Internet+", as a new thing that combines the Internet with the traditional industries, has the strong vitality and life. The State Council issued Opinions on Strengthening and Improving the Ideological and Political Work in Colleges and Universities under the New Situation, pointing out that in order to promote the reform and innovation of the ideological and political work in colleges and universities, we should strengthen the construction of the network ideological and political work carriers, and carry out the ideological and political education by means of the expressions that the students like. Therefore, this paper will study the method innovation from the perspective of the "Internet+" environment and the university students' ideological and political education.

4 Implementation Strategies of Constructing the Ideological and Political Education Service Platform Based on the Campus Network Environment

The virtual ideological and political platform is an information platform that uses the application technology of the virtual reality and integrates the three-dimensional models, sounds, images and animations. It is devoted to disseminating the advanced technologies and theories to the majority of the ideological and political educators, making them master the scientific practice, and ultimately realizing the widespread ideological and political work in colleges and universities through the way of the education that the teachers and the students like. In the current era, the big data industry is developing rapidly. It is an urgent problem for the ideological and political educators to think about how to dig deep into the connotation values of the big data through the integrated management of the massive data on the virtual ideological and political platform and provide the reliable data impetus for the innovation of the ideological and political educational reform. The campus network is a computer network that provides the resource sharing, the information exchange and the collaborative work for the school teaching, the scientific research and the management. There are many resources in the campus network that only the school users can enjoy, such as various digital book resources of school libraries, various courseware and teaching resources of the school educational administration department, and the resources shared by campus network users and so on. You can imagine the campus network as a university campus in the cyberspace, which has the valuable resources that cannot be accessed by the external network.

Construct a positive information education platform for the ideological and political education at all levels of the schools, students, society and parents. We should grasp the trend of the times, keep pace with the development of our education, make use of the network media to disseminate the new media, actively promote the construction of the network and informationization of the ideological and political education, and build a positive information education platform for the communication and exchange between colleges, parents and students through the micro-blog, QQ, website and other platforms. The colleges now have such special websites as the ideological and political network, the psychological and ideological health network, the campus education safety network, and the red theory practical education network and so on. The college homepage offers such ideological and political work columns as the psychological health consultation, opens the micro-blog and establishes the blog, which gradually forms a positive information education system with the clear positioning grasp, the distinctive characteristics of the ideological and political education, the comprehensive and complementary functions, and a broader and deeper coverage.

Build a multi-level, diversified and multi-participatory practical education platform for schools, enterprises and society. Supported by the characteristic school-running mode of "school-in-factory, factory-in-school, doing-in-school and learning-in-doing", the colleges jointly build a "double base" for the professional practice and the ideological and political education with the cooperative enterprises, and build a platform for the practical education involving schools, enterprises and society. Through the "double base" construction, the corporate cultural elements are introduced, and the ideological and political education contents are organically integrated into the professional practice, so that the practice education work has achieved new results.

Build a multi-angle teaching and education platform of theories, practice and scientific researches. The colleges actively play the main channel role of the ideological and political teaching, and pay more attention to the curriculum construction of the students' ideological and moral education. Now, two courses have become the excellent courses at the college level. The innovative reform of the classroom teaching in the course of the ideological and the Moral accomplishment and Legal Basis is very obvious. The idea of combining learning with doing runs through the entire process of our teaching. The advantages of the modular teaching system are applied to the classroom teaching in a wide range to ensure and achieve the better results. We invite college leaders, off-campus experts and business leaders to give lectures on various topics. Strengthen the practical teaching, and organize students regularly to carry out the legal knowledge competitions, the social practice surveys, and visits to the practice bases and school history halls and other practical activities.

We should pay more attention to the construction of the campus education cultures and build a platform for the humanistic education. Promote the environmental and cultural education, achieve the educational effect of fine rain moistening things, and build a green, ecological, safe and harmonious campus, which have always been the goal of the colleges. The colleges are dedicated to carrying out the campus greening, beautifying the "beautiful project" and the campus culture construction, focusing on the "green", "beautiful", "cool" and "harmony" characteristics. The colleges attach great importance to the combination of the campus construction with the strengthening of the students' ideological and political education, the quality education and the creation of a civilized atmosphere, so as to build a platform for the cultural education, so that the work of the cultural education is outstanding.

5 Conclusion

In the era of the network information, more and more college students have joined the ranks of the smart-phone netizens. The powerful functions of various applications and their convenient and fast features make the mobile new media an important tool for college students to learn and communicate daily and also profoundly influence and change their thinking mode and values. Such characteristics of the network era make the current ideological and political education of college students face great difficulties and challenges. We should actively build a theoretical research platform, guide the students to participate in the ideological and political researches, well implement them, and improve the students' autonomous learning abilities.

References

1. Chen X (2017) Research on the innovative path of the ideological and political education in college students' living community under the new situation – taking Fujian Jiangxia University as an example to implement "five entries, one unity and one center". J Fujian Jiangxia Univ (04):144–145
2. Yang Y, Zhou H, Tan Y, Liu J (2017) How to operate the network public platform of colleges and universities to serve the ideological and political education better – based on the analysis of the questionnaire survey of the civil engineering college of central South University. SE Commun (12):103–104
3. Guan P, Huang Q, Wu G (2018) Innovate the network ideological and political education work relying on the platform of the changing classes in Guizhou Normal University – take Guizhou Normal University as an example. West China Qual Educ (10):151–152

4. Yang W, Wang J (2018) The relationship between the ideological and political education in medical colleges and the cultivation of the humanistic medical practicing abilities under the background of the great ideological and political education. Health Vocat Educ (11):137–138
5. Wang Y, Deng Y (2018) Exploration of college students' network ideological and political education model based on the network public platform–taking "animation students" network public platform as an example. Sci Educ Article Collects (11):105–106

Research on the Image Feature Matching Technology and Its Application in the Computer Vision System

Qing Liu(✉)

Chongqing Creation Vocational College, Yongchuan, Chongqing 402100, China
pinleqinl@163.com

Abstract. The research content of the computer vision includes the imaging equipment and the digital equipment. The ultimate goal of the computer vision is to enable the computers, like the human beings, to observe and understand the world through the vision and have the ability to adapt to the environment independently. With the rapid development of the mobile Internet technology, the digital storage devices and the computer communication devices, we should have the abilities to browse and select the image data that people are interested in from the massive image database, and extract the structured and the meaningful logic, entity and even the relational network from these unstructured data, so as to forecast the trend of the social groups and the individual behaviors.

Keywords: Computer · Visual system · Image features · Matching technology · Application strategy

The image matching technology is an important technology in the computer vision and the image analysis and processing. It can use the modern information technology and the computer technologies to complete the process of the human visual cognition and understanding. It has great practical value in the real life [1]. The computer image recognition technology can be used in many fields, including the text recognition, the face recognition, and the real-time audio and video and so on. It can extract the data information, obtain the key information and meet the needs of the relevant industries.

1 Development of the Computer Vision System

The computer vision technology is widely used in various fields, from the medical image to the remote sensing image, from the industrial detection to the file processing, and from the nanometer technology to the multimedia database, and it has become one of the important research fields of the computer science [2].

Z. Xu et al. (Eds.): CSIA 2019, AISC 928, pp. 188–193, 2020.
https://doi.org/10.1007/978-3-030-15235-2_29

1.1 The Relationship Between the Computer Vision and the Related Fields

The image processing usually transforms one image into another. The input of the image processing system is the image output, which is still the task of the image information recovery, and it is left to people to complete [3]. The main object of this study is to transform the two-dimensional images, especially for the pixel-level operations. This feature shows that the research contents of the image analysis is irrelevant to the specific contents of the image. The pattern recognition is used to identify various symbols, drawings and other plane graphic patterns. It generally refers to the common feature patterns of a class of things which are different from other things [4]. There are two recognition methods: the statistical method and the syntactic method. The computer graphics is generated by the geometric primitives such as lines, circles and free surfaces. The image plays an important role in the visualization and the virtual reality.

1.2 The Application Fields of the Computer Vision

The application fields of the computer vision mainly include the interpretation of photographs and the video data such as the aerial photographs, the satellite photographs and the video clips, the precise guidance, the visual navigation of mobile robots, the medical assistant diagnosis, the hand-eye system of the industrial robots, the map drawing, the object three-dimensional shape analysis and recognition, and the intelligent man-machine interface [5]. One of the purposes of the early digital image processing is to improve the quality of photographs by using the digital technologies, and assist in reading, distinguishing and classifying the aerial and satellite photographs. Because of the large number of the photos to be interpreted, it is hoped that there will be an automatic visual system to interpret them [6, 7]. Under this background, many aerial and satellite photo interpretation systems and methods have been developed. The further application of the automatic interpretation is to directly determine the nature of the targets, to classify the targets automatically in the real time, and to integrate them with the guidance system [8, 9]. At present, the common guidance methods include the laser guidance, the television guidance and the image guidance. In the missile systems, the inertial guidance and the image guidance are often combined, and the image guidance is used for the accurate terminal guidance.

1.3 Evaluation Methods of the Digital Images

In the process of the image communication, the transmission of images to the receiving ends will go through the process of acquisition, transmission, processing and recording. The quality of the images will be affected by the advantages and disadvantages of these technologies, that is, the fidelity of the images and the readability of the images. The traditional evaluation methods include the objective evaluation and the subjective evaluation. The objective evaluation is to measure the quality of the restored images by the errors of the restored images deviating from the original images. The subjective evaluation is to let the observer judge the quality of the test images according to the visual effect according to some pre-defined evaluation scales or his own experience, and give the quality scores, and implement the weighted average of the scores given by all the observers. The result is the subjective quality evaluation of the images.

2 Connotation and Value Cognition of the Image Feature Matching Technology

The image matching is a basic part of the image processing. It is a process of the image matching by analyzing the similarity and consistency between the sample images and the images to be matched in the features, gray levels, and structures and so on. Many fields, such as medicine, agriculture, remote sensing, machinery and artificial intelligence, are related to the image matching technology. For the image matching, this paper mainly studies from two aspects: the gray level and the feature. Among them, the matching method based on the image features has the higher accuracy and the faster speed than the matching method based on the gray level, and has better adaptability to the gray level changes, the image deformation and the occlusion.

The image matching algorithm is systematically studied. The matching algorithm is deeply analyzed and its performance is compared. The research direction of this paper is the feature-based image matching technology. An improved ORB feature extraction algorithm is proposed. This paper introduces and compares the feature extraction methods in details, including the Harris algorithm, the Fast algorithm, the SIFT algorithm, the SURF algorithm, the ORB algorithm, and the FAST-SURF algorithm, and verifies the detection effect of each algorithm through experiments. The ORB algorithm, as one of the most widely used feature extraction algorithms, has the advantages of being fast and the rotation invariance, while the SURF algorithm can make up for the lack of the scale invariance of the ORB algorithm.

Therefore, an ORB-SURF feature extraction algorithm is proposed, which makes full use of the fast superiority of the ORB algorithm and the scale invariance of the SURF algorithm, and can greatly improve the matching effect. An image matching algorithm based on the OR-SURF is proposed. The improved K-D tree search strategy is used to extract the feature points for the OR-SURF algorithm to find the nearest and the next nearest neighbors to achieve the rough matching, and the random sampling consistency algorithm is used to eliminate the mismatching points in the matching and complete the image matching. Experiments show that this algorithm not only compensates for the poor performance of the ORB algorithm in the scale transformation, but also improves the matching speed and accuracy when the scale changes occur. Moreover, the algorithm still has the strong robustness in many complex situations.

3 Analysis of the Image Feature Matching Technology in the Computer Vision System

In the computer vision target recognition, the image matching can find the image matching with the models in the images to be recognized. It is widely used in the image retrieval (such as the image recognition, the character recognition, the target recognition and the image search). Through the analysis of the similarity and consistency of the image contents, features, structures, textures and gray levels, we can find the images similar to the image features to be queried. The purpose is to find one or more transformations, so that two or more images from the same scene but from the different times, different sensors or different perspectives are spatially identical.

The stereo vision mainly studies how to use the multi-image imaging technology to obtain the distance (depth) information of objects in a scene from the multiple images. The basic method is to observe the same scene from two or more viewpoints, obtain a group of the images from different perspectives, and then obtain the position differences of the corresponding pixels in different images through the triangulation principle. Otherwise, the parallax is used to infer the spatial positions of the targets in the scene. The stereo matching is one of the most basic key problems in the computer vision and the non-contact measurement research. This technology makes it possible to obtain the depth or distance information by disparity of the image points. It can provide the useful information for the three-dimensional reconstruction, the robot navigation, and the autonomous vehicle navigation and so on. In the practical applications, because of the influence of the deformation, distortion, and occlusion and so on, the stereo matching is a morbid problem which is difficult to solve thoroughly. In the practical application, the binocular vision is closer to the principle of the human binocular vision, and it is easier to achieve in the practical application.

The method based on the correlation window uses the neighborhood similarity to match. The main problem of this method is the selection of the size of the correlation window. If the selected window is too small, the matching rate will decrease in the smooth area, and if the selected window is too large, then the disparity edge will become blurred. To solve this problem, the whole image is divided into the feature region and the non-feature region to match, and the matching strategy and solutions are given. Aiming at the problem of the large amount of the computation in the correlation window method, the redundancy calculation in the matching process is analyzed and reduced, and the execution speed of the algorithm is greatly improved. Aiming at the problem of the large search range in the large disparity image matching, a matching method based on the global estimation disparity estimation and gradient restriction is proposed to reduce the running time of the algorithm.

4 Application Strategies of the Image Feature Matching Technology in the Computer Vision System

The matching method based on the image features can overcome the shortcomings of using the image gray information for the matching, but the matching measure of the feature points is sensitive to the changes of the positions, and the calculation cost of the feature extraction method is usually high, and some free parameters and thresholds selected according to experience are needed, so it is not convenient to the real-time applications. The image transformation is an indispensable and necessary means in the image processing. Based on the analysis of the Walsh transform and the Rank transform, the matching work is carried out by using the transformed features. The Walsh transform features contain more features than the gray level information of the original images, such as zero crossing point, and positive and negative coefficient and so on.

The Rank transforms the statistics of the size relationship between the current signals and the surrounding signals, and the transformation result is the number of the signals smaller than the current signal amplitude, and matches with the Census transform distance as a constraint condition.

The computer vision is an unlocking inverse problem, that is, estimating the geometric primitives and other features from the images. Therefore, the computer graphics belongs to the image synthesis. The computer vision belongs to the image analysis. The artificial intelligence involves the design of the intelligent systems and the research of the intelligent computing. The neural network is an information processing system, which is connected by a large number of the simple processing units through the strong connections. Comparing the application of the simple gray features as the matching primitives, using these two transformed features as the matching primitives can achieve better results.

The visual-based method is an important research field in the obstacle detection, among which the effective and fast image segmentation is one of the keys. Aiming at the complexity of the threshold determination of the multi-threshold segmentation method and the time-consuming of the region growth-based method, a simple and fast segmentation method based on the feature region is proposed. The feature area is extracted by the smoothness index function, and the extracted images are simplified to the binary images, which make the segmentation more simple and fast. The segmentation area is matched by its center of the gravity and the area features, and the location and the distance of the obstacles are judged by the contour and the parallax of the matched areas. This method does not need to understand and model the targets in advance, and it has the generality and practicability in the practical application.

Because there are noise points in the video image acquisition, we need to process the noise points in order to reduce the noise. The filtering smoothness filtering process has two ways, the linear and the non-linear methods. The linear method is simple and fast for the processors, but the images after the processing will be unclear. After the non-linear processing, although it can reduce the noise points and ensure the local characteristics of the signals, its operation speed will be slower. For the frame image processing, the detection edge filling method or the corrosion expansion method is usually used. The filling method refers to the identification of the objects by the edge detection method after the detection of the targeting objects, and then to fill by the morphological flooding filling method.

Before the image difference, the background pattern needs to be determined and initialized. In order to facilitate the difference calculation of the real-time background images in the future detection, only in this way can we obtain the excellent foreground effects. In the process of the image difference, the first frame background image needs to be determined according to the designated method as the first background image, and then the background is updated according to the algorithm in the detection process. Because the cameras exist in different scenes and environments, both in the outdoor and the indoor, with the changes of the scenes, they will have an impact on the image acquisition. In order to extract the targets from the foreground images, it is necessary to use the effective means to update the background in the real time in the detection system.

5 Conclusion

The modern electronic technologies, the computer graphics and the information processing have been concentrated on the image measurement technology. For example, the sizes and shapes of the three-dimensional images can be accurately measured by the image measurement technology. The image measurement technology also has important applications in aviation fields such as the precision instruments and the remote sensing. The image features need to be processed accurately and meticulously by using the image software in computers. Generally, images are captured by computers and then extracted. The images need to be represented by digitalization, so that the images are not distorted and easy to be processed by computers. Then the image segmentation is completed by computers, and the extraction and description of the image features are realized.

References

1. Zou D, Yan F (2017) Development and application of the computer image recognition technology. Comput Prod Circ (09):122–123
2. Song W, Zhu L (2017) Research on the feature extraction and the matching technology of images. Geomat Spat Inf Technol (07):105–106
3. Tong Y, Zhang Y (2017) Image matching algorithm based on the dual detection strategy coupled with the PROSAC technology. Comput Eng Des (11):153–154
4. Wang T (2017) Research on the image matching algorithm based on the improved SURF. Technol Wind (12):124–125
5. Huang Y, Zhang F (2018) Image matching algorithm based on the differential constrained coupled triangular network constraints. J Xinjiang Univ (Nat Sci Ed) (11):138–139
6. Leo M, Medioni G, Trivedi M et al (2017) Computer vision for assistive technologies. Comput Vis Image Underst 154:1–15
7. Xu Z, Mei L, Hu C et al (2016) The big data analytics and applications of the surveillance system using video structured description technology. Cluster Comput 19(3):1283–1292
8. Rautaray SS, Agrawal A (2015) Vision based hand gesture recognition for human computer interaction: a survey. Artif Intell Rev 43(1):1–54
9. Fathi H, Dai F, Lourakis M (2015) Automated as-built 3D reconstruction of civil infrastructure using computer vision: achievements, opportunities, and challenges. Adv Eng Inform 29(2):149–161

Research on the Influences of the Invisible Environment of Language Laboratories on the Practical Teaching Based on the Network

Zhang Fan(✉)

Xinke College of Henan Institute of Science and Technology, Xinxiang 453000, Henan, China
ssme2017@126.com

Abstract. There is a close relationship between the environmental factors and the human learning. The language laboratory is an important place for the foreign language learning. Building a good learning and practice environment of language laboratories plays an important role in the foreign language teaching. Learning can never take place in a vacuum. It always takes place in a variety of environments. Some of these environments may be more conducive to the development of the cognitive, emotional and social skills than others.

Keywords: Network technology · Language laboratory · Invisible environment · Practical teaching · Impact research

The learning environment of the digital language laboratories can include the physical environment, the language environment and the psychological environment. Understanding the influence of the environment on the learning in various ways is particularly important for the language teachers and the language learners, because the better they understand the various environments in which the targeting language is located, the more likely they are to understand the use of different languages [1].

1 Problems to Be Noted in the Open Language Laboratories

To meet the individual needs and cultivate interests in learning, the autonomous foreign language learning under the network environment means that learners use the service system provided by the network environment to actively, exploratively, designedly and purposefully study according to their actual foreign language levels [2]. This learning method can greatly meet the individual needs of the students, stimulate their inherent learning potentials, and encourage the students to devote themselves to the language learning and practical communication activities. It is a very effective way to improve their foreign language proficiency in an all-round way.

Z. Xu et al. (Eds.): CSIA 2019, AISC 928, pp. 194–199, 2020.
https://doi.org/10.1007/978-3-030-15235-2_30

1.1 Focus on the Guiding Role of Teachers in the Self-regulated Learning

The network environment of the language laboratory is a huge library of learning resources. Students study here, and if there is no teacher's guidance and control, students' learning is like looking for the needles in a haystack, and it is difficult to for them to find the focus of their learning and the knowledge they need [3]. Especially some students with poor self-control abilities can easily "walk around the world" in the network environment, killing the good time. Therefore, teachers must control the open language laboratories and guide the students' learning contents and methods. For example, teachers should choose the websites for the students, propose their learning scopes, assign the learning tasks to the students, and establish the online communication and feedback of the learning information between the teachers and the students [4]. The autonomous learning of the foreign languages is the development trend of the foreign language learning in the future, and only under the guidance and supervision of the teachers can the autonomous learning be the core content of the autonomous learning.

1.2 Improve the Management Level of the Language Laboratory

Opening the language laboratory after class adds a lot of work to the foreign language teachers and the experimental managers. It puts forward higher requirements for the computer application levels of the foreign language teachers and the management abilities of the experimental personnel. Many foreign language teachers are "croaking" in their lectures, but they are reluctant to work hard on the computer operations in their non-specialty. It is necessary to improve the foreign language teachers' awareness of the importance of the students' autonomous learning and their computer skills through training [5]. With the rapid development of the educational technologies and the continuous updating of the equipment, some experimenters are getting old and unable to do their best. Schools should regularly adjust the team of the experimenters according to the actual situations, and do a good job of alternating the old and the new experimenters. The experimenters should constantly update their knowledge and learn the new technologies to meet the needs of the modern foreign language teaching [6, 7].

2 Requirements of the Invisible Environment of the Language Laboratory for the Practical Teaching

At present, most foreign language teaching models are based on the teachers' teaching of the language knowledge, which are theoretical and lack the practicality in the language application. In order to solve this problem, foreign language courses can be offered in the language training, which can effectively strengthen the students' language training in the real or the simulated scenes, train the students to master the accurate pronunciation, intonation, speed and tone from the basic stage of the language, and improve their listening and speaking abilities on the basis of consolidating the language.

The ultimate goal of the foreign language teaching is to train and cultivate the students' practical application abilities. However, the phenomenon of the tape teaching, which has been confined to a single textbook for a long time, is far from the ultimate goal of the foreign language teaching. The universal language learning system combines listening, speaking, reading and writing of the foreign languages organically. It is suitable for any network computer classrooms or the local area network. It creates a good foreign language environment, increases the students' chances of contacting and using the foreign languages, and mobilizes the students' enthusiasms for the learning foreign languages. The students with the full autonomy can easily use the learning resources and methods they are interested in, and timely carry out the personalized learning, which has a positive role in promoting the quality of our teaching. In this regard, we still need to continue our efforts to research and explore and develop, so as to make it more suitable for the modernization of the foreign language education.

The pure software language learning system can update the software in time according to the social development and the teaching needs. It has strong vitality and will never be out of date. Deficiencies include the following. A universal language learning system with "one room and multi-purposes" wastes the teaching resources if one room is dedicated. A universal language learning system with "one room and multi-purposes" saves the teaching resources if one room is multi-purposed, but is relatively complex in the equipment maintenance and the multi-course coordination, and the students have different computer skills, and the barrier rate of the equipment is greatly increased, which easily affects the teaching and causes the teaching accidents. The universal language learning system must be guided by the computers and the networks, and the rich teaching resources from the Internet are inexhaustible. There are inherent bandwidth problems and the delay problems in the Internet, which cannot provide the students with a real and real-time language learning environment in the concurrent state.

Based on the above shortcomings, the emergence of the specialized language learning system has become inevitable. The specialized language laboratory is a professional language learning environment for training the professional talents. This product is popular for its cost, reliability, pertinence and easy operation. The professional language laboratory uses the ATM network technology to overcome the inherent bandwidth and delay problems of the Internet, and can provide the students with a real-time language learning environment in a concurrent state.

3 The Influences of the Invisible Environment of the Language Laboratory Based on the Network on the Practical Teaching

The purpose of the open digital language laboratory construction is to provide the better services for the students' learning. The main advantages and significance of this open laboratory construction include the convenient language teaching and more intuitive services for our teaching and so on. The open digital language experiment teaching can make good use of the favorable conditions of the language laboratory, transform the teaching contents into animations, pictures and other forms, and create a vivid learning environment for the students. Therefore, the construction of an open language laboratory is of great help to improve the students' practical abilities. It provides a language environment for the students to experience the atmosphere and scenes of the foreign languages.

The contents of the digital teaching are very rich. Compared with the traditional teaching mode, we use the paper textbooks, chalks and blackboards. But in the digital teaching, we use the multimedia tools. The virtual environment has an important influence on the setting of the teaching environment. Due to the problems of the teaching mode and funds, we cannot meet most of the teaching needs of the students. However, through the teaching of the language laboratories, we can let the students come to the contents of the courses.

But through the language labs, we can do it. Through the virtual environment, we can create an English atmosphere. Let the students feel that they are surrounded by the foreigners. In such a teaching environment, students will be forced to use English, and they will unconsciously improve their English levels. If some students learn the small languages, we can convert the virtual mode into the language mode of this small language, which not only provides the learning environment for the students, but also saves time for the students. For other majors, this model is also applicable. For example, in the history teaching, we can also use this tool. Put the students in a certain historical period, just like traversing, and let them experience the knowledge. This mode is a revolutionary storm for our teaching. It completely subverts the traditional teaching mode, makes up for the vacancies in the traditional teaching, and serves our teaching well.

4 The Invisible Environment of the Language Laboratory Based on the Network for the Practical Teaching Mechanism

The project of virtualization and the research and practice teaching enable the students to have a certain understanding of the server virtualization technology, to improve the students' practical and innovative abilities, to fully mobilize the students' interests, and to make the students take the initiatives in their practical teaching, thus realizing the students' autonomous learning, and creating the team cooperation for our research and exploration. The learning atmosphere not only expands the students' knowledge, but also improves the students' employment abilities, and has received the good teaching effect. The purpose of the construction of the practical teaching system of the language in and out of schools is to create the real language learning situations for the students and enable them to acquire the multi-level, multi-angle and three-dimensional real language teaching, so as to improve their comprehensive language application abilities and their cross-cultural communication abilities, and to make use of the foreign languages to exchange information orally and in writing effectively in their future work and social exchanges.

It is necessary to establish the principle of "Taking the cultural atmosphere as the platform, the language practice as the core and the innovative practice as the carrier" in the practical teaching system of the language within and outside the schools. It is necessary to establish the horizontal practical teaching consisting of "language training course + cultural experience course + second classroom activity" and the three levels of "basic level, improving level and innovative level". A foreign language experimental teaching system which combines the vertical practical teaching with the foreign language teaching is established. The composition of the horizontal practical teaching system meets the practical teaching requirements of combining the language and the culture, the in-class and the out-of-class in the foreign language learning. The composition of the vertical practical teaching system meets the practical teaching requirements of combining "foreign language + specialty, and specialty + foreign language" in the foreign language learning.

For the development of the language laboratories, the study in class and out of class is equally important. In the ordinary English teaching, the key is to make full use of the valuable resources of the language laboratory. The practical teaching consists of "language training course + cultural experience course + second classroom activities". The modern educational resources are used to increase the contents of the auxiliary teaching and the rich and diverse teaching methods. Three practical teaching links, namely, the language basic training, the language national cultures and the language practice, are developed to stimulate the students' interests and enthusiasms in their learning and cultivate the students' autonomous learning abilities, and then improve the students' comprehensive abilities to use the foreign languages.

To learn a foreign language well, we should not only master the language knowledge, but also understand the cultural background knowledge of the language country. The second language learning involves the second culture learning, and it is also a process of the cultural transfer and self-adjustment to adapt to the new cultural environment. The cultural experience course is to take the students as the main body and integrate them into the cultural atmosphere of the language country through the form of practice. And therefore, they can experience different cultures personally. We can offer the practical courses such as the situational teaching, the on-site teaching and the simulated teaching so as to enable the students to understand the cultures of the language country in an all-round way, to form a holistic grasping and to avoid the fragmentation of their cultural cognition. The learning of the cultural experience course can not only enable the students to feel the cultures through the language, but also can edify the language through the cultures, and integrate the language learning and training into the cultural atmosphere. At the same time, it can broaden the international horizon, grasp the differences between the different cultures, and carry out the cultural comparison, so as to improve the overall qualities and enhance the cross-cultural communication abilities.

The center adopts various means such as the situational simulation teaching and the online network guidance to promote the all-round development of the students' vocational skills and language abilities, and forms a multi-dimensional experimental teaching mode of "combining the laboratory with the classroom, the distance teaching with the independent learning, the situational simulation with the field training". The laboratory teaching and the classroom teaching integration center can make full use of the multimedia technologies, the computer network technologies and the digital technological interactive mode, give full play to the advantages of the traditional classroom communication, let the students in the two language learning environments of the language laboratory and the traditional classrooms, train their language basic skills, and improve their language comprehensive abilities and professional factors. The web-based language autonomous learning center and the curriculum platform have been set up in the distance learning and autonomous learning integration center. Teachers' remote guidance and the online question answering have been arranged, which provide the students with the rich language learning resources, as well as the fast and convenient learning conditions.

5 Conclusion

The language learning, and especially the foreign language learning, has limited classroom learning time and is occupying most of the time by teachers' explanations. Therefore, the classroom time alone cannot meet the requirements of the students' learning and using of the foreign languages. The second classroom activities are the important practical link to make up for the shortcomings, and we can vigorously carry out the international cultural and art festivals, the student societies, the spoken language corner, the speech contests, the classical Chinese and foreign language reading, the foreign language volunteers and other activities, which are designed to meet the practical needs of the students' language input and output. Foreign teachers and the part-time teachers are also equipped to provide the guidance to form a perfect second classroom activity system, which can mobilize the students' enthusiasms and initiatives in their learning. Combine the foreign language knowledge in the classroom teaching, further improve the knowledge structures and train the students' abilities to comprehend and innovate the foreign language knowledge.

References

1. Wang X, Wang B (2018) Construction and application of the trinity practice teaching platform in the transition and development of the newly-built undergraduate colleges. Exp Technol Manag (11):100–101
2. Jiang X, Jiao Y (2018) Promoting the innovative development of the practice education through the ideological and political education in colleges and universities in the new era – "high-end forum on the ideological and political education practice in colleges and universities in the new era". Sch Party Build Ideol Educ (11):154–155
3. Jin J, Li X, Deng D (2018) Construction of the PLC experimental platform and practical teaching system based on the OBE concept. Technol Wind (11):105–106
4. Zhang L, Jiang B, Li X, Li D, Liu K, Zhang L, Guo L, Liu H, Cui H (2018) Research on the community-hospital-community practice teaching model for the undergraduate nursing students guided by the professional certification. Health Vocat Educ (11):122–123
5. Zhang W, Wang J, Sun H (2018) Exploration of the application of the mathematical model in the urban rainwater course and practice teaching under the background of the engineering education certification. Educ Teach Forum (11):142–143
6. Bao W, Zhou H, Xie F (2017) An optimal design of web based online English teaching by using struts framework. Bol Tec/Techn Bull 55(10):58–63
7. Bao W, Zhou H, Lu W, Xie F (2016) The system of knowledge management using web based learning and education technology. Comput Syst Sci Eng 31(6):469–473

Research on the Motion Control and the Coordination Method of the Autonomous Mobile Robots

Li Ning(✉)

Guangxi Technological College of Machinery and Electricity, Nanning 530007, Guangxi, China
fqdlsec@yeah.net

Abstract. With the continuous development of the automation technology in the modern society, the robotic equipment has been more and more widely used in all aspects of the current society. The autonomous mobile robot is one of the most advanced modern robotic devices, which has obvious characteristics of the intelligence and automation. In the practical application process of the autonomous mobile robot, in order to make its functions better realized and its functions better developed, it is necessary to design its motion control reasonably so as to better serve the current social production and people's life.

Keywords: Autonomous mobile robot · Motion control system · Coordination method · Design analysis

1 Introduction

With the development of the new era, the automatic mobile robot is a kind of the robot equipment with the high intelligence level in the current society. For the autonomous mobile robot, the motion control system is a very important component, and it can play a decisive role in the realization of its functions and the effective play of various roles [1]. Therefore, it is of great significance to design the motion control system of the autonomous mobile robots scientifically and reasonably. In this paper, the design and implementation of the motion control system for the autonomous mobile robot are analyzed.

2 The Motion Control and Coordination Technology of the Autonomous Mobile Robots

2.1 Open Modular Control System Architecture

The distributed CPU computer structure is adopted, which is divided into the robot controller (RC), the motion controller (MC), the photoelectric isolation I/O control board, the sensor processing board and the programming teaching box. The robot controller (RC) and the programming teaching box communicate through the serial

Z. Xu et al. (Eds.): CSIA 2019, AISC 928, pp. 200–205, 2020.
https://doi.org/10.1007/978-3-030-15235-2_31

port/CAN bus. The main computer of the Robot Controller (RC) completes the functions of the motion planning, interpolation, position servo, main control logic, digital I/O, and sensor processing and so on, while the programming teaching box completes the display of the information and the input of the keys [2].

2.2 Modular Hierarchical Controller Software System

The software system is based on the open source real-time multi-task operating system Linux, and adopts the hierarchical and modular structure design to realize the openness of the software system. The entire software system of the controller is divided into three layers: the hardware driver layer, the core layer and the application layer. The three levels are faced with different functional requirements and corresponding to different levels of the development [3]. Each level of the system is composed of several modules with the opposite functions. These functional modules cooperate with each other to achieve the functions provided by this level.

2.3 Robot Fault Diagnosis and Safety Maintenance Technology

It is the key technology to ensure the safety of the robots to diagnose and maintain the faults of the robots through various kinds of the information.

2.4 Networked Robot Controller Technology

At present, the application engineering of robots is developing from a single robot workstation to a robot production line. The networked technology of the robot controllers is becoming more and more important. The controller has the functions of the serial port, the field bus and the Ethernet [4]. It can be used for the communication between the robot controllers and between the robot controllers and the host computers, and is convenient for monitoring, diagnosis and management of the robot production lines.

3 Design of the Motion Control System Based on STM 32

The embedded controller has the advantages of the simple structure, easy control and relatively low costs, which is more and more popular. ARM is an open source environment, which can be used for the reference and the abundant research technology. STM32 MCU has the characteristics of the high cost performance, the good performance, the low cost and the low power consumption, which is very suitable for the data processing and the control unit [5]. And the rich multi-functional motor control interface is convenient to realize the precise control of various driving motors. So STM32 MCU is chosen as the core of the control system to design the control system.

3.1 Basic Composition of the Control System

The autonomous mobile control system of the vehicle-like robots includes two parts: (1) completing the motion control function of the robots and (2) sensing the autonomous environment. The key technologies include the motion control technology, the sensor technology, the robot vision technology, the multi-sensor information fusion technology, the wireless network communication technology and the system integration technology which integrates the sensor, the control and the communication.

The robot autonomous mobile control system is a typical hardware component. The whole structure includes the upper computer with the industrial computer and the lower system with the STM32 chip as the control core. The upper computer continuously sends the instructions and the feedback signals from the coder to the lower computer, which controls the specific actions of the mobile robots. At the same time, the state information and the sensor data of the robots are fed back to the upper computer, including the vector speed of the robots, various kinds of the sensor information, and the video information and so on. The upper computer adopts the industrial computer with the strong processing power, which is not described in details in this paper.

3.2 Architecture of the Control System

The architecture is the physical framework of the robot and the carrier of the intelligent control system, including the deliberate, the reactive and the hybrid. The reflective architecture has the strong overall control ability, but the poor responsiveness. The disadvantage of the reactive architecture is that it has the poor management ability for the global tasks. The advantage of the reactive architecture is that it has the good robustness/robustness and the strong dynamic response ability. The hybrid architecture combines the characteristics of the above two architectures, which has both the high dynamic response ability and the basic strong system planning ability, especially suitable for the autonomous mobile robots in the uncertain environments. The robot autonomous migration needs the strong task management ability and the dynamic environment response ability to ensure that the decision-making tasks can be completed in various uncertain environments. In view of this functional requirement of robots, the control system of robots is designed under the framework of the hybrid system, so as to achieve the goal of the simple structure, the good transformation resistance and the strong expansion ability.

3.3 Robot Control System Architecture

The function of the STM32 control system belongs to the prudent level, in which the task planning module has the highest level and is responsible for the realization of the human-computer interaction and the remote server interaction. The sensor management module monitors the sensor data to ensure the normal operation of the sensor. The sequence generator and the execution monitoring module are the transitional layers, whose functions are to decompose the tasks and supervise the execution of the actuators. The behavior control module completes the functions of the behavior layer. The robot perception module mainly collects and processes the sensor data, including the data directly transmitted to the behavior database, as well as the information transmitted to the STM32 control system for the environmental mapping. The actuator is responsible for the final completion of the work. The overall structure of the control system of the robot adopts the multi-threaded software working mode, which reduces the limitation of the process of the deliberation on the overall response ability of the robots.

4 SLAM Models and Principles of the Robot Autonomous Migration

The environmental localization and mapping can be seen as the key to the realization of the autonomous mobile robot technology. The simultaneous localization and mapping (SLAM) of mobile robots can be described as follows. When a robot begins to move in an unknown environment, its initial position is unknown, and in the course of its movement, it can continuously locate itself according to position estimation and the map information, and build the incremental maps on the basis of its own localization to realize the robot's self-determination. On the basis of the self-localization, an incremental map is built to realize the autonomous localization and navigation of the robots. The localization and the mapping are carried out simultaneously and complement each other.

The SLAM problem of robots includes two aspects. One is the self-localization and the state control in the process of the mobile robot and the other is the information collection of the environment in which the robot is located. When a robot moves in an unknown environment, it continuously perceives its environment through a variety of sensors and camera equipments, which include detecting the roadblocks, detecting the landmarks, collecting the environmental image information and locating its own environment.

4.1 Study of the Motion Control with the Exoskeleton Robots as an Example

Although the process of the human gait can be determined by the human motion analysis, because of the particularity of the exoskeleton service object, the safety and the comfort of the users must be guaranteed while meeting the efficiency. When people walk, they also have to combine the road conditions, such as the uneven bottom, the upper and the lower steps and other factors, and must adjust the external skeleton system control procedures. In order to make the exoskeleton meet different application environments, it is necessary to have certain flexibility when working.

Research on the Control Scheme

The exoskeleton users studied in this paper refer to those who have certain muscle strength and the clear consciousness. The common exoskeletons use the seat belts to bind the key parts of the users to the external skeletal scaffolds, and drive the movement of the user's joints through the rotation of the upper exoskeleton joints. The exoskeletons usually adopt a fixed abnormal gait, but this is not suitable for different pavement and different body types of the users.

If users with different usage environments and different body shapes can capture the starting and the ending signals of each action intention, the exoskeleton control system can be quickly and accurately reflected, and will no longer be affected by the inability of the preset pace to adapt to different external conditions. So we design the control scheme.

Exoskeleton Control System
The step control signal of the exoskeleton control system described in this paper is input by the pressure sensor and the angle sensor.

The angle sensor is used to detect the exoskeleton angles. Fixed on the user's upper body and the external skeleton, the inclination degree of the users and the external skeleton is determined by determining the angles of the upper body, and is transmitted to the controller in the real time.

The pressure sensor is a sensitive device which transforms the measured strain into an electrical signal. When the user's lower limbs are walking, the muscles of the thighs and legs will make the corresponding actions. The pressure sensor receives the actions and transmits them to the control system. When we are walking, the heel and the toe bones on the foot sole of the stressed leg play the main load-bearing role of the human body. The pressure sensor receives the action and transmits it to the control system.

Motion Control Modes
As is mentioned above, people's pace can be divided into the simple force on both legs, the force on one leg and the toe bone, the force on one leg, the force on one leg and the heel, and the force on both legs. When the user's legs are stressed, the heel and the toe bone on the sole of the foot bear the main gravity of the body, so in this case, the pressure sensor on the sole will feed back greater pressure. When the user exerts one leg to lift a foot, the muscle of the thigh will first give a pressure signal to the corresponding pressure sensor, and the signal will be transmitted to the controller to make the root of the controller. According to the input pressure, the motor speed is controlled and the thigh is lifted. At this time, the sole and the heel leave the ground first, so the pressure sensor value at the heel first reduces to 0, reaching the one-leg and the toe bone stress step. According to this method, the entire walking process of the user can be controlled and detected. The angle sensor is transmitted to the user's current angle of the control system in the real time. The controller can judge whether the user will fall down or not. If the user inclines too much, the exoskeleton forces some actions to adjust the user's posture in order to prevent the user from falling down and causing injuries.

4.2 Path Tracking Control Method for Autonomous Mobile Robots

The goal of the path tracking control is to enable the robots to track the geometric path stably and efficiently to reach the designated location. According to the control rate, this paper calculates the appropriate linear speed and the angular speed, so that the mobile robots can reach the targeting position stably and efficiently according to the recorded path coordinate points, and can stabilize at the position point. After the interruption of the control signal, the mobile robot starts the autonomous return mode. The control unit receives the sensor data, substitutes it into the kinematics model, and calculates the current pose of the robots. In the course of the path recording, the last path coordinate point recorded is the first point of the path tracking. At the same time, the robot continuously receives the sensor data, calculates the current position of the robot, determines whether the targeting point has arrived, and continues to track the next point until the communication is restored or the initial position is returned. The key of the path tracking control is the design of the control rate. This paper requires the mobile robot to be stable and efficient in the process of returning. For the linear speed of the mobile robot, it is also required that the speed be faster when it is far from the targeting point and slower when it is near the targeting point. The course angle is the key to control the mobile robots to reach the targeting point. It requires that the calculated angle can be rotated accurately and reliably. So the PID control algorithm is used to control the course angle of the mobile robots.

5 Conclusion

In a word, the autonomous mobile robots are more and more widely used in all aspects of the current society. In order to make them play a better role and realize their functions better, the operation control system should be reasonably designed. Through the reasonable design of the hardware and the software of the system, the system functions can be better realized, and the system design can get more ideal results, and then make the system more effective, so that the mobile robots can be better applied and developed.

References

1. Jin X, Wang H, Zong W, Wang P, Wang C (2018) Research status of the obstacle avoidance technology for autonomous mobile robots. Transducer Microsyst Technol 37(05):5–9
2. Li Y, Wang S, Yu N (2018) Research and implementation of the SLAM and the path planning method for mobile robots based on the RGB-D information. CAAI Trans Intell Syst 13 (03):445–451
3. Yu N, Wang S, Xu C (2017) A RGB-D-based method for the autonomous exploration and mapping of the unknown indoor environment for mobile robots. Robot 39(06):860–871
4. Li H (2017) Robot autonomous positioning and navigation system and its application. Strateg Emerg Ind 40:16–17
5. Wang Z, Zhang G, Wu H, Ou X, Hu X (2017) Design of the library mobile robot system integrating the autonomous roaming and the remote monitoring. J Huaqiao Univ (Nat Sci Ed) 38(03):391–396

Research on the Motion Simulation System Based on Association Rule Mining

Zhanke Ma(✉)

Institute of Physical Education, Ningxia Normal University, Guyuan 756000, Ningxia, China
zhiyu_zhang@126.com

Abstract. The motion control is indispensable in the industrial control. The physical prototype is often used as the controlled object in the design. With the rapid development of the computer and the modern integrated manufacturing system technology, the development of the low-cost and short-cycle virtual prototype is gradually replacing the traditional physical prototype, showing greater and greater effect in the design and manufacturing process. The purpose of this design is to provide a general simulation platform for debugging the motion control system, and to control the virtual prototype of the universal controller of the elevator and refueling vehicle with the help of the simulation platform. The simulation platform has the characteristics of versatility, flexibility, operability, intuitive display and iconicity.

Keywords: Association rule mining · Motion simulation system · Sports injury · Injury factor · Interest model

The real-time simulation technology can complete the performance evaluation of the control algorithm without the participation of the controller prototype. The dSPACE system is a good application platform for the research on the real-time simulation technology. It provides the real-time control mode to meet the control requirements of different systems. Based on the MATLAB, a real-time simulation control model for the direct torque control of the permanent magnet synchronous motor (PMSM) with double-closed loop of the speed and the current is established [1].

1 Connotation and Stage Analysis of the Association Rule Mining

In 1993, Allegra, et al. first proposed the concept of the association rules, and gave the corresponding mining algorithm AIS, but the performance was poor. In 1994, they established the theory of the item set lattice space and proposed the famous Apriori algorithm based on the above two theorems [2]. So far, Apriori is still widely discussed as a classical algorithm for mining the association rules. Many researchers have done a lot of researches on mining the association rules. The mining process of the association rules mainly consists of two stages. The first stage is to find all the frequent item-sets

Z. Xu et al. (Eds.): CSIA 2019, AISC 928, pp. 206–211, 2020.
https://doi.org/10.1007/978-3-030-15235-2_32

from the data set, and the second stage is to generate the association rules from these frequent item-sets [3].

In the first stage of the association rule mining, we must find all the large item-sets from the original data set. The frequency means that the frequency of a project group must reach a certain level relative to all records [4]. The frequency of a project group is called support. Take a 2-itemset with two items A and B as an example, we can get the support of the project group containing {A, B} by formula (1). If the support is greater than or equal to the set minimum support threshold, then {A, B} is called the frequency item-set. A k-item set that satisfies the minimum support is called a frequent k-item set and is generally expressed as a Large K or Frequent K. The algorithm generates Large k + 1 from the Large k item set until a longer frequent item set can no longer be found.

The second stage of the association rule mining is to generate the association rules. Generating the association rules from the frequent item sets is to use the frequent k-item set of the previous step to generate the rules [5]. Under the condition of the minimum confidence, if the confidence obtained by a rule satisfies the minimum reliability, the rule is called the association rules. For example, the reliability of the rule AB produced by the frequent k-item set {A, B} can be obtained by formula (2). If the reliability is greater than or equal to the minimum reliability, it is called the association rule [6–8].

In the risk assessment for the sports injuries, each evaluation expert selects the sports injury risk assessment index set and uses the language variable to carry on the appraisal, and carries on the comprehensive analysis to each evaluation expert's appraisal results through the triangular fuzzy soft set method, to obtain the sports injury risk assessment index weight separately, and obtain the order of the injury risks according to the index weight [9, 10]. The correlation degree between the risk factors and the target determined by this method is poor, and there is a big deviation between the actual results and the evaluation results. A method of the sports injury risk assessment based on the association rules is proposed.

The data mining is the process of extracting the valuable knowledge from the large amounts of data by the machine learning. Apriori association rules are used to analyze and mine the association indexes which have influence on the sports injury. According to the statistical results of the historical injury data and the calculation method of the association rules, the weighting coefficients of the individual state variables in the athletes' comprehensive injury state variables are determined, and the scores of the single state variables in the athletes' comprehensive injury state variables are calculated. The experimental results show that the proposed method is basically consistent with the open questionnaire Pareto analysis.

2 Risk Assessment Method of Sports Injuries Based on the Association Rules

The motion control (MC) is a branch of the automation that uses devices commonly known as servos such as hydraulic pumps, linear actuators or motors to control the position or speed of machines. The application of the motion control in the field of robots and CNC machine tools is more complex than that in the specialized machines, because the latter has a simpler form of motion, often referred to as General Motion Control (GMC).

The data pretreatment method was used to process the injury data of the high-level athletes in different sports events. Aiming at the initial index set of sports injury risk assessment, the Apriori association rules were used to analyze and mine the related indexes which had influence on sports injuries, and the main risk states of sports injuries were classified and sorted out, to build a comprehensive risk status for athletes. According to the statistical results of the historical injury data and the calculation method of the association rules, the weighting coefficients of the individual state variables in the athletes' comprehensive injury state variables were determined, and the scores of the individual state variables in the athletes' comprehensive injury state variables were calculated. Because the association rules cannot distinguish the continuity numerical data of the historical sports injuries, it is necessary to discretize the statistical historical sports injury data, and classify the statistical sports injury data by the normalization method.

Using the method of the Apriori association rules analysis, in the first stage, the frequent iterations of all the items in the injury data set of the athletes are searched by iteration layer by layer. In the second stage, the strong association rules satisfying the lowest confidence are mined from the frequent concentration of the injury risk factors. In order to find out the relationship between the injury data given the minimum support and the confidence threshold, the association relation is used to mine the injury risk factor frequent item data set of the athletes' injury risk factor data set, and the association relation between the frequent item sets of sports injury risk factors is analyzed by introducing interests. The degree model is used to get the important association rules.

According to the confidence of the association rules of the comprehensive state variables of the athletes' injuries, the confidence of a certain state variable reflecting the state of the sports injuries can be calculated. The confidence of a certain state variable reflecting the state of the sports injuries can be compared with that of the weight coefficients of the state quantities of a certain state variable reflecting the state of the sports injuries. Similarly, we can calculate the weight coefficients of the individual variables in other states which lead to the sports injuries in the training and competition stages, and obtain the objective weight coefficients of the individual state variables, such as the physical quality and the psychological quality. The variable weight theory is used to describe the equilibrium performances of the comprehensive evaluation of the sports injury states, and the formula is used to calculate the scores of each individual state in the comprehensive state variables reflecting the sports injury states.

3 Research on the Motion Simulation System Based on the Association Rule Mining

There are still some shortcomings in the application of the Apriori algorithm. Through the corresponding improvement, it is found that the association rules obtained by the classical Apriori mining algorithm contain more redundant rules. After the improvement of the Apriori algorithm, there are no redundant rules and no rules missing in the calculation results of the Apriori algorithm. It is simple and clear through the analysis of the experimental data. It is found that the execution efficiency of the improved Apriori algorithm is lower than that of the classical Apriori algorithm. Through the simulation experiment platform of the motion control system designed by MATLAB/GUIDE (graphical user interface design), the modeling and simulation of various speed control systems can clearly and definitely simulate various speed control experiment items and output various waveforms for our analysis. The making of the test platform makes the students understand the system better and master the entire knowledge of the system analysis, the mathematical modeling and the simulation. The analysis of the simulation waveform makes it easier for the students to understand the transition process of the motion control system, such as speed, current, voltage and torque, so as to deepen the theoretical knowledge and play an auxiliary role in our teaching.

The control system simulation technology uses the ground simulation equipment to study the dynamic performance of the aircraft control system. The simulation equipment is composed of the computer and all kinds of the physical simulation equipment. It can simulate the aircraft, the control system and various flight environments. According to the nature of the models, the simulation of the control system can be divided into three categories: the mathematical simulation, the semi-physical simulation and the full-physical simulation. The full physical simulation is the most realistic, but in the development process of the control system, the three kinds of the simulation are complementary to each other.

After R. Agrawal and others put forward the problem of the mining association rules, a number of the effective algorithms for the mining association rules have been greatly developed in the past few years. Up to now, its main research directions are: mining algorithm based on the data dimension involved in rules, mining algorithm based on processing variable categories in rules, and other association algorithms and so on. The contents of the constraint-based rule mining constraints can be as follows: (1) Data constraints: Users can specify which data to mine, but not all data at certain times. (2) Specify the dimensions of the mining: Users can specify which dimensions of the data are mined and which levels are on those dimensions. (3) Rule constraints: You can specify which types of rules are required, introduce a template concept, and use it to determine which rules are interesting. If a rule matches an included template, it is interesting, but if a rule matches a restricted template, it is recognized as lack of interests.

It refers to the association rules involving two or more variables, according to whether the same dimension is allowed to repeat, and the multi-dimensional association rules can be divided into the inter-dimensional association rules and the mixed dimension association rules. The dimension association rules and the mixed dimension association rules also need to consider different field types, namely, the category data and the numerical data. For the category data, the general association rule algorithm can handle, but for the numerical data, it is necessary to convert these data into the category data. At present, most of the association rule mining algorithms are based on the relational database or the transactional database. It is very meaningful to design and apply the association rule mining algorithms to other types of the databases, such as the object-oriented database, the data warehouse, the text data, the graphics and image data, and the multimedia data and so on.

With the rapid increase of the data volume in the data mining and the application of the large-scale parallel computing in the data mining, due to the reasons of the mining system itself, the parallel data mining process tends to coarse-grained mining, and cannot achieve the arbitrary degree of the parallelism. At present, there are still some problems to be solved in the parallel data mining: the increasing amount of data, the increasing dimension, the data location, the data asymmetry, the dynamic load balancing, data distribution and the index scheme of the multi-table database, the incremental method, the parallel database management system and the file system.

4 Conclusion

The data mining can also be called the knowledge discovery in the database. It refers to the entire process of extracting some information and knowledge which people do not know beforehand but have certain potential application values in a large number of the fuzzy, incomplete and random data. It is an important step to discover the knowledge in the knowledge mining. Nowadays, the Apriori algorithm has been applied in many fields such as the data analysis, the high-level decision-making and the business intelligence. The classical Apriori algorithm needs to use the database to scan many times to generate a large number of the candidate sets. Therefore, the classical algorithm usually produces more redundant rules, so it is based on the association rules by the data mining algorithm. It is very important to propose the improved Apriori algorithm to extract the association rules better.

References

1. Zhang Y, Xie Y, Chen C, Zhang C, Zhuang Y (2016) Retrospective analysis of the pharmacological action characteristics of the real world compound matrine injection in the treatment of the malignant tumors based on the association rule Apriori algorithm. China J Chin Materia Med (12):100–101
2. Xu Z, Zheng C, Zhang Y, Zhang J, Guo Y, Li Q, Hu D, Gao X (2017) Optimization of the operating parameters of the coal-fired power plant desulfurization system based on the improved association rule algorithm. Proc CSEE (08):109–110
3. Alai A, Tian X, Xiao Q, Yang Y, Yan C, Zhang W, Akbar G, Cao M (2018) Application of the association rules and the logistic regression in the data mining of the metabolic syndrome in Uygur health examination population. Mod Prev Med (04):130–131
4. Li W, Lin S, Zhou Y, Ma P, Yang M (2018) Research progress in the reliability evaluation of the complex simulation systems. Sci Sin Inf (07):153–155
5. Zhang H, Zhao L, Deng H (2018) Establishment and modeling of the virtual simulation system for the relative gravity measurement. Geospatial Inf (10):144–145
6. Zhao J, Guo Y, Xiao X et al (2017) Multi-step wind speed and power forecasts based on a WRF simulation and an optimized association method. Appl Energy 197:183–202
7. Moreno MN, Segrera S, López VF et al (2016) Web mining based framework for solving usual problems in recommender systems. A case study for movies' recommendation. Neurocomputing 176:72–80

8. Jung H, Yoo H, Chung K (2016) Associative context mining for ontology-driven hidden knowledge discovery. Clust Comput 19(4):2261–2271
9. Wang Y, Shao L (2017) Understanding occupancy pattern and improving building energy efficiency through Wi-Fi based indoor positioning. Build Environ 114:106–117
10. Zhang J, Williams SO, Wang H (2018) Intelligent computing system based on pattern recognition and data mining algorithms. Sustain Comput: Inform Syst 20:192–202

Research on the Optimizing Process of the Basic Image Processing Algorithms

Huie Zhang(✉)

Xi'an Siyuan University, Xi'an 710038, Shaanxi, China
luding1656486@163.com

Abstract. The image processing mainly includes the image compression, the image filtering, the image sampling, the image segmentation and the image analysis. At present, the image processing has an important role in many practical fields, such as the image recognition and the face recognition and so on. Its technical core is the image processing. With the continuous expansion of the image processing scale in these areas and the continuous improvement of the real-time performance requirements, how to improve the performance of the image processing algorithms has become a current research hotspot.

Keywords: Basic image · Processing algorithm · Optimization process · Image filtering

Aiming at the realization of the real-time and the efficient image processing tasks, an open-loop tracking algorithm for moving the objects based on the template matching in the video images is taken as an example to evaluate its tracking performances based on the prototype algorithm, and the multi-level optimization process of the algorithm is introduced in details. The compressive sensing is widely used in the field of the image compression [1]. Its compression process includes the following steps. The original image is observed randomly. After the observation, the compressed data is the compressed data of the images.

1 Basic Image Processing

The image processing is the technology of analyzing, processing, and managing images to meet the visual, psychological and other requirements. The image processing is an application of the signal processing in the image domain [2]. At present, most of the images are stored in the digital form, so in many cases, the image processing is the exponential image processing. Next, we will briefly introduce the classical algorithms in the field of the digital image processing.

1.1 Introduction of the Image Processing Principles and Procedures

The binocular vision is widely used in many fields because of its low cost, the convenient use and the high efficiency. The computer vision system mainly consists of

Z. Xu et al. (Eds.): CSIA 2019, AISC 928, pp. 212–217, 2020.
https://doi.org/10.1007/978-3-030-15235-2_33

three steps: the image acquisition, the image processing and analysis, and the output and display [3]. After the refinement, it can be divided into the following steps.

The binocular camera calibration is based on an effective camera imaging model. The internal and the external parameters of the camera are determined by the experiments and calculations, and then the set the positions of the object surface points and the relationship between the coordinates of the corresponding projection points can be established correctly. This is an indispensable key step in the computer binocular vision system [4]. The image acquisition is the information source of the binocular vision system. After the camera captures the scene, it converts it into the digital signal and generates the two-dimensional graphics. On this basis, the three-dimensional graphics are formed. However, the light conditions and the performances of the camera have a great influence on the quality of the acquired images.

The image processing technology will process the input images by the processing algorithm, preprocess the original data, and reduce the interference of the useless information and other information, so as to improve the image qualities and enable the images to be further analyzed. The purpose of the image matching is to find the projection points of the same scene on the left and the right image planes, to obtain the position relationship and the parallax, which is the most important step in the binocular vision system.

After the image matching and the disparity are obtained, the depth information of the objects can be calculated according to the triangulation method, so that the three-dimensional coordinates of the surface points of the objects can be obtained [5]. In the entire process, the image matching is the most basic problem in the field of the binocular vision, and it also lays the foundation for the subsequent image analysis and understanding. However, because the binocular cameras take pictures in different scenes, different angles and positions, and are affected by the light conditions, each image has great differences, such as the gray level, the scene location, and the resolution and so on. The image matching is to find the invariant features of these factors, and then match the two images according to these features.

1.2 Analysis of the Factors Affecting the Image Processing

When a binocular camera acquires an image, due to the different camera equipment, the sufficient light and the shooting angle, the image will be distorted to a certain extent. The errors caused by these distortions will be magnified in the image matching, which will lead to the wrong matching results. Firstly, when the geometric changes such as the translation and the rotation occur, all the pixels of the two images will be displaced, which has a great impact on the image matching. Rotation is due to the difference in the perspective of the camera equipment, and the relative relationship between the two images rotated. The second is the light conditions. When shooting the same scene, the difference of the light conditions will also lead to the image differences. For example, when the light is strong, the average gray level of the image is high, which will produce shadows. In the case of the insufficient light, the resolution of the acquired image will be reduced, resulting in the partial occlusion. These differences make the image matching very difficult. The third is the influence of the sensor noise. The sensors convert the acquired information into the electrical signals, so when acquiring the images, the electromagnetic radiation, the sensor devices and the switching devices and so on will all produce certain imaging noise. The system will discretize and encode the signal for the computer processing, but in this process, there will be the quantization errors, which will affect the image matching.

1.3 Research Status of the Image Processing Algorithms

Specifically, it can be divided into two categories: based on the image gray level and based on the image features. Comparing the two algorithms, the matching algorithm based on the gray level is to count the average gray level, the gray histogram, the average absolute value, the square deviation and the covariance of the image. According to these characteristics, matching has the advantage of the high accuracy. But there are also problems such as the large amount of the calculation and the low time efficiency, so it cannot meet the needs of our real life. The feature-based image matching method is to extract the information containing various features in the images, and the amount of the data is relatively small, so the matching efficiency is higher, and it can meet the needs of the time-to-time image processing. Specifically, the image matching algorithm based on the gray level is to analyze the gray level information of the image, calculate the similarity between the images, and then find the best image matching. That is to say, the feature selected by this method is the gray level of the image.

2 Analysis of the Key Problems in the Image Processing Optimization

2.1 Establishment of the Optimization Model

In the process of the image processing, in order to get the best result, it is often necessary to establish the most appropriate optimization model. The optimization model is established according to the different tasks of the image processing. Before that, scholars at home and abroad have established a large number of the image processing models. The effect of this model in dealing with the transmitted and reflected light intensity images is almost the same as that of the human visual system. Florea, et al. have established a more detailed linear piecewise logarithmic image processing model, and Kim has established the number of the image restoration. Pohl proposed a multi-image fusion model. Therefore, for different image processing tasks, these image processing models can be established.

2.2 Selection of the Optimal Algorithms

According to the different image processing problems, choosing the best optimization algorithm is a very critical issue for the performance of the image optimization. At present, the traditional optimization algorithms and the evolutionary algorithms have their own advantages and disadvantages. The traditional optimization algorithms can be roughly divided into three categories: the heuristic algorithm, the enumeration algorithm and the search algorithm. The heuristic algorithm finds the optimal solution or the approximate optimal solution by a heuristic rule. Although the efficiency of the solution is high, the heuristic rules found are not universal and are not suitable for other problems. The enumeration method requires the discretization of the continuous functions, so it can never get the optimal solution, because the search space is very large and the efficiency is very low. Although the search algorithm cannot guarantee the optimal solution, the quality and efficiency of the solution can be balanced by using the heuristic knowledge. The evolutionary algorithms can automatically discover the characteristics and rules of the environment according to the changes of the environment. Searching by means of the population can search the multiple regions in the solution space. However, as an artificial intelligence method, the evolutionary algorithm has shortcomings in the theoretical research and the algorithm implementation, and the slow convergence speed in practice. The PSO algorithm is relatively simple, and only a small number of the parameters need to be adjusted, which is relatively easy to achieve, and can effectively solve the multi-objective optimization problem. If PSO uses the idea of the evolutionary algorithm for reference, it can design a better multi-objective algorithm.

2.3 Design of the Optimal Algorithms

The uniform design is a scientific and effective method for optimizing the parameters of various optimization algorithms. When the model is complex, the number of the uniform design experiments is less and the uniformity is better. The uniform design transforms the parameter allocation problems of the algorithm into a multi-level, multi-factor and experimental design problem, which can optimize the parameter allocation. The simple uniform design is used to determine the operation parameters of the optimization algorithm to avoid the blind attempts and improve the scientificity of the optimization algorithm. The uniform design method is used to set the parameters of the optimization algorithm. The uniform design table is established and the operation range of the parameters of the optimization algorithm is determined. The test calculation is completed according to the uniform design table. The optimum parameters are selected according to the calculation results.

3 Analysis of the Several Key Algorithms in the Image Processing

3.1 Image Enhancement Technology

The image enhancement algorithm in the image processing is to highlight the main part of the image processing, so that the image sharpness and resolution are higher than the outstanding effect of the entire picture. The modern technology is usually used to reduce the gray of the images or enhance the colors of the images to achieve the effect of the image enhancement. For example, in a landscape picture, in order to achieve the prominent close-range image in the image processing, the noise resolution effect around the image can be removed first, and then the prominent part of the image can be mixed. Generally, the gray degree of the image is processed. The gray value of 3×3 neighborhood pixels around the prominent part of the image can be reduced, which can realize the roles of the overall image stereo enhancement of the gray level of the image and the image sharpness enhancement. On the other hand, the implementation of the image enhancement technology can also improve the overall image effect by moderately adjusting the colors of the images. The adjustment of the colors is generally about 1%–5% increase in the color brightness, 10% increase in the color sensitivity and 10%–20% increase in the color matching coordination, but the increase of the colors of the images should also be proportional to the resolution of the images, to avoid blurring the images by adjusting the colors of the images.

3.2 Image Edge

The image edge is a commonly used form of the image processing. This image processing technology can improve the image softness. Combining with the modern image processing technology, the Robert edge processing technology is mainly used. The Robert edge detection operator uses the difference between two adjacent pixels in the diagonal direction to detect the gradient amplitude, and finally achieves the imaging processing of the image edge and the image main body. The processing technology that improves the image effects, levels and heights has an impact on the image edge processing effect, including the interaction of the adjacent resolving value of the image, further achieving the diagonal ladder rendering effect between the image processing structure and the image overall picture, increasing the stereoscopic effect of the image, and improving the overall image processing effect of the softness and the image clarity. On the other hand, the operation of the image edge algorithm also includes the change of the image angle of the color resolution in the image edge processing. Usually, the image edge effect changes from zero degree to 90°, or the change of the color distribution step color resolution, the change of the color resolution and the concentration of the image effect of the main body inside the image can be greatly changed by the image edge.

3.3 Threshold Calculation Method

The threshold calculation is a method of the image processing from the perspective of the human vision. This method of the image processing can meet the needs of people to appreciate the images to the greatest extent. According to the visual imaging effect of the image processing, the recognition ability and the resolution ability of the human visual angle are guaranteed. The threshold calculation method is a common image processing method in the modern image processing, and its application in our life is comprehensive. The pulse degree algorithm is also a common processing method in the modern image processing. The pulse degree is the main technical form of the single laser hitting pulse imaging light wave. The image processing calculation of the pulse imaging must be based on the frequency of the image imaging. Therefore, the impulsivity algorithm must implement the comprehensive planning of the modern overall impulse effect map, and reasonably carry out the image imaging rate of various blowing processes in the image processing, that is, to reasonably carry out the imaging frequency of each frame of the image, to increase the pulse speed, and to improve the effect of the image processing sharpness.

4 Conclusion

The modern electronic technology, the computer graphics and the information processing have been concentrated on the image measurement technology. For example, the size and the shape of the three-dimensional images can be accurately measured by the image measurement technology. The image measurement technology also has important applications in the aviation fields such as the precision instruments and the remote sensing. The image features need to be processed accurately and meticulously by using the image software in computers. Generally, images are captured by computers and then extracted. The images need to be represented by digitalization, so that the images are not distorted and are easy to be processed by computers, and then the images are segmented by computers to realize the extraction and description of the image features.

Acknowledgment. Foundation Project: Shaanxi Natural Science Foundation Project: 17JK1076.

References

1. Ma W, Wang X (2016) Robot vision image processing based on the smooth denoising block K-means algorithm. J SSSRL 12:51–53
2. Zhang H, Cao B, Liu W (2017) Hyperspectral image compression with the multi-band and the look-up table prediction correction. Laser Infrared 9:101–102
3. Gu H, Zhang J (2018) Application of the outlier detection algorithm based on the deviation degree in the cluster analysis. J Jiamusi Univ (Nat Sci Ed) 7:132–133
4. Zhao Y, Zhu Q, Hu R, Qu X (2018) Hybrid recommendation algorithms based on the self-coding machine and clustering. Microelectron Comput 11:93–94
5. Huang Y, Zhang F (2018) Image matching algorithm based on the differential constrained coupled triangular network constraints. J Xinjiang Univ (Nat Sci Ed) 11:138–139

Research on the Privacy Security and Defense Strategies on the Internet of Things Platform

Qian Bian[1(✉)], Zhenduo Wang[1], Yingqiang Wang[1], and Weigang Zhang[2]

[1] School of Telecommunications, Xi'an Siyuan University, Xi'an 710038, Shaanxi, China
sncv06@yeah.net
[2] Chang'an University, Xi'an 710064, Shaanxi, China

Abstract. With the maturity and development of the Internet of things technology, some security and privacy protection issues have emerged slowly. At present, various networks such as the mobile communication network, the wireless sensor network and some terminal devices still have many technical shortcomings, and the privacy security of objects is also threatened. Therefore, this paper takes the security and privacy protection of the Internet of things as the research direction, and analyzes the privacy protection model technologies and methods of the Internet of things.

Keywords: Internet of things platform · Privacy security · Defense strategy · New technology

1 Analysis of the Technical Characteristics and the Connotation of the Internet of Things Platform

The Internet of things (IOT) is widely used in the network convergence through the pervasive computing, the intelligent sensing and the recognition technology. It is also called the third wave in the development of the world information industry after the computer and the Internet. The Internet of things is an application expansion of the Internet. It is more a business and application than a network [1]. Therefore, the core of the development of the Internet of things is the application innovation. The soul of the development of the Internet of things is the Innovation 2.0 with the user experience as the core.

The Internet of things is a new concept emerging in recent years. It can realize the communication of the information and is closely related to the privacy of the state, enterprises and individuals. According to the survey, many countries in the world have shown great interest in the Internet of things technology [2]. The aim is to connect the Internet of things with the research work by researching the related technologies, so as to help people reduce their working pressure and improve their working efficiency. In a sense, the Internet of things is a perceptual information system and a virtual information space dominated by the wireless. The intelligent processing methods are ubiquitous and can be used in all walks of life [3]. However, beyond people's

Z. Xu et al. (Eds.): CSIA 2019, AISC 928, pp. 218–223, 2020.
https://doi.org/10.1007/978-3-030-15235-2_34

expectations, there is a serious security threat in this way of the information exchange in the Internet of things, and the phenomenon of being attacked has even led to the paralysis of the system, which has caused serious losses to the country.

The Internet of things is an important part of the new generation of the information technology, and also an important stage of the development in the era of "informatization". As the name implies, the Internet of things is the Internet with the connection of things. This has two meanings. First, the core and foundation of the Internet of things is still the Internet, which is an extension and expansion of the network on the basis of the Internet. Second, its client expands and extends to any goods and objects, and carries out the information exchange and communication, that is, the things interrelated.

The Internet of things is widely used in the convergence of the networks through the intelligent sensing, the identification and the pervasive computing [4]. It is also called the third wave of the development of the information industry in the world after the computers and the Internet. The Internet of things is an application expansion of the Internet. It is more a business and application than a network. Therefore, the application innovation is the core of the development of the Internet of things, and the Innovation 2.0 with the user experience as the core is the soul of the development of the Internet of things.

The Internet of things is not only an important stage of the development in the information age, but also an important part of the new generation of the information technology [5]. It contains two meanings. First, the Internet is the core and foundation of the Internet of things and the Internet of things is the extension and expansion of the Internet. Second, the users of the Internet of things extend to any items, and we exchange information between items.

2 Research on the Privacy Security in the Internet of Things Platform

In the process of the information transmission and exchange, there will be a lot of confidential and highly sensitive private information. Only when a platform can fully respect the personal privacy and gain the user trust can it enhance the customer experience. Security, privacy, trustworthiness, non-compulsion and ease of use are the key factors for the maturity of the Internet of things platform.

In order to protect the information privacy in the collection and communication, we must first achieve the confidentiality attributes in the security. Therefore, using the mature cryptography technology is the basis of the privacy protection. For the privacy protection of the network communication, it is necessary to mention the protection technology in the process of the sensor data collection. Therefore, when the server collects the real-time sensor data, it must adopt some special means, such as changing the frequency of sending the packets, changing the size of the network packets, sending the virtual packets, and protecting the acquisition time and the location information. Even if the data is acquired in the transmission process, it is impossible to determine the contents of the specific data, so as to achieve the effect of the privacy protection. At the same time, in the process of the privacy protection of the Internet of things, we cannot ignore some environment-aware data, such as the occupancy and the memory usage and so on, because these data can easily infer the contents of the actual data.

Another mature way in the field of the privacy protection is to use the anonymous network transmission when it is impossible to distinguish the different devices or the different user data, and the stolen data will lose its significance. Anonymity is of great significance to the privacy protection, which can ensure that the information will not leak the privacy of source information when it is sent to a third party. However, when the privacy protection is done, the principle of the security protection cannot be ignored. Otherwise, it is easy to cause confusion in the system. The difference between the privacy protection and the security is that the third party can judge objectively whether an operation has damaged the data source or its owner, but the premise of the privacy is to respect the user's settings.

3 Privacy Security Vulnerabilities in the Internet of Things Platform

As the foundation of the Internet of things architecture, the RFID system is also the core technology of the Internet of things. The security of the RFID system directly affects the application security of the Internet of things technology in the industry. As the basis and the core module of the Internet of things, there are many system security vulnerabilities in the RFID system, which are mainly manifested in the following aspects. There are defects in the RFID tag structure. Because of its limited structure and low economic costs, it is difficult to guarantee a certain degree of the security and the protection capability of the RFID tags. Currently, the market price of the commonly used RFID tags is only 10 cents. Its internal structure only includes about 5000 logic gates, and only a small number of the logic gates can support the security functions.

In fact, the implementation of the complex encryption algorithm requires 4000 logic gates, so the simple structure of the RFID tags doomed to have great shortcomings in the privacy information protection. There are the security defects in the communication network channel. In the process of the Internet of things and its application, the wireless network communication channel is mainly used to complete the information transmission, which often provides opportunities for the illegal users or hackers. Because of its simple structure, it is easy for hackers to forge the readers, thus illegally obtaining the tag information, thus tampering with, modifying and destroying the data information in the information system of the Internet of things.

The RFID tags are the important part of the information system of the Internet of things. They store a large amount of the information with business values, which has great temptation for the illegal users and hackers. Once the information is leaked or tampered with, it will cause disastrous consequences. Common methods of attacking the RFID system include the data stealing and tracking. If the data information between the RFID system and the RFID readers is not encrypted and protected, illegal users may use devices that can eavesdrop on the readers to obtain the data information of the tag and the business information of the system. This attack method can obtain the business information of the system without directly accessing the tag.

There is the attack between the humans. Hackers can use the man-in-the-middle attack to intercept the interactive data between the tags and the systems based on the data stealing and tracking. Illegal users access the reader devices between the tags and the systems. After stealing the tag identity authentication information, they tamper with and destroy the data information, and then transmit the reader data. The entire process is reasonable. There is also the cloning and deceptive attack. After stealing the identity authentication information of the tags, this attack obtains the legal identity of the tags, and then sends the acquired data information to the readers in a short time, or sends a large amount of the data information at a time, which causes the reader system to fail to process and cause the system to paralyze. In addition, the cloning attacks can also be real-life. The intercepted data information is now written to several illegal labels.

There is the attack of the service denial. This attack method is similar to the denial of services of the Internet. By placing a signal interference source in the same frequency band in the area between the label and the system communication, the system will be paralyzed due to the influence of the normal communication between the label and the readers, or the retail of goods will be destroyed, or the transaction process will be destroyed, or it will send a large amount of the garbage data in a short time, causing the system congestion, resulting in the system paralysis, unable to provide the normal services. There is the inactivated tag attack. Illegal users may steal the goods by making the inactivated devices that inactivate the labels before trading.

4 Analysis of the Privacy Defense Strategies on the Internet of Things Platform

In view of the risks existing in the application process of the Internet of things, we can mainly establish the corresponding defensive measures from the collection and processing of the data information and the functional procedures of sensors. The anonymous risk prevention measure is mainly to fuzzify the collected information in the application of the Internet of things, to reduce the frequency of the sensitive information. Reducing the identification of the specific information in the aggregation of a large number of the data information not only hides the property functions of the devices in the Internet of things, but also anonymously processes the user's various operations and their usage traces, which can make the original information difficult to trace and can greatly improve the user's security.

4.1 Common Defensive Measures

The first is the encryption-based risk prevention measure. This method, mainly through various forms of the passwords the users set up, prevents the privacy leakage. Based on the various encryption conditions of the data, the cryptographic programs can hide the original data very effectively. The encryption defense measure can ensure the security of the data without affecting other programs. The second is the network protocol risk prevention measure. This method, mainly through restricting the use permissions of the specific network, protects the sensor data and the location information of the Internet of things, which, to a large extent, can reduce the probability of the intrusion into the system due to the human factors, to create a secure network environment for the normal operation of the Internet of things.

4.2 Improve the Ability of the Information Security to Protect the Privacy

To protect the personal privacy from infringement, it is necessary to protect the personal privacy and the information security. For example, when an individual inquires about the location of a nearby hotel through his mobile phone, he relies on the location of the user at that time, but the user does not want to expose his geographical location, so as to protect himself from being tracked. At the technical level, we need to study the key technologies of the Internet of things in order to ensure the confidentiality and authenticity, and then solve the problem of the protection and trust of the user privacy.

4.3 Improve China's Laws to Protect the Personal Privacy and the Information Security

First, we need to define the scope of the personal privacy protection. How to define the public dissemination and use of the personal information and the protection of the privacy rights is a problem that needs to be solved in the legislation and policy-making. With the development of the Internet of things, more and more information is regarded as the personal privacy information. For example, in terms of the personal financial information and other information reflecting the personal credit status, opening these businesses must be authenticated by the real name of the user information, and how to protect the personal information from disclosure is worth studying. The personal records are one or a group of the information identified by the name or the certificate number. The personal record information may include all the personal records related to the personal education, work, and economic status and so on. Some information may not be confidentiality on the surface, but may only record the situations at that time, such as the micro-blog, and the album and so on. But if the interested person makes use of the information defect of the Internet of things to carry out the "human flesh search", all the personal privacy related to the parties themselves can be exposed to the public and given to the parties, causing embarrassment and harassment. Therefore, the personal privacy should also include the information about the personal records.

4.4 Perfect the Personal Information Collection

We will improve the system for the legal collection and use of the personal information. Protecting the personal privacy and the information security should focus on the collection and use of the personal records. Without the collection and dissemination of the personal information, there is no violation of the personal privacy. At present, our country has the Personal Information Protection Law, which prohibits any group to disclose the personal information to others without the personal consent. The Personal Information Protection Law will better protect the personal privacy. Compared with the traditional network, the information security, the network security, the data security and even the national security issues brought about by the development of the Internet of things will be more prominent. In the future development of the Internet of things, we should strengthen the security awareness, put the security first, strengthen the construction of the legal systems and technical means, and promote the improvement of the laws and regulations on the information security, the intellectual property infringement, and personal privacy protection and so on.

5 Conclusion

In the 21st century, with the rapid development of the Internet in our country, although the Internet of things in our country started relatively late, the pace of its development is not slow. At the same time, we have made some gratifying achievements. However, there are still many problems that need to be solved. The development of anything is a double-edged sword. Because the development of the Internet of things in China is late, the technologies are not very advanced, and therefore there are also great shortcomings in the privacy security. This paper studies how to use the anonymity method to improve the confidentiality, reduce the complexity of the encryption method and reduce the communication overhead of the routing protocol method and so on.

Acknowledgment. (1) The 13th Five-Year Plan of Shaanxi Education Science in 2018, Project No.: SGH18H526. (2) Scientific Research Program of Shaanxi Education Department, Project No.: 18JK1099. (3) Shaanxi Province Education Science 13th Five-Year Plan in 2017, Project No.: SGH17H446.

References

1. Zhang X (2017) The development status and countermeasures of the Internet of things. China Comput Commun (Theor Ed) (12):191–192
2. Wang Y, Xu Q (2017) Research on the privacy data security for the Internet of things. Comput Program Skills Maint (12):155–156
3. Xue R, Peng C, Huang X, Liu J, Yu Y (2018) Introduction to the new privacy-oriented technologies and the cryptographic algorithms. J Softw (07):108–109
4. Xu Y (2018) Framework analysis of the privacy protection and the information security enhancement model based on the Internet of things. Mod Inf Technol (09):119–120
5. Sun D (2018) Research on the application security and privacy of the Internet of things. Educ Forum (11):125–126

Research on the Security Analysis and Management of the Network Information System Based on the Big Data Decision Making

Weigang Liu(✉)

Tianjin Modern Vocational Technology College, Tianjin 300000, China
lh5ztc@yeah.net

Abstract. In the era of the big data, no matter whether it is the government industry or the financial industry or even all walks of life, they are facing the threat of the information security changes. In this situation, to deal with information security problems, we need not only to enhance the awareness of the information security protection, but also to establish and improve the information security safeguarding system and deepen the reform of the information security technologies. The problem of the big data information security cannot be ignored. When the enterprise users choose the services provided by the cloud platform to fulfill their business needs, they can build the information security mechanism by establishing and perfecting laws and regulations, establishing the information security complaint and service center, doing well the defensive measures, and strengthening the vulnerability scanning, to ensure the information security.

Keywords: Big data decision-making · Network information system · Security analysis · Management research

With the emergence of a large number of the data from all walks of life around the world, there is a higher requirement for the real-time and validity of the data processing and analysis, which promotes the rapid development of the big data technology. The big data is a relatively new concept [1]. In a sense, it is a collection of many new technologies, including some new analysis techniques, storage databases, and distributed computing and so on. Compared with the traditional technologies, its structure, information types and working modes have undergone qualitative changes. Many new technologies are integrated together. Faced with the huge amount of data to be processed by the big data technology, the following new challenges are brought to us in the field of the information security [2].

1 Time Characteristics of the Big Data Decision Making

With the rapid development of the information technology, the Internet of Things, the cloud computing, the mobile Internet and other emerging technologies make a variety of computers, sensors and mobile devices and so on continuously create exponential

Z. Xu et al. (Eds.): CSIA 2019, AISC 928, pp. 224–229, 2020.
https://doi.org/10.1007/978-3-030-15235-2_35

growth of information, which includes not only the human, but also all kinds of things, and the growth rate is also accelerating [3]. Our world has quietly entered the era of the "big data". Such a huge change has made the conventional computing tools we used before unable to meet various new challenges, which has also attracted the attention of the industry, the science and technology circles and the governments.

The big data refers to the large scale of the data involved, which cannot be intercepted, managed, processed and organized into the information with more positive purposes for various decisions within a reasonable time through the current mainstream software tools. Generally speaking, it has four characteristics: the large amount of data, many types of data, fast generation and processing speed, and low value density. The big data is considered by the U.S. government as "the new oil in the future", and its ability to use it will become one of the manifestations of the comprehensive national strength of the country in the future, and will also be one of the core assets of the country. However, with the further concentration of the data and the increase of the information quantity, the change of the processing methods has brought us some new problems in the information security [4].

The core of the big data technology has changed from the traditional storage and transmission of the information to the mining and application of the information, which brings great changes to the business model of the whole world, and its potential application values will bring about the new huge markets. Faced with the ubiquitous data, new requirements are put forward for the information security. With the progress of the technologies, it will inevitably bring about the rapid development of the information security industry. The industry chain related to the big data technology will also usher in a new development period [5–7]. It can be seen that the importance of the big data technology in the future development of the science and technology and many opportunities it will bring have attracted the worldwide attention.

The big data is a large amount of the data produced in people's daily life [8]. It cannot be measured by the traditional tools, but it has a far-reaching impact on the government organs and the enterprises, and plays a vital role in the decision-making. Since 2012, many countries have started the data analysis programs, and hope to get some hidden information through the big data on the Internet. Many senior officials believe that an important symbol of a country's overall national strength is its data processing ability, and the future victories are in the hands of the data owners. For some enterprises, by collecting some data, they can make more scientific and effective decisions in line with the development lines of enterprises. However, how to effectively manage the massive data and information has gradually become a problem of concern.

2 Major Problems Facing the Security of the Big Data Decision-Making Network Information System

The development of the big data cannot be separated from the vigorous development of the Internet. At the same time, the Internet security has always been the focus of our attention. If we do not pay attention to it, it will cause a series of the fatal blows to the generation, development and growth of the big data. Compared with the traditional data, the big data has higher commercial values and the huge data volumes, so it is easier to become the target of hackers and viruses. At present, the architecture of the big data database system is analyzed, which mainly presents the characteristics of "parallel concurrency". For the operation of the client applications, different nodes are used to meet the requirements of the controlling, communication and other aspects. In order to complete the processing of all the data nodes in the authentication process, there is great difficulty. Therefore, in practice, we often adopt an open management mode that does not restrict the user access, which will cause a lot of potential security problems.

Analyzing the traditional database management environment, it has a high success rate in the implementation of the real-time and matching detection at a single time point. However, considering the objective conditions in the large data database environment, it is impossible to achieve the real-time detection for such a continuous and complex situation in the APT attack detection. At the same time, considering the low value density of the big data, it is difficult to carry out necessary security protection work for a certain value point. In this way, there will be many potential viruses and Trojan horse attacks in the big data, which undoubtedly poses serious challenges to the network information system security of the network operators, enterprises and institutions.

In the security management of the network information system, the problems of the network vulnerabilities cannot be ignored. The focus of attacks is often the vulnerabilities in the operating system, the application software vulnerabilities, the network protocol vulnerabilities and the security management vulnerabilities, which will lead to the information leakage and become the target of attacks. These situations are very common. There are various kinds of errors in the design of the software and the hardware, resulting in the impairment of the information integrity, availability and confidentiality, which will lead to hidden dangers of the network information system security. The system backdoor, the SQL injection, the cross-site scripts, the weak passwords, the phishing and the unencrypted login, these are the common system vulnerabilities, and also the focus of attacks. Taking into account a large amount of data stored in the big data system, the risk of the leakage of the information has greatly increased. For example, in the process of the online shopping, consumers fill out the related information, including names, addresses and mobile phone numbers. If the personal information is leaked, it will seriously affect their normal life and work, and the severe cases can cause the huge economic losses. In the related network fraud, it is because of the influence of the third-party trading platform disclosure of the personal data information. Therefore, in the era of the big data, we should pay full attention to the threat of the leakage of the information resources.

The distributed data storage is easy to be attacked. The big data storage is the distributed storage. Obviously, the distributed storage can effectively solve the storage problem of the big data, and its technical requirements are low. However, the low security is a major drawback of the distributed data storage technology. In this case, hackers can easily access the data, which poses a threat to the information security of the Internet users. There are hidden dangers of the information leakage in the data processing process of the big data platform. In order to improve the efficiency of the data processing, the big data platform usually carries out sampling, acquisition, analysis and processing of the relevant data. All the above operations need to comply with the relevant provisions of the network information processing, so that there are no hidden dangers to the information security in each link. Nevertheless, the problem of the external network crossing existing in the process the distributed computing information transfer is inevitable, which has become an important way of the information leakage in the big data platform. Therefore, with the increase of the data volume, the safety management measures for the data protection need to be adjusted at any time. The big data information management risks that the storage brings: the specialty based on the data types and structures, and the storage spaces of the big data platform are increasing. The massive increase of the data volume and the rapid increase of the data types make the data storage structures to be changed. Therefore, the concurrent process of the big data application platform has great risks, which increases the difficulty of the data management and brings huge security risks to the related work.

3 Network Security Protection Strategy for the Big Data Decision Making

3.1 Enhance the User Account Security

At present, the computer users register and use different user accounts in different websites or software clients, many of which are interrelated and mutually verifiable. Therefore, in order to improve the security of the data and accounts, it is necessary to set the accounts and passwords complex and not easy to crack. Secondly, the user names and passwords that are set in each site or client are distinguished from each other, and do not set the similar accounts and passwords. Lastly, classify the accounts and passwords. Set corresponding passwords for the accounts of different levels, and carry out the separate management of the important accounts and passwords, as well as the regular replacement, so as to ensure the security of the accounts.

3.2 Install the Firewall and the Anti-virus Software

In order to use the network resources safely in the context of the big data, the security of the computer network can be improved by using the firewall technology and the virus anti-killing technologies. The so-called firewall technology is, according to the specific security standards, to real-time detect the data packets transmitted by the network system. If suspicious data packets are found, alarm or stop them in time, so as to effectively shut out the illegal data packets and ensure the data security of the computer users. In addition, the anti-virus software and the virus monitoring software should be actively used to monitor and eliminate viruses in our daily operation. The hidden dangers of the computer viruses should be eliminated by means of the periodic upgrade of the software and the periodic anti-virus scanning.

3.3 The Use of New Technologies

Using the big data mining, analysis and processing, improve the level of the intrusion detection technology, and detect whether the monitoring network and the computer system are abused or the precursor of the intrusion. Through the data mining and statistical analysis of the big data, the rule model of the intrusion detection is formed. In the process of the system operation, the normal operation of the system is judged, and the active monitoring mechanism is formed. Under the background of the big data, the application of the file encryption and the digital signature technology in the network security can effectively improve the security level of the computer network. Among them, the file encryption technology is the main technology used to prevent the secret data from being stolen, destroyed or intercepted, and also to improve the security and confidentiality of the information systems and the data. The main purpose of the data signature technology is to encrypt the data stream in the transmission.

3.4 Strengthen the Privacy Data Query Authority

In the context of the big data, through the analysis and processing of the massive data, carry out the data mining of the associated information in each site node especially for the computer users, and we can obtain a large amount of the user's privacy data. Once there is the malicious use of the data, it will cause a very bad effect. For this reason, the big data technology and the related data processing platforms should restrict the display of the detailed data by the big data users. The analysis and query of the big data technology should mainly focus on the analysis and processing of the statistical data, so as to avoid the abuse of the big data technology at the technical level. The development of the big data era brings opportunities and new challenges to enterprises. The data information of enterprises is the core of the enterprise development. The business data needs to be paid attention to and protected. Therefore, monitoring the network traffic, ensuring the data security and preventing the hacker intrusion are the core of the enterprise data security. We can start from two aspects. (1) Strengthen the structure and standardization of the data storage, so that it can resist certain attacks and damage. (2) Strengthen the protection and backup of the data, so that the data can be attacked as far as possible to reduce losses, while recovering in a very short time, so that the work of enterprises will not be affected. The data security is guaranteed, and the whole industry can develop faster and better. The network information security management and the enterprise development are inseparable, so enterprises should do a good job in the information management, so that enterprises can be more prosperous in the era of the big data.

4 Conclusion

With the progress of the technology, users' privacy data becomes more and more transparent in the computer network, and the network security becomes more and more important. For all departments and individuals using the computer networks, abiding by the relevant rules and regulations for the network uses is an important link to realize the network security. The state and the relevant organizations should timely conform to the development of the new technologies, constantly revise and improve the laws and regulations of the computer network under the background of the big data, and construct the last barrier of the computer network security with laws and regulations.

References

1. Meng Z, Xiao L, Hu F (2017) Theoretical path and application of the big data-driven educational changes – an analysis of the first china education big data development forum. J Distance Educ (03):112–113
2. Wang L, Lu X, Wang X (2017) Coupling analysis and mechanism innovation of the targeted poverty alleviation operational mechanism driven by the big data – based on the cases of Guizhou and Gansu. J Public Manag (07):144–145
3. Feng Y, Zhong J, Li X, Zhou S (2017) Research and practice on the cooperative innovation training of the advanced technical talents with the big data – taking the cooperative innovation training of the computer full-time professional degree postgraduates and undergraduates as an example. China Educ Technol (06):154–155
4. Chen G, Wu G, Gu Y, Lu B, Qiang, W (2018) Research and application challenges driven by the big data in the context of the management decision-making – paradigm changes and research directions. J Manag Sci China (07):120–121
5. Zhou Q (2018) Multi-layer affective computing model based on emotional psychology. Electron Commer Res 18(1):109–124
6. Zhou Q, Luo J (2015) Artificial neural network based grid computing of E-government scheduling for emergency management. Comput Syst Sci Eng 30(5):335–343
7. Zhou Q (2016) Research on heterogeneous data integration model of group enterprise based on cluster computing. Clust Comput 19(3):1275–1282
8. Peng C, Hu C, Chen X (2018) Research on the data-based performance audit model under the big data environment – taking the performance auditing of intelligent parking planning and management project in X city as an example. Audit Res (03):140–141

Research on the Security Risks and the Preventive Strategies of Android Smart-Phones

Zhenduo Wang(✉), Qian Bian, and Xinzhi Tian

Xi'an Siyuan University, Xi'an 710038, Shaanxi, China
chlliusjtu@163.com

Abstract. With the advent of the era of the mobile Internet, Android smart phones have been widely used because of their powerful information processing abilities and convenient and fast characteristics. However, Android smart phones not only bring great convenience to people, but also face various security risks. Android smart phones store the address books and account passwords in the mobile phones. Once the important information is lost or leaked by the virus, the consequences are unthinkable. Based on the comprehensive analysis and study of various security risks faced by the Android smart phones, the corresponding preventive measures are put forward in order to ensure the security of the Android smart phones to some extent.

Keywords: Android system · Android smart-phone · Security risk · Preventive mechanism · Application strategy

China has become the largest Android smart-phone market, and in the Android smart-phones, the personal information of the mobile phone users is stored, and it becomes particularly important to ensure the information security of the Android smart-phones. Because the Android smart-phones are more open than the traditional PC clients, the security threats faced by the Android smart-phones in the application process are more diverse and complex [1]. If we want to apply this system better, we must pay attention to the detailed identification of its security risks and formulate the specific prevention strategies.

1 Security Performances of the Android Smart-Phones

With the development of the network technology and the arrival of the information age, the Android smart-phone applications are increasingly widespread [2]. They not only provide convenience for people's life, but also bring some potential safety hazards. Combined with the security problems existing in the Android Smart-phone application process, this paper puts forward the practical defensive measures, which can improve the security of the Android smart-phones, reduce the occurrence of the leaks, and ensure the privacy security of all users [3].

Because of the severity of the Android fragmentation, the security issues remain unavoidable. From the proportion of the Android version, its updating speed is

Z. Xu et al. (Eds.): CSIA 2019, AISC 928, pp. 230–235, 2020.
https://doi.org/10.1007/978-3-030-15235-2_36

obviously insufficient. On the other hand, the Android devices are very vulnerable to the malware because they can install the third-party applications at will. In fact, even Google Play, the official store of Google, often comes out with news of the undercarriage of the application with the Trojan horse virus, which shows that there are many difficulties in keeping the Android phones away from the malicious programs [4].

The Trojan horse is called the Trojan horse. It is a deceptive document. The major difference between the Trojan horse and the virus is that the Trojan horse does not replicate itself like the virus, and its purpose is not to destroy the memory of the mobile phones to destroy the information, but to carry out the remote control and steal the passwords and other user sensitive information. In this sense, the Trojan horse is actually a "thief". The mobile phones attacked by the Trojan horse virus can be remotely controlled without ringing the bell or displaying any information as long as they carry batteries [5].

It specially refers to the software and hardware destruction of the mobile phone system software [6, 7]. The common way of the destruction is to delete or modify the important system files or the data files, resulting in the loss of the user data or the system being unable to start normally, so that it must be repaired. The means of the transmission are generally through the network download and the PC copy. The typical virus is the Skull series [8, 9]. It can replace the system applications and make them impossible to apply. Once infected, all the application icons are replaced with the skeleton and the Cross Skull icons. And the icon is no longer associated with the application, and we can only answer the calls, but the Internet, SMS, MMS, photography and other functions are all lost.

Viruses used to be the patent of computers, but with the continuous development of the smart phone hardware and the continuous progress of the mobile communication technology, viruses have also opened up new areas on the mobile phones. Nowadays, there are two kinds of the mobile phone viruses. One is the virus that infects the mobile phone operating system, and the other is the virus that collects the memory information of the mobile phones. The mobile viruses, from the simple Bluetooth attacks to the spread through MMS and GPRS networks, have become more and more diverse. Users must understand their transmission and attacking methods in order to effectively prevent and cure them.

2 Cause Analysis of the Android Smart-Phone Security Risks

The personalized and diversified applications provided by the mobile Internet have greatly enriched people's lives. However, the security of the mobile smart terminals is also facing severe challenges. The spam messages, the mobile viruses, the malware and other phenomena are becoming more and more rampant, which have brought great trouble to the users of the smart terminals. The number of the mobile virus samples is showing the explosive growth.

There are several main reasons for the frequent security problems of mobile applications. First, the openness of the Android operating system has led to the in-depth study of the system. Attackers can implant the malicious code by tampering with or counterfeiting the official applications. Secondly, the App market is a mixed application of dragon and fish. For the IOS system, it is relatively difficult to find the system vulnerabilities by using the top-down control application. However, the Android system application is not censored by the App market. It relies on the user group defense to solve the problem. Let the users score and comment on the applications, and let the users see how many users there are. If enough users criticize a piece of software, Google will remove it from the market, and users who install the application will have security risks.

Thirdly, all kinds of the applications have control over the security rights, privacy rights and other rights. About 70% of customers do not care about the application's rights when installing the App applications. They install the App applications by default, which leads to the information leakage. The Android apps on the market, when downloaded, will clearly list the system permissions required for the apps, such as the access to mobile phone numbers or sending and receiving the text messages. If users do not agree to grant these permissions to the apps, they cannot install the apps. This is the most important and basic protection measures. Basically, the apps cannot be like the viruses. It's the Android's system feature that steals the personal data or sends and receives the text messages behind your back. Everything it must be with the user's own consent, but the problem is that the malicious apps will try their best to entice you to grant the permissions.

Fourthly, the user's security awareness is weak, and the bad usage habits lead to the leaks. The random login to the WLAN in unfamiliar areas, the prison break, scanning the two-dimensional codes, all of these behaviors contain too many unsafe factors. The WLAN in unfamiliar areas can implant malicious code on the network devices, resulting in the user information leakage. The prison break technology can destroy the program signatures and enable the Android smart-phones to run the applications through the unofficial channels, thus undermining the protection provided by the system. The first point is that freezing the code signatures causes the platform to be exposed to the malware and the other problem is to allow many applications without signatures to run in a non-isolated environment with the root-level privileges. The prison break patches also allow the memory to be written and executable, making the data execution protection technology ineffective. For the two-dimensional codes, there are potential threats such as the malicious URL access, and the malicious program download and so on.

3 Security Risk Prevention Strategy Analysis for the Android Smart-Phones

With the advent of the era of the mobile Internet, smart phones have been widely used because of their powerful information processing ability and the convenient and fast characteristics. Open the settings of the automatic detecting and updating of the system, update the operating system in time, and repair the vulnerabilities, which can reduce the security risks of the mobile phones. There are numerous vulnerabilities and defects in the smart-phone operating system, and the Android, the IOS and other mainstream operating systems will be upgraded after a period of time, so we should always pay attention to the update information issued by the official website of our mobile phones, timely update the operating system, and patch the vulnerabilities. Download the official version of the payment, tools and game mobile apps from the regular application store to ensure the security and reliability of the downloaded apps. Pay attention to the types of the privileges prompted by the system when installing the applications, and be vigilant about some privileges that are not related to the applications. If a smart-phone has to grant the above permissions to download an app for the system functions, read the number of the downloads of the apps as well as the scores and comments, and confirm that tens of thousands of people have used the apps without filing a victim complaint before agreeing to the permission and downloading the apps.

Attach importance to the security of the mobile payment. The mobile payment is convenient and fast, which is favored by people. Criminals aim at the field of the mobile payment and steal the users' money. The mobile payment security should be vigilant against leaks and Trojans, and we should protect the mobile payment security mainly from the following two aspects. Develop good habits, and do not carry out the improper operations. Set a safe and reasonable password. When surfing the Internet in public places, especially when it involves the financial payment, do not connect the free WIFI from the public and unknown sources, in order to avoid the data being subjected to being stolen away. Distinguish the similarities and differences between the phishing websites and the official websites. Don't click on the unfamiliar links easily. Don't fill in the personal information in the unknown website interface. Beware of the malicious software attached to the two-dimensional code. Set different passwords in different places, and regularly modify the password of the social accounts in order to prevent the "database collision". Set the upper limit of the one-day transfer quota for the online banking and the mobile banking to reduce losses. Install and update the security protection software in time, such as the Tencent mobile phone housekeeper, and the 360 security guards and so on, and open the harassment interception function, to effectively intercept the junk information. Open the financial protection and account protection function, regularly check the virus and clean up the garbage, and timely upgrade the mobile phone security management software to the latest version.

Establish a unified third-party application store security audit and supervision mechanism to ensure that every application in the store is safe and reliable. The Android users can browse, download and purchase the applications from the official Google Play or the third-party stores (Amazon, Ahead, Android, and Pea Pod and so on). However, there is no security audit and regulatory mechanism in the third-party Android market in China. Malicious developers can upload their apps to the third-party application stores for confusing the household vision, resulting in the explosive growth of a lot of the hooligans and malicious APP.

Improve the authorization mechanism of the Android operating system, and refine the strategies to control the granularity, and further enhance the security and reliability of the Android system. The Android system uses the "all or nothing" mechanism. That is, when the Android users download and install the APP, they need the user's consent to grant all the application permissions to install the APP. Otherwise, if any application for permission is not authorized by the user, the installation will not be possible. After installing the APP, Android users have no idea how to use the application permissions. They can't set the dynamic permissions according to their location and time, which makes them vulnerable to the malware attacks. If the user installs the APP, the permissions of the dangerous level will be listed on the screen. Allowing the users to grant only part of the permissions can also ensure the normal work of the APP, and allow the users to restrict the application access to the sensitive information without uninstalling the program.

To build the Android system-level intelligent decision support, users only need to select the security level of the platform, and the rest of the decision-making work is done by the system. The installation authorization of the applications depends entirely on the users, and most users do not have the professional knowledge, which greatly reduces the security of the system. According to the survey data of China Internet Information Center, 44.4% of the users will carefully check the authorization instructions when downloading and installing the application software, and 40.7% of the users will not carefully check the authorization instructions, while 14.9% of the users say that they are "hard to say". That is to say, most of them have the risk of the blind authorization. In the Android users' information disclosure, mobile phone contacts are the focus of the information disclosure. Once they are leaked, they will encounter the spam messages and the fraudulent words and so on, which even bring the direct economic losses.

Strengthen the awareness of the safety precautions and improve the ability of the self-protection. When choosing the mobile phones, choose the well-known brands, which have a relatively complete security protection system, and refuse the emulational mobile phones. When using the mobile phones, we should set up the password of the boot-up and lock the screen. When maintaining the mobile phones, we should supervise the entire process to prevent people from installing the secret hardware and software. When using the cloud services reasonably, we should not upload the personal privacy and the sensitive information to the cloud platform. Encrypt the uploaded data to ensure that the information will not be easily cracked after loss. Turn off the mobile phone's geolocation function when no location service is needed.

4 Conclusion

At present, the Android and the IOS dominate the smart-phone systems, and Android has become a popular mobile phone system. Unfortunately, Android is also facing the greatest security risks, which has become a concern for many mobile phone users. From 2014 to 2016, the IOS vulnerability level remained relatively stable, but the total number of the malicious software in the Android lineup increased by 105% in 2016. It should be pointed out that the domestic Android users face greater risks. The lack of the user security awareness and the fact that some manufacturers are willing to sacrifice the security to keeping costs low make the Android users in Asia more vulnerable to the attacks. Therefore, in the current Android smart-phone application process, we must start from the potential security risks and explore the reasonable prevention strategies.

Acknowledgement. This project is subsidized by Special Scientific Research Program (18JK1104) sponsored by Shaanxi Education Department in 2018 and "Research on the Intrusion Detection Technology Based on Android", a scientific research project at the school level of Xi'an Siyuan University in 2019.

References

1. Hu Q, Wei G, Liao F (2017) Application of the mobile learning mode based on the smart-phones in classroom-assisted teaching of analytical chemistry. Sci Educ Art Collects (12):122–123
2. Pang Y, Mei D, Fan W, Chen X (2017) Study of the effect of the exercise management mode of using the smart-phones to record the footsteps on the blood sugar in patients with type-2 diabetes mellitus. China Med Equip (12):105–106
3. Wang T, Yang X, Ren W (2017) Design and research of the app systems for smart phones in the expressway service area based on the LBS technology. Energy Conserv Environ Prot Transp (12):133–134
4. Li Z, Hu Q, Feng Y, Song Y, Li X, Ding Y (2017) Measurement of the rigid body moment of inertia by smart phones. Phys Eng (12):124–125
5. Liang Y, Niu X (2018) Influence of smart phones on the college English classroom teaching in the context of "Internet+" and its coping strategies. Learn Wkly (11):138–139
6. He D, Chan S, Guizani M (2015) Mobile application security: malware threats and defenses. IEEE Wirel Commun 22(1):138–144
7. Li Y, Dai W, Ming Z et al (2016) Privacy protection for preventing data over-collection in smart city. IEEE Trans Comput 65(5):1339–1350
8. Cho J, Kim D, Kim H (2018) User credential cloning attacks in android applications: exploiting automatic login on android apps and mitigating strategies. IEEE Consum Electron Mag 7(3):48–55
9. Azfar A, Choo KKR, Liu L (2016) Android mobile VoIP apps: a survey and examination of their security and privacy. Electron Commer Res 16(1):73–111

Research on the Simple Algorithms for the Logistics Network Layout in the Mobile Network E-Commerce

Fan Ming(✉)

Xi'an Eurasia University, Xi'an 710000, Shaanxi, China
j363wg@yeah.net

Abstract. The e-commerce is the dominant strategy of the digital business management, which has the characteristics of guidance, innovation and sociality. Based on this, by exploring the relationship between the e-commerce and the modern logistics distribution and service activities, and combining the analysis of its status, roles, and impact, we can give full play to the advantages of the intelligent management and promote the purpose of the social and economic cycle development. In the layout activities of the e-commerce, the rational use of the simple layout algorithm is not only conducive to the improvement and development of the e-commerce system, but is also an important factor for the development of the e-commerce.

Keywords: Mobile network · E-commerce · Logistics network layout · Simple algorithm

1 Overview of the Mobile E-Commerce

With the popularity of the e-commerce, the innovation of the logistics management is becoming more and more important. With the rapid development of the online shopping mode, the e-commerce logistics has become a new form of the modern logistics industry [1]. The development of the e-commerce logistics requires that the distribution center, as its foundation, play an increasingly important role.

The mobile e-commerce is the use of the mobile phones, PDA and portable computers and other wireless terminals for the B2B, B2C or C2C e-commerce [2]. It combines the Internet technologies, the mobile communication technology, the short distance communication technology and other information processing technologies perfectly, so that people can carry out various business activities at any time and anywhere [3]. Achieve the shopping transactions, the electronic payment, various business activities, the financial activities and the related integrated services at any time, anywhere, online and offline.

Compared with the traditional e-commerce through the computer platform, the mobile e-commerce has a broader user base. At present, the number of the Internet users in China is 350 million, while the total number of the mobile phone users in China is 710 million, of which 360 million are the mobile Internet users [4]. With the issuance of the 3G licenses and the gradual popularization of the new generation of the

Z. Xu et al. (Eds.): CSIA 2019, AISC 928, pp. 236–242, 2020.
https://doi.org/10.1007/978-3-030-15235-2_37

mobile communication technology, it is expected that the number of the mobile broadband users will exceed the number of the fixed broadband users in 2011, which shows that the mobile e-commerce has a broader market prospects. The development of the mobile Internet application and the wireless data communication technology has laid a solid foundation for the development of the mobile e-commerce.

With the gradual popularization of the mobile e-commerce applications such as the mobile payment, the mobile e-commerce is slowly integrating into our lives [5]. As we all know, the e-commerce activities and the logistics management activities are closely linked, so the integration of the mobile e-commerce and the logistics management is an inevitable trend.

2 Analysis of the Characteristics of the Logistics Network Layout Structures of the Mobile Network Electronic Commerce

2.1 Structure of the Single Core Node

In the logistics network, there is usually only one core node in the single-core node structure, and the core node is mainly responsible for the logistics activities. At the same time, it effectively and timely handles the distributed logistics information, and assumes the dual functions of the logistics center and the distribution center. The core nodes can perform many functions in the logistics, and have been widely used in some small and medium-sized enterprises [6].

2.2 Interactive Structure of the Double Core Nodes

The so-called dual-core node interactive structure specifically refers to the logistics network containing two core nodes, namely, the distribution center and the logistics center. This kind of the structure ensures that the distribution center and the logistics center can effectively carry out the circulation of the goods, and the corresponding information flow and logistics are both the bidirectional circulation.

2.3 Single Item Structure with the Double Core Nodes

In the logistics network, the single structure of the dual core nodes generally has two core nodes, namely the logistics center and the distribution center. Among them, the distribution center is mainly responsible for providing the commodity distribution services to the users, while the logistics center is mainly responsible for the suppliers' logistics services, and the logistics center is responsible for distributing goods to the distribution centers.

2.4 Multi-core Node Structure

The multi-core node structure has three or more core points in the logistics network, and the consumers and the suppliers are responsible for the corresponding services by the core nodes, and the logistics information platform is responsible for the transmission of the logistics information, so that individuals and enterprises can better grasp the dynamic information of the products.

3 Analysis of the Influencing Factors of the Logistics Network Layout in the Mobile Network E-Commerce

3.1 Logistics Network Structures of the Mobile Network Electronic Commerce

Generally speaking, the outlets of the express enterprises in the same city can be basically divided into three levels (Fig. 1): the high-level outlets (the urban distribution center), the intermediate outlets (the business offices, and the urban transit sites), and the low-level outlets (the business halls, and the intra-city outlets). The first-level network mainly undertakes the tasks of the international, inter-city and part of the same-city express delivery, such as transfer, sorting and assembly, with the huge business volume. The second-level network contacts the first and the third-level network, and also undertakes the sorting and assembly tasks, with more than the first-level network, and the business volume is lower than the first-level network. The third-level network contacts the courier directly downward, and the second-level network contacts the couriers directly. The second level network is mainly responsible for the same city delivery, and to receive and dispatch to the customers, with the largest number and the most dispersed distribution.

3.2 Analysis of the Factors Influencing the Distribution of the Logistics Outlets in the Mobile Network E-Commerce

The logistics service quality, the service efficiency and the cost of the mobile e-commerce affect the survival of the express delivery enterprises. The service quality is related to the management levels and concepts of the express enterprises, the quality of the employees, and the efficiency of the express delivery and so on. The service efficiency depends not only on the management level of the express enterprises, but also on whether the network layout of the express enterprises is reasonable or not. The cost composition of the express delivery includes various equipment costs and personnel, the employee salary, freight, site rent, and maintenance and management expenses and so on. It can be seen that the layout of the express outlets is very important for the development of the express enterprises.

Generally, the layout of the mobile network e-commerce logistics network is mainly affected by the traffic, population, economic development levels, site rent, policies and regulations. Specifically, different levels of the network have different impacts because of their different functions and business volumes. The transportation affects the layout of the express delivery network in the form of the transportation cost, the transportation equipment and the personnel costs and the service efficiency. The first-level network is most dependent on the traffic factors, and needs to maintain the convenient links with the city's external large flow, the high-speed traffic lines, and the city's internal main traffic lines. Therefore, it is generally located in the hub of the urban internal and external traffic links. The secondary network points require the higher traffic accessibility and convenience in the location selection, so as to evacuate the vehicles quickly and speed up the distribution speed. Therefore, the urban internal traffic trunk line and the vicinity of the hub nodes become a more rational layout location. Because of the connection of the terminal customers, the third-tier network points are mostly dispatched by the small trucks or the electric vehicles, so the traffic factors have little influence.

4 Simple Algorithms for the Logistics Network Layout in the Mobile Network E-Commerce

With the rapid development of the economy and the intensification of the market competitions, the logistics is not only the third source of the profit, but also the representative of the core competence of enterprises. It is a strategic resource that can bring the competitive advantages to the enterprises. As a logistics enterprise, its own development can not only improve the profits of the enterprises, but also has great significance for the economic development of the modern society. In the past, many logistics outlets neglected the roles of the market rules, so they had blindness and repeatability, which made the operation of some logistics outlets unsatisfactory. Therefore, how to scientifically study and analyze the location and the layout of the logistics outlets is very important to improve the business income and the development scale of the logistics companies.

4.1 Simple Algorithm

There are many decision-making problems in our daily life. In the process of comparing, judging, evaluating and making decisions on various schemes, subjective factors account for a considerable proportion. The analytic hierarchy process (AHP) has certain advantages in solving such problems. The simple algorithm method is a practical analysis method which can effectively deal with the multi-scheme comparison and selection. Its characteristic is that it combines the qualitative and the quantitative methods to solve the complex problems systematically and hierarchically. Firstly, according to the nature and the general objective of the problem, it is decomposed into different structures, namely the hierarchical problem. Then, according to the relationship and membership among the factors, the factors are combined in different levels, and finally a multi-level structural model is formed. The simple algorithm is simple and flexible to use. Since it was introduced into China in 1982, it has been widely used and valued in the logistics network planning, and the scientific research evaluation and so on, because of its characteristics of dealing with various decision-making factors qualitatively and quantitatively. The basic idea of the analytic hierarchy process (AHP) is to rank the attributes or the influencing factors of the research objects according to their importance, and then to determine the optimal schemes after the comprehensive analysis.

4.2 The Model Solving Steps in the Analytic Hierarchy Process

The analytic hierarchy process (AHP) decomposes the decision-making problems into different hierarchical structures according to the order of the general objectives, the sub-objectives at each level, and the evaluation criteria and the specific alternatives. Then, by solving the eigenvectors of the judgment matrix, the priority weights of each element at each level to a certain element at the upper level can be obtained, and the weighting method can be added at last. The final weight of each alternative to the overall objective is merged hierarchically, and the one with the largest final weight is the optimal one.

The steps of the analytic hierarchy process to solve the problems are as follows. Establish a multi-level progressive structural model for the various elements constituting the decision-making problem. Compare the elements of the same level (level) with those of the higher level as criteria in two ways, to determine their relative importance according to the evaluation criteria, and to establish a judgment matrix accordingly. Calculate and confirm it. The relative importance of each factor is determined, and the basis of the scientific decision-making is provided for the decision-makers by calculating the comprehensive importance.

4.3 Establish the Hierarchical Structure Model

When using the simple algorithm to solve the location problems of the e-commerce logistics network in the mobile network, firstly, by organizing and hierarchizing the problems, a structural model suitable for the hierarchical analysis is constructed. In this structural model, complex problems are decomposed into combinations of different elements. These elements are divided into several groups according to their attributes, forming the different levels. Elements of the same level as the criteria play a dominant role in some elements of the next level, while it is also dominated by the upper level. The hierarchical structure model is generally divided into three levels. The highest level is the target level, for the location of the distribution centers, and the decision-making goal is to select the best distribution centers. The middle level is the base level of the decision-making, and for the location of the distribution centers it is the main factors affecting the location, such as the social, political and economic factors analyzed above. The bottom level is the decision-making plan, and also the alternative address.

4.4 Construct the Judgment Matrix

Let a layer have n factors, X = {X1, X2, … Xn}. To compare their impact on a criterion (or the target) at the upper level, determine the proportion of the criterion in that level relative to a certain criterion (i.e., rank the influence degree of n factors on a daily target at the upper level). The simple algorithm uses the method of the pairwise comparison to get two elements Ai and Aj for the criterion CK, and the scales are used in the theory of the simple algorithms to assign the important values.

4.5 Consistency Check

The key of the analytic hierarchy process is to determine the weight of each index, so the method of the pairwise comparison is often used. However, it is unrealistic to fully satisfy the consistency of the pairwise comparison matrix. The average random consistency index is obtained by calculating the eigenvalues of the random judgment matrix repeatedly and calculating the arithmetic mean. RI is related to the order of the judgment matrix. The larger the general order is, the greater the likelihood of the consistent random deviation will be. When checking whether the judgment matrix has the satisfactory consistency, the consistency index CI must be compared with the average random consistency index RI, and the consistency ratio of the test number CR (Consistency Ratio) is obtained. When CR0.1 is used for the judgment matrices of more than three orders, the judgment matrices must be corrected appropriately to maintain a certain degree of consistency.

4.6 Select the Best Plan

With the steady development of China's economy, the e-commerce and the information technology have also made considerable progress, which have also greatly promoted the development of the logistics industry. In addition to the express industry, many enterprises have begun to develop their own logistics system. The logistics industry has become more and more important in our daily life. Enterprises in the logistics sector have become more and more important. The input in the logistics of enterprises is also increasing. As far as the express delivery industry and the e-commerce enterprises are concerned, their logistics system covers almost all the regions and cities in the country, and involves many levels from the top to the bottom. As the basic unit of the logistics system, the logistics outlets play a very important role. It is necessary to study the layout of the multi-level and the multi-dimensional logistics outlets. In order to save the resources and minimize the logistics costs, the research on the logistics network layout can also make the coverage of the transit warehouse and the logistics flow more reasonable and improve the logistics efficiency.

5 Conclusion

In this paper, the contents and characteristics of the e-commerce and the impact of the e-commerce on the logistics system are studied, and the idea of building the urban logistics distribution center system based on the e-commerce is put forward. By analyzing the simple algorithms, the layout of the distribution center under the e-commerce environment is optimized. By using a simple algorithm, the weights of each evaluation index can be calculated accurately, and then each index can be scored. By synthesizing the scores of each alternative point, the location with the highest score can be selected as the optimal scheme.

References

1. Tang Y, Lu C, Hou J (2016) Simple algorithm for calculating the relative permeability from the production data of five-point pattern. Pet Geol Oilfield Dev Daqing (02):75–76
2. Tang X, Yang L, Shi J, Chen L, Zhang J, Ji P (2015) Optimizing the distribution of the cold chain logistics system for the fresh agricultural products. Logist Technol (03):12–13
3. Guo Y, Hu Y, Liu J, Liang X, Liang W (2016) Study of the logistics network layout scheme of the rural E-commerce in Guangxi – taking Sanjiang County of Guangxi as an example. Mod Bus Trade Ind (05):89–90
4. Wang H, Li M (2018) Layout research of the automotive parts sales store based on anylogic. Software (09):21–22
5. Zhang H, Wan H, Quan H (2018) Simple algorithm for the distribution of the e-commerce logistics network based on the carbon footprint. Logist Technol (09):155–156
6. Zhou Q (2018) Multi-layer affective computing model based on emotional psychology. Electron Commer Res 18(1):109–124

Research on the Sports Biomechanics Modeling of the Human Motion Technical Movements

Jiafa Li(✉) and Huihui Du

Aba Teachers University, Wenchuan 623002, Sichuan, China
wangfei843684@126.com

Abstract. The motion display of the sports biomechanics is based on the image analysis of the sports biomechanics. It can organically integrate the complex human-computer dialogues into a system, and feedback quickly in the form of the three-dimensional animations. It can directly show the results of the image analysis to the athletes and the coaches, and directly serve the actual sports. Constructing the appropriate human body model and choosing the appropriate three-dimensional computer programming technology are the keys to the three-dimensional computer display of the biomechanics movement technology of the actual movement.

Keywords: Human motion · Technical action · Sports biology · Mechanical modeling

1 Analysis of the Development Status of the Sports Biomechanics

With the development of technologies, biomechanics uses the advanced motion capture technology to collect and analyze the human motion data, but these methods cannot obtain the muscle strength. In order to calculate and analyze the key muscle forces of the human motion, it is necessary to simulate and model the human motion. There are many methods of the simulation and modeling at present. This paper compares the current mainstream modeling software LifeMOD and OpenSim from the method of modeling, and reveals their respective characteristics, so that we can make the best choice in accordance to the actual need in the simulation and modeling researches in the future [1].

The functional anatomy and the dynamic anatomy played an important role in the early formation of the sports biomechanics theories. The basic idea of this subject is to regard our human body as an organic entity [2]. All the experimental and theoretical starting points correspond to the human body units. The research direction of this subject focuses on the relationship between the human body's morphological structures and the functional abilities, and tries to reveal the coordinated movement of the human body from the perspective of the "functional group". The working mechanism of this work is to study the working status of the bone joints and the muscle groups, and to establish their own theoretical systems [3].

Z. Xu et al. (Eds.): CSIA 2019, AISC 928, pp. 243–248, 2020.
https://doi.org/10.1007/978-3-030-15235-2_38

Another characteristic of the theoretical system of biomechanics is that the mathematics and the mechanics are the main parts. Because the research goal of this subject is to reveal the laws of the human mechanical movement in sports, in order to ensure the universality and logic, it needs the mature models in other disciplines to prove it [4]. The closest to its subject research direction is mechanics in physics, and the research on mechanics cannot be carried out without the support of mathematics, so the two disciplines become the means of the proof to study the laws of the human motion.

The sports biomechanics plays an important role in the sports teaching. It drives the principles of the sports teaching mechanics, strengthens the students' scientific and reasonable exercise, promotes the diversification of the sports teaching process and enriches the teaching contents [5]. The exercise biomechanics refines the movement in the process of the physical education, which makes the teaching contents of the physical education simple and easy to understand, and enhances the coordinated development of the students' physical shapes and qualities and their physical fitness [6–8]. Sports biomechanics can closely combine the anatomy, movement laws and the movement trajectory, which plays a decisive role in the process of the sports teaching.

Biomechanics is a combination of the principles and methods of biology and mechanics. It is a science that recognizes the laws (theorems) of the life process and uses them to maintain and improve the human health. At the same time, biomechanics applies the mechanical principles to medicine and biology to provide services for the theorization and accuracy of medicine and biology. Sports biomechanics is a subject which combines the principles of mechanics and biology with the laws of the human motion in sports. It mainly involves the analysis and diagnosis of the motor techniques, the prevention of the sports injuries, the exploration of the rules of the sports training, the design and improvement of the sports equipment and the adaptability of the nerves, muscles and bones to the sports.

Sports biomechanics, as an applied subject with the strong practicality, has attracted the attention and application of many P.E. teachers because of its concise and rigorous mechanical analysis of the sports technical movements. For example, as a golden key to the sports science, sports biomechanics is an important scientific means and method for us to carry out the sports scientific researches, and also a passport for the innovation of China's sports science and technology to enter the new century. It entrusts the viewpoints and methods of biology and mechanics to all kinds of the movement techniques in sports, makes the complex sports movement techniques based on the basic laws of biology and mechanics, and quantitatively describes them in the form of mathematics, biology and the principles of sports technology, so as to raise the sports technology from the qualitative analysis to the quantitative analysis.

2 Background of the Sports Biomechanics Modeling of the Human Motion Technical Movements

Aiming at the low accuracy of the human motion modeling in the current key frame motion data acquisition methods, a human motion modeling method based on the motion biomechanics is proposed. Firstly, the biomechanical data analysis and interpolation reconstruction of the human motion are carried out, and then the motion state equation is constructed to capture and characterize the key data of the human motion modeling. Finally, the simulation results show that the method improves the accuracy of the human motion analysis, and has good reconstruction ability for the human motion such as walking, jumping and rollover. The results are reasonable.

The human motion is a complex biomechanical system. The modeling and analysis of the human motion will have the important application value in guiding the sports training, the computer game development, the virtual reality simulation, and the film and television special effects performances and so on. The process of the human motion modeling is the process of the data analysis and extraction of the information of the human motion mechanics. By capturing the data of the human motion biomechanics, the feature compression, the information retrieval and reconstruction, the human motion process can be synthesized and edited to achieve the purpose of the human motion skeleton reconstruction and the human body researches. The motion modeling method is also of great significance in guiding the sports rehabilitation training.

In the traditional methods, the human motion modeling mainly uses the key frame information extraction method, combined with the motion image analysis, to achieve the human motion reconstruction. For example, using the curve simplification method, the human motion process is regarded as a key feature point tracking process of a motion trajectory curve. The bone segment curve of the human motion is a behavior trajectory in the high-dimensional space. The hierarchical curve method has achieved some results in the behavior reconstruction, but the traditional method needs to set the resolution threshold between the key frames. In the presence of the large disturbance of the motion characteristics, the error of the human motion reconstruction is large, and the effect of the human motion modeling is poor. Aiming at the low accuracy of the human motion modeling, a human motion modeling method based on the motion biomechanics is proposed.

The establishment of the hip biomechanics model is one of the most important tasks in the human body modeling of the sports biomechanics. There are many different opinions from the establishment of the human muscle mechanics model to the simplification of the muscles around the hip joint to the establishment of the final hip mechanics model. The main task of this project is to create a simple muscle model, to simplify the muscle around the hip joint to the greatest extent, and to establish and test the model reasonably. In the study of the lower limb mechanics model, the research of the hip joint mechanics model is less than that of the knee joint mechanics model. The previous models of the hip joint are mainly based on the physiological experiments. The parameters in the model have no clear mechanical significance. The determination of the parameters is related to physiological experiments, which brings great difficulties for the future research. Establishing a reasonable mechanical model of the human hip joint can provide some reference and convenience for the follow-up studies.

Exercise is one of the most basic functions of the human body. The motor system performs the human motion functions, including bones, joints (bone connection) and skeletal muscles. The human motion system is connected by all the bones and joints of the entire body to form a skeleton, which forms the human body's framework and endows the basic forms of the human body. The skeletal muscle adheres to the bone, contracting and relaxing under the control of the nervous system, changing the positions and angles of the distraction bones and generating the movement. Sports biomechanics is a mechanical problem related to the human motion. At present, the motion biomechanics experts can accurately record the three-dimensional motion trajectories of the markers pasted on the human body through the optical motion capture systems (such as VICON), the high-speed cameras and other advanced equipment, and analyze the motion characteristics of different links and joints in various forms of the human motion. According to the characteristics of different forms of the human motion, it is an important content of sports biomechanics to establish the corresponding mechanical model of the rigid body system, solve the corresponding mathematical equations and carry out the computer simulation researches.

3 Biomechanical Modeling Mechanisms of the Human Motion Techniques

3.1 Reconstruction of the Modeling Mechanics

In the process of the motion, the postures of the motion will change constantly. It is necessary to reconstruct the posture information mechanically by the human motion modeling. When the state of the human body is not accelerated, the information captured by the motion is expressed as follows: fi = 1, 2, …, n denotes the four-element vectors in the process of the human motion, and n denotes the sequential positions transmitted back through the sensors worn on the human body. In the formula, Dp (om. rm) is expressed as the position error of the human body relative to the world coordinate system in the course of the movement. Du (om.rm) is expressed as the difference between the joint speed, and u is expressed as the step length in the renewal. According to the above experiments, it can be concluded that the difference of the joint velocity occupies a very small proportion. In this paper, we set it as 1. The reconstructed biomechanics formula of the human motion modeling is obtained as follows. d (omi, rmi) is the parameter of the stability of the first frame of the original motion state under the dynamic acceleration motion. wi is expressed as the weighted value, which is the human motion. The weight of joint j in the stable state represents the importance of joint j.

3.2 Structural Definition of the Human Data Model

The data of the same structure are defined and stored, which is convenient for the biomechanical calculation. The establishment of the data structures needs to include the parameters such as the moment of the inertia and the center of the mass and so on. The joints that generate the motion in each link are clear. Foots, hands and the spine are the special organs of the motion in the human motion. Here, the right foot is described as an example, and its data structure is described and stored.

The hand joint motion is analyzed. From the second finger to the fifth finger, each joint has a degree of freedom, which is also called the bending movement. The base of the finger joint includes two degrees of freedom, such as the rotation and the bending. The thumb usually bends, and also carries on the abduction movement. The palm moves in the translation and rotation. There are 3 degrees of freedom in each movement and 27 degrees of freedom in this hand. The spinal motion is analyzed. The spine is composed of one sacrum, 24 vertebrae, one coccyx and 23 intervertebral discs, with the ligaments and the joints connecting them. The range of the spinal motion is limited, but its range of the motion is relatively large. The spine can flex and extend around the frontal axis, the circle and gyration around the vertical axis, and the lateral flexion around the sagittal axis.

The foot movement is analyzed. In the ankle joint, the main movement core of the foot is closely connected with the ankle joint, while the two feet are basically symmetrical. Foot is a skeletal 26 projects, and the foot movement can be divided into four parts: heel, toe, middle foot, and ankle joint. And each part has a degree of freedom. The center of the heel movement is the center rotation of the posterior inferior part of the ankle, and the center rotation of the lower part of the tibia, and the center rotation of the middle part of the foot is between the ankle and the toe tip.

In order to carry out the finite element analysis of biomechanics, the first problem to be solved is the model problem, but this is a difficult problem. There are many kinds of the current modeling software, but because of the complexity of the anatomical structures and the arbitrariness of the surface, the current top-end CAD software cannot design the anatomical structure model that meets the needs. Another method may be to use the traditional reverse engineering method, which is to scan the real anatomical structures, and then use the reverse software for the three-dimensional reconstruction. However, it is unavoidable that the model is inaccurate because only the external contour of the anatomical structures can be obtained by this method, while the internal structures cannot be helped. Furthermore, for the organizations with the complex structures, there is no way to reconstruct them by using the reverse methods. Because the finite element model cannot be obtained, the biomechanical finite element analysis of the anatomical structures cannot be carried out well for a long time.

At present, the main interfaces provided by Mimics are patran, nastran, abaqus, fluent and ansys. Some people may have the idea that these finite element software can be read into the Mimics output file to carry out the finite element analysis. This is a wrong idea. Firstly, Mimics' three-dimensional reconstruction model is a surface mesh model, so it cannot be read into the finite element software for the direct finite element analysis. On the contrary, the Mimics output surface model is read into the finite element software mainly to do the volume mesh generation.

4 Conclusion

The simulation results show that the accuracy of the human motion analysis is improved, and the reconstruction ability of the human motion such as walking, jumping and rollover is good. By capturing the biomechanical data of the human motion, the feature compression, the information retrieval and reconstruction are carried out to realize the synthesis and editing of the human motion process and achieve the goal of the human motion reconstruction.

References

1. Hong Q (2016) The influence of the speed of the middle and long distance running on the performances under the perspective of the sports biomechanics. J Hunan City Coll (Nat Sci Ed) (03):75–76
2. Zhao F (2017) Biomechanical characteristics of the knee joint muscles of sprinters. J Weifang Univ (12):17–18
3. Fang L, Yu S (2018) Design and experiment of the shoulder biomechanical simulator in vitro. J Donghua Univ (Nat Sci Ed) (08):56–57
4. Wu W, Zhang T, Chen L, Deng Z, Yan Z (2018) The biomechanical effects of the vertebral tunnel technology on the upper endplate of C4 vertebral body were studied using C2-T1 finite element model. J Chongqing Med Univ (09):129–130
5. Bao C, Meng Q, Yan M (2018) Analysis of the biomechanical characteristics of the ankle joint in the basketball players' quick stop and take-off. Chin J Appl Mech (11): 131–132
6. Zhou Q, Lou J, Xie F, Liu Q (2011) The method research on optimum selection of agricultural machinery. Int J Digit Content Technol Appl 5(6):337–338
7. Xie J, Luo J, Zhou Q (2016) Data mining based quality analysis on informants involved applied research. Clust Comput 19(4):1885–1893
8. Zhou Q (2018) Multi-layer affective computing model based on emotional psychology. Electron Commer Res 18(1):109–124

Research on the Sports Teaching System Based on the Data Visualization

Zhanke Ma(✉)

Institute of Physical Education, Ningxia Normal University, Guyuan 756000, Ningxia, China
zhiyu_zhang@126.com

Abstract. A very important factor to judge whether a country is strong or not is to see whether its sports are strong or not. Compared with other countries, some of China's sports teaching activities are strong while others are weak. With the development of our science and technology, the development of the sports teaching activities should also keep pace with the times. Whether the visualization of the sports data can help the sports teaching activities achieve the desired results is worth watching. The data and information are also the basic concepts of the system engineering, and the information theory is one of the theoretical foundations of the system engineering. Therefore, the research on the development and application of the data and information in our teaching activities is an important part of the innovative research on the new teaching activities.

Keywords: Data visualization · Physical education · System construction · Modernization · Big data

The sports teaching environment has an important impact on the sports teaching effect, so it is necessary to set up a good sports teaching environment to provide conditions for the realization of the sports teaching objectives. In the past, scholars mostly focused on the teaching methods and the teaching models in our sports teaching, and less on the related researches on the sports teaching environment [1]. Therefore, it is necessary to adopt a new research method to conduct the comprehensive researches on the environment of the physical education.

1 Dependence Between the Data and the Information

From the big data structural models, we can see the relationship between the information (the information platform) and the data (the database). The objective world consists of three elements: materials, energy and information. The data and information is an objective existence, which is neither the material nor the energy. It is necessary to study the relationship and interaction between the data and the information, between the elements, between the locals, between the local and the global, between the global and the system, and between the system and the environment, all of which are realized by exchanging, processing and applying the information platforms and the databases [2]. The evolution of the systems, the generation of the overall characteristics, and the

Z. Xu et al. (Eds.): CSIA 2019, AISC 928, pp. 249–254, 2020.
https://doi.org/10.1007/978-3-030-15235-2_39

emergence of the high-level artificial intelligence applications need to be understood from the point of view of the data and the information.

The big data can be divided into the broad sense and the narrow sense. The big data in the broad sense refers to the massive and complex data sets that cannot be extracted, stored, searched, shared, analyzed and processed by the existing software tools. In a narrow sense, the big data refers to the information with experience, knowledge, intelligence and values acquired from the massive and complex data after collection, extraction, mining, processing and analysis. The existence of the information does not transfer to the existence of the subject (such as people, things and equipment). Even if the subject does not exist, the information can exist. The information objectively reflects the reality of an objective thing. The information can be the data, records, documents, works and images in the form of material carriers, or the ideological concepts, ideas, religions, beliefs and visions and so on [3]. The information has nothing to do with the four-dimensional space, and the information can transcend the constraints of the distance, space and time, and can exist independently and permanently. In a sense, the information belongs to the "fifth dimensional space", that is, the concept of the "virtual space". The information can be accepted and utilized subjectively, and will affect people's thinking and guide people's actions. Information, unlike other resources, is not consumed as the material resources, such as water, coal and electricity and so on [4]. Therefore, information is an inexhaustible resource.

The data and the information are two interdependent basic elements in the information systems. The data is a record symbol produced in all activities. The data can be processed and combined to express the characteristics of the real objective world (such as objects, attributes, events, states and activities). The information is the processed data, and is the interpretation and meaning of the data, and the data is the raw material and carrier of the information [5]. Therefore, the data and the information are independent and closely related.

2 Connotation and the Related Work of the Data Visualization

There are many different definitions of the data visualization. Broadly speaking, the data visualization includes more mature scientific visualization and the younger information visualization. In addition, it also includes the knowledge visualization developed on this basis. In a narrow sense, the data visualization refers to the use of the computer graphics knowledge and the image processing technology, the interactive graphics and the images to show the theories, methods and technology of the data [6, 7]. Its visualization object is only the narrowly understood data. Combining with the self-developed cloud classroom teaching platform, the following educational and teaching data are visualized: the construction of curriculums and classrooms, the construction of resources in our teaching, the participation of the teachers in the cloud classrooms and the students' online learning behavioral data.

The data visualization classification is a key link in the visualization research. The classification of the data visualization can guide people to design a reasonable visualization scheme, and promote the visualization technology from the theoretical research to the practical application stage. The data visualization can be divided into four types: the scientific computing visualization, the data visualization, the information visualization and the knowledge visualization.

The geometry-based data visualization technology realizes the visualization of the multi-dimensional data through projection and transformation in geometry. The standard 2D/3D technology is one of the most widely used visualization technologies in the process of the data visualization. The commonly used visualization graphics are the bar graph, the pie graph and the geometric projection technique and so on. This paper classifies it as one of the geometric techniques. There are two kinds of the geometry technology: the geometric transformation and the geometric projection. The geometric transformation technology is a mapping transformation technology to represent the multi-dimensional data with the two-dimensional graphics points, lines and surfaces and so on. The geometric projection technology is to represent the data to be visualized by means of the geometric projection.

3 Pressure Analysis of the Sports Teaching System Based on the Data Visualization

In order to promote the development of our physical education, this paper combines the physical education with the cite space, carries out a visual analysis of the physical education, analyzes the spatial distribution of the physical education research (cooperative institutions, the amount of articles, and the author's cooperation), as well as the key research points and frontiers, and sorts out and analyzes and summarizes the development of the physical education. Through the cite space software, we can make a systematic study of the physical education, and show the development process of the physical education in our country in the past 30 years through the data atlas, so that researchers can have a clearer understanding of the development process of the physical education in our country. Through summing up the key points of the physical education in various stages, this paper aims to find out the teaching systems suitable for the development of the physical education in our country.

Data are produced every day and exist objectively. It is extremely difficult for an individual to play a role. First of all, in terms of the current large amount of data, the collection rate is very low, and it is difficult to carry out an objective and effective collection. Second, the individual data and other education data, the social behavior and other data can play a role in China, and only a small number of the famous schools in Peking University and Tsinghua University have established the data center, but there is a lack of the effective management, thus reducing the benefits. Therefore, in the process of collecting, managing and using the data, we should conduct the relevant guidance and supervision.

The development of the big data brings convenience to people, but there are also some problems. The most important thing is the leak of the college students' personal privacy, and especially the leak of the personal information, and the family information and so on, which brings great anxiety to the students. If the database established by the schools lack the good network maintenance, it will make many lawless elements have the opportunity to steal the students' information for illegal activities, to bring economic and life problems to the students. For example, some time ago in Linyi of Shandong Province, the quasi-university students who were cheated of his university tuition suddenly died of heart disease, and such a shocking example has to be reflected.

The establishment and development of the big data is accompanied by the continuous development of the information technology. The massive information of the big data needs the corresponding professional personnel to maintain and operate. However, there is a shortage of the talents related to the big data in China. Only a few professions have opened this major, including Peking University, Fudan University and Zhongnan University. In the past two years, the professionals have been trained. Because of the lack of the professionals in the professional field, the difficulty of using the big data in China is increasing.

We should innovate the traditional teaching methods, so that we can play the role of the big data. The most important thing is to enrich the teaching and organizational forms of the physical education. For example, in the traditional classroom learning, teachers often use the teaching-demonstration learning method, and the students acquire knowledge from the teachers. But in the context of the big data, teachers should learn to understand the significance of using the big data processing, change the ways of our teaching, and establish an effective open PE classroom, so that students can obtain and learn the resources through a variety of ways.

4 The Sports Teaching System Based on the Data Visualization

The sports data visualization is an important technology in the sports news and the sports competitions. The sports data can be divided into the one-dimensional statistical data of sports attributes, the multi-dimensional data of sports attributes and the space-time attributes. There are mainly sports data news, sports data professional analysis and visual analysis system. This paper summarizes and concludes the basic methods used in the sports data visualization. From the perspective of the data space-time, there is the visualization based on the technical statistical data, the visualization with the combination of the technical statistical data and the spatial data, the visualization with the combination of the technical statistical data and the temporal data, and the combination of the technical statistical data and the temporal and spatial data. From the point of view of the players, there is the visualization of the individual players and the visualization of the multiple players.

The low-dimensional data contains the one-dimensional linear data, the two-dimensional data with two attributes and the three-dimensional data with the three-dimensional attributes. The one-dimensional data can be divided into two types: the quantitative and the non quantitative. In the quantitative type, the point mapping on the directed line segment is used, while the non-quantitative type is generally displayed with the color blocks. For the visualization of the two-dimensional data, this paper divides it into two kinds according to the different visual channels: the location-based method and the color and area based method. The location-based methods map the two dimensions of the exponential data to the positions to represent their differences. The typical statistical graphs include histograms, polygons, and simple scatter plots. The color and area based methods refer to the mapping of the one dimension of the data to colors and the other to areas. These methods include the pie charts and the Blue Tinger Rose charts.

The high-dimensional data has many independent attributes. How to present the multiple attributes in the two-dimensional space is a difficult problem for the high-dimensional data visualization. The visualization of the high-dimensional data needs to be dimensionalized first. There are many kinds of the dimensionality reduction methods, which can be divided into two categories: the linear and nonlinear. The linear methods include MDS (multi-scale analysis), NMF (non-negative matrix decomposition) and PCA (principal component analysis). The nonlinear methods include LLE (local linear nesting) and ISOMAP (isometric feature mapping). For the high-dimensional data, the platform adopts two kinds of the data visualization methods: the geometry-based method and the icon-based method. The geometric graphics-based methods mainly include the scatter matrix, the parallel coordinates and the parallel set visualization. The icon-based methods display the high-dimensional data by having a number of easily recognizable feature icons, each feature mapping a dimension. The more famous ones are the Chernoff Faces and the Radar Chart.

The multilevel data are essentially the tree structured data, and there are both the juxtaposition and the subordination among the nodes. The non-root node has only one parent node, and the parent node has one-to-many relationship with its children. The relationship between the nodes can be divided into the brotherhood and the father-son relationship. The hierarchical data visualization methods implemented in this teaching platform mainly includes the node linking method and the space filling method. The method of the node linking uses the lines to represent the parent-child relationship between the nodes, and points to represent each data entity, and recursively divides the rectangles in a nested way. Each rectangular area represents a data entity, and the sub-region of the rectangle represents its children. The relational data, unlike the hierarchical data, does not have a fixed top-down or bottom-up hierarchical relationship, reflecting the direct or the indirect relationship between the data, and the expression is more freely. The visualization of the relational data is the visualization of the graphs. The platform is based on the edge and the node visualization. In order to avoid the overlapping of the large numbers of nodes, the force guide placement method is used.

This series of the data visualization tutorials comprehensively introduces the significance and practical application of the data visualization, from the popularization of the data visualization science to the actual scene in which the data visualization is used with different difficulty, to the study of the excellent cases in the visualization and the sharing of the visual learning methods, so that the audience and users interested in data visualization can have quick access to and understand the beauty of the data visualization. Based on the magic mirror, the first big data visualization analysis tool in China, this course introduces the usage of various conventional and advanced visualization graphics.

5 Conclusion

The data and the information have become the symbols of today's society, and the Internet+, the big data, the cloud computing, the Internet of Things, the artificial intelligence and other new technologies are based on the data and information technology and the information applications. The data and the information knowledge have become the first driving force of the social development, and the data and information in our social production and life has played an irreplaceable role, and will continue to expand the connotation and extension of the information. The information technology applications have penetrated into the development of our government affairs, the social and people's livelihood and other industries in all aspects. The information platform and the database system will play an immeasurable role in the process of the comprehensive construction of the sports teaching system.

References

1. Fan J (2016) Application of the big data in our physical education. Phys Educ (14):181–183
2. Liu G, Liu H (2016) On the influence of the big data age on the physical education and teaching research. J Yangtze Norm Univ (4):121–122
3. Shi Q, Yu N (2015) The influence of the big data technology on the physical education in primary and secondary schools. Contemp Sport Technol (26):245–246
4. Xie C (2016) Research and thinking on college physical education classroom teaching in the age of the big data. Course Educ Res (35):106–107
5. Zhang R (2018) Playing the games according to the data – talking about the roles of the visual design in sports. Art Panor (05):154–155
6. Bao W, Zhou H, Wang L, Xie F (2016) The system of knowledge management using web based learning and education technology. Comput Syst Sci Eng 31(6):469–473
7. Bao W, Zhou H, Xie F (2017) An optimal design of web based online English teaching by using struts framework. Bol Tec/Tech Bull 55(10):58–63

Reverse Digital Repair and Evaluation Analysis of Injection Mold Face Defects

Jun Fan[1,2](✉)

[1] School of Mechanical and Electronic Engineering, Suqian College, Suqian 223800, China
1979888068@qq.com
[2] Precision Inspection and Reverse Engineering Technology Research Center of Suqian City, Suqian 223800, China

Abstract. It is limited to the repair and the manufacture of the injection mold contours, ignoring digital repair technology and sample evaluation technology. For the first time, this paper combines parameter repair technology, virtual evaluation technology, rapid prototyping technology and sample evaluation technology. In this paper, aiming at the surface defects of injection mold, the mold surface defect model is reconstructed reversely, of which, the high-precision and rapid reverse modeling method is explored. Quantitative analysis of the feature data of surface defects, and parametric repair, which quantitative technology is studied, and the defect area of the mold surface are done. By means of mold face analysis technology and rapid prototyping technology, digitized evaluation and materialized assessment of the repair effect are carried out respectively, to realize the comprehensive evaluation and finalization of the mold sample for defects repair. The new comprehensive technology means has important practical significance and application value.

Keywords: Reverse engineering · Injection mold · Face defects · Digital repair · Precision control

1 Introduction

At present, there are more researches on reverse repair technology of molds [1–8], less research on reverse repair technology of mold surface [9–11], and less research on reverse repair technology of injection mold [12, 13]. As the core feature of the mold process structure, the mold surface quality directly determines the quality of mass production products. In this paper, the digital repair technology of mold surface defect based on reverse engineering is studied, the method of high precision reverse modeling is explored, the technology of quantitative parametric repair of mold surface defect is studied, and the quick comprehensive evaluation of the repair effect of mold surface defect is realized, which provides a kind of high precision and rapid repair for mold surface defect.

Z. Xu et al. (Eds.): CSIA 2019, AISC 928, pp. 255–264, 2020.
https://doi.org/10.1007/978-3-030-15235-2_40

2 Model Reverse Reconstruction

The process of model reverse reconstruction mainly consists of four stages, namely, sample entity analysis stage, point cloud data acquisition stage, model digital reconstruction stage and model precision analysis stage.

2.1 Sample Entity Analysis

Injection mold sample, as shown in Fig. 1. The region I is a curve defect to be repaired, the region II is a surface defect to be repaired, and the region III is a structure defect to be repaired.

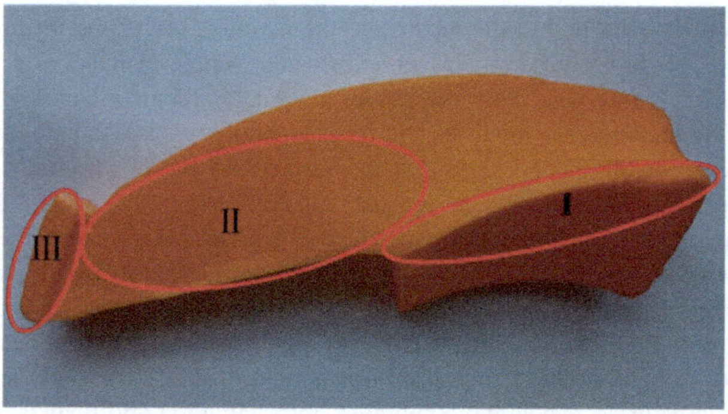

Fig. 1. Injection mold sample

2.2 Point Cloud Data Acquisition

Data Acquisition Equipment
AUTOS 3D structure optical scanner is shown in Fig. 2.

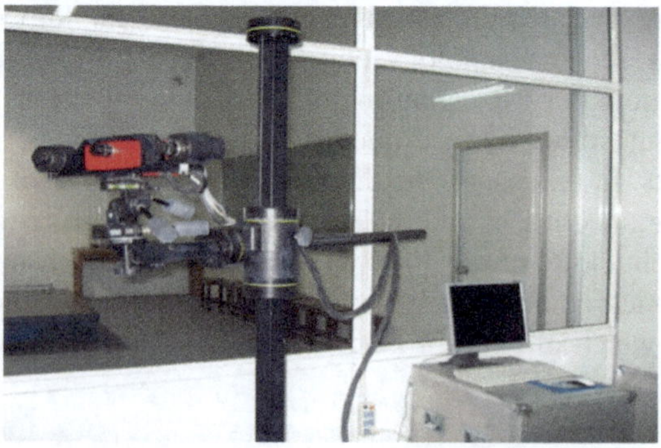

Fig. 2. AUTOS 3D structure optical scanner

Sample Data Acquisition

The point cloud of the injection mold to be repaired is shown in Fig. 3. The data acquisition and the data preprocessing of point cloud are carried out.

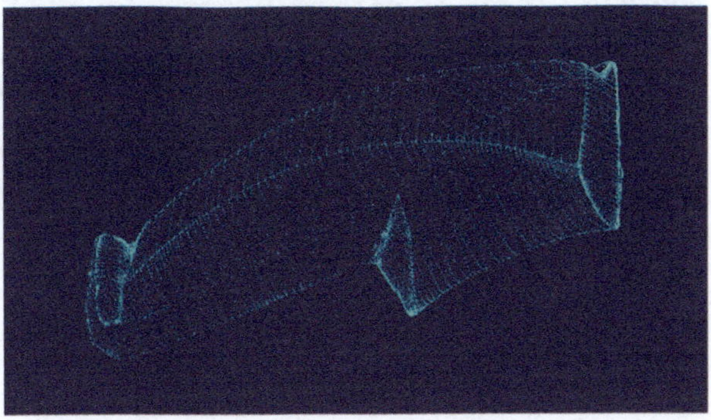

Fig. 3. The injection mold point cloud

2.3 Model Digital Reconstruction

Combined with the point cloud data of the injection mold to be repaired, the model of the injection mold is reconstructed, as shown in Fig. 4.

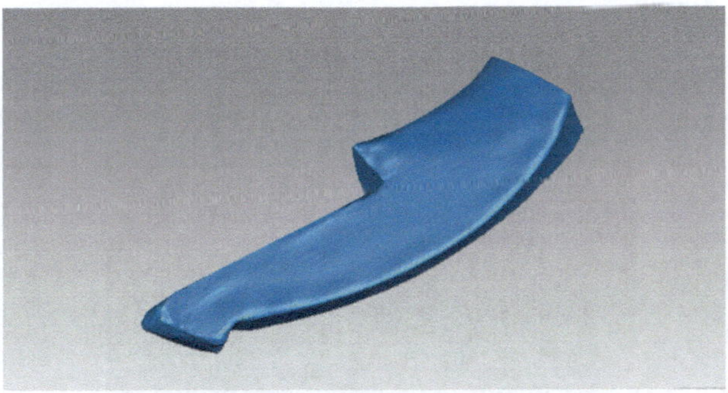

Fig. 4. The model digital reconstruction

2.4 Model Accuracy Analysis

The precision of the digital model reconstruction of the injection mold is analyzed, as shown in Fig. 5, which accuracy is controlled at 0.1 mm overall.

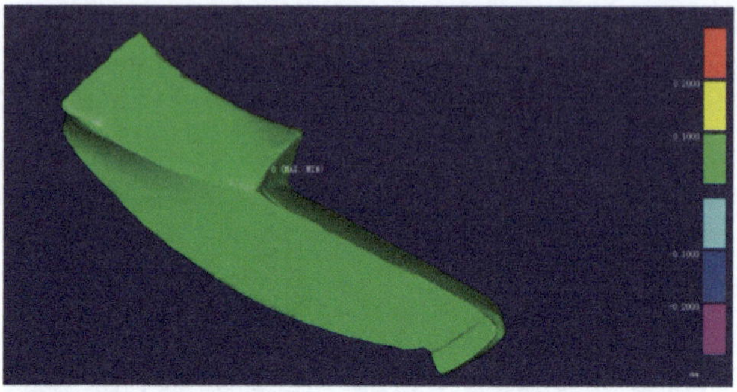

Fig. 5. The model accuracy analysis

3 Profile Defect Analysis

The profile defect analysis process is mainly composed of three stages. They are the profile curve defect analysis stage, the model profile defect analysis stage and the sample profile defect analysis stage.

3.1 Profile Curve Defect Analysis

The defects position of curve 1 in the injection mold area was analyzed, as shown in Fig. 6. There is one defect in the curve 1, that is, the circular transition is incomplete.

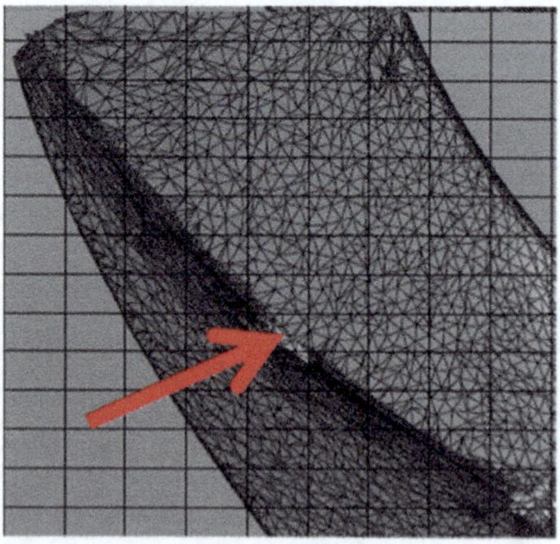

Fig. 6. The defects position of curve 1

3.2 Model Profile Defect Analysis

The mold surface point cloud is subjected to data processing and reverse reconstruction. In the reverse modeling of injection molding products, there is no smooth transition in the partial mold surface. The mold surface profile is uneven, has collapsed.

3.3 Sample Profile Defect Analysis

The overall profile defect and the local profile defect of the injection mold sample are completely consistent with the integral profile defect and the local profile defect in the reverse reconstruction model.

4 Model Digital Repair

The process of model digital repair is mainly composed of three stages. They are the point cloud shape digital repair stage, the solid modeling digital repair stage and the model repair accuracy analysis stage.

4.1 Point Cloud Shape Digital Repair

The point cloud digital morphology comparison analysis of the injection mold surface defect is shown in Fig. 7. The white phantom is the defect profile of the injection mold, and the blue outline is the digital shape profile constructed according to the white phantom.

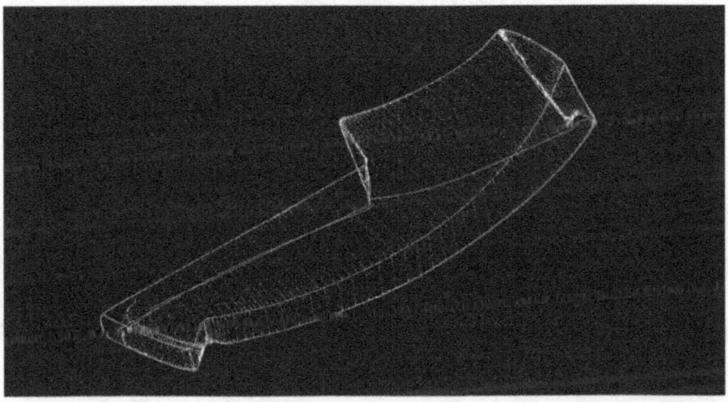

Fig. 7. Comparative analysis of point cloud digital forms

4.2 Solid Modeling Digital Repair

The digital reverse reconstruction model of the point cloud after the injection mold surface repair is carried out. The injection mold surface repairs the surface effect, as shown in Fig. 8. On the basis of the surface repairing effect of the injection mold surface of Fig. 8, the reverse reconstruction model of the digital repair of the mold surface is shown in Fig. 9. With reference to Fig. 4, the overall profile defect and the local profile defect of the injection mold sample were repaired to some extent.

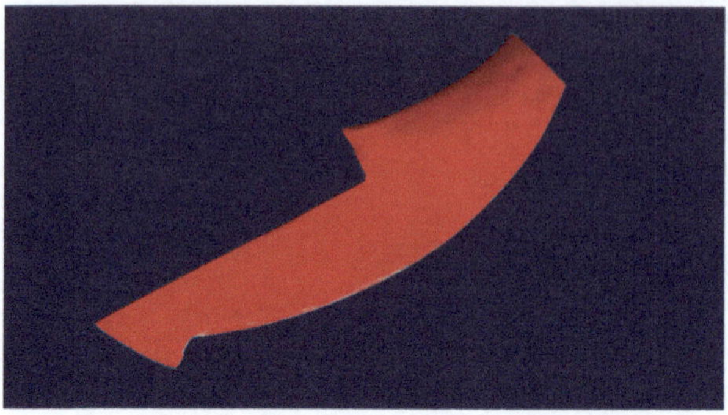

Fig. 8. Injection mold surface repair effect

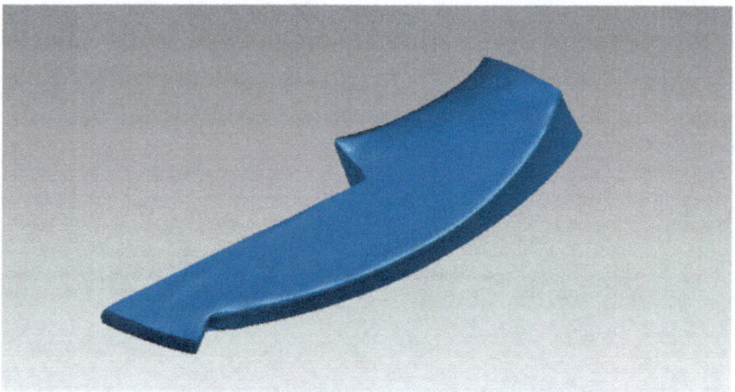

Fig. 9. Reverse modeling of the surface digital repair

4.3 Model Repair Accuracy Analysis

The accuracy analysis of the mold profile repair is shown in Fig. 10. The accuracy of the mold profile after repair is controlled at 0.1 mm.

Fig. 10. Accuracy analysis after mold surface repair

5 Comprehensive Evaluation Analysis

The comprehensive evaluation analysis process mainly includes five stages. They are profile curve evaluation analysis, model profile evaluation analysis, repair accuracy evaluation analysis, process effect evaluation analysis, and sample profile evaluation analysis, respectively.

5.1 Profile Curve Evaluation Analysis

Optimization effect analysis of mold surface defect curve 1, wherein, the repair of profile curve 1, as shown in Fig. 11. The curve 1 of mold surface defect region has been repaired to some extent.

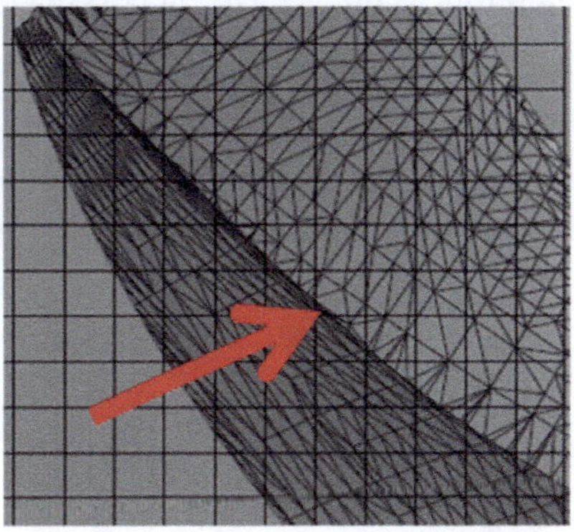

Fig. 11. The repair of profile curve 1

5.2 Model Surface Evaluation Analysis

The defect repair model in Fig. 9 shows a smoother transition of the local die surface than those in Fig. 4.

5.3 Repair Accuracy Evaluation Analysis

The precision of the model before and after the defect repair was analyzed. The model accuracy analysis table before and after repair of mold surface defects is drawn, as shown in Table 1. The average precision of the model after repair is 0.081 mm, the average precision before repair is 0.086 mm, the average precision after repair and the average precision before repair are all higher than 0.1 mm, which is the basic precision requirement of reverse development. Both the model after defect repairing and the model before defect repairing meet the precision requirement of reverse development. Compared with the model before defect repairing, the defect of mold surface has been repaired to some extent.

Table 1. The model accuracy analysis table

	Positive limit (mm)	Negative limit (mm)	Variation range (mm)	Average value (mm)
Before repair	0.078	−0.083	0.161	0.081
After repair	0.086	−0.087	0.173	0.086

5.4 Process Effect Evaluation Analysis

The surface process fitting optimization of the model before and after the repair of the mold surface was carried out precisely. A process deviation analysis table before and after the repair of the mold profile is drawn. The process deviation analysis table before and after the mold profile repair is shown in Table 2. The average deviation of the model before repair is 0.1016 mm, the average deviation after repair is 0.0980 mm, and the average deviation before repair is higher than that after repair. The above data show that the process deviation of the model after repair is better than that before repair, and the model after repair is more suitable for the use of injection mold.

Table 2. Process deviation analysis table

	Mean deviation range (mm)	Mean deviation (mm)
Before repair	0.2031	0.1016
After repair	0.1963	0.0980

5.5 Sample Surface Evaluation Analysis

Mold Repair Sample

The overall profile defect repair of the mold sample is shown in Fig. 12. The overall profile defect and the local profile defect of the injection mold sample have been repaired to some extent.

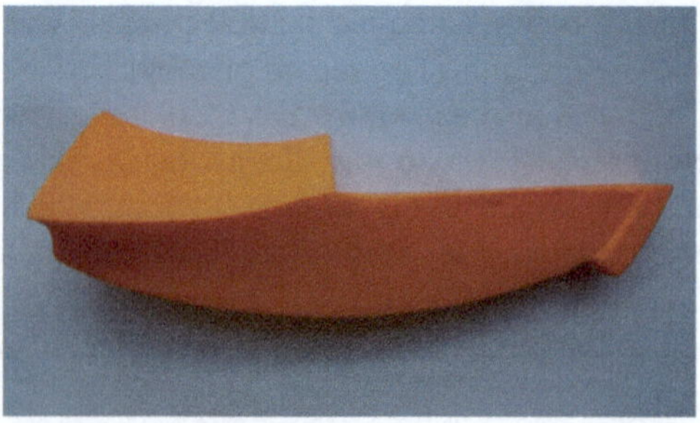

Fig. 12. The overall profile defect repair of the mold sample

Repair Surface Assessment
In Fig. 12, the defect of mold surface curve became more continuous. The mold defects in the area are smoothed and repaired, and the process becomes more accurate. The mold face structure defects were repaired and the structure became more complete. In summary, compared with Figs. 1 and 12, both integral and local surface defects of the injection mold sample have been repaired to a certain extent.

6 Conclusions

This paper provides a new comprehensive technical method for high-precision and rapid repair of mold surface defects. The implementation of the method can closely combine the repair work of mold profile defects with reverse engineering technology, precision measurement technology, computer aided technology and rapid manufacturing technology to improve the accuracy and efficiency of mold repair, greatly reduce the cost of the mold and extend the service life of the mold.

Acknowledgment. This work is supported by Qing Lan Project for Jiangsu Provincial University training target funding of excellent young backbone teachers (No. 2016), Engineering Technology Research Center Construction Project for 2016 Provincial Enterprise of Jiangsu Province (No. BM2016285).

References

1. Zhang J, Rao X, Fan L (2011) Study and application on mould repairing based on reverse engineering. Modul Mach Tool Autom Manuf Tech (7):76–79 (in Chinese)
2. Li J, Zhang D, Yao F, et al (2014) 3D remodeling of mould's broken area based on reverse engineering. Modul Mach Tool Autom Manuf Tech (9):33–34 (in Chinese)
3. Urbanic RJ (2015) A design and inspection based methodology for form-function reverse engineering of mechanical components. Int J Adv Manuf Technol 81:1539–1562
4. Wang H, Zhou J, Zhao T et al (2016) Springback compensation of automotive panel based on three-dimensional scanning and reverse engineering. Int J Adv Manuf Technol 85:1187–1193
5. Xú S, Anwer N, Mehdi-Souzani C et al (2016) STEP-NC based reverse engineering of in-process model of NC simulation. Int J Adv Manuf Technol 86:3267–3288
6. Zhang X, Li W, Adkison KM et al (2018) Damage reconstruction from tri-dexel data for laser-aided repairing of metallic components. Int J Adv Manuf Technol 96:3377–3390
7. Zhang X, Li W, Chen X et al (2018) Evaluation of component repair using direct metal deposition from scanned data. Int J Adv Manuf Technol 95:3335–3348
8. Cai P (2018) Based on reverse engineering of automobile covering parts mould repairing. Mod Manuf Eng (6):153–156 (in Chinese)
9. Liang B, Ma L, Li Y (2014) Application of the reverse engineering technology in the repair of the pump front sheath mould. Die Mould Technol (5):45–46 (in Chinese)
10. Li Y, Lv Y, Ni M (2017) Reverse design method of blade die surface. Forg Stamp Technol 42(9):140–145 (in Chinese)

11. Fan J, Wang L (2013) Precision analysis of parametric modification of the die face based on reverse engineering. Appl Mech Mater 385–386:121–124
12. Matsumoto M, Yang S, Martinsen K et al (2016) Trends and research challenges in remanufacturing. Int J Precis Eng Manuf-Green Technol 3(1):129–142
13. Jhavar S, Paul CP, Jain NK (2016) Micro-plasma transferred arc additive manufacturing for die and mold surface remanufacturing. Miner Metals Mater Soc 68(7):1801–1809

Sedimentary Facies and Paleoenvironment Significance of Hole BZK01 in the Yangtze River Delta Plain Since Pliocene Based on Big Data Analytics

Zhongyang Lin[1,3], Xianglong Jin[1,2(✉)], Zhaohui Zhu[3], Shaojun Pan[3], and Xudong Zhao[3]

[1] Faculty of Earth Resources, China University of Geosciences, Wuhan, China
zhihuo705253@163.com

[2] Key Laboratory of Submarine Geosciences, Second Institute of Oceanography, SOA, Hangzhou, China

[3] Institute of Geology and Mineral Resources, Zhejiang Institute of Geological Survey, Hangzhou, China

Abstract. The analysis of the sediment ^{14}C dating, grain sizes, sporopollen and micropaleontology in the BZK01 borehole in the Yangtze River Delta Plain was carried out. The sporopollen assemblages, the sedimentary environment evolution, the sedimentary facies and the paleoclimate characteristics revealed by the borehole since the Pliocene were discussed in combination with the gamma logging and the lithologic stratigraphy, and the quaternary multiple stratigraphic division of the borehole was established. The results show that the study area has undergone many environmental changes since Pliocene due to the influence of the primitive topography, the sea level changes, the land-sea changes, the paleoclimate changes and the Neotectonic movement. From bottom to top, the drilling holes reveal that the lithology and the strata are bedrock Changhe Formation, the Neogene-Quaternary sedimentary Jiaxing Formation, the Qiangang Formation, the Dongpu Formation, the Ningbo Formation and the Zhenhai Formation in turn. The sporopollen assemblage can be divided into 16 sporopollen belts. According to the periodic variation of the paleoclimate cycle during the warm, the humid and the cold and the dry periods, BZK01 drilling reflects that in the total area there are seven paleoclimatic cycles. The multi-stratigraphic division and correlation of the boreholes revealed that 151.9–278.0 m is the Pliocene-Early Pleistocene period of the alternating fluvial and lacustrine deposits, and there are two climatic cycles; 83.5–151.9 m is the Middle Pleistocene of the fluvial-lacustrine-shallow lacustrine depositional environment, which includes two climatic cycles. 15.4–83.5 m is the Late Pleistocene and sinks. From bottom to top, the sedimentary environment is the estuarine-fluvial-tidal-flat-littoral-fluvial facies, corresponding to Wangdian-Hangzhou transgression period, including two climatic cycles. 0–15.4 m is Holocene of the estuarine-fluvial-lagoon facies sedimentary environment, corresponding to Fuyang transgression, including one climatic cycle.

Keywords: Yangtze River Delta Plain · Quaternary · Sedimentary facies · Paleoenvironment · Paleoclimate

Z. Xu et al. (Eds.): CSIA 2019, AISC 928, pp. 265–276, 2020.
https://doi.org/10.1007/978-3-030-15235-2_41

1 Introduction

The Yangtze River Delta is an alluvial plain before the Yangtze River enters the sea. During its development and evolution since Quaternary, it has been affected by the neotectonic movement, the paleoclimate and the sea level fluctuation, as well as the supply of the terrigenous materials. As a result, the Quaternary strata and the sedimentary environment have changed frequently and the abundant information of the regional sedimentary environment evolution has been recorded [1, 2]. The climate and the sea-level changes in the Yangtze River Delta since the last glacial period have attracted much attention from scholars. Some hot issues include the number and extent of the transgression, the rate of the sea-level changes since the last glacial melting and other events [3–7]. The granularity, the sporopollen and the paleontological analysis methods have been widely used in the Paleoenvironment research. It is an effective and reliable method to judge the transgressive layer by using the presence, abundance and assemblage characteristics of the foraminifera and the ostracoda. A lot of work has been done by predecessors in the Yangtze River Delta Plain and many important achievements have been achieved [8–15], focusing on the Yangtze River. Palaeoenvironments of the Delta and its adjacent areas during the Holocene period have been studied in depth [16–18], but the related studies since the Pliocene, and especially those exposed to the bedrock, are relatively weak, which may be due to the limitations of the dating resolution accuracy and the borehole sample collection.

The BZK01 borehole selected in this paper is located in the southern flank of the Yangtze River Delta Plain. It has a complete and continuous Quaternary stratigraphic sedimentary sequence since Pliocene. It can well represent the Quaternary transgression and the paleoclimate changes in the Yangtze River Delta Plain area [19–21]. Based on the systematic analysis of the sporopollen, foraminifera and Ostracoda fossils in the sediments of BKZ01 borehole, and the control of the sedimentary facies division and the establishment of the chronological framework by combining the grain size and the dating data, the authors studied the evolution characteristics of the paleovegetation, the paleoclimate and the sedimentary environment in the plain area of the Yangtze River Delta, in order to provide evidence for the stratigraphic division and correlation of the Quaternary period in this area, which further enriches the Quaternary stratigraphic data of the Taihu Basin in the Yangtze River Delta Plain since Pliocene.

2 Sample Collection and Research Methods

Borehole BKZ01 (30° 55′36 N, 120° 35′16 E) is located in Nantangcun, Shengze Town, Wujiang District, Suzhou City, Jiangsu Province (Fig. 1), west of Taihu Lake and south of Hangjiahu Plain. It belongs to the Yangtze River Delta region where transgression and regression have occurred many times, and is sensitive to the climate changes. The drilling depth is 284.0 m, of which the bedrock is below 278.0 m. After the cores are cut longitudinally from the center, trimmed and lithologically described, the sediment dating, the grain size content, the sporopollen content analysis and the biological identification (foraminifera and ostracoda) samples are collected. The sampling interval is generally 1 m, and the sampling depth of the particle size is 0–278.0 m, and the sampling depth of the sporopollen, foraminifera and Ostracoda is 1.0–227.0 m.

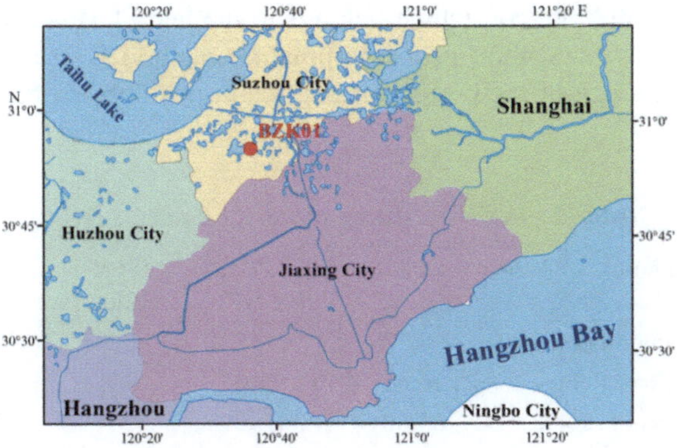

Fig. 1. BZK01 borehole location map

The AMS^{14}C dating of three sediment samples containing the carbon debris was carried out by Guangzhou Institute of Geochemistry, Chinese Academy of Sciences, using the sealed packages. The dating methods and the corrected age calculation can be seen in the references [22, 23].

That is to say, 50 g samples were taken from each sample, and after the acid-base treatment and the heavy liquid flotation, 264 samples containing spores were taken for the particle size analysis, with the references for the sample pre-treatment [23, 24]. Mastersizer 2000 laser particle size analyzer was used for the measurement. The sporopollen analysis was carried out by the routine sporopollen analysis method [25]. The flotation product of the pollen was prepared and identified under the microscope, and the percentage content was counted and calculated. 139 samples of the foraminifera and the Ostracoda were collected at a sampling interval of 1 m. The sediment samples were dried at 60 °C. The dried samples were soaked in water for 1–2 days. After the dispersal, the samples were screened with 0.063 mm and 0.125 mm pore sizes respectively. The sediments larger than 0.125 mm were the main body of the foraminifera and the ostracoda, and the sediments were the main body of the foraminifera and the Ostracoda. The species identification and the quantitative statistics were carried out under Nikon E 200 biomicroscopy (calculated with 100 g dry samples) [26]. The particle size, the sporopollen analysis and the identification of the foraminifera and the Ostracoda were completed in the Department of Geosciences, Sun Yat-sen University.

3 Results

3.1 Chronological Sequence

According to the annual data measured by AMS^{14}C method, as is shown in Table 1, the corrected age at the hole depth of 8.30 m is 7937–8009 a BP, which belongs to the Holocene epoch. The corrected age at the hole depth of 26.50 m is 13029–13136 a BP, which belongs to the late Pleistocene epoch. And the corrected age at the hole depth

83.60 m is 38745–39756 a BP, which belongs to the late Pleistocene epoch. Based on the data of the past years, the linear interpolation method is used to interpolate and extrapolate, and the age of the corresponding depth position is obtained.

Table 1. AMS^{14}C dating results of borehole BKZ01

Sample number	Dating substance	Depth (m)	^{14}C Age (yr BP)	Error value (yr)	Corrected ages (yrcal BP, 95%)
BZK01-C1	Plant debris	8.30	7145	25	7937–8009
BZK01-C2	Plant debris	26.50	11210	30	13029–13136
BZK01-C3	Whole rock organic matter	83.60	34740	190	38745–39756

3.2 Sporopollen

132 sporopollen samples were collected and 129 sporopollen types were identified, including 61 woody plant pollens, 49 herbal pollens and 19 fern spores. The sporopollen assemblage was dominated by herbs, followed by woody plants, with 68.68% and 25.41% sporopollen assemblages and 5.91% fern spores respectively. The pollens of the woody plants are mainly as follows: Pinus, Dec Quercus), Ere Q, Betula/Carpinus, Ulmus, Tsuga, Altingia/Liquidambar and Moraceae and so on. The main pollen types of herbaceous plants are as follows: Gramineae, Artemisia, Chenopodiaceae, Thalictrum, Polygonum, Labiatae, Cruciferae, and Rosaceae, followed by Scrophulariaceae, Ranuncula, Caryophyllaceae, and Primula and so on. The aquatic plants pollen is the Typha, with occasionally the wet plants Cyperaceae and other molecules. Fern spores are mainly composed of the Monolete spores, Polypodium, Pteris, Microlepria/Adiantum/Dicranopter and Lycopodium. According to the changes of the sporopollen types and the characteristics of the sporopollen assemblage, pore BZK01 can be divided into 16 sporopollen assemblage belts from old to new.

P1: 277–213 m

The total sporopollen concentration was low, and the average concentration was only 8.8 grains per gram. The pollen content of the herbaceous plants accounted for the absolute predominance, with an average content of 80.3%, of which Chenopodiaceae and Artemisia were the most abundant. The woody plants had an average content of 18.2%, mainly coniferous pollen such as Pinus.

P2: 211–203 m

The total sporopollen concentration was slightly higher, averaging 106 grains per gram. The pollen content of the herbaceous plants is still dominant, with an average of 64.7%, mainly cattail and gramineae, followed by Chenopodiaceae and Artemisia. The average pollen content of the woody plants is 29.3%, mainly coniferous plants such as Pinus.

P3: 201–109 m

The total sporopollen concentration was low with an average concentration of 10.5 grains per gram. The pollen content of the herbaceous plants is still dominant, with an average content of 73.5%, mainly Artemisia, followed by Chenopodiaceae and Gramineae. The average pollen content of the woody plants is 20.7%, mainly coniferous plants such as Pinus.

P4: 107–103 m

The total sporopollen concentration was slightly higher, with an average concentration of 97.3 grains per gram. The pollen content of the woody plants was slightly dominant, with an average content of 57.1%. Pinus and other coniferous plants were the main pollen species, with a small amount of the hemlock and the honeysuckle. The average pollen content of the herbaceous plants was 19%, mainly Artemisia.

P5: 101–97 m

The total sporopollen concentration was low with an average concentration of 55.3 grains per gram. The herbaceous plants had a slightly dominant pollen content of 45.9%, mainly in Gramineae, followed by Artemisia and Chenopodiaceae. The ferns had an average content of 29.4%, mainly in monospora, followed by Polypodia. The woody plants had an average pollen content of 24.7%, mainly in coniferous plants such as Pinus.

P6: 95–91 m

The total sporopollen concentration was high, with an average concentration of 1867.7 grains per gram. The dominant pollen content of the herbaceous plants was 59.9%, mainly Gramineae, followed by Artemisia. The average pollen content of the woody plants was 36%, mainly deciduous oak and evergreen oak.

P7: 89–85 m

The total sporopollen concentration was low, with an average concentration of 18.7 grains per gram. The dominant pollen content of the herbaceous plants was 57%, mainly Artemisia and Chenopodiaceae, followed by Gramineae. The average pollen content of the woody plants was 28%, mainly coniferous plants such as Pinus. The average content of ferns was 15%, mainly Phoenix fern.

P8: 83–67 m

The total sporopollen concentration was higher, with an average concentration of 2233.2 grains per gram. The pollen content of woody plants was slightly dominant, with an average content of 47.4%, mainly coniferous pollen of evergreen oak and pine, followed by deciduous oak, herbaceous pollen content of 38.3%, mainly Artemisia and Gramineae, fern average content of 14.3%, mainly single seam spore.

P9: 65 m

The total sporopollen concentration was 12 grains per gram. The pollen of woody plants and the herbaceous plants accounted for 50% respectively, mainly for Quercus deciduous and Artemisia.

P10: 63–43 m

The total sporopollen concentration was higher, with an average concentration of 2026.4 grains per gram. The pollen content of the woody plants was slightly dominant, with an average content of 52.4%, mainly coniferous pollen of Pinus, followed by evergreen oak and deciduous oak, and herbaceous pollen content of 29.9%, mainly Gramineae.

P11: 41–37 m

The total sporopollen concentration was slightly lower, with an average concentration of 112.7 grains per gram. The herb pollen content was slightly dominant, with an average of 46.4%, mainly in Gramineae and Chenopodiaceae, followed by Artemisia. The woody pollen content was 38.6%, mainly in coniferous plants such as Pinus.

P12: 35–27 m

The total sporopollen concentration was higher, with an average concentration of 2751.2 grains per gram. The pollen content of herbaceous plants was dominant, with an average of 63.5%, mainly in Gramineae and Chenopodiaceae, followed by Artemisia, while that of the woody plants was 35.2%, mainly in deciduous oak and evergreen oak.

P13: 25–21 m

The total sporopollen concentration was low, with an average concentration of 152.7 grains per gram. The pollen content of the woody plants decreased, and the average content was about 33.2%, still dominated by Pinus. The pollen content of the herbaceous plants increased slightly, and the average content was about 53.8%. The main herbaceous plants were rich in Liliaceae, Gramineae and Cruciferae.

P14: 19–15 m

The total sporopollen concentration was high, with an average concentration of 6385.3 grains per gram. The pollen content of the herbaceous plants was dominant, with an average of 83.4%, mainly in Gramineae, followed by Cruciferae, and 14.1% in woody plants, mainly in coniferous plants such as Pinus.

P15: 13–9 m

The total sporopollen concentration was slightly lower, with an average concentration of 414.7 grains per gram. The pollen content of the herbaceous plants was dominant, with an average of 87.7%, mainly in Chenopodiaceae and Artemisia, followed by Gramineae, while that of the woody plants was 11.3%, mainly in coniferous plants such as Pinus.

P16: 7–1 m

The total sporopollen concentration was very high, with an average concentration of 30982.3 grains per gram. The pollen content of the herbaceous plants still accounts for the majority, with an average content of 65.8%, mainly in Gramineae, while that of the woody plants is 27.5%, mainly in coniferous plants such as Pinus.

3.3 Foraminifera

A total of 139 the foraminifera samples were analyzed and 22 species of the foraminifera belonging to 17 genera were identified, including 19 benthic foraminifera and 3 planktonic foraminifera. Each of the 100 g sediment contains 0–32146 foraminiferal shells. According to the depth distribution characteristics of the foraminifera, seven foraminifera horizons are divided and the genesis of the borehole strata is inferred.

F1: 227–87 m

No foraminifera was found.

F2: 85–70 m

Most of the samples had the foraminifera, but the abundance was low. The highest number of the foraminiferal shells per 100 g sediment was 152, and the abundance of most samples was less than 100. The highest simple differentiation was 8.

F3: 69–57 m

The abundance and differentiation of the foraminifera increased compared with the previous period. The highest number of the foraminiferal shells per 100 g sediment was 1901. The simple differentiation can reach 13. Ammonia beccar II, Nonion akitaense and rotelphidium granosum were the main types of the differentiation. Other genera and species only appear sporadically. Planktonic foraminifera appeared in some horizons.

F4: 55–29 m

Foraminifera are rare. Sporadic foraminifera fossils are found in only a few horizons. The number of the foraminiferal fossils is less than 100/100 g, of which Ammonia beccar II is dominant.

F5: 28–27 m

The abundance and differentiation of the foraminifera reached the highest value of the foraminifera. Each of the 100 g sediment contains 5535–32146 foraminiferal shells. The main genera and species are the Ammonia beccar II and the Protelphidium granosum, with a small number of the planktonic foraminifera.

F6: 25–17 m

No foraminifera and Ostracoda were found.

F7: 16–2 m

There are abundant foraminifera in two horizons: 16–15 m and 7–5 m. At 16 m, the highest abundance of the foraminiferal shells was 4038 per 100 g sediment, and the Ammonia beccar II was dominant.

3.4 Mussel-Shrimp

A total of 132 Ostracoda samples were analyzed. 9 genera and 9 species of the Ostracoda were identified. The Ostracoda had few genera and species and the low abundance. There were 0–305 Ostracoda shells per 100 g sediment. According to the depth distribution characteristics of the ostracods, they can be divided into five horizons:

O1: 277–117 m

Ostracoda fossils are rare. Sporadic Sinocytheridea impressa is found in only one horizon at the bottom.

O2: 115–113 m

A small amount of the Ostracoda fossils with the low abundance and differentiation are found. The maximum number of the Ostracoda shells per 100 g sediment is not more than 305 petals, and the Sinocytheridea impressa is still dominant.

O3: 111–77 m

There are almost no ostracods. Two Sinocytheridea impressa were found in only one horizon (depth 95 m).

O4: 75–73 m

A small number of the ostracods were found, and their abundance and differentiation were low. There were no more than 31 ostracodal shells per 100 g of the sediments, mainly Sinocytheridea impressa. The Cytheromorpha acupunctata, Sinocythere dongtaiensis and Pontocythere spatiosus only appeared sporadically.

O5: 71–1 m

Ostracoda fossils are not abundant, only in four horizons (depth 53 m, 27–25 m, 15 m and 7 m), and no Ostracoda fossils are found in other horizons. Cytheromorpha acupunctata, Sinocytheridea impressa, Neosinocythereelongata and so on appeared in a small amount.

4 Paleoenvironment Reconstruction

The lithological photos of BZK01 hole are shown in Fig. 2. According to the lithological stratification characteristics and the stratigraphic unit division basis of the study area, the drilling holes are successively the bedrock Changhe Formation carbonaceous mudstone (Ech, 278.0–284.0 m), the Neogene-Quaternary sedimentary Jiaxing Formation, the Qiangang Formation, the Dongpu Formation, the Ningbo Formation and the Zhenhai Formation from bottom to top.

4.1 Jiaxing Formation

All the continental sediments are divided into three sections according to the rules of the grain size and the lithology. The hard soil layer is the top, and the sand and the gravel are the bottom, and the thickness is 126.1 m. 151.9–167.8 m is the third member of the Jiaxing Formation (al-lN-Qp_1j^3) of the alluvial-lacustrine origin. Its thickness is 15.9 m. Its lithology is dominated by the sub-clay, some silt and fine sand. 167.8–202.2 m is the third member of the Jiaxing Formation (al (pl) N-Qp_1j^3) of the alluvial (flood) origin, with a thickness of 34.4 m. Lithologic silt, sub-clay and gravel occur alternately. 202.2–231.3 m is the second member of the Jiaxing Formation (al-lN-Qp_1j^2) of the alluvial-lacustrine origin. Its thickness is 29.1 m, and its lithology is dominated by the sub-clay. 231.3–239.0 m is the second member of the Jiaxing Formation (al (pl) N-Qp_1j^2) of the alluvial (flood) origin. Its thickness is 7.7 m, and its lithology is dominated by the fine sand. 239.0–278.0 m is the first member of the Jiaxing Formation (al-lN-Qp_1j^1) of the alluvial-lacustrine origin. Its thickness is 39.0 m and it is mainly the subclay.

4.2 Qiangang Formation

It is marked by the hard soil layer and is divided into two sections. The second section contains the marine sediments besides the hard soil layer, and the first section is the floodplain clay with the silt, 68.05 m thick. 83.85–87.5 m is the second member of the Qiangang Formation (al-lQp_2q^2) of the alluvial-lacustrine origin, with a thickness of 3.65 m and the clayey sandy soil. 87.5–94.1 m is the second member of the Qiangang Formation (lhQp_2q^2) of the lacustrine-marshy origin, with a thickness of 6.6 m, clay mixed with sandy soil, brown grey-yellow grey, soft plastic and well-developed bedding. 94.1–141.7 m is the first member of the Qiangang Formation (al-lQp_2q^1) of the alluvial-lacustrine origin, with a thickness of 47.6 m. The subclay is dominant and some sandstones are interbedded. 141.7–151.9 m is the first member of the alluvial Qiangang Formation (alQp_2q^1), with a thickness of 10.2 m. The lithologic medium sand, sub-clay and fine sand are dominant.

4.3 Dongpu Formation

The hard soil layer is the grouping mark, with the thickness of 20.9 m. 62.6–66.1 m is the alluvial-lacustrine Dongpu Formation (al-lQp_3d), with a thickness of 3.5 m and the clayey sandy soil. 66.1–83.5 m is alluvial-marine Dongpu Formation (al-mQp_3d), with a thickness of 17.75 m and interbedded with the clay and the sandy soil.

4.4 Ningbo Formation

The hard soil layers are the grouping marks, and other main sediments are the marine, the alluvial and marine clay, the clay and the sandy soil interbedded and the sandy soil, with a thickness of 47.2 m. 15.4–26.0 m is the second member of the Ningbo Formation (al-lQp_3n^2) of the alluvial-lacustrine origin. Its thickness is 10.6 m. It has a high content of the clay, carbonaceous and more iron-manganese oxides. 26.0–34.9 m is the second member of the Ningbo Formation (m$Q_{p3}n^2$) of the marine origin, with a thickness of 8.9 m. The silty clay is sandy, gray and fluid-plastic, with a large number of the very thin or irregular clays. 34.9–37.3 m is the second member of the Ningbo Formation (al-mQp_3n^2) of the alluvial-marine origin, with a thickness of 2.4 m, sub-sandy soil, dark gray, micro-bedding development. And a single layer of 4–6 mm is the mainly clay, with a large number of the mica fragments visible. 37.3–39.1 m is the first member of the Ningbo Formation (al-lQp_3n^1) of the alluvial-lacustrine origin, with a thickness of 1.8 m. The sandy soil is interbedded with the clayey, grey-yellow, sandy soil slightly-medium-dense, soft-plastic and bedding developed. 39.1–54.1 m is the first member of the Ningbo Formation (mQp_3n^1), which is of the marine origin. Its thickness is 15.0 m. The clay is the sandy soil. There are more bivalve shell fragments at 46.4–46.5 m, and a few calcareous nodules can be seen locally. 54.1–62.6 m is the first member of the Ningbo Formation (al-mQp_3n^1) of the alluvial-marine origin, with a thickness of 8.5 m and clay.

4.5 Zhenhai Formation

8.6–12.5 m is the hard soil layer, overlying the marine silty soil, underlying the alluvial and marine clay with the sandy soil, with the thickness of 15.4 m. The bedrock exposed by drilling is the carbonaceous mudstone of the Changhe Formation. 0–8.6 m is the second member of the Zhenhai Formation (m$Qhzh^2$) of the marine origin. Its thickness is more than 8.6 m, and 0–0.9 m is the miscellaneous fill, and the lower part is the silty clay, and 2.6–2.8 m is the sandy soil. 3.5–3.65 m is peat, and contains a small amount of the Semi-Carbonized plant residues, and 8.3 m is a Gastropoda about 1 cm long. 8.6–12.5 m is the first member of the Zhenhai Formation (al-l$Qhzh^1$) of the alluvial-lacustrine origin, with a thickness of 3.9 m and mainly subclay. 12.5–15.4 m is the first member of the Zhenhai Formation (al-m$Qhzh^1$) of the alluvial-marine origin, with a thickness of 2.9 m and clayey sandy soil.

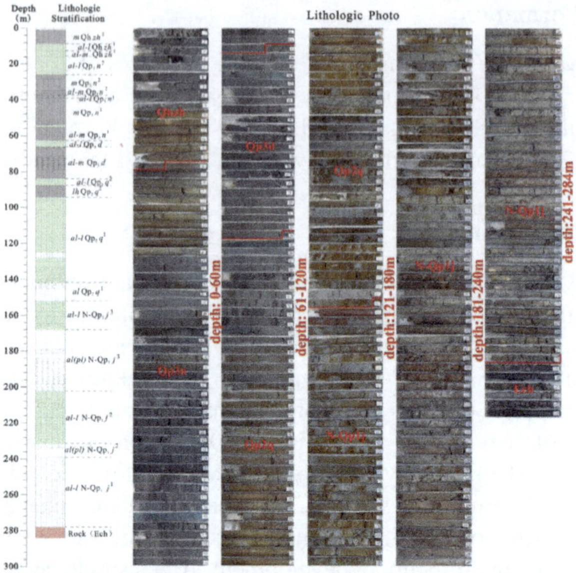

Fig. 2. Lithological stratification and stratigraphic division of BZK01 hole

5 Conclusion

(1) The corrected age of the BZK01 borehole at 8.30 m is 7937–8009 a BP, which belongs to Holocene. The corrected age at 26.50 m is 13029–13136 a BP, which belongs to the late Pleistocene, and the corrected age at 83.60 m is 38745–39756 a BP, which belongs to the late Pleistocene.

(2) According to the characteristics of the sedimentary facies, the micropaleontology and the borehole lithology, it is inferred that the bottom-up strata of the BZK01 borehole are the bedrock Changhe Formation and the Neogene-Quaternary sedimentary Jiaxing Formation, the Qiangang Formation, the Dongpu Formation, the Ningbo Formation and the Zhenhai Formation. The Pliocene and the Early Pleistocene strata are the Jiaxing Formation, all of which are the continental sediments. They are divided into three sections according to the rules of the grain size and the lithology. The hard soil layer is the top, while the sand and the gravel are the bottom. The thickness is 126.1 m. The Middle Pleistocene stratum is the Qiangang Formation, marked by the hard soil. It is divided into two sections. The second section contains the marine sediments besides the hard soil, and the first section is the floodplain clay with silt. The thickness is 68.05 m. The Late Pleistocene strata are the Dongpu Formation and the Ningbo Formation. The hard soil strata are the grouping marks. The other main sediments are the marine, alluvial and marine clay, clay and sandy soil interbedding and sandy soil, with the thickness of 21.25 m and 47.2 m respectively. The Holocene stratum is the Zhenhai Formation, and 8.6–12.5 m is the hard soil, which is covered with the marine silty soil and the underlying alluvial and the marine clay with sandy soil, with a thickness of 15.4 m.

The bedrock exposed by drilling is carbonaceous mudstone of the Changhe Formation.

(3) The sporopollen assemblage can divide boreholes into 16 sporopollen belts and 7 climatic cycles. The early Pliocene and the early Pleistocene correspond to the sporopollen P1, P2 and P3 belts, and the lower part is the river-lake sediments, while the sporopollen assemblage reflects less vegetation and cold and dry climate. The middle part is the river sediments with the herbaceous pollen content dominant, reflecting that the vegetation type should be temperate grassland and the climate is warm and humid. The upper part belongs to the river-lake sedimentation. Accumulation and sporopollen assemblage reflects less vegetation, cold and dry climate. The late stage of the Middle Pleistocene corresponds to the sporopollen zone P3 and P4–P7, including the sedimentary environment of the fluvial-lacustrine-shallow-lacustrine-fluvial facies. The climate type from bottom to top reflects the temperate continental climate, warm and humid subtropical monsoon climate, warm and humid subtropical monsoon climate, and cold and humid subtropical monsoon climate. The late Pleistocene corresponds to the zone P8–P14, which undergoes the climatic cycles of warm and humid, cold and dry, warm and humid, and cold and dry. The Holocene corresponds to the 15–16 belt of the sporopollen assemblage, which is deposited in the fluvial and lagoon facies, reflecting the process of cold, dry-warm and humid climate change.

Acknowledgment. Foundation Projects: China Geological Survey "Geological Survey of Zhejiang Marine Economic Development Demonstration Zone (Jiaxing) Urban Agglomeration" (1212011220008) and "Zhejiang 1/50000 Regional Survey Project of Jiaxing, Pinghu, Jiashan, Shengze and Xinfeng Towns" (1212011220538).

References

1. Li J, Yulan Z (2006) Palynological and paleoenvironmental studies of the late neogene sediments in the eastern margin of the Yangtze river delta. J Micropaleont (01):70–76
2. Liu Z et al (2014) Genesis and spatial characteristics of the marine-terrestrial sedimentary soils in the Yangtze Estuary. J. Undergr Space Eng (01):102–108
3. Wang Z et al (2008) A preliminary study of the late quaternary chronostratigraphic framework and two transgressions in the Yangtze river delta. J Palaeogeogr (01):99–110
4. Liu D, Ye Y (2005) Relative sea level rise and the land subsidence in the Yangtze river delta. J Geol Hazards Environ Prot (04):400–404
5. Sun Q et al (1997) Impact assessment of the sea level rise on the Yangtze river delta. Resour Environ Yangtze River Basin (01):59–65
6. Zheng Q (2015) Review of the Holocene sea level reconstruction in the Yangtze river delta. Shanghai Land Resour (01):80–85+94
7. Zhu J, et al (1994) The impact of the sea level rise on the Yangtze river delta and adjacent areas. Geogr Sci (02):109–117+199
8. Zhang Y (2005) Palynological and environmental studies of the late quaternary sediments in the front area of the Yangtze river delta. J Ocean (Chin Ed) (04):82–87

9. Zhao B (2013) Late quaternary sequence characteristics in the front of the Yangtze river delta and responses to the post-glacial sea level fluctuations and sediments, Shanghai Land Resour (02):50–54+63
10. Bai X, et al (2016) Magnetic characteristics of the tidal flat sediments in the Holocene strata of the Yangtze river delta and their paleoenvironmental significance. J Sediment (06):1165–1175
11. Liu J, et al (2007) Paleovegetation and paleoclimate since the Holocene on both sides of the Qiantang river and the Hangzhou bay. Earth Sci Front (05):235–245
12. Zhong S (2012) Late quaternary calcareous nannofossils of core Dy03 in the Yangtze river delta and their sedimentary environmental significance. Acta Micropalaeontologica Sinica 29(2):137–144
13. Zhang R et al (2011) Palaeoenvironmental evolution of subaqueous Yangtze delta inferred from sedimentary records. Mar Geol Quat Geol 31(1):1–10
14. Miao WD, et al (2016) Stratigraphic division of NB5 core in the Yangtze delta area and its environmental change information. Geology in China
15. Miao Y et al (2015) Late quaternary pollen records from the Yangtze river delta, east China, and its implications for the Asian monsoon evolution. Arab J Geosci 8(10):7845–7854
16. Zhao XT, et al (2017) Reviews on the research of late Cenozoic geology and environment of the Yangtze river delta area. J Geomech
17. Wu SJ (2008) Nanjing and Nanjing, sedimentary characteristics of the palaeochannel and environment evolution during the Holocene in the NW Taihu plain, Yangtze river delta. J Strat
18. Su J et al (2017) Postglacial sequence stratigraphy and sedimentary environment evolution of the Yangtze river subaqueous delta. J Palaeogeogr
19. Li B, et al (2011) Quaternary sedimentary records and paleoenvironment response in the Yangtze river delta. Quat Sci (02):316–328
20. Wei T, et al (2006) Distribution characteristics of the geochemical elements in the quaternary sediments of the Yangtze estuary and their palaeoenvironmental significance. Quat Sci (03):397–405
21. Wu YH (2013) Sedimentary characteristics and environmental implications of PJ core in the South Wing of the Yangtze river delta. J Guangxi Teach Educ Univ (2013)
22. Zhuang C, et al (2013) PRD16 pore microorganism records and paleoenvironment reconstruction in the Pearl river delta since the late pleistocene. J Sun Yat-Sen Univ (Nat Sci Ed) (3):128–132+138
23. Wu J, et al (2014) Palaeoenvironmental significance of the late quaternary microfauna in the eastern pearl river delta. J Sun Yat-Sen Univ (Nat Sci Ed) (6):45–52
24. Xue MA, et al (2015) Grain-size analysis of the quaternary sediments in Borehole ZKA4 in The Yangtze river delta and its paleoenvironment and paleoclimate implications. J Strat
25. Chen J, et al (2009) Paleoclimate changes and its coupling with sea level reflected by the sporopollen records since 120kaB.P. in the Yangtze river delta. Mar Sci (02):68–73
26. Liu C, et al (2008) High resolution ostracodal records and the late quaternary paleoenvironment reconstruction of PRD05 hole in the pearl river delta. J Paleogeography (3):313–322

Simulation Analysis of Dynamic Characteristics of AC Motor Based on BP Neural Network Algorithm

Shuang Wu and Jian Liu(✉)

School of Electrical Engineering and Automation, Wuhan University, Wuhan, China
weixiong619@126.com

Abstract. The difficulty of traditional AC speed regulation has been hindering the development of AC motors, and vector control technology can make AC motors obtain the control characteristics of DC motors. Based on this, this paper establishes the voltage output characteristic model of AC motor dynamic system using BP neural network optimized by genetic algorithm, and trains some measured data as training samples of BP neural network optimized by genetic algorithm. The research shows that the neural network algorithm is used to estimate the rotational speed, and a high-speed and high-precision control system without speed sensor based on vector control is realized. The maximum synchronization error of the AC motor at start-up is reduced by 78.77%, and the maximum synchronization error is reduced by 20.88% in the case of sudden change in speed. The results show that compared with the traditional controller, the neural network controller has better adaptability and robustness to the model and environment, and effectively improves the control result of the system and achieves the expected purpose.

Keywords: BP neural network · PID control · Simulation analysis

1 Introduction

Control theory has gone through two stages of development: classical control theory and modern control theory [1]. Now it has entered the stage of development of non-linear control theory and intelligent control theory [2, 3]. Because BP neural network has excellent fitting ability for any differentiable function mapping, it is widely used in automatic control, identification of non-linear systems, control and fault diagnosis of systems [4]. In 2005, research on collaborative P-based virtual environment research collaborative robot systems was proposed [5]. After that, someone proposed PASSIM: research on the development process of multi-agent system based on simulation [6]. At the same time, modern aircraft is developing towards multi-power and full-power. It has become a trend to replace hydraulic and pneumatic energy with electric power, and the electrical system of aircraft will be more complex. The advanced technology of CNC machine tools is one of the measures of the level of industrialization in a country [7]. At present, CNC machine tool manufacturers mostly use the spindle AC drive system based on vector control. The structure of the three-phase AC asynchronous

Z. Xu et al. (Eds.): CSIA 2019, AISC 928, pp. 277–286, 2020.
https://doi.org/10.1007/978-3-030-15235-2_42

motor is not complicated, the price is low, and the maintenance is convenient. If the advanced control technology is used above, it can have good static and dynamic performance [8]. It can be affirmed that the introduction of artificial neural networks into the field of motor control is an inevitable development trend [9]. This paper is an improved intelligent PID control method which is applied to the permanent magnet synchronous motor speed controller by neural network. It is an improvement and optimization of traditional PID control [10].

2 BP Neural Network Speed Estimation Vector Control System

Under the deviation coupling control structure based on the traditional BP neural network PID, the synchronization error between each motor is large, and the time to restore stability when the speed is abrupt is also long. Therefore, based on the above analysis of the synchronization error between motors in different control environments, Table 1 and Fig. 1 are obtained:

Table 1. Analysis of maximum synchronization error under different control structures

Motor number	Traditional structural deviation	Improve structural deviation
Motor 1 and electricity 2 (rad/s)	66.45	56.42
Motor 1 and electricity 3 (rad/s)	78.27	84.32

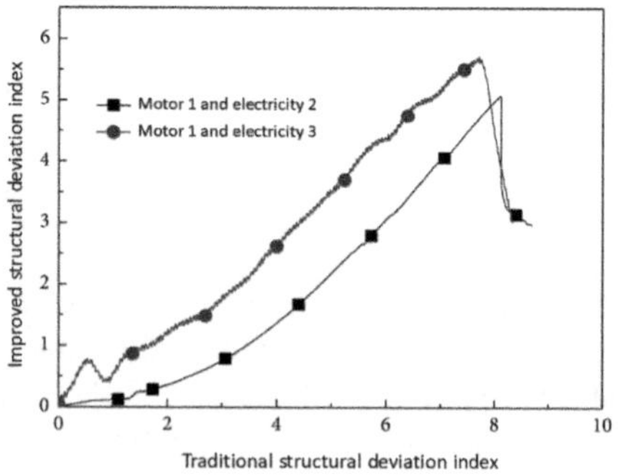

Fig. 1. Analysis of maximum synchronization error under different control structures

The essence of BP neural network algorithm can be divided into two parts: 1. the forward propagation of input information and 2. the reverse transmission of calculation error. The forward propagation of input information refers to the process of computing input information layer by layer from input layer, hidden layer to output layer, and the state of neurons in each layer will only affect the next layer of neurons. The state has an impact. In the multi-motor synchronous control system, in order to intuitively evaluate the effect of the multi-motor synchronous control, the following error and synchronization error can be used to investigate the two parameters. The ultimate goal of multi-motor synchronous control system is to make these two parameters close to zero in the shortest time. In addition to network structure, the ability and effectiveness of network problem solving depends to a large extent on the activation function adopted by the network. The BP algorithm transforms the mapping problem of neural network learning input output into a nonlinear optimization problem, and uses the gradient descent algorithm in optimization to correct the network weight with iterative operation to achieve the minimum mean square error between the network output and the expected output. Turn. The improved BP neural network algorithm is used to design the speed PI controller. Its function is to enhance the system's ability to resist load disturbance and suppress speed fluctuation. In the vector control system, the output of the speed loop is the torque current. Under the constant load state, the electromagnetic torque can be controlled to control the speed. The controller typically uses a PID algorithm to control the object. And for the same initial weights and thresholds, the training results of the network are unchanged. By introducing the genetic algorithm, it is to optimize the optimal initial weight and threshold of the network.

In order to more accurately reflect the synchronism of each motor in the multi-motor synchronous control system, this paper also analyzes the following errors of each motor in the multi-motor control system. It can be seen from Fig. 1 that the time for the two motors to achieve stability under the improved control structure is improved to some extent compared with the conventional control structure, and Table 2 and Fig. 2 are more intuitively embodied.

Table 2. Analysis of following errors in different control structures

Motor number	Traditional structural deviation	Improving structural deviation
Motor 1 and electricity 2 (rad/s)	0.358	4.61
Motor 1 and electricity 3 (rad/s)	6.732	5.88

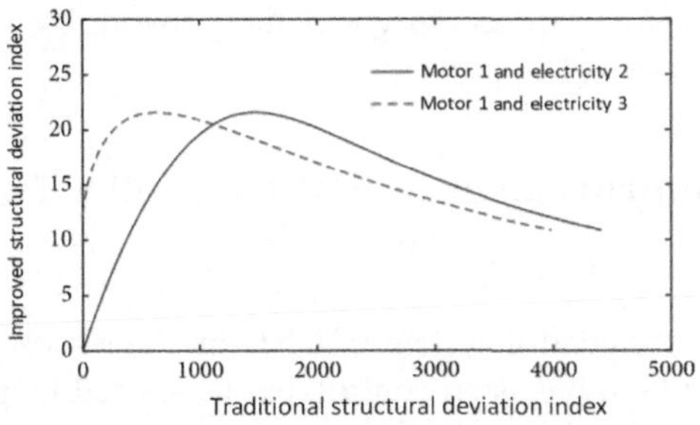

Fig. 2. Analysis of following errors in different control structures

Table 3 and Fig. 3 below are the comparison of the training results of two kinds of neural network BP algorithm in the application of fault diagnosis of aircraft AC asynchronous motor. Obviously, the improved BP algorithm is superior to the standard BP algorithm in terms of learning speed and training accuracy.

Table 3. Comparison of training characteristics of two BP neural networks

Precision esD	Standard BP algorithm number	Improve BP algorithm number of times
0.02	62	74
0.01	96	69

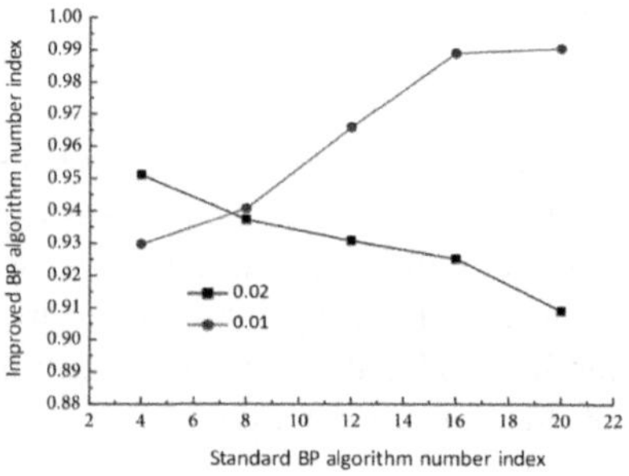

Fig. 3. Comparison of training characteristics of two BP neural networks

In the research of multi-motor synchronous control system, researchers have proposed many control algorithms. Among them, digital PID control is the most common control method in the production process, and has been widely used in metallurgy, machinery, chemical industry and other fields application. The so-called neural network control, that is, neural network-based control or simply neural control, refers to the use of neural network in the control system to model complex nonlinear objects that are difficult to accurately describe, or to act as a controller, or to optimize Calculate, or perform reasoning, or troubleshooting. It is mainly for the speed of network convergence. Improving the convergence speed of the network by scientifically assigning initial weights is the focus of improvement.

3 Design and Improvement of PID Controller for BP Neural Network

In the closed-loop vector control system of CNC machine tool spindle, the fuzzy PID algorithm can achieve high precision control, but the closed-loop speed feedback is an indispensable condition for the realization of closed-loop speed control. However, the traditional speed sensor has some problems in the spindle system, such as difficult installation, high maintenance cost and poor stability under specific conditions. For the multi-motor synchronous control system, the selection of control structure directly affects the control accuracy. This chapter analyses and studies the deviation coupling control structure in the multi-motor control structure. The role of speed compensator in the deviation coupling control structure is very important. For the sinusoidal brushless DC motor, more precise rotation is needed. The structure of position sensor of sub-position signal is more complex, so square wave brushless DC motor is more widely used. The closed-loop control in vector control aims to achieve high-speed and high-precision control effects. Closed-loop control is required in various occasions where the performance of the control system is required. Only through closed-loop control can the system respond quickly and achieve accurate torque. control. The basic idea of BP neural network based on genetic algorithm optimization is as described above. The initial weight and threshold of BP neural network cannot accurately obtain this deficiency. By using genetic algorithm to optimize its optimal initial weight and The threshold, so that the optimized BP network can better predict the sample. The fixed speed signal compensation gain only deals with the moment of inertia between the motors. When the load changes greatly, because of different motor parameters and electromechanical time constants, it is easy to cause large fluctuations in the rotational speed, and it takes a long time. Time to eliminate this fluctuation.

The step response curves obtained by the ordinary PID design method and the BPNN design method are compared, and the results are shown in Fig. 4. At the same time, the performance indexes obtained by the two design methods are compared. The results are shown in Table 4.

Table 4. Speed controller PI design method and its corresponding performance index

Speed controller PID design method	Overshoot	Adjustment time
Conventional PID	36.12%	1.564
Improved BP neural network algorithm	33.27%	0.153

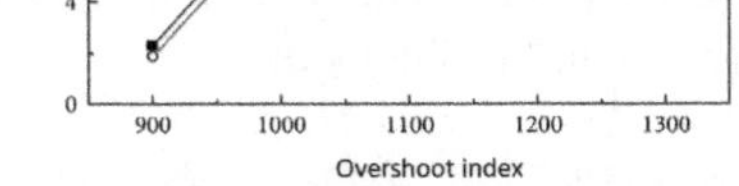

Fig. 4. Speed controller PI design method and its corresponding performance index

In all kinds of control algorithms, control is one of the earliest developed control strategies. Control has been widely used because of its intuitive, simple calculation, easy implementation, good robustness and high reliability. However, although the traditional PID controller is the most concise and widely used control, its control effect can not meet the requirements of the control system in this paper, because the three parameters of PID can not be adaptively changed according to the size of the error after being determined by debugging. In order to make the residual between the predicted value and the expected value of BP network as small as possible when predicting the test sample. In the multi-motor synchronous control system, the disturbance encountered by any motor in the operation process is unknown. Once a motor in the system is disturbed by the outside world, it will take a certain amount of time to eliminate the influence of disturbance on the multi-motor system. It is especially suitable for deterministic control systems that can establish accurate mathematical models, so it is generally necessary to know the transfer function and mathematical model of the controlled system. In fact, the selection of the number of hidden layer nodes is more or less related to the requirements of the problem, the number of input nodes and the number of output nodes, and the changes of the hidden layer nodes will directly affect the internal expression of the network. The improved BP neural network algorithm is used for optimization. In order to avoid overshoot, the penalty function is adopted. Once overshoot, the overshoot is used as the optimal index, and no overshoot is achieved. As with DC motor control, the difference between the desired torque current component and the actual feedback torque current component, i.e., the error, reflects the speed change characteristic.

One of the simplest PID control systems consists of a PID controller and a controlled object. Where N is the actual input signal, W is the output signal after the control system is set, and D is the real-time error signal. It can be expressed as:

$$HWt = \frac{\sum_{i=1}^{N} D_i(x)}{N} \tag{1}$$

Because the control process is based on the sampling of control data, the deviation of sampling time can only be calculated, so the integral and differential operations can not be directly carried out by using the formula, and the discretization of the w deviation of the upper formula is needed to be used. Do the following transformations:

$$h_f(x) = \begin{cases} 1 & \sum_{t=1}^{T} (\log 1/\beta_t) h_t(x) \geq \frac{1}{2} \sum_{t=1}^{T} \log 1/\beta_t \\ 0 & \sum_{t=1}^{T} (\log 1/\beta_t) h_t(x) < \frac{1}{2} \sum_{t=1}^{T} \log 1/\beta_t \end{cases} \tag{2}$$

The sampling period needs to be small enough to achieve higher control accuracy. The discretized w expression can be obtained:

$$D(p_1) = A \cdot w_1 \left[1 - \frac{p_1}{v_1} \left(\frac{p_1 w_1 + p_2 w_2 - m(1-r)}{\frac{w_1}{v_1} p_1{}^2 + \frac{w_2}{v_2} p_2{}^2} \right) \right] \tag{3}$$

The control object of this paper is permanent magnet synchronous motor. The increment of control quantity is particularly important in the control process. So incremental PID control algorithm is chosen in this paper. The recurrence relations are as follows:

$$R_t(p_{1t}, Q_t) = p_{1t} \cdot \min(I_t + Q_t, D_t) - (p_{0t} \cdot Q_t + C_t \cdot AI_t) + R_{t-1} \tag{4}$$

Unsupervised Hebb learning algorithm: Hebb learning is a correlation-based learning algorithm. When two interconnected single neurons receive excitation and are excited together, then the connection weights of the two single neurons and the incentives they receive the product is in a proportional relationship. As shown below:

$$AI_t = \frac{(I_t + Q_t)}{2} \cdot \frac{(I_t + Q_t)}{D_t} \tag{5}$$

Supervised learning algorithm: Teacher signal is introduced based on Hebb learning algorithm, which mainly changes the I stimulation of one of the neurons into teacher signal. The learning rules are as follows:

$$w(t) = w_2 + (w_1 - w_2)\frac{T - t}{T} \tag{6}$$

PID control is widely used in various industrial processes due to its simple structure, robustness to model errors and easy manipulation. It is a control method for mature technology in control systems. In theory, BP neural network has no limit on its input parameters, but since not all indicators in the sample are the same, this may cause the order of magnitude of some variables in the original sample to be very different. When a motor in the system is in an unsynchronized state, but the control system of this motor has not yet had time to react, but the PID controller has started to work, that is, before the control amount changes, the PID controller has already I started to adjust. The method first turns the operator's experience into a fuzzy rule, and then blurs the input signal. The input of the fuzzy rule is the blurred signal, and the fuzzy reasoning is completed according to the artificially set fuzzy rule, and finally output to the controlled object. For neural networks with multiple output variables, it is necessary to normalize the original output variable data. In this way, the following errors among the motors in the multi-motor synchronous control system can be quickly eliminated by detecting the speed of other motors in the system, and the system tends to be stable. With the continuous development of technology in related fields, the requirements for control systems are getting higher and higher. In the actual process, the traditional control system design still uses the trial and error method to solve, so the accuracy of the traditional control will inevitably be affected and restricted to a certain extent.

The completion of the coordinate transformation can achieve the decoupling control, but also the core magnetic field orientation of the vector control technology. This paper is based on the rotor field orientation, that is, the y-axis and the rotor magnetic field are assumed to be in the same direction, and the n and qm axes are rotated at the synchronous angular velocity:

$$
\begin{aligned}
o_k &= f(net_k) \quad net_k = \sum_{j=0}^{m} w_{jk} y_j \\
y_j &= f(net_j) \quad net_j = \sum_{i=0}^{n} v_{ij} y_j
\end{aligned} \tag{7}
$$

Because the small resistance voltage drop does not affect the stability of the system at high speed, the current model also needs to be observed at this time, except that the observed values are not included in the system and filtered through the low-pass filter. At low speed, the observed values are filtered out by the high-pass filter using the voltage model method and the observed values by the voltage model method at low speed, so that both of them can be smoothed through Ferry. The Y in the upper form can be transformed into:

$$
\begin{aligned}
\delta_k^o &= (d_k - o_k)(1 - o_k)o_k \\
\delta_j^y &= \left[\sum_{k=1}^{l} (d_k - o_k) f'(net_k) w_{jk}\right] f'(net_j) = \left(\sum_{k=1}^{l} \delta_k^o w_{jk}\right)(1 - y_j)y_j
\end{aligned} \tag{8}
$$

According to the repeated simulation sampling, before the neural network starts working, the sample quantity is 700 sets of data, and the weights are trained offline, and then the speed estimation experiment is performed. The simulation error curve of the network is shown in Figs. 5 and 6:

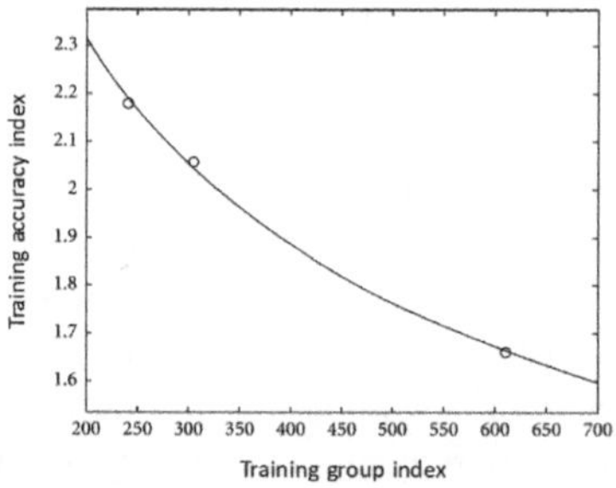

Fig. 5. Neural network training error curve

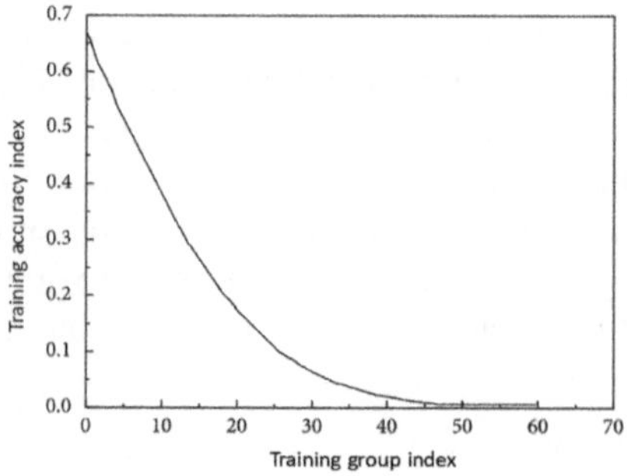

Fig. 6. Improved neural network training error curve

The output of the PID control is the given signal of the torque. By controlling the size of P, the electromagnetic torque K can be controlled. When the load is constant, the torque current can be controlled to control the speed. The input of the PID controller is the deviation between the expected speed F and the actual speed B. After the speed regulator realizes the control through the formula, it can be obtained that:

$$p_u(f) = C\sum_{i=1}^{n} K(\left\|\frac{f - z_i}{h}\right\|^2)\delta[b(z_i) - u] \tag{9}$$

The PID controller is a linear controller that is based on a given value M. The control deviation is formed with the actual output value R. The formula is:

$$T = \frac{M}{R} = \frac{i\eta_e M_e}{r} \tag{10}$$

The proportion, integral and differential of deviation are combined with M in the above formula, and then the control quantity is formed by linear combination to control the controlled object. The control law is as follows:

$$W = \alpha(\beta(\frac{E^2{}_{i-current}}{E^2{}_{i-init}}) + (\mathbf{1}-\beta)\frac{d_i}{d_{\max}})) \tag{11}$$

Increasing control effect can reduce errors and improve accuracy, but reduce stability. On the contrary, in order to ensure stability and limit control effect, the accuracy of control is reduced. The expression of discretization is as follows:

$$E\left[d^2_{toCH}\right] = \int\int(x^2 + y^2)\rho(x, y)dxdy = \int\int r^2\rho(r, \theta)drd\theta \tag{12}$$

A high-speed and high-precision vector control system based on neural network speed estimation is established. The core of the system is fuzzy PID speed closed-loop control and neural network speed estimation. Firstly, each module of multi-motor synchronous control system is modeled separately, including improved BP neural network PID controller, permanent magnet synchronous motor and improved speed compensator. The synchronization performance of multi-motor synchronous control system under the traditional BP neural network PID deviation coupling control structure and the improved BP neural network deviation coupling control structure is analyzed and compared through simulation experiments. The combination of neural network and traditional control can effectively solve the difficulty of determining the uncertainty of control parameters and the uncertainty of environmental disturbance, because it has the ability to approximate arbitrary nonlinear functions, and the structure and learning algorithm are simple and clear. Through off-line training, the parameters such as the weight of the neural network are determined, and then the network can be used for the estimation of the speed under the working phase. Finally, the estimated speed is obtained through a given input, and the closed-loop speed control is performed. According to the BP neural network PID controller and the speed compensator in the design deviation coupling control structure, the simulation platform of three motor control systems is built in the Matlab/Simulink environment.

4 Conclusions

With the continuous improvement of industrial automation, multi-motor synchronous control systems play an increasingly important role. Aiming at the problem that the initial weights and thresholds of BP neural network can not be accurately obtained and have a great impact on the training of the network, the genetic algorithm is introduced to optimize the optimal initial weights and thresholds of the network, so that the BP neural network can obtain better training results, and make the network obtain higher prediction accuracy and good results. Generalization ability. The application of neural network control in AC motor control is mainly studied. AC motor is a non-linear, strong-coupling control object, and its parameters will change with the change of external factors. Only using neural network for speed identification, if more advanced algorithms can be used for speed identification, there will be more insights.

References

1. Meng L, Liu Z (2016) Fault simulation and diagnosis for vector control of AC motor drive. IEEJ Trans Electr Electron Eng 11(5):618–623
2. Li T, Yang SL, Pan HL (2014) Simulation of temperature compensation of pressure sensor based on PCA and improved BP neural network. Adv Mater Res 846–847(5):513–516
3. Rehmani MH, Viana AC, Khalife H et al (2013) SURF: a distributed channel selection strategy for data dissemination in multi-hop cognitive radio networks. Comput Commun 36 (10–11):1172–1185
4. Paudel S, Elmtiri M, Kling W et al (2014) Pseudo dynamic transitional modeling of building heating energy demand using artificial neural network. Energy Build 70(2):81–93
5. Meng J, Yang EH (2013) Interactive encoding and decoding based on binary LDPC codes with syndrome accumulation. IEEE Trans Inf Theory 59(5):3068–3103
6. Huang DY, Lin TW, Hu WC et al (2013) Gait recognition based on Gabor wavelets and modified gait energy image for human identification. J Electron Imaging 22(4):043039
7. Hu X, Zeigler BP (2005) A simulation-based virtual environment to study cooperative robotic systems. Integr Comput Aided Eng 12(4):353–367
8. Sivashanmugam K, Miller JA, Sheth AP et al (2005) Framework for semantic web process composition. Int J Electron Commer 9(2):71–106
9. Cossentino M, Fortino G, Garro A et al (2008) PASSIM: a simulation-based process for the development of multi-agent systems. Int J Agent-Oriented Softw Eng 2(2):132–170
10. Guan Q, Wang L, Clarke KC (2005) An artificial-neural-network-based, constrained CA model for simulating urban growth. Am Cartogr 32(4):369–380

Simulation and Optimization of the Scenic Beauty of Green Plants in the Landscape Design

Xiang Pan(✉) and Qing Lin

Sichuan Agricultural University, Chengdu 610000, Sichuan, China
peili490749@126.com

Abstract. Plants are the most important part of the landscape architecture, and their configuration is of important and practical significance. From the point of view of the region, the season and the collocation, this paper introduces the principles of the plant collocation in the landscape design, summarizes the main methods of the plant collocation in the landscape design, and provides references for maximizing the practical utility of the landscape design. Chinese people pay attention to the unity of nature and man. At the beginning of any design, people-orientation and serving people should be taken into account. In the allocation of the garden plants, this factor should be fully taken into account so that nature and man can complement each other and coexist harmoniously.

Keywords: Landscape design · Green plants · Scenic beauty · Optimization simulation

The purpose of the landscape design is to create a relaxed and happy natural environment and improve people's quality of life. With the acceleration of the social development, the ecological environment has also been greatly damaged. Therefore, from the perspective of development, improving the ecological environment is one of the important tasks of our mankind [1]. Plants are the regulators of nature. The allocation and utilization of plants in the landscape design is not only related to the success of the landscape, but also to the development of cities.

1 The Connotation of the Landscape Design

The landscape design is a huge and complex comprehensive discipline, which integrates the theories of sociobiology, cultural anthropology, art, architecture, contemporary science and technology, history, psychology, regional science, nature and geography and so on, and they interact with each other [2]. The landscape design is an ancient and new discipline. In a broad sense, the conscious environment transformation that humans are engaged in today can be called the landscape design. It is a creative activity with the dual nature of time and space. It develops with the development of the times. Each era endows it with different connotations and puts forward new and higher requirements. It is a process of creation and accumulation [3].

Z. Xu et al. (Eds.): CSIA 2019, AISC 928, pp. 287–292, 2020.
https://doi.org/10.1007/978-3-030-15235-2_43

What is the landscape design? The landscape design refers to the creation of a relatively independent landscape with a certain social and cultural connotation and aesthetic values in a certain area. It must have two attributes. One is the natural attributes. It must be a sensible factor of light, shape, color and body, a certain spatial form, and a relatively independent object which can be easily separated from the regional morphological background [4]. The other is the social attributes. It must have certain social and cultural connotations, the ornamental functions, and the functions of improving the environment and application, which, through its connotation, can trigger people's emotions, interests, associations, empathy and other psychological reflection, that is, the so-called landscape effect.

If we understand the landscape design as a process of implementation of any problem concerning the use of the outdoor space and lands by our human beings, the ways to solve the problems and the ways to supervise the solution. The purpose of the landscape design is to create leisure and activity space for people and to create comfortable and pleasant environment. The responsibility of the landscape architects is to help people, buildings, communities, cities and human life to live in harmony with the earth [5].

In recent years, "ecological design" has always been a hot topic of concern, and also a point of doubt. The eco-design is still in its infancy in the field of the architectural design and the landscape design, and its concepts are interpreted differently. Generally speaking, there are two aspects: applying the principles of ecology to guide the design, and making the design result friendly to the environment while meeting the needs of human beings. Referring to the definitions of Sim Van der Ryn and Stuart Cown, any design form that coordinates with the ecological process and minimizes its damages to the environment is called the ecological design, which means that the design respects the species diversity and reduces the resource deprivation, maintains nutrition and water circulation, maintains the quality of plant and animal habitats, and contributes to the improvement of the human settlements and the ecosystem health.

The author holds that "The ecological design is to inherit and develop the experience of the traditional landscape design, follow the principles of ecology, build the multi-level, multi-structure and multi-function scientific plant communities, and establish a new order of human, animal and plant association, so as to achieve the ecological beauty on the premise of the minimal damage to the environment. The unity of the scientific beauty, the cultural beauty and the artistic beauty will create a clean, beautiful and civilized landscape environment for our mankind".

2 Scenic Beauty of the Green Plants in the Landscape Design

At present, the "ecological design" of the landscape is not yet mature and in the transitional period. It needs the clearer concepts, the solid theoretical basis and the clear principles and standards, which need further research and exploration and continuous practice. The scenic beauty evaluation (SBE) is a psychophysical model evaluation method proposed by Daniel et al. This method is based on the classification evaluation method, which allows the testers to score different scenery pictures, and eventually get the beauty score of the picture samples. The landscape values in this method are not judged by a few experts, but are based on the public aesthetics, so it can reflect the aesthetic values of a landscape more objectively. The process of the scenic beauty evaluation is divided into three parts. Firstly, select the samples, measure the public aesthetic attitudes, obtain the scenic beauty scores and standardize the data. Secondly, decompose the landscape of the evaluation samples and determine the weight of each factor. Thirdly, establish the model of the relationship between the scenic beauty and each landscape factor, and measure the relationship between the landscape elements and the scenic beauty.

This study randomly selected the samples of the typical garden plant clusters constructed by Palm Eco-Town Development Co., Ltd. in the residential areas, the commercial areas and the public green areas all over the country. Samples were taken by the same photographer using the same camera in sunny weather. Thirty-two plant cluster photos were selected from 90 primary samples, as the final survey samples. The sample evaluation adopts the webpage survey to obtain the public aesthetic attitudes, namely the beauty scores. Thirty-two samples were randomly arranged on the page without restricting the test population. The beauty of the seven-point system was used as the measurement criterion, respectively dislike very much, dislike, dislike not very much, general, like a little, like and like very much. The corresponding scores were 1, 2, 3, 4, 5, 6 and 7. A total of 265 online questionnaires were received, of which 257 were valid. The landscape factors of the plant clusters are divided into 10 project indicators and 19 measurement indicators. Among them, the morphological diversity, the color diversity, the elevation level and the species diversity are the quantitative indicators, while the rest are the qualitative indicators, which are quantified into three levels. The detailed elements are decomposed in Table 1. The scenic beauty model is constructed by using the multivariate quantitative model I, and the scenic beauty prediction model is established by using the multivariate linear regression program Backward in the SPSS 19.0 statistical software.

The plant allocation should be based on the needs of the landscape design, using the plant materials such as arbors, shrubs and grasses, combining with the ecological habits and the biological characteristics of the plants themselves, to allocate different kinds of plants to each other in order to synthesize the plant space that can express certain artistic conception. The plant landscape design should ultimately solve the plant selection and the plant allocation. For the layout of the plant landscape, trees are used as the background images and skylines, and the shrubs are used to divide the space and guide the vision, and the ground cover and lawn are used as the landscape background. The planting design often arranges the typical plant clusters at the important nodes, which are generally divided into four layers on the elevation level. The proportion of the evergreen and deciduous species is about 1:3. In addition, it should conform to the basic aesthetic principles of the color changes, shapes and the texture contrast. In the design and construction, except for the important nodes, the other nodes can adopt the single-layer and the two-layer planting structures to increase the diversity of the plant landscape. It is not necessary to follow the hierarchical structures of the typical plant groups. In the strategy of the group planting of the garden plants, a successful garden planting can be constructed as long as the interlayer structure on the facade is guaranteed and the transition is natural, and the planting density and species on the plane are reasonable, and the complex plant stacking is avoided, and the color contrast is obvious in the aesthetic expression, and the plant morphology is changed, and the ground cover line is clear and the lawn is blank, so as to build a successful garden plant group.

3 Simulation and Optimization of the Scenic Beauty of the Green Plants in the Landscape Design

All things will come to themselves, with deep purples and bright reds. There are many kinds of plants that are available for the selection and allocation in the spring gardens. How to maintain the plant diversity and emphasize the sustainability is very important. The layered allocation should be based on the florescence and colors of the flowers and trees. Plants with longer florescence should be in the first two layers of the hierarchical allocation, and they should also have a considerable advantage in quantity.

In summer, it's very hot. The most important thing to give the visitors in summer is the fresh green shade. Some shade trees should be planted beside the seats, water banks and public facilities where tourists can rest, so as to cover the sun. Obviously, the shade trees in gardens should not be shrubs, but trees with tall crowns and good shade. There are also subtle differences in the green of the shade trees, from shade to depth, from light to dark, and planting different trees separately will reflect the different hues. The green of the shade trees in gardens is only a feature, not a bright spot. The green of the aquatic plants is more attractive to tourists.

The autumn plants are not luxuriant in spring and summer. They are the transitional period of most plants. Most of the garden plants are facing withering. The plant allocation needs to be cautious and prudent. We should not introduce tree species at will. We should pay more attention to the plant habits and the overall environment of the garden. Scientific selection of the tree species will help beautify the gardens. Some plants in autumn are not suitable for planting in clusters, and some other style leaves can be planted alone. For example, the leaves of the deciduous tree of Chinese tallow trees overlap each other, forming a diamond shape. The leaves are very beautiful. After a frost in autumn, the leaves are red and gorgeous, and a tree becomes a scene. Liquidambar formosana is also a kind of the red-leaf tree. It is also a deciduous tree. Its leaves are palmately trilobed and its edges are serrated. When the leaves are red, the layers of trees are dyed, and it has the reputation of "frost leaves are red than February flowers". There are fewer varieties of flowers that can be selected for the gardens in autumn, but chrysanthemums are unique among many flowers. Autumn chrysanthemum is a perennial herb, which is cool and cold-resistant. It is one of the four famous gentlemen in flowers. It has the high ornamental value and is one of the important choices of the garden plants.

Winter gardens have been facing bottlenecks in the plant allocation, because winter gardens lack the charm of autumn, not to mention the colorful and gorgeous spring and summer, and few flowers and trees are available. Besides paying attention to green, we also need to pay attention to the physique effect of the tree species. The cold-resistant trees such as the pine and the cypress are often used as the ornamental trees in gardens. Their deciduous leaves alternate in an orderly manner, which seems as if they are not deciduous as a whole, giving people the feeling of evergreen in winter and summer.

Confucius once said, "When the weather is the coldest every year, we know that most of the other plants are withering, and only pines and cypresses stand tall and do not fall." A large number of the pine and cypress trees make up for the green defects in gardens and enhance the appreciability of winter gardens. When it comes to the winter flower viewing, it is natural for the plum blossom blooming alone in the cold winter. In the gardens, the plum blossom can be planted centrally to form a plum blossom garden with the unique artistic conception. The landscape greening in winter can also rely on the snow scenery. The vegetation wrapped in the silver-wear can make the tourists have the unlimited poetry and reverie.

Optimize and adjust the plants that are not suitable for the growth to make them scientific and reasonable. Because of many reasons such as the soil, climate and environment and so on, some plants are not suitable for the local growth and the effect is very poor, so we should optimize and adjust the situation and re-plant the landscape plant varieties suitable for the growth. The unreasonable planting of plants, and especially those with serious potential safety hazards, should be adjusted in time to avoid serious consequences. The design of the landscape plants requires not only aesthetics, but also safety. For example, the tall trees, and especially plants with well-developed roots, such as Ficus macrophylla, cannot be too close to the buildings and walls. Otherwise it is very easy to damage the houses and walls. For example, deciduous plants should be planted in front of doors and windows of houses, so that they can get light in winter and shade in summer, and have certain seasonal changes. The landscape plants should also be optimized and adjusted to meet the existing functional requirements for the changing building environment. Because the application function has changed, the original plant landscape does not meet the requirements of the current environmental function, so after changing the application function, the requirements of the plant landscape have changed accordingly. This requires the rectification and adjustment of the original plant landscape. Remove the unsuitable plants and increase the suitable plant varieties.

4 Conclusion

At the beginning of the design, in order to achieve the desired landscape effect, it is necessary to ensure a certain density. However, with the continuous growth of plants, the original density has changed greatly, which not only destroys the original landscape effect, but also is not conducive to the normal growth of plants. At this time, the density should be readjusted so as to make plants grow properly. Replace the withered, distorted or even dead plants to maintain the beautiful landscape effect. Because of the damage and death of the plant diseases and insects in the process of the plant growth, it has a serious impact on the beautiful landscape. In order to maintain the original landscape effect, it is necessary to replace the plants with poor morphology or even death in time.

Acknowledgement. This study was sponsored by the Youth Foundation of Education Department of Sichuan Province, China, "Evaluation and Sustainable Development Plan of Sichuan Salt Road Ancient Town Based on World Heritage Standard" (Grant No. 16ZB0041) and the Social Science Research Base Foundation of Education Department of Sichuan Province, China, "Evaluation and Sustainable Development Plan of Sichuan Minority Villages' Cultural Landscape Based on World Heritage standards" (Grant No. JGYQ2015011).

References

1. Wu J (2017) Construction practice and planning of the ecological agricultural demonstration parks based on the landscape renovation – taking sima mining area as the research object. Chin J Agric Resour Reg Plan (04):111–112
2. Liang Y, Chen T (2017) Exploration on the design concept of the landscape park in northern Duisburg, Germany – based on the perspective of the ecological restoration in Huangshi mining area. J Hubei Polytech Univ (Hum Soc Sci Ed) (11):121–122
3. Miao Y, Chang X (2017) A preliminary study of the landscape design of the waterfront unused land in Kunshan Urban fringe area under the concept of "Urban Double Repair" – a case study of the ecological restoration of riverside green space along Wusong river in Kunshan. Chin Hortic Abstr (11):132
4. Wang M (2018) Exploration and practice of the characteristic urban leisure waterfront landscape design – taking the design of the waterfront landscape engineering on both sides of Liuyang river (Weichuan River-Sanyuangong section) in Liuyang city as an example. Mod Hortic Educ Ref (10):133–134
5. Yin X, Cao L (2018) Street landscape improvement model based on the historical context and the spatial status – design experiments on Kaifeng governing hall street. Popul Lit Art (11):105–106

Simulation of the Hydraulic Support System and the Constant Pressure Liquid Supply Based on Intelligent Systems

Mei Sun[1,2](✉) and Jianping Tan[1]

[1] Central South University, Changsha 410000, Hunan, China
xjp6667@163.com
[2] Changde Vocational and Technical College, Changde 415000, Hunan, China

Abstract. The hydraulic support is the most important and key supporting equipment in the fully mechanized mining face. In the process of the redesign, the hydraulic support provides enough safe space for the production of the working face from the angle of ergonomics, and is the main equipment to promote the advance of the working face equipment. With the progress of the science and technology, the fully mechanized face is developing toward the electro-hydraulic control and unmanned working face, which requires the higher and higher performance of the equipment, and especially the hydraulic support. The performance of the hydraulic support mainly depends on the hydraulic technology.

Keywords: Hydraulic support system · Simulation technology · Steady pressure supply · Pressure fluctuation

In order to improve the following characteristics of the support to the shearer, the mathematical simulation model of the support hydraulic system was established by AMESim based on the hydraulic principle of the support system in the working face. By simulating the process of the hydraulic support operation, the law of the action of the flow rate on the speed and the pressure of the support operation is analyzed and summarized, and the idea of the stable-pressure fluid supply adapted to the support operation is put forward.

1 Research Background of the Hydraulic Support System Simulation and the Pressure-Stabilized Fluid Supply

The hydraulic system of the hydraulic support forms a multi-stage parallel hydraulic supply and control network from the emulsified pump station to each hydraulic cylinder of each hydraulic support [1]. The system network includes the main hydraulic supply system and the hydraulic system of the support. The diameter of each cylinder, the flow rate of the emulsion pump, the diameter and length of the rubber hose, the size and specifications of the joint and the control valve are different. These parameters are mainly reflected in the resistance of the circuit, which will ultimately affect the speed of

Z. Xu et al. (Eds.): CSIA 2019, AISC 928, pp. 293–298, 2020.
https://doi.org/10.1007/978-3-030-15235-2_44

the bracket. How to reasonably design the configuration of the pipeline system according to the requirement of the support movement speed or to analyze and calculate the flow characteristics and the support movement speed according to the existing system analysis is very urgent to design and analyze the high-pressure emulsified fluid supply system and the hydraulic support hydraulic system of the automated and intelligent fully mechanized mining face [2]. The actual demand is also the content of this paper.

The hydraulic support is one of the most important equipments in the fully mechanized coal mining face. With the development of China's coal mining industry, the demand for the hydraulic support is increasing, and its performance requirements are also increasing. The hydraulic technology is the technical basis of the hydraulic support. Its performance determines the working performance of the support. Its dynamic performance directly affects the performance of the whole support [3].

Firstly, the basic composition, the working conditions and the common basic circuit of the hydraulic support system are introduced. Secondly, taking a certain type of the hydraulic support as an example, the hydraulic system is designed and selected, including the determination of the support types and the basic parameters, the design of the hydraulic cylinder, the design of the anti-slip hydraulic system, and the design of the large flow system. The pressure loss of the system is checked and calculated, and the rationality of the system design is verified theoretically [4]. Then, the mathematical model of the main components in the hydraulic system of the hydraulic support, the force balance equation between the hydraulic cylinder and the load and the continuity equation of the hydraulic cylinder are established, which lays the foundation for the establishment of the simulation model from the theoretical level. Finally, the simulation model of lifting and pushing the control system of the hydraulic support is established, and the dynamic characteristics of the system are analyzed. In the research of the control system, it is very important to establish the mathematical model of the system [5]. It is an indispensable and powerful tool for designing the system, analyzing the system performance and improving the system structure [6].

Through the theoretical analysis of the characteristics of the process of the pressure-stabilized fluid supply, a mathematical formula for calculating the pressure-stabilized fluid supply flow adapted to the operation of the support is derived. The formula shows that the pressure-stabilized fluid supply flow is mainly determined by the action types, the action stroke, the action quantity and the pressure setting of the unloading valve under the same hydraulic system of the support. Finally, the simulation model validates the technology of the steady-pressure supply, and compares the effect of the rated and steady-pressure supply schemes during the operation of the hydraulic support. The results show that the technology of the steady-pressure supply not only ensures the fast execution of the support action, but also reduces the fluctuation of the system pressure and reduces the pressure impact of the hydraulic system of the support. Based on the MSC.EASY5 software, the hydraulic control system of the hydraulic support is simulated and modeled. The lifting and falling pillars, the sliding and sliding frame, the overflow load and the unloading impact conditions of the hydraulic support are simulated and analyzed. The inner pressure of the front and the rear pillars under different loads is studied, and the time of lifting the frame of the hydraulic support is calculated.

The research results show that the hydraulic system design of the support is reasonable and the simulation model of the hydraulic system is correct.

2 Characteristics of the Hydraulic Support and the Hydraulic System of the Hydraulic Support

The hydraulic support is powered by the high-pressure liquid of the emulsified liquid pump station. By using the hydraulic control system, the support and its auxiliary devices can complete supporting, falling column, moving frame, pushing conveyor and protection actions, thus realizing the mechanization of the support work. In addition to the main working conditions of lifting the pillar support and unloading and moving the pillar, the working conditions of the hydraulic support also include the auxiliary working procedures such as adjusting the frame, protecting the wall, the side protection, falling down and anti-skid, such as the relative movement between the internal parts of the support and the adjacent support. These auxiliary processes can work independently or at the same time with some of the main processes, except that the adjustment must be carried out during the removal process.

During the support process of the hydraulic support, the support relationship between the support top beam and the roof of the working face will change due to the closure of the control valve, that is, from the original active support to the later passive support. Therefore, the support process can be divided into the active support (initial support stage) and the passive support (bearing stage). When the hydraulic support starts to drop the column, due to the sudden opening of the hydraulic control one-way valve, the force of the main spool is unbalanced, resulting in the phenomenon of the transient impact. Only when the high-pressure liquid in the column is unloaded (i.e. unloading the impact) can the support fall normally. Therefore, unloading the impact is regarded as an independent basic condition when studying the dynamic characteristics of the hydraulic support. The main operating procedures of the hydraulic support play a decisive role in the performance of the support, in which the unloading drop pillar seriously affects the support status of the support, and the sliding and the moving of the support are related to the speed of the support and the support of the lifting pillar, which will determine the performances of the support.

The hydraulic system of the support in the fully mechanized mining face mainly consists of the pumping station, the liquid supply, the return pipeline and the end support and the intermediate support connected in parallel in the liquid supply and the return pipeline. The layout of the supply and return pipelines includes the simple single-line supply, the circular supply, the single-line branch supply, the trapezoidal network supply and the half-face supply.

The hydraulic system of the hydraulic support has the following characteristics. The hydraulic system is huge and has many components. The hydraulic support is laid along the full length of the mining face with a large length (up to 200 m). In the hydraulic system, there are a large number of pillars (80–1000) and jacks (80–1500), as well as a large number of the safety valves, the hydraulic control one-way valves, the control valves, and a large number of the high-pressure hoses and pipe joints and so on, so the entire system is intricate. The sealing and reliability of the components in the system have a great influence on the work of the support.

The working pressure is high. The hydraulic support in the working face support roof requires a greater supporting force, and the initial supporting force is related to the working pressure of the pump station, generally 10–35 MPa. After the initial support, the piston chamber of the column is closed, and the hydraulic pressure can reach 40–60 MPa at the working resistance. Therefore, the hydraulic components are required to have the sufficient high-pressure strength.

The distance of the liquid supply is long and the pressure loss is large. The working fluid of the hydraulic support's column and the jack is supplied by the pump station located in the downward groove of the working face. The hydraulic energy needs to be transported over a long distance. The pressure loss is large. Especially when moving the frame and pushing the conveyor, the working fluid of the support's hydraulic system has a large capacity to circulate, so the main road is required to have enough passage flow section.

The working environment is bad, humid and dusty, with the limited working space, and the stope conditions often change. The inspection and repair is inconvenient, requiring the reliable hydraulic components and the long working time. There are the high requirements on the hydraulic components. The working fluid of the hydraulic support is the emulsified liquid, and the water accounts for about 95%, so its viscosity is low, and the lubricity and the rust resistance are not as good as the mineral hydraulic oil. Therefore, the material of the hydraulic components is required to be good, with the high precision, the good rust and corrosion resistance.

3 Simulation of the Hydraulic Support System and the Constant Pressure Liquid Supply

With the development and popularization of computers, the digital simulation method is formed based on the modern control theory. The method first establishes a mathematical model describing the state of the hydraulic system, and then calculates the dynamic and stable characteristics of the hydraulic components or the systems by the computer simulation.

The mathematical model for the dynamic analysis and the design of the support hydraulic system is established on the basis of the force and motion equation, the fluid mechanics and the thermodynamics equation in the system composed of components and elements. The factors such as the compressibility of the oil, the friction characteristic and the resistance of the unsteady moving parts in the pipeline should also be taken into account. Although there are some uncertainties and the approximation to a certain degree, compared with the classical control theory, it can reflect the real situation of the system more comprehensively and describe the state changes of the whole working process of the system.

To optimize the hydraulic system of the hydraulic support, we should not only consider the need of the system itself, but also take economic benefits as the basis. The imported hydraulic components are expensive and have the high design cost. This is not in line with the current situation of China's coal mining industry. Therefore, we must improve the accuracy of the domestic hydraulic components and strengthen the research and development of the new sealing materials.

Open the AMESim software and enter its sketch interface mode. Call the components in the mechanical library, signal the control library and the hydraulic library provided by the system to build the AMESim model of the hydraulic system of the mine hydraulic support, and carry out the simulation analysis. The names of the components have been marked on the diagram. Click on the preferred sub-model button to make the AMESim automatically select the simplest and the most suitable sub-model for each component. If you need to select another sub-model, you can double-click the model and select the sub-model you need. All the components of the hydraulic system can be automatically allocated by the software.

In order to minimize the leakage of the hydraulic system of the hydraulic support, the selection of the seals is the foundation, and the influencing factors of the hydraulic system, such as the design and processing of the sealing grooves, should be considered comprehensively. In terms of the pollution of the hydraulic oil, effective filtering measures should be taken to cut off the source of the pollution to the hydraulic system, strengthen the control of the pollution sources, and take full consideration of the optimal design of the leakage. The optimal design of the hydraulic system of the hydraulic support aims to improve the working performances of the hydraulic support, enhance the stability and reliability of the hydraulic system, reduce the probability of the hydraulic system failure, and thus improve the mining efficiency.

In order to reduce the joints of the pipeline, the integrated design method is adopted as far as possible. By reducing the loss of the pressure and the link of the failure, this design method makes the hydraulic system of the hydraulic support more sensitive, so as to improve the working efficiency, and also make the pipeline layout of the hydraulic support more concise and neat. The operation efficiency of the support plays a decisive role in the efficiency of the work. The diameter of the supporting column is large. Therefore, when designing the hydraulic system of the support, the flow through the support column and the jack should be increased as much as possible.

If the load of the working object can be determined, the working pressure of the whole hydraulic support system determines the economy and rationality of the system. The size of the actuator is determined according to the working pressure. If the working pressure is higher, the pressure level of the actuator will be correspondingly higher, and the corresponding equipment cost will be higher. Generally, the working pressure of the hydraulic support is 20 MPa to 32 MPa. When designing the hydraulic system, we should have a detailed understanding of the pressure bearing range of each component and take all the measures to ensure that the system pressure of any part of the hydraulic system in the working face is within the allowable range.

The control modes of the hydraulic support include the manual control mode, the one-way adjacent frame control mode, the two-way adjacent frame control mode and the automatic control mode, such as the grouping program control mode, the pilot program control mode and the remote control mode. From the control of the frame to the automatic remote control, its selection mainly depends on such factors as the safe operation, the reliable operation, the quick operation and the convenient maintenance. As is mentioned above, the coal seam dip is one. Under such coal seam conditions, it is easy for the coal blocks and the slam stones to roll down, which poses a great threat to the personal safety of the workers. In order to ensure the safety in the production, this type of the support adopts the one-way adjacent frame control mode.

4 Conclusion

The hydraulic support is one of the most important equipments in the fully mechanized coal mining face. Its investment accounts for about the total investment of the complete equipment in the fully mechanized coal mining face. Its function is not only to support the roof and maintain the safe working space, but also to push the conveyor and the shearer. The hydraulic transmission technology is the technical basis of the hydraulic support, and its performance determines the working performances of the support, and its dynamic performance directly affects the performances of the whole support. At present, the hydraulic system design of the hydraulic support relies on the experience and the analogy method, and then check whether the design is reasonable through the prototype. This not only takes a lot of time and energy, but also has the huge risks. Through the simulation analysis of the hydraulic system, unreasonable points can be found in the early design stage and the loss can be reduced.

References

1. Li Z (2017) Research and analysis on the main supply and the return fluid system of the high mining height hydraulic support. Mech Manag Dev (10):130–131
2. Sun J, Zhang G, Zhao Q, Xie H (2018) Research on the anti-slip mechanism and the safety protection device of the large inclined hydraulic support. Coal Eng (07):181–182
3. Cao L, Guo H, Wang T, Li P, Li H (2018) Diagnosis of the valve core stuck in the hydraulic system of the hydraulic support column. J Shandong Univ Sci Technol (Nat Sci Ed) (09):188–189
4. Liang B, Wang Y, Wu X, Su F (2018) Development and application of the tracked hydraulic support loading machine for the fully mechanized mining face. Coal Sci Technol (10):151–152
5. Zhou Q (2018) Multi-layer affective computing model based on emotional psychology. Electron Commer Res 18(1):109–124
6. Zhang Y, Chen J (2018) Strength analysis and structural optimization of the ZY6400/21/45 hydraulic support base. J Henan Inst Eng (Nat Sci Ed) (12):15–16

Study of the Chinese-English Translation of Computer Network Terms

Fenfen Zhu(✉)

Lushan College of Guangxi University of Science and Technology, Liuzhou 545006, Guangxi, China
1959278975@qq.com

Abstract. With the continuous development of the computer science, the computer network terms play an increasingly important role in the understanding of computer English. This paper makes a general study of the Chinese translation of the English computer network terms by analyzing the characteristics of the computer network terms, the comparison of the English and Chinese word-formation, and the translation principles and methods and so on. Although the history of the computer development is short, it has a wide range of applications. The computer terminology has its own characteristics. The computer network terms produce a large number of the new words through infiltration, derivation and transformation, which is a significant feature of it.

Keywords: Computer · Network terminology · English translation · Language environment · Cultural connotation

The computer network terms, in terms of their meanings, are mostly monosematic. Most computer network terms can be translated literally, so the literal translation is the most common and effective translation method in the process of translating the computer network terms.

1 Analysis of the Particularity of the English-Chinese Translation of the Computer Network Language

Translation is a highly individualized and creative work. The Chinese translation of the computer network terms can be flexibly applied according to the actual situations by means of the literal translation, the free translation, the transliteration, the borrowing, the synthesis, the transliteration and the abbreviation [1]. In order to translate the terminologies comprehensively, thoroughly and accurately, translators must follow the basic principles of the terminology translation: accuracy, simplicity and standardization. In terms of the translation principles, there are different opinions among different people.

For example, some scholars advocate "accuracy, readability and transparency", while others advocate "singleness, simplicity and standardization", and still others believe in "accuracy, simplicity and standardization" and so on [2]. The author tends to "accuracy, simplicity and standardization", because "conciseness" need not be

Z. Xu et al. (Eds.): CSIA 2019, AISC 928, pp. 299–304, 2020.
https://doi.org/10.1007/978-3-030-15235-2_45

discussed. The terminology of the original texts is not concise, and why should "translation" be brevity? However, simplicity is a very important principle, because the meaning of a word and the translation of a word also conform to the international and the national standards. A term refers to only one concept, and a concept is expressed in only one term [3].

The computer network terminology also belongs to the category of the scientific and technological terminology. Therefore, in the process of translating the computer network terminology, translators should strictly follow the translation criteria of the scientific and technological terminology [4–6]. Firstly, the translation of the computer network terms must follow the principle of accuracy. Only by grasping the meanings of the terms accurately can the original meanings be faithfully conveyed. Secondly, the translation of the terms into Chinese should also have the principle of conciseness. "If the translated name is too long or the translated explanation is too much, it will not conform to the characteristics of the simplicity of names in the Chinese culture". Secondly, the principle of memorability in the translation of the scientific and technological terms is also a criterion that must be followed. If we translate "IR C" in the computer field into "Internet Relay Chat Room" literally, it will appear that the translation is too long and hard to remember.

2 Analysis of the Common Methods of the Terminology English Translation

Generally speaking, there are two strategies for the terminology translation. One is the "foreignization" strategy, and the other is the "domestication" strategy [7]. Usually, the path of the terminology translation is foreignization before domestication, not vice versa [8]. This is different from the literary translation because the terminology translation considers accuracy, simplicity and standardization rather than the recognition, acceptance and convenience of the targeting language readers, which is what the literary translation has to pay attention to. In the process of the translation, these two strategies are embodied in our transliteration and the free translation, both sounds and meanings, and the image translation. Next, I will briefly talk about these translation methods.

The transliteration, as its name implies, is based on the pronunciation of the original words, which will be translated according to the Chinese Pinyin. The transliteration is the most obvious sign of foreignization and the most commonly used method in the initial stage of our translation. In addition, the transliteration is also a popular method of the terminology translation in the world. It is conducive to the international exchange of the science and technology and the unification of the terminology translation.

The so-called free translation is to reproduce the original meanings with the most appropriate and natural words in Chinese according to the meanings of the original words. In order to better grasp and apply the "free translation", it is not only necessary but also beneficial to analyze the formation of the terms. The following four situations need to be considered. The synthesis method: It is a characteristic of the computer network terminology. Even if more than two words are used to synthesize a compound word, the meaning of this compound word is a combination of the two original meanings. The affix method: It is the most basic word formation in the computer network terminology, that is, to form a new word by adding the affixes (including the prefixes and the suffixes). When translating into Chinese, it is enough to translate according to the meanings of the affixes and the roots. The patchwork: It is to cut off two or more words and turn them into a new word. To translate this kind of the vocabulary, merge the original meanings. The abbreviation: It is to take the initial letters of several words to form a new word. When translating, we can translate each word according to its original meanings. When translating, both the transliteration and the free translation are used, usually half to half. Most of these terms consist of the proper nouns. In the computer network terminology, there is often a term consisting of an alphabetical form plus a word. There are two situations in translating such terms.

Another very important principle of the terminology translation is to respect the history, "established by the people through long social practice" and "name follows owner". As for the terms that have been translated in history and have been widely accepted, even if they are indeed flawed by the present standards, the author believes that they should be "customized" without spending a lot of the manpower, the material resources and the time to discuss whether it is right or not. For example, the term Marilyn Monroe does not need to be translated into "Marilyn Monroe". Similarly, Confucius must not be translated into Kongzi, but be reverted to Confucius. The zinc plate in English does not need to be translated into "galvanized iron sheet". There is no need to change to "laptop" at all in the translation of the laptop computer.

Similarly, the terminology New York cannot be translated into "New York", nor can New Zealand be changed to "New Zealand". If we do not adhere to this principle, and do not maintain the existing terms which are widely accepted by the society and are relatively easy to use, then it is conceivable that the translation and application of the terms will be confused. However, it must be pointed out that the author does not oppose the revision of the terms with obvious errors and shortcomings, but emphasizes that we must adhere to a scientific concept of the time and space, and not rashly deny a definite translation in history. What need to be discussed and revised is those terms that have just appeared, with only the trial translation, but have not yet been translated.

3 Analysis of Notices in the English-Chinese Translation of the Computer Network Language

With the continuous development of the computer science, the computer network terms play a more and more important role in the understanding of the computer English. Although these terms are short in length, they have a strong information function and are the starting point and cornerstone of the computer research. Therefore, attention should be paid to the translation of the computer network terms. However, due to the constraints of the region, the environment, the cultures, the professional knowledge, the self-condition and the ability of the recipient to understand and bear, the translator is either unique, pleasing the audience, or also tends to translate hard, which makes the translation of many terms open to discussion.

The mistranslation of the computer network terms involves a wide range of the phenomena, and its manifestations are diverse. There are many reasons for the problem. However, the main reasons lie in the translator's subjective disconnection, poor use of the dictionaries and unfamiliarity with the relevant professional knowledge. This requires the translators to fully recognize the seriousness of these errors, understand their causes, read the professional dictionaries and learn more about the relevant professional knowledge in the process of their translation, so as to correct erroneous translation methods, ensure the accuracy and appropriateness of the computer network terminology translation and improve the quality of our translation.

4 Analysis of the Translation Strategies for the English-Chinese Translation of the Computer Network Terms

For the computer network terminology, this paper points out the process of the terminology translation from eight aspects: the nominal and adjective modification, the orthogonal and ironic order, the monosemy and conciseness, the conciseness and rationality, the Chinese characteristics and conciseness, the transliteration and the free translation, the analogy and adaptability, the translatability and untranslatability. Contradictions exist in the process of the development of everything and run through the entire process of the development of everything. They depend on each other, penetrate each other and transform each other under certain conditions. Generally speaking, the existence of any contradiction is only relative and temporary, and there is no absolute contradiction.

Due to the particularity of the computer English, the traditional translation methods such as the literal translation, the free translation and the transliteration sometimes fail to translate the terminology accurately and concisely. When the traditional translation methods are not competent, "zero translation" may be a safer and wise choice. The concept of "zero translation" was put forward by Professor Qiu Maoru. He believes that "zero translation" refers to the translation of the ready-made words from the unused words into the source language, which contains two meanings:

First, the words in the source texts are intentionally untranslated, and second, the ready-made words in the unused words are translated into the words in the source texts. In the Chinese translation of the computer network terms, many zero translation methods are adopted, such as "CPU, LCD, and DOS" and so on. These English terms may be more convenient for everyone to remember than "CPU", "LCD", and "Disk Operating System". Another example is "BIOS-system". After the literal translation, the Chinese name should be "basic input and output system". If the literal translation is chosen, the translation will become cumbersome and fluent, which cannot embody the principle of the terminology translation concise and easy to remember. The zero translation method makes the translation much simpler, and "BIO S" is also the computer professionals' consistent abbreviation of "basic input and output system", so that the translation meets the professional requirements and is easy for readers to accept. There is also "Excel, Access, PowerPoint and Word" software, which cannot be replaced by the corresponding Chinese language in the process of the Chinese translation, and the English terms of this office software are familiar to everyone. In this case, copying the original texts is undoubtedly the wisest way.

Starting from the importance of the computer network terminology, this paper probes into the translation principles of the computer network terminology, and analyses the translation methods of the computer network terminology through the specific examples. It is difficult to use the principles of "faithfulness, expressiveness, elegance" or in the "spirit and form" in the translation of the computer network terminology, instead of using the literal or the free translation as a general requirement. On the basis of familiarity with the computer knowledge, translators should consult dictionaries and the related literature to avoid looking at the texts for meanings. Terminology has the characteristics of monosemy only in a specific field. Therefore, the understanding of the terminology semantics must start from the connotation of the whole text, so as to accurately translate the meanings of the terminology and avoid taking out meanings out of the contexts.

At present, there are many English-Chinese dictionaries about computers, such as the English-Chinese Computer Dictionary, the English-Chinese Computer, the Communication Dictionary, and the New English-Chinese Computer Words Dictionary and so on, which provide us with a lot of help in our study and work. When we encounter the terminology which is difficult to determine in the process of our translation, we must consult several dictionaries more, and we must not allow exhaustive translation, so as not to cause the mistranslation, such as ease, modem, printer, Photoshop, drive and run and so on. In addition, although some computer network terms are not included in any dictionary, as a translator, we cannot give up simply because we can't find a dictionary or two without finding the terms we want to find. Because there is a problem of the uniform naming of terms, if everyone has their own understanding and translation of the same term, it will bring great inconvenience to our communication. For example, "computer" can be translated into "computer", "emblem machine" and "personal computer" in Chinese, but now in most cases the name "computer" is commonly used, which is more appropriate.

5 Conclusion

To translate the computer network terms well, we need to base on the particularity of the language. Therefore, according to this particularity, this paper draws the following conclusions. The computer network terminology needs to be changed according to the changes of the times and the development of the science and technology, and the computer network terminology needs to be adopted according to the wishes of the audience. The English-Chinese translation of the computer network terminology needs to adopt the transliteration and the free translation, and we should combine the methods to express the contents of the terms more appropriately.

References

1. Hu F, Liu L (2015) Exploring the methodology of the Chinese and the western translation studies and approaches in the study of the translation skills – comments on the comparative study of the Chinese and the western translation studies in the philosophical dimension. J Central South Univ For Technol (Soc Sci Ed) (02):145–146
2. Zhao J, Cui H, Hua F (2015) Translation practice report on how we got to now: six innovations that made the modern world. J Beijing Univ Chem Technol (Soc Sci Ed) (21):255–256
3. He G, Hu S, Long L (2016) Literal translation, free translation, combination of transliteration and free translation or the transliteration plus annotation? – A brief analysis of the translation skills of the health-preserving terms in the yellow emperor's canon of internal medicine. J Chongqing Norm Univ (Philos Soc Sci Ed) (12):321–323
4. Bao W, Zhou H, Xie F (2017) An optimal design of web based online English teaching by using struts framework. Boletin Tecnico/Tech Bull 55(10):58–63
5. Bao W, Zhou H, Lu W, Xie F (2016) The system of knowledge management using web based learning and education technology. Comput Syst Sci Eng 31(6):469–473
6. Bao W (2014) Research on college English teaching based on intercultural communication in engineering universities. World Trans Eng Technol Educ 12(3):191–195
7. Li Z, Lan X (2017) Integration of the "explicit" and "implicit" translation teaching for the non-English majors – reflections on enhancing the translation abilities of the non-English majors. J Xihua Univ (Philos Soc Sci Ed) (05):111–112
8. Guan R, Luo W (2018) The enlightenment of the English translation practice of handbook of international conference on MTI translation teaching – taking the English translation of handbook of China-ASEAN information port forum 2016 as an example. J Chongqing Univ Educ (01):150–151

Study of the Soybean Imitation Cheese Based on Intelligent Systems and Big Data Analytics

Zhiguo Zhang(✉) and Qi Zeng

College of Food Science and Engineering, Qilu University of Technology (Shandong Academy of Sciences), Jinan 250353, Shandong, China
lixiao694370507@163.com

Abstract. The imitation cheese is the product with the cheese characteristics which can meet the specific requirements by using the non-milk-derived protein and fat or the full substitution of the milk protein and the milk fat. This paper introduces the production principles, processes and key control points of the imitation cheese with the soybean as main raw material, summarizes the researches and development of the soybean imitation cheese in recent years, and forecasts the market prospect of the soybean imitation cheese in China.

Keywords: Soybean imitation cheese · Functional characteristics · Key control points · Prospects

Cheese is a fresh or fermented mature dairy product made from milk, skim milk, cheese or a mixture of these products by curdling and separating the whey. It is a kind of food with rich nutrition and uniquc flavor. It is known as the pearl on the crown of the dairy industry. It occupies a very important position in the western diet structure. Cheese is rich in nutrients. Besides the high-quality protein, it also contains sugar, organic acid, calcium, phosphorus, sodium, potassium, iron, zinc, magnesium and other trace mineral elements, as well as the fat-soluble vitamin A, carotene and water-soluble vitamin B_1, vitamin B_2, vitamin B_6, vitamin B_{12}, nicotinic acid, pantothenic acid, biotin and other battalions [1, 2]. The traditional cheese made from the cow's milk has a high cost and does not conform to the consumption concept and the dietary habits of the traditional dairy products in China, which restricts the development of the traditional cheese in China. Therefore, the research and development of the soybean cheese imitation has great market potential and practical significances.

1 Imitation Cheese

The imitation cheese is usually defined as the milk fat and/or the milk protein substituted by some or all the non-milk sources of raw materials. It is generally a product of the vegetable-derived fat or the protein substitution. Generally speaking, based on the raw materials and the manufacturing technology, the imitation cheese is mainly divided into three types: the analogue cheeses or cheese analogues, the filled cheeses and the vegetable protein cheese [3, 4]. At present, scientists have systematically and thoroughly studied the effects of the soybean protein, casein, sodium caseinate, milk

Z. Xu et al. (Eds.): CSIA 2019, AISC 928, pp. 305–310, 2020.
https://doi.org/10.1007/978-3-030-15235-2_46

protein, corn starch and other substitutes for the protein in cheese on the imitation cheese [5–7], and the effects of the vegetable oil/animal fat instead of the milk fat on the microstructure and texture of the imitation cheese [8, 9]. The advantages and disadvantages of the imitation cheese made from different protein and fat sources were systematically expounded. Researchers have also studied the effects of the raw materials and processes on the functional properties of cheese, such as microstructure, rheology, texture and flavor, and the internal relationship between them. These studies provide guidance and theoretical basis for the diversified development of the cheese products [4, 10].

2 Soy Imitation Cheese

2.1 Brief Introduction to the Soybean Imitation Cheese

The soybean imitation cheese is a more comprehensive imitation cheese which has been studied extensively and comprehensively at present. It belongs to the vegetable protein cheese in the imitation cheese. It is mainly a kind of the cheese product which is obtained by completely or partially replacing the milk protein and the milk fat in the milk cheese with the soybean protein and the soybean fat (the soybean hydrogenated oil) [4]. The production process of the soybean cheese is similar to that of the traditional dairy cheese. Mainly using the soybean as the raw material, through grinding, filtering, boiling and other processes, inoculate the lactic acid bacteria to ferment, solidify the soybean protein in the soybean milk, and then after the pressing and discharging of the whey, ripening and other processes, the cheese product produced has the good color, aroma and texture. It has the good flavor and the unique functional properties [11]. Compared with the traditional fermented soybean food in China: tofu, it has the advantages of the low salt content, the good safety, the strong function, and the rich and comprehensive nutrition [12].

2.2 Research Status of the Soybean Imitation Cheese at Home and Abroad

The concept of using the soybean instead of part of the milk to make the soybean cheese was put forward by Y. D. Hang and H. Jackon in 1967. Y. D. Hang and others used the pure soybean milk as the raw material, compared the characteristics of the cheese coagulation obtained by the three coagulation methods: the acid-lowering coagulation, the CaSO4 coagulation and the lactobacillus fermentation coagulation. They discovered that the cheese clots have the advantages of the low moisture content and the good texture compared with other two curd methods, but the curd time is longer. Garro et al. [14] studied the sugar metabolism in the soymilk by the three fermentation strains, the streptococcus thermophilus, the lactobacillus casei and the lactobacillus fermentation. The results showed that there was no significant difference in the sucrose metabolism among the three strains. They could metabolize 60–75% sucrose in the soymilk, but only the lactobacillus fermentation could metabolize 85% of the stachyose. Atia et al. [15] optimized the processing technology of the mixed soybean cheese, determined that the mixed soybean cheese with 5% soybean protein had better ripening and sensory characteristics, and analyzed its microstructure.

Shao [16] and others introduced the preparation method of the vegetable protein cheese. Using the soybean protein as the raw material, the vegetable protein cheese by fermentation of the soymilk with the lactobacillus casei and the streptococcus lactis was produced. Wang [17] and others used the soybean as the raw material, and through grinding, filtering and boiling processes, they inoculated the lactic acid bacteria fermented soybean milk, and after the protein precipitation, and after pressing, blanking and other processes, the soybean cheese products with good color, aroma and texture were obtained by inoculating the hairy mould and seasoning and ripening. He [18] and others used the soybean protein isolates to replace the enzymatic coagulation casein to produce the mozzarella cheese. The effects of the soybean protein isolate addition, the heating temperature and the stirring speed on the qualities and characteristics of the imitation cheese were studied.

3 Key Control Points in the Production of the Soybean Imitation Cheese

In the process of the soybean cheese production, the key control points, such as the selection of the starter, the selection of the enzymes for production, the addition of salts and the selection of the technological conditions, are very important for the characterization of the flavor, the texture and the properties of the soybean cheese.

3.1 Fermentation and Rennet

The choice of the starter and the chymosin has an important influence on the texture and flavor formation of the soybean cheese. The fermentation agents include not only the lactic acid bacteria, but also the post-ripening fungi (such as yeast and mold). The microflora on the surface and inside of the cheese, together with the lactic acid bacteria and their metabolites, together with the processing technology and the internal and the external conditions, endowed the unique structures and flavor characteristics of the soybean cheese. The acid produced during the fermentation can reduce the PH in the soybean milk and make the soybean protein precipitate, to achieve the coagulation effect [13]. The addition of the chymosin can also improve the coagulability, water retention, chewiness and elasticity of the products.

Zhang [14] used the lactobacillus plantarum as the starter, and the glutamine transaminase (MGT) as the rennet, to produce the soybean cheese, and used the fungi isolated from Huizhou Mao tofu to ferment after ripening, which improved the sour vegetable flavor after the fermentation of the lactobacillus plantarum. The soybean cheese had good texture and flavor. Feng [19] fermented the soymilk imitation cheese with the streptococcus lactis, the streptococcus lactis and the leuconostoc lactis, and used the chymosin MAXIREN 150 to produce the soymilk imitation cheese. The products had good cheese flavor and no distinct bean odor. The nutrient composition had no significant difference with the pure cow cheese, which effectively reduced the production costs of the cheese. Li [20] and others used the streptococcus thermophilus and the lactococcus bulgaricus (ratio 1:1) as the starter to produce the soymilk mixed cheese, and the produced soymilk-milk mixed the fresh cheese has the good color and the moderate hardness.

3.2 Addition of Salts

The addition of salts can act on the non-enzymatic coagulation stage, including the calcium acetate, calcium chloride, calcium sulfate, and magnesium chloride and so on. At present, there are three main theories about the mechanism of the salt ions on the coagulation. (1) Based on the salting-out theory, the cations in salt and the negatively charged amino acid residues on the surface of the heat-deformed soybean protein. In combination, the electrostatic repulsion between the protein molecules decreases to form gels. Because the water and ability of salts are stronger than that of the proteins, when salt is added, the competition for the surface hydration layer of the protein molecule leads to the decrease of the protein stability and the formation of gels. (2) Based on the characteristics of the protein isoelectric point sedimentation, adding salt ions can reduce the PH value of the soybean milk. When the PH reaches about 6, the soybean milk begins to coagulate. (3) A large number of the carboxyl groups contained in the soybean protein can bind to the bivalent cations, such as Ca2+ and Mg2+, which are dissociated from the salt coagulants, and form a protein metal ion bridge to form the protein gels.

The salt additions and types play an important role in the curd milk of the soybean protein, which can directly affect the texture characteristics of the products.

The soybean milk fermented by the starter has more diverse biological and functional properties. The research of Yang et al. [1] shows that the fermented soybean juice has the stronger ability to scavenge 1,1-diphenyl-2-picrylbenzene callus (DPPH) free radicals. The protease produced by the fermentation strains can hydrolyze the soybean protein into the small molecular active peptides and the amino acids. A lot of studies have shown that the soybean protein peptide has the biological functions of reducing the cholesterol [21], antioxidant [22], and improving the immunity [23]. The fermented soybean food contains more oligosaccharides, including the natural oligosaccharides (such as raffinose oligosaccharides, mainly the raffinose, the stachyose, the pilose and the osmanthus and so on), as well as the oligosaccharides, sucrose, fructooligosaccharides, galactoligosaccharides, isomaltooligosaccharides and oligosaccharides produced during the fermentation. These oligosaccharides in the fermented soybean food, also known as the bifid factors, cannot be digested and absorbed by humans, but can increase the beneficial bacteria such as the bifidobacteria in the intestine and reduce the harmful bacteria, thus improving the digestive system function, lowering the blood pressure and the serum cholesterol, reducing the toxic products, enhancing the immunity and delaying the senility and other physical functions.

3.3 Process Conditions

In the production process of the soybean imitation cheese, the soybean water ratio, the fermentation temperature, the fermentation time and other technological conditions also have a great impact on the texture and flavor of the soybean imitation cheese. The improper soybean water ratio will lead to the loose or hard cheese texture and the poor taste, which will have a negative impact on the subsequent ripening process of the cheese. The fermentation temperature can make the starter play the best role, avoid the waste of the raw materials, and produce the good texture and unique flavor substances of the cheese. The appropriate technological conditions are indispensable for the production of the soybean imitation cheese products with good qualities and flavor.

4 Research Significances and Prospects of the Soybean Imitation Cheese

Cheese is a kind of the dairy food with rich nutrition and unique flavor. However, the development of the cheese in our country has been greatly restricted, and the market acceptance is very low. The main reasons for this situation are as follows. 1. The high cost of the cheese production leads to the high price of the traditional cheese products. 2. The flavor of the traditional cheese is not acceptable to our people. In the face of this situation, the soybean imitation cheese came into being at the historic moment. Firstly, using the abundant soybean milk resources in China to replace part of the milk, the cost of producing the soybean cheese is greatly reduced, and the traditional flavor of the milk cheese can be retained, and the complementarity of the protein and the amino acid contained in the soybean and milk can be realized for the human growth, and development, various physiological activities and the tissue repair provide nutrients [24]. In addition, the soybean milk contains no cholesterol, and the unsaturated fatty acids such as the linoleic acid and the α-Linolenic acid are high in content. They have the functions of softening the blood vessels and preventing the cardiovascular diseases [25]. The soybean cheese is rich in protein, fat, vitamins, calcium, phosphorus and the essential amino acids, fatty acids and other nutrients, and the nutritional value is very high.

According to China's natural resources, population and development potential of the dairy animal husbandry, the way of combining the cheese products with the soybean protein conforms to the basic national conditions of our country. The mixed cheese not only reduces the production costs of the cheese, alleviates the contradiction of the insufficient milk source, but also suitably changes the taste of the cheese and makes it suitable for the consumption habits of our people. The mixed soybean cheese will certainly improve the current situation of the insufficient high-quality protein in the dietary structure of Chinese people, provide the new channels for the development and utilization of the soybean resources in China, and provide new opportunities for the development of the related foods.

Acknowledgement. This research is supported by Shandong Provincial Natural Science Foundation, China (ZR2016CM38).

References

1. Yang JH, Mau JL, Ko PT et al (2000) Antioxidant properties of fermented soybean broth. Food Chem 71(2):249–254
2. Song J (2012) Development of the soybean compound cheese. Agric Technol 32(4):208
3. Guinee TP, Carić M, Kaláb M (2004) Pasteurized processed cheese and substitute/imitation cheese products. Cheese Chem Phys Microbiol 2:349–394
4. Zheng Y, Mo B, Liu Z et al (2014) Research progress of the imitation cheese. Food Ferment Ind 40(2):174–179
5. Aljewicz M, Grażyna C, Kowalska M (2011) Cheese-like products, analogs of processed and ripened cheeses. Zywnosc Nauka Technologia Jakosc 18(5):16–25
6. Hoffmann W, Hinrichs M, Johannsen N et al (2005) Use of different acid caseins in analogue low-moisture mozzarella. Milchwiss_Milk Sci Int 60(40):395–398
7. Ye A, Hewitt S (2009) Phase structures impact the rheological properties of rennet-casein-based imitation cheese containing starch. Food Hydrocollids 23(3):867–873
8. Cunha CR, Dias AI, Viotto WH (2010) Microstructure, texture, colour and sensory evaluation of a spreadable processed cheese analogue made with vegetable fat. Food Res Int 43(3):723–729
9. Jana AH, Upadhyay KG, Solanky MJ (2005) Quality of Mozzarella cheese analogue made using different fats. J Food Sci Technol-Mysore 42(6):497–500
10. Bachmann HP (2001) Cheese analogues: a review. Int Dairy J 11(4):505–515
11. Garro MS, Valdez GFD, Giori GSD (2002) Application of conductimetry for evaluation of lactic starter cultures in soymilk. J Food Sci 67(3):1175–1178
12. Qi Z (2012) Study of the processing technology of soybean cheese. Nanjing agricultural university
13. Hang Y, Jackon H (1967) Preparation of soybean cheese lactic starter organisms: I. General characteristics of the finished cheese. Food Technol 21(7):95
14. Garro MS, Valdez GFD, Oliver G (1998) Growth characteristics and fermentation products of Streptococcus salivarius subsp. thermophilus, Lactobacillus casei and L. fermentum in soymilk. Zeitschrift für Lebensmitteluntersuchung und -Forschung A, 206(1):72–75
15. Atia M, Xia W, Zhang G (2004) Effect of soy protein supplementation on the quality of ripening Cheddar-type cheese. Int J Dairy Technol 57(4):209–214
16. Shao W, Li A (2000) Development of the vegetable protein cheese. China Dairy Ind 28 (6):3–4
17. Wang J, Wang C, Han C et al (2002) Development of the Chinese cheese – bean cheese. Food Sci Technol 1:21–22
18. He S, Li X, Wu P (2010) Study of the technology of simulating mozzarella cheese with soybean protein isolate. China Dairy Ind 38(6):16–19
19. Feng J (2003) Study of the technology of the mixed cheese with soybean milk and cow milk. Northeast agricultural university
20. Li Y, Li R, Wang J et al (2014) Optimization of the processing technologies of the soymilk-milk mixed fresh cheese. Jiangsu Condiment Subsid Food 2:12–15
21. You X (2000) Functional fermentation products. China Light Industry Press, Beijing
22. Chen HM, Muramoto K, Yamauchi F (1995) Structural analysis of antioxidative peptides from soybean. beta.-Conglycinin. J Agric Food Chem 43(3):574–578
23. Yamauchi F, Suetsuna K (1993) Immunological effects of dietary peptide derived from soybean protein. J Nutr Biochem 4(8):450–457
24. Ji Y, Zhang C, Liu Y et al (2005) High-yield cultivation techniques of a new high-protein soybean variety (line) Lianke 928. Mod Agric Sci Technol: J Seeds Seedl 12:42–44
25. Peng Y, Zhao M (2003) Study of the soybean protein replacing milk fermentation. Sci Technol Food Ind 24(1):17–19

Study of the Structural Renewal of "Snow Bud" in Shutao Company Based on Intelligent Systems and Big Data Analytics

Yiming Xu, Siyang Gu, and Pinwu Li(✉)

Sichuan Agricultural University, Chengdu 611130, Sichuan, China
deyi48456115@126.com

Abstract. The improved "Snow Bud" packaging mainly includes the medium packing and the external packaging, with the cuboid as the main structure, and the paper as the main material. The inner packaging uses the tin foil paper bag. Because of the thin thickness of the carton used and the problems in colors, patterns, character design, structures and other aspects, the purpose of the improvement is to use the new structures to enhance the practical functions of the packaging and to explore the application of the new materials in the tea packaging.

Keywords: Shutao · "Snow Buds" · Packaging structure · Packaging materials

1 Summary of Shutao "Snow Bud" Tea Packaging

1.1 The Cultural Origin of Tea

The so-called "tea culture" is a general term of various concepts, beliefs, emotions, and love and hatred between people, and between people and nature with tea as a carrier expressed in the process of discovering and utilizing tea as foods, drinks, gifts and sacrifices by the Chinese people living in the land of China. It is also usually embodied in the four aspects of sociality, mass, nationality and regionalism. As Mr. Lu Xun once said, "Having good tea and being able to drink good tea is a kind of the pure happiness".

The tea culture in China has a profound foundation. China is the home of tea. There are six kinds of tea, green tea, black tea, Oolong tea, yellow tea and white tea, with twenty tea-producing provinces and 80 million tea growers, so that we are the real big tea-producing country. At the same time, different varieties of tea can give people different tastes, and people can feel different tea cultures, so when designing the tea packaging, we should customize it according to the characteristics of the tea varieties [1].

1.2 Functions of the Tea Packaging

Men depend on clothes, while horses on saddles. Whether a man is in high spirits depends on how well he dresses. Whether a product is attractive or not depends on its packaging. With the continuous development of the social economy, the dependence

Z. Xu et al. (Eds.): CSIA 2019, AISC 928, pp. 311–318, 2020.
https://doi.org/10.1007/978-3-030-15235-2_47

on the commodity packaging is increasing in the process of the commodity sales, and the design of the commodity packaging is also getting more and more attention. The packaging design usually refers to the selection of the appropriate packaging materials according to the characteristics of the products, specific environment and user preferences, and the use of certain design techniques, through scientific and reasonable conception, to shape the external shape of the packaging, in order to achieve better expected goals. At the same time, it is also an important product image design language, which can convey the designer's feelings and express the corresponding artistic aesthetic feelings by means of the design. It can be said that the packaging design is the process of fully expressing the designer's aesthetic interests and design skills.

1.3 Background of the "Snow Bud" Packaging Design

After more than half a century of development, Shutao Company has developed into a modern tea industry enterprise integrating the tea planting, production, processing, sales and research and development. There are nearly 50 thousand mu tea bases with good ecological environment and excellent tea varieties in Emei Mountain, Qingcheng Mountain, Ya'an Mingshan and Pujiang of Chengdu. The development model of the unified planning, the unified planting, the unified management and the unified picking is adopted for the quality management, and the authoritative certification of the pollution-free agricultural product bases has been passed.

The "snow bud" is collected from the company's own production base, where the environment is good, mist-shrouded, and the winding paths are quiet, and is carefully made by selecting the early spring green tea. The shape of the strands is fine and uniform. The soup is bright yellow and green, with a long lasting fragrance and refreshing taste.

1.4 Analysis of the Current Situation of Shutao "Snow Bud" Packaging

In view of the characteristics of the tea itself, such as the moisture susceptibility, the odor absorption and the fragility, the requirements of the tea packaging are more stringent. A good tea packaging can not only increase the share of the sales, but can also enhance the brand images.

The existing packaging of the selected "Snow Bud" uses the carton as the main material. On the basis of 300 g specifications, the cuboids (the length and the width are equal while the height is not) as the large packaging, and the small cuboids (the length, the width and the height are not equal) as the inner packaging, with the tinfoil paper as the small packaging to form the overall structure. The carton is not thick enough, and the moisture proof is not good, and it looks as if it is of very poor quality. It can be seen that there are many problems in it. On the whole, its shape is similar to the packaging of a chocolate milk tea. It has the poor identification and is easy to mislead the consumers. In addition, the number and the specifications of the internal packaging and the medium packaging are not consistent enough, with the lack of the material selection and application, resulting in many shortcomings in its practical and environmental protection. So for the existing packaging, a survey of some consumers around was done, and more than 90% of the per capita think that the packaging still has a lot of room for improvement. Therefore, this study takes Shutao "Snow Bud" as the object, aiming at the product positioning area, to improve its product packaging, and we strive to achieve the cultural connotation, the appropriate packaging, the green packaging, the simplicity, the elegance, the simplicity and the fashion.

2 Psychological Strategies in the Commodity Packaging Design

Today's economy is called the eyeball economy. If we can attract consumers' eyeballs, the brand will be more popular with the consumers. Therefore, when designing the product packaging, we must consider it in conjunction with the consumers' psychology. On the one hand, the commodity packaging acts on the consumers' senses through the appearance images, such as colors, shapes, texts and patterns, as well as the cultural connotations, idea purposes and emotional sentiment composed of these elements, so as to make them pay attention to the products, stimulate their emotions towards the products, and ultimately produce the purchasing behavior. On the other hand, the psychological and behavioral characteristics of the consumers and their laws are prerequisites for the packaging design. That is, the packaging design of the commodities must meet the characteristics of the consumers' psychological needs. From the perspective of the consumer psychology, in people's memory, the proportion of the language information and the image information is 1:1,000, which shows that it is better to use the specific visual images to promote the goods than the abstract concepts [2].

3 Study of the Packaging Concepts and Improvement of Shutao "Snow Bud"

In the twenty-first century, the design was diversified, and various design styles are flourishing. With the "functionalism" as the center, the internationalism packaging style, which emphasizes the standardization and regularization of the production, has been tired of by people. People need new ideas, and they need to pursue a higher level of aesthetic style [3].

For the tea packaging design, the design positioning, the visual performance and the structural materials are the three basic elements of the overall design thinking of the tea packaging [4, 5]. Nowadays, the relationship between tea and the modern life is increasingly close. In sales, the tea packaging not only plays a very good role in the quality of tea, but also disseminates the tea cultures. In recent years, we have been pushing forward the old and bring forth the new of the tea packaging. Designers are required to have a deeper understanding of the cultural research of the Chinese tea packaging design, pay attention to the combination of the cultural taste and the marketing, and reflect the unique styles and forms of the Chinese tea packaging [6].

In view of the cultural connotation of Shutao "Snow Bud" itself, in the packaging design, the cultural connotation of the product is integrated on the basis of considering its own characteristics, so that the consumers can not only understand its information, but can also deeply feel its cultural connotation in the process of choosing and purchasing, so that they can accurately interpret the product information at the same time, which can enhance the market competitiveness of Shutao "snow buds" products.

3.1 Packaging Design of Shutao "Snow Bud" – Materials

Although there are many controversies about the green tea packaging on the market, and they think that these designs have been overflowing and have no new ideas, I think it is not a good thing for the brand to be not particularly loud green tea, so it is not necessarily a good thing to let it take the most routine route. Therefore, the improved design, based on the "green", adheres to the principle of simplicity, elegance, simplicity and low carbon, uses the environmental protection materials of composite tin foil, takes simplicity and elegance as the design theme, highlights the unique cultural connotation of the "Snow Bud" and increases its market value [7].

The quality of the tea packaging materials is directly related to each link of the tea storage, the quality assurance, the transportation and the sales. It occupies an important position in the entire packaging industry and is an important basis for developing the packaging technology, improving the packaging quality and reducing the packaging costs. Therefore, the correct understanding and selection of the packaging materials is one of the key technologies of the tea storage and preservation, and is also related to the cleanliness, hygiene, beauty and practicality of the products. The material selection of the tea packaging should pay attention to the characteristics of the oxidation, adsorption, moisture absorption and fragility of the tea itself. The material selection should be appropriate to avoid the resulting changes in the tea quality. Paper is the main packaging before the improvement. Its advantages include protecting the products, changing the structures, easy for processing, printing, low cost, and easy to make composite materials and recycling. So when improving, we intend to continue to use the paper materials. However, due to the insufficient thickness of the carton before the improvement, the tea cannot be well preserved, so this aspect has been strengthened. The tin foil was used in the inner packaging before the improvement. Considering its fragility and lower packaging grade, the polyethylene plastic bags with lower costs were finally chosen. The materials are as follows. The inner package is made of the embedded polyethylene plastic bags, and the cuboid stereotactic carton is used in the middle packaging, while the rectangular packing box is used in the outer packing. The entire package is equipped with an overall foam model to be embedded in the middle package, and a red silk layer is added on the foam, and the composite paper is used in the handbag.

3.2 Packaging Design of Shutao "Snow Bud" – Structures

Product Styles and Specification Design

The structure of the improved packaging is mainly the cuboid configuration, which is relatively common, and is more convenient in the production, design, and printing and so on. Therefore, the cuboid configuration is still used in this improvement. However, considering the layout of other aspects of the improved design, it was decided that the cuboid outer packing with equal length and width should be changed into the cuboids with different lengths, widths and heights instead, and the inner packing should be changed into the cuboids with equal length and width. The packaging specification before the improvement is 300 g, which is divided into three packages and three inner packages. Considering the deteriorating nature of the tea, it is decided to adopt the inner packages of 3 g bags to meet the one-time brewing. At the same time, the inner packages will be increased to six, each containing 10 identical inner packages. In order to facilitate the consumers to carry, and also to increase the levels of the packaging, a handbag is adding on the basis of the improvement. Under the principle of commodity, epoch, simplicity and fashion, CorelDRAW, Photoshop modeling and mapping software were used to improve the original packaging design, including the inner packaging (3 g/bag), the inner packaging (30 g/bag), the outer packaging box (180 g/box), and the gift handbag.

Structural Design of the Inner Packaging

The inner packaging before the improvement is a vacuum way to preserve the tea after wrapping the tea with the tinfoil, which makes the tea lumpy and inconvenient to use. So after changing the capacity of the inner package, the vacuum pumping method is replaced by the pumping nitrogen.

The nitrogen extraction packaging is an effective method for the storage of the famous tea in recent years. It is widely used. It extracts the air from the packaging, forms a “vacuum” state, then fills the inert with the nitrogen gas, and finally seals. The nitrogen extraction and charging can not only reduce the oxygen contents in the package to about 1%, but can also prevent the reaction between tea contents and the oxygen, thus preventing the aging and deterioration of the tea [7].

The inner packaging material is a socket polyethylene plastic bag, and the net content of the bag is 3 g/bag, which can meet the best consumption of the single brewing tea for the consumers. The length and the width of the bag are 100 mm and 70 mm respectively. The model diagram of the three-dimensional structure is shown in Fig. 1.

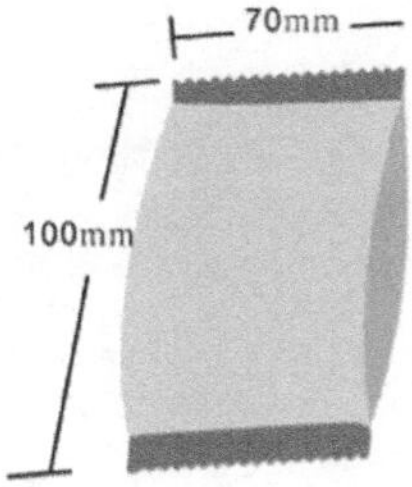

Fig. 1. The stereoscopic structural model of the inner package

Structural Design of the Medium Package

This time, the packaging material is basically the paper, and each carton can hold 10 inlaid polyethylene plastic bags. Its specifications are 100 mm, 100 mm and 70 mm in length, width and height respectively. It is cuboid in the three-dimensional configuration, and its three-dimensional structure model is illustrated in detail in Fig. 2. The geometric packaging structure not only reflects the rigor of the products, but also saves space as much as possible. The inner packaging box can be placed as much as possible in the outer packaging box to avoid the excessive packaging, and the supporting force of the corners can be used to protect the inner products from being damaged in the process of the transportation, storage and sale.

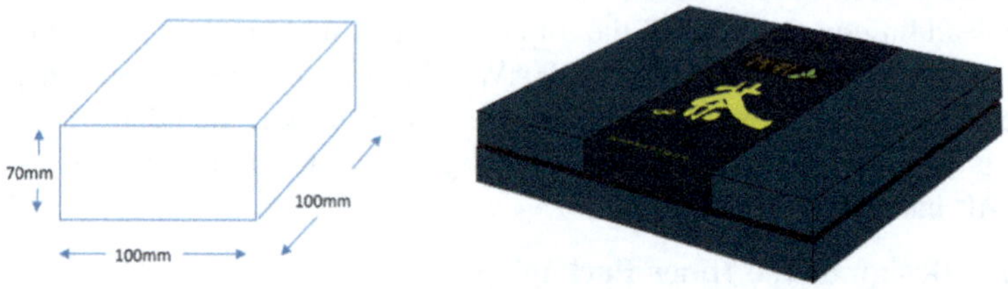

Fig. 2. The stereoscopic structural model of the medium package

Structural Design of the Outer Packaging

The choice and design of the structure should not only satisfy the engineering bearing capacity and the protection function, but should also take into account the aesthetic needs and the control of the packaging cost, and we should also pay attention to the practicability of the packaging and how to make better use of the packaging space. At present, with the continuous updating of materials, the structure of the packaging is more free and diversified. The improved outer packing still chooses the improved composite carton. But because of the stability of the foam used for fixing the middle packaging before, the foam with greater density has been selected in the improved design, and the quality of the silk fabric has also been increased. Its length, width and height are 350 mm, 250 mm and 150 mm respectively, which are the cuboid stereo configuration. The stereo structure model is shown in Fig. 3 in details.

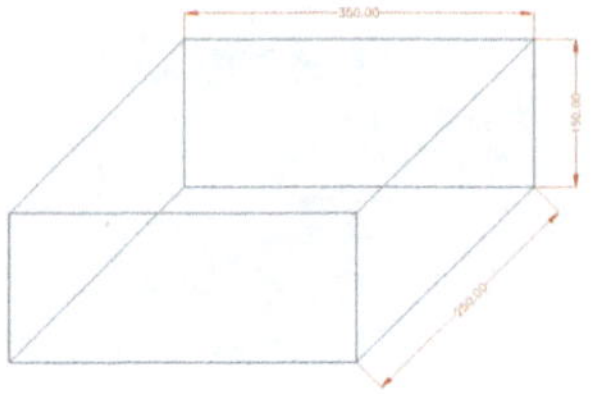

Fig. 3. The stereoscopic structural model of the outer package

Structural design of the handbag packaging

Shutao "Snow Bud" gift-type handbags should have the characteristics of light weight, green and environmental protection, which can be easily carried by the purchasers. Its material is paper, and its specifications are 355 mm in length, 155 mm in width and 255 mm in height respectively. Its three-dimensional structural model is illustrated in details in Fig. 4.

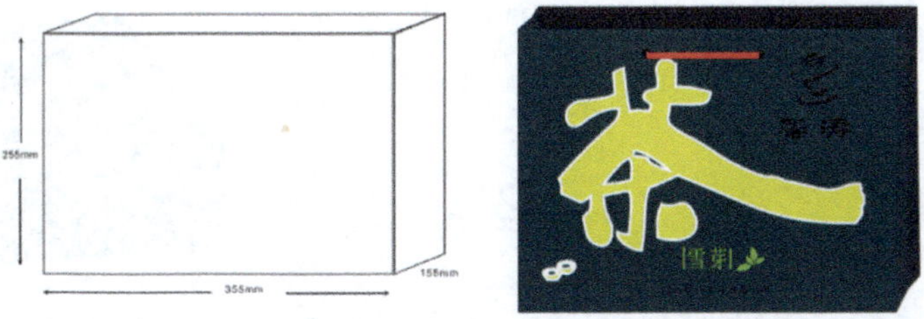

Fig. 4. The stereoscopic structural model of handbags

3.3 Cost Budget of the Packaging Design

According to the requirements of the improved design of the tea packaging, through the online trading information of the B2B e-commerce packaging customization enterprises on the Alibaba website, we can find out the trading price of all kinds of the materials in the improved design of the tea packaging. Detailed information can be found in Table 1. The specific trading prices of each packaging material fluctuate with the changes of the customized quantity. Therefore, the cost of the improved design of "Shutao" packaging can be calculated preliminarily: the unit cost price of the gift packaging is 9.39 Yuan.

Table 1. The price of each packaging type when it reaches a customized quantity and the information of the material suppliers

Packing types	Price (Yuan/PCS)	Custom quantity (PCS)	Suppliers
Gift inner packaging	0.04	50000–99999	Youyou Bag Factory in Qian Ku Town of Cangnan County
Gift medium packaging	0.25	10000–99999	Jiufang E commerce Trading Company in Yiwu
Gift outer packaging	7.5	$\geq$ 10000	Shenzhen Weichengfa Paper Packing Co., Ltd.
Gift handbags	1.6	2000–4999	Yiwu Chaiyun Packaging Co., Ltd.

4 Comparative Study of the Packaging Before and After the Improvement

In terms of the morphology, more than 70% of the people chose the improved one. Most people think it looks more stable than before. Rational use of the internal silk also increases the packaging grade. In terms of the texts, 56% of the candidates who have been improved are richer in the typesetting, and simpler and clearer before the improvement. In terms of the materials, 74% of the people choose the latter. Although both of them are made of the paper as the main material, they are better in the thickness and the material quality than before, and the rational use of the silk and foam has also become an important reason for people to choose it (Fig. 5).

Fig. 5. (a) Packaging before the improvement, (b) Packaging after the improvement

Acknowledgment. Project: Research Center for Development and Communication of the Agricultural Brands with Sichuan Characteristics "Research on the Roles of the Regional Cultural Symbols in the Construction of Sichuan Rural Tourism Brands" CAB1067.

References

1. Liao X (2017) On the embodiment of the cultural characteristics in the Chinese tea packaging design. Mod Bus Trade Ind (17):114
2. Luo Q, Sun H, Zhou Y (2012) Interaction between the commodity packaging and the consumer psychology. Packag J (1):72–74
3. Li C (2015) Study of the relationship between the consumer psychology and the experiential design. Art Sci Technol (5):130
4. Yuan E (2004) Application and research on the consumer psychologies in the packaging design. Packag Eng (1):105–107
5. Ma F (2002) Perception, understanding, imagination and emotion – on the psychological activity process of sports aesthetics. Sports Lit Hist (5):38
6. Han T (2012) Talking about the characteristics of contemporary packaging design. J Taiyuan Urban Vocat Coll (12):161–162
7. Wang S (2006) The overall design thinking of the tea packaging. Guangdong Tea (3):20–21

A Brief Analysis of the Literature Network Resource Mining and Data Development

Jianguo Gao(✉)

Mianyang Teachers' College, Mianyang 621000, Sichuan, China
scchuanyanhan@126.com

Abstract. The development of today's society cannot be separated from the information technology, in which the literary network data resources occupy an important position. Especially with the continuous development of the network, the literature network resources are also growing rapidly, so the problem of the network information overload is becoming increasingly prominent, and the network information mining is also developed on this basis. Then this paper mainly summarizes the data mining and the application of the literature network information resources.

Keywords: Literature · Network resource mining · Data development · Big data

1 Introduction

Compared with other data resources, the network data resources have more advantages, such as the faster update, the more convenient retrieval and the more rich contents. It has become an indispensable resource in the current social development. The data mining of the network information resources has gradually emerged and attracted people's attention [1]. The data mining of the literary network information includes not only a kind of the mining for the contents of the web pages, but also a kind of the mining for the linking patterns, as well as the information generated by the user's browsing, accessing, publishing and other related operations. The effective mining and utilization of the network information can further enhance the attractiveness of the website, attract more users, and realize the efficient use of the literary network resources.

2 Network Resource Mining and Data Development of Literature

2.1 The Significance of the Literature Network Resource Mining and Data Development

Compared with the traditional general information resources, the network information resources have their unique characteristics, with the rich information contents, the frequent changes in the information, the complex information structures and the diverse information formats [2]. Admittedly, the value of the information itself varies. The

Z. Xu et al. (Eds.): CSIA 2019, AISC 928, pp. 319–324, 2020.
https://doi.org/10.1007/978-3-030-15235-2_48

information environment we live in is constantly changing, and the information technology conditions are revolutionary to the information environment. The information environment influences, catalyzes or determines people's information needs to a certain extent.

The development and mining of the network information resources is to obtain the effective information through the reasonable cost expenditure and the scientific means from the complicated "big data". Generally speaking, the "big data" for the user behavior level is a huge data source based on the user behavior information, and is also the product of the user behavior information explosion [3]. On the premise of the profound changes in the technology of the self-Media and the mobile terminals, the big data of the user behaviors has several basic characteristics, also known as "4 V". The first is the volume (referring to the massive data of the user behavior information), and the second is the velocity (referring to the fast generation and update of the user behavior information). The third is the variety (the type of the user behavior information is complex and diverse), and the fourth is the veracity (the authentic user information based on the intelligent mining such as the self-media).

The summary of the data mining based on the big data is to extract the potential effective information we need from the multiple, large, incomplete, fuzzy, random, fragmented and scattered data [4]. These kinds of the information can be fully used in the e-commerce, the life services, the economic and trade, the academic research, the technological upgrading and many other fields.

2.2 Literature Network Resources Mining and Data Development Contents

Step 1: Identify the objects. Although the results are unpredictable, we must clearly analyze the direction of the business objects, but not blindly carry out the data mining, and then we cannot clearly understand how to process the data, so we cannot get the correct results at all.

Step 2: Data preparation. This stage is the most important stage. It takes 60%–80% of the time. Only when it is based on the detailed and accurate data will the following mining not produce the wrong results. The following will describe the specific process of the data preparation in details. (1) The data collection and filtering and input. The data acquisition channels are wide, and the data collection should be as complete as possible. We should collect the internal and the external data about this business object as far as possible. Enter the data to ensure the accuracy and integrity of the data as far as possible in the same place. (2) Data preprocessing. The pre-processing is basically to further process the data on the basis of the data collected in the previous step. At this time, the data includes not only the data just inputted, but also the existing data, which can be cleaned up, integrated, transformed, stipulated or discretized. (3) Data analysis and modeling. Through the data processing in the previous step, the data now can be directly processed after the optimization of the data. Now we analyze the existing data and convert the data into an analysis model. This analysis model is established for the mining algorithm, and the establishment of an analysis model suitable for mining the algorithm is also one of the keys to success.

Step 3: Data mining. The design of the specific data mining process involves the following areas of knowledge:

(1) Generalized knowledge: The generalized description of the category characteristics, the higher-level knowledge with the generality and macroscopicity which can reflect the common nature of the same kind of things, found from the micro-characteristics of the data, is the generalization, refinement and abstraction of the data. (2) Relevant knowledge: The knowledge that reflects the dependence or association of one event with another. If multiple correlations are found, other items can be predicted according to one item, which is generally the most computational part. (3) Classification knowledge: The characteristic knowledge that reflects the common nature of the same kind of things and the difference characteristic knowledge between different things. (4) Predictive knowledge: According to the data of the time series, the future data can be inferred from the historical and the current data, which can also be regarded as the related knowledge with time as the key attribute.

Step 4: Data evaluation. After the above three steps, through the data processing and analysis, the appropriate model building, the appropriate analysis method selection and the use of the related algorithms, we get the results. At this time, we will generally show the results by combining the visualization technology. The advantage of the visualization technology is to visualize the results directly, which is helpful to find the specific correlation and get the useful knowledge.

3 Literature Network Resource Mining and Data Development Practice

3.1 Practice of the "Digitalization" and the "Data" Based on the Literary Research

In the era of the "big data", the database construction is the foundation. Without the database, there is no way to talk about the "big data". Compared with the Chinese ancient literature, the database construction in the modern and the contemporary literature researches is obviously lagging behind [5]. As far as the current situation is concerned, the databases commonly used by the Chinese modern and contemporary literary researchers are mainly the comprehensive databases: Google books, China national knowledge internet, the full-text database of journals of the late Qing Dynasty and the Republic of China (Shanghai Library), the Hanwen Republic of China library, the newspaper database of Airu University of the Republic of China, the old newspaper database of Dacheng, and the academic literature database of Taiwan and so on. These comprehensive databases cover many disciplines and specialties. There are few professional databases of the modern and the contemporary literary disciplines. Peking University Press has developed a searchable "New Youth" database, but it can only be used on CD-ROM, not online. Chongqing Normal University is building a "Great Rear Area Literature Historical Data Database", which cannot be used at present. In contrast, there are many thematic databases of the ancient literature, such as Complete Library of Four Branches of Books, Four Series, Compilation of Historical Materials of Stone Carvings in the Past Dynasties, Ten Links, Chinese Basic Ancient Books, Ancient and

Modern Books, Longyu Hantang Classical Books Database, Complete Poetry of the Tang and Poetry of the whole Song Dynasty. In the field of the modern literature researches, although Lu Xun's research is a prominent one, so far no decent database has been built. Therefore, it is urgent for the modern and the contemporary literary researchers and the relevant departments to develop and construct the thematic databases. However, as far as the existing databases are concerned, especially the full-text databases of journals of the late Qing Dynasty and the Hanwen Republican Library, which are often used by the modern literary researchers, only complete the process of transforming the paper books into images. They can only be retrieved according to the authors, titles, sources and other factors, and basically cannot carry out the statistics and the retrieval of the full text. This is only a digital process. It cannot be called the data.

3.2 Basic Model of the Network Information Resources Development and Mining

For the development and mining of the more professional information resources, such as the academic information resources, we mainly use the network search engine and the open access mode to develop and mine. The first is a comprehensive search engine, such as Baidu, Sohu, Sina, Google, Yahoo, All the Web, dmoz, and MSN Search and so on. The second is a dedicated search engine, including the search for the academic resources, the image information, the e-mail, the phone numbers, the people, the newsgroups, and the FTP servers, such as the WW Virtual Library, the Intute, the SCIRUS (scientific and technological literature), and the Google scholar. The use of various web search engines is basically the same. According to the user's needs, you can input the search terms for the subject queries or you can query step by step from the categorized catalogue. The third is the open access resources, such as Open Access (OA), a movement launched by the international scientific and technological circles, academia, publishing circles and information dissemination circles to promote the free dissemination of the scientific research results through the Internet. Anyone can get all kinds of the documents free of charge through the cover technology. The open access is not a technical issue, but a conceptual and cultural issue.

The diversified and pluralistic information needs to be realized by means of the data mining tools. In the mining for the data tools, the entire network information data are stored in the databases or the text files with the clear field definitions, which is called the structured network information data mining tool. It is mainly used for forecasting, clustering analysis, correlation analysis, time series analysis and statistical analysis. In addition, it should be pointed out that for some data information, we need to make full use of the text mining tools and the technologies to achieve it.

Big Data Feedback Creation

The big data is used to analyze the users and the market of the online literary works in the creative stage, which can provide guidance for the writers in their writing directions. Through the data collection and collation, it is found that the urban, fantastic, historical, military and other works are popular with male readers, while the female readers prefer the romantic works. The secondary elements, history, mystery and other subjects are being favored by the post-95 generation. After a long period of the development on various platforms, the network literary works have accumulated a large number of readers and the reading data, and recorded the reading preferences of the readers of different age groups. On the basis of the massive data, they can provide the objective and the reliable suggestions for the authors, directly tell the authors which contents readers are more interested in, and how to set the suspense to attract the authors. This kind of the direct writing reference increases the survival rate of the works.

The big data technology directly improves the efficiency of the content creation. Authors on various platforms upload 150 million words a day, including a large number of the obscene and the pornographic, politically sensitive, and destroying the national unity, which requires the careful and strict auditing by the staff. The use of the big data technology can quickly scan the uploaded documents and quickly locate the suspicious contents, which are more convenient and faster, and can improve the efficiency of the auditing, and shorten the publication time of the works.

Data Analysis Improves the Content Production

The big data is valuable. The company scientifically and rigorously evaluates the users and the markets, uses the big data to track and analyze various core resources of the film and television industry chain, such as the works, directors, actors and broadcasting platforms and so on, with the in-depth mining and analysis, in order to provide the decision-making basis and the resource guarantee for the production of the high-quality plays. For example, the "Nirvana in Fire" produced by Shandong Film and Television Media Group chooses to broadcast it in IQI, which is based on the statistics that IQI has more customers, the clearer picture quality and the higher user satisfaction. The "Flowers and Bones" crew quantitatively assessed the audience's likeability, the role conformity, the star exposure and the fans' expectations, and finally chose Huo Jianhua and Zhao Liying as the male and the female stars after analysis and statistics. At present, the society has entered the era of the data explosion, and the combination of the data and the IP is the general trend.

Big Data Constructs the Whole Literature Product Chain

A good literary work is not a "hammer sale". It needs to constantly enrich the characters, enhance the readability of the story, extend the product line, and make the brand continue to be extended in the continuous adaptation and operation. In order to maintain the lifeline of the relevant literature, we can draw lessons from the writing mode of the American sitcom, establish a stable team of the writers, use the relativity of the big data to analyze and screen the writers who have the basis of writing on the related subjects, conduct the training, clarify the writing requirements, or discuss collectively, and brainstorm to select more appropriate plots. Or each of them is responsible for a part, and performs each of his or her duties, and completes the content creation. This not only guarantees the quality of the works and cultivates the literary values, but also establishes a good team renewal mechanism for the authors.

4 Conclusion

With the arrival of the era of the big data, the network literary resources are becoming more and more dominant in our daily life. The level of the development and utilization of the network literary resources will directly affect its rational use. Only by effectively mining and exploiting the resources of the network literature can it become an orderly information space.

Acknowledgment. Foundation Project: Sichuan Education Department Project (18SB0256).

References

1. Wang Z, Du R (2018) New problems and regulations on the copyright protection of the network literature in the big data era. Sci-Technol Publ (11):26–31
2. Sun G (2018) Contemporary literary researches and the empirical style of studies in the age of the big data. J Xiangtan Univ (Philos Soc Sci Ed) 42(02):116–120
3. Song X, Qiao H (2017) New production mechanism and landscape of the network literature under the background of the big data. Lit Art Crit (03):100–107
4. Ding L, Zhang W (2016) Research on the strategy of the network literature production under the background of the big data. J Chang Univ Tradit Chin Med 32(03):646–648
5. Hao K (2014) Application and identification of the modern literary historical materials in the network resources. Ser Mod Chin Lit Stud (09):54–60

A Study of the 3D Animation Technology Based on the Internet

Jing Pu[1(✉)] and Xinxin Nie[2]

[1] Sichuan Agricultural University, Dujiangyan 611830, Sichuan, China
guanjuchen@126.com
[2] Jincheng College of Sichuan University, Chengdu 611731, Sichuan, China

Abstract. The rapid development of the Internet technologies has exerted profound influence on the multimedia, and the 3D animation, as an important category of the multimedia, has also been undergoing vital innovation and development adapting to the characteristics of the Internet. This paper summarizes the mainstream representation patterns of the current 3D animation on the Internet. It analyzes the advantages and defects of different techniques by appreciating and commenting on the typical works. It is hoped that the analysis and proposed development prospect will provide a useful reference for the study of the 3D animation in the field of the Internet.

Keywords: 3D animation · Internet · HTML5 · Web3D · Flash

With the comprehensive arrival of the information age, the Internet has penetrated into the human life. The Internet technology eliminates the differences in various channels of the information dissemination and makes it possible for the information to spread rapidly and freely. The characteristics of the Internet have led to the re-evolution of the multimedia technologies. As a form of the multimedia, animation has also undergone tremendous innovations, such as the FLASH animation, which is a kind of the animation innovation in the Internet era. FLASH is widely known on the Internet because of its simple production and small amount of data. It has produced many well-known animation works, such as "Little Broken Children" and "Friends under the Happy Tree" and so on. The early Internet animation was restricted by the network bandwidth and the terminal equipment, and was presented in the two-dimensional form. However, with the increasing network bandwidth and the increasing performances of the terminal hardware, the wide application of the three-dimensional animation in the network becomes possible. Compared with the two-dimensional animation, the three-dimensional animation shows the vivid visual pictures through the virtual three-dimensional space effect, and approaching the real visual effect makes the original static and planar scenes into the vivid three-dimensional images, which brings more intense sensory impact to the audience. The foreign research on the three-dimensional animation in the field of the Internet started earlier, and there are good research results [1, 2]. The domestic research on the application of the three-dimensional animation on the network started late, but we also made a lot of research results. For example, Rong Yandong analyzed the application of the WebGL-based 3D technology in the web pages [3], and Puqing analyzed the Cult3D technology to achieve

Z. Xu et al. (Eds.): CSIA 2019, AISC 928, pp. 325–331, 2020.
https://doi.org/10.1007/978-3-030-15235-2_49

the three-dimensional animation on the network [4]. Wang Dongze conducted the research on the three-dimensional animation and modeling of the virtual reality technology (VMRL) on the network [5]. This paper will summarize the current mainstream Internet three-dimensional animation technologies, and analyze the advantages and disadvantages of various technical solutions in the production and dissemination through the evaluation of some typical works.

1 Implementation of the Animation Based on HTML5

HTML5 is the latest version of HTML. It not only has the new syntactic features, but also provides the rich multimedia features to meet the needs of other technical support and plug-ins. Among them, the following three characteristics provide the basis for the realization of the three-dimensional animation:

(1) Canvas: Canvas is a rectangular area composed of the pixels, in which the lines, squares, circles and other shapes can be drawn using the canvas elements. Through the Canvas technology, animators can draw vector graphics more efficiently and flexibly, thus creating the richer and more interactive web pages that can be viewed on any browser and device that supports the HTML5 language.
(2) WebGL: WebGL is a 3D drawing API interface in the form of JAVAScript, which supports the direct rendering of the 3D contents in the Canvas without any plug-ins. Its kernel uses OpenGL ES 2.0 graphics acceleration engine, which can directly use the graphics hardware acceleration capability of the system itself for the image rendering.
(3) Audio and video: HTML5 provides the local support for the video and audio elements. Users can listen to and watch the videos without downloading the plug-ins. They can also embed the audio and video through tags or the use attributes to customize the multimedia experiences. The control attributes can complete the playback control that HTML could not do before, such as the playback, jumping and buffering and so on.

1.1 Appreciation of Works

This paper refers to the HTML5 3D technology on the network to produce a 3D computer room animation effect map, and the effect is as follows:

Fig. 1. Three-dimensional animation work based on HTML5

In the effect, the animation produced has the characteristics of the three-dimensional animation. In addition to its width and height, there is the depth. The static frame picture above shows the extension of the X, Y and Z axis from the high angle. It has the obvious three-dimensional characteristics. Compared with the two-dimensional plane picture, the model in the above picture is more stereoscopic, and can show the orientation and angles of the objects and the room. It is easier for the audience to understand the specific location distribution of the objects in the computer room (Fig. 1).

1.2 Advantages and Disadvantages

From the analysis of the typical works and the development of the HTML technology, this paper draws the advantages and disadvantages of the HTML5 technology:

Advantages: 1. There are unified standards and norms, and the mature industry alliances are responsible for the technology promotion and expansion, and the development prospects are very bright. 2. The current mainstream browsers support the HTML5 standards, and the users only need to upgrade the browsers, without the extra plug-in support, to facilitate the cross-platform animation dissemination. 3. The HTML5 technology is an open source. And the development tools are completely free, lowering the learning threshold.

Defects: At present, there are fewer 3D graphics interfaces supported by the HTML5 technology. Compared with the mainstream animation production software (such as 3DMAX), the HTML5 technology cannot achieve the advanced 3D effects. The HTML5 technology needs a certain programming basis, and is relatively unfriendly to the animation designers. Using the HTML5 technology to produce the animation has a higher learning cost, and the production and the modification of the design are also more difficult, which, to some extent, affected the animation production based on the HTML5 technology.

However, in view of the shortcomings of HTML5 technology mentioned above which is not friendly to the animation designers, domestic companies have developed the animation tools based on HTML5, of which Mugeda is more famous.

2 Implementation Scheme of the 3D Animation Based on Web3D

Web3D is the general name of the network three-dimensional realization technology. At present, the mainstream technologies of Web3D are:

(1) CULT3D: Cult3D is a new kind of the 3D network technology, which consists of three parts: the output plug-in, the designer and the browser plug-in. The Cult3D Designer is the core part to produce the three-dimensional images. It has the excellent image quality, the fast running speed and the good interaction performances. Through Cult3D, developers can build models based on sounds, links, animations, clickable JAVA codes and control their movements. In addition, it has the low hardware requirements, so even devices with the low configuration can browse them smoothly.

(2) VRML technology: It is an open and extensible virtual reality modeling language, which is the scene modeling language used to build the real-world scene models or people's fictional three-dimensional world. The VRML objects become the nodes, and the collection of the sub-nodes can form the complex scenery. The VRML technology is now applied in many fields, forming different virtual worlds, completing the two-dimensional to the three-dimensional transformation of the Internet, and giving people the unprecedented three-dimensional experience.
(3) VIEWPOINT: It generates very small file format, and the adjustable three-dimensional polygon mesh structures and the excellent transmission characteristics make it very suitable for the transmission on the Internet. In the process of downloading the 3D data, it can complete the transformation from the low-precision rough model to the high-precision complex model. It consists of many media elements such as the 3D objects, materials, animations, interactive actions and scene definition information. It has a high-quality rendering engine, which makes the rendering effect real and does not need any hardware equipment to accelerate.

2.1 Appreciation of Works

As is shown in Fig. 2, the animation works create the car model in the 3DS MAX software, and create the interactive car scenes through the development the software based on the Cult3D technology, and finally synthesize the media in Director, so as to realize the use of the virtual reality technology to present the three-dimensional demonstration space. Users can view the car from any direction, or they can dynamically change the colors of the car by clicking a button.

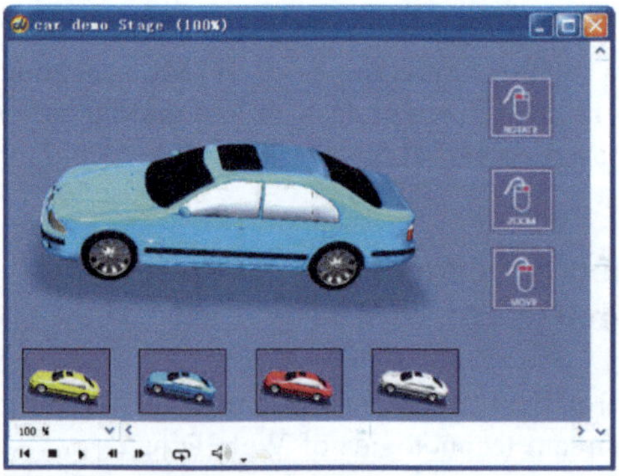

Fig. 2. Three-dimensional animation work based on the Cult3D technology

2.2 Advantages and Disadvantages

Based on the current development status of Web3D and the evaluation of the typical works, it is concluded that Web3D has the following advantages: 1. Web3d supports the image engine rendering, integrates the processing of various special effects, and shows the excellent results. 2. Web3D can support the importation and production of the model originals generated by various graphic production tools (such as 3DMAX) through the plug-ins, with rich production materials.

But the biggest disadvantage of Web3D is that there is no uniform technical standard, which makes it difficult to popularize the Web3D-related technology, with the high costs. Meanwhile, the skills of the animation producers are limited by the specific technology, and the learning threshold is high, so it is difficult to expand the technology. In addition, because there is no uniform standard, every manufacturer needs to develop their own plug-ins to meet the requirements of their animation technologies, and the users must install the corresponding plug-ins and require the plug-ins to be compatible with the platforms in order to get a good experience, and such requirements are too harsh. The above shortcomings seriously affect the promotion and dissemination of the three-dimensional animation works based on the Web3D technology.

3 Implementation of the Flash-Based Three-Dimensional Animations

As the two-dimensional animation software, FLASH itself does not support the direct three-dimensional animation production, but it can produce animations with the three-dimensional effect through some special techniques or plug-ins. The main ways to realize the three-dimensional animations are as follows:

(1) ActionScript is used to achieve the three-dimensional effect. It simulates the 3D coordinates by calculating the 3D coordinates and then by perspective. The basic technical principle is to transform the three dimensional coordinates into the two-dimensional coordinates through the algorithm, and then calculate the coordinates of the epitomes FLASH, so as to achieve the three-dimensional effect.

(2) The three-dimensional effect of the FLASH animations is realized with the assistance of the third-party 3D software. Using the third-party software of cool3D, Swift 3D, 3D MAX, Maya, Dimension and other three-dimensional rendering software, the materials with the three-dimensional effect produced by them are imported into FLASH, and the three-dimensional FLASH animations are used and produced.

(3) It can be achieved through the FLASH's own component library. The animated character's body is decomposed into different components, and the body's different dynamics and different angles are drawn and stored in the components one by one. In the process of the animation production, we can use different body components to realize the dynamics and angles of the whole character. This method greatly reduces the workload for the large animated cartoons. This way of production is to use the dimension principle of the three-dimensional animation to produce the two-dimensional animations, to reduce the workload, but it is not the real three-dimensional effect.

3.1 Appreciation of Works

Figure 3 shows a model of an animated character, which is made by using the Flash plug-in COOL3D. It is a true vector three-dimensional animation. It not only interprets the three-dimensional characteristics from the three dimensions, but also conforms to the characteristics of vectors, enlarges without distortion, and the data volume is small and is easy to spread. From the observation of the work, we can find that although the model is a three-dimensional model, there is no real effect produced by the conventional three-dimensional software (such as 3DMAX), and the visual effect is closer to the vector graphics.

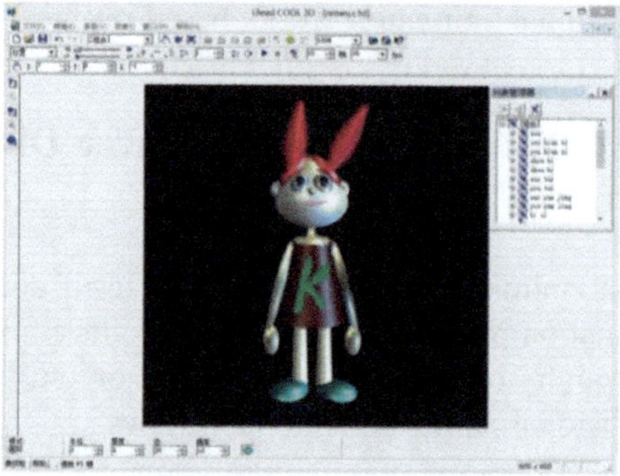

Fig. 3. Three-dimensional animation works based on the Flash plug-in

3.2 Advantages and Disadvantages

Based on the evaluation of the typical works and the current development of the FLASH technology, its advantages include: 1. The Flash technology is very mature, rich in the animation materials and the components resources. 2. The amount of the animation data generated is small, and the production cost is low, and it is convenient to spread under various network conditions.

But Flash itself is the two-dimensional animation software, which can only achieve the simple three-dimensional animation effect through the plug-ins, and cannot complete the complex three-dimensional animation effect. Flash's biggest problem is that its technology is not the open source, and has been abandoned by Aodbe, and there is no possibility of its further expansion. At present, the mainstream mobile operating platforms do not support Flash. With the increasing mobility of the Internet, the future of the Flash technology is very gloomy.

4 Summary and Prospect

When the network bandwidth and the terminal hardware capabilities are no longer limited, the three-dimensional animations will inevitably get better development because of the realistic and cool characteristics. It can be predicted that in the near future, the status of the three-dimensional animations and the two-dimensional animations will be reversed. This paper sums up the three main three kinds of the 3D animation implementation technologies, HTML5, Web3D and Flash. Taking the typical works as an example, it expounds their production techniques and aesthetic characteristics, and analyses the advantages and disadvantages of each technology, which provides a useful reference for the further study of the development of the three-dimensional animations on the Internet. Based on the analysis of this paper, the author thinks that the technology of the three-dimensional animations must evolve towards the goal of standardization and cross-platform.

Acknowledgment. Foundation Project: The scientific research project funded by the Research Center of Aesthetics and Aesthetic Education, Sichuan Key Research Base of Social Sciences (15015).

References

1. Lim S-B, Lee H-J, Jin W-Y, Shim S-M (2017) CSS3 extensions for setting web content in a 3D view volume and its stereoscopic 3D display. Comput Stand Interfaces 50:65–75
2. Zhou Q, Luo J (2018) The study on evaluation method of urban network security in the big data era. Intell Autom Soft Comput 24(1):133–138
3. Rong Y (2015) Application of the WebGL-based 3D technology in the web pages. Inf Secur Technol 08
4. Pu Q (2005) Realization of the interactive network three-dimensional animations based on the JAS development technology of Cult3D. J Eng Graph 3
5. Wang D (2005) Research on the VRML three-dimensional modeling and animation technology based on Web3D. Sichuan University

Analysis of the Information Service Levels of the Agricultural Economy in the Environment of the Internet of Things and the Cloud Computing

Meng Yue(✉)

Sichuan Agricultural University, Chengdu 610000, Sichuan, China
jinhong_xu@yeah.net

Abstract. China is accelerating the pace of the rural economic reform, and is also facing the new challenges of the agricultural economic management. Therefore, in the process of development, we need to use the information technologies to manage and maintain the sustainability of agriculture. The rapid and stable development status quo and the application of the information management in each link can increase the speed of the agricultural development in China. The cloud computing, as a new network technology platform, is an additional product of the distributed processing and the parallel processing of the modern physical computing in the process of the mutual promotion and development. It is an efficient channel for the network information transmission.

Keywords: Internet of things · Cloud computing environment · Agricultural economy · Information service level

With the rapid development of the modern cloud computing, the cloud computing has become a hot word in the whole IT industry. Strengthening the construction of the agricultural information service is an important means of promoting the circulation of the agricultural products and the development of the rural economy in China [1]. From the perspective of promoting the development of the rural economy in China by the agricultural information services, this paper puts forward the current situations of the construction of the rural information services in China, in order to promote the expansion of the agricultural information channels, enhance the level of the farmers' agricultural knowledge, improve the economic efficiency of the agricultural products in the market circulation, and also promote the prosperity and development of the agricultural economy, so as to be conducive to promoting the progress of the rural society and accelerating the pace of the rural construction [2].

Z. Xu et al. (Eds.): CSIA 2019, AISC 928, pp. 332–338, 2020.
https://doi.org/10.1007/978-3-030-15235-2_50

1 Overview of the Internet of Things and the Cloud Computing

The network cloud computing platform is composed of a large number of the simple PC registered users, the total terminal data service system, the transportation channels and the nodes. This huge information processing system needs the special maintenance, operation and upgrade personnel to manage and protect, and also has a complete computer in the total terminal data service system structural and maintenance capabilities [3]. The Internet of things is an important part of the new generation of the information technology. Its English name is "The Internet of things". As the name implies, "The Internet of Things is the Internet with the connection of all things". The Internet of things (IOT) is defined as a kind of the network that connects any item to the Internet according to the agreed agreement through the radio frequency identification (RFID), the infrared sensors, the global positioning systems, the laser scanners and other information sensing devices which is used for the information exchange and communication to realize the intelligent identification, positioning, tracking, monitoring and management of items [4]. It is a network that extends and expands on the basis of the Internet. Compared with the traditional Internet, the Internet of things has the distinct characteristics of the terminal diversification, the intelligent management and the strong network inclusiveness.

The cloud computing is an Internet-based computing method. The core idea of the cloud computing is to uniformly manage and schedule the computing resources connected through the network, forming a huge computing resource pool to provide the users with the on-demand allocation, which we call "cloud". The resources in the cloud seem to be infinitely expandable to the users, and can be accessed at any time, used on demand, expanded at any time, and paid for by use. This feature is often referred to as the use of the IT infrastructure like the hydropower.

The cloud computing in a narrow sense refers to the mode of the delivery and use of the Internet infrastructure, through which resources (hardware, platform and software) are acquired in an on-demand and scalable manner. The cloud computing in the broad sense refers to the mode of the delivery and use of services, which means to obtain the required services through the network in an on-demand and scalable manner [5]. This kind of services can be related to the software and the Internet, or any other services. In this way, the shared resources on the network can be provided to the computers and other terminals on demand. It aims to integrate several relatively low-cost computing entities into a perfect system with the powerful computing power through the network [6].

2 Characteristics of the Agricultural Information Service in the New Period

2.1 Information Is Growing Explosively

With the development and construction of the Internet of things, the main subjects of providing the information are not simply the human beings, and anything may become the source of the information release. The information explosion is a description of the rapid development of the information in recent years, describing the speed of its development as an explosion sweeping the whole earth. The information explosion is reflected in four aspects in the field of the agricultural information. Firstly, there are more information released by the government and less information released by the farmers themselves. Secondly, there is more information in various types of the publicity and advertising, which plays a less substantive guiding role for the farmers. Thirdly, there are more theoretical information and less actual data. Fourthly, there are more comprehensive information and less classified information. The problems of the information overload and the ambiguity of the direction bring obstacles to the application of the information service.

2.2 High Degree of the Digitalized Information

The super-large-scale storage of the cloud computing system and the all-round perception function of the Internet of things make the digitization of the agricultural information higher and higher. Through the websites or the professional systems, farmers can quickly find the relevant information they need. The digital information has the following characteristics: the multimedia as the content feature, various types of the information resources, multi-level information service functions and fast updating speed and strong timeliness, the powerful retrieval system, and no limit by the time and space. The development of the multimedia technology promotes the development of the digital information, which makes the digitization of the agricultural information not only reflected in the informatization of words, but also the more direct ways such as images and videos more easily accepted by the farmers. The higher the digitalization degree of the information is, the better the development of the agricultural information service will be.

2.3 Terminal Diversification

The information acquisition and presentation is a very important part of the information services. Traditionally, as the terminal equipment of the information presentation, TVs, computers and broadcasting and so on are used. With the further development of the chip technology, the network bandwidth has multiplied. A large number of the new terminals such as mobile phones and tablets have been applied in the field of the information services. The terminal equipment of the information acquisition has also changed from the traditional manual inputs and scanners to the new terminals such as sensors, cameras and pick-up heads.

2.4 Strong Individualized Information Demand

Driven by the explosive growth of the information and the digitization of the information, the information service platform has accumulated a large amount of information. How to filter out the personalized information which is in line with the user's needs is very important. In the field of the agriculture, the demand for the information varies from regions to times and from individuals to individuals, which is embodied in the contents and forms of the information.

2.5 The Costs of the Information Services Have Decreased

The costs of the information service can be divided into two parts: the hardware cost and the information application cost. In recent years, with the decrease of the costs of the electronic components and the implementation of a series of policies of benefiting farmers in the countryside organized by the government, the price of computers and other information presentation equipment is also decreasing year by year. But for some farmers with lower incomes, the cost is still too high to afford and it cannot be popularized. The information dissemination methods such as televisions and radios are relatively fixed in time, but they are not targeted enough to meet the needs of the individual farmers. In addition, there are still more complex computer operations, difficult to get started. For the farmers whose general cultural quality is not high, its application is a difficult problem. With the emergence of the cloud computing system, the complex operations and a large number of the information processing are put on the cloud platform. Users only need a simple terminal, such as mobile phones, to query the information they need.

3 Information Service Mechanisms of the Agricultural Economy in the Environment of the Internet of Things and the Cloud Computing

3.1 Improve the Construction of the Database

The database construction of the agricultural information resources is the main form of the large scale and efficient development and utilization of the agricultural information resources. In view of the slow pace of the agricultural informatization development in China, it is better for the government departments to set up a fixed special fund control department for the agricultural informatization, which will be the more mature digital agricultural technology. This kind of the high-tech investment of the information capitals and equipment should be carried out according to the economic situation of the region. For those areas without the conditions to achieve the above objectives, we can strengthen the construction of the agricultural information network and the information platform of the local and the county agricultural departments, make full use of and integrate the rural information resources, and vigorously promote the integration of telephones, televisions and computers as the key points. The agricultural comprehensive information services can solve the problem of the information blockade in the rural areas, and realize the economic promotion of the new socialist countryside through the information supply.

3.2 Improve the Information Service Model of the Agricultural Economy

The government departments should set up the relevant agricultural economic information promotion websites, provide the current agricultural information throughout the region through the websites, regularly broadcast the agricultural science and technology programs, and distribute the agricultural science and technology books and periodicals free of charge. In the aspect of the agricultural information collection, the agricultural commissions should fully collect the local agricultural related information, and at the same time collect and organize the basic agricultural information of the whole region. The collection of the information should ensure its rapidity, stability and practicability, so that it can be transmitted to the agricultural producers as quickly as possible by means of the website dissemination. The commercial information and the technical information of the agricultural enterprises have certain confidentiality, because the information is directly related to the competitions of enterprises. But if paid information supply is adopted, it will be possible to turn the enterprise information into a propaganda and profit-making means for the marketization of the agricultural enterprises.

3.3 Multi-angle Agricultural Information Services

In order to further expand the means of the information services and the service space, the database information can be transplanted to the short message platform and edited into the agricultural short messages, and the agricultural short message services can be developed through the customization and the universal distribution of the farmers. We can also make full use of the advantages of the "one-stop" system of the Ministry of Agriculture and the wide coverage and influence of the online exhibition hall, do a good job of the information online and publishing, publicize the characteristic advantages of the agricultural products and leading processing enterprises in an all-round way, and actively explore the market space for the agricultural products. In addition, the subscription of the agricultural information technology journals, the establishment of the advanced agricultural technology training courses, and the demonstration of the technicians' actual operations are also the traditional agricultural technology extension modes, which have the function of the information services and are all the feasible agricultural information service modes.

The business design mainly includes two core functions. The first is the remote diagnosis service. Aiming at the scientific guidance of the production process and the scientific diagnosis and treatment of the crop diseases, the platform supports the remote diagnosis and consultation services of the agricultural experts, traces the diagnosis process, and supports the organic combination of the accurate diagnosis with the off-line laboratories, shortens the intermediate link of the traditional agricultural diagnosis, and ensures the high-efficiency and convenience, to solve all kinds of the problems in the peasant household production. The second is the remote monitoring business. It includes collecting and early warning of the production environment parameters, the real-time video surveillance of the production environment, controlling the equipment through the remote instruction and improving the production environment. Monitoring the businesses can minimize the hidden dangers in the production and improve the ability to prevent the crop epidemics.

3.4 Informatization of the Agricultural Management

In the field of the agricultural management, the intelligent agriculture planning has designed the "Integrated Information Service Platform for Agricultural Product

Industry", aiming at "opening all links of the agricultural industry chain by means of informationization, and helping the farmers to complete the whole business process quickly and safely", and the platform will be followed up before, during and after the production, serving farmers as a whole. By registering the information of the planting varieties, areas, locations and expected prices on the platform, farmers can obtain the resources and the information provided by the platform, including the whole process of the production, processing, warehousing, logistics, sales and agricultural materials. Thereafter, farmers can receive orders from merchants at home as long as the follow-up operations are carried out according to the requirements of each link of the platform. According to the orders, send information to the warehousing and logistics companies, so that the agricultural products can be transported to the merchants and the market in a timely manner through logistics.

After the agricultural products are packed into the warehouse, label each box (record the weight, variety, grade, origin, farmers, output, use of chemical fertilizers, pesticides, diseases and insect pests and other information). When entering the warehouse, there will be the temperature, humidity and carbon dioxide sensors in the warehouse, which will have the real-time record of the quantity of the agricultural products into and out of the warehouse. After the warehouse receives the orders from the farmers, it organizes the delivery according to the orders, and at the same time, it orders the logistics company to let the logistics company's vehicles arrive at the warehouse location at the prescribed time. After the goods are loaded, the warehouse managers, farmers and merchants can check the vehicle's driving position and the vehicle's internal environment through the terminals.

The central platform is open to the society and involves a wide range of the social forces in the agricultural information services. At the same time, it forms a window in the rural market and attracts the social institutions to enter. When it comes to the remote management, the center-to-region is realized through the internet, with the real-time connection between the center platform and the regional platform, and between the cloud and the end. The whole system is integrated, which can realize both the remote management of the businesses and the remote management of the system. In the terminal experience, at present, it is mainly the computers, mobile phones, integrated computers, and multi-functional televisions and so on. With the integration of the three networks and the development of the Internet of things and the mobile Internet, more terminals will emerge. At the same time, with a unified central platform and the integrated services, the service functions and roles of the terminals will be greatly enhanced.

4 Conclusion

Under the environment of the Internet of things and the cloud computing, the information service business mode of the whole agricultural economy has gradually extended from a relatively single hotline telephone mode to a full range of the three-dimensional services, such as the hotline service, the short-messaging service, the intelligent terminals, the e-commerce, the traceability, the positioning, the micro-blog, and the smart agriculture and so on, in order to better serve the farmers and increase the farmers' income, and at the same time solve the current problems of the limited scale, the terminal distress, the insufficient impact, the scattered resources, the difficult coordination and the lack of standards in the field of the agricultural services in China. In the future, a perfect vertical platform service system can be built in a relatively short time with the mature deployment methods.

References

1. Zhou Q, Lou J, Xie F, Liu Q (2011) The method research on optimum selection of agricultural machinery. Int J Digit Content Technol Appl 5(6):337–344
2. Liu S (2017) On the informatization construction of the agricultural economic management. Friends Farmers Get Rich (06):142–143
3. Ping W (2018) Sustainable development of the agricultural economic management informatization. Agric Technol (03):124–125
4. Qiao X (2018) Application of the informatization in the agricultural economic management. China's Strat Emerg Ind (04):110–111
5. Yang X (2018) Current situations and countermeasures of the agricultural economic management in the new period. China South Agric Mach (07):154–155
6. Zhang J, Wang J (2018) Problems and countermeasures in the informationization of the agricultural economic management. Mod Mark (09):157–158

Application of the Architectural Animation Virtual Technology in the Landscape Architecture Design

Chunyan Zhu(✉), Yizhi Jiang, Yuanyuan Zhang, Yujia Zhang, and Ying Xie

Sichuan Agricultural University, Chengdu 611130, Sichuan, China
yingcan260272943@163.com

Abstract. With the continuous development of the virtual technology, the field of the architectural animation is increasingly using in the simulation of this virtual reality. The virtual reality technology of the architectural animation can make us have a deeper experience and a more real feeling of the architecture, and has the broad application prospects. The display information of the landscape design needs the virtual animation technology to better simulate the three-dimensional scene. Using the web platform, we can develop the virtual animation technology to complete the requirement of the landscape visualization.

Keywords: Architectural animation · Virtual technology · Application mechanism · Landscape architecture · Design scheme

For the architectural animation, it is the best method and the most significant way to realize the architecture visualization. With the advent of the Internet era and the rapid development of the information technologies, the three-dimensional design technology in the field of architecture has been improved and developed [1]. The virtual technology of animation makes use of the computer technologies and the image technologies to directly and vividly display the architectural design drawings and the text data in the form of images through the computer technologies on the screens.

1 Overview of the Virtual Technology of the Architectural Animation

1.1 Concept of the Virtual Technology of the Architectural Animation

The architectural animation is to be able to show the activities related to the architecture with the activities of the architectural films. For animation, it is to apply the animation technology to the field of architecture. The animation virtual technology is to express the creative intention of the architectural design by using the three-dimensional graphics on the platform of computers. Usually, it provides the dynamic space for the audience to experience. In the design and utilization of the animation virtual technology, a series of the related technologies such as graphics, iconography, lens, lighting

Z. Xu et al. (Eds.): CSIA 2019, AISC 928, pp. 339–344, 2020.
https://doi.org/10.1007/978-3-030-15235-2_51

and stereo technology are used [2]. The perfect combination of the film technology and the animation technology forms the dynamic demonstration, the virtual reality and the decision-making analysis and a series of technologies to show the architectural design works. The architectural animation is mainly generated by the computer technology. The virtual three-dimensional building environment is produced by the computer technology [3]. According to the actual needs and the realistic terrain conditions, the virtual three-dimensional scenes are created, and the required building models are arranged using the three-dimensional schemes.

1.2 The Practical Significance of the Virtual Technology of Architectural Animation

Using the three-dimensional animation, the real object to be displayed can be virtualized, and the complex information will be displayed directly. The architectural animation virtual technology combines all kinds of the technologies in the field of the design organically, which brings good observability, appreciation and interaction to the customers, makes the customers understand the products more accurately and clearly, and also improves people's working efficiency to a certain extent [4]. The virtual reality technology has the characteristics of perception, interaction, autonomy and existence in the process of its application. The existence of the perception ensures that the virtual reality technology can realize the perception function of things in the process of its application. However, due to the technical constraints, the virtual reality technology at this stage only has several aspects of the perception abilities of the visual, tactile and auditory perceptions. The existence of the perception performance can realize the simulation of the design site, and the simulated scene is very lifelike, effectively helping people experience the feeling of immersion [5]. Interactivity mainly refers to that when applying the virtual reality technology, designers can rely on the virtual scenarios to get the real inspiration and achieve the vulnerability repair of the design schemes. As for the autonomy, it mainly refers to the autonomous normative action according to the physical laws of the objects when applying the virtual reality technology.

2 The Impact of the Three-Dimensional Virtual Technology on the Architectural Animation Technology

With the development of the science and technology, the virtual reality technology is more and more widely used in our life. The virtual reality technology can artificially arrange the environment and facilitate the communication and exchange between the professional and the non-professional people. In the process of its application, it can effectively combine the computer technologies with the network technologies or the image technologies, and adopt the mode of the image integration. Today, many kinds of technologies are integrated. This is a multi-solution model. The virtual reality technology develops rapidly in the modern society and is applied more and more widely.

2.1 The Feasibility of the Virtual Reality Technology Applied in the Field of the Architectural Animation

The virtual reality technology is more and more widely used in our daily urban construction, and has brought considerable social and economic benefits. The virtual reality technology can effectively display the new buildings and the new environment of the city in the three-dimensional view, which facilitates the masses to understand the changes of the city, and constructs an all-round three-dimensional city model. In this model, people use the virtual technology to communicate, which is also a way to obtain the information, facilitating the masses to query the building information.

2.2 The Virtual Reality Technology Is Widely Used in the Urban Planning

The virtual reality technology is widely used in the urban geographic system and the land resource planning query system. This query mode is basically based on the situation of the database. The traditional display of the geographic information system is mainly digital, which is not sustainable. If the virtual reality technology is effectively combined with your building animation technology, you can insert the data of the traditional geographic information system into the virtual reality technology. This kind of the real data can simulate the original environment to the greatest extent, and then use the computer technologies and the virtual reality technology so that the entire geographic information system (GIS) is displayed in three dimensions. Only in this way can it bring real feelings and experiences to the audience, and give people great visual impact and shock. This technology of the virtual reality is more acceptable, and it has more important practical significance for displaying the external images of the city and the urban planning.

2.3 The Virtual Reality Technology Is Widely Used in the Scenic Area Planning

If we use the three-dimensional animation in the scenic spots to make the reality virtual, we can make the scenic spot scenery in the three-dimensional form show in front of people, so as to deepen the appreciation of the scenic spots for the masses, and play an irreplaceable important role in expanding the popularity of the scenic spots and attracting more tourists. This kind of the scene display is more authentic and three-dimensional. With the help of the computer technologies and the Internet technologies, it can play a good publicity effect for the scenic spots. The virtual reality technology can also be used for the traditional building management data acquisition, and then establish a three-dimensional database to protect the original cultural relics. This precise data preservation has important and practical significance for the future restoration and reconstruction of the cultural relics. This cultural heritage can be preserved in our memory forever in the form of the data.

3 Innovative Design Methods of the Landscape Architecture

3.1 Use the Scientific and Rigorous Methods to Implement the Innovative Design

The gardening can either follow the ancient methods or draw lessons from the western forms of expression, neither of which should be excluded. It is an inevitable trend to combine the past with the present, to make the past serve the present and to make the foreign serve the present. This paper probes into the history of gardening, both at home and abroad, their aesthetic thoughts, and the historical and cultural conditions. We should inherit the tradition, absorb the essence, take the advantages of the West, make up for the shortcomings of the Chinese gardens, integrate the connotation of the Chinese cultural thoughts and the western modern concepts to create the modern gardens with Chinese characteristics, and design them in a rigorous manner along the context of the national cultures. The pure imitation and duplication are often immature. It is not advisable to copy and patch up the western and the classical gardens without knowing anything about them.

3.2 Carry Out the Innovative Design Using the Agricultural Industry Method

In the fringe areas of most large and medium-sized cities, besides a large number of the artificial and semi-natural environments, such as the residential buildings, factories, university towns and theme parks and so on, which have been developed, there are also a large number of the rural environments. We can create a rural environment that not only provides certain production functions, but also satisfies the needs of viewing and resting in terms of the village itself and the rural environment, combined with the agricultural industry. The environment places needed for the leisure vacation, such as the sightseeing agricultural parks and the urban agricultural parks, can meet the natural needs of the urban residents to return to natural pastoral areas and close to nature. At the same time, it can let the cities into the countryside and integrate the countryside into the cities, so as to achieve the urban-rural integration.

3.3 Innovative Design Using the Modern Technologies

In the modern landscape design, its connotation and extension have been greatly expanded and deepened. The application of the landscape elements such as concrete, glass and stainless steel is relatively prominent in the hard landscape. Concrete can not only replace the traditional hard landscape, but also have higher plasticity. The creative performance of the glass reflection, refraction, transmission and other characteristics let us move between reality and illusion. The concise and elegant shape of the stainless steel can let us appreciate the beauty that the traditional gardens do not have. In the soft landscape, the introduction of a large number of the thermoplastic and synthetic fibers, rubber and polyester fabrics is adding luster to the appearance of the gardens, and even fundamentally changing the appearance of the traditional landscape.

With the rapid development of the modern lighting technology, a new type of the landscape, the night landscape, emerged. The light sources with different colors and lamps with different effects bring our visual and psychological feelings into a dreamlike and illusory state of confusion. As we all know, both the classical gardens and the modern landscapes are inspired by nature, and the natural landscapes are always changing. The change of seasons and the rise and fall of vegetation often make the most beautiful moment of the natural landscape fleeting. The classical gardens can only be "let nature take its course" basically.

4 The Setting of the Virtual Animation in the Landscape Design System

The virtual animation plays an important role in the landscape design system. Visitors can understand the landscape design in an all-round way. Through the use of the camera tools in the 3ds max, they can switch lenses at will and view the landscape in an all-round way, reflecting the momentum and styles of the gardens. The virtual animation technology makes use of the scene changes in the landscape design to facilitate the visitors to understand the surrounding environment of the gardens. Some well-designed birds, animals, and the sun shuttling in the clouds and so on are added to the animation to foil the atmosphere and fabricate a variety of the beautiful scenery atmosphere.

To achieve these effects, we first need to adjust the camera's animation. In the scene, we usually use the target camera to complete the corresponding animation requirements, bind the camera's path constraints, and control its "XYZ" axis as the default controller of the "position" and "rotation" parameters. The controller can control the parameters of the X, Y and Z axes separately and modify them separately. It can edit the camera animation more carefully. The path constraint controller is an editing modifier for the "position" parameter. The controller can restrict the average distance of the object moving along one or more paths and take photos. The settings of the camera animation parameters can be more precise.

After completing the basic settings of the camera animation, we need to use the dynamic tools to simulate the physical scenes of the objects in the landscape design to set up the animation, which means that we can create the animation effects close to the real world, and can easily modify the animation. The dynamics tools use the physical simulation scenarios to set the animation, rather than the traditional key point method. This is a new concept of the animation settings. When the simulation scenario is set up, the object will automatically complete the animation according to its own physical properties. The dynamics tools support the fully integrated rigid and soft dynamics, the material simulation and the fluid simulation. After setting the physical attributes in the scenes, we need to animate some characters in the scenes. This will help to enhance the vividness of the entire landscape design and make the virtual scenes more humanized. We can adapt the skeleton models by creating a group of the skeleton animations, and then determine the location of the skeleton according to its shape. And simplify them properly, and finally complete the creation of the skeletons. When creating the skeleton, we should also take into account the types of the calculator, as well as the complexity of the analysis of the model animation and other factors.

The visualization technology of the landscape design planning has always been more important. The virtual animation technology can better show the designer's creativity in the landscape design, and play a guiding role in the future gardening. The application of the virtual animation technology in the scenery, architecture and animation can better show the anticipated landscape design planning. Because the scenery and the architecture need a lot of the texture mapping and the mapping, it is bound to require the higher computer hardware, but with the development of the science and technology, this problem has been well solved, so that the development of the virtual animation technology has a technical platform support.

Nowadays, many universities and institutions are researching and applying the virtual animation technology. It can be said that the virtual animation technology can be well integrated with the landscape design and other industries. It makes the process of the landscape design become controllable and the design drawings become visual. The virtual animation combines sounds, images, animations, and technical interaction, which makes visitors more immersive into the designed scenes of the landscape.

5 Conclusion

With the emergence and development of the virtual technology of the architectural animation, the virtual technology of the architectural animation has higher advantages than the architectural model and the plane perspective in the aspects of reality. The architectural animation virtual technology incorporates the manifestation of the films and the animation technologies. When it is applied in the landscape architecture, the effect of the project can be simulated dynamically. The landscape architecture animation virtual technology can achieve the three-dimensional rendering of the scenery, the bird's-eye view, the shuttle, the free-operation mode of the long and short range lens, and can also show the light and the shadow art performance of the landscape architecture, such as the day and night alternation, and the seasonal changes.

References

1. Chen H (2014) Realistic perception of the modern landscape planning, design and practice. China Forest Ind (16):157–159
2. Liu Y (2015) On the present situation of the modern landscape garden planning and design and the relevant countermeasures. Xiandai Hortic (11):119–121
3. Xu D, Li W, Meng X, Hu Y, Zhang M (2016) Teaching reform and practice of the landscape architecture design preliminary course in agricultural and forestry colleges – taking northeast forestry university as an example. Guangdong Landscape Archit (06):131–132
4. Liu S, Lu R, Wang X (2017) Parametric landscape architecture design methodology and practice – taking Guixi eco-park landscape of Huancheng eco-district of Chengdu as an example. China Landscape Archit (05):116–117
5. Qian J, Wu X (2018) Research on the embodiment of the natural education in the modern landscape architecture design – taking the planning and design of changle national forest germplasm park in Zhejiang province as an example. J Chin Urban Forest (05):102–103

Application of the Big Data Technology in the Chain Operation Management

Xiaohua Tang(✉)

Sichuan Technology and Business College, Dujiangyan 611830, Sichuan, China
qilinkai86@163.com

Abstract. The chain operation of enterprises has been gradually improved in today's society, and now it has already begun to have an excellent management mode. In the operation of the chain enterprises, the corporate headquarters plays a decisive role. It is the flag of all the subsidiaries and sub-organizations. Therefore, it is necessary for the corporate headquarters to play an excellent management function and bring the excellent corporate cultures into play to the chain enterprises. This paper, based on the big data technology, analyzes its application strategies in the chain operation management.

Keywords: Chain operation · Management work · Big data technology · Application

1 Introduction

The chain operation management has its particularity, which is different from the general independent individual operation. Its corresponding management should always focus on the business objectives of the chain enterprises [1]. The era of the big data is testing the ability of the enterprises to use the data. For the traditional enterprises with the non-Internet attributes, there are still many limitations in building the idea of the big data management, including the reconstruction of the market supply relationship, the imbalance of the resource allocation in each link, and the immaturity of the industrial chain, and so on. Among them, the most critical is that the traditional business management model cannot effectively collect, mine and analyze the data, which leads to the poor market adaptability and the lagging services [2]. Therefore, the advent of the era of the big data requires the enterprises to reverse the traditional management ideas.

2 Understanding the Characteristics of the Chain Operation

In order to further analyze the application of the big data technology in the chain operation management, it is necessary to understand the characteristics of the chain operation. The chain operation is a special modern business management mode, which is usually composed of the enterprise headquarters and a number of the subsidiaries or sub-organizations. In the chain operation, the headquarters makes the overall planning, transfers the characteristics of the enterprises to each of the subordinate companies, and

Z. Xu et al. (Eds.): CSIA 2019, AISC 928, pp. 345–350, 2020.
https://doi.org/10.1007/978-3-030-15235-2_52

ultimately makes every agent enterprises under its banner develop synchronously with the headquarters [3]. Therefore, the chain operation has six characteristics: the six unifications: the unified procurement, the unified distribution, the unified logo, the unified marketing strategy, the unified price, and the unified cancellation. But in the category of the chain operation, there is a term often heard, that is, the integrated purchase and distribution. As the name implies, the chain operation enterprises unified the purchase, and then separate the sales [4]. In order to reduce the cost of the purchasing, several sales departments unite to centralize the purchasing and maximize the profits of the purchasing department by increasing the quantity of the single batch purchasing, but this is not the characteristic of the chain operation. Because there is no requirement of the close relationship between the distribution departments, it can only be said that it is a form of the organization with the common interests. As the headquarters of the chain enterprises, more important than the unified procurement and distribution is that it also has the comprehensive management planning, the commodity development, the business model development and research, the business guidance, the staff training and other aspects of work [5]. Because the headquarters of the enterprises will be the flag of the entire chain enterprises, all the subsidiaries and sub-organizations are in fact to continue their excellent aspects in order to guarantee the enterprise characteristics and the business styles can be fully reflected in all the chain stores. Let all the chain stores carry forward the excellent corporate cultures and carry it out in the hearts of the consumers. Therefore, by contrast, the distribution stores that only have the function of the integrated distribution cannot be regarded as the chain headquarters. Their ideology is low-level, and the content is extremely limited. They can only achieve the unified procurement and distribution, and cannot play a decision-making role. Therefore, the unified distribution and distribution is by no means a chain operation.

3 Defects of the Chain Enterprise Management Model Under the Background of the "Big Data"

3.1 Defects of the Chain Enterprise Management Model Under the Concept of the "Big Data"

Firstly, the core of the big data is the data management, that is, to guide the management of the production, marketing, finance and other aspects of the enterprise management with the data standards. The enterprise information construction is the basis of the data management, but not the core element. The ultimate goal is to achieve the "data resource" through the data management.

Secondly, the key to the big data is the data quality. Once the big data is "resourced", it means that it has evolved into an important resource of the social concern, the enterprise competition and the government management. The quality of the data directly affects the quality of the big data management, and then affects the level of the business intelligence of the enterprises. "Business Intelligence" refers to the methods, technologies and tools that the enterprises can quickly analyze the data. At present, the theory of the "Business Intelligence" has not been popularized in the modern enterprise management mode of our country.

Thirdly, the ultimate goal of the big data is the construction of the data ecosystem. If we take the technology development, the information security, the theory perfection and other factors into account, its realization cannot rely on an enterprise or institution, let alone simply regard it as a "network". On the one hand, the production space of the big data, the expanding objects, sensor devices, smart mobile terminals, network providers, and data service providers and any of the endpoints involved in the Internet can generate the data. On the other hand, the enterprises have different demands for the big data contents, and need to further strengthen the data segmentation and the construction of the correlation system, which needs to set up more data leading roles.

3.2 Chain Business Managers Do Not Attach Enough Importance to the Commercial Value of the Big Data

Many business managers believe that they can cope with the "data impact" caused by the information explosion as long as they are well prepared for the informationization. This management mode based on the idea of the confrontation not only ignores the commercial values of the data mining and utilization, but also exposes the irrational adherence to the traditional management mode of enterprises. The business value of the big data can be understood through the supply chain competition model. On the one hand, the competition of enterprises in the market economy is inevitable, but the dominant factors are not the resources, channels and wealth, but the talents, innovation and technologies. On the other hand, the competition of enterprises in the market economy is relative. The "weapon of competitions" is no longer a simple product, but all the business relations related to the products, including all the supply and distribution elements in the upstream and downstream of the enterprises, and especially the disappearance of the estrangement between the enterprises and the consumers in the Internet economic environment. The public is the decision-making body of the enterprise managers, from whom we can get more valuable data. In this case, the financial data can only reflect the stage and limitations of the situation, and cannot play a long-term role in the enterprise management, and nor can it provide more targeted and valuable decision-making basis.

4 Ways to Ensure the Application Efficiency of the Big Data Technology in the Chain Management

4.1 Construction of the Intelligent Management Model for the Chain Enterprises Based on the Big Data

Construct the Big Data Culture

In the highly informationized and Internet-based modern economic society, the data has been regarded as a strategic resource, and the competition for the ownership and dominance of the data has become a new situation of the commercial competitions. The intelligent management in the era of the big data must be based on the big data, strengthen the talent training of the big data mining, attach importance to the IT planning at the top level, improve the sharing utilization rate, standardize the enterprise operation with the data, promote the enterprise insight, the industry insight and the user insight with the big data, realize the decision support, optimize the operation, reduce the costs, precise the marketing, enhance the enterprise security, promote the innovation and change the business models.

Operation Mechanisms of the Intelligent Management System for the Chain Enterprises
Under the background of the big data, the framework model of the enterprise intelligence management is defined from two dimensions: the enterprise group intelligence (including the enterprise cultures, the enterprise systems, the enterprise concepts, and the personnel structures and so on) and the man-machine integrated intelligence (including the value chain management, the marketing management, the financial management, the operation management, the information management, the knowledge management, the innovation management, and the opportunity management and so on), and is finally embodied as the follows four application levels: the decision data source, the decision information, the decision knowledge and the decision conclusion. The information and the values contained in the data are presented by the expert system, the decision support system and the knowledge base system.

Intelligent Management System Architecture of the Traditional Chain Enterprises
The innovation of the application of the enterprise intelligent management system includes the three dimensions of the customer insight, the product design, and the precision marketing, mainly based on the description of the product data and the user behavior data. The traditional customer segmentation of the large groups cannot reflect the individualized needs and the private customization requirements in the Internet era. The concrete customer needs and the small market segmentation form the discrete core customer groups of the enterprises. The large-scale and real-time data reduces the time for the enterprises to acquire the user data, in order to improve the efficiency of the enterprise management and the user experience, and ultimately achieve the insight into the customers. Secondly, the innovation of the product design is based on the analysis and positioning of the user behavior data and the product data. It discovers the recessive and the diversified psychological needs of the users, designs and develops products or services to meet various needs, and realizes the value proposition of the enterprises. The users' consumption time is fragmented and they are accustomed to the multi-screen consumption. The marketing channels of the products or services need to be further integrated. With the help of the information platform and the technology equipment based on the multi-screen interaction, the users' consumption habits and consumption processes can be tracked in time to complete the accurate marketing of the multi-channel integration.

4.2 Chain Enterprise Information Security Management System in the Big Data Environment

Planning Phase
The main tasks are to prepare for the safety management, to set up the safety management organizations, to establish the framework of the safety management system of

the big data and the safety management process strategy, to formulate the scope of the safety management, and to implement the safety responsibility to people. Specific work includes: providing the organizational guarantee for the safety management, setting up the effective security organizations, such as the security committees, assigning the roles, defining the authority and implementing the relevant responsibilities to all kinds of the personnel to ensure the smooth progress of the management. Secondly, hold the safety management meetings to plan the overall objectives of the enterprise information security management system in light of the actual situations of the enterprises. The main goal of the cloud security management is often to ensure the effectiveness of the security technology, the availability of the security products, the rationality of the security operation and the security performances, and to achieve a balance between the high security and the high performance, so as to provide guidance for the formulation and inspection of various management measures.

Implementation Phase

Investigate and analyze the security gap of the enterprises, identify the security vulnerabilities and risks, formulate the specific management plans, and clarify the contents of the security management from the physical security, the network security, the host security, the application security and the data security. The safety management covers all fields of the information security, including the risk assessment management, the safety certification management, the security strategy formulation, the planning of the management measures, the emergency plan establishment, the operation specification formulation, the environmental and physical safety management, the system development safety management, the operation and business safety management, the organization safety management, the safety awareness training, the safety education training, and the emergency response. Enterprises can choose the key contents to manage in the field of the information security management and establish the complete information security management measures and rules in order to ensure that the information security management has rules to follow. They can take action according to the formulated countermeasures. They should first apply the security management in the laboratory environment, and then the technologies and the management initiatives are put into practice.

Inspection Phase

Mainly through the daily inspection, the internal audit and review, the automatic control procedure alarm, the improvement of the field analysis and other measures, check whether the management measures are effective, whether they meet the safety management standards, and whether they meet the requirements of the laws and regulations, and record the inspection results, as the basis for the next stage of processing.

4.3 Improve the Management Decision-Making Ability

Firstly, in the era of the big data, the chain enterprises should constantly improve their ability of the internal data analysis and the management decision-making technologies. At present, the data analysis technology applied in enterprises is mainly based on the cloud computing. Simply put, the cloud computing is a basic tool for the data processing. Its greatest role is to carry the data, manage a large amount of the data collected by the enterprises, and process the data with differences, so as to reduce the difficulties in the process of the enterprise data analysis. Practice has proved that after the application of this technology, enterprises have been effectively promoted in the data information collection, and the data information analysis and evaluation. In addition, with the support of the cloud computing, a lot of the data information can be displayed in a specific form, and be applied to the information display by enterprises, in order to reduce the difficulty of the users in the information understanding and increase the user's recognition of the enterprises.

Secondly, the rapid increase of the data information in the era of the big data leads to the low efficiency of extracting the information values. How to find the data information that has a positive impact on the decision-making in a large amount of the information is an urgent problem to be solved in the era of the big data. In the process of the information screening, the traditional ways and methods are not applicable, so new methods must be developed and new technologies must be introduced. At present, the most valuable method is the knowledge discovery method, which is outstanding in the information value extraction and can shorten the application time of the information screening to a certain extent. Of course, this method is not perfect in the actual application process, and it is difficult to effectively grasp the relationship between the data, and it is difficult to respond to some implicit information in time, so the knowledge discovery technology in the future still has a large space for development, and it is worth the further exploration and improvement by the relevant personnel.

5 Conclusion

In a word, the application of the chain operation in China is not long. The chain enterprises are still the new things in some towns. The theoretical research and the case study on the financial management of the chain operation in the related enterprises are limited. Therefore, Chinese enterprises must strengthen management in the process of using the chain operation mode, especially for the application of the big data technology in the chain operation management, and constantly improve the application efficiency, so as to make the chain operation create more possibilities for the enterprises to improve their economic benefits.

References

1. Sun J (2018) How to achieve the high-level enterprise management from the perspective of the chain management. Mod Econ Inf (20):60
2. Zhang B (2017) Research on the innovation of the catering enterprise management based on the chain operation. Mark Mod (15):125–126
3. Zhang D (2015) Design and implementation of the chain enterprise management information system. Mark Mod (27):30–31
4. Wu X (2015) Design and implementation of the chain enterprise management information system. Mark Mod (23):22–23
5. Li J, Wang J (2015) Research on the catering chain enterprise management system based on the supply chain. Mod Econ Inf (08):122

Application of the Computer Vision Technology in the Image Feature Extraction

Yunhui Qu(✉)

Computer Teaching and Research Section, Xi'an Medical University, Xi'an 710054, Shaanxi, China
haibincong@126.com

Abstract. The image feature extraction is separated from the computer vision and the image processing. Computer is used to analyze and process the image information, and then to determine the invariant features of the image, and then to process the actual problems with the extracted features. Based on the basic principles and theories of the image processing technology, this paper carries out the application analysis of the computer vision technology in various fields and implements the comparative analysis of its methods.

Keywords: Computer vision technology · Image features · Extraction technology · Application research

As we all know, computers do not know pictures, but only know numbers. In order to enable computers to "understand" images and thus have a real sense of "vision", this chapter will study how to extract useful data or information from images, and obtain the "non-image" representation or description of images, such as values, vectors and symbols [1]. This process is the feature extraction, and the representation or description of these "non-image" extracted is the feature. With these numerical or vector features, we can teach the computers how to understand these features through the training process, so that the computers have the ability to recognize images.

1 Computer Vision Technology

The computer vision technology is a new technology developed under the application of the computer technologies. It is mainly used to study the macroscopic or the explicit functions of the computer simulation organisms [2]. The application of this technology will involve many disciplines, such as computer science, neurobiology, artificial intelligence, pattern recognition and image processing. The comprehensive application of the multidisciplinary technologies makes the computers have the ability to "perceive" the surrounding world, which is the core of this technology [3].

The characteristics of the computer vision technology lie in that, firstly, it can complete the detection of the tested person without touching the tested person, and secondly, the application fields and objects of this technology are very wide, and it can complete the detection of the ultrasonic, microwave and infrared rays which are difficult to observe by human beings under the application of sensitive devices. Last but

Z. Xu et al. (Eds.): CSIA 2019, AISC 928, pp. 351–356, 2020.
https://doi.org/10.1007/978-3-030-15235-2_53

not least, the technology also breaks through the limitation of the long-term work in the visual observation, and can observe the objects for a long time.

2 Overview of the Image Features

Features are the corresponding (essential) characteristics or characters of one class of objects that are different from those of other classes or the set of these characteristics and characters. It is characterized by measuring or processing the data that can be extracted [4]. For an image, each image has its own characteristics that can be distinguished from other kinds of images, some of which can be intuitively perceived as the natural features, such as brightness, edge, texture and color, while others need to be transformed or processed, such as moments, histograms and principal components.

We often combine several or more characteristics of a class of objects to form a feature vector to represent the class of objects. If there is only a single numerical feature, the feature vector is a one-dimensional vector, and if it is the combination of n features, it is an n-dimensional feature vector. This feature vector is often used as input to the recognition system [5]. In fact, an n-dimensional feature is a point in the n-dimensional space, and the task of the recognition and classification is to find a partition of the n-dimensional space [6].

For example, in order to distinguish three different species of Iris tectorum Maxim plants, we can choose their petal length and petal width as the characteristics, so that a two-dimensional feature can represent a plant object, such as (5.1, 3.5). If the sepal length and the sepal width are added, each Iris tectorum Maxim plant object is represented by a four-dimensional feature, such as (5.1, 3.5. 1.4, 0.2). The image recognition is actually a process of classification. In order to identify the category of an image, we need to distinguish it from other different categories of images. This requires that the selected features can not only describe the images well, but can also distinguish different types of the images well.

We hope to select those image features which have little difference between the same kind of images (smaller intra-class distance) and the large differences between different kinds of images (larger class distance). We call them the most discriminative features. In addition, the prior knowledge plays an important role in the feature extraction. How to rely on the prior knowledge to help us select the features is also a continuing concern.

Generally speaking, the feature extraction should be analyzed in details and its evaluation criteria are subjective. However, there are still some general principles that can be followed as our guidance in the feature extraction practice. The following are summarized. Features should be easy to extract. In other words, we can't pay too much to get these features. Of course, this should be weighed against the classification ability of the features. The selected features should be insensitive to the noise and the irrelevant transformations. For example, to identify the license plate numbers, the license plate photos may be taken from various angles, while we are concerned about the contents of the letters and the numbers on the license plates, so that it is necessary to obtain the descriptors that are insensitive to the geometric distortion and transformation, so as to obtain the features of the invariant rotation or the projection distortion.

3 Application Analysis of the Computer Vision Technology in Various Fields

3.1 Application in the Industrial Fields

The industrial production requires high quality of products. The application of the computer vision technology in the industries mainly concentrates on the following three aspects: the product shape and size detection. For the manufacturing industry, whether the shapes and sizes of the products are qualified or not will directly affect the roles of the products in the actual application process. The application of the computer vision technology can detect the geometric features of products, such as roundness, position and shape. The second is the detection of the missing parts of products. In the process of the production line operation, the computer vision technology can accurately detect whether there are rivets, screw and other parts missing in the production process, and whether there are impurities in the production process. The computer vision technology realizes the effective detection of the product surface texture, roughness, scratches, cracks and other aspects.

3.2 Application in the Field of the Agricultural Production

The application of the technology in the agriculture is mainly concentrated on the following two aspects: the prediction of diseases and insect pests. The key link of predicting and forecasting is to establish the recognition system of the computer vision technology for all insects. There are two main methods used in the digital modeling of the insect image recognition system. One is to detect the edge of the insect pests by the mathematical morphology, and then extract the characteristics of the insect pests. The other is to extract the basic information of the insect perimeters, areas and complexity from the binary images of the insects. This information establishes a template library for the insect pests to achieve the fuzzy decision analysis of insects. The second is the monitoring of the crop growth. The common method is to use the non-contact monitoring system under the computer vision technology to continuously monitor the relevant factors, such as light, temperature, humidity, and nutrient solution concentration and so on, under the growing environment of the crops, and then to judge the growth trend of the crops.

3.3 Application in the Forestry Production

The application of the technology in the forestry production is mainly focused on two aspects: the pesticide spraying and the forest cone collection. As far as the forestry pesticide spraying is concerned, the conventional pesticide spraying methods can easily cause the loss of a large amount of pesticides, which can not only fail to achieve the purpose of preventing the forest pests, but can also waste a lot of the manpower, material resources and financial resources. The application of the computer vision technology can get the specific dosage and the exact location of the pesticide application through the real-time analysis of the target images of the pesticide application. The pesticide application under the guidance of this technology has played a great role in the pesticide application. As far as collecting the forest cones is concerned, the operation of collecting is always very difficult. The main methods used in our country are collecting by the manual use of the professional tools, collecting by the aerial working vehicle and collecting by shaking and picking machines under the use of the mechanical equipment. Both methods have certain security and efficiency problems. The application of the computer vision technology can obtain the specific positions of the cones by the image acquisition of the cones which need to be collected, and then complete the cone acquisition with the use of the professional manipulators. The technology not only saves a lot of labor, but also greatly improves the picking efficiency.

3.4 Application in the Detection of the Agricultural Products

The agricultural products are greatly affected by the natural environment in the production process, so the agricultural products will not only produce the quality differences, but also cause great differences in colors, sizes, shapes and other appearances. Because most of the agricultural products are classified into different grades when they are sold, the computer vision technology is applied to the detection of their colors and shapes and sizes, which effectively achieves the purpose of the detection of the agricultural products. The detection of the appearances and sizes not only improves the efficiency of classifying and grading the agricultural products, but also reduces the damage to the products to a large extent. By detecting the colors of the watermelon and other agricultural products, we can accurately judge whether it is mature or not and effectively avoid the mistakes under the manual operation.

4 Application of the Computer Vision Technology in the Image Feature Extraction

With the rapid development of the mobile Internet technologies, the digital storage devices and the computer communication devices, images transmitted through the social network applications such as Face-book, Twitter, Instagram and the instant messaging applications such as WhatsApp and WeChat continue to grow rapidly. How to browse and select the image data that people are interested in from the massive image database, and how to extract the structured and meaningful logic, entity and even relationship network from these unstructured data, so as to predict the trend of the social groups and the individual behaviors, is a new challenge and a new opportunity in the era of the big data on the Internet.

4.1 Smoothing Filtering Processing

Because there are noise points in the video image acquisition, we need to process the noise points in order to reduce the noise. The filtering process of the filtering smoothness has two ways, the linear and the non-linear. The linear method is simple and fast for the processor, but the image after processing will be unclear. After the non-linear processing, although it can reduce the noise points and ensure the local characteristics of the signal, its operation speed will be slower.

4.2 Image Padding

For the frame image processing, the detection edge padding method or the corrosion expansion method is usually used. The padding method refers to the identification of the objects by the edge detection method after the detection of the target objects, and then padding by the morphological flooding filling method. The corrosion expansion of the images is mainly due to the performances of the cameras.

4.3 Real-Time Background Updating

Before the image difference, the background pattern needs to be determined and initialized. In order to facilitate the difference calculation of the real-time background images in the future detection, only in this way can we obtain the excellent foreground effect. In the process of the image difference, the first frame background image needs to be determined according to the designated method and designated as the first background image, and then the background is updated according to the algorithm in the detection process. When updating the entire image, the main processes are: judging and reading whether the image is the first frame; transforming the image processed by Opencv into a single channel gray value; processing the real-time collected image with the Gauss smoothness to remove the noise; and finally using the morphological filtering to process the noise.

When detecting the moving objects, the satisfactory foreground tracking results can be obtained only if the accuracy is ensured in the detection process. This process is divided into two steps. The first step is to segment the binary image, and the second step is to pre-process the image and fill it fully to ensure the integrity of the foreground image. Among them, the foreground image extraction is mainly divided into the following steps. First, the foreground image and the background image are differentiated, and then the difference image is binarized, and then the foreground image edge in the background is detected, and the image is filled according to the contour. Because cameras exist in different scenes and environments, both the outdoor and the indoor change of the scenes will have an impact on the image acquisition. In order to extract the target from the foreground image, it is necessary to use the effective means to update the background in real time in the detection system.

5 Conclusion

The visual technology is used in the complex environment. Because of the changes of the illumination, the environmental factors in the scenes have a great impact on the performances of the video acquisition equipment. The environmental factors will reduce the quality of the acquired image information, and there are unavoidable noises in the image, which will have a great impact on the detection of the moving objects and the image acquisition. After acquiring the video frame images, the data need to be preprocessed, such as the smoothness filtering, the image filling, and the image background updating and so on.

References

1. Huang X, Bi J, Zhang N, Ding X, Li F, Hou F (2017) Application of the computer vision technology in agriculture. Agric Sci Technol (11):129–130
2. Yin Q (2017) Report on the new media conference on the "one belt and one way" international cooperation summit forum – examination of the media literacy of the VR visual communication and the new media technology and the information service team. J Int Commun (06):119–120
3. Jia Z, Fu L, Yang X, Shi C, Wang H, Zhou J, Wang Y (2018) Advances in the application of the machine vision technology in the sensory detection of the aquatic foods. Food Sci (10):155–156
4. Tang H, Zeng Z (2018) Image technology, visual presentation and modern China: comments on Wu Guozhong's "having maps on the left and history books on the right – a home library and there is a word in the painting: a study of the modern Chinese pictorial journals (1874–1949)". J Int Commun (07):126–127
5. Zhou Q (2016) Research on heterogeneous data integration model of group enterprise based on cluster computing. Clust Comput 19(3):1275–1282
6. Hu F (2018) Application and research of the "micro course" in the design classroom teaching in colleges and universities – taking the visual communication major of engineering and technology college of Hubei university of technology as an example. China New Telecommun (06):100–101

Chinese Entity Synonym Extraction from the Web

Xiuxia Ma[1(✉)], Xiangfeng Luo[1], Subin Huang[1], and Yike Guo[1,2]

[1] School of Computer Engineering and Science, Shanghai University, Shanghai, China
ma930912@i.shu.edu.cn, {luoxf,huangsubin}@shu.edu.cn
[2] USA Department of Computer, Imperial College London, London, UK
y.guo@imperial.ac.uk

Abstract. Entity synonyms play an important role in natural language processing applications, such as query expansion and question answering. There are three main distribution characteristics in texts on the web: (1) appearing in parallel structures; (2) occurring with specific patterns in sentences; and (3) distributed in similar contexts. These characteristics are largely complementary. Existing methods, such as pattern-based and context-based methods, only consider one characteristic for synonym extraction and ignore the complementarity among them. For increasing accuracy and recall, we propose a novel method that integrates the three characteristics for extracting synonyms from the web, where Entity Synonym Network (ESN) is built to incorporate synonymous knowledge. To further improve accuracy, we treat synonym detection as a ranking problem and use the Spreading Activation model as a ranking means to detect the hard noise in ESN. Experimental results show our method achieves better accuracy and recall than the state-of-the-art methods.

Keywords: Synonym extraction · Entity Synonym Network · Spreading activation

1 Introduction

Entity synonyms play a vital role in many natural language processing applications [1, 2]. The distribution of synonyms in the web has three main characteristics: (1) appearing in parallel structures such as tables and lists; (2) occurring with specific patterns in sentences; (3) distributed in similar contexts. Existing methods mostly consider one characteristic of them to extract synonyms. Pattern-based methods [3, 4, 6–8] infer the relation of two words by analyzing specific patterns mentioning them, which are more accurate and interpretable. But they may lead to low recall, because synonymous words may not be co-mentioned with any pattern. Context-based methods [9–11, 13–15] assume words which often appear in similar contexts are semantically similar. The methods may cause higher recall but lower accuracy, since most synonyms will appear in similar context, but semantically related words may also share similar context.

These characteristics are largely complementary, but few researchers attempt to integrate them. In this paper, we propose a novel extraction method by combining them.

Z. Xu et al. (Eds.): CSIA 2019, AISC 928, pp. 357–363, 2020.
https://doi.org/10.1007/978-3-030-15235-2_54

To incorporate the outputs of the three characteristics, Entity Synonym Network (ESN) is built. Then we treat synonym detection as a ranking problem. Inspired by the study of human cognition process, a 'human-like' ranking method based on the Spreading Activation model [5] is proposed for ranking entities in ESN.

2 Related Work

There exist mainly two kinds of methods: the pattern-based and context-based methods. The pattern-based methods infer the relation of two words by analyzing the local contexts mentioning both of them. Hearst et al. [3] are pioneers to apply patterns like "X such as Y" to discover hypernym-hyponym relations. Simanovsky et al. [4] proposed a method to calculate confidence for extracted patterns. Wang et al. [6] proposed a pattern construction approach to extract verb synonyms. Batista et al. [7] researched bootstrapping process for relation extraction and calculated the confidence of extracted patterns and tuples. Hu et al. [8] treated synonym extraction as a sequential labeling problem and train CRF model to extract synonyms. The context-based methods regard words sharing similar contexts may be synonyms. Based on the assumption, words are usually represented with distributional feature vectors. Hagiwara et al. [11] treat synonym detection as binary classification and train a classifier to predict candidate pairs by constructing distributional features. Recently some work has focused on utilizing low dimensional word embeddings [10, 13] for identifying synonyms. Olivier et al. [14] proposed a method to turn distributional thesauri into word vectors for synonym extraction. Leeuwenberga et al. [9] trained word embeddings by using both the continuous bag-of-words and the skip-gram model for synonym extraction. Nguyen et al. [15] employed word embedding technique to encode relational phrases and then cluster the phrases.

The above methods either consider pattern-based or context-based characteristic of synonym distribution. These characteristics are largely complementary, but there have been few attempts to combine them, causing low recall or accuracy.

3 Proposed Method

As encyclopedias in the web cover topics in nearly every imaginable domain, we collect web pages as corpus from Chinese encyclopedias, including BaiduBaike and HuDongBaike. Firstly, we acquire the entries within a certain domain from encyclopedias and crawl web pages of these entries. Firstly, we extract candidate pairs by utilizing parallel and pattern characteristics. After inputting a target entity, we expand its candidates by context characteristic. Secondly, we build ESN (Entity Synonym Network) to combine the extraction results. Finally, we use the Spreading Activation to rank entities in ESN. Details will be given in the following sections.

3.1 Chinese Entity Synonym Extraction from the Web

Candidates Based on Parallel Characteristic. Many web pages on encyclopedias include parallel structures, called infoboxes, which are tabular summaries of attributes. The infobox may list its synonyms with specific attribute names appearing. Some of them are shown in Table. 1. We crawl web pages with infoboxes, and match HTML tags, where attribute names appear, to acquire synonyms.

Table 1. Attribute names which imply synonyms

Attribute names	别名/alias,又名/also named as,中文学名/Chinese name, 俗称/commonly named as, 其他名称/other name, 别称/alias ,简称/abbreviated as…

Candidates Based on Pattern Characteristic. While infoboxes contain much valuable structured knowledge, they suffer from incompleteness. Therefore, pattern characteristic is utilized to improve coverage of synonym extraction. We observe synonym relations are usually expressed with specific cue words, such as "又叫 (also called as)" and "简称 (abbreviated as)". We markup sentences with 1000 synonymous pairs, and then extract patterns between terms and their synonyms. This paper uses an entropy-based feature selection method to get cue words. $P = \{p_1, p_2, \ldots p_n\}$ denotes a set of patterns between synonym pairs. $W = \{w_1, w_2, \ldots w_m\}$ represents all verbs and nouns in P. For word w_i, we use the following formulas to calculate its entropy E_{w_i}, where L and R indicate a set of words that appear in the left neighbors and the right neighbors of $w_{i.}$

$$p(lw|w_i) = \frac{\text{count}(lw, w_i)}{\sum_{lw' \in L} count(lw', w_i)} \tag{1}$$

$$p(wr|w_i) = \frac{\text{count}(wr, w_i)}{\sum_{wr' \in R} count(wr', w_i)} \tag{2}$$

$$E_{w_i} = -\left(\sum_{lw \in L} p(lw|w_i) \log p(lw|w_i) + \sum_{wr \in R} p(wr|w_i) \log p(wr|w_i)\right) \tag{3}$$

The larger the value of E_{w_i}, the more important the word w_i. Top ranked words are selected as original cue words. Since computation of word similarities via low-dimensional vectors has been proven efficient, we train similar words of original cue words by the word2vec model of Mikolov [13]. Finally, we obtain more than 200 cue words. According to these cue words, soft patterns like "X 又名(also called as) Y" are constructed. Receive a single target entity as input and search for strings of the other entity as candidate synonyms.

Candidates Based on Context Characteristic. Words sharing similar contexts are likely to be synonyms. We firstly represented words with low-dimensional vectors by the trained word2vec model. Synonymous relations extracted by the parallel characteristic are represented as an undirected graph $<W, S>$, where $W = \{w_1, w_2, \ldots w_n\}$ is a vocabulary and edges $(w_i, w_j) \in S$ indicate synonym relation. We then adopted an effective retrofitting method [10] to improve word vectors. The aim is to learn the vector which is both close to the initial vector and its neighbors in the graph. For a given entity e, the top 10 most similar words are obtained by retrofitted vectors. For each similar word w_i, its similarity value with e is recalculated by using (4).

$$\mathrm{score}(\mathrm{e}, \mathrm{w_i}) = \frac{\mathrm{cosine\_similarity}(\mathrm{e}, \mathrm{w_i})}{\sum_{\mathrm{w_j} \in \mathrm{Top10}} \mathrm{cosine\_similarity}(\mathrm{e}, \mathrm{w_j})} \tag{4}$$

This will give words that have a high cosine similarity compared to other words a high score. Finally, we select top 5 ranked words as final candidates of the given entity.

3.2 Building Entity Synonym Network (ESN)

To combine the candidates extracted by the above three characteristics, we propose to build Entity Synonym Network (ESN) for a given entity.

$$\mathrm{ESN}(e) = <E, W> \tag{5}$$

Where E is a set of nodes, W is a set of undirected weighted links composed all edges $(u, v) \in E \times E$.$\mathrm{w}_{u,v}$ can be calculated as,

$$\mathrm{w}'_{u,v} = \partial \times \mathrm{pl}(u, v) + \beta \times \mathrm{cw}(u, v) + \gamma * \mathrm{cx}_{(\mathrm{u,v})}.score \tag{6}$$

$$\mathrm{w}_{u,v} = f(\mathrm{w}'_{u,v}) = \frac{1 - e^{-w'_{u,v}}}{1 + e^{-w'_{u,v}}} \tag{7}$$

Where u and v denote word u and v. $\partial, \beta, \gamma \in (0, 1)$ indicate the accuracy parameter of each characteristic. $pl(u, v)$ denotes the frequency of pair u↔v extracted by parallel characteristic. $cw(u, v)$ denotes frequency extracted by patterns. $\mathrm{cx}_{(\mathrm{u,v})}.score$ denotes the similarity score between pairs trained by context characteristic. $f(x) = \frac{1 - \mathrm{e}^{-x}}{1 + e^{-x}}$ is a normalization function.

3.3 Synonym Ranking Based on the Spreading Activation

Based on the Spreading Activation model [15], the ranking process for candidates of a given entity in ESN begins with a source node. We choose the target entity as the source node. The entities in ESN initially have an activation value 0, except for the source node has 1. This activation spreads through the entities that are directly connected to the target entity in a series of iterations. In each iteration, every entity propagates its activation to its neighbors. The process ceases when it reaches nodes that are far away from the target entity. For each activated node, the activation value is computed by (8).

$$A_{\mathrm{i}}^{t} = A_{\mathrm{i}}^{t-1} + \sum_{j} \lambda \times \mathrm{A}_{j}^{t-1} w_{i,j} \tag{8}$$

We denote the activation of A_{i} in t^{th} iteration as A_{i}^{t}. The activation value A_{i} for the node i is calculated by a function of activation value A_j of its neighbors j and the edge weights $\mathrm{w}_{i,j}$. The decay factor λ is generally set 0.75, which means the association strength between nodes decreases when propagation distance increases. The activation value of each node is a numerical value that represent the activated probability.

4 Experiment

4.1 Dataset

Chinese encyclopedias in the web, such as BaiduBaike (https://baike.baidu.com/) and HuDongBaike (https://www.baike.com/), which are like Wikipedia but larger than Chinese Wikipedia, cover topics in nearly every imaginable domain. Entries in encyclopedias include "tags", which means domains they belong to. According to "tags", we crawl web pages in the food domain as datasets. A set of entries that have the tags "食品或食物 (food)" is collected. Finally, we crawled 245679 web pages from BaiduBaike and 130707 web pages from HudongBaike.

4.2 Evaluation Metrics

In many applications, it is more desirable to have a subset of high-quality synonyms as output. The quality of output can be evaluated by accuracy and recall. However, it is difficult to know the universal set of synonyms for each entity, and hard to report the traditional recall number. Similar to the case [12], we use average number of synonyms per entity as an alternative for recall. (1) *Accuracy: number of true synonyms divided by the total number of synonyms; (2) Avg#Syn: average number of synonyms per entity.*

4.3 Experimental Result

As it is not feasible to analyze all entities, we randomly sample different number of test entities N (N = 100, 200, 300, 400, 500) to analyze the experimental results. We empirically set $\partial = 1$, $\beta = 0.8$ and $\gamma = 0.4$ to build ESN for entities. Since each entity gets a set of candidates sorted by activation value, we set appropriate threshold to obtain a subset of synonyms per entity. In Fig. 1, we conduct comparison with the following methods. (1) "parallel", "pattern" and "context" represent the extraction results by parallel, pattern and context characteristics; (2) Batista [7]: a bootstrapping method for relation extraction, which can be applied to synonym extraction by treating synonym pairs as seeds; (3) Hu [8]: a self-supervised Chinese synonym extraction method.

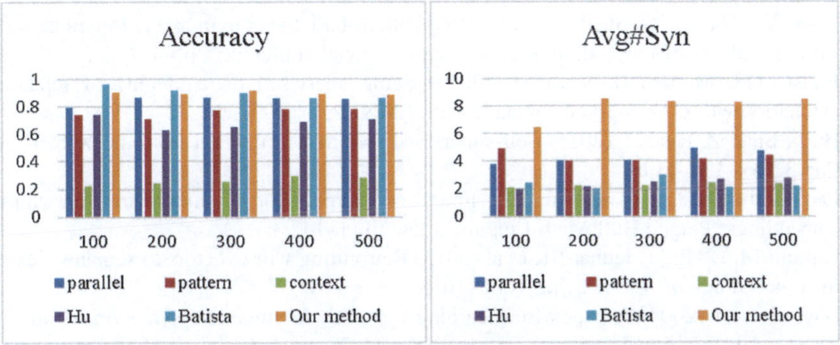

Fig. 1 Comparison of our method with other methods

Figure 1 shows that our method by combining three characteristics achieves better results than method considering single characteristic and Hu [8]. Compared with the method of Batista [7], when the number of test entities is less than 200, the accuracy of our method is slightly lower. However, when increasing the number of sampling entities, our method greatly improves recall and can also achieve better accuracy. The results reveal that our method can improve the performance of synonym extraction.

5 Conclusion

This paper has integrated three characteristics of synonym distribution in corpus. For the first characteristic, we identified entity synonyms listed in structured tables from Chinese encyclopedias. For the pattern characteristic, we proposed to employ information entropy and word vector for getting cue words of patterns. For the third one, we combined retrofitted word vector and relative cosine similarity to identify synonyms. This paper proposed ESN (Entity Synonym Network) to combine the outputs of these characteristics. To further improve accuracy, we innovatively used the Spreading Activation model as a ranking means for detecting noise in ESN. The experimental results have shown that our method can achieve higher accuracy and recall than other methods.

References

1. Krishnan A, Deepak P, Ranu S et al (2018) Leveraging semantic resources in diversified query expansion. World Wide Web 21(4):1041–1067
2. Hristovski D, Dinevski D, Kastrin A et al (2015) Biomedical question answering using semantic relations. BMC Bioinf 16(1):1–14
3. Hearst MA (1992) Automatic acquisition of hyponyms from large text. In: 14th International conference on computational linguistics, pp 539–545
4. Simanovsky A, Ulanov A (2011) Mining text patterns for synonyms extraction. In: Database and expert systems applications, pp 473–477
5. Pelegrina AB, Martin-Bautista MJ, Faber P (2013) Contextualization and personalization of queries to knowledge bases using spreading activation. In: Flexible query answering systems, pp 671–682
6. Wang W, Thomas C, Sheth A et al (2010) Pattern-based synonym and antonym extraction. In: Proceedings of the 48th annual southeast regional conference, p 64
7. Batista DS, Martins B, Silva MJ (2015) Semi-supervised bootstrapping of relationship extractors with distributional semantics. In: EMNLP, pp 499–504
8. Hu F, Shao Z, Ruan T (2015) Self-supervised synonym extraction from the web. J Inf Sci Eng 31(3):1133–1148
9. Leeuwenberg A (2016) A minimally supervised approach for synonym extraction with word embeddings. Prague Bull Math Linguist 105(1):111–142
10. Faruqui M, Dodge J, Jauhar SK et al (2015) Retrofitting word vectors to semantic lexicons. In: Proceedings of NAACL, pp 1606–1615
11. Hagiwara, M (2008) A supervised learning approach to automatic synonym identification based on distributional features. In: Meeting of ACL, pp 1–6

12. Chakrabarti K, Chaudhuri S et al (2012) A framework for robust discovery of entity synonyms. In: Knowledge discovery and data mining, pp 1384–1392
13. Mikolov T, Sutskever I, Chen K et al (2013) Distributed representations of words and phrases and their compositionality. Adv Neural Inf Process Syst 26:3111–3119
14. Ferret O (2017) Turning distributional thesauri into word vectors for synonym extraction and expansion. In: International joint conference on natural language processing, pp 273–283
15. Nguyen NT, Miwa M, Tsuruoka Y (2015) Identifying synonymy between relational phrases using word embeddings. J Biomed Inf 56:94–102

Mining Friend Relationship Existing in Microblog Based on Building Relationship Description Dictionary

Nana Wang[1], Shunxiang Zhang[1](✉), and Zheng Xu[2]

[1] School of Computer Science and Engineering, Anhui University of Science and Technology, Huainan 232001, China
287758594@qq.com, sxzhang@aust.edu.cn
[2] Guangxi Key Laboratory of Cryptography and Information Security, Guilin, China
xuzheng@shu.edu.cn

Abstract. The character/person relationship is difficult to effectively extract because of microblog characteristics such as massive content, low value density etc. To solve this problem, this paper proposes a microblog friend relationship mining method based on building relationship description dictionary. First, the semantic similarity calculation method of the combination with HowNet and Tongyici Cilin is proposed. Based on the two friend relationship description Dictionary, a more accurate friend relationship description Dictionary is obtained. Then, add core predicates, dependency syntax, and semantic analysis methods when working with text data, the core predicate can determine the entity pair, dependency syntax can describe the relationship between words before and after words, semantic analysis can further mining text information. When combine use them can achieve the mining of friend relationship from microblog. Experimental results show that the friend relationship mining performance is significantly improved.

Keywords: Friend relationship · Core predicate · Dependency syntax · Semantic analysis

1 Introduction

With the rapid development of social networks, the huge number of users makes microblog become the largest virtual community in China. Due to the massive content and simple operation of social networks, more and more users choose to interact with others through registration microblogs, and there are lots of social relationships among characters in microblog text. So mining microblog friend relationship can better accelerate the development of social networks.

In recent years, the relationship between characters has extremely important application value in the field of microblog market, advertising, network public opinion, etc. Making the social relationship among microblog characters become a hot research topic in recent years. For example, using Chinese online resources to construct Chinese character relationship extraction system [1], using dependency-kernel-based SVMs [2],

Z. Xu et al. (Eds.): CSIA 2019, AISC 928, pp. 364–372, 2020.
https://doi.org/10.1007/978-3-030-15235-2_55

extracting frames from the information in the text document of the clause [3], using Information Gain Value create template to realize automatic extraction of Web character relationship [4], developing the NERC system and express its characteristics with simple words of various types [5], using the DLA subspace learning algorithm [6], constructing a semantic relationship tree and proposing a convolution tree kernel [7], etc.

To solve the problem mentioned above, we propose a microblog character relationship mining method based on building relationship description dictionary. The major contributions of our work include two aspects: (1) we effectively extend relationship description dictionary. (2) We mine the friend relationship from the microblog text data. Experimental results show that the friend relationship mining performance is significantly improved.

The rest of the paper is organized as follows: Sect. 2 builds the relationship Dictionary. Section 3 mining friend relationships in microblog text data. Section 4 is experimental evaluation of the proposed mining approach. Conclusions are given in Sect. 5 (Fig. 1).

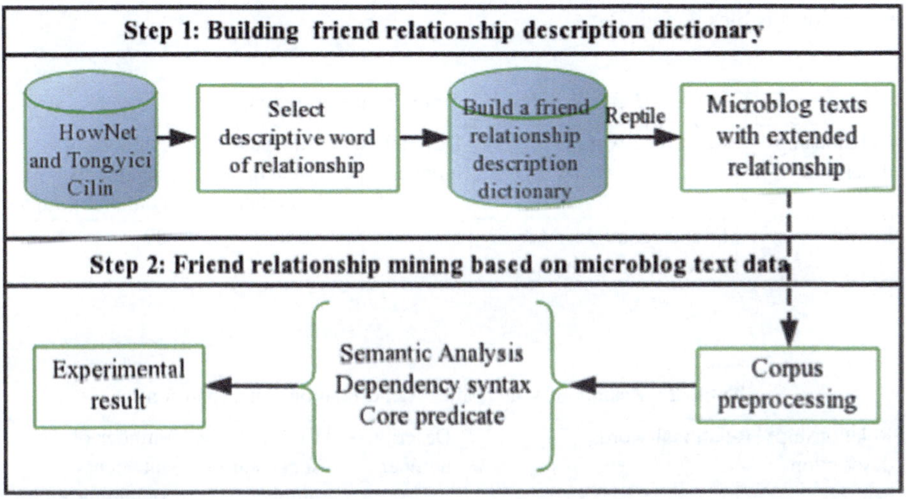

Fig. 1. The mining process of friend relationship

2 The Expansion of Friend Relationship Description Dictionary

2.1 The Preprocessing of Microblog Text Data

Before the microblog relationship mining, this paper uses the LTP software of Harbin Institute of Technology to preprocess the microblog text data. It mainly includes deleting "@" symbols in the microblog text, deleting "#" symbols and the contents of symbols, converting the traditional characters in the microblog text into simplified characters, filtering the emoji, and so on.

2.2 The Building Description Dictionary of Friend Relationship

In order to improve the accuracy of friend relationship mining, this paper proposes a method for calculating the semantic similarity of words in combination with HowNet and Tongyici Cilin. And the value of similarity is 0–1. For any words W_1 and W_2, according to their distribution in HowNet and the synonym word forest, the similarities are calculated as S_1 and S_2 respectively; then, weights λ_1 and λ_2 are assigned to S_1 and S_2, and the weights satisfy $\lambda_1 + \lambda_2 = 1$, According to the formula 1, the semantic similarity of the words of the combination with HowNet and Tongyici Cilin is calculated.

$$S = \lambda_1 S_1 + \lambda_2 S_2 \tag{1}$$

By experimenting with the above method, in this paper, $S = 0.7$ is set as the threshold, and the friend relationship descriptive words are compared before and after expansion (Tables 1 and 2).

Table 1. Results of pre-expansion search

Relationship description	Relational words	Descriptive number	Corpus statement number	Number of statements containing tuples
Friend	Friends/good friends/ mates/pal/acquaintances	5	248	89

Table 2. Results of search after friend relationship expansion

Relationship description	Relational words	Descriptive number	Corpus statement number	Number of statements containing tuples
Friend	Friends/good friends/ mates/pal/acquaintances/ old friends/colleagues/ close friends/old pals/ girlfriends/boyfriends	11	523	234

Based on the detailed discussion above, a simple algorithm can be devised to help analyze how this article extends the friend relationship Dictionary. The detailed process of the mining algorithm is shown in Algorithm 1.

Algorithm 1: Word similarity calculation algorithm

Input: words W_1 and W_2

Output: Word similarity value S of W_1 and W_2

```
1: The Similarity calculation method of HowNet
2: for (input words W1 and W2)
3:     do (calculate)
4:         {similarity is S1}
5: end if
6: The Similarity calculation method of Tongyici Cilin
7: for (input words W1 and W2)
8:     do (calculate)
9:         {similarity is S2}
10: end if
11: for (weights λ1 and λ2 are assigned to S1 and S2, and the weights
    satisfy λ1 + λ2 = 1)
12:     do (Calculate according to Formula 1)
13:         {similarity is S}
14: end
```

Algorithm 1 mainly includes three judgments (i.e., three selection sentences).

(1) The first judgment, the steps 1–5, is used to calculate the similarity between the words W_1 and W_2 as S_1 with the method of HowNet.
(2) The second judgment, the steps 7–10, is used to calculate the similarity between the words W_1 and W_2 as S_2 with the method of Tongyici Cilin.
(3) The third judgment, the steps 11–14, is used to Calculate according to Formula 1, and get similarity is S.

3 The Mining of Friend Relationship

3.1 The Determination of Core Predicates

Core predicates play an important role in undertaking entity relationships and acquiring entity boundaries. A large number of experimental data show that the distance between the entity and the core predicate in the statement is significantly different from other predicate distances. So it can indirectly reflects the existence of a relationship characteristic between entities. Figure 2 shows, the sentence "Zhang Yong and Li Jun become close friends, they get along very well and often play together". In Fig. 2, ROOT indicates that the root node is the core predicate "become", predicates also include "getting along" and "playing". It is calculated that the average distance between entities and the core predicate is 3, and the average distance from other predicates is 5.

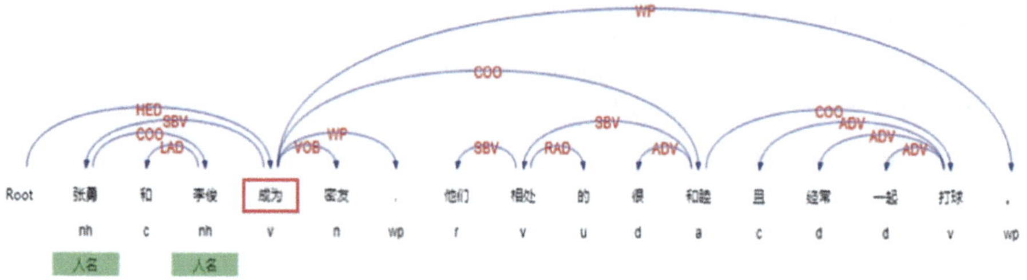

Fig. 2. The instance for core predicate

3.2 Dependency Parsing

Dependency parsing is to determine the syntactic structure of a sentence or the dependencies between words in a sentence. Mainly by analyzing the collocation between words. In the sentence, the elements such as "subject predicate object" and "fixed complement" are used to obtain the connection between the components. Because the friend named entities in the statement will appear as a phrase structure in the dependency structure, the friend relationship characteristics between entities in the same sentence will be reflected in a certain dependency relationship. This paper takes the core predicate and the dependency syntax as a syntactic feature of entity relation extraction. Figure 3 shows, the sentence "Ma Yun, speaking in Israel, said that he and Bill Gates became a very close friend". And "Ma Yun" and "Bill Gates" represent character entities, "Israel" means a place name entity, it also forms preposition-object relationship with "being", the core predicate is "become", Get the entity relationship pair: become (Ma Yun, Bill Gates).

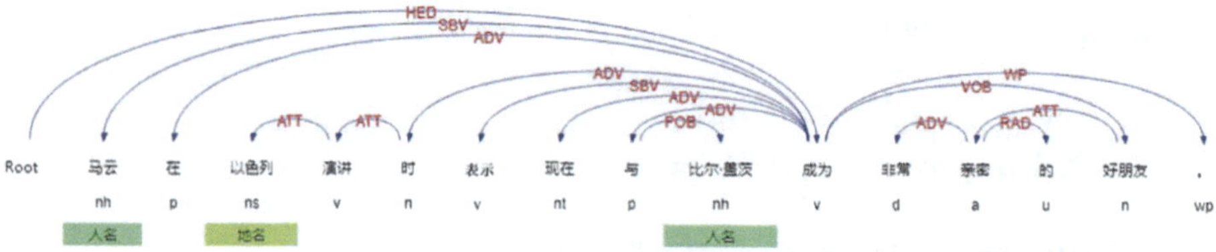

Fig. 3. The instance of dependency syntax

3.3 Dependency Parsing

Semantic analysis refers to the prediction of a formal representation that can express the meaning of a sentence based on the sentence's syntactic structure and the meaning of each real word in the sentence. And semantic role labeling is a shallow and commonly used semantic analysis technology [8]. Figure 4 shows, "Zhang Qiang often interacts with Wang Juan on microblog, and they become friends through mutual powder". The predicate in this sentence is "interactive", the agent of the action (A0) is "Zhang Qiang", and the influencer of the action (A1) is "Friend", this sentence indicates that the influence of the action implementer is to become a "friend".

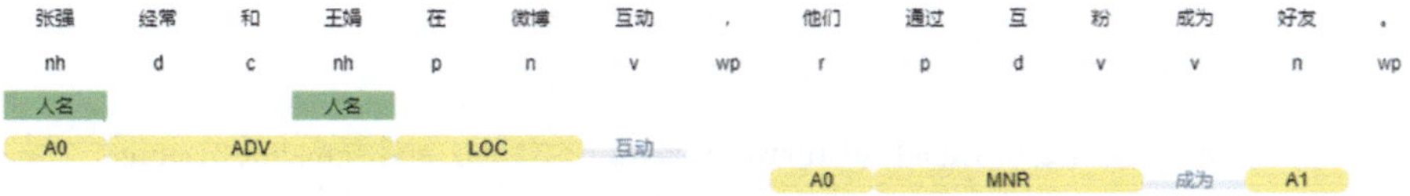

Fig. 4. The instance of semantic analysis

3.4 The Combination of Syntax and Semantic Feature

Based on the traditional character relationship extraction, this paper adopts the language processing technology of LTP to improve, and then obtain more sentences with syntactic and semantic features. The specific methods are as follows:

(1) Sentence structure. Select the microblog text of the friend relationship pair in the statement.
(2) Core predicates. According to the position of the entity and the core predicate in the sentence, this paper calculates the distance between entities and core predicates and other predicates, and the distance between the entity and the core predicate is taken as the feature of the relationship extraction. When there are more than one pair of entities in the sentence, two different entity positions closest to the core predicate are selected as the object of this study.
(3) Dependency parsing. Obtain the dependency syntactic relations between each pair of entities in the corpus.
(4) Semantic analysis. After the above operations, the language processing technology of LTP is selected to perform semantic analysis on the sentence, and the analysis result is taken as the feature extraction feature.

Figure 5 shows, the sentence "Boss Wang and Boss Zhang have become best friends of each other and often cooperate". Using traditional methods in terms of word segmentation, part of speech, contextual context, based on the characteristic word "cooperation", it is inferred that "Boss Wang" and "Boss Zhang" belong to the cooperative relationship; Based on the traditional character extraction relationship, this paper adds the core predicate, dependency syntax, semantic analysis and other extraction features, and infers that the core predicate ROOT is "become". The two characters entities are "Boss Wang" and "Boss Zhang" respectively. Through analysis, the relationship between the entity and the relationship is expressed as a friend relationship.

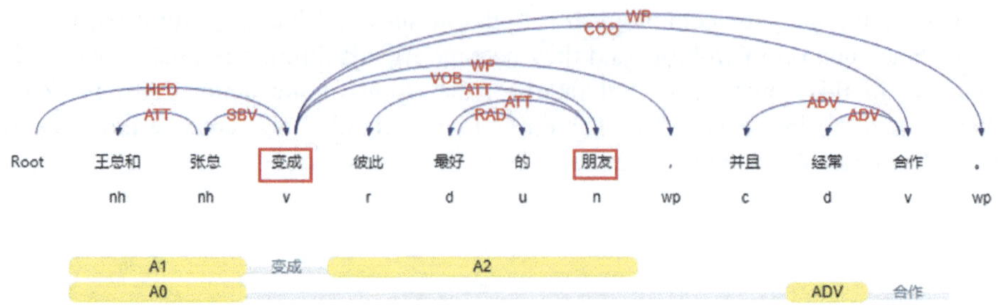

Fig. 5. Instance of friend relationship based on syntax and semantic features

This chapter first preprocesses the microblog text data, then use the LTP to determine the core predicate, dependency parsing analysis, semantic analysis and other operations on the text. The results show that the accuracy of friend relationship recognition is significantly improved.

4 Experiments

In this section, we present our experimental evaluation method and the experimental results of the proposed approach.

(a) Evaluation method

The experiment of evaluating indicators mainly include three aspects which are the accuracy rate P, the recall rate R and the F_1 index, the specific formula is as follows:

$$P = \frac{C}{T} \times 100\% \tag{2}$$

$$R = \frac{C}{S} \times 100\% \tag{3}$$

$$F1 = \frac{2 \times P \times R}{P + R} \times 100\% \tag{4}$$

Where C is the number of instances of a class that were correctly extracted; T is the number of instances actually extracted; S is the number of related instances in the experimental data set.

(b) Experimental analysis

Experiment: getting the microblog about the topic "Friends" from the microblog, randomly extract 2000 microblog comments under the topic of this microblog to analyze the friend relationship, and calculate the accuracy, the recall rate and the $F1$ value of the friend relationship between the characters in the relationship. The experiment was divided into four groups: A, B, C, and D, corresponding to core predicates and dependency syntactic groups, semantic analysis and dependency syntactic groups, core predicates and semantic analysis groups, and semantic analysis, dependency syntax, and core predicate groups, the specific experimental data are shown in Table 3.

Table 3. Comparison of microblog data statistics results of the topic "Friends"

Groups	P(%)	R(%)	F1(%)
A	68.36	66.43	67.38
B	68.75	66.88	67.80
C	67.97	66.65	67.30
D	69.05	66.94	67.98

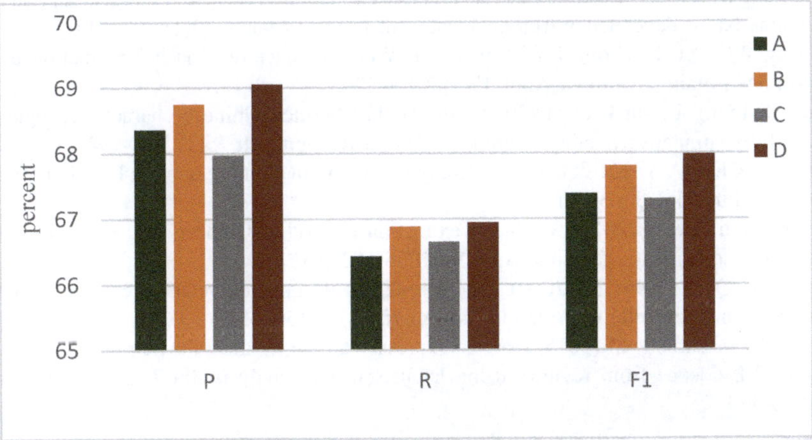

Fig. 6. Microblog statistics of the friend relationship

From Table 3 and Fig. 6, the microblog statistics of the topic "Friends" can be obtained. The accuracy rates of experimental results in group A, group B, group C and group D were 68.36%, 68.75%, 67.97%, and 69.05%, respectively. Where the difference between group A, group B and group C was not obvious. After the experimental characteristics of group A, group B and group C were comprehensive, the accuracy rate P was significantly higher than that of group A, group B and group C. The effect is more significant. This paper shows that the method proposed in this paper can provide effective help in the mining of microblog friend relationship.

5 Conclusions

Microblog text has features such as short text and non-standard grammar. In order to min friend relationship, this paper proposes a microblog friend relationship mining method based on building relationship description dictionary. The experiment prove that the method can effectively min the friend relationship. In the future, the friend relationship identification method can be combined with deep learning in a variety of fields, such as entity relationship mining, sentiment analysis, user recommendation and so on.

Acknowledgement. This Research work was supported in part by 2018 Cultivation Project of Top Talent in Anhui Colleges and Universities (Grant No. gxbjZD15), supported by Guangxi Key Laboratory of Cryptography and Information Security (No. GCIS201719).

References

1. Pan Y, Blebu L, Isha B, Yang J et al (2015) Remote supervision of character relationship extraction using Chinese online resources. Small Comput Syst 36(4):701–706
2. Choi M, Kim H et al (2013) Social relation extraction from texts using a support-vector-machine-based dependency trigram kernel. Inf Process Manag 49(1):303–311
3. Huang WC, Xu L, Xiong LY et al (2016) Web character relationship extraction based on information gain. J Comput Appl 33(8):2286–2293
4. Tao D, Liang L, Jin L et al (2014) Similar handwritten Chinese character recognition by kernel discriminative locality alignment. Pattern Recogn Lett 35(1):186–194
5. Guo JY, Chen P, Yu ZT et al (2016) Entity extraction in Chinese domain based on multi-core fusion. Chin J Inf Sci 30(1):24–29
6. Li HG, Wu X et al (2013) A relation extraction method of Chinese named entities based on location and semantic features. Appl Intell 38(1):1–15
7. Zhou G, Qian L, Fan J (2010) Tree kernel-based semantic relation extraction with rich syntactic and semantic information. Inf Sci 180(8):1313–1325
8. Ahmad SN, Laroche M (2015) How do expressed emotions affect the helpfulness of a product review? Evidence from reviews using latent semantic analysis. Int J Electron Commer 20 (1):76–111

Remaining Useful Life Prediction of Power MOSFETs Using Model-Based and Data-Driven Methods

Jinjing Wu[1(✉)], Zheng Xu[2], and Xiao Wei[3]

[1] The Third Research Institute of Ministry of Public Security, Key Lab of Information Network, Security of Ministry of Public Security, 339 Bisheng Road, Shanghai, China
xyzl.xt@163.com

[2] Guangxi Key Laboratory of Cryptography and Information Security, Guilin, China

[3] Shanghai University, Shanghai, China

Abstract. Prognostics and health management has become an advanced engineering technology in avionics systems which can implement condition monitoring and reduce unnecessary downtime. A prognostic application to power MOSFETs is developed in this paper. Firstly, failure mechanism of the power MOSFETs under power cycling aging tests is analyzed. Then, the drain-source on-state resistance is considered as a leading precursor of failure as it exhibits a decaying trend. Finally, a degradation model is established to predict the remaining useful life based on Kalman filter and LS-SVM, respectively. Several results are analyzed to demonstrate the feasibility and effectiveness of these methods.

Keywords: Prognostics and health management · Remaining useful life · Least square support vector machine · Kalman Filter

1 Introduction

Power semiconductor devices such as MOSFETs (Metal Oxide Field Effect Transistors) are essential components of electronic and electrical subsystems in many fields including photovoltaic power generation, smart grid and radar systems [1]. Due to the harsh operating conditions, these devices often suffer from mechanical stress, thermal overstress and electrical overstress like over voltage or over current which may cause performance degradation or a failure. For mission critical systems, it is extremely important to avoid such failures. Thus, development of prognosis for power devices has been one of the most challenging frontiers to improve reliability, reduce downtime, and prevent failures of these systems [2]. The technology of prognostics is based on the analysis of failure modes, detection of early signs aging and fault conditions, which are then correlated with a damage propagation model and suitable prediction algorithms to complete remaining useful life (RUL) estimation [3].

In order to monitor the health condition or predict the RUL of power devices, identification of the characteristic parameters which indicate failure precursors has

Z. Xu et al. (Eds.): CSIA 2019, AISC 928, pp. 373–381, 2020.
https://doi.org/10.1007/978-3-030-15235-2_56

received considerable attention in recent years [4]. The collector-emitter voltage and turn-off time of discrete IGBTs are identified as health indicators which are affected by die-attach degradation as a result of thermal overstress [5]. Meanwhile, it is stated that on-state resistance (R_{on}) changes due to its dependence on junction temperature and can be used as a precursor of failure because of the difference in coefficients of thermal expansion (CTE) among various material constituents of the chip and the package [6]. Zheng presented a method by constructing the hybrid model of buck circuit and recursive least squares (RLS) algorithm to obtain the on-state resistance of power MOSFET [7]. Wu proposed a nonlinear dual-exponential degradation model for MOSFETs and developed a method to predict the degradation state based on a strong tract filter [8]. A non-interfering measurement technique based on spread spectrum time domain reflectometry (SSTDR) which may cost a lot has been developed to identify the aging level of power MOSFETs inside a live converter circuit by building a model between the SSTDR output and on-state resistance [9]. The PCoE at Ames Research Center has conducted the accelerated aging experiments to establish an exponential empirical model of MOSFET on-state resistance and implemented the RUL prediction by Kalman Filter [10]. However, it is a model-based methodology which means the prediction accuracy depends on the accuracy of the established model.

This paper presents a data-driven method by LS-SVM (Least square support vector machine) and a model-based method by Kalman Filter (KF) to develop a comparative study of the RUL prediction for power MOSFETs. Both methods are able to provide valid results. Prognostic performance metrics are employed to evaluate and compare the algorithms.

This paper is organized in the following manner. Section 2 describes Failure Mechanism and Precursor of Power MOSFETs. In Sect. 3, frameworks of prognostic application for Power MOSFETs by Kalman filter and LS-SVM are presented. Several results are analyzed in Sect. 4. The conclusion of this paper is presented in Sect. 5.

2 Failure Mechanism and Precursor of Power MOSFETs

Accelerated aging experiments provide more degradation information which is different from accelerated life tests (ALT). Prognostics require not only analysis of run-to-failure data but also information about the degradation process. Therefore, it is necessary to record in-situ measurements of the observable parameters in the accelerated aging process.

Failure of semiconductor devices can be categorized into two groups: (a) chip related failure, and (b) packaging related failure. The reasons for chip related failures are electrical overstress, electrostatic discharge, latch up, charge effects and radiation effects. The packaging related failures arise from the difference in coefficients of thermal expansion (CTE) among various material constituents in the component's packaged structure. Packaging failures are of mainly two types, bond failures and die-attach degradation. For a MOSFET, conditions such as latch-up, thermal run away, or failure to turn on due to loss of gate control, are considered as failure modes. The parameter on-state resistance (R_{on}) is found to be the most significant aging factor in power MOSFETs which mainly arise from die-solder layer failures.

A typical MOSFET structure showing the resistances of each layer and region from source to drain terminals is as shown in Fig. 1. In a MOSFET, Ron is the total resistance between the source and the drain during the on-state which determines the maximum current rating and loss, and can be expressed as in Eq. (1).

$$R_{on} = R_{N^+} + R_{CH} + R_A + R_J + R_D + R_S \tag{1}$$

Where, R_{N^+} is the resistance of the source region with N+ diffusion, R_{CH} is the resistance of the channel region, R_A is the accumulation resistance, R_J is the resistance of the JFET region, R_D is the drift region resistance, R_S is the resistance of the substrate region.

R_{on} increases with an increase in temperature as a result of mobility reduction of the charge carriers. This mobility reduction is attributed to increased scattering of charge carriers with temperature. Therefore, the variation in R_{on} is considered a precursor to failure due to its dependence on junction temperature.

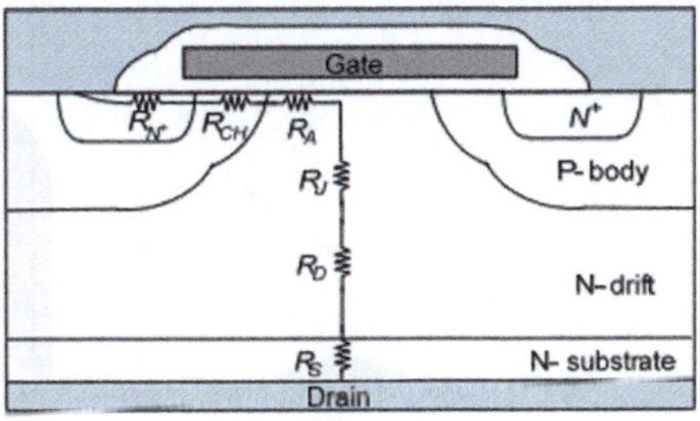

Fig. 1. Typical MOSFET structure with resistance components

3 Framework of Prognostic Application for Power MOSFETs

In order to conduct degradation tests, several power MOSFETs (IRF520Npbf) with a TO-220 package suffer from the power cycling leads to thermo-mechanical stresses, while the drain current I_D and the drain to source voltage V_{DS} are recorded as the device is in the aging regime. The on-state resistance is computed as the ratio of V_{DS} and I_D during the on-state of the square waveform. The degradation dataset of R_{on} in this paper derives from reference [8]. Compared to the initial value of a power MOSFET, the end of life (EOL) is defined as the time at which the change of on-resistance increases to 0.05 Ω, and it is considered that a failure has occurred. The changes of Ron for device #36 are illustrated in Fig. 2. It can be seen that this process grows exponentially as a function of time. Based on least square fitting method, an exponential curve is modeled as shown in Fig. 2 and Eq. (2).

$$R_{on}(t) = \alpha \cdot \exp(\beta t) + R_0 \tag{2}$$

Where t is time, α and β are model parameters, R_0 is the initial value of on-state resistance. Defining the parameters α and β be time dependent parameters, then the derivative of Eq. (2) is given by,

$$R_{on}(t) = (R_{on}(t) - R_0) \cdot \beta \tag{3}$$

The forward difference method is used to approximate the time derivatives in order to discretize the model in Eq. (3). The first step in the process is

$$R_{on}(k+1) = R_{on}(k) \cdot (1 + \Delta \cdot \beta) - R_0 \cdot \Delta \cdot \beta \tag{4}$$

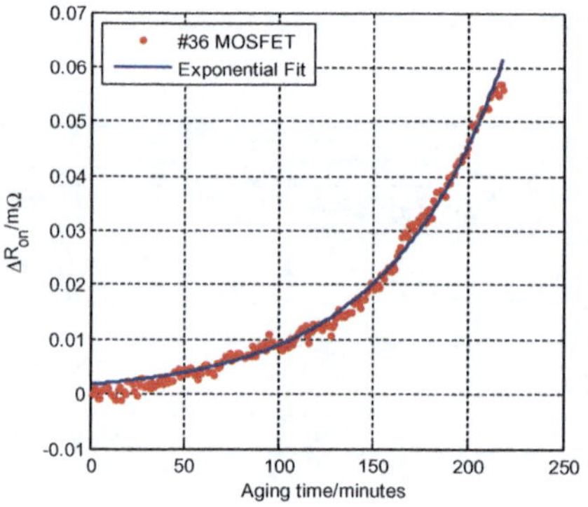

Fig. 2. Exponential Fit of Ron changes for device #36

3.1 Model-Based Method by Kalman Filter

Kalman Filter (KF) is a widely acknowledged optimal state estimator assuming a Gaussian distribution through minimizing the mean square error (MSE) of the estimates considering the errors in the measurements and the model.

$$x_k = Ax_{k-1} + Bu_k + w_k \tag{5}$$

$$z_k = Hx_k + v_k \tag{6}$$

Where, w_k and v_k denote the process and measurement noises, respectively. The noise terms are assumed to be white and they have Gaussian distribution with zero-mean and the following covariance Q_k and R_k

$$Q_k = E\left[w_k w_k^T\right] \tag{7}$$

$$R_k = E\left[v_k v_k^T\right] \tag{8}$$

The estimation error covariance P_k is defined as the following equation.

$$P_k = cov[x_k - \hat{x}_k] \tag{9}$$

The KF is defined for a discrete system as given in Eqs. (5) and (6). The time updates of the KF for the priori state estimate and priori error covariance estimate are given as

$$\hat{x}_k' = A_k\hat{x}_{k-1} + B_k u_k \tag{10}$$

$$P_k' = A_k P_{k-1} A_k^T + Q_k \tag{11}$$

The measurement update involves the calculation of optimal Kalman gain, posterior state estimate, and posterior error covariance, which are expressed as

$$K_k = P_k' H_k^T \left(H_k P_k' H_k^T + R_k \right) \tag{12}$$

$$\hat{x}_k = \hat{x}_k' + K_k \left(z_k - H_k \hat{x}_k' \right) \tag{13}$$

$$P_k = (I - K_k H_k) P_k' \tag{14}$$

At every time step, the new state is estimated in a recursive manner using the previous posterior estimate that is found through the time and measurement updates. The flowchart of RUL prediction by KF is as shown in Fig. 3, D represents the exponential model. The filter algorithm is used to track the state of health and the degradation model is used to make predictions of remaining useful life. RUL is computed as the time between time of prediction t_p and the time at which the forecasted state crosses the failure threshold value.

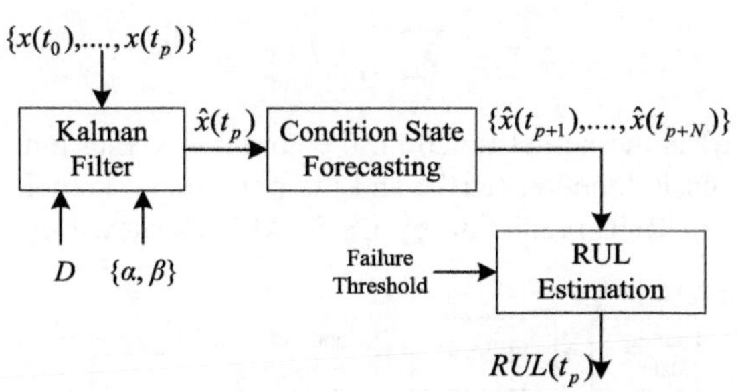

Fig. 3. The flowchart of RUL prediction by KF

3.2 Data-Driven Method by LS-SVM

Support vector machine (SVM) is a machine learning algorithm proposed based on structural risk minimization principal. However, SVM takes heavy computation and long training time. Least squares support vector machine (LSSVM) which is introduced by Suykens and Vandewalle improves the traditional SVM algorithm by adding a least square term in the cost function, which significantly reduces the computation complexity. Given a training data set of N points $= \{x_i, y_i\}, i = 1, 2, \ldots, N$, with the input data $x_i \in R^n$ and output data $y_i \in R$. The decision function can be defined as:

$$y(x) = w^T \varphi(x) + b \tag{15}$$

Where, $\varphi(x)$ denotes the nonlinear function that maps the input space to a high dimension feature space, w is the weight vector and b is the bias term.

For the function estimation problem, the structural risk minimization is used to formulate the following optimization problem:

$$minJ(w, b, \varepsilon) = \frac{1}{2}\|w\|^2 + \frac{1}{2}c\sum\nolimits_{i-1}^{n} \varepsilon_i^2 \tag{16}$$

Subjected to the equality constraint

$$y_i = w^T \varphi(x_i) + b + \varepsilon_i \tag{17}$$

Where, c is the regularization constant and ε_i is the i-th estimation error.

To derive the solutions w and ε, the Lagrange function is constructed

$$L(w, b, \varepsilon, \eta) = J(w, b, \varepsilon) - \sum\nolimits_{i-1}^{n} \eta_i \left[w^T \varphi(x_i) + b + \varepsilon_i - y_i\right] \tag{18}$$

Where, η_i is the introduced Lagrange multiplier.

According to the Karush-Khun-Tucker conditions, the finally result into the LSSVM model for function estimation can be described as:

$$g(x) = \sum\nolimits_{i-1}^{n} \eta_i K(x, x_i) + b \tag{19}$$

Where, $K(x, x_i)$ is the kernel function, the common kernel functions are the linear kernel, the radial basis function (RBF) and the polynomial kernel.

The flowchart of RUL prediction by LS-SVM is as shown in Fig. 4.

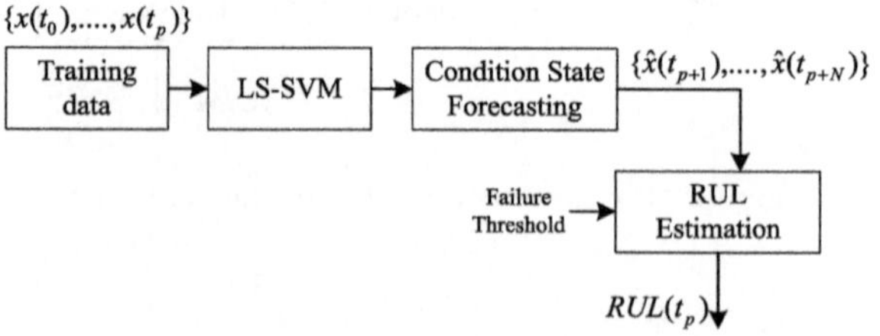

Fig. 4. The flowchart of RUL prediction by LS-SVM

4 Remaining Useful Life Prediction Results

This section presents the results of the two algorithms implemented. RUL predictions for device #36 are made at t_p: 170, 175, 180 min into aging. Subtracting the time when the prediction was made from the time when the predicted increase in resistance crosses the failure threshold (0.05 Ω) gives the estimated remaining component life. The prediction step starts after t_p and the end of life for device #36 from Fig. 2 is 202.4 min. The estimation error and relative accuracy (RA) are presented as performance metrics. RA is defined as

$$RA = 100 * \left(1 - \frac{|RUL^{*} - RUL'|}{RUL^{*}}\right) \tag{20}$$

Where RUL^{*} is the measured value and RUL' is the predictive results.

Figures 5, 6 and 7 illustrate the health state estimation trajectories based on KF and LS-SVM. Also, the RUL can be calculated and as shown in Table 1. As the prediction time closes to the end, the model-based method by KF has a more accuracy than the data-driven method by LS-SVM. However, both of them get the satisfied results and demonstrate the feasibility and effectiveness.

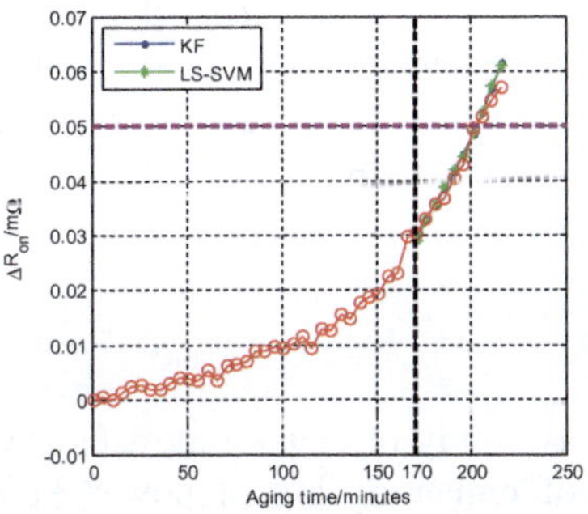

Fig. 5. RUL estimation at t_p = 170 min based on KF and LS-SVM

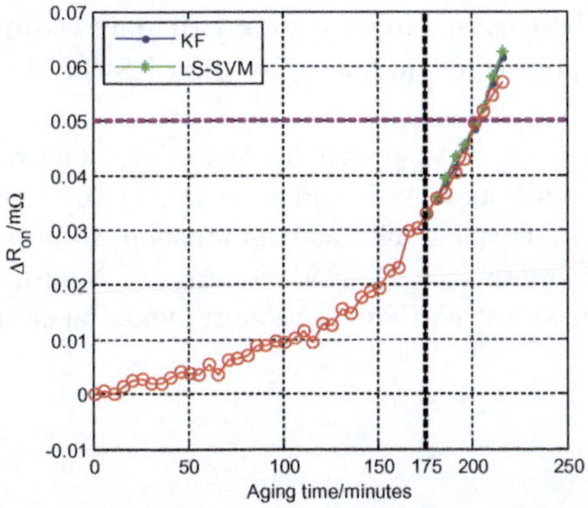

Fig. 6. RUL estimation at t_p = 175 min based on KF and LS-SVM

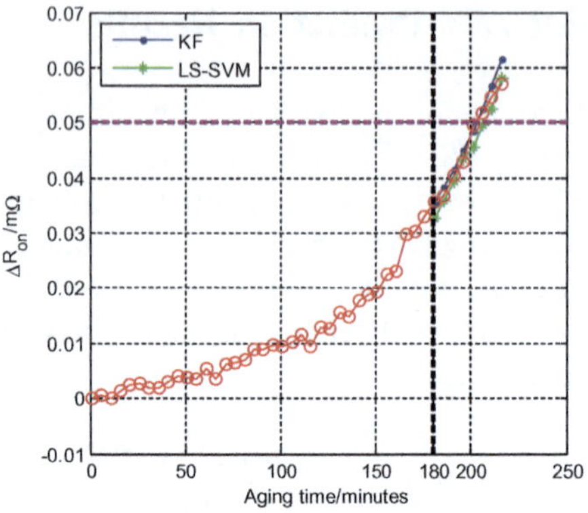

Fig. 7. RUL estimation at t_p = 180 min based on KF and LS-SVM

Table 1. Results of the RUL prediction based on KF and LS-SVM

t_p (minute)		RUL'	RUL^*	Error	RA
170	KF	32.6	32.4	0.2	99.38
	LS-SVM	32.8		0.4	98.77
175	KF	27.8	27.4	0.4	98.54
	LS-SVM	26.7		0.7	97.45
180	KF	22.1	22.4	0.3	98.66
	LS-SVM	26.7		4.3	80.80

5 Conclusion

This paper presents on a case study of employing data-driven and model-based techniques for the prediction of remaining life of power MOSFETs. The model-based method should establish a monotonous change model like an exponential process in terms of the on-resistance degradation data set. The data-driven method does not need a given model and it depends on the relationship between these time series data. Kalman filter and LS-SVM can both gain a high accuracy of state estimation. However, the KF shows more obvious superiorities and stability than LS-SVM during RUL estimation.

Acknowledgement. This work is supported by Domestic Cooperation Project of Shanghai Science and Technology Innovation Action Plan in 2017 (No. 17595800900), supported by Guangxi Key Laboratory of Cryptography and Information Security (No. GCIS201719), and supported by Key Lab of Information Network Security of Ministry of Public Security (The Third Research Institute of Ministry of Public Security) under grant no. C18607.

References

1. Celaya JR, Saxena A, Saha S et al (2011) Prognostics of power MOSFET. In: IEEE 23rd international symposium on power semiconductor devices and ICs (ISPSD), San Diego, USA, 23–26 May, pp 60–163
2. Song Y, Wang B (2013) Survey on reliability of power electronic systems. IEEE Trans Power Electron 28(1):591–604
3. Yang S, Xiang D, Bryant A et al (2010) Condition monitoring for device reliability in power electronic converters: a review. IEEE Trans Power Electron 22(11):2734–2752
4. Sonnenfeld G, Goebel K, Celaya JR (2008) An agile accelerated aging, characterization and scenario simulation system for gate controlled power transistors. In: IEEE AUTOTESTCON, Salt Lake City, USA, 8–11 September, pp 208–215
5. Patil N, Celaya J, Das D et al (2009) Precursor parameter identification for insulated gate bipolar transistor (IGBT) prognostics. IEEE Trans Reliab 58(2):271–276
6. Celaya JR, Saxena A, Wysocki P et al (2010) Towards prognostics of power MOSFETs: accelerated aging and precursors of failure. In: Annual conference of the prognostics and health management society, Portland, USA, 10–16 October, pp 1–10
7. Zheng X, Wu L, Guan Y, Pan W, Wang G (2013) Research on parameter identification of switching power supply power MOSFET. Meas Control Technol 32(6):119–123
8. Wu L, Guan Y, Li X et al (2015) Anomaly detection and degradation prediction of MOSFET. Math Prob Eng 2015:1–5
9. Nasrin MS, Khan FH, Alam MK (2014) Quantifying device degradation in live power converters using SSTDR assisted impedance matrix. IEEE Trans Power Electron 29(6):3116–3131
10. Celaya JR, Saxena A, Kulkarni CS et al (2012) Prognostics approach for power MOSFET under thermal-stress aging. In: Reliability and maintainability symposium (RAMS), Reno, USA, 23–26 January, pp 1–6

Application of Improved Grey Predictive Control Adaptive PID Control Algorithms in Temperature Control

Yinjun Zhang and Mengji Chen(✉)

School of Physics and Mechanical and Electrical Engineering, Hechi University, Yizhou 546300, Guangxi, People's Republic of China
18107786221@163.com

Abstract. In view of the characteristics of the large time delay system of temperature, this paper presents an adaptive PID control algorithm which improves the grey predictive control, and applies this algorithm to temperature control. In this paper, the optimized initial conditions are used to improve the model, and the model is used as a prediction model. Then the predicted results are replaced by the measured values of the controlled object. This can not only overcome the problem of temperature lag, but also apply the optimized performance indicators to the PID controller, and realize the optimal control of adaptive PID. The simulation results show that the method has good adaptability and robustness to temperature control.

Keywords: Grey prediction · PID control · Pure delay

1 Introduction

In most industrial control systems, time-lag process exists to varying degrees. In large time-delay systems, time-delay will seriously affect the stability and dynamic characteristics of the system. Moreover, when the controlled object is particularly complex and time-varying, it is difficult to achieve predictive compensation control by establishing an accurate model of the object. Grey prediction can be predicted based on a small amount of information. It does not need to master the prior information about the structure of the controlled object model and the control experience data. The advance step can be revised online. It can overcome the time-varying characteristics of the object model and has strong adaptability, so it is very suitable for real-time control of industrial processes. But at the same time, because the grey model is based on the data of the sequence itself to find rules for prediction, sometimes there will be large prediction errors. In recent years, some achievements have been made in the research of grey model, such as the original data processing [1, 2], the modification of initial conditions [3], the construction of new background values [4–6], and the improvement of time response function [7]. On the basis of comprehensive analysis of existing research literature, this paper proposes an improved model which optimizes both background values and initial conditions, and further improves the prediction accuracy of the model. At the same time, combining grey prediction with adaptive PID control,

Z. Xu et al. (Eds.): CSIA 2019, AISC 928, pp. 382–387, 2020.
https://doi.org/10.1007/978-3-030-15235-2_57

using grey predictor and introducing quadratic performance index into the setting process of PID controller, the weighting coefficient is modified according to the negative gradient direction of performance index, and the optimal control law of adaptive PID is realized. The simulation results show that the method is feasible and has strong adaptability and robustness for large time-delay systems, and various control performances are obviously improved.

2 Adaptive PID Control Based on Improved Grey Prediction Model

Combining grey prediction with adaptive PID control, using grey predictor, using historical data of process output, the output value of future step is predicted, and the predicted value is compared with the expected set value to get the deviation. As the input of adaptive PID control, the output of controller is designed according to the PID control law, so that the delayed controlled quantity can advance and reverse. By mapping to the controller, the controller can move ahead and realize "pre-regulation", thus reducing the overshoot and accelerating the adjustment process, and eliminating the influence of time delay on the control quality of the system. The overall structure of the system is shown in Fig. 1.

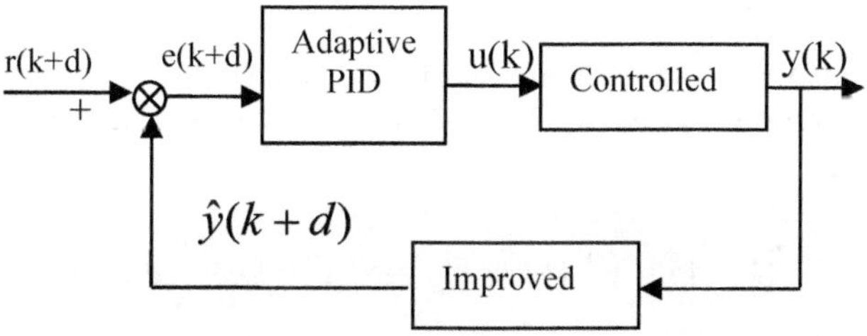

Fig. 1. Gray predictive adaptive PID control system

3 Improved Grey Prediction Model

The general process of $GM(1,1)$ modeling is: $Y^{(0)}$ is primitive non-negative data sequence: $Y^{(0)} = (y^{(0)}(1), y^{(0)}(2), \cdots, y^{(0)}(n))$, A cumulative generation operation (AGO) is performed on $Y^{(0)}$, and the 1-AGO sequence is obtained $Y^{(1)} = (y^{(1)}(1), y^{(1)}(2), \cdots, y^{(1)}(n))$, where

$$y^{(1)}(k) = \sum_{i=1}^{k} y^{(0)}(i),\ k = 1, 2, \ldots, n. \tag{1}$$

Neighbor Mean Generation of Sequence $Y^{(1)}$ is performed, and the Neighbor Mean Generation Sequence $Z^{(1)}$ of Sequence $Y^{(1)}$ is obtained, where

$$z^{(1)}(k) = 0.5[y^{(1)}(k) + y^{(1)}(k-1)],\ k = 1, 2, \ldots, n. \tag{2}$$

Thus we can get the grey differential equation of $GM(1,1)$

$$y^{(0)}(k) + az^{(1)}(k) = b, \tag{3}$$

The corresponding whitening equation is

$$\frac{dy^{(1)}(t)}{dt} + ay^{(1)}(t) = b \tag{4}$$

Where, it is called development coefficient, which is grey action. It can be obtained by least square method.

The solution of Eq. (4) is

$$y^{(1)}(t) = (y^{(1)}(1) - \frac{b}{a})e^{-at} + \frac{b}{a} \tag{5}$$

The time response sequence of the corresponding Eq. (3) is

$$\hat{y}^{(1)}(k+1) = [y^{(0)}(1) - \frac{b}{a}]e^{-ak} + \frac{b}{a},\ k = 2, 3, \ldots, n, \tag{6}$$

The cumulative generation operation of sequence $\hat{y}^{(1)}$, i.e. the inverse operation of cumulative generation, is recorded as IAGO, and the predictive sequence can be obtained. Where

$$\hat{y}^{(0)}(k+1) = \hat{y}^{(1)}(k+1) - \hat{y}^{(1)}(k) = (1 - e^{a})[y^{(0)}(1) - \frac{b}{a}]e^{-ak},\ k = 2, 3, \ldots, n. \tag{7}$$

Formula (6) shows that the prediction accuracy of the model $GM(1,1)$ depends on: (1) the values of a and b, and the values of a and b depend on the structure of the original sequence and background value $Z^{(1)}$; (2) the selection of the initial conditions of the grey differential equation model, the original $GM(1,1)$ model takes $\hat{y}^{(1)}(1) = y^{(1)}(1)$ as the initial conditions. Reference [5] According to the exponential characteristics of A-grey model, using the method of integral in the interval [k, k + 1], makes the

$$z^{(1)}(k) = \frac{y^{(1)}(k) - y^{(1)}(k-1)}{\ln y^{(1)}(k) - \ln y^{(1)}(k-1)},\ k = 2, 3, \ldots, n. \tag{8}$$

The background value is optimized. According to the new information priority principle, a $GM(1,1)$ model with $y^{(1)}(n)$ initial conditions is proposed in reference [1].

$$\hat{y}^{(1)}(k+1) = [y^{(1)}(n) - \frac{b}{a}]e^{-a(k-n+1)} + \frac{b}{a} \tag{9}$$

According to formula (10), if the $k + d$ time is predicted, then the accumulated data is restored to get the restored data, and the $k + d$ time is predicted as follows:

$$\hat{y}(k+d) = [y^{(1)}(n) - \frac{b}{a}]e^{-a(k+d-n+1)}(1 - e^{a}) \tag{10}$$

The two methods mentioned above can improve the accuracy of $GM(1,1)$ independently and completely independently. In this paper, using these two improved methods, an improved $GM(1,1)$ model is proposed, which optimizes the background value and initial conditions simultaneously, and improves the prediction accuracy of $GM(1,1)$ model.

4 Adaptive PID Control Algorithm

The PID discrete control formula is

$$u(k) = K_p e(k+d) + \frac{K_p}{T_i}\sum_{j=1}^{k+d} T_s e(j) + k_p T_d \frac{e(k+d) - e(k+d-1)}{T_s} \tag{11}$$

Where T_s is sample period, k is sample number, K_p is proportionality coefficient, T_d is differential time, T_i integration time, $e(k+d)$ is error.

$$e(k+d) = r(k+d) - \hat{y}(k+d), \tag{12}$$

as

$$u(k) = u(k-1) + \Delta u(k), \tag{13}$$

In order to obtain an adaptive form, the formula (13) is written as

$$\Delta u(k) = \sum_{i=1}^{n} w_i(k)x_i(k) \tag{14}$$

where $n = 3$,

$$w_i(k+1) = w_i(k) + \Delta w_i(k+1) \tag{15}$$

$$\begin{aligned} &x_1(k) = e(k+d),\ x_2(k) = e(k+d) - e(k+d-1), \\ &x_3(k) = e(k+d) - 2e(k+d-1) + e(k+d-2), \end{aligned} \tag{16}$$

5 Simulation Analysis

In order to verify the effectiveness of this method, a large-lag system model in reference [8], such as mode (19), is used as the simulation object, $T_s = 1$ s, Given input $r(t) = 1(t)$, Modeling Dimension of Grey Predictor $n = 5$, Prediction step number $d = 10$.

$$G(s) = \frac{e^{-10s}}{1.5s + 1} \tag{17}$$

The control quality of conventional PID control, grey predictive PID control and adaptive PID control based on improved grey predictive model proposed in this paper are compared. The simulation results obtained by MATLAB are shown in Fig. 2.

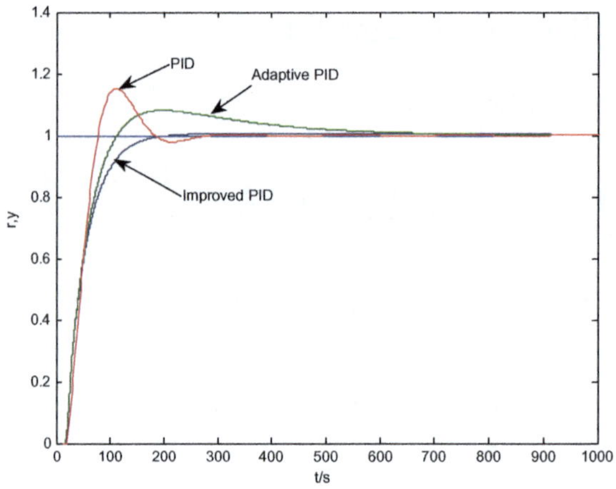

Fig. 2. Step response comparison of control method

As can be seen from Fig. 2, grey predictive control can effectively reduce system overshoot and shorten adjustment time. The adaptive PID control based on improved grey predictive model proposed in this paper combines the characteristics of PID control and grey predictive control, which makes the system have good dynamic characteristics. Compared with PID control, it can significantly reduce system overshoot and system oscillation, and make the system more rapid. It converges to the target value; compared with the general grey predictive control, the proposed control algorithm can make the system converge faster.

6 Conclusions

This paper presents an adaptive PID control based on Improved Grey Prediction model, which combines grey prediction model with adaptive PID control. The improved *GM* (1,1) model, which optimizes both background values and initial conditions, is used as the prediction model. The prediction accuracy of the grey model is improved and the time delay is overcome. The adaptive PID controller based on gradient optimization is used to realize the optimal control of the adaptive PID. The simulation results show that the algorithm can effectively control the time-varying large-time-delay system in industrial process, and it is easy to implement. It has good application prospects.

Acknowledgments. The work was supported by the Hechi University Foundation (XJ2016ZD004), Guangxi teacher Capability enhancement project (2018KY0495), Hechi university Youth teacher Foundation (XJ2017QN08) and was supported by the Projection of Master Foundation (2017HJA001, 2018LG004).

References

1. Dang Y, Liu S, Liu B (2012) Models for initial conditions. Chin Manag Sci 2:132–135
2. He B, Meng Q (2012) Research on the extension method of grey prediction model. Syst Eng Theor Pract 9:137–140
3. Wang Z, Peng X, Dai Q (2015) Improvement and application of grey prediction model based on initial value correction. Chongqing Inst Technol 21(10):81–84
4. Tan C (2017) Background value construction method and application of the model (1). Syst Eng Theor Pract 4:98–103
5. Wang Z, Wu C (2013) Improved model and its application. Pract Underst Math 33(9):20–25
6. Li J, Dai W (2016) New method and application of background value construction of model based on interpolation and Newton-Cores formula. Theor Pract Syst Eng 10:122–125
7. Liu B, Liu S, Zhai Z, Dang Y (2014) Optimizing the time response function of the model. China Manag Sci 11(4):54–57
8. Han M, Guo W, Wang J (2015) Adaptive predictive PID controller for large delay systems based on RBF network. J Instr Instr 26(10):1039–1043
9. Hongjun L, Han P, Wanye Y (2014) Adaptive grey predictive PID control for time-varying large-lag systems. J North China Electr Power Univ 31(5):44–47
10. Li C-Y, Huang T-L (2014) Optimal design for the grey prediction PID controller for power system stabilizers by evolutionary programming. In: Proceedings of the 2004 IEEE, international conference on networking, sensing control, Taipei, Taiwan

Application of Improved Grey Predictive Control in Alcohol Temperature Control

Yinjun Zhang and Mengji Chen(✉)

School of Physics and Mechanical and Electrical Engineering, Hechi University, Yizhou 546300, Guangxi, People's Republic of China
782723236@qq.com

Abstract. Temperature in the fermentation process directly affect wine quality and yield. Commonly used in the actual production of ordinary PID control of temperature, method is simple, convenient, the most widely used. But for the fermentation process nonlinear, time-varying uncertainty and large delay, it is difficult to establish accurate mathematical model, PID control algorithm to control the quality is not very high, especially in error as the basic adjustment items, derivative action is only apparent in the system deviation function, is after the control, it cannot very well overshoot suppression system.

Keywords: Grey prediction · PID control · Temperature control

1 Introduction

Alcohol is by microorganisms through glycolysis (EMP) pathway generated by decomposition of glucose, a yeast metabolites. Temperatures are affecting microbial growth and survival of one of the major environmental factors. A great influence on the fermentation. Temperature on the growth of microorganisms in the fermentation of specific performance: ① affect enzyme activity. ② affect cell membrane fluidity. ③ affect the solubility of substances [1, 2]. Temperature in the fermentation process directly affect wine quality and yield. Commonly used in the actual production of ordinary PID control of temperature, method is simple, convenient, the most widely used. But for the fermentation process nonlinear, time-varying uncertainty and large delay, it is difficult to establish accurate mathematical model, PID control algorithm to control the quality is not very high, especially in error as the basic adjustment items, derivative action is only apparent in the system deviation function, is after the control, it cannot very well overshoot suppression system [3, 4]. Therefore, to be the appropriate choice of programs to control the temperature of the fermentation process. In response to these problems, based on improved gray PID temperature control method.

System information is completely clear, said white systems, information systems completely unclear, said black system [5, 6]. Information section identifies some of the system is not clear, said Gray. Gray system method is to estimate the uncertain part of the establishment of gray, to compensate for the uncertainty part to reduce its impact, thereby improving system control performance and increasing robustness [7, 8].

Z. Xu et al. (Eds.): CSIA 2019, AISC 928, pp. 388–393, 2020.
https://doi.org/10.1007/978-3-030-15235-2_58

2 Grey PID Algorithm Design

The first step: using PID control, the controller starts the process, first the gray portion of the estimator uncertain model parameters to establish GM (0, N) model to estimate where PID control algorithm is:

$$u_p = u(t) = k_p e(k) + k_i \sum_{k=1}^{n} e(k)T + k_d de(k) \tag{1}$$

Gray estimator uncertain part of the model parameters to establish GM (0, N) model specific algorithm is:

(1) create the original discrete series $x_i^{(0)}(k)$, where
(2) $i = 1,\ 2,\ \cdots,\ n;\ k = 1,\ 2,\ \cdots,\ N;\ N \geq n;$

$$\left.\begin{array}{l} x_1^{(0)} = \left(x_1^{(0)}(1)\quad x_1^{(0)}(2)\quad \cdots \quad x_1^{(0)}(N)\right) \\ x_2^{(0)} = \left(x_2^{(0)}(1)\quad x_2^{(0)}(2)\quad \cdots \quad x_2^{(0)}(N)\right) \\ \qquad\qquad\qquad\qquad \vdots \\ x_n^{(0)} = \left(x_n^{(0)}(1)\quad x_n^{(0)}(2)\quad \cdots \quad x_n^{(0)}(N)\right) \end{array}\right\} \tag{2}$$

(3) calculating a cumulative generation (1-AGO) series, in which: $i = 1, 2, \cdots, n;$ $k = 1, 2, \cdots, N;\ N \geq n;$

$$\left.\begin{array}{l} x_1^{(1)} = \left(x_1^{(1)}(1)\quad x_1^{(1)}(2)\quad \cdots \quad x_1^{(1)}(N)\right) \\ x_2^{(1)} = \left(x_2^{(1)}(1)\quad x_2^{(1)}(2)\quad \cdots \quad x_2^{(1)}(N)\right) \\ \qquad\qquad\qquad\qquad \vdots \\ x_n^{(1)} = \left(x_n^{(1)}(1)\quad x_n^{(1)}(2)\quad \cdots \quad x_n^{(1)}(N)\right) \end{array}\right\} \tag{3}$$

(4) calculating data matrix B, it must be reversible (i.e.); if reversible, it should adapt to increased N, until reversible [9, 10].

$$B = \begin{bmatrix} x_1^{(1)}(2) & \cdots & x_n^{(1)}(2) & 1 \\ x_1^{(1)}(3) & \cdots & x_n^{(1)}(3) & 2 \\ \vdots & & \vdots & \vdots \\ x_1^{(1)}(N) & \cdots & x_n^{(1)}(N) & N-1 \end{bmatrix} \tag{4}$$

calculation of the discrete series $D^{(0)}(k)$, $k = 1, 2, \cdots, N$.

$$D^{(0)} \triangleq \begin{pmatrix} D^{(0)}(1) & D^{(0)}(2) & \cdots & D^{(0)}(N) \end{pmatrix}^T \tag{5}$$

(5) Computing $(1 - AGO)$ Discrete Sequence $D^{(0)}(k)$, $k = 1, 2, \cdots, N$.

$$D^{(1)}(k) \triangleq \sum_{l=0}^{k} D^{(0)}(l) \tag{6}$$

$$D^{(1)} \triangleq \left(D^{(1)}(1), D^{(1)}(2), \cdots, D^{(1)}(N)\right)^T \tag{7}$$

The estimated uncertainty part of the gray model parameter vector.

$$D^{(1)}(x,t) = V_1 x_1^{(1)} + V_2 x_2^{(1)} + \cdots + V_n x_n^{(1)} + f^{(1)} \tag{8}$$

$$\hat{V}^T = (B^T B)^{-1} B^T D^{(1)},\ \hat{V}^T = \begin{pmatrix} \hat{V}_1 & \hat{V}_2 & \cdots & \hat{V}_n & \hat{f} \end{pmatrix}^T$$

Step 2: Press the estimated parameters plus compensation control, the estimator stops working, gray PID control algorithm is:

$$u = u_p + u_c.\ \text{Among them: } u_c = -\left[\sum_{i-1}^{n} \hat{V}_1 x_i + \hat{f}\right] \tag{9}$$

Hardware Platform

The control algorithm in the ARM9 series S3C2410 processor platform. Using C++ programming algorithms, because the program is too long, here describes only the main program, the main program is as follows:

```
// TODO: Write code for the behavior of the application here.
double simTime=0.001;      // Sampling time
int ctype=2;               //"1" is conventional PID and "2" is grey
PID
greyPIDcontrol gpc(nNumRow,nNumColoumn); // Get the con-
structor
CMatrix data=gpc.initDataMatrix(simTime);// Establishing Primi-
tive Discrete Sequences
gpc.selectControllerType(ctype);
CMatrix x1=gpc.AGOmatrix(data);      // Computation of a Cumu-
lative Generation (1-AGO)
                                     Discrete Sequencex1(k)
CMatrix B=gpc.dataMatrixB(x1);  //Computational Data Ma-
trixB(x1)
gpc.data_Dk(data,simTime);      // Computing Discrete Sequence
D(k)
gpc.AGO_D1k(gpc.D); // Computing (1-AGO)//Discrete Sequence
D1(k)
gpc.paramaterVectorEstimation(B);// Estimating parameter vector
V of grey model with uncertain
                                 part bD (x, t)
for(int i=0;i<2000;i++)
gpc.greyCompensationControl(simTime,i);// Grey Compensation
Control
fprintf(fp,"%f\t%f\t%f\t%f\t%f\t\n",i*simTime,gpc.r[i],gpc.x[0],gp
      c.r[i]-gpc.x[0],gpc.u); // Getting Control Algorithms U(t)
```

3 Simulation

Extraction temperature fermentation plant model as a simulation object [5]:

$$G(s) = \frac{160}{s(s+35)} \tag{10}$$

The transfer function into a state equation of the form:

$$\dot{x} = Ax + bu + bD(x,t) \tag{11}$$

In the formula, $A = \begin{bmatrix} 0 & 1 \\ 0 & -35 \end{bmatrix}$, $b = \begin{bmatrix} 0 \\ 160 \end{bmatrix}$.

Second-order transfer function of the object, the number of iterations optional $N \geq n+1$, so desirable $N = 3$.

Plus interference check $D(x,t) = V_1 X_1 + V_2 x_2 + f$, Take interference parameters $\hat{V} = [5,5,5]$, $k_p = -35$, $k_i = 0$, $k_d = -5$. The gray PID control, after three sampling times, to obtain the interference parameter estimates V = [4.7824 5.0031 5.0363], Figs. 1, 2 and 3 were estimated using the gray compensation is not PID control (PID CONTROL) and gray PID control (grey PID control) of the tracking error, the tracking error change rate, and u(t).

From the simulation results can be seen in Figs. 1, 2 and 3 Grey PID compensation than ordinary PID control added after the error becomes small, and the changes in uniform and no larger overshoot; while improving system robustness and speed of

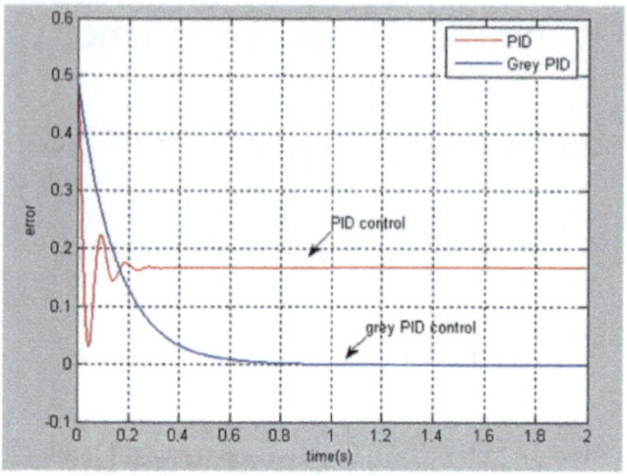

Fig. 1. Comparative tracking error curve

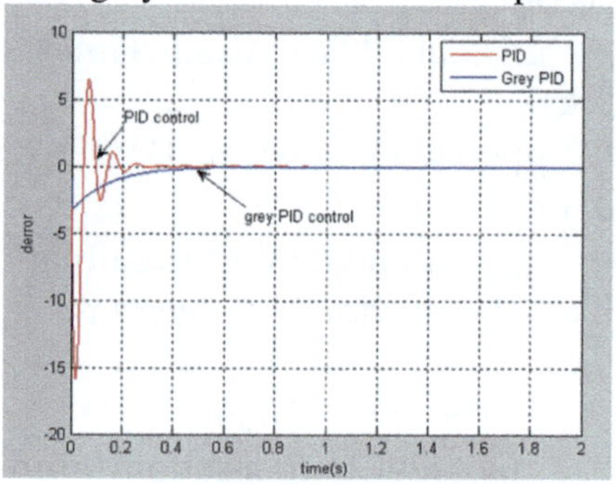

Fig. 2. Comparison tracking error change rate curve

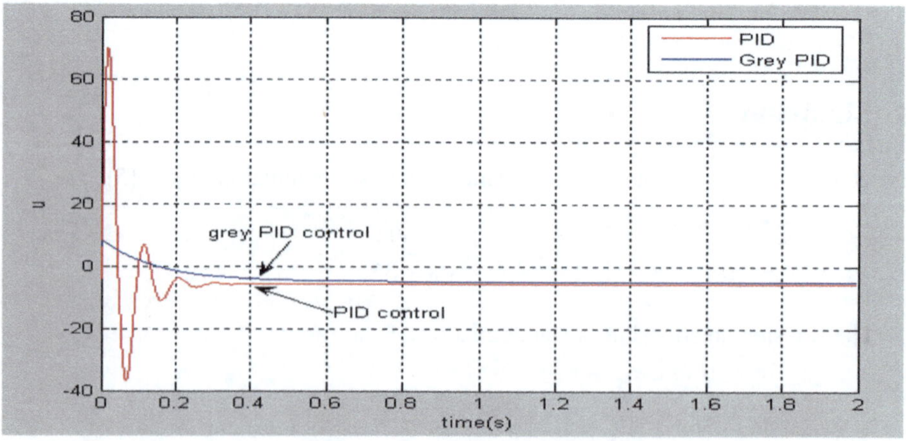

Fig. 3. $u(t)$ compares the curve

response, and received better system tracking. Reduce errors and increase robustness, that is well controlled fermentation temperature may be within the specified range, greatly improved the quality of the wine and the drinker.

4 Conclusions

PID control method based on gray wine fermentation process to achieve higher accuracy of the temperature requirements, without changing the hardware devices in the case of wine through the temperature control to improve the quality and yield.

Acknowledgments. The work was supported by the Hechi University Foundation (XJ2016ZD004), Guangxi teacher Capability enhancement project (2018KY0495), Hechi university Youth teacher Foundation (XJ2017QN08) and was supported by the Projection of Master Foundation (2017HJA001, 2018LG004).

References

1. Zhou H (1996) Different fermentation temperature on the impact of higher alcohols generated. Making Sci Technol 4:8–17
2. Kayai I, Tan N, Atherton DP (2006) A refinement procedure for PID controllers. Electr Eng 88(3):215–221
3. Dostal P, Bobal V, Vojtesek J (2009) Adaptive temperature control in a tubular chemical reactor. In: 17th mediterranean conference on control and automation, Thessaloniki, pp 1209–1214
4. Xu W (2003) CAO column. control theory and design, 1st edn. Shanghai Jiaotong University Press, Shanghai pp 74–82
5. Hangzhou Press (2006) Advanced P1D control and MATLAB simulation, 1st edn. Electronic Industry Press, Beijing p 96–99
6. Su C, Lin J, Lin H (2004) Fuzzy PID control and MATLAB. J Comput Appl 4:51–55
7. Liu B, Liu S, Zhai Z, Dang Y (2014) Optimizing the time response function of the model. China Manag Sci 11(4):54–57
8. Han M, Guo W, Wang J (2015) Adaptive predictive PID controller for large delay systems based on RBF network. J Instr Instr 26(10):1039–1043
9. Liu H, Pu H, Yao W (2014) Adaptive grey predictive PID control for time-varying large-lag systems. J North China Electr Power Univ 31(5):44–47
10. Li C-Y, Huang T-L (2014) Optimal design for the grey prediction PID controller for power system stabilizers by evolutionary programming. In: Proceedings of the 2004 IEEE, international conference on networking, sensing control, Taipei, Taiwan

Human-Simulation Logical Predictive Controller Design

Mengji Chen and Yinjun Zhang(✉)

School of Machinery and Electronic Information Engineering, Hechi University, Yizhou, Guangxi, China
18107786221@163.com

Abstract. Simulating the thinking characteristics of human beings, a new type of intelligent controller is proposed: human-like logic predictive controller. The controller combines the characteristics of Pan-Boolean algebra-based logic controller and model-based predictive controller. It is a multi-valued logic hybrid dynamic system. The simulation results of MATLAB show that the controller performs well in model matching and can still operate satisfactorily in model mismatch. Compared with other types of AI controllers, the controller has simple structure, clear physical background and clear mathematical concepts. It is easy to popularize and apply in the field of industrial control.

Keywords: Human-Simulation · Pan-Boolean algebra · Logical control · Predictive control

1 Introduction

In the era of classical control theory and modern control theory, it is difficult to describe some systems which are difficult to be described by general physical and chemical laws. There are many influencing factors, and there are cross-coupling systems among them. The establishment of their models is very complex. Some production processes lack appropriate testing means, or testing devices cannot enter the tested area, which makes it impossible to establish systems [1, 2].

The multi-valued logic controller based on Pan-Boolean algebra is an intelligent controller [3–8] proposed by Mr. Zhang Nanlun. Its characteristic is that it does not need the concrete model of the object and the setting parameters are simple. There is no interaction between the parameters of the PID controller, nor the problem that the output of the fuzzy control is difficult to understand intuitively. It is a kind of controller that simulates human control behavior. Literature [9] has tested its performance in many aspects and considered that it is superior to fuzzy control and PID control. However, logical controllers have inevitable weaknesses: fluctuation.

Inspired by the thinking characteristics of human beings in control, this paper proposes a humanoid logic predictive controller, which combines the advantages of logic controller and predictive controller. Its characteristic is that it simulates the control behavior of experienced teachers, overcomes the fluctuation problem of logic controller, enhances the overall performance of the system in case of model mismatch, and improves its robustness. Matlab simulation proves this point.

Z. Xu et al. (Eds.): CSIA 2019, AISC 928, pp. 394–398, 2020.
https://doi.org/10.1007/978-3-030-15235-2_59

Carefully observing the behavior of the boiler-burning master in controlling the water temperature of the boiler, we can find that the control behavior has at least three characteristics: Active open-loop system; Strong time-varying system; Experience in the characteristics of the system;

2 Principle and Implication of Predictive Controller

The original model can be obtained by many methods, such as shock response, and can be divided into one-step prediction and multi-step prediction according to the number of prediction steps. At this time, it can be considered as open-loop control, if the model is completely accurate and there is no error. Because of the error between the original model output and the actual object system output, the above open-loop model needs to be modified. In predictive control, the output error feedback correction method, i.e. closed-loop prediction, adds the output error to the predictive model. This is the online correction stage, or feedback correction. At the same time, in order to avoid the jump of control function, an exponential reference trajectory is set up, which is called softening trajectory. The softening degree of the trajectory is determined by the parameter alpha. The larger the alpha, the better the flexibility and robustness, but the slower the response speed. On-line optimization is the core of predictive control, i.e. using known information, the time-varying control function u (K) can be obtained to minimize the error of output signal and input control.

The prediction model can input $\{u(k+j-1)|j=1,\ldots$ according to the historical information of the controlled object $\{u(k-j), \{y(k-j)|j>1\}$ and future input $\{u(k+j-1)|j=1,\ldots M\}$, predicting the future output of the object $\{y(k+j)| J=1,\ldots P\}$.

A closed-loop predictive output equation is set up.

$$y_p(k+i) = y_m(k+i) + [y(k) - y_m(k)]$$

Reference trajectory equation:

$$y_r(k+i) = \alpha^i y(k) + \left(1 - \alpha^i\right)w$$

$$y_r(k) = y(k)$$

Then the objective function is optimized:

$$J_{\min} = \sum_{i=1}^{p} [y_p(k+i) - y_r(k+i)^2]q_i$$

It can be seen from the equation that the essence of predictive control is to minimize the mean square error between the actual output and the set control input.

3 Contrastive Simulation Research

In order to study the performance of the humanoid predictive controller, several different controlled objects are designed and compared in MATLAB environment.

The controlled object is:

$$G(s) = \frac{3}{2s^2 + 3s + 4}$$

The two poles are s1 = −0.7500 + 1.1990i and s2 = −0.7500 − 1.1990i respectively. The system is stable. If the influence of random noise is not considered, the control parameters of phase plane zoning are selected as follows:

$K'_{4+} = 10,\ K'_{3+} = 8,\ K'_{2+} = 3,\ K'_{+} = 1,\ K'_0 = 0,\ K_{-} = -1,\ K'_{2-} = -3,$ $K'_{3-} = -8,\ K'_{4-} = -10$, The derivative $\dot{e}_0$ of deviation e_0 and deviation is taken as values: $e_0 = \pm 0.1\quad \dot{e}_0 = \pm 0.1$. The dynamic process of phase plane partition control is completed by the interaction of nine control forces, and finally stabilizes under the action of K_0^. Closed-loop PID control parameters P = 3, I = 2, D = 1. The system output under the above control parameters is shown in Fig. 1.

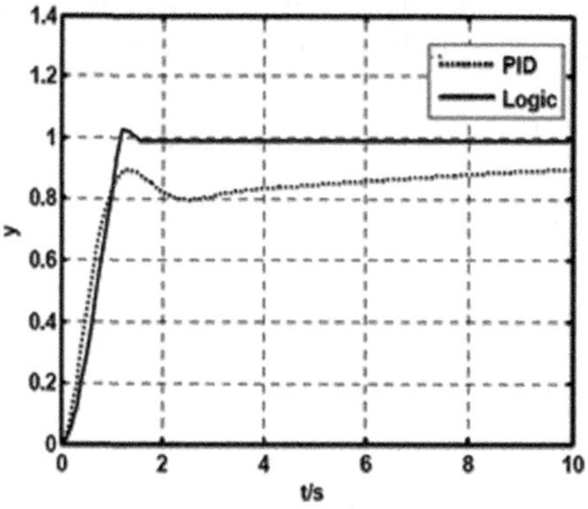

Fig. 1. Comparison of PID control and logic control

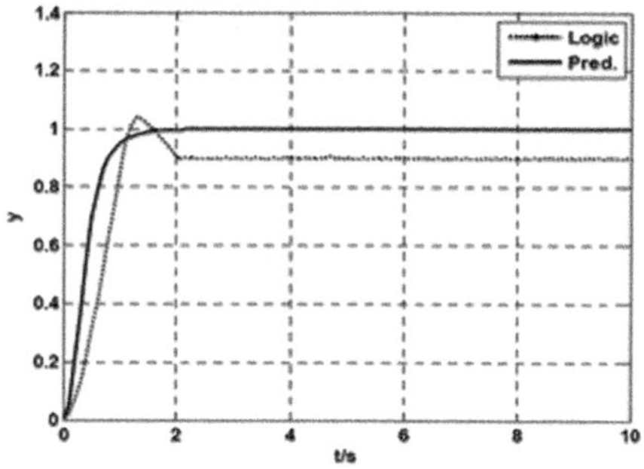

Fig. 2. Contrast between logical control and model-matched predictive control

In the figure, the continuous line is the result of logical control. Compared with PID, logic control is better than PID control. Of course, if you zoom in further, you will find that there are many small oscillations in logical control. This is the main disadvantage of logical control. Because logic control is a time-varying multi-valued switch control. Similarly, the model is input into the predictive model of predictive control, and the results of predictive control are compared. As shown in Fig. 2.

As can be seen from Fig. 2, the predictive control results are optimal when the model matches perfectly. This is not surprising, because predictive control itself is a time-varying optimal control. The output of the proposed human-like logic predictive controller is shown in Fig. 2 when the model matches perfectly. As can be seen from Fig. 3, the output of the humanoid logic predictive controller is basically the same as that of the predictive controller. This is because when the system runs in the allowable range, the switcher of the humanoid logic predictive controller will switch the control to the U (t) end of the predictive controller. When the system runs normally, it depends on the experience of the master to maintain a better control effect.

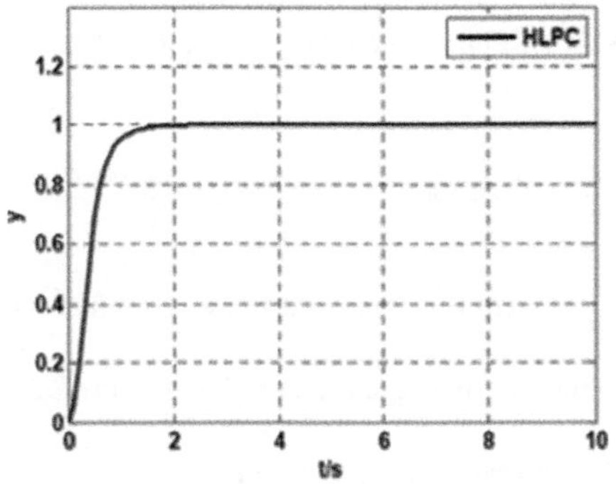

Fig. 3. Human-like logic predictive controller (model matching)

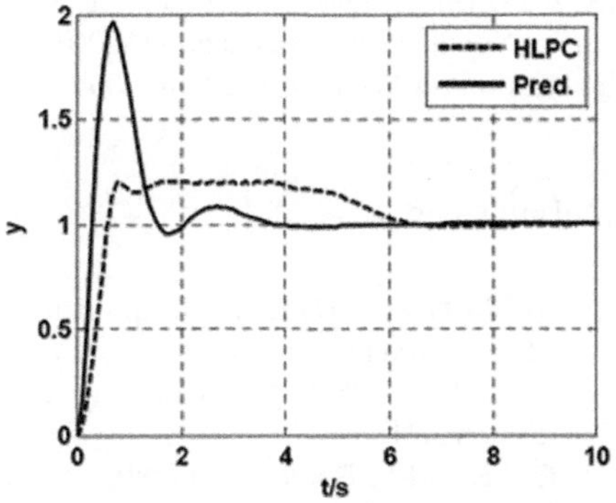

Fig. 4. Controlled output under model mismatch

Next, the performance of human-like logic predictive controller is studied when the predictive model and the actual model are mismatched. Change the controlled object to:

$$G(s) = \frac{9}{2s^2 + 3s + 4}$$

Without changing any parameters of the controller, the output of predictive control and humanoid logic predictive control is shown in Fig. 2. In the figure, the thick line is the output of the step response of the humanoid logic controller, while the thin line is the output of the predictive control. As can be seen from Fig. 4, in the case of model mismatch, the predictive control oscillates a lot, while the analog logic control oscillates a little. Maintain within a certain range of errors. This is because the logic controller is forced to control beyond the allowable error range and force the output to return to a reasonable range.

4 Conclusion

The paper presents a new intelligent control model, the humanoid predictive controller. This controller combines the Pan-Boolean algebra-based multi-valued logic controller and predictive controller model, which are used to simulate the three characteristics of human when dealing with control problems. Through the simulation of the simulation system by Matlab, it is found that the controller can operate satisfactorily in the case of model mismatch, while maintaining the advantages of predictive control in the case of accurate model. This controller has simple structure, clear mathematical concept and close control strategy to human thinking. It is believed that it can be widely used in industrial process control.

Acknowledgement. The work was supported by the Hechi University Foundation (XJ2016ZD004), Guangxi teacher Capability enhancement project (2018KY0495), Hechi university Youth teacher Foundation (XJ2017QN08) and was supported by the Projection of Master Foundation (2017HJA001, 2018LG004).

References

1. Saridis GN (1997) Hierarchical intelligent control system for urban and freeway traffic.TR-EE-8 l, Purdue University
2. Tu Y, Li Z (1994) A new design method of human-like intelligent controller. J Autom 20 (5):616–621
3. Camacho EF, Bordons C (1999) Model predictive control. Springer, London
4. Rossiter JA (2003) Model-based predictive control: a practical approach. CRC Press, Boca Raton
5. Huiping Z, Bo D et al. (2006) Application and development of modern control theory in process industry. J Beijing Inst Petrol Chem Technol 14(3)
6. Li Z, Liu J (2006) Robust control, control and decision-making for saturated constrained systems 21(6):641–645
7. Zhang N (2005) New control principle. National Defense Industry Press, Beijing
8. Nanlun Z, Guoting Q, Lingli J (2004) 9.5-state controller and its simulation analysis. Papers of the fifth world congress on intelligent control and automation. Zhejiang University Press, pp 2570–2574
9. Xu H, Yu H, Qiu G, Zhang N (2005) 27.9-state controller and its third-order object simulation analysis. J Syst Simul 35(6):767–769
10. Lin W, Feng H et al (2007) Dynamic control model and logic rule control of piezoelectric ceramic actuators. Nanotechnol Precis Eng 5(1)

Iterative Learning Control for Uncertain Time-Delay Reaction Diffusion System

Yinjun Zhang[1,2] and Mengji Chen[1(✉)]

[1] School of Physics and Electrical Engineering, Hechi University, Yizhou 546300, Guangxi, China
782723236@qq.com

[2] Aeronautics and Astronautics Engineering Institute, Air Force Engineering University, Xi'an 710038, Shan'xi Province, China

Abstract. The work is connected with the development of stability theory methods of iterative learning control for uncertain time-delay reaction diffusion system, This research works out the specific control law for the system, and proves the robustness, and the convergence of uncertain time-delay reaction diffusion is constructed by the iterative learning control via L^2 norm condition. A numerical example illustrates the effectiveness of the proposed method and a simulation shown its effectiveness. It allows us to obtain new interesting results both for control theory and for applications, and also for knowledge theory as a whole. It is important that the new algorithm expands the reaction diffusion system scope of application.

Keywords: Iterative learning control · Reaction diffusion system · Uncertain time-delay

1 Introduction

Iterative learning control (ILC) can be applied to the systems which perform the required task repeatedly. It improves the tracking performance of the system by utilizing the repetition of the task as an experience to improve the control quality. One typical example is of a robotic manipulator performing a designated task in the production industry. Iterative learning control is formulated by Arimoto et al. [1, 2] in 1984. Since then, there have been some great deal of efforts by the researchers to synthesize a better iterative learning control scheme relating to the robotics applications. The Kuramoto–Sivashinsky equation is a model of reaction–diffusion systems [3, 4]. This equation is related to various pattern formation phenomena accompanying the appearance of turbulence. It is well known as a model of chaos in partial differential equations. This nonlinear dissipative fourth-order partial differential equation also describes incipient instabilities in a variety of physical and chemical systems. For examples, such as falling liquid films [5, 6], unstable flame fronts and interfacial instabilities between two viscous fluids [7]. Motivated by this, we study stabilization of the zero steady state solutions of the reaction–diffusion systems. Stabilization of the Kuramoto–Sivashinsky equation has been studied. Adaptive control for reaction–diffusion systems has been considered for the Burger's equation.

Z. Xu et al. (Eds.): CSIA 2019, AISC 928, pp. 399–406, 2020.
https://doi.org/10.1007/978-3-030-15235-2_60

This paper is concerned with nonlinear boundary stabilization of the generalized Kuramoto–Sivashinsky equation by ILC. The iterative learning controller is constructed by the concept of high-gain nonlinear output feedback and the error mechanism of the unknown parameters. In the control system the robustness and the convergence of the system state to zero will be guaranteed.

2 System Description

Consider the following reaction-diffusion system:

$$\begin{cases} \frac{\partial u(x,t)}{\partial t} = D\Delta u(x,t) + [A(t) + \delta A(t)]u(x,t) + B(t)u(x,t-\theta) \\ y(x,t) = C(t)u(x,t) + G(t)u(x,t-\theta) \end{cases} \tag{1}$$

System (1) satisfies the following initial condition:

$$u(x,t) = \mathrm{r(x,t)}, \mathrm{(x,t)} \in \Omega \times R^{+} \tag{2}$$

$$u(x,t) = 0 \text{ or } \frac{\partial u(x,t)}{\partial N} = 0, \mathrm{(x,t)} \in \partial\Omega \times [-\tau, +\infty] \tag{3}$$

where $(x,t) \in \Omega[0,T]$, Ω is sufficiently smooth and an open bounded domain in R^q. $u \in R^n, \mathrm{y} \in R^m$ are state vector and output vector respectively; Δ is the Laplace diffusion operator; $D = diag\{d_1, \cdots, d_n\}$; Time-delay $\theta \geq 0$; N is unit normal vector; For Ω domain, the vector function $\mathrm{a}(x) = \mathrm{col}(\mathrm{a}_1(\mathrm{x}), \mathrm{a}_2(\mathrm{x}), \cdots, \mathrm{a}_n(\mathrm{x}))$, we mark $\|\mathrm{a}(x)\| = \{\mathrm{a}^{\mathrm{T}}(\mathrm{x})\mathrm{a}(\mathrm{x})\}^{1/2}$, $\nabla \mathrm{a}(x) = \mathrm{col}(\nabla \mathrm{a}_1(x), \cdots, \nabla \mathrm{a}_n(x))$, $\|\mathrm{a}(x)\|_{\mathrm{L}^2} \triangleq \left\{\int_\Omega \|\mathrm{a}(x)\|^2 \mathrm{dx}\right\}^{1/2}$. We mark the solutions of Eq. (1–3):

$$u(x,t) \triangleq u(x,t,r) = \mathrm{col}(u_1(x,t,r), u_2(x,t,r), \cdots, u_n(x,t,r))$$

For the system (1), supposed its expected output is $y_d(x,t)$, its corresponding input is $u_d(x,t)$, and its output equation is $\mathrm{y_d}(x,t) = C(t)u_d(x,t) + G(t)u_d(x,t-\theta)$. We look for a control input sequence $\{u_k(x,t)\}$, let $\lim_{k\to\infty} u_k(x,t) = u_d(x,t)$. The P-type iterative learning control algorithm is

$$u_{k+1}(x,t) = u_k(x,t) + \Gamma e_k(x,t+\theta) \tag{4}$$

where $e_k(x,t) = y_k(x,t) - y_d(x,t)$; Γ is gain matrix.

3 Convergence and Robustness Analysis

Theorem 1. If the gain of Eq. (4) and initial condition satisfy Eq. (5), then the algorithm determined by Eq. (4) is convergence in $t \in [-\theta, T-\theta]$, and $\lim_{t\to\infty} \|e_k(x,t)\|_{L^2} = 0$, $\forall t \in [0,T]$.

$$\|\mathrm{I} - \mathrm{G}(\mathrm{t})\Gamma\|^2 \leq \rho \in [0, 1) \tag{5}$$

Proof. According to (4), we have

$$e_{k+1}(x,t) = \overline{e}_k(x,t) + \overline{C}_k(x,t) \tag{6}$$

where $\overline{e}_k(x,t) = (I - G(t)\Gamma)e_k(x,t)$, $\overline{C}_k(x,t) = -C(t)(u_{k+1}(x,t) - u_k(x,t))$, so

$$\left(\overline{e}_k^T(x,t) + \overline{C}_k^T(x,t)\right) \leq 2\rho\|\overline{e}_k(x,t)\|^2 + 2b_c\|\overline{u}_k(x,t)\|^2 \tag{7}$$

mark $\|\overline{u}_k(x,t)\| = \|u_{k+1}(x,t) - u_k(x,t)\|, b_c = \max_{0 \leq t \leq T}\left\{\|C(t)\|^2\right\}, b_\Gamma = \max_{0 \leq t \leq T}$ $\|\Gamma\|^2$. Integral (7) about x on Ω, we have

$$\|e_{k+1}(\cdot,t)\|_{L^2}^2 \leq 2\rho\|e_k(\cdot,t)\|_{L^2}^2 + b_c\|\overline{u}_k(\cdot,t)\|_{L^2}^2 \tag{8}$$

Then $u^T(x,t)$ is left multiplied to the two sides of system (11), we have

$$\begin{aligned} &u^T(x,t)\frac{\partial u(x,t)}{\partial t} \\ &= u^T(x,t)D\Delta u(x,t) + u^T(x,t)[A(t) + \delta A(t)]u(x,t) + u^T(x,t)B(t)u(x,t-\theta) \end{aligned} \tag{9}$$

because $\frac{\partial}{\partial t}[u^T(x,t)u(x,t)] = 2u^T(x,t)\frac{\partial}{\partial t}u(x,t)$, so

$$\begin{aligned} \frac{\partial}{\partial t}\int_\Omega u^T(x,t)u(x,t)dx = 2d_i &\int_\Omega u_i(x,t)\Delta u_i(x,t)dx \\ &+ 2\int_\Omega u^T(x,t)\frac{1}{2}\left[A^T(t) + A(t)\right]u(x,t)dx \\ &+ 2\int_\Omega u^T(x,t)\delta A(t)u(x,t)dx + 2\int_\Omega u^T(x,t)B(t)u(x,t-\theta)dx \end{aligned} \tag{10}$$

Using divergence theorem and boundary condition (3), we have

$$\int_\Omega u_i(x,t)\Delta u_i(x,t)dx = -\int_\Omega [\nabla u_i(x,t)]^2 dx, i = 1, 2, \cdots, n \tag{11}$$

By Poincare inequality, we have

$$\int_\Omega [u_i(x,t)]^2 dx \leq h^2 \int_\Omega [\nabla u_i(x,t)]^2 dx, i = 1, 2, \cdots, n \tag{12}$$

where $h = l/\sqrt{m}$.

As $\alpha+\beta< -\lambda-b$, it is easy to proof the positive integer ε and k satisfy the followings inequality. $2(\lambda+\alpha-h^{-2}\varepsilon)+b+\beta+k<0$ and $\varepsilon<\min_{1\leq i\leq n}\{d_i\}$. Note that $g(t)\triangleq 1+\frac{(b+\beta)e^{\tau t}}{2(\lambda+\alpha-h^{-2}\varepsilon)+b+\beta+t}$ is continuous in interval(0,k). For the ε, by Eqs. (11) and (12), we have

$$d_i\int_\Omega u_i(x,t)\Delta u_i(x,t)dx+h^{-2}(d_i-\varepsilon)\int_\Omega u_i^2(x,t)dx\leq-\varepsilon\int_\Omega[\nabla u_i(x,t)]^2dx \quad (13)$$

Substituting Eq. (13) into Eq. (10), we can get

$$\begin{aligned}&\frac{d}{dt}\int_\Omega\|u(x,t)\|^2dx\\&\leq 2\int_\Omega u^T(x,t)\left\{\frac{1}{2}\left[A^T(t)+A(t)\right]-h^{-2}diag(d_1,d_2,\cdots,d_n)+\left(\alpha-h^{-2}\varepsilon\right)I\right\}\\&-2\varepsilon\int_\Omega[\nabla u_i(x,t)]^2dx+\|B(t)\|\int_\Omega\|u(x,t-\theta)\|^2dx\end{aligned} \quad (14)$$

Let $y(t)=\int_\Omega\|u(x,t)\|^2dx$, then integral (14) about t, yields

$$\begin{aligned}&y(t)\leq y(0)e^{\left[2\left(\lambda+\alpha-h^{-2}\varepsilon\right)+b+\beta\right]t}\\&-2\varepsilon\int_0^t e^{\left[2\left(\lambda+\alpha-h^{-2}\varepsilon\right)+b+\beta\right](t-s)}\int_\Omega[\nabla u_i(x,s)]^2dxds\\&+\int_0^t e^{\left[2\left(\lambda+\alpha-h^{-2}\varepsilon\right)+b+\beta\right](t-s)}\|B(t)\|\int_\Omega\|y(x,t-s)\|^2ds\end{aligned} \quad (15)$$

Taking a positive integer $q\in(0,k)$, then e^{qt} is left multiplied to the two sides of Eq. (15), we have

$$y(t)e^{qt}\leq y(0)e^{\left[2\left(\lambda+\alpha-h^{-2}\varepsilon\right)+b+\beta+q\right]t}-\frac{(b+\beta)e^{q\theta}}{2(\lambda+\alpha-h^{-2}\varepsilon)+b+\beta+q}\sup_{-\theta\leq s\theta}\{y(s)e^{qs}\} \quad (16)$$

as $2(\lambda+\alpha-h^{-2}\varepsilon)+b+\beta+q<0$, so

$$\sup_{-\theta\leq s\theta}\{y(s)e^{qs}\}\leq y(0)\left\{1+\frac{(b+\beta)e^{q\theta}}{2(\lambda+\alpha-h^{-2}\varepsilon)+b+\beta+q}\right\}^{-1}\leq y(0)[g(q)]^{-1} \quad (17)$$

Obviously,

$$y(t)\leq\sup_{-\theta\leq s\theta}\{y(s)e^{qs}\}e^{-qt}\leq y(0)[g(q)]^{-1}e^{-qt},t\geq 0 \quad (18)$$

Substituting Eq. (18) into Eq. (15), yields

$$y(t) \le y(0)\left\{1 - \frac{[g(q)]^{-1}(b+\beta)e^{q\theta}}{2(\lambda+\alpha-h^{-2}\varepsilon)+b+\beta+q}\right\}e^{-q\theta} -$$

$2\varepsilon \int_0^t e^{[2(\lambda+\alpha-h^{-2}\varepsilon)+b+\beta+q](t-s)} \int_\Omega [\nabla u_i(x,s)]^2 dxds$, when $t > k > 0$,

$$y(t) + 2\varepsilon \int_{t-k}^{t} e^{[2(\lambda+\alpha-h^{-2}\varepsilon)+b+\beta+q](t-s)} \int_\Omega [\nabla u_i(x,s)]^2 dxds \le y(0)[g(q)]^{-1} e^{qt} \quad (19)$$

For the Eq. (19), we use intermediate value theorem, zoom inequality, we have

$$y(t) + 2\varepsilon e^{[2(\lambda+\alpha-h^{-2}\varepsilon)+b+\beta]k} \int_\Omega \|\nabla u_i(x,\eta)\|^2 dx \le y(0)[g(q)]^{-1} e^{qt}, \eta \in [t-k, t] \quad (20)$$

The inequality (20) implicates

$$\sup_{-\theta \le s\theta} \{y(s)\} + 2\varepsilon e^{[2(\lambda+\alpha-h^{-2}\varepsilon)+b+\beta]k} \sup_{t-k \le s \le t} \left\{ \int_\Omega \|\nabla u_i(x,\eta)\|^2 dx \right\} \le y(0)[g(q)]^{-1} e^{-q(t-\theta)}$$

So $lim_{t\to\infty} \|u(x,t)\|_{L^2} = \lim_{t\to\infty} \|\nabla u(x,t)\|_{L^2} = 0$. According to (4), we have $\|u_{k+1}(x,t) - u_k(x,t)\| \le \|\Gamma e_k(x, t+\theta)\|$ then

$$\|u_{k+1}(\cdot, t-\theta) - u_k(\cdot, t-\theta)\|_{L^2}^2 \le b_\Gamma \|e_k(\cdot,t)\|_{L^2}^2 \quad (21)$$

Substituting inequality (21) into inequality (14), zoom the inequality, yields

$$\begin{aligned} \|\nabla \overline{u}_k(\cdot,t)\|_{L^2}^2 &\le \|\nabla \overline{u}_k(\cdot,0)\|_{L^2}^2 e^{ht} \\ &- 2d \int_0^t e^{h(t-s)} \|\nabla \overline{u}_k(\cdot,t)\|_{L^2}^2 ds + g b_\Gamma \int_0^t e^{h(t-s)} \|\overline{e}_k(\cdot,s)\|_{L^2}^2 ds \end{aligned} \quad (22)$$

Let λ be large enough and $\lambda > h$, $e^{-\lambda t}$ be left multiplied to the two sides of inequality (22),

$$\|\nabla \overline{u}_k(\cdot,t)\|_{L^2}^2 e^{-\lambda t} \le \delta r^k + \frac{g b_\Gamma}{\lambda - h} \|e_{k+1}(x,t) - e_k(x,t)\|_{(L^2,\lambda)}$$

So

$$\|\nabla \overline{u}_k(\cdot,t)\|_{(L^2,\lambda)}^2 \le \delta r^k + \frac{g b_\Gamma}{\lambda - h} \|e_k(x,t)\|_{(L^2,\lambda)} \quad (23)$$

For inequality (8), we use iterative method to get

$$\|e_{k+1}(\cdot,t)\|_{L^2}^2 \le (2\rho)^k \|e_0(\cdot,t)\|_{L^2}^2 + \sum\nolimits_{i=1}^{k-1} (2\rho)^{k-i-1} 2b_c \|\bar{u}_k(\cdot,t)\|_{L^2}^2 \tag{24}$$

e^k is left multiplied to the two sides of inequality (24), we have

$$(\|e_{k+1}(\cdot,t)\|_{L^2}^2 e^k)e^{-\lambda t} \le \|e_0\|_{(L^2,\lambda)}^2 + \sum\nolimits_{i=1}^{k-1} (2\varepsilon\rho)^{k-i-1} \varepsilon \left[2b_c\delta(\varepsilon r)^i + \frac{2b_c g}{\lambda - h} b_\Gamma \|e_i\|_{(L^2,\lambda(\varepsilon))}^2 \right] \tag{25}$$

Let $F(\varepsilon,\lambda) = \frac{2\varepsilon\rho}{1-2\varepsilon\rho}\left[\frac{2\varepsilon g b_c}{\lambda - h}\right]$, then according to (25), we have

$$\left(\|e_k(\cdot,t)\|_{L^2}^2 e^k\right) e^{-\lambda t} \le \|e_0\|_{(L^2,\lambda)}^2 + \sum_{i=1}^{k-1} (2\varepsilon\rho)^{k-i-1} 2\varepsilon b_c \delta(\varepsilon r)^i + F(\varepsilon,\lambda) \sup_{1\le i\le k} \|e_i\|_{(L^2,\lambda(\varepsilon))}^2$$

then

$$\|e_k(\cdot,t)\|_{L^2}^2 \le \|e_0\|_{(L^2,\lambda)}^2 + 2b_c\delta\varepsilon + F(\varepsilon,\lambda) \sup_{1\le i\le k} \|e_i\|_{(L^2,\lambda(\varepsilon))}^2$$

So

$$\sup_{1\le i\le k} \|e_i\|_{(L^2,\lambda(\varepsilon))}^2 \le \frac{\|e_k(\cdot,t)\|_{L^2}^2 + 2b_c\delta\varepsilon}{1 - F(\varepsilon,\lambda)} \tag{26}$$

Hence $\|e_k(\cdot,t)\|_{L^2}^2 \le \varepsilon^{-k} e^{\lambda t} \sup_{1\le i\le k} \|e_i\|_{(L^2,\lambda(\varepsilon))}^2 \le \varepsilon^{-k} e^{\lambda t} \frac{\|e_k(\cdot,t)\|_{L^2}^2 + 2b_c\delta\varepsilon}{1-F(\varepsilon,\lambda)}$ as $\varepsilon > 1$, so $lim_{k\to\infty} \|e_k(\cdot,t)\|_{L^2}^2 = 0, t \in [0,T]$.

4 Simulation Analysis

For system, we give a numerical example to verify the effectiveness of the proposed method.

$$D = \begin{bmatrix} 1 & 0 \\ 0 & 1 \end{bmatrix}, A = \begin{bmatrix} sint & 0 \\ 0 & cost \end{bmatrix}, B = \begin{bmatrix} e^{-t} & 0 \\ 0 & e^{-t/2} \end{bmatrix}, C = \begin{bmatrix} 1 & 0 \\ 0 & 1 \end{bmatrix},$$

$$G = \begin{bmatrix} 0.7 & 0 \\ 0 & 0.7 \end{bmatrix}, r = 0.1$$

When $\delta = 1, \theta = 0.5$, if we choose learning gain $\Gamma = \begin{bmatrix} 0.8 & 0 \\ 0 & 0.8 \end{bmatrix}$, then it is easy to meet the condition of theorem 1. The simulation figure shown in Figs. 1 and 2 is ordinary reaction diffusion system. According to Fig. 3, we know the iterative learning algorithm improve the reaction diffusion system convergence speed that shown in Fig. 3.

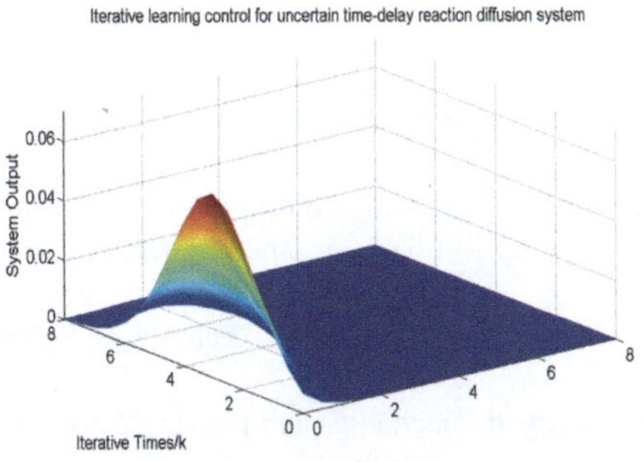

Fig. 1. Iterative learning control for uncertain time-delay reaction diffusion system

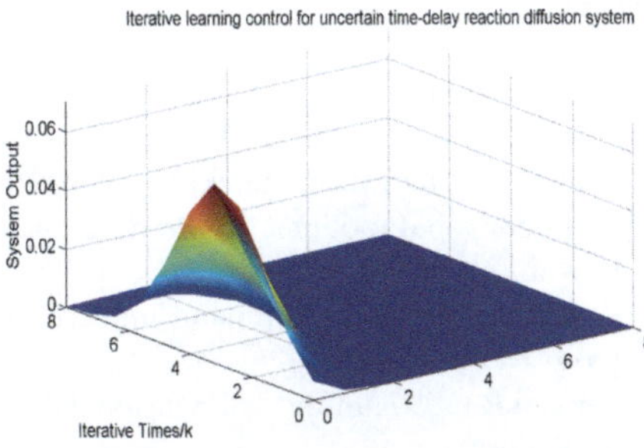

Fig. 2. Ordinary uncertain time-delay relay reaction diffusion system

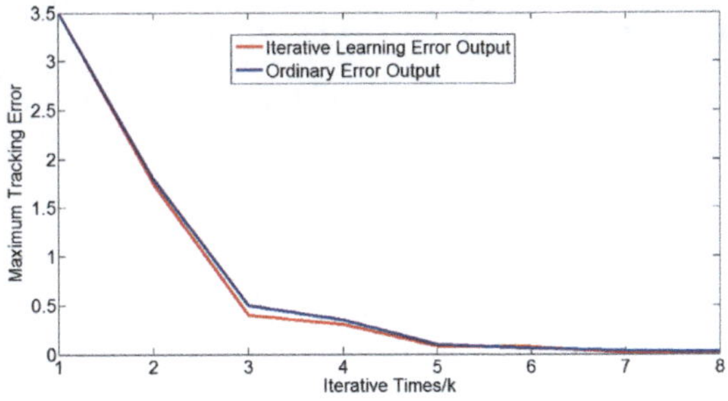

Fig. 3. Maximum tracking error curve comparison

5 Conclusion

We have investigated the iterative learning control for uncertain time-delay reaction diffusion system by nonlinear boundary. Under the existence of bounded deterministic disturbances, the robustness and convergence of uncertain time-delay reaction diffusion are constructed by the iterative learning control to obtain high-gain nonlinear output feedback of the unknown parameters. In the whole control system the robustness and the convergence of the system state to zero is guaranteed.

Acknowledgments. The work was supported by the Hechi University Foundation (XJ2016ZD004), Guangxi teacher Capability enhancement project (2018KY0495), Hechi university Youth teacher Foundation (XJ2017QN08) and was supported by the Projection of Master Foundation (2017HJA001, 2018LG004).

References

1. Arimoto S, Kawamura S Miyazaki E (1984) Bettering operation of roboties by leanring. J Robot Syst 1(2):123–140
2. Arimoto S, Naniwa T, Suuzki H (1990) Robusntess of P-type leaning control with a forgetting factor for robot motions. In: Proceedings of the conference on decision and control. honoluu Hawaii, pp 2640–2645
3. Kuramoto Y (1978) Diffusion induced chaos in reactions systems. Prog Theor Phys Suppl Progr 64:346–367
4. Temam R (1997) Infinite-dimensional dynamical systems in mechanics and physics. Springer, New York
5. Christofides PD, Armaou A (2000) Global stabilization of the Kuramoto-Sivashinsky equation via distributed output feedback control. Syst Control Lett 39:283–294
6. Curtain RF, Zwart HJ (1995) An introduction to infinite-dimensional linear systems theory. Springer, New York
7. Hooper AP, Grimshaw R (1985) Nonlinear instability at the interface between two viscous fluids. Phys Fluids 28:37–45

Iterative Learning Control of Discrete Generalized Distributed Parameter

Mengji Chen and Yinjun Zhang(✉)

School of Physics and Electrical Engineering, Hechi University, Yizhou 546300, Guangxi, China
18107786221@163.com

Abstract. In this paper, the iterative learning control problem of discrete descriptor distributed parameter systems is studied. P type learning control law is constructed. By using the principle of iterative convergence, it is proved that this learning law can make the output tracking error of the system converge along the iterative axis.

Keywords: Iterative learning control · Discrete system · Generalized distributed parameter system

1 Introduction

Iterative learning control (ILC) is an advanced intelligent control method of systems which perform tasks repetitively over a finite time interval. Since the original work by Arimoto et al. [1] in 1984. In the last three decades, ILC has been constantly studied [2–4].

However, the discrete system has its own particularity. Therefore, many methods on continuous system cannot be directly applied to discrete systems, such as Lyapunov analysis. There are essential differences between discrete systems and continuous systems. The iterative learning of discrete systems is controlled by sampling points [5–7] and is described in the way of generalized matrices. Of course, 2D [8–10] theory can also be used to study it, but 2D theory can only study continuous systems at present, and it can not be studied for discontinuous systems. At present, many scholars have studied the iterative learning control of discrete systems. The convergence of the discrete iterative learning control, the optimal control problem, the robustness of discrete iterative learning and so on are studied. Different from the continuous system, the scope of application of the discrete system is relatively large, it has certain practical significance. For the iterative learning control problem of discrete generalized distributed parameters, there is no related research.

The main arrangement of this paper is as follows: firstly, the linear generalized distributed parameter system is discretized. Secondly, we use the knowledge of discrete iterative learning to prove its convergence. Finally, an example is given to verify the effectiveness of the proposed algorithm.

Z. Xu et al. (Eds.): CSIA 2019, AISC 928, pp. 407–414, 2020.
https://doi.org/10.1007/978-3-030-15235-2_61

2 The Description of Discrete Singular Distributed Parameter Systems

Consider the following systems:

$$E\frac{\partial z(x,t)}{\partial t} = A\frac{\partial^2 z(x,t)}{\partial x^2} + Bu(x,t),$$
$$z(x,t) = z(1,t) = 0,\ z(x,0) = \varphi(x), 0 \le x \le 1 \qquad (1)$$

where $z(x,t) \in R^n, z(x,t) \in H_0^1((0,1)\times(0,\infty))$ (sobolev space) is the state of the system. $u(x,t) \in R^m$ is the input, $E, A \in R^{n\times m}, B \in R^{n\times m}$ are constant matrices, $\phi(x) \in R^n$ is a continuous vector function, det E = 0 $\det(\lambda E + (n\pi)^2 A) \neq 0$, $rankE = r < n$. Solving Eq. (1) by using separation principle: Let

$$z(x,t) = X(t)T(x) \qquad (2)$$

where $X(t) \in R^{n\times n}, T(x) \in R^n$. Introducing Eq. (2) into Eq. (1) yields

$$\mathrm{E}X'(t)T(x) = AX(t)T''(x) \qquad (3)$$

Accord to the Ref. [11], we know the transform of system (1) is

$$E\dot{X}_n(t) = -(n\pi)^2 AX_n(t) + Bu_n(t), X_n(0) = \varphi_n, n = 1, 2, \cdots \qquad (4)$$

Because $(E, -(n\pi)^2 A)$ is regular, nonsingular matrices P and T exist, so that Eq. (4) is discrete and limited equivalent to

$$x_{n1k}(i+1) = A_{11}x_{n1k}(i) + A_{12}x_{n2k}(i) + B_1 u_{nk}(i) \qquad (5)$$

$$0 = A_{21}x_{n1k}(i) + A_{22}x_{n2k}(i) + B_2 u_{nk}(i) \qquad (6)$$

where

Suppose system (6) satisfies the following assumptions:

(1) The system is regular and A_{22} is reversible.
(2) The initial value of system is $x_{nk}(0) = x_{nd}(0), k = 1, 2, 3, \cdots$
(3) The system is controllable and reachable. For any realizable bounded output trajectory $x_{nd}(i)$, there exists a unique control input $u_{nd}(i)$ to make the following equation hold:

$$\begin{cases} x_{n1d}(i+1) = A_{11}x_{n1d}(i) + A_{12}x_{n2d}(i) + B_1 u_{nd}(i) \\ 0 = A_{21}x_{n1d}(i) + A_{22}x_{n2d}(i) + B_2 u_{nd}(i) \end{cases} i \in [1, 2, \cdots, T] \qquad (7)$$

Here we use the following iterative learning control law to analyze it

$$u_{n(k+1)}(i) = u_{nk}(i) + [L_1 L_2]\begin{bmatrix} e_{n1(k+1)}(i) \\ e_{n2(k+1)}(i) \end{bmatrix} \tag{8}$$

where

$e_{n1(k+1)}(i) = x_{n1d}(i) - x_{n1(k+1)}(i), e_{n2(k+1)}(i) = x_{n2d}(i) - x_{n2(k+1)}(i), L_1 \in R^{m \times q}$, $L_2 \in R^{m\times(n-q)}$, L_1, L_2 are gain, and meet $(I + L_1\bar{B}_1 - L_2\bar{B}_2)$ is invertible.

3 Convergence Analysis

Lemma 1. For the generalized distributed parameter systems (5) (6), if the above assumptions (1), (2), (3) are satisfied and the following convergence conditions are satisfied:

$$\|G\| \le \varepsilon < 1 \tag{9}$$

Then we can get that the algorithm is uniformly convergent, and when k approaches zero, the system control input error also approaches zero, namely $\lim_{k\to\infty} u_{nk}(i) = u_{nd}(i)$. The actual trajectory also consistently converges to the desired trajectory $\lim_{k\to\infty} x_{nk}(i) = x_{nd}(i)$. Where $G = H_1^{-1}, H_1 = I + L_1\bar{B}_1 - L_2\bar{B}_2$.

Proof. By the system (6) and the A_{22} invertible, we can obtain

$$x_{n2k}(i) = -A_{22}^{-1}A_{21}x_{n1k}(i) - A_{22}^{-1}B_2 u_{nk}(i) \tag{10}$$

By Eq. (10), yields

$$x_{n2(k+1)}(i) = -A_{22}^{-1}A_{21}x_{n1(k+1)}(i) - A_{22}^{-1}B_2 u_{nk}(i) \tag{11}$$

Introducing Eq. (11) into Eq. (5) yields

$$x_{n1k}(i+1) = (A_{11} - A_{12}A_{22}^{-1}A_{21})x_{n1k}(i) + (B_1 - A_{12}A_{22}^{-1}B_2)u_{nk}(i) \tag{12}$$

According Eq. (12), we have

$$x_{n1(k+1)}(i) = (A_{11} - A_{12}A_{22}^{-1}A_{21})x_{n1(k+1)}(i) + (B_1 - A_{12}A_{22}^{-1}B_2)u_{n(k+1)}(i) \tag{13}$$

Let $\bar{A}_{11} = A_{11} - A_{12}A_{22}^{-1}A_{21}, \bar{B}_1 = B_1 - A_{12}A_{22}^{-1}B_2$, then the Eq. (13) can be rewritten

$$x_{n1(k+1)}(i+1) = \bar{A}_{11}x_{n1(k+1)}(i) + \bar{B}_1 u_{n(k+1)}(i) \tag{14}$$

Similarly, we can get the following equations

$$x_{n1d}(i+1) = \bar{A}_{11}x_{n1d}(i) + \bar{B}_1 u_{nd}(i), x_{n2d}(i) = -A_{22}^{-1}A_{21}x_{n1d}(i) - A_{22}^{-1}B_2 u_{nd}(i) \quad (15)$$

and as

$$\Delta u_{n(k+1)}(i) = u_{nd}(i) - u_{n(k+1)}(i) \quad (16)$$

$u_{nd}(i)$ is desired control input. Introducing Eq. (16) into Eq. (8) yields

$$\Delta u_{n(k+1)}(i) = \Delta u_{nk}(i) - L_1 e_{n1(k+1)}(i+1) - L_2 e_{n2(k+1)}(i) \quad (17)$$

Define the status error

$$e_{n1(k+1)}(i) = x_{n1d}(i) - x_{n1(k+1)}(i)\ e_{n2(k+1)}(i) = x_{n2d}(i) - x_{n2(k+1)}(i) \quad (18)$$

then we have

$$\begin{aligned} e_{n1(k+1)}(i+1) &= \bar{A}_{11}e_{n1(k+1)}(i) + \bar{B}_1 \Delta u_{n(k+1)}(i), \\ e_{n2(k+1)}(i) &= -\bar{A}_{21}e_{n1(k+1)}(i) - \bar{B}_2 \Delta u_{n(k+1)}(i) \end{aligned} \quad (19)$$

$$e_{n2(k+1)}(i) = -\bar{A}_{21}e_{n1(k+1)}(i) - \bar{B}_2 \Delta u_{n(k+1)}(i) \quad (20)$$

Where

$$\bar{A}_{21} = A_{22}^{-1}A_{21}, \bar{B}_2 = A_{22}^{-1}B_2 \quad (21)$$

Introducing Eqs. (20) and (21) into Eq. (17) yields

$$\Delta u_{n(k+1)}(i) = \Delta u_{nk}(i) + (L_2\bar{A}_{21} - L_1\bar{A}_{11})e_{n1(k+1)}(i) + (L_2\bar{B}_2 - L_1\bar{B}_1)\Delta u_{n(k+1)}(i) \quad (22)$$

Merger of similar items, we can get

$$(I + L_1\bar{B}_1 - L_2\bar{B}_2)\Delta u_{n(k+1)}(i) = \Delta u_{nk}(i) + (L_2\bar{A}_{21} - L_1\bar{A}_{11})e_{n1(k+1)}(i) \quad (23)$$

Let $H_1 = I + L_1\bar{B}_1 - L_2\bar{B}_2$. From the above iterative learning law we can know H_1 is reversible, so by Eq. (23),

$$\Delta u_{n(k+1)}(i) = H_1^{-1}\Delta u_{nk}(i) + H_1^{-1}(L_2\bar{A}_{21} - L_1\bar{A}_{11})e_{n1(k+1)}(i) \quad (24)$$

because $G = H_1^{-1}$, let $\mathcal{H} = G_1^{-1}(L_2\bar{A}_{21} - L_1\bar{A}_{11})$. The Eq. (24) can be rewritten as

$$\Delta u_{n(k+1)}(i) = G\Delta u_{nk}(i) + \mathcal{H}e_{n1(k+1)}(i) \quad (25)$$

From the initial state of the system, we know that the initial state of the system at each iteration of the system is at the corresponding desired trajectory position. According to Eq. (20), we have

$$e_{n1(k+1)}(i+1) = \sum_{j=0}^{i=1} \bar{A}_{11}^{i-j-1} \bar{B}_1 \Delta u_{n(k+1)}(j) \tag{26}$$

Introducing Eq. (26) into Eq. (25), yields

$$\Delta u_{n(k+1)}(i) = G \Delta u_{nk}(i) + \mathcal{H} \sum_{j=0}^{i=1} \bar{A}_{11}^{i-j-1} \bar{B}_1 \Delta u_{n(k+1)}(j) \tag{27}$$

Taking norm of Eq. (27), we have

$$\left\| \Delta u_{n(k+1)}(i) \right\| \leq \|G\| \|\Delta u_{nk}(i)\| + \|\mathcal{H}\| \left\| \sum_{j=0}^{i=1} \bar{A}_{11}^{i-j-1} \bar{B}_1 \right\| \left\| \Delta u_{n(k+1)}(j) \right\| \tag{28}$$

Using $\lambda^i, 0 < \lambda < 1$ to multiply both sides of (28), let $\sup_{\substack{i \in [0,T] \\ j \in [0, i-1]}} \bar{A}_{11}^{i-j-1} \bar{B}_1 = b$, $\|\mathcal{H}\| = h$. we have

$$\lambda^i \left\| \Delta u_{n(k+1)}(i) \right\| \leq \lambda^i \|G\| \|\Delta u_{nk}(i)\| + \lambda^i h \left(\sum_{j=0}^{i=1} \bar{A}_{11}^{i-j-1} b \left\| \Delta u_{n(k+1)}(j) \right\| \right) \tag{29}$$

According to definition of λ norm, we get

$$\left\| \Delta u_{n(k+1)}(i) \right\|_\lambda \leq \|G\| \frac{1-\lambda}{1-(1+hb)\lambda} \|\Delta u_{nk}(i)\|_\lambda \tag{30}$$

By the convergence condition, we know that there exist the enough large λ to make

$$\|G\| \frac{1-\lambda}{1-(1+hb)\lambda} < 1 \tag{31}$$

so

$$\lim_{k \to \infty} \|\Delta u_{nk}(i)\|_\lambda = 0 \tag{32}$$

namely

$$\lim_{k\to\infty} \|\Delta u_{nk}(i)\| = \|\Delta u_{nd}(i)\| \tag{33}$$

Repeat the above λ condition for Eq. (26), then using λ^i to left multiply both sides of (26), it becomes

$$\lambda^i e_{n1(k+1)}(i+1) = \lambda^i \sum_{j=0}^{i=1} \bar{A}_{11}^{i-j-1} \bar{B}_1 \Delta u_{n(k+1)}(j) \tag{34}$$

Taking norm of Eq. (35), we have

$$\left\|e_{n1(k+1)}\right\|_\lambda = b\frac{\lambda}{1-\lambda}\left\|\Delta u_{n(k+1)}\right\|_\lambda \tag{35}$$

So we have

$$\lim_{k\to\infty} \sup_{0\le i\le T} \|e_{n1k}\| = 0,\ \lim_{k\to\infty} \sup_{0\le i\le T} \|e_{n2k}\| = 0 \tag{36}$$

According to Eqs. (36) and (41), we know that the real state of system converges uniformly the desired state on $0 \le i \le T$, namely

$$\lim_{k\to\infty} x_{nk}(i) = x_{nd}(i) \tag{37}$$

Proof is completed.

4 Numerical Simulation Analysis

For the system (1), we take the matrices as follows

$$E = \begin{bmatrix} 1 & 0 \\ 0 & 0 \end{bmatrix}, A = \begin{bmatrix} 1 & 1 \\ 2 & 1 \end{bmatrix}, B = \begin{bmatrix} 1 & 0 \\ 0 & 1 \end{bmatrix}$$

the nonsingular matrices P and T are unit matrices, the system is regular and the learning gain is $L_1 = [0.50]^T, L_2 = [-0.5 - 1]^T$, it can be calculated, It can be known that the convergence condition is true. Now suppose the desired trajectory of the system is $x_{nd}(i) = [5\sin(0.4i)i^2]^T$, the initial state is $x_{nk}(0) = [00]^T$, $u_0(0) = [00]^T$, we take $i \in [0, 30]$. The simulation of the curve shown in Figs. 1 and 2.

Figure 1 shows the tracking curve of the first component of the system at the 28th and 22nd times. We can clearly see from the Fig. 1 that the first component of the system can trace the desired trajectory very well at the 28th iteration. Although there is relatively large fluctuation at 22nd times, it does not affect the stability of the system. The Fig. 2 shows the trace of the second component at the 4th and 18th times. It is

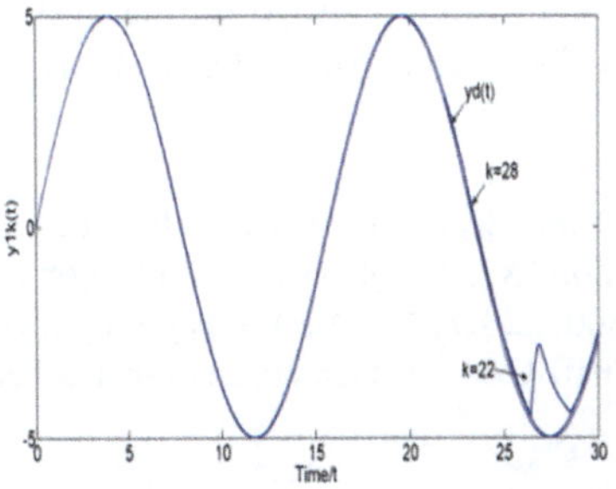

Fig. 1. The iterative output tracking curve of $x_{n1k}(k)$

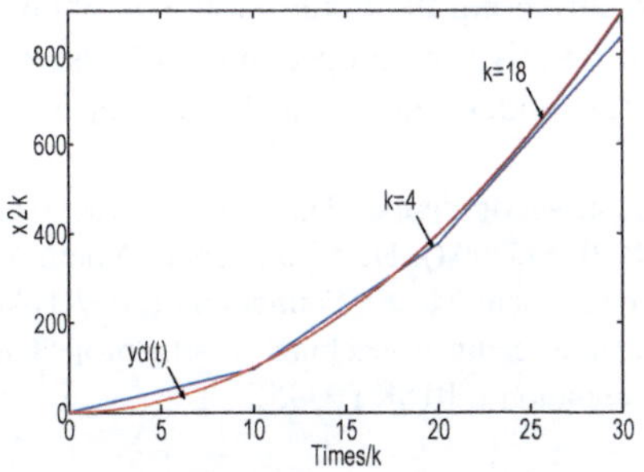

Fig. 2. The iterative output tracking curve of $x_{n2k}(k)$

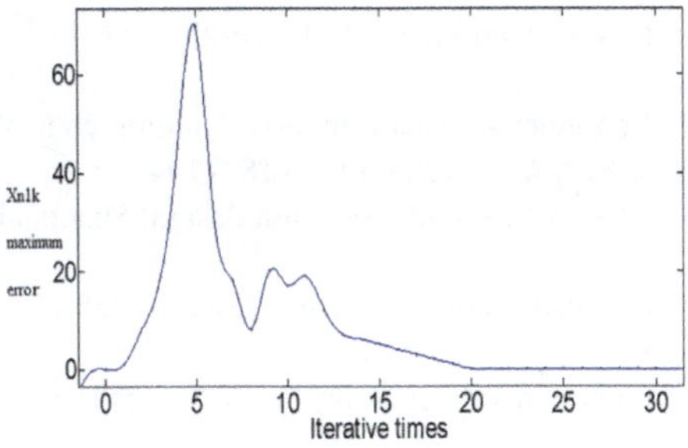

Fig. 3. Iterative error tracking curve of $x_{n1k}(k)$

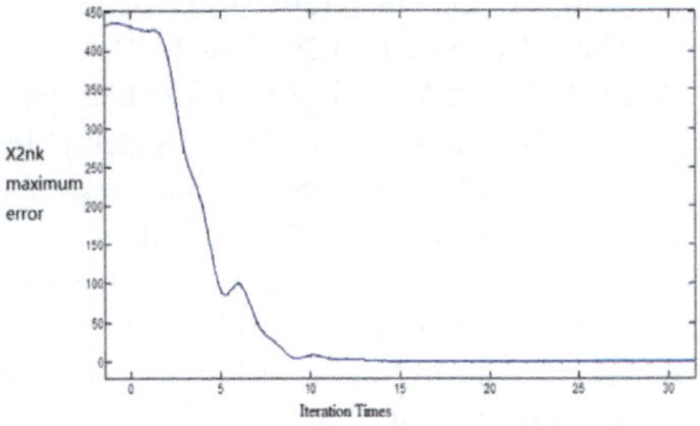

Fig. 4. Iterative error tracking curve of $x_{n2k}(k)$

clear from the figure that the second component of the system has been well tracked desired track at 18th times We know from Fig. 1 and Fig. 2 that the algorithm is feasible and effective.

Figure 3 shows the first component of the tracking error curve. From the Fig. 3, we can see that the corresponding error of the system is very small at 28th times, which is consistent with Fig. 1. Figure 4 is the same. At the 18th times, the error is very small. This is also consistent with the Fig. 2

Figures 3 and 4 illustrate that with the number of iterations increases, the error between the state trajectory and the desired trajectory of the system gradually converges to zero, it is in agreement with the above proof.

5 Conclusions

This paper presents an iterative learning control problem for a discrete generalized distributed parameter system. Under the condition of the definite boundary value solution for a class of linear discrete systems, the P-type learning law is applied to design iterative learning controller. The convergence of the output tracking error of the system along the iteration axis is obtained. We get good results. How to further design the iterative learning control of discrete generalized distributed parameter systems in other modes remains to be further explored.

Acknowledgments. The work was supported by the Hechi University Foundation (XJ2016ZD004), Hechi university Youth teacher Foundation (XJ2017QN08), the Projection of Environment Master Foundation (2017HJA001, 2017HJB001, 2018LG004), The key project of the new century teaching reform project in Guangxi (2010JGZ033), Guangxi Youth teacher Foundation (2018KY0495).

References

1. Arimoto S, Kawamura S, Miyazaki F (1984) Bettering operation of robots by learning. J Robot Syst 1(2):123–140
2. Zhang Y, Li Y et al (2015) Vector analysis for iterative learning control algorithm. J Comput Theor Nanosci 12(12):4724–4729
3. Su J, Zhang Y et al (2016) Singular distributed parameter system iterative learning control with forgetting factor with time-delay. Int J u- e-Serv Sci Technol 9(7):182–194
4. Liu P (2011) Nonlinear distributed parameter system robust fault detection design. Shanghai Jiao Tong Univ J 45(2):241–246
5. Wang H, Dong J, Wang Y (2015) High-order feedback iterative learning control algorithm with forgetting factor. Math Probl Eng 2015(3):1–7
6. Lan T, Yan F, Lin H (2017) Iterative learning control with forgetting factor for urban road network. J Control Sci Eng 2017(4):1–7
7. Yuan Y, Zhaoqiang GE (2010) Exponential stability of the first order singular distributed parameter systems. Syst Sci Math 30(3):315–322
8. Liu F, Shi G, Luo Y (2010) An equivalent proposition for stability by feedback of the second order singular distributed parameter system. In: International conference on advanced computer theory and engineering. IEEE, pp V3-71–V3-74
9. Benton SE, Rogers E, Owens DH (2002) Stability conditions for a class of 2D continuous-discrete linear systems with dynamic boundary conditions. Inter J Control 75(1):5–15
10. Kaczorek T (2011) Stability of continuous-discrete linear systems described by the general model. Bull Polish Acad Sci: Tech Sci 59(2):189–193
11. Liu Y (1999) Variable structure control method of structural stability for singular distributed parameter system. J South China Univ Technol 27(4):9–12

The Multi-model Predictive Control Method Research on the Outlet Temperature Control of Hot-Blast Stove

Mengji Chen and Yinjun Zhang(✉)

School of Machinery and Electronic Information Engineering,
Hechi University, Yizhou, Guangxi, China
18107786221@163.com

Abstract. It is difficult to establish the precise mathematical model of the outlet temperature of hot-blast stove in the real manufacturing, for the control system is hunting system, with uncertainty, time variation, and coupling. So it will be more difficult to guarantee the effort of control, based on the control algorithm of precise mathematical model. Aiming at these problems, a new predictive control strategy which can quickly suppress disturbances has been provided by this paper. The strategy is based on the idea of control of dynamic matrix control, and combines dynamic matrix control and the advantages of both time optimal control. It is proved by the simulation result that this strategy acts a good effect on the outlet temperature control of hot-blast stove and will quickly suppress the interference.

Keywords: The outlet temperature of hot blast stove · Dynamic matrix · Time optimal control · Interference suppression

1 Introduction

Hot-blast stove is heat exchangers of storage heat. The stove is the important equipment in the production iron process, it will be important to control the combustion quality, for air temperature will directly affect the level of production in the blast furnace. It provided heat air into stove, and is very active factor in smelting process, according to the combustion chamber temperature changes, search for optimization AFR proportion coefficient quickly, while stove is a nonlinear, large hunting system, with large delay, uncertainty, time-varying, nonlinear and coupling etc., there are many factors affecting stove, and a variety of factors restraining each other, so it is very complex and difficult to control the process and describe the mathematical model precisely. It lack of flexibility and response using traditional model method under the entire control system in the framework of the model, it is difficult competent control of complex systems. Now there are many control strategies, such as the traditional PID control [1], improved PID algorithm (fuzzy PID control [2, 3]), neural network control [4], expert control etc.

Z. Xu et al. (Eds.): CSIA 2019, AISC 928, pp. 415–419, 2020.
https://doi.org/10.1007/978-3-030-15235-2_62

In addition, the controlled target is a hunting system, now it used the Smith Predictor Control [5–9], but as what mentioned before, the system demanded a strategy that is not very high precision, but also control strategy with predictive, because the system do not require the model structure that has some prior knowledge as well as rolling optimization strategy, so predictive control can be enhanced control robust effectively, so these features can get fine control effect for the system that has a large hunting system and the lack of accurate mathematical model.

According to process demand and analysis process to identify the quality and yield of product, safe operation and economic operation, and environmental protection as the controlled parameter constitute the process control system, these parameters has decisive role and can be directly measured, so the system have more control targets, compare with single predictive control for overcoming constant value adjustment of interference has some flaws, just considered the robust of the system, but the system have more disturbances and single predictive control will ignore the controlled object interference rapid suppression capability, so both ways more difficult. Some scholars present a good balance between immunity and robust by changed online type of error correction vector, add an auxiliary unit [10] is precondition, undoubtedly it adds to the complexity of the system and it is implemented difficultly. In this paper, outlet temperature of stove model as research object, uses a dynamic matrix control based (Dynamic Matrix Control) strobe control strategy. The new control strategy has a faster response speed and rapidly suppress interference than traditional control method.

2 Control Strategy Overview

Aiming at the control target, the new strategy combines the time optimal control and dynamic matrix control their each characteristics, the core idea is to control the dynamic matrix, on the basis, utilize fast control features of the time optimal control to improve controller unpredictable disturbances rapid response capability.

The control strategy of the system in disturbed or model adaptation case response sensitivity, depending on the control mode is used to switch between the two tolerance values es. When the es is set small, the controller will be higher sensitivity to disturbances, suppression will be faster because in this case depends mainly on the time optimal disturbance suppression control to achieve, but accordingly it may reduce system stability. On the contrary, the system disturbance suppression time will be longer. Controlled target model is the formula (1):

$$\tilde{A}(q^{-1})y(t) = q^{-(d+1)}B(q^{-1})u(t) + C(q^{-1})e(t)/\Delta \tag{1}$$

Can be expressed as formula (2):

$$\gamma(t+1) = \sum_{i=1}^{n} A_{1,i}\gamma(t+1-i) + \sum_{i=1}^{m} B_{1,i}U(t-d-i) + \sum_{i=1}^{r} C_{1,i}E(t+1-i) \tag{2}$$

Where $\gamma(t), U(t), E(t)$ respectively N-dimensional column vector output, M-dimensional column vector input, and R-dimensional are independent zero-dimensional white noise sequence, d + 1 is the system delay. $A_{1,i}(i = 1, 2, K, n), B_{1,i}$ $(i = 1, 2, K, n), C_{1,i}(i = 1, 2, K, n)$ are N × N, N × M and N × R dimensional matrix. By formula (2) recursion, the system time in the future minimum variance output prediction model is formula (3):

$$\gamma = \gamma_m + G\Delta U \tag{3}$$

There

$$\begin{aligned} \gamma &= [\gamma^T(t+d+1|t), \gamma^T(t+d+2|t), \ldots, \gamma^T(t+p|t)]^T \\ \gamma_m &= [\gamma_m^T(t+d+1|t), \gamma_m^T(t+d+2|t), \ldots, \gamma_m^T(t+p|t)]^T \end{aligned} \tag{4}$$

$$\Delta U = [\Delta U^T(t), \Delta U^T(t+1), \ldots, \Delta U^T(t+p-d-1)]^T$$

$$G = \begin{bmatrix} B_{1,0} & 0 & \cdots & 0 \\ B_{2,0} & B_{1,0} & \cdots & 0 \\ \vdots & \vdots & \cdots & \vdots \\ B_{p=d,0} & B_{p-d-1,0} & \cdots & B_{1,0} \end{bmatrix}$$

P is the predicted length.

Formula (4) $\gamma_m(t+k)$ is entirely in the past control inputs and outputs determined by the following formula soon:

$$\gamma_m(t+k) = \sum_{i=1}^{n} A_{1,j}\gamma_m(t+k-i) + \sum_{i=0}^{m} B_{1,j}U(t+k-1-d-i/t) + \sum_{i=0}^{n} C_{1,j}E(t+k-i/t)$$

Next $k - 1, 2, \Lambda, p$, consider the lagging system by the following formula recursive calculated step reference trajectory p-d constitute vectors γ_r.

$$\begin{aligned} \gamma_r &= [\gamma_r^T(t+d+1), \gamma_r^T(t+d+2), \ldots, \gamma_r^T(t+p)]^T \\ \gamma_r(t+d) &= \gamma_m(t+d) \\ \gamma_r(t+d+i) &= \alpha\gamma_r(t+d+i-1) + (I-\alpha)\gamma_s \end{aligned}$$

Where the column vector of the set value γ_s, I is N × N dimensional matrix, α is the output coefficient matrix softens. Finally, if we want to outlet temperature deviation as small as possible, so the need to take the minimization of the objective function of the formula (5):

$$J = min\{(\gamma - \gamma_r)^T(\gamma - \gamma_r) + \gamma\Delta U^T\Delta U\} \tag{5}$$

Get the current system time optimal control in increments of:

$$\Delta U = (G^T G + \gamma I)^{-1} G^T (\gamma_r - \gamma_m)$$

The current optimal control law is:

$$U(t) = U(t-1) + D^T(\gamma_r - \gamma_m) = ze(k) \quad (6)$$

The matrix of first M rows of $(G^T G + \gamma I)^{-1}$ is DT, $G^T G$ is $(p-d)M \times (p-d)M$ -dimension matrix. Where Z is a constant, formula (6) is the result of the time optimal control. The basic idea of time optimal control can be described as formula (7):

$$\begin{cases} x(t) = Ax(t) + Bu(t), X(t_0) = X_0 \\ y(t) = Cx(t) \end{cases} \quad (7)$$

Assuming the target Xi was in the time optimal state, it required to seek the optimal control within the allowable control range system in the shortest possible time from the initial state X0 to Xf the performance indicators is formula (8):

$$min\, J = \int_{t_0}^{t_f} 1dt = t_f - t_0 \quad 0 \le u_t \le M \quad (8)$$

The value of the shortest controls time and the amount of control can be obtained between in (0, M). When the condition is met by equation of state of the target, the optimal solution is formula (9) within the constrained control function uj(t):

$$\tilde{u} = \begin{cases} 0 \\ M \end{cases} \quad (9)$$

From the basic idea of time optimal control can be seen, its controlled variable is not selected by performance indicator, so the controlled variable is not necessarily optimal, but it improved controller's rapid response capability using the extreme values to control and played a very fine effect. This is just to make up the shortage of the dynamic matrix in constant value adjustment of overcoming interference.

Simulation experiment and users use the same control device, and simulation similar object model with and field equipment environment using Matlab programming and debugging, and running on the real device, gradually verification.

Consider one of the following multi-input multi-output systems, from the system we know the control target is multivariable systems with lag factor. From Fig. 1, the delay in the dynamic matrix control algorithm is unpredictable disturbed to object in the time-lag target control, using DMC can be more reach a steady state quickly based on time optimal algorithm, and the estimated control effect can be get guaranteed in, the results shown in Fig. 1. Figure 2 shows, the single DMC controller interference cannot achieve rapid response regulator, this strategy improve the system noise immunity effectively.

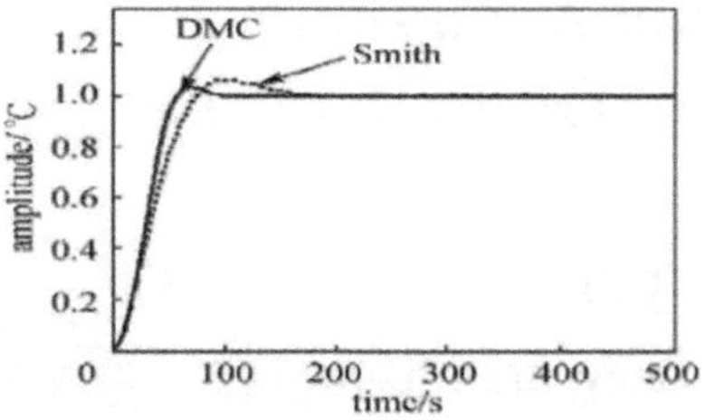

Fig. 1. Comparison of step responses

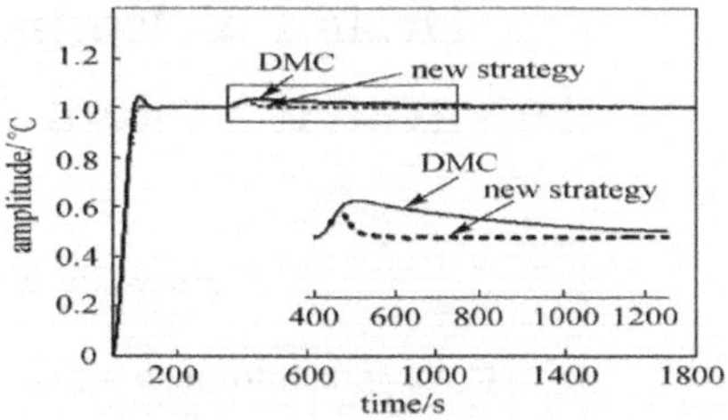

Fig. 2. Disturbance contrast

3 Conclusions

According to the characteristics of the controlled target, combination the time optimal and dynamic matrix control based on dynamic matrix control step control idea, can form a fast disturbance rejection predictive control strategy, through simulation on the Matlab and test on the real device, the strategy has good anti-jamming capability, the outlet temperature of hot-blast stove has fine control effect.

Acknowledgement. The work was supported by the Hechi University Foundation (XJ2016ZD004), Guangxi teacher Capability enhancement project (2018KY0495), Hechi university Youth teacher Foundation (XJ2017QN08) and was supported by the Projection of Master Foundation (2017HJA001, 2018LG004).

References

1. Ashoori A, Moshiri B, Khaki-Sedigh A, Bakhtiari MR (2009) Optimal control of nonlinear fed-batch fermentation process using model predictive approach. J Process Control 19:1162–1173
2. Jun D, Yongmao X (2004) Single neuron adaptive PID controller and its application. Control Eng China 11:27–30
3. XiuZhilong SH (2005) Control of temperature in biological ferment process. Control Eng china 12:30–33
4. Sharu Z (2008) Study on temperature fuzzy control in ferment process. J Tianjin Polytech Univ 27:70–73
5. Hisbullah, Hussain MA, Ramachandran KB (2003) Design of a fuzzy logic controller for regulating substrate feed to fed-batch fermentation. Inst Chem Eng 81:138–146
6. Saffer II DR, Castro JJ, Doyle III FJ (2005) A variable time delay compensator for multivariable liner processes. J Process Control 15:215–222
7. Li G, Tang Y (2018) Study on properties of forged AZ61 magnesium alloy based on fuzzy PID temperature control. Hot Working Process 47(23):130–132+136
8. Li X, Liu Y, Zhu E, Ma X (2018) Design of intelligent temperature fuzzy control PID system. Inf Rec Mater 19(11):118–120
9. Sang J (2018) Design of temperature control system based on fuzzy PID controller. Sci Technol Innov 26:153–154
10. Ge Z, Wu L (2018) Study on temperature control strategy of boiler main steam based on fuzzy PID control. Heilongjiang Electr Power 40(04):358–361

Design of Embedded Data Acquisition Integrated System Based on SQLite Database

Jin Xing and Yonghui Zhang(✉)

College of Information Science and Technology, Hainan University, Haikou, China
916941919@qq.com

Abstract. The modern data acquisition system gradually develops in the direction of strong real-time, multi-parameter, high-precision, and large-capacity. For this reason, this article has designed a kind of data acquisition system with strong real-time and can be stored in the database. The controller in this system is designed based on the LPC1769 chip and realizes multi-module plug-and-play expansion. The computer program is achieved by C# language, supported by Iocomp Component Library to build the UI interface, and completes the storage and management of large amounts of data by using SQLite Database. Data is transmitted between the computer and the controller according to the serial protocol. In order to improve the accuracy, the system also provides the ADC self-check function. The embedded data acquisition system can collect data in real time by the controller, and can provide comprehensive data with acquisition time as in the test. The storage of the data can be performed by the computer program, both the dynamic and static curve of the data can be displayed and the analysis of the data can be completed.

Keywords: Component · Embedded system · Data acquisition integrated system · SQLite database · Communication protocol · ADC self-check

1 Introduction

Data acquisition is the basic means of obtaining information. Data acquisition technology is a comprehensive application technology based on sensor technology and microcomputer technology. With the wide application of computer technology in industrial monitoring, control and management, data acquisition systems play an important role in the process of industrial automation.

2 Overall System Design

The embedded data acquisition integrated system combines the basic composition and functional requirements of the data acquisition system [1]. The upper computer program communicates with the lower computer through a custom serial protocol. The system framework is shown in Fig. 1.

Z. Xu et al. (Eds.): CSIA 2019, AISC 928, pp. 420–428, 2020.
https://doi.org/10.1007/978-3-030-15235-2_63

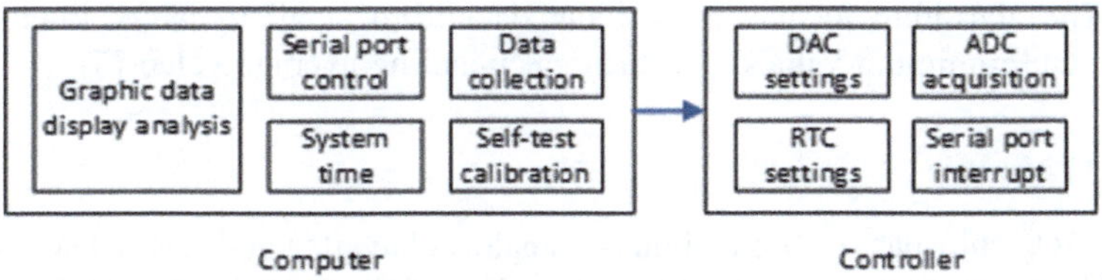

Fig. 1. System framework diagram.

3 Controller Design

The controller uses NXP's LPC1769 as the main control chip. Its operating speed can reach 120 MHz. It has rich on-chip peripherals such as RTC, DAC, UART, and interrupt controller NVIC. etc. In addition, it has large SRAM on-chip and Program ROM.

The hardware part adopts a modular design and consists of a main control board and three expansion modules. The expansion modules are: digital display module, data acquisition module and switch module. After the expansion module is inserted into the main control board, the module can be identified and has the function of the module, which is very convenient for plug and play [2].

3.1 RTC Module

LPC1769 chip provides RTC function, namely real-time clock function, which can realize the clock function with the appropriate frequency. This design uses a standard clock crystal with a frequency of 32.768 kHz. It synchronizes with the clock information sent by the host computer to provide accurate time stamp information for data acquisition [3].

3.2 ADC Module

This design uses two amplifiers to form the preamplifier circuit of the ADC module.

The role of the front-end circuit is to expand the range, and the calculation formula of the circuit is derived as follows:

In the formula, is the maximum voltage value allowed, the maximum acceptable voltage of the pin is 3.3 V, and the resistance is used for proportional calculation. According to the formula, the maximum value is approximately equal to 5 V, which means that the circuit can achieve the purpose of extended range. The specific use can still continue to expand according to the above formula, and can be verified to meet the circuit design [4].

In order to improve the sampling accuracy and data smoothness of the ADC, the system uses a median filtering method in the ADC data acquisition program. The principle of this method is to combine two different digital filters with different filtering functions: 'median value filter method' and 'arithmetic average filter method' to form a composite digital filter, or multi-level digital filter. This algorithm combines the advantages of the above two filtering methods. It can not only suppress random interference, but also filter out obvious impulse interference. The basic algorithm is as follows. The algorithm idea is to sort the data from small to large, filter out the maximum and minimum values, and then calculate the average value [5].

3.3 DAC Module

The LPC1769 chip has a 10-bit digital-to-analog converter and has a DMA control function. DMA control can make the conversion of the DAC without CPU control, thereby reducing the burden on the CPU and improving the overall conversion speed. The role of the DAC module is to provide a self-test function to the module to calibrate the ADC. The circuit design is as follows [6].

4 Serial Communication Protocol

The controller and the computer adopt serial communication. The specific communication contents include the data from the computer to the controller, the time synchronization data, the DAC calibration data, and the ADC data with the time stamp information from the controller to the computer. In order to avoid confusion, this article designed a set of communication protocols to judge. The idea of the protocol is to encapsulate the data with a fixed protocol header and protocol tail, and then elaborates on the content of the communication protocol from the lower controller to the computer [7].

First, we define the time structure, then define the data structure, and include the time structure in the data structure. In the program, the RTC time is assigned to the time structure 'Time 1', and the data sampled by the ADC is assigned to the data structure 'Data 1'. Finally, 'start' is used as the protocol header, and 'stop' is encapsulated as the protocol end. And send the data to the computer through the serial port.

5 Computer Design

The system needs to display the data collected by the controller intuitively and clearly, and realizes the timely data storage and calculation and analysis functions. Finally, the data is displayed in the computer in the form of a graphic, so the supporting computer software is designed to enable the user to get the measurement result from the upper computer directly.

5.1 Overall Design

Because this design focuses on applications, the interface strives to be elegant, concise and clear. After summarizing the various functions that need to be implemented, a reasonable layout of each function control is performed. According to functional design requirements, the computer software interface is divided into the following sections: serial port control section, data acquisition and database operating section, graphic data display and analysis section, ADC module self-test section, and system clock section. The function of each section is briefly described below [8].

5.1.1 Serial Port Control Section

The serial port control section uses a series of combo box controls to implement the setting of the serial port parameters of the computer, and uses the LED control to display the current serial port status.

5.1.2 Data Acquisition and Database Operating Section

This section consists of a character display table and five function buttons. The table uses the DataGridView control in the Winform framework to display uploaded data in real time, row by row. When there is a lot of data, a scroll bar will be automatically generated on the right side, and the data will be displayed in drop-down. The five function buttons are: Start Sampling, Stop Sampling, Clear Data, Export Data, Import Data. The first three buttons are data collection functions, because they are associated with the serial port control section, so only when the serial port is open, can the corresponding function be executed. For this reason, this design sets the prompt function. When the serial port is opened, the corresponding function can be performed. Otherwise the prompt interface will pop up. Other functions associated with the serial port also have similar prompts. The latter two buttons are associated with the function of the database. Clicking the Export Data button will call the data in the database and display it. The Import Data button will complete the data storage function.

5.1.3 Graphic Data Display and Analysis Section

This section includes a graphic display control and four function buttons. The four function buttons are: Static Analysis, Clear Display, Start Dynamic, and Stop Dynamic. Among them, the Static Analysis button implements the data extraction and the one-by-one display function for the static data in the database; the Start Dynamic button and the Stop Dynamic button perform the display control for the dynamic data being sampled; after the Clear Display button is pressed, the data curves displayed in the axis are deleted.

5.1.4 ADC Module Self-test Section

The system designed the DAC module function to perform self-calibration of the ADC module. The functional principle is to set the parameters of the built-in DAC module of LPC1769 to output different analog voltages, and connect the ADC of the controller to the DAC, and then read out the converted value of the ADC module, and finally perform comparison and calibration.

The computer provides two functions. One is a single DAC setting. The data comes from the TextBox control and is sent to the controller. For the TextBox control, its input is limited here. When the input is non-numeric or exceeds the preset range of the DAC, it will be discarded and given a prompt; Another function is the continuous DAC setting, which will start the second timer inside the computer, and then generate a random number between the input range (0–1023) of the DAC and send it to the controller.

5.1.5 System Clock Section

This part uses four digital controls to display the system time with a time synchronization button. The computer system time can be obtained in real time, and the year, month, hour, minute, and second are displayed in real time on the digital tube. The function of the Time Synchronization button is to read the time of the current host computer, and send the data package to the controller according to the agreement. The controller updates the RTC module after the data is parsed, and builds an accurate clock synchronized with the computer to provide time stamps for uploading comprehensive data [9].

5.2 Interface and Function Implementation

5.2.1 Implementation of Iocomp Control

The computer program interface design adopts the Winform framework. The advantage of this framework is that it is easy to set up a software form interface and has a variety of visual controls. Therefore, it is used during the programming of the computer software interface.

Due to the need for real-time graphics display and analysis of data, the design introduces the Iocomp external industrial control library. This control library is powerful and can provide all kinds of engineering controls with stable performance and rich features, and is designed specifically for real-time systems, so it is very suitable for the system application of this design.

5.2.2 Serial Port Control Implementation

The design idea of this part mainly refers to the implementation method of general serial port debugging assistant: the method of pulling down the combo box, setting parameters, and opening the serial port. The flow chart is shown below (Fig. 2).

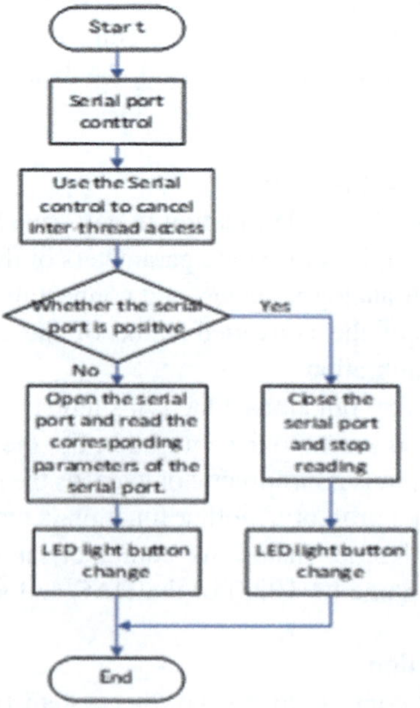

Fig. 2. Serial port control flow chart.

This uses the SerialPort control to implement and cancel cross-thread access restrictions, otherwise the program will report an error. The rest of the functions are shown in the flow chart. By judging the state of the serial port, it can be turned on and off, and the status of the LED can be indicated.

5.2.3 Data Acquisition and Database Operating Implementation

The data collection part of the flow chart is shown below (Fig. 3).

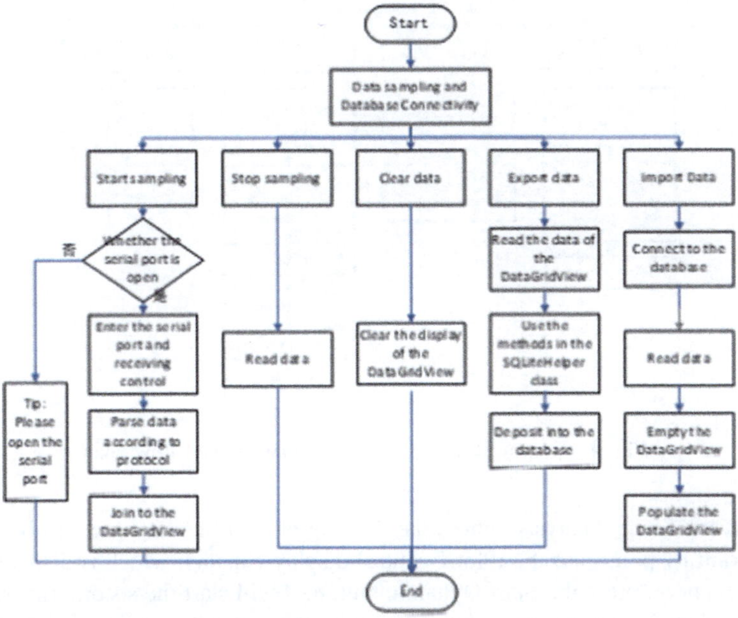

Fig. 3. Flow chart of data collection and database operation.

Press the Start Sampling button, it will first determine whether the serial port is open, 'NO' will pop up the 'Please open the serial port' prompt window, 'YES' will enter the receiving state, and the data from the controller is received in the serial port event response function of the computer, and then according to the agreement to analyze the data, the final analysis of the data will be added to the DataGridView control to be displayed; Stop Sampling button and Clear Data button, respectively, to terminate the data reception and clear the data displayed in the DataGridView control action; The function of Export Data button is to read the data in the DataGridView control and store the data in the database with the corresponding database processing function. Pressing the Import Data button reads the database, invokes the saved data in the database, and displays it after clearing the existing data in the DataGridView control.

5.2.4 Graphic Data Display and Analysis Implementation

The function of the graphical data display is implemented by the iPlotX control in the Iocomp control library. The control can process the data curve in real time, provide the function of saving and displaying, and can set the color curve, line shape, line width, coordinates, form, and horizontal axis of the data curve. The process of this part is shown in the figure below (Fig. 4).

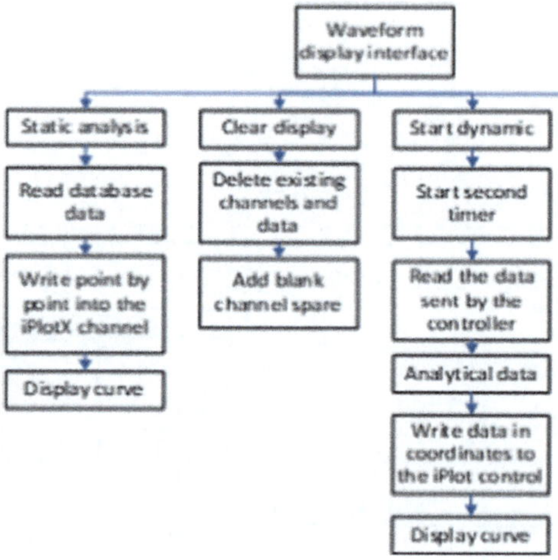

Fig. 4. Graphical data display and analysis part flow chart.

Press the Static Analysis button, the existing data in the database will be read, and wrote point by point into the iPlotX control display channel, which is displayed in the form of a curve; press the Start Dynamic button, it will start the second timer, and one second is the time cycle to read the data sent by the controller, and then parses the read data according to the protocol and writes it in the channel of the iPlotX control to complete the display; press the Clear Display button to delete the existing curve; press Stop Dynamic button, the timer will be turned off and stop receiving data [10].

5.2.5 SQLite Database Design

The purpose of this design is to collect data for small devices and make the computer work in stand-alone mode, so it is not suitable to configure the server separately. Therefore, the lightweight, green, open-source, lightweight, embedded database, SQLite database, which without external dependencies has become the best choice.

SQLite official provides API class library based on ADO.NET technology. This class library provides a large number of API functions for the SQLite database processing. It is more stable and convenient when using C# language for SQLite operation and programming.

In order to further facilitate the development, the design has also undergone secondary development. The corresponding SQLiteHelper helper class has been written, and the functions provided by the API class library have been encapsulated.

This class provides a static string variable that holds the database name. SQLite database is a file type database, which exists in the form of files in the application directory. To access the database, you need to assign value to the variable.

In addition, the helper class also provides the following static functions: ExcuteNoQuery, ExcuteQuery, ExcuteScalar, and ExcuteTable. Their functions are shown in the following table. The first three functions use the SQLiteDataReader class, and the last function uses the SQLiterDataAdatper class, which is extremely convenient for programming needs.

Access to the database is mainly to add, delete, modify, check the four operations, as an important component of the C# application program to access the database ADO. NET provides a complete implementation of the program, its ideas are divided into the following three steps:

1. use SQLiteConnection object to connect to the database;
2. establish SQLiteCommand object responsible for the execution of the database operation statement and the call of the stored procedure;
3. operate the returned result after the stored procedure is executed.

6 Conclusion

This design uses SQLite database, combines the computer software and the controller to form a complete embedded data acquisition integrated system, and the sampled data can be viewed on the graphic data display interface of the computer. Another powerful way to view is to use SQLiteExpert software to view and analyze data.

Although the data volume of the embedded system does not belong to a very large level, it also has a certain scale, and there is a need for saving and quick searching. Using the SQLite embedded database can quickly and easily meet this requirement. The system has good test results and can implement data acquisition functions. It can be applied to embedded systems with similar data requirements. It is hoped that this design can provide a reference for a large number of embedded developers.

Acknowledgments. This work was supported by CERNET Innovation Project and Innovative research topics for graduate students in Hainan Province.

References

1. Cai Z, Tan Z (2018) Design of automatic monitoring data acquisition system for IoT intelligent wireless nodes. Modern Electr Tech 41(04):183–186
2. Yang L, Liu M, Gao H, Dang C (2017) Design of high-speed data acquisition system based on FPGA and ARM. J Shanxi Datong Univ (Natural Sci Ed) 33(06):32–36+55
3. Han L, Wei M (2017) Application of SQLite embedded database in automated production workshop. Comput Era 2017(12):42–45
4. Dai L, Li Z (2017) Embedded ethernet data acquisition and control system design. Comput Meas Control 25(10):85–88

5. Yan M (2017) The design of embedded data acquisition system based on μCOSII. Electr Test 2017(07):14–15+10
6. Zhu L, Song C (2017) Comprehensive experiment design of FIR digital filtering based on DSP. Exp Technol Manag 34(01):206–209+213
7. Yan J (2016) Research on energy consumption optimization method of embedded data acquisition system based on mach. Mod Electron Tech 39(15):149–151+155
8. Xiong H, Guo X, Xie W (2014) Biodiesel remote monitoring system design based on IOT. In: International conference on human centered computing. Springer, Cham, pp 750–756
9. Liao J, Si LU (2017) Design and implementation of intelligent gateway based on android for agricultural IoT. J Agric Sci Technol
10. Du K, Chu J, Sun Z et al (2016) Design and implementation of monitoring system for agricultural environment based on WebGIS with Internet of Things. Trans Chin Soc Agric Eng

The Analysis of Structure and Freedom for Space Position Adjustment Platform

Shuguang Zhang(✉), Chenbing Guo, and Mingchen Wu

The 713th Research Institute of China Shipbuilding Industry Corporation, Zhengzhou, China
3237610382@qq.com

Abstract. In order to satisfy the space 6 degree of freedom attitude adjustment of the module during the precision installation process, a 6 degree of freedom position adjustment platform for the space 6 degree of freedom is designed. According to the technical requirements of precision installation, the six position adjustment functions required to be realized are solved, and the mechanical structure of the horizontal adjustment mechanism and the plane adjustment mechanism is put forward. The geometric mapping method is used to intuitively describe their respective working modes. To lay the foundation for kinematics analysis.

Keywords: Six degrees of freedom · Attitude adjustment · Working modalities · Clever

1 Introduction

The space 6 DOF position adjustment platform is used to accurately adjust the LRU optical module space position, so that its space position parameters meet the requirements of the installation to complete the final installation task [1–4]. The space position adjustment platform consists of two kinds of low-degree of freedom parallel mechanisms, the horizontal adjustment mechanism and the plane adjustment mechanism, and completes the adjustment work of six degrees of freedom for the LRU optical module space [5, 6]. The horizontal adjustment mechanism can realize the movement of 3 degree of freedom along the Z axis and the rotation around the X axis and the Y axis [7]; The plane adjustment mechanism can adjust the motion of the LRU optical module along the X axis, the Y axis, and the rotation around the Z axis [8]. Firstly, the function of the 6 DOF position adjustment platform is analyzed [9]. Then, carry out mechanical structure design [10]; Finally, screw theory is used to solve the freedom of horizontal adjustment mechanism and surface adjustment mechanism.

2 Functional Analysis of Spatial Position Adjustment Platform

The overall scheme design of mechanical system is the most critical part in the whole process of mechanical product design, because it directly determines the performance and quality of mechanical products. Through analyzing the work flow of the design

Z. Xu et al. (Eds.): CSIA 2019, AISC 928, pp. 429–436, 2020.
https://doi.org/10.1007/978-3-030-15235-2_64

object, the functional requirements and performance indicators that it needs to achieve are clearly defined, and then the functional requirements are decomposed, the working principles for each function are determined, and the overall design plan is formulated to serve as a guide for the next design work. The process of setting up LRU optical modules will now be carefully analyzed, which will be very useful for clarifying the function of the spatial position adjustment platform.

The installation process of the LRU optical module is as follows: (1) After rigorous functional testing in the OAB, the LRU optical module is placed on the space position adjustment platform in the clean holding box. The clean keeping box ensures that the LRU optical module is not contaminated by the external environment. They will be transported together by an automatic guide vehicle. (2) The automatic guide vehicle transports the clean retention box loaded with the LRU optical module to the designated installation location in the laser hall to achieve the rough positioning of the LRU optical module. (3) The machine vision system is accurately measured, feedback to the position parameters that need to be adjusted for the space position adjustment platform, and the position adjustment platform adjusts the six degrees of freedom position parameters of the LRU optical module space to achieve the precision requirements for precision mounting. (4) The LRU optical module is then sent to the designated installation location by the lifting mechanism to complete the installation task.

Through the analysis of the installation process of LRU optical module, we can determine the functional requirements of the spatial position adjustment platform: that is, it can accurately adjust the spatial position and attitude of LRU optical module in real time. The adjustment of position and attitude includes 3 displacement components and 3 rotation components in space, and a total of 6 degree of freedom adjustments. Its technical indicators are shown in Table 1 below.

Table 1. Technical specifications of spatial pose & orientation adjusting platform

	X moving	Y moving	Z moving	X turning	Y turning	Z turning
Travel requirements	−50–50 mm	−50–50 mm	−50–50 mm	−12°–12°	−12°–12°	−15°–15°
Precision requirements	0.1 mm	0.1 mm	0.1 mm	0.1°	0.1°	0.1°

3 Structural Design of Spatial Position Adjustment Platform

With the continuous improvement of the industrial automation level, people have put forward more and more high requirements for the performance of the mechanism such as movement performance, bearing capacity, work space, and dynamic response. Structural design based on a single basic mechanism has great limitations, and it is difficult to achieve complex and strict engineering technical standards. At this point, it is an important means for institutional innovation design to organize basic institutions in an orderly manner.

The mechanism combination is to form an ideal mechanism by organizing certain types of basic mechanism configurations according to certain logical laws. Its purpose is to improve the shortcomings of the basic mechanism configuration and better realize the special functions in engineering practice. The space 6 DOF position adjustment platform is a new type of parallel mechanism composed of two kinds of parallel mechanisms with less DOF. The horizontal adjustment mechanism and the surface adjustment mechanism cooperate with each other to complete the adjustment of the six degrees of freedom of the LRU optical module space. The three-dimensional structure diagram is shown in Fig. 1.

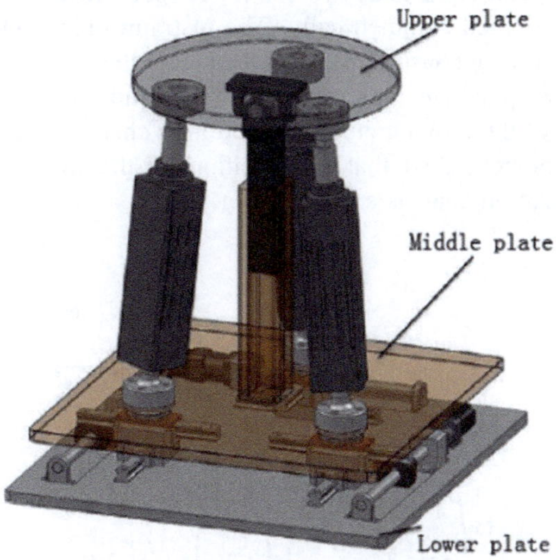

Fig. 1. The spatial pose & orientation adjusting platform

String and blending mechanism covers the dual advantages of tandem mechanism and parallel mechanism, and avoids weaknesses. It has the advantages of large working space and flexibility in tandem mechanism. It also inherits the advantages of strong bearing capacity, good rigidity, and small error. The overall structure of the space 6-DOF position adjustment platform is formed by the orderly connection of the upper support board, the middle support board and the lower support board through 6 motion branches and 1 constraint branch. Among them, the medium tray is a moving platform in the plane adjustment mechanism and is used as a fixed platform in the horizontal adjustment mechanism.

The lower tray acts as a sports base and is connected to the middle tray through three PPR motion branches to form a plane adjustment mechanism. The rotation around the Z axis and the movement along the X axis and the Y axis can be achieved with a total of 3 degree of freedom. In the horizontal adjustment mechanism, the medium support plate serves as a fixed platform and is connected to the upper support plate through 3 SPS motion branches and 1 PU constraint branch chain. The adjustment of 3 degree of freedom can be achieved along the Z axis and the rotation around the X axis and the Y axis.

3.1 Structural Design of Horizontal Adjustment Mechanism

From the institutional point of view, the horizontal adjustment mechanism can also be called a 3-SPS/PU parallel mechanism. Its fixed platform(medium tray) and moving platform(upper tray) are connected by three SPS motion branches and one PU constraint branch.

Each SPS movement branch has two ball hinges at both ends, with a moving pair in the middle and distributed symmetrically. The restraint branch chain is composed of Hooke hinges and moving constraints and is located in the geometric center of the fixed platform and moving platform. The servo electric cylinder acts as a drive to drive the relative movement of the moving auxiliary, and then changes the length of the rods of the three motion branches, so that the position of the moving platform in space changes. Its organization chart is shown in Fig. 2.

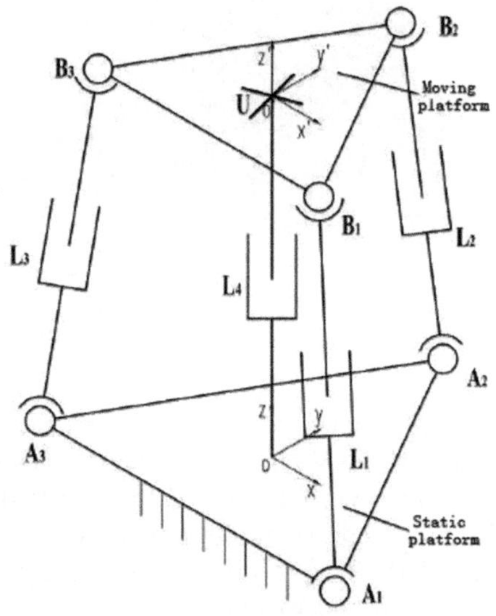

Fig. 2. Schematic diagram of 3-SPS/PU parallel mechanism

Through the linkage adjustment of the expansion of three servo electric cylinders, the length of the rods of the three motion branches can be changed, so that the moving platform of the 3-SPS/PU horizontal adjustment mechanism presents different spatial positions. There are a total of 5 typical working modes:

(1) When the electric cylinder No. 2 is elongated and the electric cylinders No. 1 and No. 3 do not work or shorten the same distance, the moving platform rotates around the X axis, as shown in Fig. 3B below; (2) When the No. 3 electric cylinder is elongated, the No. 1 and No. 2 electric cylinders shorten a certain distance according to a certain rule, and the moving platform rotates around the Y axis, as shown in Fig. 3C below; (3) When the electric cylinders No. 1, No. 2, and No. 3 work together according to a certain rule, the moving platform makes a coupling motion around the X and Y axes, as shown in Fig. 3D below; (4) When the electric cylinders No. 1, No. 2, and No. 3 extend at the same distance, the moving platform moves along the Z axis for translation, as shown in Fig. 3E below; (5) When the electric cylinders No. 1, No. 2, and No. 3 work together according to a certain rule, the moving platform makes a composite movement, as shown in the Fig. 3F below.

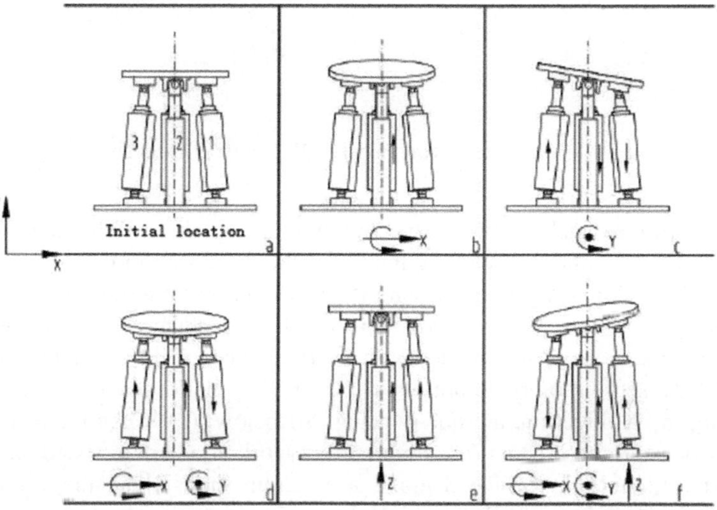

Fig. 3. 5 working conditions of 3-SPS/PU parallel mechanism

3.2 Structural Design of Planar Adjustment Mechanism

The moving platform(medium tray) and the fixed platform(lower tray) of the plane adjustment mechanism are connected by three identical PPR motion branches and are therefore also called 3-PPR parallel mechanisms. Cross guide rails form two moving pairs and can achieve relative movement perpendicular to each other. Its lower rail is fixed to the fixed platform, and the upper rail is equipped with a precision ball shaft ring through the connecting part, and then it is combined with the shaft hole of the moving platform to form a rotating pair. The servo motor acts as a driver, axially connecting the planetary gear reducer and the ball wire bar nut pair to make the sliding block output connected to the nut pair move relatively. When the relative position of the three sliders changes, the spatial position of the moving platform will also change. The organization chart is shown in Fig. 4 below.

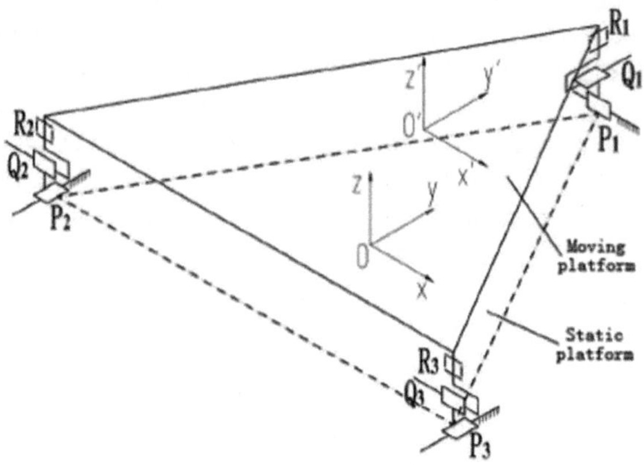

Fig. 4. Schematic diagram of 3-PPR parallel mechanism

Through further analysis of the specific structure of the 3-PPR plane adjustment mechanism and the different working modes of the three servo motors(X servo motor driving P1 slider, Y1 servo motor driving P2 slider, Y2 servo motor driving P3 slider), Five working conditions of the planar adjustment mechanism can be obtained:

(1) When the X servo motor rotates at a certain speed or in a certain direction and the Y1 and Y2 servo motors do not work, the moving platform will move at a certain speed along the X direction, as shown in Fig. 5B below; (2) When the X servo motor does not work and the Y1 and Y2 servo motors work at a certain speed and the same steering, the moving platform will move at a certain speed along the Y direction, as shown in Fig. 5C below; (3) When the X servo motor rotates at a certain speed and a certain direction, and the Y1 and Y2 servo motors work at a certain speed and the same turn, the moving platform will move at a certain speed along a certain direction in the XOY plane. As shown in Fig. 5D below; (4) When the X servo motor rotates at a certain speed and in a certain direction, when the Y1 and Y2 servo motors work at a certain speed and turn differently, the moving platform will rotate around the Z axis, as shown in Fig. 5E below; (5) When the X servo motor, Y1, and Y2 servo motors work together at the same time, the moving platform performs a composite motion in the XOY plane, as shown in the Fig. 5F below.

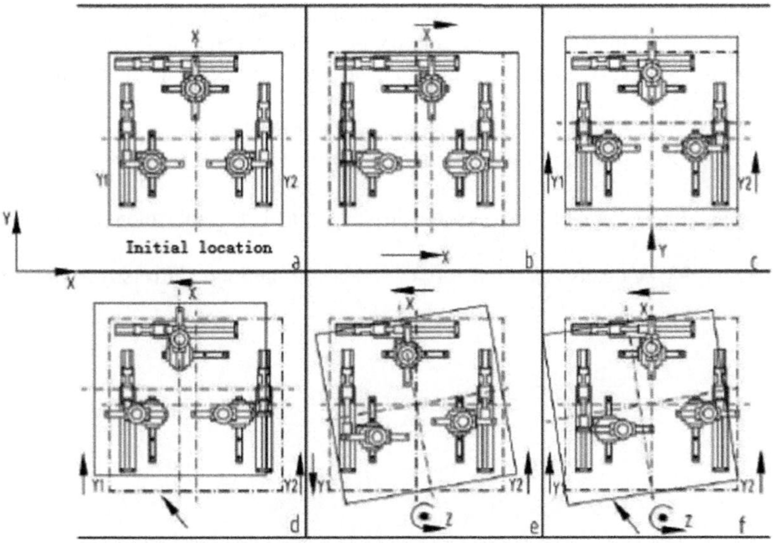

Fig. 5. Working conditions of 3-PPR parallel mechanism

4 Conclusion

Firstly, this paper analyzes the installation flow of the LRU optical module, and clarifies the function of the 6 DOF position adjustment platform to realize the 6 DOF motion of the LRU optical module. On the basis of determining the use of function and performance indexes, the advantages and disadvantages of the string and parallel robot are analyzed. A string and hybrid mechanism is proposed to realize the motion of six degrees of freedom in space. Then, the structure of the 3-SPS/PU horizontal adjustment mechanism and the 3-PPR planar adjustment mechanism are designed, and the five working modes are obtained, which lays the foundation for the next step of kinematics analysis.

References

1. Bonanno RE (2004) Assembling and installing LRUs for NIF. Proc SPIE-Int Soc Opt Eng 5341
2. Egner S (1996) Semi-numerical solution to 6/6-Stewart-platform kinematics based on symmetry. Appl Algebr Eng Commun Comput 7(6):449–468
3. Moses EI, Campbell JH, Stolz CJ et al (2003) The national ignition facility: the world's largest optics and laser system. Proc SPIE-Int Soc Opt Eng 5001:1–15
4. Liu, XJ, Wang JS, Gao F (2000) On the optimum design of planar 3-DOF parallel manipulators with respect to the workspace. In: IEEE international conference on robotics & automation, vol 4, pp 4122–4127
5. Hunt KH (1990) Kinematic geometry of mechanisms. Clarendon Press, Oxford

6. Guo XJ, Chang FQ, Zhu SJ (2004) Acceleration and dexterity performance indices for 6-DOF and lower-mobility parallel mechanism. In: ASME 2004 international design engineering technical conferences and computers and information in engineering conference, pp 251–255
7. Gough VE, Whitehall SG (1962) Universal tyre test machine. Proc Fisita Int Tech Cong
8. Stewart D (1965) A platform with six degrees of freedom. Proc Inst Mech Eng 180(1):371–386
9. Merlet JP (2006) Parallel robots (solid mechanics and its applications). Springer-Verlag, New York
10. Ball RS (1998) A treatise on the theory of screws. Cambridge University Press, Cambridge

Application of Support Vector Machine Based on Particle Swarm Optimization in Short-Term Load Forecasting of Honghe Power Network

Jing Hua, Wei Xiong(✉), Lin Niu, and Linlin Cao

Engineering College, Honghe University, Mengzi 610000, Yun-Nan, China
459353359@qq.com

Abstract. In order to predict short-term load accurately and effectively, a short-term load forecasting model (PSO-SVM) based on particle swarm optimization (PSO) and support vector machine (SVM) is proposed. The parameters of the support vector machine are regarded as the velocity and position of a particle, and the optimal support vector machine parameters are found through continuous updating of the speed and position of the example. It can overcome support vector machine algorithm's shortcoming. The model of short-term load forecasting of the Red River power grid is established according to the optimal parameters, and the model performance is simulated. Try. The simulation results show that, compared with the SVM prediction model, PSO-SVM not only speeds up the optimization speed of SVM, but also improves the precision of load forecasting, and is more suitable for the need of short-term load forecasting in regional power grid.

Keywords: Short term load forecasting · Support vector machine · Particle swarm optimization algorithm

Power load forecasting is the basis of the economic operation of the power system. Accuracy of load forecasting will directly affect quality of the power grid planning and construction. Therefore, how to use the existing data scientifically and rationally and correctly apply various load forecasting methods is an important problem for the study of the grid planning scholars and the staff of the power grid planning and design [1].

In recent years, the local economy of Honghe state has developed rapidly and the demand for electricity consumption is increasing. The power supply of Honghe Power Grid is closely related to the local economy of Honghe state.

1 Regional Load Analysis

There are twelve counties and cities in Honghe state. Because of the differences in geographical location, economic development is unbalanced, and the small hydropower resources in some cities are very rich. Some cities contain large loads with frequent load fluctuations, such as iron and steel plants, and some cities also contain high iron and electric iron impact negative load. With the conventional load forecasting method,

Z. Xu et al. (Eds.): CSIA 2019, AISC 928, pp. 437–443, 2020.
https://doi.org/10.1007/978-3-030-15235-2_65

it is difficult to make the prediction precision reach the ideal level [2]. The traditional load forecasting method is difficult to meet the requirement of prediction accuracy, while the support vector machine theory is based on the statistical learning theory, and the original structure risk minimization is used, and there is no data dimension limit. The multi group genetic algorithm is used to optimize the support vector machine, which can dynamically control the mutual conversion process between local search and global search. It has better convergence and functional ability, and can improve the accuracy of prediction and meet the precision of short-term load forecasting. Practical application shows that this method can achieve good results in load forecasting accuracy of Honghe Power Network [3].

2 Load Forecasting Based on Improved SVM

2.1 SVM Model

Set training set sample

$$f(x_i, y_i), i = 1, 2, \cdots, n, x \in R^a, y \in R \tag{1}$$

In the high dimensional space, the linear regression function is:

$$f(x) = \omega x + b \tag{2}$$

The insensitive loss function of all training data is less than the error.

$$\min_{\omega} \phi(\omega) = \frac{1}{2}\|\omega\|^2 \tag{3}$$

$$y_i - \omega \cdot x_i - b \le \varepsilon, i = 1, 2, \cdots n \tag{4}$$

Into the relaxation variable, (2), (3) change into

$$\left\{\begin{array}{l} \min \phi(\omega, \xi) = \frac{1}{2}\|\omega\|^2 + c\sum_{i=1}^{n}(\xi_i + \xi_i^*) \\ y_i - \omega x_i - b \le \varepsilon + \xi_i \\ \omega \cdot x_i + b - y_i \le \varepsilon + \xi_i^* \\ \xi_i \cdot \xi_i^* \ge 0 \end{array}\right\} \tag{5}$$

In formula (5), C is the penalty coefficient. By constructing the Lagrange function, the duality function can be obtained as follows [4]:

$$\min \frac{1}{2}\sum_{i=1}^{l}\sum_{j=1}^{l}(\alpha_i - \alpha_i^*)\left(\alpha_j - \alpha_j^*\right)(x_i^T * x_j) + \sum_{i=1}^{l} y_i(\alpha_i - \alpha_i^*) - \varepsilon\sum_{i=1}^{l}(\alpha_i + \alpha_i^*) \tag{6}$$

The constraints are as follows:

$$\sum_{i}^{n} \left(\alpha_i - \alpha_i^*\right) = 0, \alpha_i, \alpha_i^* \in [0, c] \tag{7}$$

The nonlinear regression problem is mapped to high-dimensional space, and radial basis function is selected as kernel function.

$$k(x, x_i) = \exp\left(-\|x - x_i\|^2/(2\mu^2)\right) \tag{8}$$

Bring the kernel function into the inner product of (8) form, and get the regression function of SVM.

$$f(x) = \sum_{i}^{n} \left(\alpha_i - \alpha_i^*\right) K(x_i \cdot x) + b \tag{9}$$

The research shows that compared with other types of kernel functions, the RBF function has fewer parameters and better performance than the other types of kernel functions. Therefore, the radial basis function is used to model SVM, which is defined as follows [5]:

$$K(x_i, x_j) = \exp\left[\frac{\|x_i - x_j\|^2}{2\sigma^2}\right] \tag{10}$$

The size of SVM parameters determines the accuracy of prediction and classification accuracy. Therefore, selecting the optimal parameters of SVM is a key problem. In this paper, particle swarm optimization (PSO) is used to optimize SVM parameters.

2.2 Particle Swarm Optimization Algorithm

Particle swarm optimization (PSO) is a group based stochastic optimization technique, which is inspired by the simulation study of bird population aggregation [6, 7]. PSO algorithm can describe every possible rout as a particle, the speed of all the particles flying at a certain level are described by the position and speed of each particle. The optimal positions of individual particles and the whole population are pbest and gbest respectively, and the updating formula of velocity and position of particles is:

$$v_{i,d}^{k+1} = \omega \cdot v_{i,d}^{k} + c_1 rand() \cdot \left(pbest_{i,d}^{k} - x_{i,d}^{k}\right) + c_2 rand() \cdot \left(gbest_{d}^{k} - x_{i,d}^{k}\right) \tag{11}$$

$$x_{i,d}^{k+1} = x_{i,d}^{k} + v_{i,d}^{k+1} \tag{12}$$

In the formula, the weight of the inertia is the learning factor and the random number between the particles is the velocity and position of the D dimension of the particle in the K iteration, and the position of the individual extreme value of the particle I in the D dimension, and the position of the group in the global extremum of the D dimension.

From (11) (12), the current position of the particle is determined by the position of the previous moment and the current speed, and the current velocity is determined by the velocity of the previous moment, the individual extremum and the global extremum. In the calculation, the local optimal solution can be escaped by changing the individual extremum indirectly and changing the global extremum directly, and the local optimal solution can be searched in other regions [8, 9]. Finally, the global optimal solution can be found.

2.3 Optimization SVM Parameters by PSO Algorithm

Short-term load forecasting accuracy based on SVM and the parameters of SVM are closely related. In this paper, PSO algorithm are used to select above parameters of SVM and to establish the optimal model of power load forecasting. The following steps are as follows:

(1) data collection: collecting short-term load data and forming SVM learning samples;
(2) determine the range of values and set the relevant parameters in the PSO algorithm.
(3) initialize the velocity and position of particles in particle swarm according to the range of parameters [10, 11].
(4) refer to the training data, the fitness value can be get from the mean variance, train the model through the cross validation method, obtain each particle's fitness value and make it in the minimum, it also can obtain the initial global extreme value gbest.
(5) Update velocity and position of particles according to formula (11) and (12).
(6) Calculate each particle's fitness value in the new position. If the fitness value of the particle is better than the fitness value, it is updated to a new position. If the fitness of the particle is better than that, it is updated to a new position [12, 13].
(7) Judge the optimal position of the particle's fitness value whether in the maximum value of goal value or not or the number of iterations. If the stopping condition is satisfied, the optimal parameter combination of the SVM is output.
(8) According to the parameters which combined of optimum, the model of a short-term load forecasting based on SVM is established. If we do not meet the return step (4), we will carry out the training again.

3 An Example Analysis

The data used in this paper are load data of a certain day in Honghe Power Grid.

We choose 12 h data as training set, train the model and established the short-term load forecasting model, than choose latter 12 h data as test set to verify model prediction effect, after verify, we should compared prediction value with actual value and then apply the mean relative error evaluation of the mean square error of the model.

The formula is as follows:

$$MAPE = \frac{1}{N}\sum_{i=1}^{N}\left|\frac{x(i) - x'(i)}{x(i)}\right|$$
$$MSE = \frac{1}{N}\sum_{i=1}^{N}(x(t) - x'(t))^2 \tag{13}$$

The experiment is implemented on the MatlabR2013a platform using the Libsvm 2.98 toolbox, and the minimum error of 10-fold cross validation is used as the criterion of SVM parameter selection. The SVM parameter search interval is set to:

$$c \in [1, 100], \mu \in [0.001, 0.1], \sigma \in [0.001, 100]$$

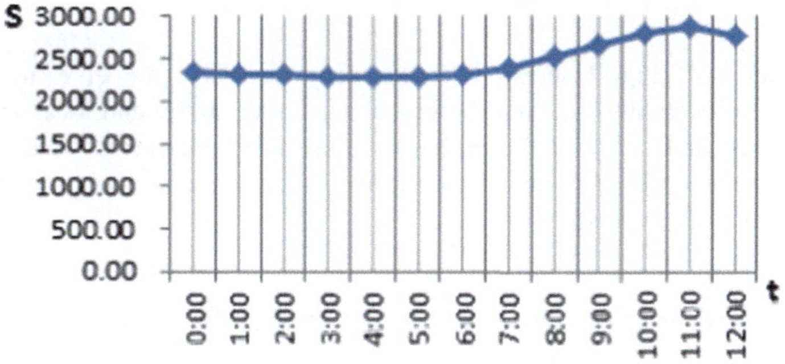

Fig. 1. Load data for the first 12 h

Convergence speed is improved and shorten the training time, the short-term load sample should be normalized, the maximum value is found in the original data, and then all data can be removed to the value less than 1, which is convenient to run in the Matlab software (Fig. 1).

$$y = \frac{x}{x_{\max}} \tag{14}$$

In the formula, for the ordinary load, for the maximum load.

The particle swarm size is 20, the maximum iteration number is 500, and the parameter optimization process is shown in Fig. 2.

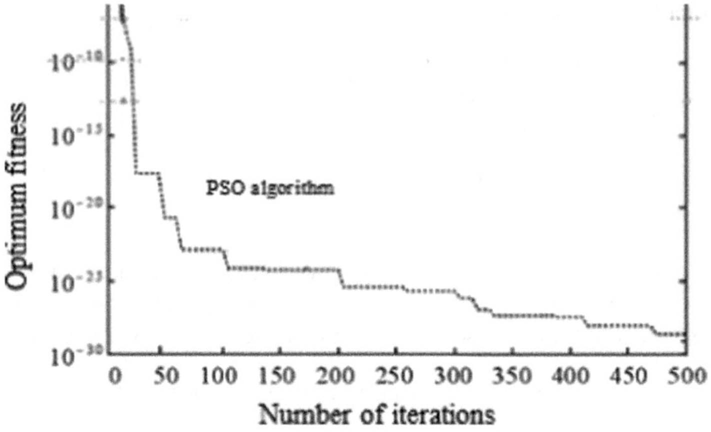

Fig. 2. Fitness curve of PSO

The analysis of Fig. 2 shows PSO not only maintains the diversity of particle swarm, but also takes into account the global searching ability and local search ability of the particles, and can find the global optimal SVM parameters faster and obtain higher learning precision. In Table 1 the mean square error and relative average error of the simulation results are shown.

Table 1. Comparison of short-term power load forecasting performance of different forecasting models

Forecasting models	MAPE	MSE
SVM	13.65	52.34
PSO	6.77	14.96

4 Conclusion

SVM in short-term load forecasting of power system, if the parameter selection is not reasonable, it will have a directive impact on the accuracy of load forecasting. In view of the shortcomings of the slow convergence rate of the SVM algorithm and the local extremes, particle swarm optimizations (PSO) are used to optimize the short-term power load forecasting of SVM. The simulation results show that the improved PSO-SVM short-term power load forecasting model has the advantages of stable output, good convergence and small prediction error. This prediction method has certain reference significance for the power management department and enterprise decision-making in the red and power grid.

Acknowledgements. Science Research Foundation of Yunnan Provincial Department of Education (2015Y453).

References

1. Lu N, Zhou J, He Y (2010) Application of particle swarm optimization neural network model in short term load forecasting. Power Syst Prot Control 38(12):65–68
2. Li J, Li Y, Li H et al (2011) Multi variable grey neural network model for non point source pollution load forecasting. J Northwest Agric For Univ (Nat Sci Ed) 29(3):229–234
3. Zhang Y, Yang Y (2009) Application of improved GA-BPN in short term power load forecasting. Comput Eng Appl 45(13):223–226
4. Wang J, Liu B, Feng Y (2008) Application of ant colony neural network in short term load forecasting. 29(7):1797–1799. 1837
5. Zeng M, Feng Y, Liu D (2008) Multi model combined electricity price forecasting based on evidence theory. Chin J Elect Eng 28(6):84–89
6. Huang W-Y, Tong M-M, Ren Z (2009) Nonlinear combination forecasting method of gas emission based on SVM. J China Univ Min Technol 38(2):234–239
7. Zhao Y, Gao J, Liu Y (2010) Fault diagnosis based on ant colony optimization neural network. J Beijing Inf Sci Technol Univ 25(2):45–48
8. Zhang L, Jiang Y, Dong L (2010) Crack opening ant colony neural network prediction model based on. Hydroelectric Power 36(5):31–33
9. Sheng R, Zheng H, Jing H (2010) Application of MMAS and rough set in fault diagnosis of bearing. Vibr Test Diagn 30(2):128–131
10. Yang J, Sun X, Che H et al (2009) Rolling force prediction of cold continuous rolling mill of neural network based on ant colony algorithm. Steel 44(3):52–55
11. Li P, Yan Y, Zheng W (2010) Short term load forecasting based on quantum neural network with evidence theory. Power Syst Prot Control 38(16):49–53
12. Li W, Zhang S (2010) Fault classification based on improved evidence theory and fusion of multiple neural networks. J Mech Eng 46(9):93–99
13. Yang J, Huang H (2008) Early fault diagnosis method for aero engine based on evidence theory. J Aerosp Power 23(12):2327–2331
14. Yang Z, Zhou J (2009) A turbine fault diagnosis method based on D-S evidence theory. Hydropower Energy Sci 27(2):152–154
15. Kang C, Xia Q, Liu M (2007) Power system load forecasting. China Electric Power Press pp 196–199

Research and Implementation of Security Vulnerability Detection in Application System of WEB Static Source Code Analysis Based on JAVA

Hui Yuan(✉), Lei Zheng, Liang Dong, Xiangli Peng, Yan Zhuang, and Guoru Deng

Information and Communication Company of Hubei State Grid Corporation, No. 197 Xu Dong Street, Wuhan, Hubei, China
882148@qq.com

Abstract. With the rapid development of Internet technology, Web applications are widely used in all walks of life, and their security requirements are increasing. Unfortunately, at present, the development of Web security technology still lags behind the development of Web application technology itself. The Web application itself and its operating environment are still relatively fragile, and its operating environment is easily forged or modified, making Web applications gradually become malicious. The object of the attack is frequently attacked. This paper investigates and analyzes the common vulnerabilities in web applications, deeply studies the basic characteristics of these vulnerabilities, and understands the principles and solutions of vulnerabilities. The static analysis method is used to analyze the vulnerabilities, and the static analysis methods are used to solve the security vulnerabilities in the Java web project.

Keywords: Static source code · Vulnerability detection · Pattern matching · Data stream analysis

1 Introduction

With the increasing number of Web applications, the code size of software has increased dramatically, which makes it more and more difficult for traditional manual testing to analyze the security vulnerabilities in the system, and the efficiency of manual testing is not high. Simple manual testing has not been able to keep up with the speed of software development. This article has conducted in-depth study on common vulnerabilities in Web applications. These vulnerabilities include some security vulnerabilities, such as SQL injection and XSS vulnerabilities. There are also vulnerabilities that affect system stability, such as resource shutdown and buffers. Problems such as overflow, and the various forms of vulnerability have been studied for their various forms of production and common solutions. In-depth study of static analysis related analysis methods, including fault-based pattern matching for character flow, abstract syntax tree, intermediate code analysis methods, and data flow analysis methods, after studying these methods to summarize this series of methods Advantages and

Z. Xu et al. (Eds.): CSIA 2019, AISC 928, pp. 444–452, 2020.
https://doi.org/10.1007/978-3-030-15235-2_66

disadvantages, in order to reduce the system's false negative rate, and finally came up with a comprehensive analysis method for vulnerability detection.

2 Common Vulnerabilities in Web Applications

2.1 SQL Injection Vulnerability Attack

SQL injection vulnerability attack refers to inserting malicious code into database operation statements through the external interface of the database to control the database and even the operating system. It is a common and harmful vulnerability in Web application systems.

The cause of the SQL injection vulnerability is that developers lack the knowledge of Web security or the experience of security programming. The code written is not perfect. The common problems are:

(1) Combine multiple SQL instructions by means of string concatenation, resulting in the same transmission channel for the data and control structure;
(2) The account permissions used by the application to connect to the database are too large;
(3) The user input has not been thoroughly verified and filtered;
(4) The database is open to functions with excessive permissions; and so on.

SQL injection vulnerabilities mainly include:

(1) GET type. The parameters in the URL link can be maliciously extended to become an injection vulnerability. (2) POST type. Injection vulnerabilities exist where the user interacts with the database, such as dialog boxes, search boxes, etc. entered by the user. User input may contain malicious code that is submitted to the background as part of the SQL statement. (3) Cookie injection type. If the server does not process and filter the data submitted by the browser and use the data directly for database interaction, a cookie-type injection vulnerability may occur.

2.2 XSS Attack

XSS attacks also result from insufficient filtering of user input by the Web application. Attackers use website vulnerabilities to inject malicious HTML code and client-side Java Script code into web pages. When users browse these pages, the browser will execute malicious code, steal user cookies, and conduct session hijacking or phishing spoofing., causing users to suffer losses.

The main reason for the XSS vulnerability is the same origin policy of the Java Script language and the Web browser. The Java Script language allows the Web server to send executable code to the browser, which poses a security risk to the Web application. The XSS vulnerability is that an attacker crosses the same-origin policy and uses a Java Script script to control the browser to obtain session cookies and so on for further attacks. XSS vulnerabilities are generally divided into the following categories:

(1) Reflective XSS

The principle of XSS attack is shown in Fig. 1. The attacker designs a URL link that contains malicious code, enticing the user to access it, and once the link is opened, the malicious code is executed. Since it can only be triggered once by a click, it is called a reflection type. Such attacks are usually hidden in the search bar of a web page, at the login portal, etc., to steal client cookies or phishing scams.

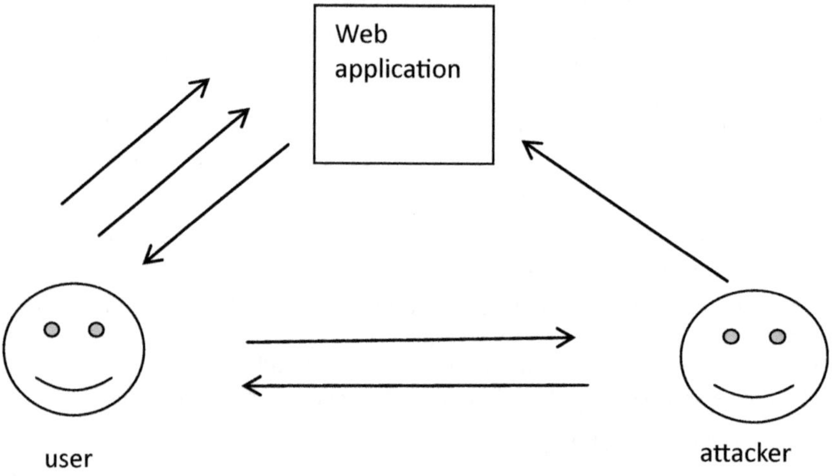

Fig. 1. Example of XSS attack principle

(2) Storage type XSS

Storage XSS, also known as persistent XSS, allows attackers to upload or store malicious code to the server in advance, hiding it in interactions such as comments, message boards, and logs. As long as the browser accesses the page containing the malicious code, the malicious code is executed without the user clicking on a specific link.

(3) DOM type

XSS's document object model (DOM) is a W3C organization's recommended process for extensibility.

The standard programming interface for markup languages, independent of browser, platform, and language. The DOM can be thought of as the output page of Java Script. The DOM-based XSS is also a vulnerability in the Java Script code, which is formed by modifying the DOM node of the page. It differs from reflective XSS and storage XSS in that: DOM type XSS code does not require direct participation of server parsing response, and browser-side DOM parsing can trigger XSS, and the attack occurs entirely on the client.

3 Vulnerability Monitoring Technology Classification

According to different analysis objects, vulnerability detection technology can be divided into two categories: source-based vulnerability detection technology and object code-based vulnerability detection technology.

For some software that can get source code (such as open source projects), you can use source-based vulnerability detection technology to find potential vulnerabilities by analyzing the source code. A typical example is vulnerability detection for the system. However, the source code of most software, especially commercial software, is difficult to obtain, so source-based vulnerability detection technology cannot be used, and only object-based vulnerability detection technology can be used. Security vulnerabilities appear in dynamic page programs. Dynamic pages are difficult to obtain source code or only partially source code. Therefore, vulnerability detection technology can be classified as vulnerability detection technology based on object code.

Vulnerability detection based on object code involves many aspects such as compiler, instruction system, and executable file format. It can be divided into white box analysis, black box analysis and gray box analysis.

White box analysis uses reverse engineering to convert the target program to binary code or to restore part of the source code. However, in general, it is difficult to completely convert the target program into readable source code. Especially when the original author uses the scrambling and encryption measures, it is very difficult to use white box analysis.

Box analysis is a method of controlling the input of a program and observing the output without reverse engineering the target program itself. It can aggregate some context-sensitive and meaningful code, reduce its complexity, and finally analyze the function module to determine whether there is a vulnerability. However, the process of black box analysis requires the analyst to have a high level of technology, otherwise it is difficult to find exploitable vulnerabilities in a short period of time.

The combination of white box analysis and black box analysis is gray box analysis. This method can reduce the difficulty compared with white box analysis, and the relative black box analysis can improve the analysis hit rate and analysis quality.

4 JAVA-Based WEB Static Source Code Analysis Vulnerability Detection System Design

The implementation of this paper divides the whole system into three layers. The first layer is the interaction layer, which can also be called the view layer. The second layer is the logic control layer, which is mainly responsible for the logic control of the whole system. The third layer is the vulnerability detection layer. The system is divided into three layers. The first one is in line with the MVC development model, which better matches the Spring MVC framework. The second is to make the system modular and convenient to implement. Figure 2 is a diagram showing the overall architecture of the system.

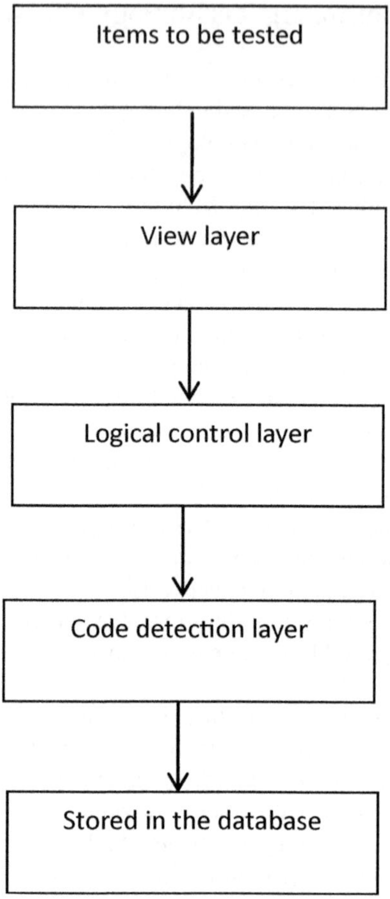

Fig. 2. System overall design architecture diagram

(1) View layer: The main task of this layer is to provide an interactive interface for the inspectors. The functions implemented include submitting the items to be tested and viewing the project analysis results. Here, JSP+CSS is used to implement the specific functions. JSP (Java Server) is the Java server page, a main dynamic web page technology standard, which is based on the traditional HTML static web page. CSS is a cascading style sheet, which is used to format the text style of the JSP page.
(2) Logical control layer: The main function of this layer is responsible for the logical control management of the whole system, that is, responsible for the forwarding of business logic, and is not responsible for the specific processing of business logic. The advantage of this is that it can separate the logic from the business. For

the system, the business is code detection. The detection includes multiple methods and preprocessing stages and subsequent report analysis stages. The logical relationship between these stages is controlled by this layer, when the underlying code scanning and report analysis and other code occur. When changing, it is not necessary to modify the code of the logic layer. This is also the application of MVC. This layer corresponds to the controller layer of Spring MVC.

(3) Code detection layer: This layer is the focus of this system. The code detection layer is composed of task trigger, code preprocessing module and analysis modules of four different processes. This layer corresponds to Spring MVC, which is Service layer, mainly business. The implementation details of the logic. The task trigger is an intermediate control module between the logic controller and the code analysis module. The function implemented by this module is that after submitting the project to be tested, the logic controller hands the authority to the task trigger, and the task trigger goes. Trigger the real code detection module, the task trigger controls when to start the detection of four different modules, wait for the report to complete the summary report and other column functions, simply, the four detection modules will start the thread pool respectively. The function of the thread in the thread, when to start these threads, the return value of the thread result is the task trigger.

The code preprocessing module is to unify the submitted code, including the coding style of the unified code and the compilation process. The confidence level proposed in the previous section of this paper is designed as follows: due to different analysis methods for the rules the definition is different, but different analysis methods analyze the same vulnerability to increase the confidence in the report. Therefore, the source code is formatted first, and the file path+file name+code line is used to locate the specific the code vulnerability is used as the source of the vulnerability. The location of the different analysis methods is the same. If this vulnerability is detected, this can increase the confidence.

The implementation of the code detection module is based on some open source detection tools, lexical analysis using JAVACC, grammar analysis by means of PMD, intermediate code analysis and data flow analysis based on: FindBugs, the process of implementation is first to write matching rules, corresponding to the matching rules to achieve the detector, Then the detector is added to the detection tool. When the task trigger triggers the detection task of the current module, an available thread is obtained in the thread pool, and then the script that executes the corresponding tool is called, and the thread is executed, that is, After the module detection is completed, the return value is obtained and summarized into an analysis report, and finally stored in the database for storage, which is used for querying the page display.

5 Static Analysis Vulnerability Detection System Test Analysis

In order to test the performance of our vulnerability detection system, we first need to set up a test environment. This article uses the system to deploy on the cloud server. The client accesses the test system through the chrome browser.

After the analysis of the detection results of the vulnerability detection system, it can be concluded that the network mall system contains many types of vulnerabilities. Here are several major vulnerabilities in the form of monitoring, as shown in Fig. 3. According to the above test process, this vulnerability detection system has well completed the design goals and implemented the detection of static code.

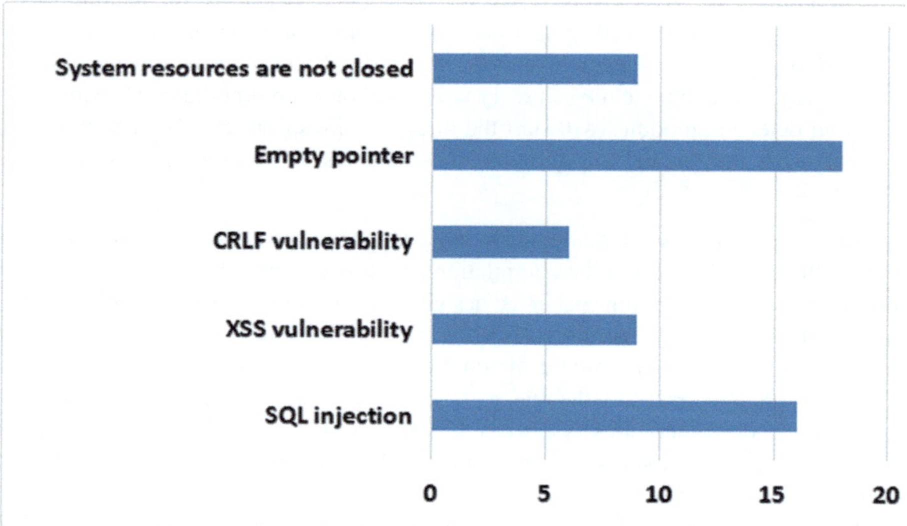

Fig. 3. Vulnerability detection system results comparison

These vulnerabilities have not been falsely reported after manual verification. This is also consistent with the high confidence level of the vulnerability detection results proposed in this paper. Although the confidence level is not reduced, the concept of confidence is in practical use. Still has practical significance. Tests were performed using FindBugs, PMD, and vulnerability detection systems, respectively. Figure 4 below summarizes the test results.

This is a non-practical measure of the fixed number, when the IP (M, A, P) value is larger, the rate of false positives will be lower, it is easy to conclude that the vulnerability detection system than other analysis tools of the false positives are lower. Of course, the table also calculated the false alarm rate, from the data can be seen that the false alarm rate is higher than PMD, compared with FindBugs reduced, this does not come to a specific vulnerability detection system in the false alarm rate changes.

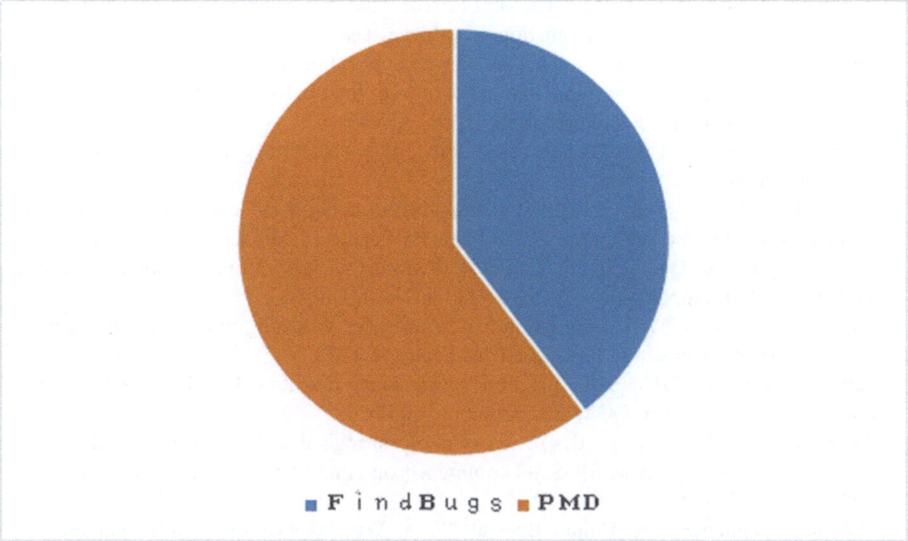

Fig. 4. Multi-tool comparison test results

6 Conclusion

Due to flaws in the software itself. This paper mainly solves these software security vulnerabilities from the perspective of static code analysis. The main research content is to solve the security vulnerabilities in the current Web system by using comprehensive analysis method. In the process of writing this paper, Java Web development technology is used to develop and implement the vulnerability detection system. The performance of the system is verified by a series of tests, Finally, the effectiveness of the system in reducing the rate of leak detection is verified by comparing the system with other test systems.

References

1. Katkar AS, Kulkarni RB (2012) Web vulnerability detection and security mechanism. Int J Soft Comput Eng 2(4):237–241
2. Antunes N, Vieira M (2015) Assessing and comparing vulnerability detection tools for web services: benchmarking approach and examples. IEEE Trans Serv Comput 8(2):269–283
3. Yuan L, Zhao W, Dan W et al (2015) A XSS vulnerability detection approach based on simulating browser behavior. In: International conference on information science and security. IEEE, pp 1–4
4. Yu F, Alkhalaf M, Bultan T et al (2014) Automata-based symbolic string analysis for vulnerability detection. Form Methods Syst Des 44(1):44–70
5. Mohammadi M, Chu B, Lipford HR et al (2016) Automatic web security unit testing: XSS vulnerability detection, pp 78–84

6. Mohammadi M, Chu B, Lipford HR et al (2016) Automatic web security unit testing: XSS vulnerability detection. In: Automation of software test. IEEE, pp 78–84
7. Bhor RV, Khanuja HK (2017) Analysis of web application security mechanism and attack detection using vulnerability injection technique. In: International conference on computing communication control and automation. IEEE, pp 1–6
8. Li Y-J, Ren HM (2016) SQL injection vulnerability detection system based on selenium. Modern Comput
9. Kumar R, Indraveni et al (2014) Automated session fixation vulnerability detection in web applications using the set-cookie HTTP response header in cookies, pp 351–354
10. Qian L, Zhang T, University A X (2015) Research on XSS Vulnerability Detection and Defense in Web Application. Nat Sci J Harbin Norm Univ
11. Antunes N, Vieira M (2016) Designing vulnerability testing tools for web services: approach, components, and tools. Int J Inf Secur 16:1–23
12. Luo MY, Ling J (2015) SQL injection vulnerability detection based on sequence value comparison of Webpage DOM tree. Comput Eng Des
13. Trinh MT, Chu DH, Jaffar J (2014) S3: a symbolic string solver for vulnerability detection in web applications. In: ACM SIGSAC conference on computer & communications security. ACM, pp 1232–1243
14. Rafique S, Humayun M, Hamid B et al (2015) Web application security vulnerabilities detection approaches: a systematic mapping study. In: IEEE/ACIS international conference on software engineering, artificial intelligence, NETWORKING and parallel/distributed computing. IEEE, pp 1–6
15. Rajeh W, Jin H, Zou D (2017) Saudi cloud infrastructure: a security analysis. Sci China Inf Sci 60(12):122102
16. Thome J, Shar LK, Bianculli D et al (2017) An integrated approach for effective injection vulnerability analysis of web applications through security slicing and hybrid constraint solving. IEEE Trans Softw Eng PP(99):1
17. Alsaleh M, Alomar N, Alshreef M, Alarifi A, Al-Salman A (2017) Performance-based comparative assessment of open source web vulnerability scanners. Secur Commun Netw 2017(3):1–14
18. Chai A (2017) Design and implementation of dynamic and efficient web crawler for XSS vulnerability detection. In: International conference on machinery, materials and computing technology
19. Nunes P, Medeiros I, Fonseca J et al (2017) On combining diverse static analysis tools for web security: an empirical study. In: Dependable computing conference. IEEE, pp 121–128
20. Nunes P, Medeiros I, Fonseca JC et al (2018) Benchmarking static analysis tools for web security. IEEE Trans Reliab PP(99):1–17

Application of Deep Learning Algorithm Based on Multimodal Regularization in Detection Robot Grasping

Shunan Liu[1(✉)], Xiaoli Wang[1], and Haidong Huang[2]

[1] Changchun University of Science and Technology, Changchun, Jilin, China
327334223@qq.com
[2] Jilin University, Changchun, Jilin, China

Abstract. This paper uses the depth learning method to settle the question of robot detection and capture in the RGB-D view of the sight comprising the object. A large number of candidates can be evaluated by using a two-step cascade system, which is faster and more effective than manual design. There are two deep-step cascade system network, a top value of the first detection system is reappraised by the second system. The first meshwork runs faster because of its fewer functions and can prune a large number of candidate fetches effectively. The second network is more powerful, so the speed is relatively slow, and the top-level detection value of the first system can be re-evaluated. To deal with multimodal input effectively, e propose a weighted structure regularization method in view of multichannel group regularization. The experimental results indicate the depth learning arithmetic can effectively improve the performance of RGBD robot grasping data sets, And the new objects can also be captured successfully.

Keywords: Robot fetching · Deep learning · RGB-D multimodal data

1 Introduction

Robot grasping involves problems in perception, planning and control. At present, people turn the perceptual problem into a detection problem, and it is necessary to use the angle of view of the object taken from the camera to infer the range of the top position where the robot gripper is placed [1]. Due to the use of 3D data, new input modes (such as RGB-D cameras) need to be merged, manual design is no longer applicable when robots interact with the new environment. The deep learning approach is able to extract useful functionality directly from the data to perform various tasks, we need to apply depth learning to the problem of robot grasp to get the best grasp of a given object to complete the grasp task.

Due to the need to process multiple input data modes(audio, video, image data, and even RGB-D information), with regard to some new associations of patterns and assignments, structured regularization method is applied to multi-pattern feature learning algorithm. In addition, in order to reduce the computational cost, A two level cascade system based on depth learning is proposed [2].

Z. Xu et al. (Eds.): CSIA 2019, AISC 928, pp. 453–461, 2020.
https://doi.org/10.1007/978-3-030-15235-2_67

2 Deep Learning System

The detection process of the two-level cascade detection system is shown in Fig. 1. The detection network of the two-stage system uses different features obtained by the deep learning algorithm, which saves time and effort compared to manual design. For an image of an object to be grabbed, the underlying feature set is used to detailed rake through latent squareness to get a suit of top-level candidate squareness, and then a stronger, deeper meshwork to find the highest of these candidates. A sorted rectangle that produces an optimum grab for a given target [3].

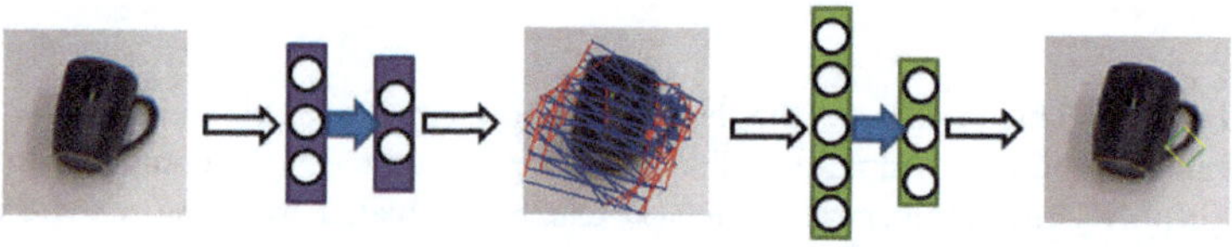

Fig. 1. Description of two-stage detection process

In the robotic gripping system shown in Fig. 2, the RGB-D picture of the spectacle of the object to be captured by the second-level cascade detection system is used to search for a large number of candidate candidates for the grab space, by extracting the corresponding A set of original features for color, depth, surface normals, and as a depth network input, the potential crawls are scored by a first-level, smaller depth network, and the top-ranked crawl candidate sets are fed to Level 2 in a larger and deeper network to generate a first-rate ranking crawl.

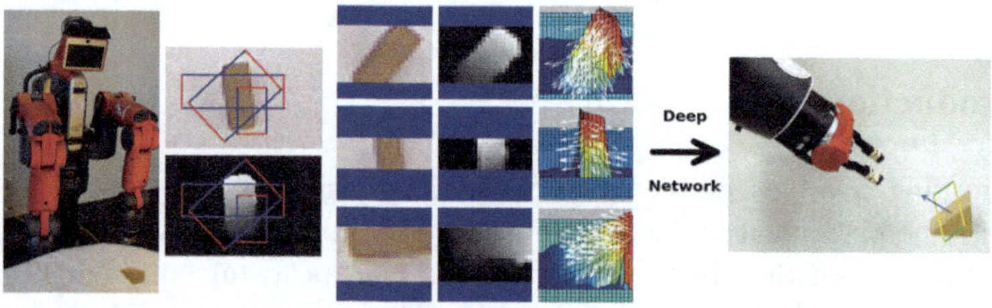

Fig. 2. Detects and performs fetching

2.1 System Model

The chance of detecting the squareness G(t) captured by the robot is simulated by the stochastic variable $\hat{y}^{(t)} \in \{0, 1\}$, and the prediction that G(t) indicates that it is to be grasped is determined by the feature $x^{(t)} \in R^N$. The left side of Fig. 3 shows a deep network with two hidden layer units h[1] and h[2], and each layer has two units K1 and K2. The characteristics of second hidden layers are used to predict whether the crawling is feasible [4]. On the right side, the logical classifier is pre-trained by the prediction of $P(\hat{y}^{(t)}|x^{(t)};\Theta)$. Each floor ℓ has a set of weight coefficients W [ℓ] to map inferent features to the hidden layer. Therefore the parameters of the mould are Θ = {W [1], W [2], W [3]}. Use the same weight to project the hidden unit output into the reconstruction of the input to form the output:

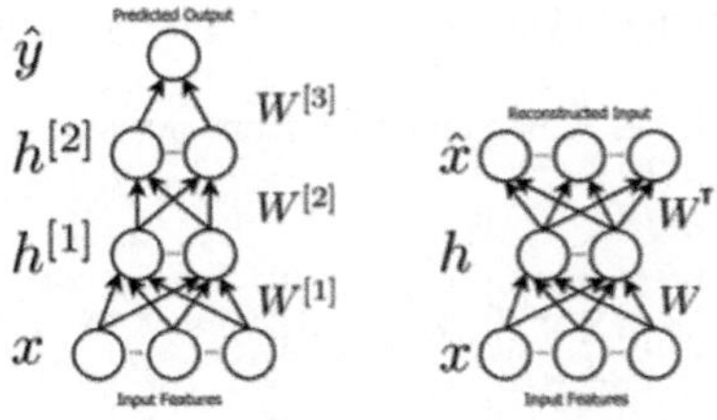

Fig. 3. Deep network and automatic encoder

$$\begin{aligned} h_j^{[1](t)} &= \sigma(\sum_{i=1}^{N} x_i^{(t)} W_{i,j}^{[1]}) \\ h_j^{[2](t)} &= \sigma(\sum_{i=1}^{K_1} h_i^{[1](t)} W_{i,j}^{[2]}) \\ P(\hat{y}^{(t)} = 1|x^{(t)};\Theta) &= \sigma(\sum_{i=1}^{K_2} h_i^{[2](t)} W_i^{[3]}) \end{aligned} \tag{1}$$

The position, direction and size of the best grab rectangle of an object are denoted by G, as follows:

$$G^* = \arg\ \max P(\hat{y}^{(t)} = 1|\phi(G);\Theta) \tag{2}$$

The identification accuracy of the system is optimized by the obtained parameter Θ, The φ function gives appropriate input data t = 1,... by extracting x(t) ∈ RN and y (t) ∈ {0,1}, and M represents a rectangle G. The first phase of the two-stage learning method uses the unsupervised features to recover the hidden layer weights W [1] and W [2]. The disciplinal process using a variant of the sparse autoencoder (SAE) (shown on the right side of Fig. 4) is to avoid overfitting. The sparse penalty function activated by hidden units is defined as g(h). Where λ controls its weight. f(W) is a regularization function, β is a weighted term, Reconstruct $\hat{x}(t)$ through x(t), the initial implicit layer weights obtained by SAE are as follows:

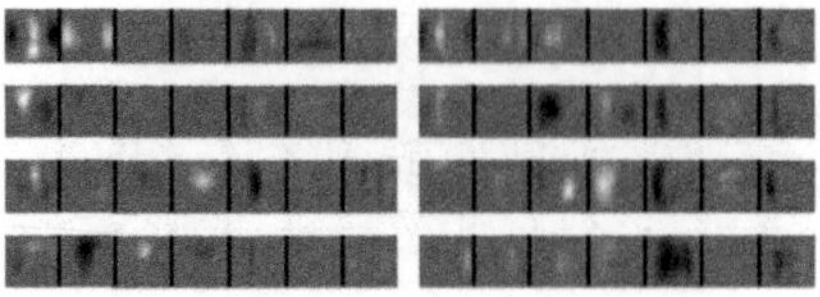

(a) Active grip corresponding to features

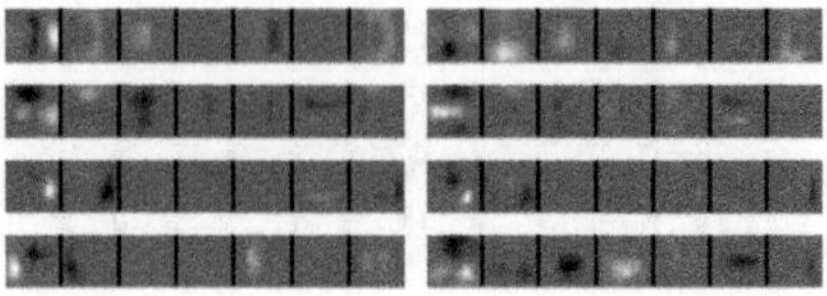

(b) Corresponds to features of negative grip

Fig. 4. Characteristics from mastery data

$$\begin{aligned} W^* &= \arg \quad \min \sum_{t=1}^{M} (\left\| \hat{x}^{(t)} - x^{(t)} \right\|_2^2 + \lambda \sum_{j=1}^{K} g(h_j^{(t)})) + \beta f(W) \\ h_j^{(t)} &= \sigma(\sum_{i=1}^{N} x_i^{(t)} W_{i,j}) \\ \hat{x}_i^{(t)} &= \sum_{j=1}^{K} h_j^{(t)} W_{i,j} \end{aligned} \tag{3}$$

The algorithm initializes W [1] to rebuild x, and then reconstructs h [1] by repairing W [2] from W [1]. The classifier weight W[3] and the fine-tuned hidden layer weight W [1] can be obtained in the second supervised stage of the learning algorithm and identified by W[2]. Use the logarithmic likelihood of the data to normalize the implicit layer weight:

$$\Theta^* = \arg \quad \max \sum_{t=1}^{M} \log P(\hat{y}^{(t)} = y^{(t)} \middle| x^{(t)}; \Theta) \ - \beta_1 f(W^{[1]}) - \beta_2 f(W^{[2]}) \tag{4}$$

In the 24 × 24 pixel image of the receiving field suitable for the network, RGB-D subimages contained in each rectangle is scaled to the aspect ratio, and 24 × 24 7 = 4032 input features are extracted from the feature values of the seven channels. The seven channels include three channels representing image intensity and color, a deep channel for representing images, and the final three being the X, Y, and Z ingredients of the external normal calculated based on the depth passageway [5]. Since the feature is calculated after the image and fixture alignment, it is always consistent with the splint.

2.2 Data Pretreatment

Whitening information is essential in-depth learning of multimodal data input. The PCA method will mask most of the depth data [6]. Therefore, for multimodal data, individual each depth patch and obtain the minimum value to reduce the average value of each feature, and divide by the standard deviation of all feature combinations to expand the data of each channel [7].

In multi-mode Settings, input data x from R different modes. The modal matrix S is R × N binary matrix, The members of the visible unit x i in the particular mode r are represented by the element S r, I, $\psi_{\mathrm{r}}^{(t)} = \sum_{i=1}^{N} S_{r,i}/(\sum_{i=1}^{N} S_{r,i}\mu_i^{(t)})$ defines the scaling factor for schema r. If $\mathrm{x}_i^{(t)}$ is masked then $\mu_{\mathrm{i}}^{(t)}$ is 1, otherwise 0. The scaling factor of case I is: $\psi_{\mathrm{i}}^{(t)} = \sum_{r=1}^{R} S_{r,i}\Psi_r^{(t)}$.

Each value of X can be scaled simply according to the corresponding scaling factor, for example $x_i^{\prime(t)} = \psi_i^{(t)} x_i^{(t)}$. However, sparse autoencoders penalize squared errors, so linear scaling x should be scaled twice. The scaled x' is used as in item to the network, but because of the initial x penalty reconstruction, scaled only after calculating the squared error:

$$W^* = \arg \min \sum_{t=1}^{M} (\sum_{i=1}^{N} \psi_i^{(t)} (\hat{x}_i^{(t)} - x_i^{(t)})^2 + \lambda \sum_{j=1}^{K} g(h_j^{(t)})) \tag{5}$$

Use the zoomed visible input to redefine the hidden units:

$$\mathrm{h}_j^{(t)} = \sigma(\sum_{i=1}^{N} x_i^{\prime(t)} W_{i,j}) \tag{6}$$

In practice, the scale factor needs to be limited to the maximum value of a certain value c, such as $\Psi_r^{\prime(t)} = \min(\Psi_r^{(t)}, c)$.

3 Structured Regularization for Feature Learning

Structural regularization is a set of weights in a higher-order regularization function, and the regularization set of each hidden unit is represented by different modes. Such as, a grouped p-norm will be used to:

$$f(W) = \sum_{J=1}^{k} \sum_{r=1}^{R} (\sum_{i=1}^{N} S_{r,i} \left| W_{i,j}^{P} \right|)^{1/P} \tag{7}$$

In the formula, $\mathrm{S}_{\mathrm{r,\ i}}$ is 1, indicating that characteristic i is part of group r; otherwise, it is 0. Due to the formation of high value weights in the high-order group, the added value is less than the low value weight. Therefore, taking p → ∞, the regularization of this group is worth the infinitely great norm:

$$\mathrm{f}(W) = \sum_{j=1}^{K} \sum_{r=1}^{R} \max_i S_{r,i} \left| W_{i,j} \right| \tag{8}$$

In practice, because infinite norm is non-differentiable, it is hard to optimize the gradient, so logarithm and exponent are taken as the maximum norm for approximate differentiation. In experiments, the first level of weight generated by this regularization function is concentrated in fewer patterns for each feature. When the sparse pattern is induced by the β value, the algorithm will get duplicate features, indicating that the maximum norm may be over-constrained. L0 norm can be used to solve this problem [8].

$$\mathrm{f}(W) = \sum_{j=1}^{K} \sum_{r=1}^{R} \mathrm{II}\{\max_{\mathrm{i}} S_{R,I} \left| w_{i,j} \right) > 0\} \Big| \tag{9}$$

II is an index function. When the parameter is true, it is 1, otherwise it is 0. Using $\log(1 + x^2)$ to approximate L0 norm, the weight of the pattern with non-zero maxima is not limited. The regularization function can directly encode the number of patterns used by each weight. Figure 4 shows the characteristics of unsupervised learning from our group regularization depth learning algorithm. We discuss these characteristics and their impact on robot grasp.

4 Experiment

4.1 Data Set

The experiment uses an extended version of the Cornell grabbing dataset containing 1035 images that can grab objects [9]. Each image in the data set contains an equal number of graspable and non-graspable rectangles. The unsupervised learning algorithm is validated by five crossover experiments. Provides 4032 input functions per network with seven 24 × 24 pixel channels as inputs. Depth learning algorithm is used to configure 200 hidden units for the first layer and the second layer of deep network, which takes about 30 min. The first depth network of the two-channel detection system searches through 200 units in detail, and then rearranges 50 hidden units to find 100 highest ranked rectangles.

4.2 Detection Indicators

For detection, the highest ranked rectangle of each method is compared to the ground truth rectangle set for each image. The two indexes of "point" and "rectangle" were used for judgment. Point measurement: if the distance between the predicted rectangular center is lower than that of the actual rectangular center on the ground, it is considered that the grabbing is successful. However, this indicator ignores the fetching direction and may overestimate the judgment result [10]. Rectangle measurement: the direction error between the top-ranked grabbing rectangle G and the real rectangle G* on the ground is no more than 30°, then the success is judged [11]. The remaining sets are evaluated by intersecting common bounding boxes, namely Area(G ∩ G*)/Area (G ∪ G*). The ground truth squareness is the whole space of covering the object.

4.3 Results and Discussion

In-depth study of robot grab gauging. Figure 4 displays the characters obtained during the unsupervised stage using the depth learning algorithm, which are highly correlated with positive and negative grip cases. Each property includes seven alleyways - In turn are depth, Y, U and V picture channels, and normal components of X, Y and Z surfaces. The upright edge corresponds to the clamping plate. The feature shows the non-zero weight of the depth channel, indicates the correlation between the depth and grasp, and the structural regularization method eliminates the weight of many modes of the feature. The feature lacks the weight of the channel corresponding to the color and can be grabbed for objects of different colors.

The three-dimensional mesh of the depth channel with four features is shown in Fig. 5, and its positive (upper) negative (lower) correlation with the grasp of the rectangle is strongest. The coordinates of X and Y correspond to the position of the deep net acceptance domain, and the Z coordinates correspond to the weight of the depth alleyway at each position. The correlation between positive (upper) and negative (lower) corresponds to the case in which grip is possible and non-grip respectively. Depth learning algorithm can also get the characteristics of cases that can be grasped and cases that can not be grasped closely without any supervised information. In the positive correlation feature, the first two handles are centered with a convex region, and the second two are handles with a circular edge. The negative correlation features are the valley between the ridge vertical to the cramp frame plane and clamper plate. Because the data used is semi-graspable and semi-graspable, the algorithm can also get useful representation of the task on the head.

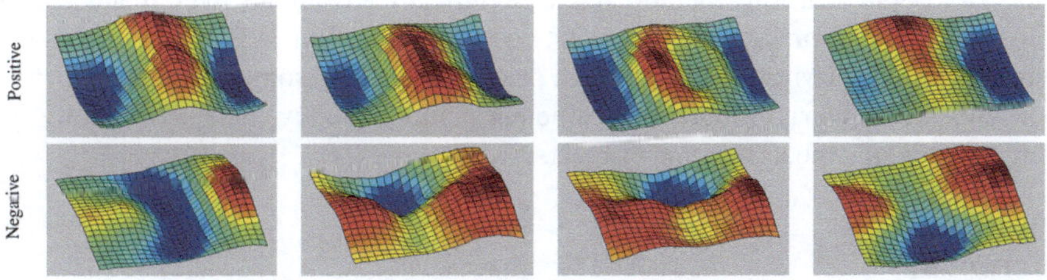

Fig. 5. Learned 3D depth features

Regularization of Multimodal Groups. Multichannel bevy regularization has higher test precision than uncomplicated L1regularization term, and can avoid over fitting of network when evaluating new objects. Figure 6 shows the use of multi-channel group regularization and L1 regularization methods for grabbing. The former can carry out effective grasping. The latter applies only to certain modes (such as the depth channel of sunglasses and enamel bottles, and RGB channel of forfex). In the figure, green and yellow are the grabbing rectangles of the group regularization method, and blue and red are the networks practiced only by uncomplicated L1 regularization. Upper: RGB picture, bottom: depth groove, blue and green edge corresponding to fixture. Figure 6 reveals that the group regularization network can better integrate all patterns of information. And It is more robust to noise and obliterated data in the belief channel.

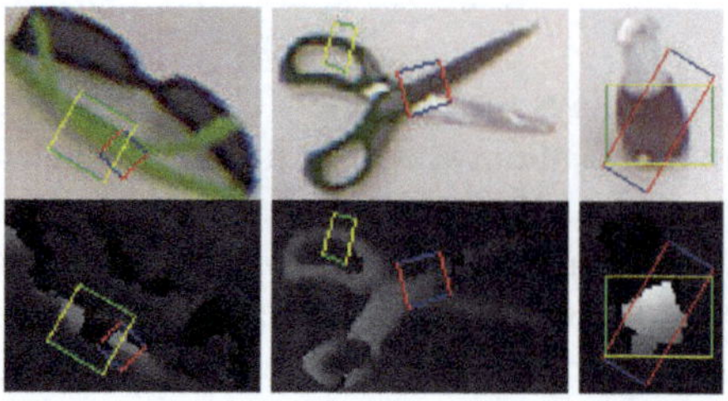

Fig. 6. Improvements from group regularization

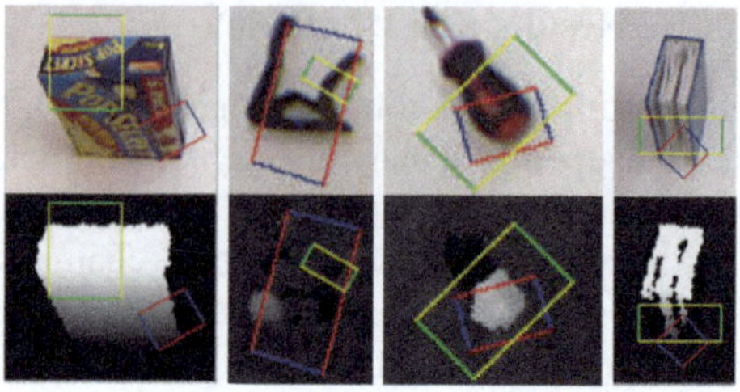

Fig. 7. Improvements from two-stage system

Two-stage Detection System. The two-stage detection system has strong computational efficiency and accuracy, and the number of network rectangles can be reduced by 1000 times. The average computing time using MATLAB was reduced from 24.6 s per image to 13.5 s per image, which improved 2.4% compared with single channel network detection performance. Figure 7 shows the capture of a two-stage system (Expressed in green and yellow) and a single-level system (Displayed in blue and red). It may be feasible for the robot to grasp by single-stage system selection, but the rectangle chosen by the two-level system can be grasped successfully.

5 Conclusion

This paper proposes a deep learning method based on RGB-D data to detect robot's grasp. The depth learning method has a significant advantage over manual design, saving time and effort. Moreover, the two-stage detection system based on multi-channel group regularization method can produce better grasping detection function of robot, and has higher computational efficiency and accuracy, and successfully grasp different types of objects and new objects.

References

1. Osadchy M, LeCun Y, Miller M (2007) Synergistic face detection and pose estimation with energy-based models. JMLR 8:1197–1215
2. Coates A, Ng AY (2011) Selecting receptive fields in deep networks. In: NIPS
3. Collet Romea A, Martinez Torres M, Srinivasa S (2011) The moped framework: object recognition and pose estimation for manipulation. IJRR 30(10):1284–1306
4. Collobert R, Weston J, Bottou L (2011) Natural language processing (almost) from scratch. JMLR 12:2493–2537
5. Ekvall S, Kragic D (2007) Learning and evaluation of the approach vector for automatic grasp generation and planning. In: ICRA
6. Girshick R, Donahue J, Darrell T, Malik J (2014) Rich feature hierarchies for accurate object detection and semantic segmentation. In: CVPR
7. Hinton G, Salakhutdinov R (2006) Reducing the dimensionality of data with neural networks. Science 313(5786):504–507

8. Hyvärinen A, Hoyer PO, Inki M (2001) Topographic independent component analysis. Neural Comput 13(7):1527–1558
9. Jiang Y, Moseson S, Saxena A (2011) Efficient grasping from RGBD images: learning using a new rectangle representation. In: ICRA
10. Jiang Y, Amend JR, Lipson H, Saxena A (2012) Learning hardware agnostic grasps for a universal jamming gripper. In ICRA
11. Jiang Y, Lim M, Zheng C, Saxena A (2012) Learning to place new objects in a scene. IJRR 31(9):1021–1043

Research and Improvement on the Linkage of 2D and 3D Based on Multi-touch Technology

Youjian Yu[1(✉)], Xiangyi Meng[2], and Zihui Chen[1]

[1] School of Control and Mechanical Engineering, Tianjin Chengjian University, Tianjin 300384, People's Republic of China
yujianneau@163.com
[2] School of Art and Design, Tianjin Tianshi College, Tianjin 301700, People's Republic of China

Abstract. As more mature technology, multi-touch appears strong interactive experience. This paper introduced multi-touch interactive technology into the linkage of 2D and 3D. Its natural and efficient features was adopted to solve the existing problem in virtual reality interaction. OTSU algorithm was applied to low the error rate of multi-touch recognition. Then defined gestures generic class and the designed linkage framework realize the linkage of 2D and 3D in multi-touch mode and promote the interaction experience.

Keywords: Multi-touch control · Man-machine interaction · The linkage of 2D and 3D

1 Introduction

With the development of technology, virtual reality has been widely used in various fields. For increased need, the interactive mode of virtual reality will develop in the direction of facilitation [1]. There are problems with difficult operation and difficult control in the traditional 3D interaction. But it shows the lack of sense of reality for 2D interaction. The linkage between 2D and 3D can solve the lack of spatial effects for 2D interaction and reduce disorientation for 3D interaction.

Multi-touch technology is one of the well-developed interactive technologies in recent years [2]. This paper introduces multi-touch technology to virtual reality. The natural and efficient features of touch control interaction are used to solve the difficult problem of virtual reality interactive operation. Based on the multi-touch technology, the improvement of the linkage between 2D and 3D can greatly increase the operability and availability of interaction.

2 The Linkage Between 2D and 3D Data Models

2D floor plan and 3D scene respectively show in the two monitors. Horizontal multi-touch screen shows for 2D interactive roam, and so on. The vertical screen shows 3D virtual scene in real time. The mechanism of the response each other is used to realize

Z. Xu et al. (Eds.): CSIA 2019, AISC 928, pp. 462–468, 2020.
https://doi.org/10.1007/978-3-030-15235-2_68

the combination of 2D and 3D coordinates, and the interaction of 3D scene controlled by 2D plane. It means that each of 2D coordinates is mapped to the coordinates of 3D scene. That is to say, ID of object's name and coordinate is unique.

For the linkage between 2D and 3D, geographical coordinates in 2D and 3D correspond each other. Each of spatial object models in 3D scene corresponds with the layer data in 2D. System operations of 2D and 3D have synchronicity to be consistent based on coordinate each other. Namely, 2D and 3D can operate zooming, translating, picking up, etc. [3].

2.1 Operation Synchronization

Coordinate mapping between 2D plane and 3D space means that coordinates of 2D layer correspond to spatial location of 3D. For interactive manipulation, synchronous changes of 2D and 3D are implemented with trigger mechanism. It is also the essence of the linkage between 2D and 3D in the visual level. Synchronous operation mainly includes visual domain synchronization and interactive browsing synchronization.

(1) Visual domain synchronization

2D plane in the multi-touch platform moves synchronously with the 3D scene. 3D scene changes with the movement of the observation direction in the 2D plane. When the position of a viewpoint is defined in the 2D plane, 3D scene will automatically switch to the position of the viewpoint.

(2) Interactive browsing synchronization

While a user roams 3D virtual scene, the corresponding position and visual field will be showed in 2D plane of multi-touch platform. Platform can track viewpoint position of 3D virtual scene in real-time and send the location of current coordinate to 2D scene from 3D scene. Receiving the coordinate values, 2D plane will plot the roaming path. Zooming and roaming in 2D plane, 3D scene can follow the actions in 2D scene synchronously, so as to realize response [4].

3 The Linkage Between 2D and 3D Based on Multi-touch Technology

3.1 Multi-touch Technology

Multi-touch technology replaces the common operating interface with usability and intuition and makes human-computer interaction abandon the traditional operation of keyboard and mouse. Then users don't need to be trained professionally and can directly do interoperate. It simplifies users understanding and study of the software and equipment and improves the synergistic effect of human-computer interaction and naturalness. As the combination of software and hardware, Multi-touch technology mainly includes the laser plane (LLP), frustrated total internal reflection (FTIR), ToughtLight, Optical Touch hardware scheme and so on. They are classified into capacitance and infrared reflectance by contact mode. FTIR is adopted by the recognition accuracy, as shown in Fig. 1.

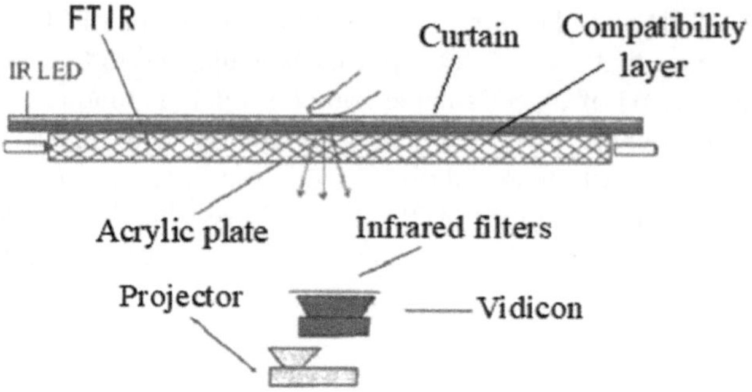

Fig. 1. FTIR principle diagram

As a kind of basic optical phenomenon, FTIR shows the infrared reflects the inside of an acrylic plate that can produce a full internal reflection. After the plane of acrylic plate is touched by fingers, threshold density of infrared media will be changed. It will lead to the failure of inner circle reflection and the light overflow in the touch point. Touching the plane, vidicon can detect the bright spots and send their image information to computer in real-time. Contact recognition software detects these images to do real-time identification and tracking and gets key information data such as the number, ID and coordinates of contact points in the camera window. Then the packet is decoded by the driver and sent to the response software.

3.2 Identifying Touch Point

The infrared camera is used to capture the original image. Removing the noise, the background model is built with the average pixel. Image noise points are further removed with smoothing processing. Then the satisfactory convex hulls are got by outlining. Finally, touch points within normal range are obtained by removing too big and small angular points. It is shown in Fig. 2.

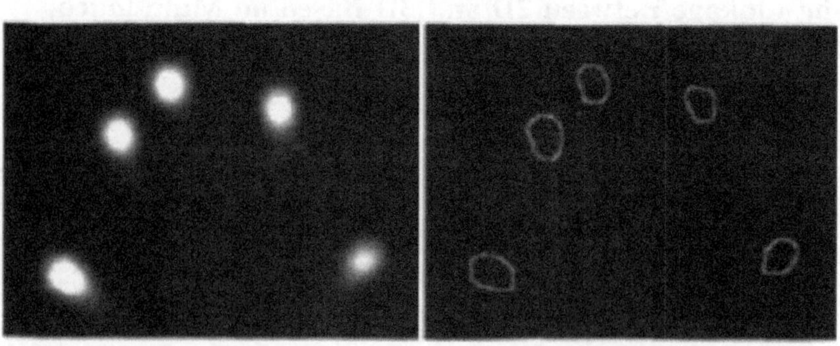

Fig. 2. Outline describing diagram

3.3 Improving the Recognition Accuracy of Contact Points with OTSU

Maximum class difference (OTSU) algorithm is used to obtain the adaptive threshold to improve the extraction accuracy of contact points. OTSU algorithm has a good segmentation effect for obvious single peak image and is suitable for image recognition of small contact objects [5]. Thus, OTSU algorithm can solve the binary problem of contact point segmentation in the optical multi-touch control system. OTSU can maximize interclass variance of foreground and background and do image binarization.

In the optical multi-touch system, the proportion of contact target is small in the whole touch screen. The overall gray value of the difference image is mainly influenced by the background gray. When the gray mean of the whole image is divided, the segmentation threshold is not the best. Based on the iterative principle, the image is divided into equal sub-modules to increase the proportion of histogram containing the grayscale information of contact points. Then the maximum class difference method is used to do threshold segmentation [6].

The gray value of the contact target is much higher than that of the background target by analyzing difference image. For contact recognition, the whole image will be segmented by OTSU algorithm to obtain the gray histogram containing the contact area. The grayscale graph of the difference image is assumed that conforms to probability density function of a certain normal distribution. Set standard deviation respectively as σ_a and σ_b.

$$\begin{cases} \sigma_a = \left\{ \sum_{i=0}^{r} [i - \mu_a]^2 P_i \backslash \sum_{i=0}^{r} P_i \right\}^{1/2} \\ \sigma_b = \left\{ \sum_{i=T+1}^{L-1} [i - \mu_b]^2 P_i \backslash \sum_{i=T+1}^{L-1} P_i \right\}^{1/2} \end{cases}$$

When the contact target and background image are completely separated, the gray distribution image should meet:

$$\mu_b - \mu_a > \alpha(\sigma_a + \sigma_b) \tag{2}$$

where a is about 2.

When the threshold value is segmented, target class A and background class B will not satisfy Eq. (2). Then the segmentation needs to continue until the optimal segmentation effect is obtained. The results of the two segmentations are shown in Fig. 3.

Figure 3 shows the optimal threshold can be automatically generated with the continuous iteration of the solution process in the self-adaptive threshold of OTSU algorithm. OTSU algorithm can not only segment the original difference image efficiently, but also reduce the process computation and improve the operability and adaptability of touch control system.

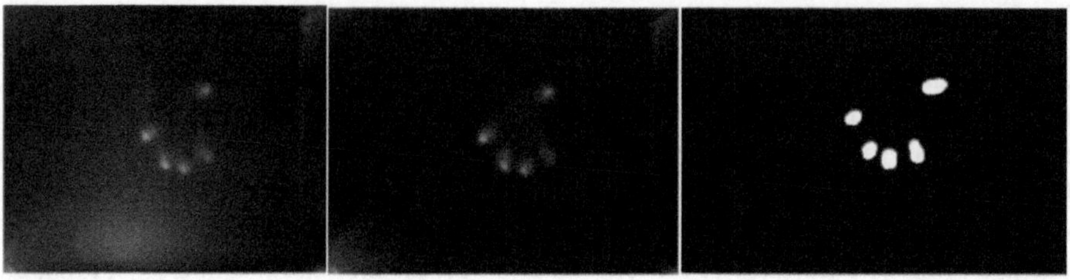

(a) The original image of band segmentation; (b) the first segmentation result; (c) the second segmentation result

Fig. 3. Segmentation results

3.4 The Gesture Design of Multi-touch Platform

Corresponding functional gestures are set in the touch control platform to realize the linkage of 2D and 3D [7]. Analyzing task mapping of mainstream gestures and combining with the basic operation types of the system, a set of operation gestures and corresponding task mapping are designed. The gesture functions are shown in Table 1.

Table 1. The gesture mapping table

Name	Mapping	Legend
Taping with single	Open the response function and select the target and scene	
Taping with two fingers	Specify the scene and make the same amount of magnification	
Pressing with single	Open the context menu	
Dragging with single	3D scene movement	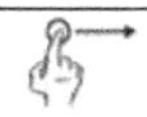
Moving up and down with parallel two fingers	Viewpoint fluctuation control	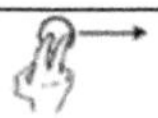
Moving up and down with parallel three fingers	Focus on the presentation and restoration of the target	

Through function interface layer and touch platform driving development layer, function interface layer oriented universal touch objects is designed. Then other platforms only need to do secondary modification for functional interface layer according to the driver development package of different touch platforms, without changing the mapping function interface in the system (Fig. 4).

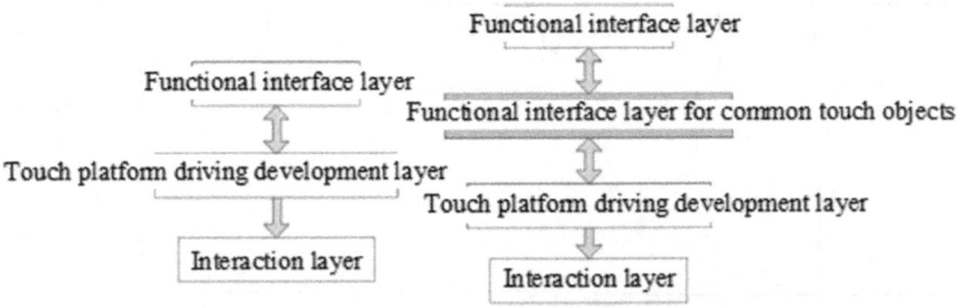

Fig. 4. Diagram of the general interface structure

According to characteristic of different touch screen driving development packages, there are the common parts of each driving development package. A set of universal touch object interface class-IES_MultiTouchInterface is extracted, which is used to encapsulate driver program in various touch platforms, as shown in Fig. 5.

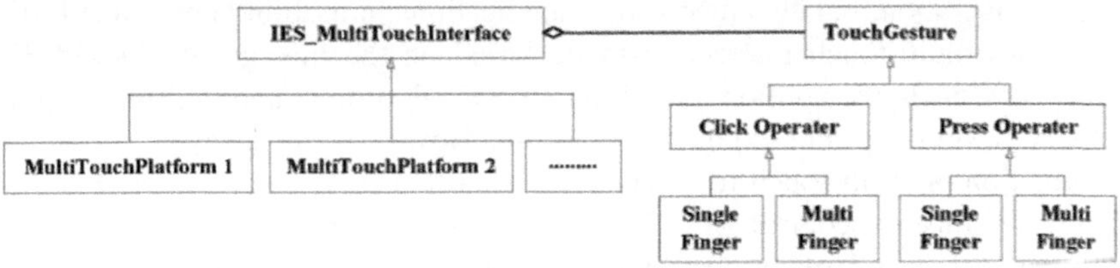

Fig. 5. The interface of universal touch object

3.5 The Linkage Control Design of 2D and 3D

Two independent subsystems of 2D and 3D system constitute the linkage of 2D and 3D through the message communication mechanism [8]. In design, there exists the independence of the two systems and interoperation between two systems, such as creating an article in 2D plane and appearing corresponding 3D article in the same coordinates of 3D scene at the same time. In addition, zooming, roaming and other interactive operations in 2D plane will change corresponding points in 3D scene. Then the design framework of this system is shown in Fig. 6.

SKYLine software provides the 3D data production, management and browsing and so on. The reasonable scheduling data can satisfy the functional requirements [9]. Based on SKYLine, scene data browsing and display are realized. Combined with the corresponding standardized system, 3D model symbol library is developed and connected to 3D system. Then the coordinate information of 2D system displays in platform. Combined with 3D display platform, interactive roaming and scene picking can be realized through corresponding message mechanism.

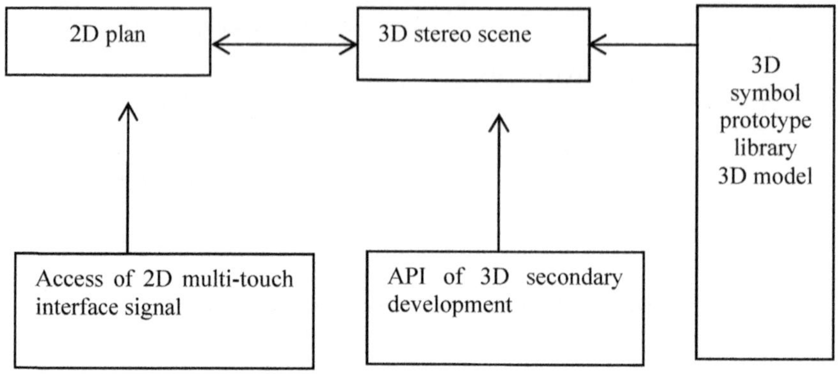

Fig. 6. System architecture diagram

4 Conclusion

The linkage of 2D and 3D in multi-touch platform is not only easy to accept in the interactive operation, but also meets most applications in usability and effect [10]. This paper shows multi-touch interactive mode based on the linkage of 2D and 3D, the accuracy of touch recognition based on OTSU algorithm and utility category for secondary developing and extending platform. However, the delay is caused by the transformation of multi-touch from 2D to 3D coordinates. It will be improved in future studies. The linkage control of 2D and 3D based on multi-point touch, roaming, picking, zooming and other operations reduce the operation difficulty of users and improve the sense of experience.

References

1. Tao X, Li T (2012) History and development of virtual reality technology. China New Commun 23:41
2. Wang F (2012) Design and implementation of a large-scale multi-person multi-touch control system and software. PLA Information Engineering University
3. Cheng H, Song L, Cao J et al (2010) Research on the linkage technology of two-dimensional and three-dimensional GIS. Zhejiang Water Conserv Sci Technol 3:31–32
4. Wang C, Li X (2008) Design and implementation of urban planning information system based on two-dimensional GIS and three-dimensional simulation. Urban Surv (3):10–13
5. Lin M (2016) An improved threshold segmentation algorithm emphasizing trough information. Comput Syst Appl 25(8):162–165
6. Zhang J (2014) Infrared image threshold segmentation based on Otsu. Xiangtan University
7. Wang Q, Peng Z, Liu L (2003) An adaptive method of image segmentation. Trans Beijing Inst Technol 23(4):521–524
8. Wan Y, Bian F (2008) Design and implementation of inter active 2D&3D GIS architecture. Geomat World 6(2):48–52
9. Zhang G, Heng X, Ling Y, Lao S (2010) Interaction gesture analysis and design based on multi-touch surface. Appl Res Comput 27(5):1737–1739
10. Wang X, Chen G, Wang D, Wang C (2012) Research on multi-touch gesture analysis and recognition algorithm. Comput Sci 39(s1):522–525

Simulation of QoS Routing Algorithm Based on SDN Framework

Zhaochenzi Zhang and Yanfang Fu(✉)

Xi'an Technological University, Xi'an, China
149481698@qq.com

Abstract. Specific to the increase of Internet dataflow at the present stage and the higher requirements for service quality (QoS), the Lagrangian relaxation algorithm is introduced into the SDN framework. In this design, new traffic scheduling system has been realized in the design, this system takes bandwidth and time delay as the cost, selects the optimal path. Secondary development of RYU controller is conducted, and Mininet simulation network topology is used to verify the system function.

Keywords: QoS routing · Mininet simulation · SDN architecture · MCMP algorithm

1 Introduction

SDN refers to a software-defined network, belonging to a new type of network innovation architecture, and also a way to realize network architecture, as well as a kind of implementation model of network virtualization, which adopts OpenFlow technology, the core technology, separating network equipment control surface from data control surface, and then the realization of the control of network traffic is able to be available, make the network intelligent. On account of the continuous development of internet technology, network environment is gradually developing to complexity, the transmission traffic type presents diversity. Under the complex network environment, traditional process mode is unable to cope with the multiple transmission flow type, and can make network architectures difficult to change.

2 Introduction of SDN Network Framework

2.1 SDN Network Framework and Interface

2.2 Application Plane

In this study, the main interaction between the application plane and the control plane is through the interface, and in the application plane, multiple programming languages can be used, at the same time, multiple programs and logic can be operated, to improve SDN's high adaptability and programming ability, the main way of representation is to support the application.

Z. Xu et al. (Eds.): CSIA 2019, AISC 928, pp. 469–478, 2020.
https://doi.org/10.1007/978-3-030-15235-2_69

2.2.1 Control Plane

The realization of the interaction between the application plane and the control plane is mainly through the interface-based approach. Controller constitutes the control plane, the controller can be one or more, and these controllers are responsible for maintaining the network and continuously implementing the topology, and then routing algorithm is adopted and passes the calculated data to the switch through the interface [1].

2.2.2 Data Plane

The switch is an important configuration of the data plane. The main function of the data plane is the management and collection of network link state, which needs to be realized through the controller and issue OpenFlow message, another function of which is to forward the data. The switch needs to forward the message according to the relevant rules issued in the controller, and the message control needs to be realized through the controller, rather than control by means of network packets transmitting (Fig. 1).

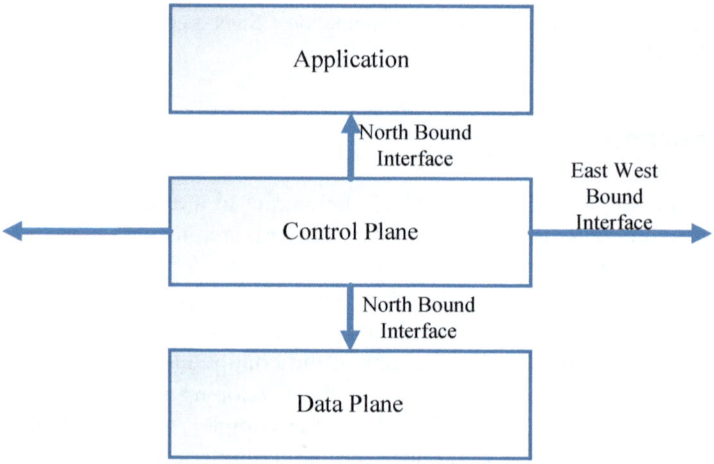

Fig. 1. Classical SDN architecture

2.3 SDN Simulation

2.3.1 Mininet Simulation System

When performing simulation, it is necessary to use Mininet for verification. Mininet is a kind of simulation tool for SDN network topology and network equipment transfer. Primarily due to the limited quantity of related resources in ordinary computers, the large-scale network topology cannot be proceeded. In view of this situation, Mininet can be used. Mininet can exist in the single host, thus simulating the topology network type and forwarding function. In addition, Mininet also possesses higher scalability and supports a variety of software - defined network components, which is also of certain portability. Therefore, it is widely used in simulation [2].

2.3.2 Ryu Controller

SDN application layer is divided into three types: SDN management application manages and controls the SDN frame through managing Restful; SDN cloud application manages and controls the network by combining Restful API with the Neutron component of OpenStack; SDN user application manages and controls the SDN frame through user-defined Restful API or RPC.

Many applications developed in the Python language are built in the Ryu framework layer, mainly including topology discovery, tenant isolation, firewall and other components, whose core functions include flow table issue, topology discovery, etc. (Fig. 2).

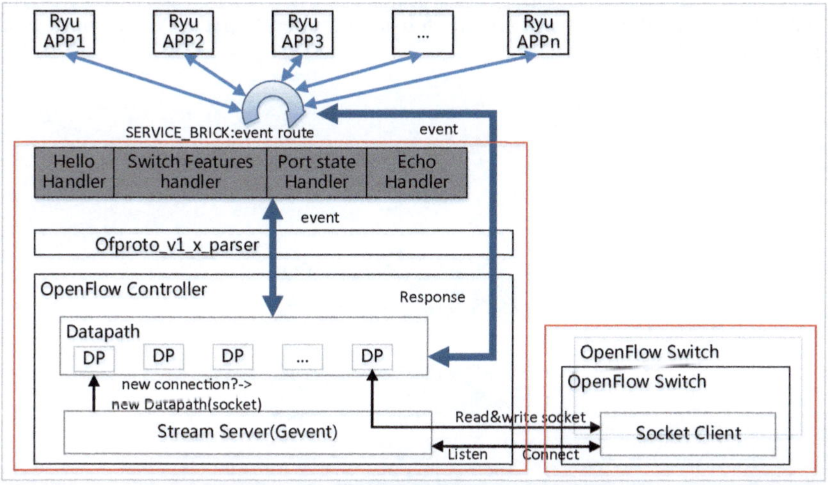

Fig. 2. Ryu internal operation mechanism

3 Research on Multi-restriction Multi-path Algorithm Based on SDN Architecture

3.1 Description of MCMP Algorithm

MCMP (Multi Constrained Multi Path) is also known as multi-restriction multi-path algorithm, when analyzing the MCMP algorithm, the entire network can be regarded as directed graph, $G = (V, E)$, among which, the letter V mainly represents various stages in the network, while the letter E represents various links in the network. $n = |V|$, $m = |E|$, these two formulas represent the number of different network nodes and network links. In normal conditions, the bidirectional transmission means is adopted in links in the network, so $e = (u, v)$ can represent the link between node u and node v, and each link has the following features, time delay $d(e)$, bandwidth $b(e)$ [3].

The design and application principle of MCMP algorithm is to fully guarantee the reliability of the network and minimize the impact of the switch failure on the network. In the actual design process, appropriate path selection can be made on the basis of the actual situation:

(1) Under the circumstances of the perfect network and the reasonable special requirements proposed by the user, that is to say, if the network conditions permit, the MCMP algorithm can be adopted to find the related path that fits the user's needs [4].
(2) Under the circumstances that the network situation is poor and there is no path to meet the user's requirements, comprehensive consideration is needed, and then in multiple restricted conditions, [5] the path that best meets the user's requirements is selected. When choosing other paths, it is required to ensure that the chosen path does not intersect with the switch node. The algorithm flow chart is shown below (Fig. 3).

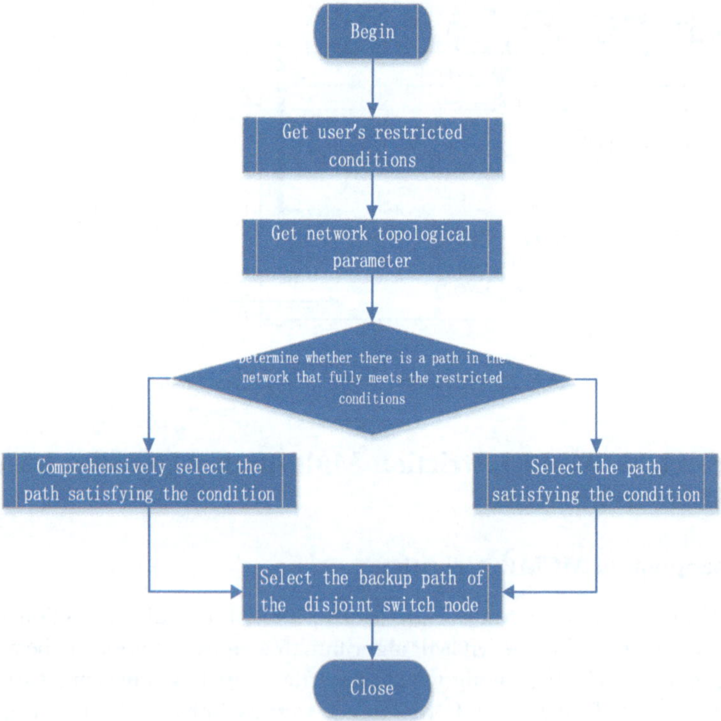

Fig. 3. MCMP algorithm flow chart

Finally, in order to ensure the path reliability, the improved LBA algorithm should be used on the basis of the selected links.

3.2 Path Selection Method

3.2.1 To Select the Path that Fully Meets the Conditions

In the path selection, if the user gives two restricted conditions, maximum time delay Dmax and minimum bandwidth B, then the path to be searched is:

$$L = \min\{c(p) : p \in P(s,\ t),\ d(p) <= \mathrm{Dmax},\ b(p) >= \mathrm{Bmin}\} \tag{1}$$

Among which, P(s, t) is a set of paths from source node s to destination node t, c(p) is the total cost of the link.

The idea of Lagrangian relaxation algorithm is adopted in this research. Suppose:

$$c\lambda = c + \lambda^* d \tag{2}$$

Among which, c = Bmin − b(l), d = d(l) − Dmax, if the calculated results of c and d of a certain link are less than 0, it indicates that this link does not meet the limit condition requirements, so the path needs to be abandoned to ensure that both c and d are positive values [7].

$$L(\lambda) = \min\{C\lambda, (p) : p \in P(s.\ t)\} - \lambda^* \mathrm{Dmax} \tag{3}$$

Therefore, for any lower bound that $\lambda >= 0$, $L(\lambda)$ is L, it is proved that:

$$\begin{aligned} L(\lambda) &= \min\{c\lambda(p) : p \in P(s.\ t) - \lambda^* \mathrm{Dmax} \\ &= \min\{c(p) : p \in P(s,\ t),\ d(p) <= \mathrm{Dmax},\ b(p) >= \mathrm{Bmin}\} + \lambda^* d - \lambda^* \mathrm{Dmax} \\ &= L + \lambda^* d - \lambda^* \mathrm{Dmax} \\ &= L + \lambda^* (d - \mathrm{Dmax}) \\ &\le L \end{aligned}$$

To get closer to L, the maximum value of $L(\lambda)$ needs to be searched,

$$L^* = \max L(\lambda)(\lambda >= 0) \tag{4}$$

so in this case, L^* is the approximate solution to L.

Making the best of the Lagrangian relaxation algorithm, the optimal λ, of the setting source node and destination node can be found out, and then the best solution can be obtained.

Observe and analyze whether there are paths in the network that fully meet the conditions, and make judgment. The main means of judgment is to use only broadband as the limited condition, and select the path PC, or use time delay as a limited condition, and then choose path Pd. If the path that meets the user's demand cannot be selected under the single limited condition, then it is judged that the user satisfaction path in the current network path does not exist, so the comprehensive routing algorithm should be adopted [8].

If the only used single parameter routing exists in the network, the path can satisfy the condition, then it indicates that there may be a path that satisfies both constraint conditions in the network. The Lagrangian relaxation algorithm is employed for the calculation [9].

3.2.2 Multiple Limiting Conditions Are Integrated to Select the Path

When the network state is poor or the user limited conditions are more, there is no fully qualified path in the network path, aiming at such circumstance, it is required to comprehensively consider each limit parameter suggested by the user, and select the path that best meets the conditions in terms of the condition. In the selecting course, AHP algorithm is needed to analyze the constraint conditions proposed by users. [10] AHP algorithm is a kind of decision analysis method, which mainly decomposes the big target into several small ones, and divides the single level into multiple levels, then analyzes the effect of realizing the small targets, so as to solve the problem. In the case of complex network conditions, user demands are no longer simple, and there will often be situations where user restricted condition and paths do not match. Therefore, AHP algorithm can be adopted to select the most appropriate path for users.

4 Implementation and Simulation of QoS Routing Algorithm Based on SDN Architecture

4.1 MCMP Module Design

4.1.1 Data Monitor Module

In the implementation of MCMP algorithm, broadband and delay conditions in the network should be acquired first, while the link discovery module and topology module are responsible for connection conditions of multiple switches, so it is unable to probe the detailed parameter information in the link. Yet when the QoS data monitoring module collects parameters index, it is by means of issuing OpenFlow messages.

4.1.2 Routing Calculation Module

Routing calculation module is the most core module in the whole experimental platform, whose main function is to write the MCMP algorithm, and in the memory module, it can also acquire the user's specific limited conditions and the related parameter information of the link, and it can also get the network topology in the topology management module. After obtaining certain data, it will calculate according to the user's limited conditions and combined with the topology situation, and then the path that meets the user's needs can be selected. Moreover, the routing calculation module can also launch all the calculation results into the switch [11].

4.1.3 Memory Module

The memory module mainly has two functions, one is to store user restrictions, the other is to store various information detected in the data monitoring module. The memory module mainly uses MySQL database to store existing data. In addition, the routing module in the Floodlight controller is used to obtain information of the user restrictions, and then link resources. After obtaining the corresponding resources, the network status information will be read and various data will be calculated [3] (Fig. 4).

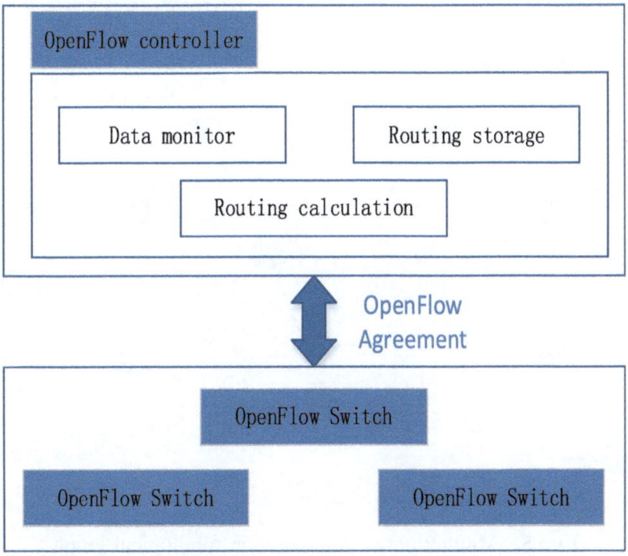

Fig. 4. Schematic diagram of MCMP architecture

5 Experimental Platform Building

When building the experimental platform, the custom topology in the Mininet simulation platform we shall be first built. After the completion of the construction, users only need to script on the basis of the provided network topology requirements.

As can be seen from Fig. 5, the custom topology generates 9 switches, 8 virtual hosts, virtual hosts h1, h2, h3 connect with virtual switch s1, virtual host h4 connects with virtual switch s2, virtual host h5, h6, h7 connect with virtual switch s6, and virtual host h8 connects with virtual switch s7.

After writing, run the Ryu controller, the current network simulation topological structure on the Ryu network visualization platform cam be seem, as shown in Fig. 5:

Next, the Ryu controller on the basis of building an automatic network topology is needed to run. First, the data monitoring module needs to be used to obtain the corresponding data, and to obtain the data accuracy. Then, the data monitoring module is used to monitor broadband data and delay data, and the corresponding data is calculated by issuing the following message. Through experimental analysis, it can be found that the data monitoring module has a high accuracy of information retrieval, and the simulation and accuracy of the experiment are obtained on the new basis of data accuracy.

Finally, the flow table information is required to be obtained. It is necessary to judge the routing effect of MCMP, set certain parameters for the path that meets the constraint condition and the path that does not meet the constraint condition simulation analysis

The parameters shown in Tables 1 and 2 are obtained through experiments:

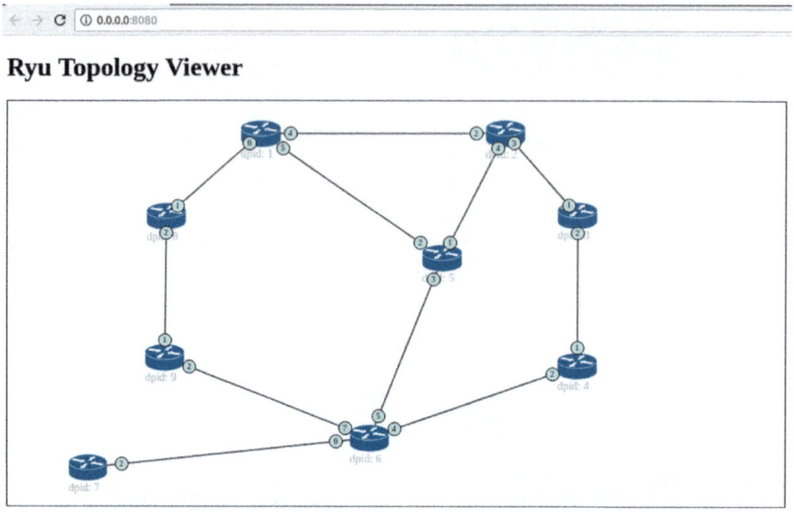

Fig. 5. Mininet network simulation topological graph

Table 1. Original link bandwidth

Bandwidth (Mbps)	S1	S2	S3	S4	S5	S6	S7
S1	–	500	–	400	500	–	–
S2	500	–	600	–	–	–	–
S3	–	600	–	500	–	–	600
S4	400	–	500	–	500	600	400
S5	500	–	–	500	–	400	–
S6	–	–	–	600	400	–	600
S7	–	–	600	400	–	600	–

Table 2. The link bandwidth measured by the data monitoring module

Bandwidth (Mbps)	S1	S2	S3	S4	S5	S6	S7
S1	–	493	–	496	488	–	–
S2	493	–	598	–	–	–	–
S3	–	598	–	490	–	–	597
S4	406	–	490	–	496	590	396
S5	488	–	–	496	–	399	–
S6	–	–	–	590	399	–	594
S7	–	–	597	396	–	594	–

Comparing the two sets of data, it can be found that the data obtained by the data monitoring module is slightly smaller than the actual bandwidth of the link. The reason for this phenomenon is that the maximum bandwidth of the link is adopted in the process of setting link bandwidth, while the network transmission of data packets in the actual network topology will occupy a certain network bandwidth, so the measured value is less than the maximum value.

In accordance with the simulation data, the link state information of the data monitoring module is basically correct.

According to the experimental platform to obtain the policy data, the algorithm simulation is strictly conducted based on the path selection method mentioned above. The analysis to the simulation results is mainly to compare the difference between MCMP algorithm and the other two algorithms, and the comparison content is the accuracy of selection path and link recovery time respectively.

(1) Comparison of path selection accuracy: The MCMP algorithm can be used to select the path that meets the constraint conditions for users who meet the restricted conditions in the network, while the application of AHP and GA will lead to instability. The main reason for this situation is that the main goal of AHP and GA is to select a relatively good path under multiple conditions of comprehensive link, but the restriction conditions are neglected, so it is impossible to select a path that meets users' needs.
(2) Link recovery time comparison: When the switch breaks down, the effect of the link disjoint algorithm is the same as that of the conventional passive routing, the main reason is that the link disjunction algorithm only guarantees the disjoint between the backup path and transmission path, but does not guarantee the node-disjoint between the backup path and the main transmission path. While the MCMP algorithm can ensure that the selected switch node does not intersect the backup path, so the recovery time is very stable.

This shows that the MCMP algorithm can satisfy various requirements and restricted conditions of users, and the algorithm path that meets the constraints of users can be accurately selected for the first time, both in terms of time delay and broadband. The MCMP algorithm can also solve the problems such as slow link recovery and so on.

6 Conclusion

Along with the continuous development and progress of SDN network, users' requirements for the network are gradually increasing. Users hope that the relevant data can be propagated in the path under certain restrictions. While the SDN network is characterized by centralized control and flexible management etc., therefore, specific to such kind of SDN network, a multi-restricted multi-path QoS algorithm, MCMP, is proposed, and the main objective of this algorithm is to select a satisfactory path for the user based on the user constraints and network conditions.

Acknowledgments. This paper is supported by the industrial research project of Shaanxi science and technology department (No. 2016GY-088), the fund of school of computer science and engineering of Xi'an Technological University (No. GSYSI2016012) and the educational reform fund of Xi'an Technological University (No. 16JGY25).

References

1. Chen Z (2017) Research and implementation of QoS control technology in SDN network. University of Electronic Science and Technology of China
2. Gao L (2017) Extensibility study on spatial information network control plane based on SDN. Beijing University of Posts and Telecommunications
3. Zhan, ZH (2016) Research on data center routing strategy and security authentication based on SDN. Anhui University
4. Lawson S (2013) Network heavy hitters to pool SDN efforts in OpenDaylight project. Network World, 8 April 2013. www.networkworld.com/news
5. Open_vSwitch, Open_vSwitch database schema, DB Schema 7. 6. 0
6. Xu K, Liu H, Liu J, Shen M (2011) One more weight is enough: toward the optimal traffic engineering with OSPF. In: 2011 31st international conference on distributed computing systems (ICDCS). IEEE, pp 836–846
7. Zhang M, Liu B, Zhang B (2009) Multi-commodity flow traffic engineering with hybrid MPLS/OSPF routing. In: 2009 IEEE global telecommunications conference, GLOBECOM 2009. IEEE, pp 1–6
8. McKeown N, Anderson T, Balakrishnan H et al (2008) OpenFlow: enabling innovation in campus networks. ACM SIGCOMM Comput Commun Rev 38(2):69–74
9. Solberg M, Silverman B (2017) OpenStack for architects. Packt Publishing Ltd.
10. Chirammal HD, Mukhedkar P, Vettathu A (2016) Mastering KVM virtualization. Packt Publishing Ltd.
11. Ivanov K (2017) KVM virtualization cookbook. Packt Publishing Ltd.
12. Subramanian S, Chowdhury CD (2015) OpenStack networking cookbook. Packt Publishing Ltd.

The Research on Core Development Technology of Security Big Data Application Platform

Gao Chai(✉)

Political College of National Defence University, Xi'an, China
124974610@qq.com

Abstract. Big data is the product of the rapid expansion of Internet data, and the potential of big data analysis has also been stimulated. Many companies will use the advantages of the new technology of big data to enhance their competition. This paper mainly studies the application of big data technology in the security work of large events. The main of this article, through the use of large data technology for integrated excavation, and through the use of application terminals for information collection in large-scale event sites, the identity of the personnel as the excavation entrance, the excavation of the identity of the collected information, crime background information, network relations, etc. With the corresponding software development technology to demonstrate the results of business classification, so as to construct a large data security platform. This paper uses Oracle data warehouse technology, combined with big data technology to construct all kinds of business data into large data layer after extraction, conversion, and filtering, to study and analyze the core development technology of security big data application platform. Using Web development technology to realize face recognition and vehicle recognition.

Keywords: Security · Big data · Data warehouse · Develop

1 Data Warehouse Concept

Data warehouse is the structured data environment of decision support system (DSS) and on-line analysis application data source. The data warehouse studies and solves the problem of obtaining information from the database. The characteristics of the data warehouse are theme-oriented, integrated, stable, and time-varying. The data warehouse was proposed by Bill Inmon, the father of the data warehouse, in 1990 [1, 2].

2 Data Warehouse Features

The data warehouse is not a so-called "large database" when the database has already existed in a large number of places, in order to further excavate data resources and for decision-making needs. The purpose of the construction of the data warehouse is to serve as the basis for the front-end query and analysis. Due to the large redundancy, the storage of W is also large. Data warehouse has several characteristics [3].

Z. Xu et al. (Eds.): CSIA 2019, AISC 928, pp. 479–486, 2020.
https://doi.org/10.1007/978-3-030-15235-2_70

Efficient. The data warehouse analysis data is generally divided into day, week, month, season, year, etc. The data requirements for the daily cycle are the most efficient, requiring 24 h or even 12 h. The customer can see yesterday's data analysis. Bad data warehouses often have problems, obviously not [4].

Data quality. The data warehouse provides a variety of information that requires accurate data, but because the data warehouse process is usually divided into multiple steps such as data cleaning, loading, query, and display, the complex architecture triggers more layers, so it is easy to lead to data distortion. May cause the customer to analyze the wrong decision, causing the loss [3].

Extensibility. Considering the extensibility of the next 3–5 years, some large data warehouse system architecture design is complex, so that the future does not need to spend too much money to rebuild the data warehouse system to ensure stability. It is mainly reflected in the rationality of data modeling. There are more middle layers in the data warehouse scheme, so that the massive data flow has enough buffer to prevent it from running when the data volume is much larger. It can be seen that the data warehouse technology can awaken the data accumulated by the enterprise for many years, not only to manage these massive data for the enterprise, but also to explore the potential value of the data. Broadly speaking, the decision support system based on data warehouse is composed of H components: data warehouse technology, online analysis and processing technology and data mining technology, in which data warehouse technology is the core of the system. The following will introduce the main technologies of modern data warehouse and the main steps of data processing around data warehouse technology, and discuss how to use these technologies to help with operational maintenance [1, 2].

3 Data Warehouse Establishment Steps

In the first step, design and define metadata, including data source's data type, structure of definition and cube, structure of movement, structure of process flow, and even set up of schedule. Cheeky.

In the second step, the target dimension, cube, poping, etc. generated by OWB compilation are deployed to the target Oracle database (in the case of 11gR2, it can also be deployed to the OC4J-based J2EE server). The specific performance is to create a tag table, create a variety of plsq1 package, and so on. Cheeky.

The third step is to call and execute plsq1 package through process flow, schedule, or manual, to actually load the data into the data warehouse. Cheeky.

The first two steps are generally repeated in the development test environment until the program is determined to be the most popular. Almost the same meaning as writing code. Once the design is completed, the H step is performed every once in a while. How to query and use the data in the data warehouse is not the time we need to consider now. These can be used by OBISE, OBIEE, and even other company products [5].

4 The Data Extraction Study of Informatica PowerCenter

4.1 Total Comparison

The condition is that the source table and the destination table contain unique fields that identify the record, such as the primary key "CUSTOMER_KEY". The method is as follows:

Reads a record from the source table, compares the corresponding field to the target table field, and updates the target table corresponding record if the source table record exists; If the record does not exist in the target table, the record is written to the target table to achieve the purpose of increasing the number of updates. The specific implementation is shown in Fig. 1.

Fig. 1. Full table comparison process

Lookup_custom_k node (find), query the record from the target table according to the primary key "CUSTOMER_KEY", and return the output "CUSTOMER_KEY_boolean" according to the input record, indicating whether the input record exists in the target table. Update if the record is in the destination table, insert if the record does not exist; Inertorupdate node (update policy), enter expression:

IIF(ISNULL(CUSTOMER_KEY_boolean),DD_TNSERT,DD_UPDATE)

There is a problem: When each record is extracted, it must be compared with the target table, and the system is expensive; The total table comparison method has low requirements and wide application to the data table, but the update speed is slow and the efficiency is low. The source table deletion can not be monitored, that is, a record of the source table was deleted but the record still exists in the target table [6, 7].

4.2 Time-Stamp Approach

The time-stamp approach is shown in Figs. 2 and 3.

Filter record: CREATE _ TIME < sydate and CREATE _ TIME & GT; = $$ LastUpdateTime
UPDATEPARAMETER(expression node)
Updated mapping variable value: SetMax Vareable($$LastUpdateTime, CREATE _ TIME)

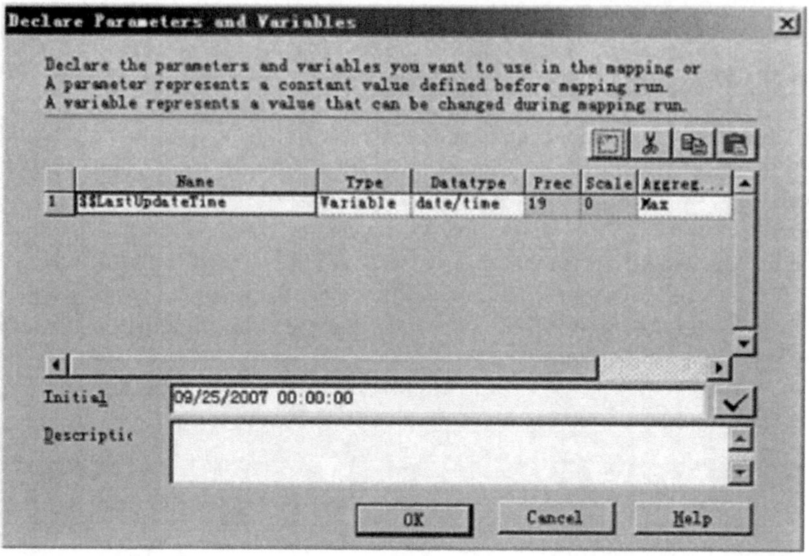

Fig. 2. Variable settings

Fig. 3. Filter settings

Problems: Fast speed, only supports insertion operations; Unable to monitor source table deletion and update; If the source data record is updated, but the timestamp field is not updated, the record maintains the old record in the target table; If the source data record is updated but the timestamp field is not updated, the record remains in the target table.

4.3 Trigger Mode

The condition is that the source table and the destination table contain the primary key that identifies the record. The method is as follows:

The flow chart of data extraction is shown in Fig. 4.

Step 1: Create a sequence(common object) in the source database.

```
create sequence seq_bx start with 1 increment by 1;
```

Step 2: Create functions(shared objects) in the source database.

```
create or replace function changeStatue(Var_table_name IN
varchar2,var_pk_column IN int)
return varchar2 IS
var_update_sq1 varchar2(1000):
BEGIN
var_update_sql:='update '||var_table_name||' set statue=1 where id='
||var_pk_column:
execute immediate var_update_sql:
COMMIT;
return'ok' ;
EXCEPTION
WHEN OTHERS THEN
Return 'error' ;
END;
```

Step 3:Create a monitoring table("customer_bx") on the source database related table(such as "customer", its primary key "c_id")to record data changes in the table.

```
create table customer_bx(
id int not null,
c_id int not null,
operation varchar(10) not null,
statue int not null,
CREATE_TIME date not null,
primary key(id)
):
```

c_id:

operation;insert、update、delete,

statue:Status code, initially 0, updated to 1 When this record is imported into the target library

CREATE_TIME:

Step 4: Create a trigger to monitor changes in the source table(such as "Custer")

```
create trigger tjfx_trg_ customer_ after_insert
after insert on customer for each row
begin
insert into customer_ bx values(sec}bx.nextval,:new. C_ID,' insert', 0, sysdate);
end trigger_ customer_ after_insert;
create trigger trigger_ customer_ after_ update
after update on customer for each row
begin
insert into
values (seq_bx. nextval,' C_ I D', :new. C_ID,' update', 0, sysdate);
end trigger_ customer_ after update;
```

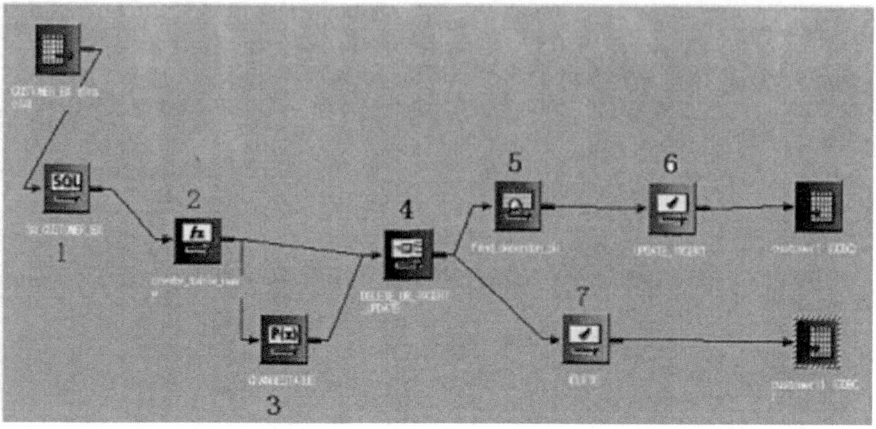

Fig. 4. Data extraction flow chart

5 Caller Image Match

Face recognition, the business processing method is to push the local image to the face server for comparison and return the matching program (0 to 100 values, the higher the match). The main code fragment of the program is as follows:

```
    CHCNetSDK.NET_VCA_FIND_PICTURECOND pi=new
    CHCNetSDK.NET_VCA_FIND_PICTURECOND();
    pi.lChannel=CHANNE;
    pi.struStartTime=start;
    pi.struStopTime=stop;
    CHCNetSDK.NET_VCA_FACESNAP_MATCH_ALARM_LOG
alarmLog=new
    CHCNetSDK.NET_VCA_FACESNAP_MATCH_ALARM_LOG();
    try
    {
    int
iflag=CHCNetSDK.NET_DVR_FindFaceMatchAlarm(server.m_lUserID,ref pi);
    if(iflag==-l)
    {
    iLastErr=CHCNetSDK.NET_DVR_GetLastError();
    Thread.Sleep(1000);
    }
    int       xxx=CHCNetSDK.NET_DVR_FindNextFaceMatchAlarm(iflag.out
alarmLog);
```

```
while(xxx=1002)//1002
{
xxx=CHCNetSDK.NET_DVR_FindNextFaceMatchAlarm(iflag.out   alarmLog);
}
while   (xxx=1000)
{
uint   SnapFacePicID=alarmLog.struSnapInfoLog.dwSnapFacePicID;
uint   BlackListPicID=alaraiLog.struBlackListInfoLog.dwBlackListPicID;
uint
RegisterlD=alarmLog.struBlackListInfoLog.struBlackListInfo.dwRegisterlD;
CHCNetSDK.NET_VCA_FACEMATCH_PICCOND   picc=new
CHCNetSDK.NET_VCA_FACEMATCH_PICCOND();
picc.dwSnapFacelD=SnapFacePicID;
picc.dwBlackListFacelD=RegisterlD;
picc.dwBlackListlD=BlackListPicID;
picc.byRes=new byte[100000];
CHCNetSDK.NET_VCA_FACEMATCH_PICTURE   pict=new
CHCNetSDK.NET_VCA_FACEMATCH_PICTURE();
pict.pBlackListFace=new   IntPtr();
pict.pSnapFace=new IntPtr();
```

6 Call Vehicle Identification

Vehicle identification, input license plate number, call the vehicle identification server database query, based on the query results (vehicle color, frame number, license plate information for information verification), the implementation code is as follows:

```
DataTable   dt=new   DataTable   ();
//connectionString
NpgConn=new   NpgsqlConnection   (connectionstring);
NpgConn.Open ();
DataSet ds =new DataSet ();
String sqlStr="select*from hps_passvehicleinfo   p   left   join   hps_gateinfo g
on p.gate_id=g.gate_id
Where g.CARNO="+ License plate number+";
NpgsqlDataAdapter objAdapter=new NpgsqlDataAdapter (sql,NpgConn);
objAdapter.Fill(ds);
dt=s.Tables[0];
```

7 Conclusion

At the beginning of the era of big data, the research results of large data from all walks of life are beginning to show results. It is important to use smart data in the integrated construction of security, defense, management, and control. The application of big data will be a necessary way to crack down on illegal crimes and comprehensive social management. Through the design of this paper and the practical development of the large data security application platform, Oracle data warehouse construction ideas and related technologies, based on the research and design, the system is developed in practice. The large-scale data security application platform basically realizes the identification and control of personnel and vehicles in various scenes (security gates, road passes, and public security patrols) during large-scale events. However, in the course of system practice, how to improve the use of experience and data impact speed in each application terminal is a place to be optimized. In the practical development of the system, more in-depth research is needed on. net technology, Oracle database technology, and popular UI technology applications.

References

1. Gopalakrishnan Nair TR, Selvarani R (2011) Defect proneness estimation and feedback approach for software design quality improvement. Inf Softw Technol 54(3):274–285
2. Unhelkar B, Murugesan S (2010) The enterprise mobile applications development framework. IT Prof Mag 12(3):33–39
3. (2010) Vocational education of communications and information technology conference (2010 NVCIC)
4. (2016) Over drive announces a series of "Big Data" re-ports. Spolanka. http://libraris.wright.edu/noshelfrequired/2012/04/11/overdrive-announces-a-series-of-big-data-reports
5. Sageman M (2004) Understanding terror networks. University of Pennsylvania Press
6. Williams K, Wu J, Choudhury SR et al (2014) Scholarly big data information extraction and integration in the CiteSeerx digital library. In: 2014 IEEE 30th international conference on data engineering workshops (ICDEW)
7. Jiang D, Chen G, Ooi BC et al (2014) epiC: an extensible and scalable system for processing big data. In: Proceedings of the VLDB endowment

Research on Parameter Optimization of Tracked Vehicle Transmission System Based on Genetic Algorithm

Lingjun Wei[1,2(✉)], Haiou Liu[1], Huiyan Chen[1], Ziye Zhao[1], Yi Xu[3], and Hongyan Zhang[3]

[1] School of Mechanical Engineering, Beijing Institute of Technology, Beijing 10081, China
wljyal66@126.com

[2] Beijing Vocational College of Transportation, Beijing 102618, China

[3] China North Vehicle Research Institute, Beijing 100072, China

Abstract. The transmission system of tracked vehicle has great influence on the fuel economy and power performance of the whole vehicle. Its drivetrain parameter design is a multi parameter, multi-objective and nonlinear optimization problem, which is an important part of vehicle design. Based on the theoretical analysis of the optimization parameters and evaluation target, established the Smulink simulation model and normalization of tracked vehicle transmission system, according to the basic principle of genetic algorithm, computer simulation technology and optimization theory is applied to vehicle power transmission system parameter optimization design, the transmission ratio and the main reduction ratio as design variables, to guarantee the basis of fuel consumption as the target function to measure the tracked vehicle fuel economy, to the acceleration time and gradient optimization mathematical model of vehicle transmission system with a target shoe as constraint conditions. By comparing the optimization scheme and the original scheme, the traction characteristics and fuel economy are greatly improved. In the scheme, the objective function is greatly improved by improving the transmission ratio parameters of tracked vehicles, which proves that the optimization design method of tracked vehicles transmission system using genetic algorithm has been greatly improved.

Keywords: Genetic algorithm · Tracked vehicle · Drive system · Parameter optimization

1 Introduction

The optimization problem has always been an important research topic in the field of Engineering technology. The problem it studies is to discuss the best plan in many schemes. Optimal matching of power transmission system is an important method to improve fuel economy and traction characteristics of tracked vehicles. Through the matching of power transmission system, the performance requirements of tracked vehicles under different structure, volume, mass, environment and working conditions can be realized [1].

Z. Xu et al. (Eds.): CSIA 2019, AISC 928, pp. 487–495, 2020.
https://doi.org/10.1007/978-3-030-15235-2_71

Action characteristics and fuel economy [6]. Genetic Algorithms (GA) is an efficient global search algorithm based on genetic theory and natural selection, which combines the random information exchange mechanism of chromosomes within the population and survival rules of the fittest in the process of biological evolution [2]. GA provides a general framework for solving nonlinear systems, multi models, and multi-objective optimization problems [3]. Because the modified algorithm only requires the optimized function to be computable and does not require continuity and differentiability, it is widely used in automatic control, combinatorial optimization, function optimization, machine learning and other technical fields [4].

The basic function of the tracked vehicle transmission system is to transmit the power from the engine to the driving track, producing the driving force, so that the tracked vehicle can move forward or backward at a certain speed [5]. The optimization matching problem of the transmission system of tracked vehicle is actually to make the engine work in the optimum working area regularly By choosing appropriate transmission system parameters and forms (transmission form, gear number, transmission ratio interval, transmission ratio, drive axle form and speed ratio), according to the conditions and requirements of the vehicle. The best tr

2 Establishment of Powertrain Simulation Model

In order to analyze the dynamic torque characteristics of the transmission system, the power transmission system model is established by using the normalized equivalent method as shown in Fig. 1 and the power transmission structure diagram as shown in Fig. 2. According to the principle that the kinetic energy, potential energy and loss of the transformed system remain unchanged, the original parameters such as the moment of inertia, torsional stiffness and damping coefficient of the rotating parts are transformed into the transmission shaft according to formula 1 [7]. There are two steps to establish the model: first simplify the original system, obtain the rotational inertia, torsional stiffness and other parameters on the lumped mass-elastic axis, and then normalize them [8].

$$\begin{cases} J^* = J/i^2 \\ k^* = k/i^2 \\ c^* = c/i^2 \\ i = n/n' \\ M^* = M/i^2 \end{cases} \tag{1}$$

Formula: k, c, J, n and M are the torsional stiffness, damping, rotational inertia, rotational speed and torque of a component before conversion (original system); K*, C*, J*, n*, and M* are the torsional stiffness, damping, rotational inertia, rotational speed and torque of a component after conversion (equivalent system); I is the actual transmission ratio of the shafting.

Through the analysis of simplified mechanics module, we can get the mathematical expression of bifurcation mechanical model [9]:

$$J_n^* \ddot{\theta}_n + k_{n+1}^*(\theta_n - \theta_{n+1}) + c_{n+1}^*(\dot{\theta}_n - \dot{\theta}_{n+1}) + k_{n+2}^*(\theta_n - \theta_{n+2}) + c_{n+2}^*(\dot{\theta}_n - \dot{\theta}_{n+2})$$
$$= k_n^*(\theta_{n-1} - \theta_n) + c_n^*(\dot{\theta}_{n-1} - \dot{\theta}_n) \tag{2}$$

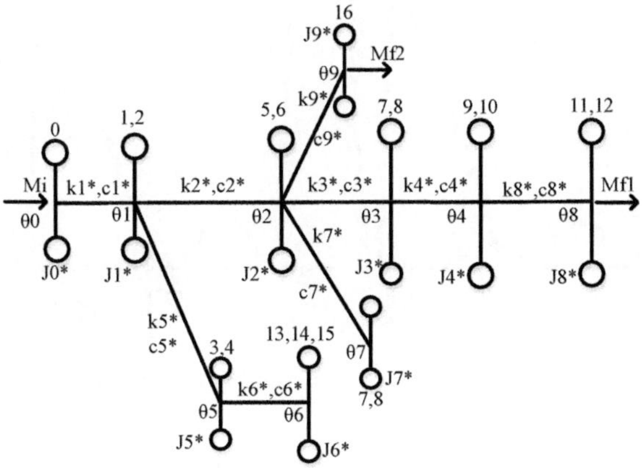

Fig. 1. Normalized equivalent system model

The Smulink simulation model is built on the basis of normalized model and transmission structure, as shown in Fig. 2.

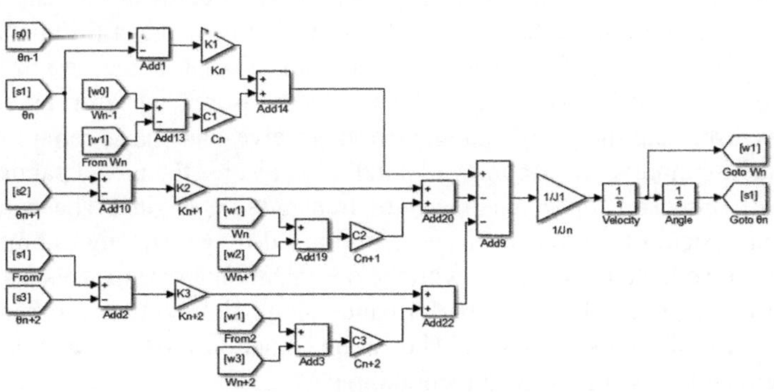

Fig. 2. Smulink simulation model

Figure 3 is a comparison of ideal velocity curves and simulation results for tracked vehicles in 60 s time.

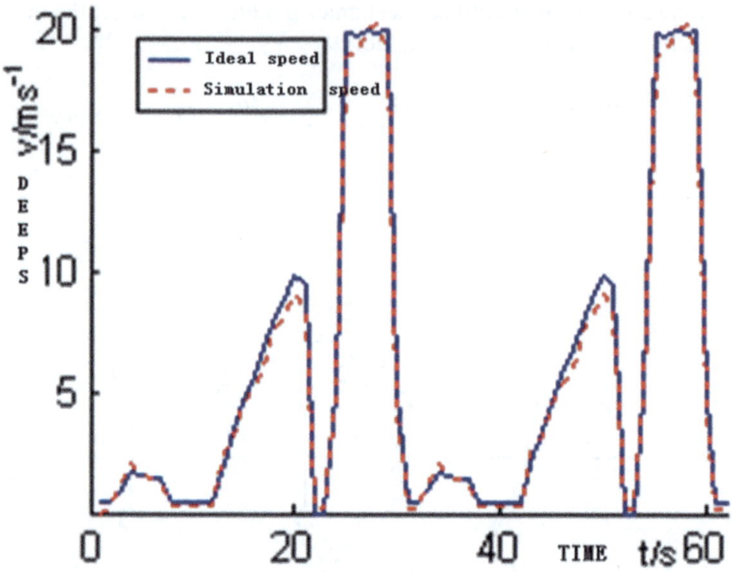

Fig. 3. Ideal velocity curve and speed curve of tracked vehicle

The normalized model of tracked vehicle established in this paper is credible, and the simulation results reflect the actual vehicle response characteristics more truly.

3 Design and Implementation of Genetic Algorithm

Through the analysis of Fig. 3, the speed characteristic curves of the real vehicle test results are similar to those of the simulation results, but there are still some differences. The main reasons for the discrepancy are: normalization of some design parameters during modeling [10]; Secondly, it is affected by measurement error, assembly error and process error, and the design parameters themselves also have certain uncertainty [11]. In order to improve the accuracy of simulation results, the model parameters must be modified. Based on the given power of transmission engine, The parameters of transmission system of tracked vehicle are optimized by genetic algorithm [12].

In order to facilitate the interface with the optimization program. This paper designs a non-interface of tracked vehicle performance simulation software to run, which is realized by an m file mine-no-gui.m. The calling function format is as follows:

[error-code-resp] = mine-no-gui(action,input)

The parameters of the call function are as follows:

(1) action

Action denotes the execution of functions and is represented by strings, such as modify in Table 1 and other five command actions.

Table 1. Actions and instructions of function execution

Action	Function
Initialize	Initializing Matlab workspace
Modify	Modification of transmission system parameter values
Grade-test	Simulation and calculation of climbing gradient
Accel-test	Acceleration simulation calculation
Drive-cycle	Instruction simulation cycle

(2) input

"Input" is the input parameter of a function and is a structure. The commonly used format is input. [field]. [subfield], where the field variable corresponds to the action in the table, and each field variable contains a different number of subfield variables below it, i.e. the parameters of each part of the specific transmission system or the cycle parameters of the working condition, for example, to be modified. Gearbox transmission ratio can be set as follows:

input.[field].[subfield] = {'gb_ratio'} input.modify.value = {x(1), x(2), x(3), x(4)}

Among them, X(1), X(2), X(3) and X(4) are the transmission ratios of four gears.

4 Optimization Model

4.1 Setting the Objective Function

The basis of parameter optimization for tracked vehicle powertrain is to determine the optimization objective function [13]. These objective functions have combined the inherent characteristics of the engine and chassis of tracked vehicles with their actual driving conditions, can quantitatively reflect the matching degree of the power transmission system of tracked vehicles, can reflect the exertion degree of the engine power and economy, and can indicate the engine corresponding to the actual driving conditions of tracked vehicles. The difference between the working conditions and the ideal working conditions can indicate the potential and possible ways to improve the driveline [14]. In short, the establishment of the objective function is a prerequisite for parameter correction, so that the similarity between the simulation results and the real vehicle test results can be quantified. The objective of model optimization is to optimize fuel economy on the basis of guaranteed throughput. The fuel economy is calculated by taking the cycle conditions shown in Fig. 4 as an example [15]. That is:

$$\min f(X) = com\_FE \tag{3}$$

In the formula, *com_FE* is the cycle condition in Fig. 4, and the fuel consumption of loader, L/h, is calculated by simulation.

4.2 Optimization Variables

Among the transmission parameters of tracked vehicles, the transmission ratio is the main parameter affecting the matching between engine and transmission system [16]. Tracked vehicles in the performance of the task, as long as the operating conditions remain unchanged, the gearbox shift unchanged, in many cases, its traction characteristics are sacrificed to ensure its economy [17]. In view of this situation, the gearbox of tracked vehicle is set to four gears, and the four optimization variables are the transmission ratio of four gears, i.e.

$$X = [i_{q1}, i_{q2}, i_{q3}, i_{q4}]^T = [x_1, x_2, x_3, x_4]^T \tag{4}$$

The upper and lower bounds for variables optimization are shown in Table 2.

Table 2. Upper and lower bounds of optimal variables

Block	i_{q1}	i_{q2}	i_{q3}	i_{q4}
Lower bound of optimal variable	4.920	2.602	1.415	0.749
Upper bound of optimal variables	3.644	1.924	1.046	0.553

4.3 Building Constraint Functions

Tracked vehicle transmission system parameters must meet the minimum requirements for the transmission system, these requirements not only to ensure its performance in special operating conditions, but also to consider its structure and workmanship [18]. Different design variables have different constraints. Considering the basic requirements of the maximum dynamic factor, the maximum gradient and the maximum speed of the tracked vehicle, the constraints are constructed as follows:

(1) $t_1 \leq 32.7$ In the form, t_1 is 0–25 km/h acceleration time (s);
(2) $t_2 \leq 22.6$ In the form, t_2 is 10–20 km/h acceleration time (s);
(3) $\alpha_1 \geq 53.1$ In the form, α_1 is the maximum gradient (%) when I block.

5 Optimization Results and Analysis

This paper takes the fuel economy as the objective function on the basis of ensuring the dynamic performance of tracked vehicles. Using the genetic algorithm toolbox in MATLAB, the appropriate functions and constraints are compiled to optimize the transmission ratio of tracked vehicle gearbox. The group size $n = 35$, crossover probability $p_c = 0.92$, mutation probability $p_m = 0.08$, After 60 generations of evolution, a relatively perfect convergence curve of transmission ratio optimization can be obtained, as shown in Fig. 5.

The optimization results are as follows: x = [4.27741, 2.25742, 1.24732, 0.64319]. The optimization results are processed as follows: x = [4.277, 2.257, 1.247, 0.643].

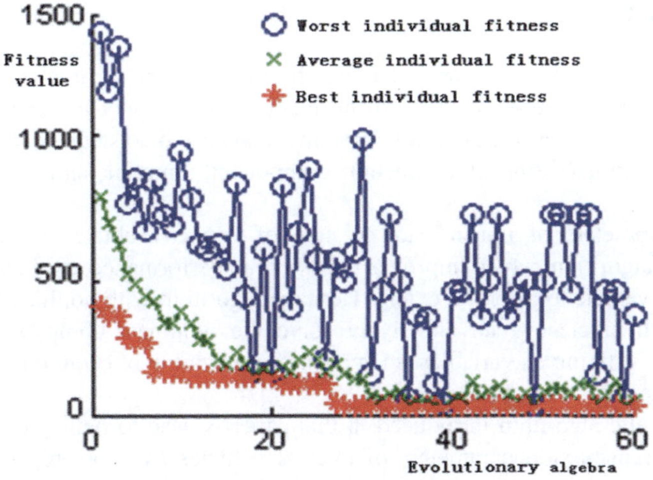

Fig. 4. Convergence curve

The final optimization results of the genetic algorithm are shown in Table 3, which shows that the fuel economy of the vehicle is much better than that of the original vehicle., which is reduced from $15.18\,L * h - 1$ to $13.55\,L * h - 1$. In the traction characteristic index, the 0–25 km/h acceleration time was reduced by 1.1 s; the first gear climbing gradient increased from 53.1% to 59.4%.

Table 3. Optimized results by genetic algorithm

Project				Original car	Optimization results
Target value com_FE/$(L \cdot h^{-1})$				15.18	13.55
Design variable	i_{q1}			2.155	4.277
	i_{q2}			0.578	2.258
	i_{q3}				1.247
	i_{q4}				0.643
Constraint	Acceleration time(s)	0–25 km/h	t_1	32.7	31.6
		10–20 km/h	t_2	22.6	20.9
	(%)I gear climbing		α_1	53.1	59.4

By comparing the optimization scheme with the original scheme, the traction characteristics and fuel economy have been greatly improved. By improving the transmission ratio parameters of tracked vehicles, the objective function is improved obviously, which shows that the optimization design method of tracked vehicle transmission using genetic algorithm is better.

6 Conclusion

(1) In this paper, many factors affecting the performance of tracked vehicles are considered comprehensively, and the mathematical model of optimal design of automotive power transmission system is analyzed systematically. Based on normalization of objective function component, Smulink simulation model is built.

(2) The parameters of transmission system of tracked vehicle are optimized by genetic algorithm, which improves the dynamic performance and fuel economy of tracked vehicle to a certain extent. Heritage algorithm with nonlinear, parallelism and high efficiency can effectively solve the nonlinear complex optimization problem with mixed variables, such as the optimization of economic and dynamic parameters of tracked vehicles.

(3) The genetic algorithm introduced in this paper is used to optimize the design of power transmission parameters of tracked vehicles can not only make a quick optimization choice for the parameter matching of the transmission system, but also make a more accurate prediction of the power performance and fuel saving effect of the matching scheme of the transmission system at the design stage. This avoids the blindness of the designer in the design stage and reduces the time of product development and development. It has a certain value in the research and development of tracked vehicles.

References

1. Ashish KB, Anil K, Girish KS et al (2015) Performance study of evolutionary algorithm for different wavelet filters for satellite image denoising using sub-band adaptive threshold. J Exp Theor Artif Intell, 1–25
2. Scott DW (2015) Multivariate density estimation: theory, practice, and visualization. Wiley, Hoboken
3. Sun J (2005) Pro/engineerWildfire mechanical design course. Tsinghua University Press, Beijing
4. Zhang J, Yang J (2006) Assembly design and motion simulation and Pro/E realization. National Defense Industry Press, Beijing
5. Ying JL (2005) MATLAB genetic algorithm toolbox and application. Xi'an Electronic and Science University Press, Xi'an
6. Yin B, Sun Y (2004) Application of penalty function method in optimal matching of automobile transmission system. Mech Des Manuf 6:4–5
7. Deb K, Pratap A, Agarwal SA (2002) Fast and elieist multi-objective genetic algorithm: NSGA-II. IEEE Trans. Evol. Comput. 6(2):182–197
8. Wu G, Chen H (2009) Multi-objective optimization of hybrid electric vehicle parameters based on genetic algorithm. Automobile Eng 31(1):60–64
9. Xu Z, Chen Z (2004) Recent progress in basic theory research of genetic algorithm 4:91–114
10. Li L (2008) Application of multi-objective optimization strategy based on genetic algorithm. Jiangnan University, Wuxi
11. Chen Q (2007) Advanced electric vehicle technology. Chemical Industry Press, Beijing
12. Yu ZS (2010) Automobile theory. Machinery Industry Press, Beijing

13. Mu Z, Cao W, Zundi H (2010) Optimum design of lateral stability of railway vehicles based on ISIGHT. Locomotive Rolling Stock 30(2):19–21
14. Qian L, Wu W, Zhao H et al (2005) J Agric Mach 36(9):5–8
15. (2010) Orthogonal optimization and application of comprehensive parameters of hybrid bus power system. J Hefei Univ Technol (Nat Sci Ed) 33(3):341–345
16. Yu WW (2006) Automobile design. Machinery Industry Press, Beijing
17. Yan F, Hu F, Tian S et al (2009) Parameter optimization design of automotive power transmission system based on multi-objective genetic algorithm. Automobile Technol 12:20–23
18. Cao J, Lee W, Cheng HS (2009) Experimental and numerical investigation of combined isotro-pickmatic hardening behavior of sheet metals. Int J Plast 25:942–972

Adaptive Discrete-Time Sliding Mode Control of Brushless DC Motor Servo System for Unmanned Surface Vehicles

Zhongjiu Zheng, Ning Wang(✉), and Hong Zhao

Center for Intelligent Marine Vehicles, College of Marine Electrical Engineering, Dalian Maritime University, Dalian, China
zhengzhongjiu@163.com

Abstract. Considering the diverse effect on the electric steering gear servo system of unmanned surface vehicles (USV) caused by the external disturbances and parameter perturbations, a novel discrete adaptive global sliding model control scheme (AGSMC) is proposed in this paper. By introducing a nonlinear switch function, the system trajectory stays on the designed sliding manifold at the initial time. On the basis of discrete nonlinear adaptive reaching law, linear extrapolations are adopted to estimate the system uncertainty dynamics to ensure the robustness of the control system. Simulation results show that, in contrast with other discrete sliding mode control methods which are based on the exponential approach law, the proposed AGSMC schemes achieve enhanced tracking precisions with reduced control consumptions where chattering is significantly suppressed.

Keywords: Electrical servo system · Adaptive control · Sliding-model control · Discrete sliding-mode

1 Introduction

During the last decades, significant improvements have been achieved in the brushless DC motor due to the development of the novel magnetic materials. The motor torque, response speed and the power mass ratio have all been improved, therefore the DC motor is used in the electric steering gear servo system of USVs. Meanwhile, the complex process environments for electric steering gear servo system result in strict performance requirements. However, considering the modeling errors, parameter perturbations and uncertainty factors such as disturbance moments, further improvements of the electric steering gear follow-up system become very difficult [1–3]. Hence, it is essential to focus on the research on the control technology of the electric steering gear servo system for USVs.

In recent years, sliding-mode control (SMC) [4, 5] has been widely used in motor control. However, the insensitivity against uncertainties of the system is guaranteed only in its sliding mode. In other words, effective improvements will be achieved if we can reduce the reaching time to the sliding surface. And shortening arrival time has become an important research content of variable structure control.

Z. Xu et al. (Eds.): CSIA 2019, AISC 928, pp. 496–504, 2020.
https://doi.org/10.1007/978-3-030-15235-2_72

According to the artificial regulation of the trajectory, the reaching law approach [6, 7] improves the transient quality of the approaching movement to shorten the reaching period of the sliding surface, making it widely applicative in industry. In this methods, the upper bound of the uncertainties are introduced to ensure the reachability of the sliding mode [8, 9], making the estimated accuracy serious constraints in industry application. Besides that, another design approach called global sliding mode method makes the trajectory of the system stay on the sliding manifold at initial time via construction of the nonlinear switch function. The reaching process of sliding-mode control is eliminated while the robustness is enhanced, resulting in widespread attentions and deep researches [10, 11]. Approaching method combining together with nonlinear switch function design can assure the global robustness of the controller in stable conditions.

For the problems mentioned above, a novel adaptive sliding-mode reaching law is proposed in this paper. Unlike some existing sliding-mode control schemes, the reaching velocity varies in the light of the distance between the sliding manifold and the trajectory of the system. Moreover, the uncertainty terms are calculated on-line by using the linear extrapolation method, which is confirmed to achieve precise estimated accuracy with low computational complexity. Thus, enhanced global tracking performance is realized using the adaptive robust control with reduced control consumption where the chattering phenomenon is not obvious.

2 Model Derivation

The electrical follow-up system is actually a closed loop angular position servo system, which is mainly composed of a controller, a driver, a DC servo motor, a speed reducer and a potentiometer. The basic structure of the system is shown in Fig. 1.

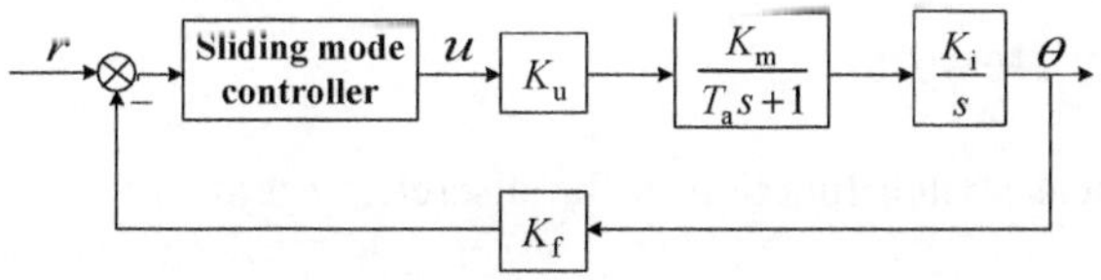

Fig. 1. Structure of the DC electric servo system.

Here, the position command, control input and the output are denoted by subscripts r, u and θ respectively. K_u is the drive amplification, while K_m and T_a are the magnification and time constants of the DC motor, K_i and K_f represent the transmission coefficients of the harmonic reducer and the feedback loop. Let $K_a = K_u K_m K_i K_f$, the state equation:

$$\dot{\boldsymbol{x}} = \boldsymbol{A}\boldsymbol{x} + \boldsymbol{b}u \tag{1}$$

Where,

$$\boldsymbol{A} = \begin{bmatrix} 0 & 1 \\ 0 & -\frac{1}{T_a} \end{bmatrix},\ \boldsymbol{b} = \begin{bmatrix} 0 \\ \frac{K_a}{T_a} \end{bmatrix},\ \boldsymbol{x} = \begin{bmatrix} x_1 \\ x_2 \end{bmatrix}$$

Moreover, $\boldsymbol{x}$ denotes the state vector of the system; $x_1 = \theta$ describes the output angle while $x_2 = \dot{\theta}$ describes the angular speed.

Due to the discrete character of the computer control, the system (1) should be processed with discretization as below

$$\mathbf{x}(k+1) = \mathbf{F}\mathbf{x}(k) + \mathbf{g}u(k) \tag{2}$$

Where,

$$\boldsymbol{F} = \begin{bmatrix} 1 & T_a(1 - e^{-\frac{T}{Ta}}) \\ 0 & e^{-\frac{T}{Ta}} \end{bmatrix},\ \boldsymbol{g} = \begin{bmatrix} K_a[T + T_a(e^{-\frac{T}{Ta}} - 1)] \\ K_a(1 - e^{-\frac{T}{Ta}}) \end{bmatrix}$$

Considering the Inevitable uncertainties, the equation is changed to

$$\boldsymbol{x}(k+1) = (\boldsymbol{F} + \Delta\boldsymbol{F})\boldsymbol{x}(k) + (\boldsymbol{g} + \Delta\boldsymbol{g})u(k) + \boldsymbol{d}(k) = \boldsymbol{F}\boldsymbol{x}(k) + \boldsymbol{g}u(k) + \delta(k) \tag{3}$$

Where, $\Delta\boldsymbol{F}$ and $\Delta\boldsymbol{g}$ denote the parameter perturbation matrix, and $\boldsymbol{d}(k)$ denotes the external disturbance vector. Here all the perturbations and disturbances are considered as an uncertainty term which is represented by $\delta(k)$

$$\boldsymbol{\delta}(k) = \Delta\mathbf{F}\mathbf{x}(k) + \Delta\mathbf{g}u(k) + \mathbf{d}(k) \tag{4}$$

3 Controller Design

Choose a nonlinear switch function of the discrete system as

$$s(k) = \mathbf{c}\mathbf{x}_\mathbf{e}(k) - \mathbf{c}\mathbf{P}(k)\mathbf{x}_\mathbf{e}(0) \tag{5}$$

Assume that $r(k)$ and $dr(k)$ are the desired output angle and angular velocity of the DC motor, the expected state vector $\boldsymbol{x}_d$ and the error vector $\boldsymbol{x}_e$ are demonstrated as

$$x_d(k) = [\,r(k) \quad dr(k)\,]^T \tag{6}$$

$$\mathbf{x}_\mathbf{e}(k) = \mathbf{x}_\mathbf{d}(k) - \mathbf{x}(k) \tag{7}$$

In addition, the parameters in (5) are shown as

$$\boldsymbol{c} = \begin{bmatrix} c_1 & 1 \end{bmatrix}^T,\ c_1 > 0$$

$$\boldsymbol{P}(k) = \begin{bmatrix} \beta_1^k & 0 \\ 0 & \beta_2^k \end{bmatrix},\ |\beta_1| < 1,\ |\beta_2| < 1$$

As shown in (6) that at initial time,

$$s(0) = \mathbf{c}\mathbf{x}_\mathbf{e}(0) - \mathbf{c}\mathbf{P}(0)\mathbf{x}_\mathbf{e}(0) = 0 \tag{8}$$

The trajectory stays on the sliding manifold $s(k) = 0$, resulting in the elimination of the reaching process.

The reaching law is

$$s(k+1) - s(k) = -qTs(k) - \varepsilon T\ \mathrm{sgn}(s(k)),\quad q > 0,\ \ \varepsilon > 0,\ \ qT < 1 \tag{9}$$

In the above discrete reaching law, if ε is decreased, the chattering of control is relatively reduced, but the reaching speed cannot be ensured when ε is tiny. Therefore, the ideal value of ε should be time-varying, that is when $|s(k)|$ is far away from the sliding manifold, ε should be set relatively large, and when $|s(k)|$ is close to the manifold, ε should be relatively small. In consequence, ε is selected as:

$$\varepsilon = |s(k)|/l,\ l > 0 \tag{10}$$

According to [12], the value of will decrease only if $|s(k)| > \varepsilon T/(2 - qT)$, which indicates that ε should satisfy the following requirements:

$$\varepsilon < (2 - qT)|s(k)|/T \tag{11}$$

Once T is selected such that

$$T < 2l/(1 + lq) \tag{12}$$

The inequality (11) is satisfied. When $l = 1/q$, T should be set as $T < 1/q$ which is also shown in (9).

The stability analysis under the (9) and (10) is established:

$$\begin{aligned} &[s(k+1) - s(k)]\mathrm{sgn}(s(k)) \\ &= [-qTs(k) - \frac{|s(k)|}{l}T\mathrm{sgn}(s(k))]\mathrm{sgn}(s(k)) \\ &= -(q + \frac{1}{l})T|s(k)| < 0 \end{aligned} \tag{13}$$

$$\begin{aligned}&[s(k+1)+s(k)]\text{sgn}(s(k))\\&=[(2-qT)s(k)-\frac{|s(k)|}{l}T\text{sgn}(s(k))]\text{sgn}(s(k))\\&=(2-qT-\frac{1}{l}T)|s(k)|>0\end{aligned} \tag{14}$$

It shows that the control system is stable. Moreover, once the identical equation holds that $s(k)\equiv 0$, the global sliding-mode robust control can be achieved, which indicates that the system is not sensitive on the disturbances during the whole process. In addition,

$$\begin{aligned}s(k+1)s(k)&=(1-qT)s^2(k)-\frac{|s(k)|}{l}Ts(k)\text{sgn}(s(k))\\&=(1-qT-\frac{1}{l}T)s^2(k)\end{aligned} \tag{15}$$

Because the sampling period is very small, the inequality $T<l/(1+lq)$ holds in general. At this point, the system trajectory will not pass through the sliding surface $s(k)=0$, but will approach it asymptotically. Hence, the curve of the control input is smooth without chattering.

Assumption is made that $\boldsymbol{cg}\neq 0$, combining (5), (7) and (9):

$$\begin{aligned}u(k)=&-(cg)^{-1}\{c[(F-I)x(k)+\Delta P(k)x_e(k)-\Delta x_d(k)\\&+\delta(k)]-qTs(k)-\frac{|s(k)|}{l}T\text{sgn}(s(k))\}\end{aligned} \tag{16}$$

Where,

$$\boldsymbol{I}=\begin{bmatrix}1&0\\0&1\end{bmatrix}$$

$$\Delta\mathbf{P}(k)=\mathbf{P}(k+1)-\mathbf{P}(k) \tag{17}$$

$$\Delta\mathbf{x_d}(k)=\mathbf{x_d}(k+1)-\mathbf{x_d}(k) \tag{18}$$

Note that, $\boldsymbol{x}_d(k+1)$ and $\delta(k)$ are unknown, thus the control law (16) is not achievable.

In general, the dynamic characteristics of the position tracking signal and those of the parameter perturbations and external disturbances are much slower than the sampling frequency. Therefore, the unknown terms can be forecast estimated using the linear extrapolation method:

$$\mathbf{x_d}(k+1)=2\mathbf{x_d}(k)-\mathbf{x_d}(k-1) \tag{19}$$

$$\boldsymbol{\delta}(k)=2\boldsymbol{\delta}(k-1)-\boldsymbol{\delta}(k-2) \tag{20}$$

From (5) it is concluded that:

$$\boldsymbol{\delta}(k-1) = \mathbf{x}(k) - \mathbf{F}\mathbf{x}(k-1) - \mathbf{g}u(k-1) \tag{21}$$

$$\boldsymbol{\delta}(k-2) = \mathbf{x}(k-1) - \mathbf{F}\mathbf{x}(k-2) - \mathbf{g}u(k-2) \tag{22}$$

The AGSMC scheme in this paper is achieved by combining (16)–(22):

$$\begin{aligned} u(k) = &-(\boldsymbol{cg})^{-1}\{\boldsymbol{c}[\sum_{i=0}^{2}(\boldsymbol{U}_i\boldsymbol{x}(k-i)+\boldsymbol{V}_i u(k-i) \\ &+\Delta\boldsymbol{P}(k)\boldsymbol{x}_e(0)-\Delta\boldsymbol{x}_d(k-1)]-qTs(k)-\frac{|s(k)|}{l}T\mathrm{sgn}(s(k))\} \end{aligned} \tag{23}$$

Where, $\boldsymbol{U_0} = \boldsymbol{F} + \boldsymbol{I}$, $\boldsymbol{U_1} = -2\boldsymbol{F} - \boldsymbol{I}$, $\boldsymbol{U_2} = \boldsymbol{F}$, $\boldsymbol{V_0} = \boldsymbol{0}$, $\boldsymbol{V_1} = -2\boldsymbol{g}$, $\boldsymbol{V_2} = \boldsymbol{g}$

As is shown in (23), the control law is realizable since all the control variables are calculated using the information which is already known.

4 Simulation Results

The conventional discrete reaching law sliding-mode controller (RLSMC) is introduced for comparison. The state vector $\boldsymbol{x}_d(k+1)$ is estimated via linear extrapolation and the upper bound of the uncertainty term is introduced in RLSMC. The control law is:

$$\begin{aligned} u(k) = &-(\mathbf{cg})^{-1}\{\mathbf{c}[\mathbf{x_d}(k-1) - 2\mathbf{x_d}(k)] + \mathbf{cFx}(k) \\ &+ (1-qT)s(k) - (\varepsilon T + M)\mathrm{sgn}(s(k))\} \end{aligned} \tag{24}$$

The parameters of the simulation object is chosen as: $T_a = 0.8$ s, $K_a = 12$, $T = 0.001$ s while the corresponding discrete state matrix is shown as:

$$\boldsymbol{F} = \begin{bmatrix} 1 & 0.0010 \\ 0 & 0.9988 \end{bmatrix}, \boldsymbol{g} = \begin{bmatrix} 7.4970 \times 10^{-6} \\ 0.0150 \end{bmatrix}$$

The coefficients of the uncertainty terms are set as:

$$\Delta\boldsymbol{F} = \begin{bmatrix} 0 & 0 \\ 0 & 0.05\sin(kT) \end{bmatrix}, \Delta\boldsymbol{g} = \begin{bmatrix} 0 \\ 0.002\sin(kT) \end{bmatrix},$$

$$d(k) = 0.12\sin(2\pi kT)$$

The design parameters for the proposed AGSMC scheme are given as follows:
$c_1 = 25$, $\beta_1 = 0.12$, $\beta_2 = 0.12$, $q = 10$, $l = 2$
$\mathbf{P}(k) = \mathbf{0}$, $\varepsilon = 1$, $M = 0.12$

In simulation, the initial state vector of the motor is set as $[0.1°, 0.1°/\mathrm{s}]^T$. The position tracking signal of the electrical follow-up system is shown in Fig. 2.

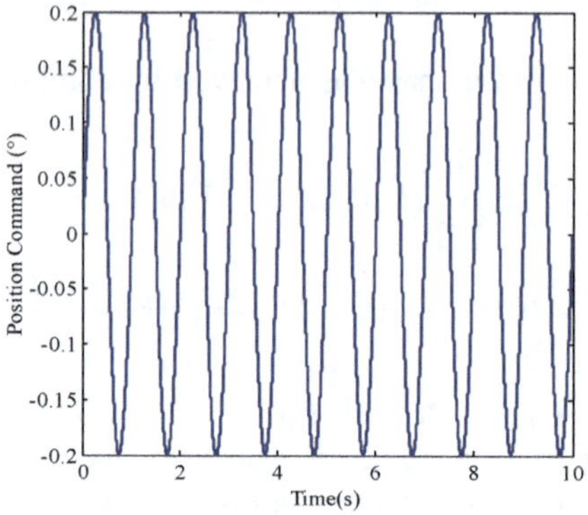

Fig. 2. Curves of position commands of the electrical follow-up system.

The tracking performances of the AGSMC and RLSMC are compared in Fig. 3.

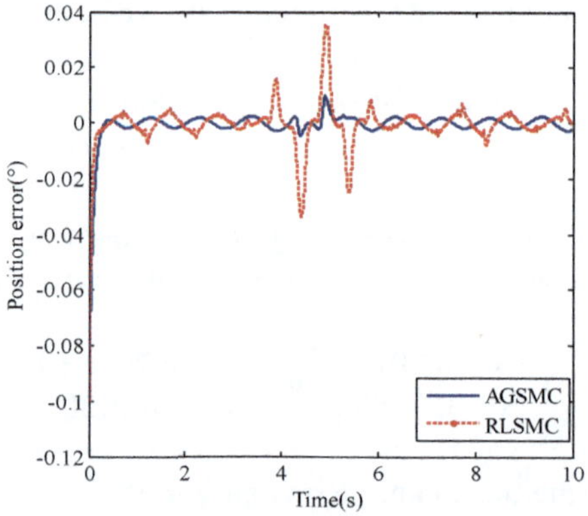

Fig. 3. Position tracking error curves under ASMC and RLSMC.

Conclusion can be made that the tracking error of AGSMC is smaller than that of RLSMC. Moreover, the tracking capability remains at the same level during the whole period, which is not affected by the parameter perturbations and external disturbances. The control input of the two scheme under this situation is shown in Fig. 4.

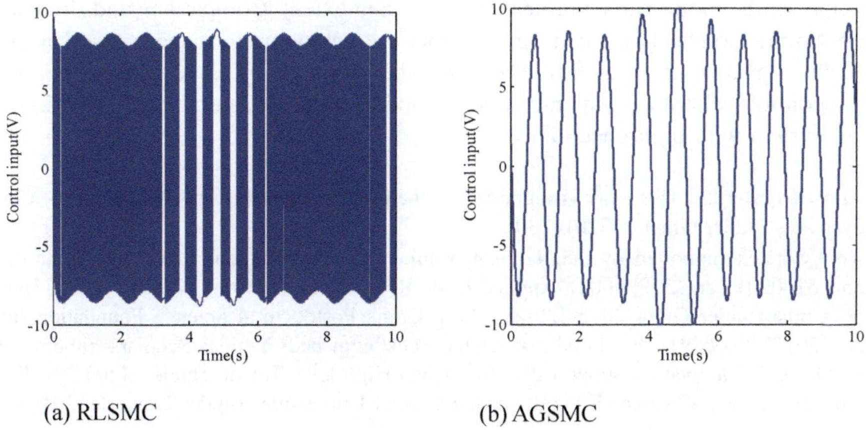

Fig. 4. Control input curves under ASMC and RLSMC.

From Fig. 4, it can be easily observed that, compared with RLSMC, the chattering of the control input has been suppressed in a large extent under AGSMC. This is mainly due to adaptive parameters in the reaching law, of which the real-time curves are illustrated in Fig. 5.

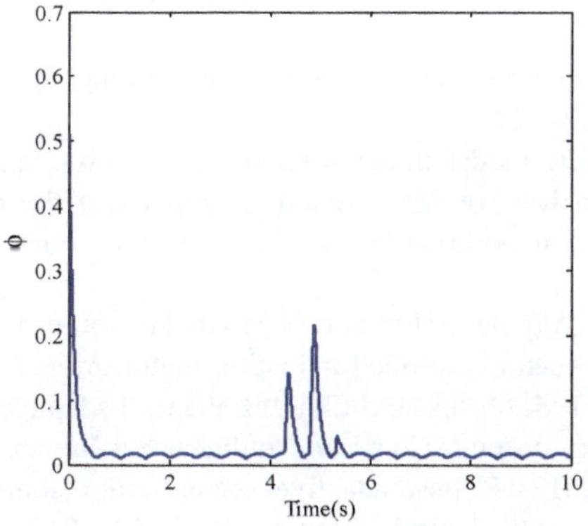

Fig. 5. Real-time curves of adaptive parameter ϕ

5 Conclusion

Considering the diverse effect on the electric steering gear servo system of unmanned surface vehicles (USV) caused by the external disturbances and parameter perturbations, a novel discrete adaptive global sliding model control scheme (AGSMC) is proposed in this paper. The key feature of the proposed scheme is the enhanced tracking property without obvious chattering control input with disturbances, so the structure of the motor can be protected to a certain extent. Comparison and simulations are performed to effectively confirm the superior tracking performance as well as global robustness against uncertainties. Therefore, the proposed AGSMC scheme is a novel servo control technology which can be applied to high precision electric steering gear servo system with high uncertainty.

Acknowledgments. This work is supported by the Fundamental Research Funds for the Central Universities (under Grant 3132018130).

This work is supported by the National Natural Science Foundation of P. R. China (under Grants 51009017 and 51379002), Applied Basic Research Funds from Ministry of Transport of P. R. China (under Grant 2012-329-225-060), China Postdoctoral Science Foundation (under Grant 2012M520629), the Fund for Dalian Distinguished Young Scholars (under Grant 2016RJ10), the Innovation Support Plan for Dalian High-level Talents (under Grant 2015R065), and the Fundamental Research Funds for the Central Universities (under Grant 3132016314)

References

1. Yan R, Liu ZH, Zhou R (2013) Application of low speed opto-electronic tracing systems based on sliding mode disturbance observer. J Beijing Univ Aeronaut Astronaut 39(6):835–840
2. Wu Y, Liu Y, Tian D (2013) A compound fuzzy disturbance observer based on sliding modes and its application on flight simulator 2013(2):388–400
3. Li Y, Sun DY (2010) Research on electro-mechanical actuator system based on discrete-time variable structure control. In: 2010 second WRI global congress on intelligent systems, pp 16–19
4. Utkin VI (1977) Variable structure systems with sliding modes. IEEE Trans Automatic Control AC–22(2):212–222
5. Utkin VI (1992) Sliding modes in control and optimization. Springer, Heidelberg, pp 1–70
6. Ghazi SE (1996) Analysis of VSC system designed via the reaching law approach. In: Proceedings of southeastcon bringing together education, science and technology, pp 170–174
7. Bayindir MI, Can H, Akpolat ZH et al (2005) Application of reaching law approach to the position control of a vector controlled induction motor drive. Electr Eng 87(4):207–215
8. Roopaei M, Zolghadri M, Meshksar S (2009) Enhanced adaptive fuzzy sliding mode control for uncertain nonlinear systems. Commun Nonlinear Sci Numer Simul 14(9–10):3670–3680
9. Li P, Zheng ZQ (2012) Robust adaptive second-order sliding-mode control with fast transient performance. IET Control Theor Appl 6(2):305–312
10. Choi HS, Park YH, Cho YS et al (2001) Global sliding mode control improved design for a brushless DC Motor. IEEE Control Syst Mag 21(3):27–35
11. Pang HP, Chen X (2009) Global robust optimal sliding mode control for uncertain affine nonlinear systems. J Syst Eng Electron 20(4):838–843
12. Gao WB, Wang YF, Homaifa A (1995) Discrete-time variable structure control systems. IEEE Trans Ind Electron 42(2):117–122
13. Li J, Ren Q (2008) Study on supply air temperature forecast and changing machine dew point for variable air volume system. Build Energ Environ 27(4):29–32 (in Chinese)

Application of Learning Trajectory Preference Algorithm Based on On-Line Iterative Feedback in Robots

Shunan Liu[1(✉)], Xiaoli Wang[1], and Haidong Huang[2]

[1] Changchun University of Science and Technology, Changchun, Jilin, China
327334223@qq.com
[2] Jilin University, Changchun, Jilin, China

Abstract. For the unstructured environment, the definition standard of the robot's trajectory varies with the user, task and environment. The trajectory of the robot is no longer a simple geometric constraint trajectory. It is necessary to study the trajectory preference problem of the robot in unstructured circumstance. Therefore, we put forward an active on-line learning skeleton frame for guiding robots to gradually understand the user's preferences for objects to be manipulated in scenario-rich environments. Users do not need to obtain the best trajectory as training data to train the robot. Our method is different from the conventional LfD method. We only need to provide the feedback type that the user expects to improve the current trajectory slightly on-line. Approaching the optimal trajectory in an iterative manner. In this paper, we apply this algorithm to the high degree of freedom robots, and compared with several common algorithms, it can obtain more ideal trajectory when environment and task change.

Keywords: Unstructured environment · Online iterative feedback · Task preference

1 Introduction

In recent years, the use of high-freedom robotic arms is largely limited to structured environments, just like the manufacturing industry needs to perform repetitive actions. In the unstructured environment, robots lack understanding of user preferences, so they can not produce ideal actions. To solve this problem, we need to study the learning preference on the trajectory of the robot with the high degree of freedom.

A key problem in operating a high degree of freedom robot is to determine a suitable mission trajectory. A suitable trajectory needs not only to be effective at geometry points, but also need to give priority to the needs of users. In the past, people's research mainly focused on imitating experts' demonstration, such as a ball cup experiment, autonomous helicopter flight, two-dimensional path planning and so on [1]. Learn from the demonstration that LfD is suitable for experts to clarify what is the best trajectory scenario. However, for different users, different tasks or different environments, the user's preference for the trajectory is ubiquitous [2]. This deems it impossible to manual geocoding in current route planners. Therefore, an algorithm is

Z. Xu et al. (Eds.): CSIA 2019, AISC 928, pp. 505–512, 2020.
https://doi.org/10.1007/978-3-030-15235-2_73

proposed to obtain the user's preference for trajectories through the user's interaction feedback in the interactive learning environment.

2 Synergistic Learning with Incremental Tickling

2.1 Robot Learning Settings

We suggest an on-line learning arithmetic that learns preferences from suboptimal contrails of users' tickling. In each procedure, tasks received by robots as inputs, and the contrail of the current estimate closest to its fractional function is output. observing users' feedback, improving the locus, and renovating the scoring function to better match users' preferences. Applying this algorithm to robots with 7-DOF arms, In the course of training, the original trail proposed by robots may not be able to achieve expected action. Thus, replace performing the tracks forthright, the user first visualizes them in the imitator and then decides on the type of feedback that they want to offer.

2.2 Tickling Mechanism

Feedback can be given by unskilled users who are not experts, so only feedback is required to be gradually increased in anticipation (relative to near optimal), and this feedback is required to satisfy the convergence of the algorithm. This is in contrast to the LfD method, which requires demonstrations of the optimal full trajectory [3]. Three feedback mechanisms are summarized.

Re-ranking. Use OpenRAVE to display track rankings on touchscreen devices, and ask users to determine if a lower ranking track is better than a top ranking one. Users observe the trajectory in their current predicted score order and click on a better trajectory than the highest ranking. Figure 1 displays the three loci of moving small sword. As couple back, the subscriber moves the contrail of level 3 to the highest place.

Fig. 1. Re-rank feedback mechanism

Zero-G. If a robot with a high degree of freedom has zero G compensation pattern, the user is allowed to modify the trajectory path by altering the robots' arm collocation by physical quantity. In such pattern, the robots' arms become ethereal and users can guide effortlessly them to requisite position. Zero G mode activates automatically by t users holding the robot's wrist. Figure 2 employing zero gravity model as a feed back way to gradually improve ballistic curve by modifying a path point [4].

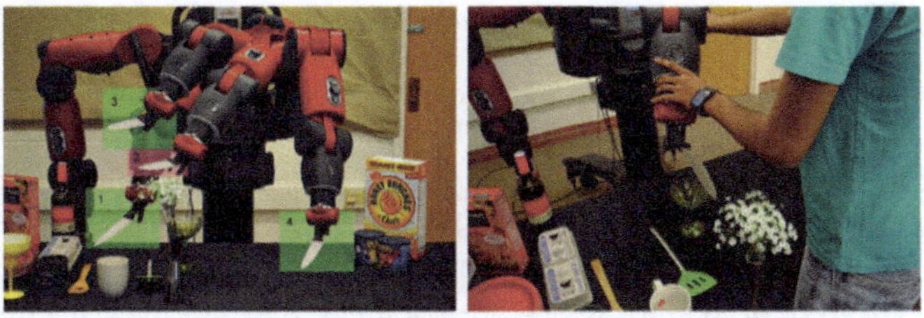

Fig. 2. Zero-G tickling

Interactive. Some robots have hardware that doesn't permit zero-gravity tickling, a seesaw Rviz-ros interphase for users to ameliorate trail by correcting path points is established. Interactive markup leads users to amend path points, and users offer feed back by adjusting it [5] (Fig. 3). The purpose of feedback is the same as zero gravity.

Fig. 3. Interactive feedback

Among the three kinds of feedback, the user will provide a slightly improved trajectory than the current trajectory, but will not display the best trajectory to the algorithm.

3 Learning Arithmetic

For each assignment, we use the following family of parameterized functions to establish users' rating function s*(x, y).

$$s(x, y; w) = w \cdot \varphi(x, y) \tag{1}$$

W is a known weight vector, $\varphi(\cdot)$ is a feature of the trajectory Y used to describe the context X. Equation 2 further divides the fractional function into two parts that only involve the object with which the trajectory interacts and the object and environment being manipulated [6].

$$\begin{aligned} s(x, y; w\,O, w\,E) &= s\,O\,(x, y; w\,O) + s\,E\,(x, y; w\,E) \\ &= w\,O \cdot \varphi\,O\,(x, y) + w\,E \cdot \varphi\,E\,(x, y) \end{aligned} \tag{2}$$

In following Sects. 3.1 and 3.2, we characterize the two conditions of $\varphi o(\cdot)$ and $\varphi e(\cdot)$.

3.1 Depicting Properties of Object-Object Interaction

The reciprocity among substances in the circumstance and objects are operated is captured in this section. Set path points of the track Y to y1, …, yn, O = {o1, ···, ok} is the object in the environment. The robot operates on the object and satisfies the space $\overline{o} \in$ O. Objects in the environment affect the points in the trajectory. For example, in Fig. 4, In environments with multiple objects, robots have two ways: 'a' and 'b' to take flowers from left to right, both of which are not ideal. Because the arms twist in 'a', but take the flowers down in 'b', this delicate preference needs to be reasoned under such constraints [7]. In a graph representing a fractional function, parameters related to object-object interaction are encoded. y1 and yn are different path points in orbit. The shaded nodes correspond to the environment and the edges represent the interactions between the nodes. Due to the proximity of O1 and O2, the path point y 3 is affected. If minimum range of impact is less than threshold value or if o k is lower than $\overline{o}$, the object o k is connected to the locus point. The edges link y j and o k are represented as (y j, o k) $\in$ E.

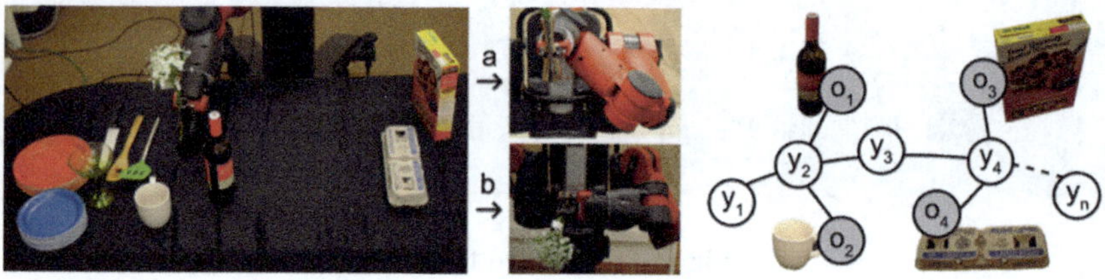

Fig. 4. A robot moves flowers

The attributes of the object are important factors that determine the quality of the trajectory, attributes are used to represent each object. For each target ok, we use a complexor $[l_k^1 \ldots l_k^M]$ of M (where, l_k^M = {0, 1}) to indicate whether the object o k has the attribute m. For example, attribute sets can be represented by [0, 1, 0, 0, 0, 0 if they are {heavy, fragile, hot, liquid, fast, and electronic} [7]. Binary variables l_p^k and l^q indicate whether $\overline{o}$ and o k have attributes q and p, respectively. For each edge of(y j, o k), below four features φoo(y j, o k) can be extracted: The shadow of the minimum interval of impact along the X, Y, Z axis and a binary variable. If O K is perpendicular to the top which is 1, otherwise 0. The fraction s O(·) is defined on this figure, as shown below:

$$S_o(x, y; w_o) = \sum_{(y_j, o_k) \in \varepsilon} \sum_{p,q=1}^{M} l_k^p l^q [w_{pq} \cdot \phi_{oo}(y_j, o_k)] \tag{3}$$

Interactions between targets with attributes q and p are captured by the weight vector wpq acquiring Eq. (2) by connecting vector wpq. If the vector of WO's location i is wUV, so the vector corresponding to the location of φO(x, y) will be $\sum_{(y_j, o_k) \in \varepsilon} l_k^u l^v [\phi_{oo}(y_j, o_k)]$.

3.2 Trajectory Characteristics

The feature φE(x, y) acquired by administering manipulations on a batch of path points contains below three genres of functions:

Robot arm allocation. Not all arm configurations can achieve the desired operational results. Not all the arm configurations have the desired effect. Using the elbow, wrist and shoulder position of the robot w. r. t., the characteristics of the robot arm structure are calculated in the cylindrical coordinate system [8]. A track is divided into three portions in time, and nine characteristics of each section are computed.

Azimuth and Time Behavior of the Manipulated Object. As time goes on, diverse sections of trajectory may have distinct demands, so it is divided into three parts, and each part of the object in the X, Y, Z direction of motion and along the vertical axis of the deviation spectrum (Fig. 5). Then the mean power spectral density (psd) of low and high frequency parts in three directions is calculated and taken as six additional characteristics. This provides a third of 9(=1 + 4 * 2) features. Add an additional feature (the maximum deviation of the object on the whole track), and get $\phi_{obj}(\cdot) \in R^{28}(= 9 * 3 + 1)$.

Fig. 5. (Left) move the cup with good (green) and bad (red) tracks. (Right) spectrum of motion in Z direction

Interaction between Objects and Environment. Capturing the time alternating quantity of perpendicular and level distances between an target and its ambient surface is achieved by this feature, which divides the locus into trisection and calculates the following contents of the object separately: (1) the least perpendicular distance from the adjoining exterior; (2) the smallest horizontal distance from the surface near it; (3) the smallest distance from the desk performing the task; (4) the smallest distance from the target position; (5) the average distance between an object and its nearest surface in horizontal and vertical directions. The purpose of plotting the time-frequency spectrum of nearest perpendicular distance is to caught the transient transformation of the distance between the object and its surroundings, and to extract six features from the time-frequency spectrum by meshing. From these features, you can see if an object rubs across the edge of the table, that is, whether there is a steep change in the vertical distance. Therefore, the feature captured from the target-surroundings interaction is $\phi_{obj-env}(\cdot) \in R^{20}(3 * 4 + 2 + 6)$. The ultimate property vector $\phi_E(.) \in R^{75}$ is acquired by uniting ϕ_{obj} and ϕ_{robot}.

3.3 Calculating the Track Ranking

In order to obtain the optimal trajectory (or the first few) of a given task in scenario X, maximize present grade function S (X, Y, W 0, W E).

$$y^* = \arg\max\ s(x, y; w_o, w_E) \tag{4}$$

There are two issues to note. First, the trajectory space is continuous and requires discretization to keep argmax within the controllable range. Two effective methods for solving this problem are discrete state space or sampling trace directly from successive room. Both methods have been studied before, however, for the high degree of freedom manipulator, the sample-based method maintains the traceability of the problem, so this method is adopted. A fast scan of the random tree (RRT) is used to sample the trace. Furthermore, for a existing set of {y(1), ..., y(n)} dispersed track, Calculation is needed, that is, the trajectory can be sorted simply by the trajectory score function S (x, y(i); w o, w e).

3.4 Learning the Scoring Function

initialization: $w_o^{(1)} \leftarrow 0, w_E^{(1)} \leftarrow 0$

for t = 1 , specimen locus of T {y (1) , . . . , y (n) }

$$y_t = \arg\max_y s(x_t, y; w_O^{(t)}, w_E^{(t)})$$

Obtain user feedback $\overline{yt}$, $w_O^{(t+1)} \leftarrow w_O^{(t)} + \phi_O(x_t, \overline{y_t}) - \phi_O(x_t, y_t)$

$$w_E^{(t+1)} \leftarrow w_E^{(t)} + \phi_E(x_t, \overline{y_t}) - \phi_E(x_t, y_t)$$

The target is to get parameters w O and w E of the grade functions (x, y; w O, w E) so as to the trajectories can be sorted according to the user's preference. To this end, the preference perceptron algorithm is adjusted according to an algorithm, which is called a trajectory preference perceptron (tpp).

4 Experimental Results

In untrained and trained cases, comparing the performance algorithms based on on-line iterative feedback with the following means:

Manual: Plan a single path according to some manually coded task-related preferences.

TPP: This is our algorithm. It is evaluated in an untrained and pre-cultivated environment.

MMP-online: It is an on-line achievement of the Maximum Profit Margin method, which can be considered a special case of the TPP algorithm, and the observed feedback is considered to be the best [9]. In each iteration, all previous feedback is used as a training instance to practice a configurable sustain vector machine (SVM). The resulting weight is used to calculate the track score for the next iteration.

Oracle-svm: The algorithm uses the experts' labels on the trajectory and uses SVM-rank for training in batch processing. This algorithm is not practical in practice because it needs to be marked on the large space of the trajectory. Only used in pre-trained settings, and only predicted once during the forecast period.

4.1 Evaluation Indicators

A data set was designed to quantitatively appraise the capability of on-line measure. According to the subjective preference of the person, an expert marked 2100 tracks (1–5 is the best) on the Likert scale [10]. These absolute ratings are only used to quantitatively evaluate different algorithms. The quality of the alignment table is quantized by normalized discounted cumulative gain (nDCG) at places 1 and 3, and they are represented as nDCG@1 and nDCG@3, respectively. Although nDCG@1 is the appropriate amount of autonomous robot to perform No. 1 track, nDCG@3 is appropriate for robot supervised sights (such as assembly lines) and reports the average nDCG value for given scalar tickling iterations.

4.2 Results

Figure 6 shows the nDCG3 and the feedback diagram for the average task of all types of activities, with the same environment and different objects on the left, the new environment and the same object in the middle, and the new environment and different objects on the right. Among them, manual programming (pink); Oracle-svm (black); beforehand drilled MMP-online (blue-); untrained MMP-online (blue–); beforehand drilled TPP (red-); untrained TPP (red–). The graph shows that the pre-trained TPP starting point has a higher nDCG3 value than the TPP, and with more feedback, the property of both arithmetics will be improved, and finally the same performance will be provided. MMP-online assumes that each user's feedback is optimal, so it accumulates many contradictory incorrect training examples over time. Pre training and untrained versions start with similar nDCG@3 values and give similar performance on them, and no TPP algorithm has the advantage in the new environment. Oracle-svm starts off a higher nDCG@3 value than other algorithms because it is beforehand exercised by expert tags on the trajectory, but for the alike cause it is not suitable for using in practice, and Oracle-svm is not recent because it needs specialist tags in the experiment set. Artificial algorithms produce poor trajectories because some preferences need to be written into the planner's code, but because some preferences are hard to specify.

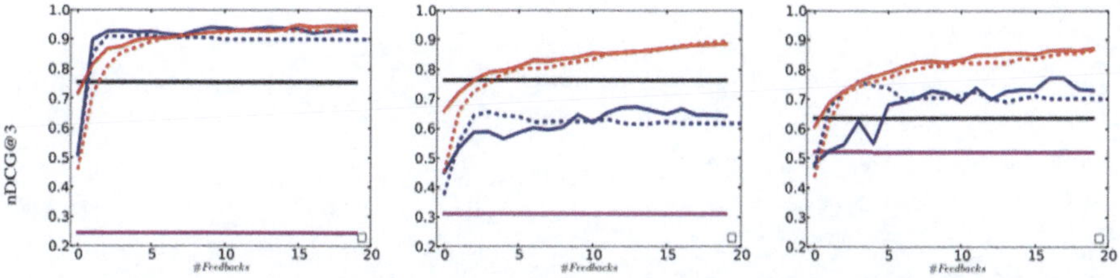

Fig. 6. Research on the generalization of objects, environment and both

5 Conclusion

In the text, a collaborative learning shell frame based on the high degree of freedom arm robot is presented, and a set of orbital characteristics are modeled based on complex scenarios and the TPP algorithm is obtained. Unskilled users provide feedback to manipulate robots to obtain ideal trajectories that satisfy users' preferences.

References

1. Abbeel P, Coates A, Ng AY (2010) Autonomous helicopter aerobatics through apprenticeship learning. IJRR 29(13):1608–1639
2. Akgun B, Cakmak M, Jiang K (2012) Keyframe-based learning from demonstration. IJSR 4(4):343–355
3. Alterovitz R, Siméon T, Goldberg K (2007) The stochastic motion roadmap: a sampling framework for planning with markov motion uncertainty. In: RSS
4. Argall BD, Chernova S, Veloso M et al (2009) A survey of robot learning from demonstration. Rob Auton Syst 57(5):469–483
5. Ratliff N, Silver D, Bagnell JA (2009) Learning to search: functional gradient techniques for imitation learning. Auton Robots 27(1):25–53
6. Van Den Berg J, Abbeel P, Goldberg K (2010) LQG-MP: optimized path planning for robots with motion uncertainty and imperfect state information. In: RSS
7. Bhattacharya S, Likhachev M, Kumar V (2011) Identification and representation of homotopy classes of trajectories for search-based path planning in 3d. In: RSS
8. Gossow D, Leeperand A, Hershberger D, Ciocarlie M (2011) Interactive markers: 3-d user interfaces for ros applications [ros topics]. Robot Autom Mag 18(4):14–15
9. Calinon S, Guenter F, Billard A (2007) On learning, representing, and generalizing a task in a humanoid robot. IEEE Trans Syst Man Cybern 37:286–298
10. Cohen BJ, Chitta S, Likhachev M (2010) Search-based planning for manipulation with motion primitives. In: ICRA

Review on Non-spherical Surface Detection Technology

Bo Yu(✉), Yan Li, Xiaodong Li, Shijun Luo, and Haibo Li

Mechanical and Electrical Engineering College,
Changchun Institute of Technology, Changchun, China
yubo745@163.com

Abstract. The paper described the importance of aspheric surface detection technology for aspheric surface, introduced three main methods for the detection of aspheric surface profile: profile measurement, interferometry and geometrical optical measurement, emphatically analyzed the diversification derivation mode of interferometry and its detection principle, in the meantime, summarized the advantages, disadvantages and applications of aspheric surface detection methods, prospected the development trend of aspheric surface detection technology.

Keywords: Aspheric surface · Face detection · Comparison and analysis · The development trend

1 Introduction

Aspheric optical element with its unique nature, popular in the optical imaging system, the use of aspherical optical components in optical system, to improve the image quality, reduce weight, improve the optical properties instruments, simplify the structure, reduce the equipment cost is of great significance [1]. Although, aspheric optical element in performance is better than that of spherical element greatly, however, because of the aspheric optical element itself makes its in machining and inspection is much more difficult than a sphere, so the application of aspheric optical element is limited by a lot. Have expert said: no high precision detection technology, manufacturing is the key to the high precision aspheric elements, therefore, aspheric component detection technology for its high quality of manufacturing is of great importance.

2 Aspherical Surface Shape Detection Technology

For a long time, the aspheric optical element processing and detection technology is the widely application of the two big problems. Aspheric components processing without detection, especially not spherical surface shape contour detection. If there is no appropriate and accurate detection method, can't work out high-precision aspheric component. Current, aspherical elements mainly adopts contour measurement, interferometry, and geometric optical measurement method to implement its surface shape accuracy of detection.

Z. Xu et al. (Eds.): CSIA 2019, AISC 928, pp. 513–520, 2020.
https://doi.org/10.1007/978-3-030-15235-2_74

2.1 Profile Measurement

Contour measurement method using contact or non-contact [2] way of measurement, a direct measurement of the aspherical vector width; And then using aspheric meridional section line equation and compare the actual aspherical Outlines, aspheric surface is obtained form the contour error, measuring principle is shown in Fig. 1.

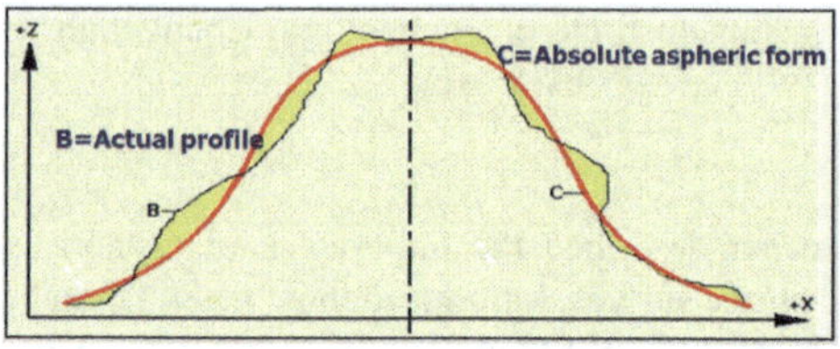

Fig. 1. Profile measuring aspherical schematic diagram

(1) Contact measurement

Contact measurement using probe direct contact with the aspherical elements, through the lateral movement probe, and record the height change, probe for aspherical surface shape contour data.

(2) The non-contact measurement

Non-contact measuring instrument, the basic structure of the same as the contact between the main difference is that measuring probe. General by optical non-contact measuring instrument measuring instrument, because the test probe is contact, therefore, the detection method can avoid scratches in the surface to be measured, at the same time, also can obtain higher measurement accuracy.

2.2 Interferometry

Interferometry is a spherical component surface shape measurement of important method. At present, a lot of interference measurement for aspheric derivative method, generally can be divided into zero measuring method and non zero measurement method. Zero measurements include: no like almost method [3, 4], compensation mirror method [5, 6] and calculation method of hologram [5, 6]; A zero measurement method that includes: part of the compensation method, band stitching method [4, 7, 8], the subaperture stitching method, the Nyquist sampling method, shear method, long waven. For zero test, due to the reference beam and the metering common optical path, can be very good to eliminate the common optical path error of interferometer, however, to achieve zero measurement, zero compensator often require high precision, and the compensator with aspheric one-to-one correspondence, therefore, generally zero test generality is poorer, the cost is higher, but the precision is higher, more for high precision detection field. Non zero test, through the implementation of part of the compensation, or special algorithm (such as: splicing method, shear method), or special receiver (such as: the Nyquist sampling method), interference or longer wavelengths (such as: long wave appearing), to enlarge the measuring range of interferometer, so as to realize the aspherical surface shape detection. This method can realize a series of aspheric component surface shape detection, has good generality, however, often resulting in loss of precision measuring range, in addition, because of the implementation in the zero state measurements, the reference light path and the metering of optical path of the road will be damaged, it will introduce some common optical path error, therefore, this method applies only to the high detection accuracy, for the field of ultra-precision, is used less.

(1) No like almost method

No like almost method to test the quadric surface, requires no additional compensation device can realize zero detection, is of high precision, but test process often need to introduce additional planar or spherical. In addition, for coaxial quadric surface, due to the existence of the center block and make the measurement error, therefore, no like almost method of measuring range is limited to a certain extent. Its detection principle as shown in Fig. 2.

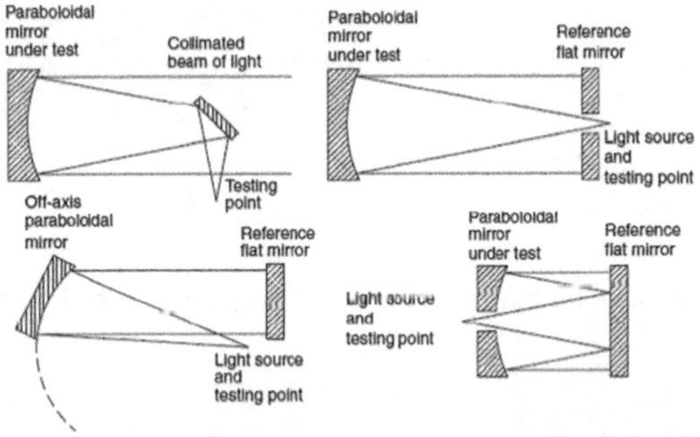

Fig. 2. No difference point method of parabola detection scheme

(2) The compensation method of mirror

Compensator method first put forward by Soviet scientists, after Couder, 'Burch, Ross, Dall, Offner and Shafer, gradually formed a variety of detection methods, including: Offner compensator the most widely used, the biggest influence, its detection principle as shown in Fig. 3, through the adoption of two pieces of lens, realize the aspherical spherical aberration and high-order aberration correction, the compensation accuracy is very high. In addition, because compensation is aspheric lens of two pieces of lens focus (near the best fitting spherical center), so, even if the size of the aspheric is bigger, the size of the compensator can also small, easy processing, this method is suitable for the detection of large diameter concave aspherical. Due to compensate the aberration of lens, however, it is often difficult to measure, so the precision of compensator requires to ensure absolute accuracy, therefore, this method has certain drawbacks in terms of practicality.

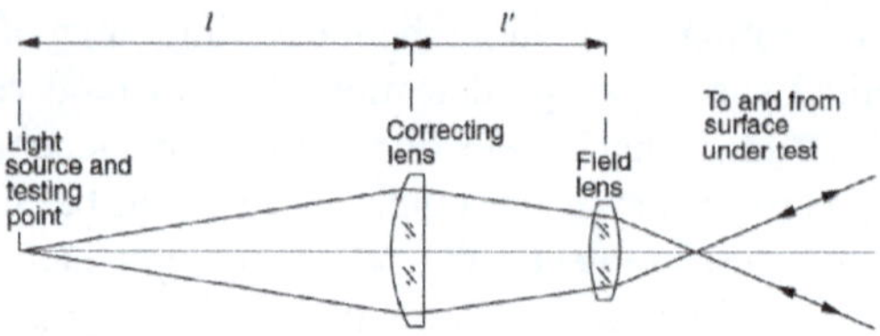

Fig. 3. Offner compensator

(3) The calculation method of hologram

Calculation method of hologram using computer to generate the required first wavefront holographic pattern, and then use processing equipment to the holographic pattern preparation to substrate generated hologram, when using specific light hologram can emersion holograms recorded wave front. Generated by this method can match the aspheric surface under test of spherical wave, so as to realize zero of aspheric surface testing, detection principle as shown in Fig. 4.

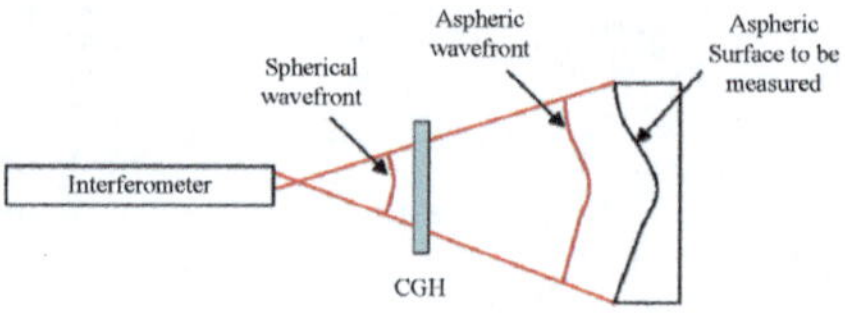

Fig. 4. Computing hologram of aspheric zero detection

(4) Band together

For rotational symmetrical aspheric surface, when aspherical along optical axis movement of interferometer, there will be a circular interference figure, as shown in Fig. 5. Although in one place can't realize the measurement of the aspheric surface, however, if you can will be "together" multiple locations of measurement data, can realize the whole aspherical surface shape measurement. For this kind of measurement method, in order to realize accurate, need to accurately measure the position of the current aspherical, therefore, this method besides phase-shifting interferometer

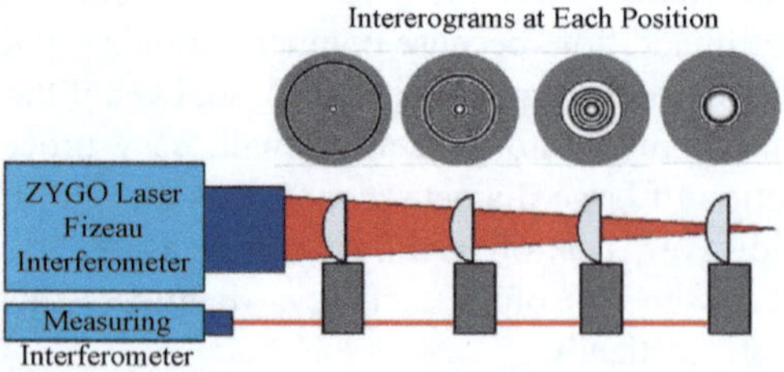

Fig. 5. Band stitching measurement principle diagram

measurement phase, often need to displacement interferometer testing aspheric interferometer with relative axial offset.

(5) The subaperture stitching

Subaperture stitching like band splicing method, and through multiple position measurements, eventually "patchwork" an outline of the aspherical surface shape. Differs between them: band stitching measurement, aspheric only along the optical axis direction; and subaperture stitching, aspheric not only along the optical axis direction, but also do translational motion and tilting movement, measuring principle as shown in Fig. 6. As each part of the sub aperture aspheric surface are measured, therefore, of subaperture stitching method has higher horizontal spatial resolution. For the steradian smaller aspherical subaperture partition, can be measured directly.

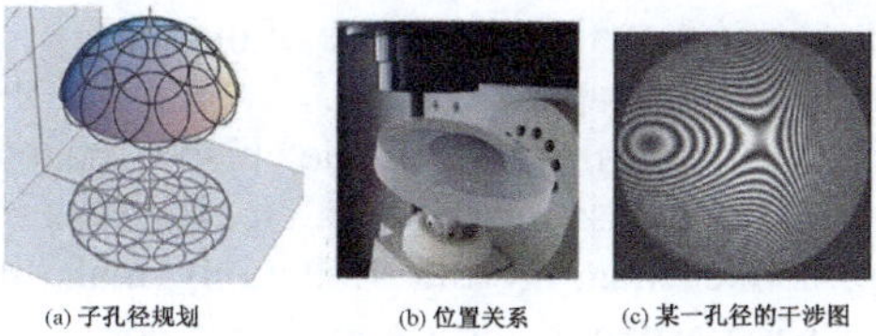

Fig. 6. Schematic diagram of subaperture stitching method

(6) The Nyquist sampling method

According to the law of Nyquist sampling, if you want to distinguish on the surface of the CCD as an interference fringe, the sample point we need at least two pixels; When the sampling frequency is lower than the Nyquist sampling frequency, interference fringe aliasing, will appear as shown in Fig. 7. If the direct measurement of aspheric surface with spherical interferometer, tend to be "the Nyquist sampling", in this case, the general solution package can't through the aspherical surface shape. The Nyquist sampling method is first adopted as shown in Fig. 8 "the Nyquist" detector to achieve the Nyquist sampling, and then use a special package solution algorithm to realize the measurement of aspheric surface shape.

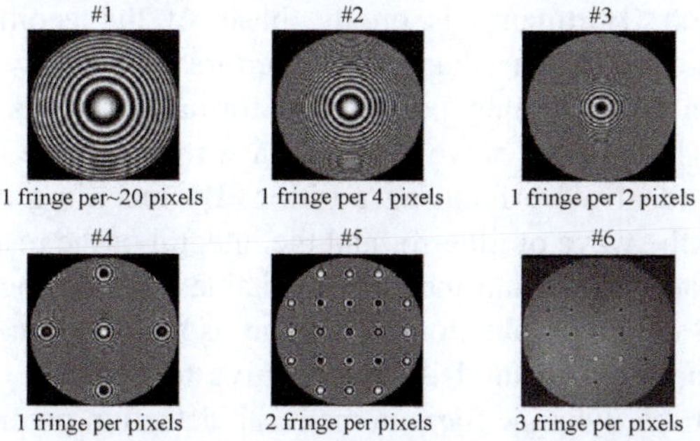

Fig. 7. Under different sampling frequency interference figure

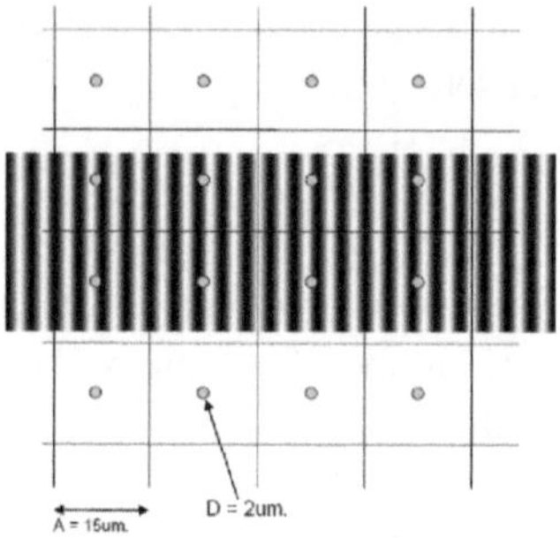

Fig. 8. The Nyquist sampling detector

2.3 Geometric Optical Measurement Method

Geometric optical measurement method is the use of the linearity of the light to achieve a method of aspheric surface shape detection. At present, the geometrical optics measurement derivative methods mainly include: blade shadow, hartmann, interval method, template method, compensation method, etc. Because of the limitation of the method in detection principle, there are big limitations, make it difficult to meet the current range, high precision aspheric surface testing needs.

(1) Blade shadow

Blade shadow method is also called foucault knife edge method, its first used in the inspection of large diameter surface shape error of optical elements, mainly is suitable for the different sizes of rotary quadratic aspherical surface shape detection. It required equipment is simple, high sensitivity, the qualitative analysis under the condition of the blade is not pay attention to, the mirror surface defects found A/20 is A easy, often used in the detection of the scene. It is not easy to scratch the mirror, fast detection and moments from the shadow figure can be found that the flaw of the mirror and the location, so it is very agile. However, it is just a qualitative method of aspheric surface testing, subjective and quantitative difficulties, the shortcomings of low sensitivity and not easy to test convex, therefore, on the applicable scope is limited to a certain extent.

(2) The method of Hartmann

Hartmann method (Hartmann) is on the basis of the geometrical optics slope measuring principle, through the diaphragm aperture sampling wave, keep a certain relationship between the sampling point wave formation, so as to realize the final inspection. If somewhere on the wave of ideal wave front tilt, the light will be focused in other places, not ideal wave focusing position. By recording the deviation vector, and then calculating the wave of tilt error and the integral of the resulting form errors of the surface to be tested. Hartmann method is suitable for the processing of the online inspection and final quality evaluation, is a large aspheric commonly used detection methods, but the hartmann method is not sensitive to external vibration and disturbance, therefore, for the dynamic measurement, its detection precision is less than the other way.

(3) The method of interval

Interval (Ronchi) method is also called the line plate method, is to use long grating, the formation of lang's figure, through processing interval graph, to realize the aspherical surface shape detection. It as the formation of stripe formed due to the diffraction and interference of graphics to explain, can use a very simple small instrument testing components of large caliber, so for aspheric surface testing, especially some large diameter concave mirrors, a significant advantage. But it is worth noting: French stripe is not directly related to the shape and aspheric surface, and is associated with the transverse aberration. Because by lang grating on each slit of light with the light of other slit interference occurs, produce a kind of have a lot of transverse shear effect of pupil, which reduces the test accuracy. Therefore, it is also a kind of qualitative detection methods, it is mainly used for coarse grinding and fine grinding surface shape detection.

3 The Aspheric Surface Testing Technology

To sum up, the testing method of aspheric are diverse, they each have advantages and disadvantages. Method of measuring directly face outline, the cost is low, but its accuracy is not high, only applies to medium precision aspheric surface testing, therefore, its limitations. Interference test can be divided into zero detection and non zero, the difference is whether need auxiliary components. Computing holographic method is an outstanding representative of zero test, its many advantages, such as: measuring precision is very high, do not need to make the reference entity, cost savings, measuring speed, wide applicable scope, etc., however, because of CGH and aspheric surface under test is one to one match between, as a result, the poor generality. Shearing interferometry and subaperture stitching are not zero, the precision of the shearing interferometry is generally low, and they need complicated mathematical processing, it has the advantage of good generality, its drawback is that need complex analysis algorithm, image processing was complicated, measurement speed is slow. When the measured surface is larger, and divided the band is large, the edge will be difficult to collect data, also can appear when stitching. It has the advantage of high precision measurement. Hartmann method is one of the geometrical optics method, the method of hartmann test environment is low, low sensitivity to external vibration and air turbulence, but, because of the limitation of this method by the aperture sampling points less, therefore, its measuring precision is low.

4 The Aspheric Surface Testing Technology

In recent years, domestic and foreign experts and scholars on the aspheric surface testing technology conducted a series of research innovation, by summing up thinking, the development direction of aspheric technology should boil down to:

(1) The precision, efficiency of direction

As the aspherical applications gradually expanding, the requirement for the quality of aspheric surface is higher and higher. High precision, high efficiency has become the primary direction of the future development. For example: some current method is of high precision, but at the expense of the efficiency, cannot meet demand.

(2) The direction of crossover and integration

This direction includes two aspects: one is the manufacturing and so on multi-disciplinary overlapping fusion; Two is the intersection of processing and testing technology, namely, through the adoption of online detection, condition monitoring and error compensation of fusion technology to improve the manufacturing accuracy, thus ensuring high quality aspheric surface.

(3) The principle of a new study

Currently, aspherical surface shape measuring principle prototype in detection base theory and prototype cost is less than from two aspects, so, if there are any new detection principle was born, and matches by hinge low cost to complete the construction of the high precision principle prototype, is a historic innovation. Based on this article reviews the research, the author will put forward in the next article personal innovative ideas.

References

1. Alessandro B, Pasquale P, Franco Q et al (2001) Fast one-dimensional profilometer with a compact disc pickup. Appl Opt 40(7):1044–1048
2. Semenov DV, Nippolainen E, Kamshilin AA (2006) Ultrahigh accuracy 3-D profilometer. Appl Opt 45(3):411–418
3. Malacara D (2007) Optical shop testing, 3rd edn. Wiley, Hoboken
4. Kim YS, Kim BY, Lee YW (2014) Design of null lenses for testing of elliptical surfaces. Appl Opt 40(19):3215–3219
5. Pruss C, Reichelt S, Tiziani HJ et al (2004) Computer-generated holograms in interferometric testing. Opt Eng 43(11):2534–2540
6. Tian C, Yang Y, Wei T et al (2011) Nonnull interferometer simulation for aspheric testing based on ray tracing. Appl Opt 50(20):3559–3569
7. Wen Y, Cheng H, Tam H-Y et al (2013) Modified stitching algorithm for annular subaperture stitching interferometry for aspheric surfaces. Appl Opt 5(13):5686–5694
8. Chen S, Li S, Dai Y et al (2008) Experimental study on subaperture testing with iterative stitching algorithm. Opt Express 16(7):4760–4765

Novel of Web Search Strategy Based on Web Page Block Granularity Analysis Algorithm and Correlation Calculation Model

Ganglong Fan[1,2] and Hongsheng Xu[1,2(✉)]

[1] Luoyang Normal University, Luoyang 471934, China
85660190@qq.com
[2] Henan Key Laboratory for Big Data Processing and Analytics of Electronic Commerce, Luoyang 471934, China

Abstract. In this paper, the running parameters of crawler nodes are dynamically adjusted to make them more manageable and configurable. The running parameters include crawling speed, crawling depth, number of crawler threads. After processing and organizing the information, the search engine provides Web information query service for users, crawls continuously on Internet, collects web information. Then the information is understood, extracted, organized and processed by the indexer, and the index database is established. The paper presents novel of web search strategy based on web page block granularity analysis algorithm and correlation calculation model.

Keywords: Web search strategy · Web page block granularity · Correlation calculation · Search engine · URL

1 Introduction

Search engine is a kind of tool that users often use when surfing the web. No doubt, every user can get the network resources they need through the search engine. The word search engine is widely used in the Internet, but it is understood differently in every region. In some European and American countries, search engines are often based on the Internet [1]. They collect web pages through web crawler programs and index every word in a web page, that is, full-text retrieval. In some Asian countries, search engines are usually web-based search services.

The development of search engines faces two major difficulties. The basic idea of the web page block granularity analysis algorithm is to divide the web page into different web page blocks through some kind of page segmentation algorithm, and exclude the links which are not related to the topic, and then deal with them further. This analysis algorithm can avoid the interference of advertising and other noise links. In the aspect of data acquisition, the system combines manual crawling and incremental crawling. The information is extracted, de-noised and structured to create the conditions for the follow-up special analysis. In practical application, the designed web crawler should be able to crawl the (Web), along a large area of hyperlinks and grab the information from the web page. Therefore, crawling hyperlinks from web pages is a

Z. Xu et al. (Eds.): CSIA 2019, AISC 928, pp. 521–527, 2020.
https://doi.org/10.1007/978-3-030-15235-2_75

necessary operation for web crawlers to expand the scope of access autonomously and automatically.

At present, there are three main performance indicators of search engine: first, the size of the search engine is considered, only when the scale reaches a certain order of magnitude, the degree of conformity of the user's search results. The web spider of the search engine must complete the information search of the target network in a relatively short time, at the same time; it can complete the feedback of the search results within a time period that the user can tolerate, and finally, the quality of the search [2]. It can remove the repeated pages, filter some useless information, and return the desired results accurately. If the crawler node is collected, it may be taken repeatedly, and if it is not collected, there may be a leak. Therefore, different ways of dealing with this problem are chosen according to different situations: discard mode, cross mode or switch mode.

The web crawler will eventually be designed to automatically read and write configuration files and execute automatically in the background. When extracting the text information of html documents, we should filter the identifiers, but at the same time, we should take care to obtain the layout information according to the identifiers and filter the useless links. Anchor text can be used as the evaluation of the content of the page and the content of the page pointed to, and can also collect some search engine cannot index files, multimedia, Images and other documents generally through anchor text and related document comments to judge the contents of the file.

2 Application Analysis of Web Page Block Granularity Analysis Algorithm

The Web resource is to use the web crawler technology to crawl the data in the Web has become a severe problem. Because of the poor scalability and fault tolerance of the traditional web crawlers, they cannot perform the task of efficiently crawling in many aspects.

Static web pages are mainly composed of html documents, which have a tree structure and clear structure [3]. The description information in a web page is generally contained in <title> <r title> node, and the web page links are generally included in <a> <a> node. Therefore, crawlers only need to grab the content of these nodes in order to capture the links and their description information, without downloading the entire page.

The system saved the initial link in the init_url file, saved some abnormal cases in the crawl.log file, saved the downloaded web page file in the page_db folder, saved the web page information file in the page_info folder, and saved the information file in the page_info file. The file name of the web page, the name of the information file and the name of the link file are saved in the link.db file.

It can be predicted that Internet information capture, mining and reprocessing will become more and more needs of people in the future, and all crawlers and related information processing tools can meet this demand. But the Internet itself, 1.0 and 2.0, is not well prepared for the crawler era. Now the parade of SEO is a strong search engine under the conditions of the impact on the structure of the site. After the advent of the reptile era, there will be a special information site on the Internet, which is provided to the crawler.

ExtJS was originally based on YUI technology, developed by developer Jack Slocum, and organized visual components by referring to Java SWING and other mechanisms, as is shown by Eq. (1), where L is from the application of CSS style on UI interface to exception handling in the process of data parsing. It can be regarded as is a rare JavaScript client technology boutique [4].

$$L(\theta) = L(x_1, x_2, \cdots, x_n; \theta) = \prod_{i=1}^{n} p(x_i; \theta) \quad (1)$$

In a crawler, the function of web crawling is similar to that of a common browser. As long as the specified URL is entered, the content of the page can be returned to the user. The process of web page capture is to access the target URL through H-bamboo P protocol and get the content of the web page. Although the basic function on accessing the Http protocol has been provided in the java.net package of JDK.

The website granularity analysis algorithm is simpler and more effective than the page granularity analysis algorithm. When a node exits in the system, there are usually two kinds of cases. One is when the control node requires the node to quit, the node notifies the other node to leave, and when the other node receives the message, it modifies the logical table and the physical table. In addition, SiteRank will not be fooled by common counterfeiting against PageRank.

Web crawling is to read the network resource specified in the URL address from the network flow and save it to the local. Similar to the use of programs to simulate the function of IE browsers, URL as the content of HTTP requests sent to the server side, and then read the server side of the response resources.

Because the firewall of the site will prevent the local IP from accessing the site, Vector space model (VSM) is used to retrieve content, and weight evaluation technique is used for statistics. In this system, the correlation degree is calculated by pattern matching module, and the roaming model is adopted to carry out the continuous retrieval in the later stage [5]. The client which is convenient to read files is mainly to filter the local documents and exchange information with the server, and transfer the URL, that needs parsing to obtain the XML documents formed by the source code of the parsed web pages.

Breadth priority is to first crawl all the initial URL pages, then select one of the linked web pages through the web page filtering strategy, and continue to grab all the URL linked web pages in the web page, so that it can be crawled layer by layer.

3 Analysis of Correlation Calculation Model

The Boolean model is easy to implement when judging the topic. At the same time, a topic is also expressed as a set of keywords [6]. In judging the relevance of a document to a topic, it is equivalent to calculating the intersection of two sets of keywords.

For the subject discriminant model based on Boolean model, the more elements are contained in the intersection, the higher the correlation with the topic is it.

ExtJS also supports a variety of JS underlying libraries, including jQuery, Prototype, in addition to YUI, allowing developers to make free choices [8]. The framework is based on pure HTML/CSS JS technology, provides abundant cross-browser UI components, and uses JSON/XML data source to develop flexibly, so that the load of the server presentation layer can be really reduced.

Crawler reads from the initial URL library and crawls pages in parallel using asynchronous I/O. After the page is downloaded, extract the hyperlink. If the hyperlink belongs to the Web site that this Crawler is responsible for collecting, join the URl collection not accessed, or store it in an intersected URL file. The batch module periodically allocates these cross URL files to the appropriate collection module, filtering to the duplicate URL, as is shown by Eq. (2), where x is again during the process.

$$\|x\|_p := \left(\sum_{k=1}^{\infty} |x_k|^p \right)^{1/p} \tag{2}$$

For the page analysis, if the node does not forward the message to the central node, the central node receives the request message and replies the reply information with all the currently running nodes to the requesting node. The initial URL seed is stored in the critical region, the web page is searched and fetched according to the breadth search operation method, and the URL is extracted and returned to the session area. The correlation degree is judged by the algorithm of judging the topic relevance degree. Then select one of the linked pages and continue to grab all the pages linked to the page [7]. If you want to get more pages, you can set r to be smaller. To get fewer pages, set the r larger.

The uniform Resource Locator (URL) is an abstract method to identify the location of the resource and the location of the resource, which is a concise representation of the location and access method of the resource available from the Internet, and uses powerful automatic web page extraction program that downloads web pages from the World wide Web for search engines and is an important component of search engines. SPider visits a site by requesting HTML documents on the site. URL is equivalent to a file name extension in the network.

At this point, the exit request message is sent to the newly selected central node, and when the new central node receives the message, it is found that the exit is the central node. It is precisely because the document is defined in the real number domain in the form of vector that all kinds of mature algorithms and calculation methods in pattern recognition and other fields are adopted which greatly improve the computability and maneuverability of natural language documents.

The specific idea based on Web URL de-duplication is: when an index page is crawled back by a Web crawler, the Web crawler will extract all the Web page URLs from the page and match them with the existing Web page URLs [8]. If the URL matches successfully, the existing local URL will be filtered out; if the matching is not found, the existing URL will be filtered out. Then, insert the URL into the "no grab URL library", waiting for the next crawler to enter.

4 Experiments and Analysis

In order to improve resource utilization, we need to extract the most important web pages as far as possible. There are many grounds for judging the importance of a web page, such as the popularity of a link (through a backlink), the importance of a link (through a certain URL function). For example, the importance of URL including. Com and home is higher than that of the web page containing. CC and map), and it is the average depth of link (judging by the depth of distance to seed), the weight of history, the quality of page, etc.

The basic workflow of the web crawler can be divided into the following steps: 1. First select some carefully selected seed URL;2. Put the URL into the URL queue to be fetched [9]. The URL, parsing DNS, is extracted from the URL queue to be fetched and the ip, of the host computer is obtained and the corresponding web pages of the URL are downloaded and stored in the downloaded web pages library. In addition, put these URL into the captured URL queue.

The node parameter is responsible for maintaining and changing the running parameters of the node. When the control node issues the control command, the communication module sends the control command to this part to modify the current running parameters of the node. Log records, responsible for maintaining the log and it is so that the management center node can monitor the crawler node running speed, crawling depth and other parameters.

Unlike Boolean models, the vector space model represents the user's query requirements and database document information as points (vectors) in vector space composed of retrieval items, and determines the similarity and select one of the linked pages and continue to grab all the pages linked to the page. Then, the query results are arranged according to the degree of similarity. After the central node receives the request message, the information of the crash node in the system table is deleted, and then the node crash exit notification message is sent to all running nodes.

First, the URL task list is established, that is the URL to be crawled. Starting from the URL task list, the Web page is crawled according to the preset depth, and whether the URL is duplicated is judged [10]. A very small number of initial URL, crawlers are needed to roam the network along the hyperlinks on the web pages according to certain rules, collect resource information, and traverse the entire web site. After that, the URL in the task list is crawled again to crawl the crawler.

This method is suitable for real-time search. For some recent hot topics, or data that changes at any time (such as stock market information), the database does not have the information of these pages. If the user is queried at this time, the relevance of the pages that have not been crawled will be judged by other pages that have been crawled.

The calculation is further analyzed by adding or reducing nodes. When a node is added, the table information is updated, and the corresponding element value in the logical table is changed to true, to add the node ID number to the physical table. Because multithreading allows for parallel transmission, even if one node fails, it will not affect the transmission with other nodes, each node has its own database to store the visited URL. When a new node is added, the repeated adoption of URL is inevitable. If the old node is migrated to the new node with the visited URL, it will cause a lot of overhead, so we can take the discard mode to deal with it.

5 Summary

For search engines, an important criterion for evaluating the efficiency of web crawlers is the cost of crawlers. The cost of crawler = the number of old pages crawled repeatedly/the number of new pages discovered, that is, the crawler should try his best to discover new pages and reduce the crawling of repeated pages, and decide that the frequency of updating a web page is related to the control of time update. In order to avoid repeatedly fetching the same page, the links that need to be stored in the database are compared with the run-queue and the completion queue in the database. The URL is sorted in descending order by the combination of domain name, link number and a Hash algorithm, so that the UR pointing to the Web resource on the same host can be assigned to the same grab queue. Other nodes are notified to add information about the new node, and other nodes update their system node table after receiving the notification message. When the crawler updates the content of the web page, he does not need to recrawl the page, but only to judge some properties of the page (such as date), and compare with the last result, if the same, there is no need to update.

Acknowledgements. This paper is supported by Henan key Laboratory for Big Data Processing & Analytics of Electronic Commerce, and also supported by the science and technology research major project of Henan province Education Department (17B520026), Key scientific research projects in Henan province universities (17A880020, 15A120012).

References

1. Zhang S, Wu G (2017) A fault-tolerant asymmetric DHT square method for dynamic heterogeneous networks. Comput Res Dev, 100–105 (2017) (in Chinese)
2. Ke Z (2014) Evaluation and future development of network search engine. J Langfang normal Univ 13(43):25–32 (in Chinese)
3. Li X, Yan H, Wang J (2015) Search engine: principle, technology and system. Science Press, Beijing (in Chinese)
4. Shen H, Pan Y (2016) A review of research on search engines. Comput Technol Dev (2016). (in Chinese)
5. Li G (2014) The present situation and development of network search engine. Enterprise Science and Technology and Development. Enterprise Science (in Chinese)
6. Huanzhe Li, Xiaoliang Liu, Daquan Guo, Fan Wang, Qiang Cao, Shumin Ma (2013) Discussion on J2EE architecture. J Hebei Acad Sci 3(4):152–156 (in Chinese)
7. Xu B, Zhang W (2013) Search engine and Information acquisition Technology. Tsinghua University Press, Beijing (in Chinese)

8. Zhang X, Pull M (2011) Research and development of intelligent search engine technology. Comput Eng Appl 3(24):265–285 (in Chinese)
9. Hongsheng Xu, Zhang Ruiling (2016) Novel approach of semantic annotation by fuzzy ontology based on variable precision rough set and concept lattice. Int J Hybrid Inf Technol 9(4):25–40
10. Zhang Y, Chen Y, Zhang G (2015) Research and development of search engine technology. Comput Eng 31(14):150–156 (in Chinese)

Electricity Consumption Prediction Model Based on Bayesian Regularized BP Neural Network

Chenghao Yuan[1(✉)], Dongxiao Niu[1], Changzu Li[1], Lijie Sun[1], and Longxiu Xu[2]

[1] North China Electric Power University, Beijing 102206, China
ychncepu@163.com
[2] China Sate Grid Xinjiang Economic Research Institute, Xinjiang 830002, China

Abstract. Increasingly complex economic backgrounds and power consumption have created more difficulties for medium and long-term load forecasting. Traditional load forecasting models such as linear regression and time series models have a good effect on linear data, but the accuracy of nonlinear electric load data in Xinjiang is poor. The BP neural network model has better processing ability for nonlinear data, but it also has the problem of poor fitting and generalization ability. Aiming at this problem, this paper establishes a medium- and long-term power load forecasting model of Bayesian regularized BP neural network. Compared with the general gradient descent method BP neural network model, it is proved that the proposed model has a good effect on the electricity consumption forecast in Xinjiang.

Keywords: Regularization method · Bayesian rule · BP neural network · Medium and long term load forecasting

1 Introduction

At present, the Chinese economy has entered a new normal, the growth rate of various macroeconomic indicators has been flat, and the growth rate of electricity consumption has also slowed down [1]. On the other hand, electric energy as a green clean energy has obvious environmental benefits compared to coal and oil. Various regions in China, including Xinjiang, have launched large-scale electric energy substitution projects, which have contributed to the increase in electricity consumption [2]. This results in power consumption data also exhibiting significant nonlinear characteristics. Traditional prediction models such as linear regression [3, 4] and exponential smoothing [5, 6] have good processing results for linear data, but have poor processing ability for nonlinear data. With the development of artificial intelligence technology, load forecasting models are increasingly considering the use of intelligent algorithms [7]. BP neural network can fully approximate arbitrarily complex nonlinear relations through signal forward propagation and back propagation of errors, and can learn and adapt to the dynamic characteristics of severely uncertain systems, thus showing good

Z. Xu et al. (Eds.): CSIA 2019, AISC 928, pp. 528–535, 2020.
https://doi.org/10.1007/978-3-030-15235-2_76

robustness when dealing with nonlinear data [8]. However, BP neural network also has the problem of poor fitting and generalization ability [9]. The regularization method corrects the neural network training performance function by limiting the network weight, thereby eliminating its over-fitting problem and improving its generalization ability [10]. The Bayesian rule uses the prior probability to correct the probability distribution and has a wide range of applications in practical engineering problems [11]. Therefore, this paper proposes to use the Bayesian regularized BP neural network model for the prediction of electricity demand in Xinjiang, and has good prediction accuracy compared with other methods.

2 Model Establishment

(1) Regularization method of neural network.

When the training samples $D = (x_i, t_i), i = 1, 2, \ldots, n$, n are the number of samples, W is the network parameter vector, under the condition given by the network structure H and the network parameter W, the neural network error sum of squares function E_D is:

$$E_D = \frac{1}{2}\sum_{i=1}^{n}\sum_{j=1}^{k}(f(x_i, W, H) - t_i)^2 \tag{1}$$

Where $f(\cdot)$ represents the actual output, and k represents the neural network output. The general regularization method adds an attenuation term E_W after the error squared function:

$$E_D = \frac{1}{2}\sum_{l=1}^{n}\sum_{j=1}^{k}(f(x_i, W,H) - t_i)^2 \tag{2}$$

Where m is the total number of network parameters, and the further total error function is:

$$F(W) = \beta E_D + \alpha E_W \tag{3}$$

Among them, the determination of super parameters α and β is the focus of regularization, controlling the distribution of neural network weight and threshold. This paper uses Bayesian regularization method to determine the selection of super parameters.

(2) Bayesian learning of neural networks.

According to Bayesian rules, the distribution function of a random variable network parameter under a given sample is:

$$p(\mathrm{W}|D, \alpha, \beta, H) = \frac{p(D|\mathrm{W}, \beta, H)p(\mathrm{W}|\alpha, H)}{p(D|\alpha, \beta, H)} \tag{4}$$

W in Eq. (4) is the weight vector, The numerator and denominator on the right side of the equal sign are the likelihood function and the normalization factor, respectively. While $p(\mathrm{W}|\alpha, H)$ represents the prior knowledge of the parameters W under the overall sample.

Under the condition of less information, the range of the prior distribution is very large. After the sample data is available, it can be transformed into a compact posterior distribution. The consistency of the weight and the network mapping can only be achieved in a relatively small range.

Assuming that the sample population and weight parameters obey the normal distribution, the likelihood function and the prior distribution function can be obtained:

$$p(D|W, \beta, H) = \frac{1}{Z_D(\beta)} \exp(-\beta E_D) \tag{5}$$

$$p(W|\alpha, H) = \frac{1}{Z_W(\alpha)} \exp(-\alpha E_W) \tag{6}$$

Where $Z_D(\beta) = \left(\frac{\pi}{\beta}\right)^{\frac{n}{2}}, Z_W(\alpha) = \left(\frac{\pi}{\alpha}\right)^{\frac{m}{2}}$

According to Eqs. (4), (5) and (6):

$$p(\mathrm{W}|D, \alpha, \beta, H) = \frac{\frac{1}{Z_D(\beta)} \exp(-\beta E_D) \frac{1}{Z_W(\alpha)} \exp(-\alpha E_W)}{p(D|\alpha, \beta, H)}$$

$$= \frac{1}{Z_F(\alpha, \beta)} \exp(-F(W)) \tag{7}$$

According to the Bayesian rule, when the posterior probability is maximized, the weight parameter is the optimal choice. As seen from Eqs. (1–7), the posterior probability maximization is equivalent to the minimization of the total error function $F(W) = \beta \mathrm{E_D} + \alpha \mathrm{E_W}$.

Then according to Bayesian rules to optimize the super parameters α and β, first find the posterior distribution:

$$p(\alpha, \beta|D, H) = \frac{p(D|\alpha, \beta, H)p(\alpha, \beta|H)}{p(D|H)} \tag{8}$$

In the above formula, the prior function $p(\alpha, \beta|H)$ has a large distribution range and can be approximated as a uniform distribution. The normalization factor $p(D|H)$ is independent of the super parameters α and β, so the posterior distribution maximum problem is transformed into the likelihood function $p(D|\alpha, \beta, H)$ maximization problem. It should be noted that $p(D|\alpha, \beta, H)$ is the normalization factor of Eq. (4), and further can be obtained:

$$p(D|\alpha, \beta, H) = \frac{Z_F(\alpha, \beta)}{Z_W(\alpha)Z_D(\beta)} \tag{9}$$

Under sufficient sample data, the posterior distribution is normally distributed. When the posterior distribution range is small, the Taylor expansion can be simplified to obtain $Z_F(\alpha, \beta)$.

Suppose that $F(W)$ has a minimum value and the corresponding weight is W_{MP}. In the vicinity of W_{MP}, $F(W)$ Taylor is expanded, and the high-order term is omitted, and further can be obtained:

$$Z_F \approx (2\pi)^{\frac{m}{2}} \left(\det\left(\nabla^2 F(W_{MP})\right)^{-1}\right)^{\frac{1}{2}} \times \exp(-F(W_{MP})) \tag{10}$$

Where $\nabla^2 F(W_{MP}) = \beta\nabla^2 E_D + \alpha\nabla^2 E_W$ is the Hessian matrix of $F(W)$ at point W_{MP}.

Then, the Eq. (10) is brought into the Eq. (9), and the maximum likelihood principle is used to determine the α and β when the maximum value of the likelihood function $p(W|D, \alpha, \beta, H)$ is satisfied, which is the optimal choice of the super parameters:

$$\alpha_{MP} = \frac{\gamma}{2E_W(W_{MP})} \tag{11}$$

$$\beta_{MP} = \frac{n - \gamma}{2E_D(W_{MP})} \tag{12}$$

In Eqs. (11) and (12), $\gamma = m - 2\alpha_{MP} tr\left(\nabla^2 F(W_{MP})\right)^{-1}$ is the number of valid parameters, and m is the total network parameter. γ is the number of parameters that play a role in reducing the error, and takes values in the range of $[0, m]$.

When calculating the Hessian matrix of $F(W)$ at point W_{MP}, the amount of computation involved is relatively large, and it can be considered to simplify by Gauss-Newton approximation:

$$\nabla^2 F(W_{MP}) \approx 2\beta J^T J + 2\alpha I_M \tag{13}$$

J in the Eq. (13) is a Jacobian matrix of E_D at W_{MP}.

(3) Bayesian regularized BP neural network training steps.

Step 1: Initial assignment of the hyperparameters α and β, we can make $\alpha = 0, \beta = 1$.Initialize network parameters based on prior distribution.
Step 2: Training with BP neural network allows the total error to be minimized.
Step 3: The Gauss-Newton approximation method is used to find the matrix $\nabla^2 F(W_{MP})$ and the number of effective parameters γ.

Step 4: The estimated values of the hyperparameters α and β are calculated by the Eqs. (11) and (12).

Step 5: Perform step 2 to step 4 again until the desired accuracy is achieved.

3 Empirical Analysis

This paper selects the data of Xinjiang Uygur Autonomous Region in China for empirical analysis, and collects electricity consumption data for Xinjiang from 2000 to 2017 as shown in Fig. 1:

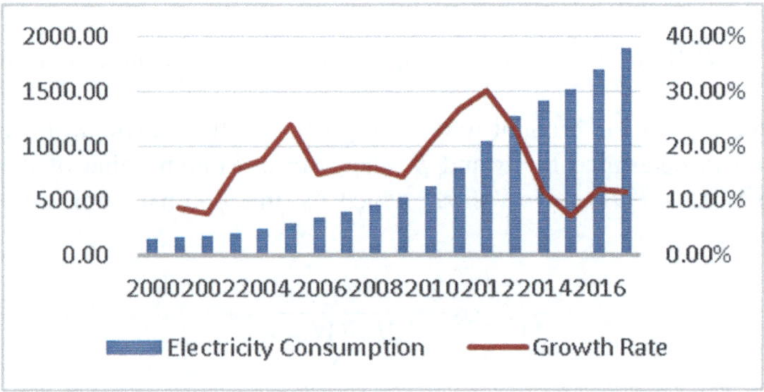

Fig. 1. Electricity consumption in Xinjiang Region in 2000–2017

The factors affecting the annual electricity consumption are mainly macroeconomic indicators. Considering the electricity consumption mode dominated by industrial electricity in Xinjiang, this paper selects three macroeconomic indicators: regional GDP, industrial added value, and annual social fixed asset investment. The summary is as follows:

The data source in the table is the website of Xinjiang Uygur Autonomous Region Statistics Bureau (http://www.xjtj.gov.cn/), where the meanings of X_1, X_2 and X_3 are regional GDP and industrial added value respectively. Investment in fixed assets with the whole society.

The collected data is divided into three parts. The data from 2000 to 2011 is the training set; the data set from 2012 to 2014 is used as the verification set; and the test set is used in 2015–2017 to test the generalization ability of the final training model.

In order to eliminate the dimension effect of each input quantity, the data needs to be dimensionless, as follows:

$$y_i = \frac{(x_i - x_{\min})}{(x_{\max} - x_{\min})} \tag{14}$$

The number of neurons in the hidden layer is difficult to determine. Too few neurons in the hidden layer will result in too little information in the training process,

Table 1. Factors affecting electricity consumption in Xinjiang region from 2000 to 2017. (Unit: 100 million yuan)

Year	X_1	X_2	X_3
2000	1365.0	422.0	612.0
2001	1485.0	450.0	706.0
2002	1598.3	473.0	812.6
2003	1875.0	571.0	1002.0
2004	2200.2	745.0	1160.0
2005	2604.0	992.0	1352.3
2006	3019.0	1218.7	1567.1
2007	3494.4	1405.1	1850.8
2008	4203.4	1790.7	2314.0
2009	4273.6	1579.9	2827.2
2010	5418.8	2105.0	3539.0
2011	6574.5	2764.1	4712.8
2012	7530.3	2929.9	6258.4
2013	8510.0	2896.0	8148.4
2014	9264.1	3179.6	9744.7
2015	9324.8	2690.0	10729.3
2016	9617.2	2440.9	9983.9
2017	10920.1	3229.1	11795.6

and the fitting effect will be poor. When the number is too large, there will be problems of poor fault tolerance and over-fitting and the error is not necessarily smaller. The number is generally determined by experience and experimentation. After many experiments, when the number of neurons in the hidden layer is 10, the prediction error is small, so the number of hidden layer neurons selected in this paper is 10.

The network has reached the fitting accuracy after 87 trainings, and the training speed is relatively fast. The forecast results of electricity consumption in 2014–2017 are shown in Fig. 2:

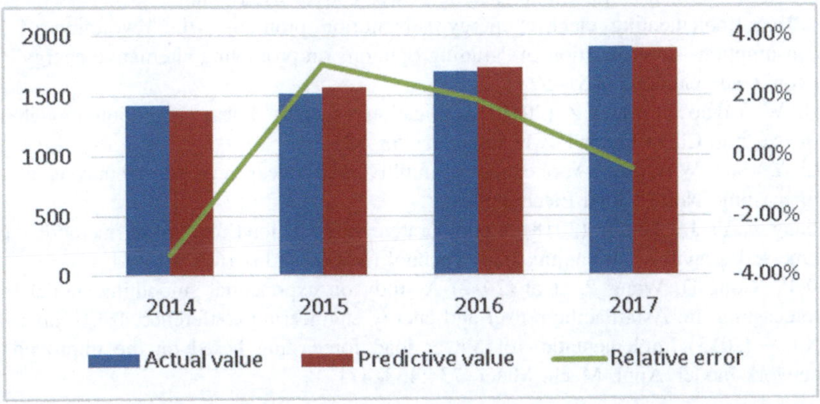

Fig. 2. Forecasting results

Suggested method is compared with the general gradient descent method BP neural network model. The comparison results can be seen in Tables 1 and 2:

Table 2. Comparison of forecasting results. (Unit: Billion kWh)

	Actual value	Suggested method		General gradient descent method BP neural network	
		Predictive value	Relative error	Predictive value	Relative error
2014	1422.63	1375.34	−3.32%	1355.86	−4.69%
2015	1522.19	1568.66	3.05%	1666.87	9.50%
2016	1704.11	1735.63	1.85%	1772.57	4.02%
2017	1900.86	1892.35	−0.45%	1968.79	3.57%
Average error		2.17%		5.45%	

From Table 2, it can be seen that the method has a smaller relative error and the prediction effect is relatively better.

4 Conclusions

This paper proposes the use of Bayesian regularized BP neural network for regional power consumption prediction, which can solve the problem of poor over-fitting and generalization ability encountered by BP neural network for predicting work. The empirical analysis of the electricity consumption data in Xinjiang from 2000 to 2017 shows that the prediction accuracy is higher than that of the general gradient descent BP neural network.

References

1. Xu Y (2017) Research on electricity demand forecast of urban power grid based on the new economic normal. North China Electric Power University, Beijing
2. (2016) Implementing electric energy substitution promoting the revolution of energy consumption—interpretation of “guiding opinions on promoting alternative energy”. China Econ Trade Guide (18):51–52
3. Qi W, Liguo S, Yufei Z (2010) Application of unary linear regression model in load forecasting. China Electr Power Educ S1:553–555
4. Li Y, Sun W (2003) Application of multivariate linear regression analysis in energy forecasting. North China Electr Power
5. Jiang Z, Ye J, Chen W (2018) A comparative study of load forecasting methods based on time series analysis. J Nanjing Insts Technol (Nat Sci Edn) 16(02):26–31
6. Ji P, Xiong D, Wang P, et al (2012) A study on exponential smoothing model for load forecasting. In: Asia-Pacific power and energy engineering conference. IEEE, pp 1–4
7. Xu Y (2015) Implementation of power load forecasting based on the improved neural network model. Appl. Mech. Mater. 734:468–471

8. Zhu H, Cui Y, Xiong H (2016) Power network load forecasting based on improved BP neural network. Mod Electron Tech
9. Li C (2018) Research on load forecasting problem based on improved BP neural network. Xi'an University of Technology
10. Yang H, Fu H (2009) Stock index prediction based on Bayesian regularized BP neural network. Sci Technol Eng 9(12):3306–3310+3318
11. Samoilin EA (2010) A method for discrimination between image signals and pulse interferences on the basis of a neural-like network and a modified Bayesian rule. J Commun Technol Electron 55(8):876–885

The Research on Teaching and Research Team Construction and Management in Vocational Education Institutions

Hong Zhang and Zhao Yao(✉)

Army Academy of Amored Forces, Changchun, China
452855684@qq.com, 15304452638@189.cn

Abstract. Faculty strength is the key factor to determine the level of education in colleges and universities. It is an effective means for colleges and universities to train their faculty, taking team building as a platform to build outstanding teaching and research teams. According to the construction of teaching and research teams in professional colleges and universities, this paper analyzes the current situation of teaching staff training in colleges and universities, discusses the connotation of teaching and research teams 'ability goals, studies the team implementation and management mechanisms, and provides theoretical guidance for institutions to carry out teaching and research team construction. It is beneficial to improve the quality of personnel training in institutions.

Keywords: Team building · Teacher strength · Culture research

1 The Present Situation of Teacher Force Construction and Talent Cultivation in Colleges and Universities

"The country is talented, the government is talented, and the industry is talented". Talent is the determinant of success [1, 2]. The teaching team and scientific research team (hereinafter referred to as the teaching and research team) are the main forces of professional colleges and universities to perform the tasks of cultivating talents, developing technology, and serving the mission of the troops [3, 4]. Taking team building as a platform to strengthen the construction of teachers, train professional leaders in disciplines, and "double and double division" faculty members to form teaching and research teams that serve the troops to safeguard the advantages and characteristics of clusters, and cultivate high-level research results [5]. It is an important measure to improve the quality of teaching and scientific research in institutions [6–9].

The construction of teachers 'strength must firmly establish the ideological understanding that "teaching ability is the core ability of colleges and teachers" and must adhere to the fundamental requirements of "focusing on actual combat and moving closer to the troops", "originating from the troops, higher than the troops, and used for the troops" to create a teaching and research team [10]. It will lay a solid foundation for the establishment of distinctive high-level arms (professional) institutions. The 21st century is the era of teams. Only by breaking through the institutional drawbacks of the management of the traditional grassroots organizations in teaching,

Z. Xu et al. (Eds.): CSIA 2019, AISC 928, pp. 536–542, 2020.
https://doi.org/10.1007/978-3-030-15235-2_77

rationally allocating resources, establishing an effective team cooperation mechanism, and building a "winning team". Let the team spirit, team culture, team management, team innovation to achieve "a combination of multiple elements, advantages". In particular, the most lacking of professional leaders in the discipline of training, in today's era, a person's power is limited, there is no perfect individual, only a perfect team. The organization structure of a team is flat, flexible and efficient. It is more conducive to people's communication, cooperation, and growth. It is more conducive to generating advantages and complementary weaknesses. It is more conducive to mobilizing initiative and creativity. It is more conducive to the passion and vitality of everyone who is moving forward. A comprehensive team force that surpasses individual talents 'abilities, which is the charm of team building.

2 Basic Characteristics of the Composition of Teaching and Research Teams

Clarity of Objectives. Setting clear goals is a prerequisite for healthy team development. For the teaching and research teams of professional colleges and universities, their goals are: to take the professional construction of disciplines as the lead, and to take teaching engineering, special construction, personnel training, scientific research and other professional colleges and universities to teach, research and research activities. Capital projects (tasks) are the goals. To conduct teaching research, gather intelligence, and serve the troops. It plays a team role in curriculum construction, information means construction, teaching model, method reform and scientific research, research on difficulties, new warfare methods, training methods, new equipment, trial of new technologies, and solution of military technical problems.

Structurally Sound. As far as the teaching and research teams in professional colleges and universities are concerned, the membership structure of the team should be determined according to different "goal" tasks, emphasizing that the team composition is "not the best, but the best". Not to mention the "high, full, and dazzling" role on the surface, a complete and benign team should be human-centered in terms of personnel composition, and they should complement each other, complement each other, and complement each other in terms of knowledge, learning edge, skills, and personality. Do not seek the best, but seek the most reasonable human resource allocation, in order to facilitate the coordination of interpersonal relationships in the team execution, reduce internal conflicts, and help improve the overall work performance of the team. Based on the principle of complementarity, it generally includes three categories of personnel, members with technical expertise in specific fields, and generally includes full-time teachers and troops of "double staff" and "double division"; Members with problem-solving and decision-making skills generally include academic leaders and teaching cadres; Members who have the skills to resolve conflicts and coordinate interpersonal relationships generally refer specifically to academic leaders or institutional personnel.

Division of Labor Collaboration. The work model of the team emphasizes that while clearly dividing the duties of its members, it also requires its members to actively participate in and coordinate their work, complement each other, and cooperate with each other. We will build a team spirit and culture, establish institutional discipline, form good communication, and build efficient team execution. We will create a kind of "gathering strength" and a comprehensive team force that surpasses individual talents 'abilities, and give full play to the overall resources and talents of the team. The three forces that make up the best team: leadership, innovation, and execution. Scientific and rational division of labor and coordination is conducive to realizing the complementarity of advantages among team members (institutions, troops, and agencies) and resolving the "short board" of individual ability qualities of team members to meet the training of "able to fight and win battles". "High-quality equipment to guarantee and protect the victory" needs talents.

Significant Performance. After a good teaching and research team has gone through different stages of team formation, agitation, standardization, and implementation, Remarkable achievements have been made in a series of construction projects aimed at "goals", such as the construction of specialty disciplines, teaching projects, key topics, key issues and problems, and the building of a team of teachers. Formed a number of "consistent with the needs of war, adapt to the development of arms, reflect the advantages of scientific research results".

3 Construction Strategy and Measures of Teaching and Research Team at Different Development Stages

Construction Phase. Accurate positioning of target and spirit. A team is a group of members that are mutually dependent to varying degrees in the process of achieving a common goal and achieving a goal. The most important thing in the establishment of the teaching and research team is to base on reality and focus on development, comprehensively analyze the resource conditions and talents of the University, and put forward feasible team building goals. Accurate positioning goals is the problem that the team is formed and team building is the first to think and solve. Any teaching and research team must "promote the main theme and strengthen the support of the spirit", "be familiar with the troops, be proficient in equipment, and understand combat", unite, self-confidence, assume responsibility, and collaborate to study and solve practical problems and contradictions that hinder the deepening of reform and innovation in education. To provide intellectual support for the realization of the goal of a strong military, "as the core spirit and goal of the team, dare to fight, be willing to contribute, be aggressive, and finally realize the value of the team.

Team Role Positioning. Forming a team does not simply gather a group of people together, but effectively complements, shares, and optimizes the reorganization of the organization's hardware strength, human resources, and personality advantages. The usual strategy for the selection and collocation of members is: Principle 1, the selection of talents will be selected. The leader of the team is at the core of the team and is the leader of the team. It may be an expert and elite in a certain field or some aspect, and at the same time it is the promoter and overall responsible person of the team. It should be a "versatile hand" with a variety of abilities. According to the content of its ability goals, the leaders of the teaching and research teams in professional institutions should generally be selected from the group of academic leaders. Principle two, member collocation by role. The team members 'collocation should be "Youdao" from the overall effectiveness of the team, "do not want the best, just for the right", five fingers clenched into a fist to fight together. Team leaders should respect the differences in the roles of team members. "Good people and talents are suitable for use". According to the different characteristics of members in the team, they should be properly matched through cooperation and agitation. Make the team a group of talents, personality, knowledge, education, age, experience, experience, etc.

Set a Code for Team Operations. First, we should do a good job of analyzing the work of team tasks. Most of the teaching and research teams are set up on a task-driven basis, do a good job of analyzing the goals and objectives of tasks, and divide the goals and projects into tasks layer by layer according to certain principles and standards. The task is broken down into a single task, and the specific task is divided into appropriate team units or members, and the appropriate responsibilities and duties of the institutions, units, agencies, or team members are established and differentiated. Specific work standards, scope and corresponding job descriptions, clear division of labor, their respective roles, so that in the team can "follow the rules". The second is to establish a team operation standard. If the team does not have a standard, it will become an unorganized and undisciplined group. Discipline is the guarantee of the achievement team. We will establish a specific and detailed management system for the operation of teaching and research teams, strictly implement their decisions, overcome difficulties, seek methods, and perform their duties. We will treat violators with no mercy, discipline every member, and replace people with the rule of law.

Stimulation Phase. Excitement period is the time period that the team runs through run-in and turbulence, and it is a special period that each team must go through. The members of the teaching and research team come from different departments and units, with different visions and requirements. After experiencing the "excitement" of the team's construction period, some potential problems are gradually exposed. The team can effectively wear and tie and successfully pass through this sensitive period. It is a test of the team's ability to integrate. The symbol of effective team "wear and tear" is that the team members are honest and cooperate with each other. Each person finds his or her position in the team. In everyone's mind, the overall goals of the team and the individual goals of the members complement each other. The cohesion of the team is taking shape, and the quality of the team's ability is steadily improved. In this sensitive period, the usual response is to build confidence. Trust is the beginning of teamwork and the cornerstone of team success. Team members advocate the establishment of "pass, help, and bring" learning organizations, through task communication and mutual cooperation, to create an internal atmosphere and learning opportunities. We should actively communicate the specific problems encountered in practical work, analyze and study the problems together, and enhance our feelings and mutual trust in the process of jointly solving specific problems. Team leaders should set up an effective communication platform, coordinate between team members in a timely manner, establish a model model, promote professional qualities, cultivate team spirit, and create a cooperative atmosphere and opportunity. The second is to avoid the "Apollo phenomenon". The position and role of the elite in the team can not be underestimated. They may be the founders, practitioners, or strong advocates of the work of the team, or they may be strong supporters of team spirit and culture. However, it does not require all team members to be elite. Elite members should be controlled within a certain range. Too many elites can not form a sense of cohesion and loyalty. Avoid the "Apollo phenomenon" in which the team's strong background, personality, and contempt for achieving team goals lead to fragmentation. The team run-in period is generally 3 to 6 months. Team members can persist through this period and will grow into the backbone of the team.

Normalization Phase. After the turbulent run-in period, the team will soon enter an efficient normative period. The team will clearly define the division of labor, coordinate and cooperate according to the responsibilities, responsibilities, responsibilities, and complementary cooperation and tacit cooperation among members. The team showed strong self-confidence and subjective initiative. The team spirit and sense of honor gradually strengthened and reflected, and achieved initial results. It was a period of cohesion and combat effectiveness. During this period, the team should implement the strategic concept of sharing and cooperation. Sharing is a moral character, but also a learning attitude. Sharing is the beginning of mutual trust and deep cooperation among team members. It is a working method of freedom, democracy, relaxed and harmonious, together, and cooperation and mutual assistance. "Cooperation" is the essence of the survival and development of the team and the core spirit of the team. Being good at cooperation is also a kind of ability. It is a kind of ability to promote the team and oneself. The team and team members who do not understand cooperation must fail.

Phase of the Implementation Period. It is the period of completing the team's mission. The team has achieved phased goals. After performance evaluation, the team has carried out necessary organizational adjustments, work integration, strength configuration, and process reengineering to achieve the goal of concentrating the superior forces of the team. The team in this period is undoubtedly the most effective high-performance team. It can use the minimum investment to achieve the maximum benefit for the team. This period is also a period when the unique team characteristics and culture are gradually formed. At the same time, players with outstanding abilities have grown into professionals who can stand alone and take greater responsibility. The team can use this opportunity to promote the metabolism of the organization and thus maintain its vitality.

In this period, the team should pay attention to two aspects of the construction: First, the execution of the ability. Bill Gates believes that "small success depends on individuals, great success depends on the team". The team's fighting power is determined by the cohesion and execution of the members. It is the result of the common behavior of the members. Execution is the foundation of the team. The team has the ability to perform as follows: full of passion, initiative, potential to be played; Strive for the best, pay attention to the details; Keep your promise and be responsible to the end; Good at learning and willing to cooperate; Perseverance, endurance; Good at innovation, dexterity. The team's execution ability is not overnight or one person can do it. We must work together to establish a set of execution modes for implementation strategies and continuously improve and improve. Second, team culture. Team culture is a kind of micro culture. It is a kind of moral norm and behavior way of spiritual strength, value, working style and member's approval that condenses and accumulates in team work. A good team culture can make the team a fighting group, with the same advancement, coordinated operations, incentives for execution, sublimation of cohesion, and hard to overcome. During the team implementation phase, the team culture is deeply rooted and can effectively promote the improvement of execution.

Rest Period. The team faces dissolution because it is about to complete its historical mission. The struggle of the past will become a good memory. During this period, the main work of the team is to summarize, document, coordinate, commend, etc. The team should further enhance emotional communication, handle good benefit sharing, improve documentation, and close the work to prepare for the formation of new teams and meet unknown challenges.

4 Teaching and Research Team Performance Management and Evaluation Measures

Teaching and research team performance evaluation is a process of internal development aspirations, target tracking, quality evaluation, internal and external environment forecasting, and regular summary analysis and comprehensive evaluation. It aims to identify the main reasons that affect the implementation of teaching and research teams and propose improvement measures. It is a process in which the team constantly challenges higher goals and sustainable development. The PDCA cycle, also known as Daiminghuan, was proposed by Dr. Deming, a quality management expert in the United States. It is a basic work procedure for quality management activities. Its main content consists of four stages and eight steps. The four stages are planning, execution, inspection, and processing. The eight steps include: (1) analysis of the current situation and identification of problems, (2) Analysis of impact factors; (3) Identification of the main factors; (4) to take measures and draw up plans; (5) Implementation of the plan for the development of measures; (6) Check the implementation of the results; (7) Analyze, summarize and revise the new implementation standards; (8) Unresolved to move into the next cycle.

The PDCA management cycle is a large ring, one ring and one ring, a small environmental protection ring, and a large ring. The stages are not clearly separated but closely linked and repeated. Each cycle goes forward and rises to a new height. There are new contents and goals. Practice has proved that PDCA management method can ensure that quality problems are solved, process control is continuously improved, and quality level is continuously improved. The dynamic, open and continuous improvement of the team's performance management during the five stages of the development of the teaching and research team determines that it can be built using the PDCA cycle.

References

1. Gao Y (2012) The team gathers strength. China Electric Power Press, Beijing
2. Jia Y (2013) Knowledge, employment and lead a team of 9 kinds of personality and team building. People's Post and Telecommunications Publishing House, Beijing
3. Chen J (2012) The team wins and wins in the wolf team. Economic Management Press, Beijing
4. Feng H (2015) Research on the management of innovation ability performance examination of university research teams. Beijing Jiaotong Univ (1)
5. Gao J (2010) An analysis of team management in higher vocational colleges. Guangxi Univ
6. Cao K (2013) In order to cultivate commanding talents who can win the war, he actively promoted the education of colleges and universities to move closer to the troops. Military
7. Li X, Li GQ, Zhao SZ et al (2012) Diode-pumped Nd:YVO4 laser passively Q-switched with graphene oxide spin coated on ITO substrate. Laser Phys 22(4):673–677
8. Du GL, Li GQ, Zhao SZ et al (2011) High slope efficiency and low threshold in a laser-diode-pumped passively Q-switched ion-implanted Nd:YVO4 planar waveguide laser with GaAs saturable absorber. Laser Phys 21(11):1880–1883
9. Chang L, Yang C, Yi XJ, et al (2012) 914 nm LD end-pumped 31.8 W high beam quality E-O Q-switched Nd:YVO4 laser without intracavity polarizer. Laser Phys 22(9):1369–1372
10. Sato Y, Taira T, Pavel N et al (2003) Laser operation with near quantum-defect slope efficiency in Nd:YVO4 under direct pumping into the emitting level. Appl Phys Lett 82 (6):844–846

Research on the Development Trend and Coping Strategies of Internet Finance

Xiaoxia Wu[1,2(✉)] and Heyu Xiao[1,2]

[1] Jilin Engineering Normal University, Changchun, China
516877066@qq.com
[2] Jilin Oilfield Senior Middle School, Songyuan, China

Abstract. With the rapid development of Internet finance, it has gradually formed a new Internet financial entity and model. Under the Internet financial environment, multiple elements in the main body have to cooperate to realize the transfer of funds from supply to demand. This will promote the development of the financial industry. In the active expansion of the development platform of Internet finance, more and more partners are participating in it, and while building the industrial ecosystem together, it also brings the development impact and challenges of the Internet era. At present, there are many problems in China's Internet financial ecosystem. This requires strengthening the construction of the Internet financial ecosystem and ensuring the sound development of China's Internet financial ecosystem. Based on the characteristics of Internet finance, this paper analyzes the risks, development trends and future coping strategies of Internet finance, and hopes to help Internet finance.

Keywords: Internet finance · Ecosystem · Development

1 Introduction

The essence of internet finance is finance. It is a financial activity based on Internet technology [1, 2]. It is a cross-border and integration of mutual network and gold, which is both a process and a result. Compared with the traditional gold-knocking industry, Internet finance has a wider range of Internet financial coverage, and it has expanded the existing market in essence, and has transformed the entire financial industry. It is the financial disintermediation and democratization product [3, 4]. As a new financial model, Internet finance has full rights and means to participate in financial activities. Under the fair environment of relative information, it fully captures the convenience and benefits brought by Jinhang Service [5]. In the active expansion of the development platform of Internet finance, more and more partners are participating in it, and while building the industrial ecosystem together, it also brings the development impact and challenges of the Internet era [6–15]. Based on the analysis of the current status of Internet finance development in China, this paper analyzes its potential risks, proposes corresponding preventive measures and analyzes the future development trend of Internet finance [16–20].

Z. Xu et al. (Eds.): CSIA 2019, AISC 928, pp. 543–549, 2020.
https://doi.org/10.1007/978-3-030-15235-2_78

2 Characteristics of Internet Finance

Internet finance can be defined as a highly open financial product and service that is dominated by the Internet industry and is based on the Internet and utilizes the latest information technology. As the intersection of the Internet industry and the financial industry, Internet finance has the following characteristics.

2.1 Virtuality

Internet financial virtualization is the biggest distinguishing feature of traditional Golden Industry, mainly reflected in virtualized financial services, virtualized service organizations and virtualized transaction media currencies. Financial business virtualization refers to the Internet file transactions, services, and offices are all electronic documents of the Internet; service organization virtualization is the establishment of e-banking and other Internet financial institutions, virtualized service activities, greatly reducing financial institutions The importance of physical construction; the virtualized transactional media currency is another upgrade of physical currency in the social development to replace the physical currency. As a media currency spread through the Internet, such as electronic data, Internet wallet, Internet cash, etc., it reflects financial activities. Credit activity in the middle.

2.2 Open Animals

The openness of Internet Gold is mainly the openness of the business environment. It shows that financial activities can break through the limitations of time and space, break the geographical restrictions of traditional financial institutions and business outlets, and attract a wider range of customers. As long as the operator opens the Internet business, anyone is likely to become a customer, and can provide services to customers at any time and in any way. At the same time, it appears in the currency of the trading medium, and can also get rid of the shackles of various forms of money, promote the internationalization of market behavior, and promote the all-round development of international financial services.

2.3 Transparency

Internet financial transparency refers to the transparency of market operations. Transparency is mainly reflected in the transmission of information, the execution of transactions, the liquidation of procedures and the formation of market prices. Internet financial transactions are fast and convenient, which greatly reduces transaction costs. Open market information greatly enhances the transparency of financial markets and plays an important role in the choice of market prices and the improvement of market regulation capabilities.

3 The Development Trend of Internet Finance

Internet finance has improved the economies of scale of financial services, improved the efficiency of financial services, and reduced the cost of information acquisition, customer management costs and information clearing costs, thereby increasing the value creation ability of financial services. Under the trend of financial enterprise Internetization and Internet corporate financialization, Internet finance has seen a new development trend.

3.1 Internet Finance Socialization will be a New Direction

According to a report released by the China Internet Information Center, Chinese mobile Internet users have become the first Internet group, and the mobile Internet has become an important platform for people's future social life. "Platformization is an important feature of the development of mobile Internet, and it is also the confidence factor for the success of mobile Internet innovation and entrepreneurship." Many financial institutions and Internet finance companies will further realize the socialization of Internet finance through social platforms such as WeChat and Weibo. From the imaginary space, mobile financial services will have a lot of convenience. In the coverage of financial services and life services, it will fully surpass the online banking services in the era of desktop Internet. With the integration of new mobile technologies such as QR code recognition and near-field payment into financial services, customers can enjoy informatized financial services anytime and anywhere, which is beyond the reach of traditional desktop Internet financial services. The socialization of Internet finance has become an inevitable trend in the development of Internet finance.

3.2 Banking Innovation will be a New Goal

With the rapid development of Internet finance, banks should use Internet thinking as a guide to accelerate the improvement of management and services, and innovate and improve online and offline services. On the one hand, banks should establish an Internet financial service platform that is different from the traditional service model, change the traditional service model, and realize the transition from "payment +credit" intermediary to "payment + credit + information" intermediary. On the other hand, we must focus on building mobile financial services. Mobile financial services is the application and development of the Internet financial service model in mobile terminals. It is expected to become the most efficient, smoothest, and best-performing banking service, and it is likely to become the mainstream service model for banks in the future. The banking industry should make full use of the Internet, cloud computing, big data and other technical means to promote the transformation of business into digital, and constantly improve the Internet financial business channel service functions.

3.3 Internet Enterprise Marketization will be the New Normal

Promoting the market-oriented operation of enterprises is the most effective means to stimulate entrepreneurial passion and tap the development potential. Under the influence of the market mechanism, a new market-oriented operation mode has been constructed, which has achieved the goal of optimizing the allocation of production factors, reducing production costs, and improving production efficiency and economic benefits. With the development of the marketization of Internet enterprises, Internet consumer finance came into being. Internet consumer finance refers to the modern financial services that customers use to make loans and purchase goods through the Internet, including housing loans and car loans. E-commerce companies such as Ali and JD.com will enter the field of online consumer finance, which will cause Internet consumer finance to usher in an industrial outbreak.

4 The Impact of Internet Finance on the Financial Industry

4.1 Accelerate the Integration of Traditional Finance and the Internet

On the one hand, the Internet can bring vast amounts of data for the financial corporations, financial enterprises make up the short board in terms of collecting customer information to better match precise customer, strengthen risk management and control, reduce operating costs, improve service quality; on the other hand, Through cooperation with financial institutions, Internet companies have taken advantage of the wealth of financial enterprises, many branches, and high consumer awareness to expand the scale of offline business and enhance the ability to operate across sectors.

4.2 Strengthen the Impact on Traditional Financial Services

The first is to divert the deposits of commercial banks. Although the third-party payment platform will not affect the total amount of commercial bank deposits, the impact on the structure and term mix of commercial banks will inevitably lead to the pressure of small and medium-sized banks. Second is the direct squeeze on the intermediary business of commercial banks. On the one hand, the rise of third-party payment platforms, especially to the traditional financial industry, is gradually squeezing the space of banks in card settlement, agency collection, agency insurance, fund sales and own online payment services; Third-party payment is gradually changing the consumption habits of financial consumers, gradually forming a habit of staying at home and reducing dependence on bank physical outlets. Once again, Internet Finance is gradually forming a direct impact on commercial banks by virtue of its advantages in information processing. Take Ant Financial Service as an example, including Alipay, Sesame Credit, Ant Jubao, Internet Merchant Bank, Ant Small Loan, Ant Financial Cloud, Yu'ebao, Lucky Treasure, Ant Flower and other products, which basically cover traditional financial "storage and loan". In the three major sections, Alipay solved the payment problem of "last mile", while ant small loan can realize 3 min application and 1 s to arrive, without any manual intervention. Finally, Internet finance's behavioral model changes in risk management will challenge the quality and effectiveness of commercial banking services and risk control systems.

4.3 Impact on Monetary Policy

Because Internet financial companies issue virtual currency, they have certain reservoirs and alternative effects on the currency, and they cannot disclose data to the public. This makes the central bank unable to accurately grasp the scale and flow of the Internet currency, thus weakening the central bank's ability to control the base currency. Increased the difficulty of currency regulation and affected the effectiveness of monetary policy.

4.4 The Impact on Financial Supervision

Due to the diversification of Internet financial risks, especially the operational risks and technological risks, it is characterized by cross-infectivity, suddenness, rapid spread and destructiveness, and the amplification effect of various risks is caused by the Internet network. These have brought challenges to financial regulation. At the same time, China's current law on Internet financial supervision is still in a blank, and the existing regulatory thinking, regulatory procedures and regulatory methods are obviously lagging behind, unable to meet the regulatory requirements in the Internet environment.

5 Research on Internet Finance Ecological Construction Strategy

5.1 Foster and Ecological Body

In every ecosystem that operates successfully, the living organisms that it constitutes play a more important role. The same is true for the Internet financial ecosystem. Only by perfecting and enriching its ecological subject, each subject can fully demonstrate its unique role in the ecosystem, and thus achieve harmonious development between each other. Promote the healthy development of the Internet ecosystem in the progress of the survival of the fittest. At the same time, it is necessary to combine the characteristics of the Internet to continuously implement segmentation and innovation of financial ecological subjects, so that financial products become more diversified, and thus constitute a good symbiotic mechanism to promote the diversified development of the Internet financial ecosystem.

5.2 Strengthen the Effective Supervision of the Financial Sector of the Internet

First, gradually improve the Internet financial supervision system. Clearly define and confirm the responsibilities and scope of the supervision department; construct a coordination organization composed of relevant departments such as the China

Insurance Regulatory Commission and the Internet Supervision, and coordinate the supervision through each other to avoid the problem of duplication and lack of supervision. Second, improve the Internet financial supervision system. Based on the existing financial regulatory system, standardize the entry threshold and business model, and formulate corresponding standards, so as to make up for the system gaps in Internet finance as soon as possible.

Finally, innovative regulatory approaches use a variety of regulatory approaches to enhance the effectiveness of regulation. Under the supervision of the Internet, regulators can use big data and cloud computing to improve their supervision; learn from the experience of international supervision and change the original single financial supervision; and encourage the relationship between innovation and risk prevention. Good treatment and moderate supervision.

5.3 Accelerate the Construction of Social Credit System

The construction of credit system is a problem that needs to be paid attention to in the development of Internet financial ecosystem. In order to ensure a good credit environment in the development of the Internet financial ecosystem, it is necessary to build a sound credit evaluation system as soon as possible. The credit system can be constructed from three perspectives, namely, three credit systems: personal consumption, commercial enterprises and capital credit. Adhere to the marketization as the basis of operation, vigorously develop credit intermediaries; formulate a mechanism for breaking the trust and punishment, reduce the default problem by raising the cost of default; build a credit evaluation labeling system, build a credit database covering the whole network according to data mobile phone and analysis, so as to be able to Achieve comprehensive resource sharing among Internet companies.

6 Conclusion

With the development of the economy, the financialization of the Internet is the trend of the times and is inevitable. On the positive side, Internet finance has promoted innovation in the traditional financial industry and facilitated people's lives. The future of Internet finance will play a greater role in profoundly changing the face of society. At the same time, we cannot ignore the risks brought by Internet finance, and we must do timely prevention and control to make Internet finance more healthy and stable.

References

1. Lu M-F, Lu S, Wang ZG, et al (2014) The development trend of new formats of inclusive finance—the calm thinking about hot issues of internet finance. J Jinling Inst Technol
2. Li J (2017) Research on the development of the third party payment standard in China—based on the development trend of internet finance. Reform. Strategy

3. Lu M-F, Xu Y-Y (2016) Strategic study on the development trend of internet finance: based on the comparison on the special rectification of internet finance and the previous successive economic rectifications. J Jinling Inst Technol
4. Ye X (2015) An analysis on new development trend of consumer finance under the background of internet finance. Credit Ref
5. Liu X (2017) The development trend of the internet consumer finance research. J Changchun Financ Coll
6. Zhao W, Gao L, Zeng T, et al (2016) The new development trend and the supervision analysis of consumer finance under the background of internet. Financ Dev Rev
7. Xu N (2015) The development trend and influence of internet finance. Credit Ref
8. Kang CY (2018) The research on the development trend of supply chain finance under the background of internet. J Changchun Financ Coll
9. Wang HP (2016) The new development trend of the finance professional teachers' practical teaching ability under the background of "internet+". Educ Teach Forum
10. Kang Y, Hu C, Hochschule Niederrhein University (2016) The business model, development trend and value creation of internet consumer finance. Explor Financ Theor
11. Ma W (2017) Development trend and international governance of cross-regional internet financial crimes based on the perspective of internationalization. J Huaqiao Univ
12. Yang F, Wang Y (2017) Traditional financial institutions internet transformation architecture analysis and its future development trend. Reform Strategy
13. Xu J, Branch HS (2016) Management countermeasures for commercial banks under the trend of "internet finance+". J Shaoyang Univ
14. Mo Y (2014) Development trends of financial industry in internet era. Financ Econ
15. Li L (2017) Study on the new trend of internet financial development and its influence. J Jiangsu Norm Univ
16. Wang L, Ding Y, University H, et al (2016) Development trend of internet finance and its risk prevention. J Huanghe S&T Univ
17. Wang YJ (2017) Research on the development model of consumer finance under the background of internet. Technoecon Manag Res
18. Dong X (2016) About the internet financial relationships with commercial banks and the analysis of development trend. Mod Indus Econ Informationization
19. Luo C, Li C-D, University ZW (2018) Research on the development of agricultural public financing model under the internet finance. Spec Zone Econ
20. Shi Z (2017) Research on the development model and trend of P2P finance. Jiangsu Sci Technol Inf

Research on the Development Trend and Protective Measures of Campus Network Security in Colleges and Universities

Deling Ran(✉)

Information Center, ShanDong University of Arts, Jinan, China
randeling@sdca.edu.cn

Abstract. With the popularization and development of the Internet, the informationization and digital construction of colleges and universities are also accelerating. As an important infrastructure for informationization and digital campus, college campus network plays a role in teaching and research, management and foreign exchange. The important tasks of the school and provide a good guarantee for the education and teaching of the school. The current campus network has spread to all departments of the school and has become the basic equipment of the entire campus. The school has higher and higher requirements for the manageability and security of the network. Therefore, it is particularly important to construct a reasonable network security system. Based on the current problems of college network security, this paper comprehensively expounds the hardware and software design loopholes of campus network security at this stage, and deeply discusses the management flaws and technical problems of campus network security problems. Based on the comprehensive improvement of the security information management level of the campus network and the establishment of a comprehensive campus network security technology system, this paper proposes a comprehensive implementation plan of firewalls and other technologies to comprehensively improve the current campus network security level.

Keywords: Network technology · Campus network · Network security · Protective measures

1 Introduction

With the popularization and development of the Internet, the informationization and digital construction of colleges and universities are also accelerating [1, 2]. As an important infrastructure for informationization and digital campus, college campus network plays an important role in teaching and research, management and foreign exchange [3, 4]. Important tasks and provide a good guarantee for the school's education and teaching [5–9]. The current campus network has spread to all departments of the school [9–15]. The campus network has become the infrastructure in campus construction. The school attaches more importance to the construction of a safe and reasonable network security system, which is mainly reflected in the management of the campus network. Sexual and safety concerns [16, 17]. For today's attacks and

Z. Xu et al. (Eds.): CSIA 2019, AISC 928, pp. 550–558, 2020.
https://doi.org/10.1007/978-3-030-15235-2_79

threats from all levels of the network, the security of the entire campus network cannot be guaranteed by a single security technology or a network device. With the development of network technology, the campus network has become complex and diverse [18–20]. The planning and construction of the campus network security system should be focused on the whole, layered construction, from the core to the terminal, and deployed everywhere, thus reducing the security risks of the entire campus network and ensuring the network security of users in the campus network. This paper proposes a comprehensive implementation plan based on firewalls and other technologies to comprehensively improve the current campus network security level.

2 Campus Network Security Features

2.1 Campus Network Environment Is Complex

Due to the limitations of the campus network, many students in the school chose to handle the off-campus agent network in order to pursue free online time and bandwidth, and link to the campus network through the small LAN of the dormitory, which makes the access mode of the campus network change. The diversity of the network allows for the presence of telecom broadband, China Unicom broadband and other operators to provide access methods; the Internet access method has also changed from a single wired connection to a wireless connection, wired connection. The campus network itself has multiple export outlets provided by the education network and network operators, and the above-mentioned private access methods for students in the school, which makes the number of campus network access ports increase, and the increase of network access ports makes the campus network's wired network face A large number of security threats. Although the emerging wireless network provides us with convenience on the use of the network, at present, due to the limitations of network infrastructure and technology, the performance of the wireless network is not very stable, and there are certain security risks and security problems.

2.2 Lack of Overall Planning for Campus Network Construction

In the initial construction of the campus network, due to the lack of comprehensive overall considerations, in the process of building the campus network, the original design plan based on the actual problems will be changed, resulting in a difference from the expected design after the completion of the project. Moreover, in the aspect of capital investment, there is a situation in which hardware contempt software is emphasized, so that the hardware cannot perform its function and value in the absence of corresponding software.

2.3 User Groups Are Dense and There Is a Peak of Internet Access

The campus network has a simple staff structure and is staffed by the school. It is mainly composed of students in the school. It is characterized by a large number of people and is active. Due to the characteristics of the school's schedule, the peak period of the Internet is lunch break, evening and rest day. During this time, the campus network will consume a lot of core bandwidth and export bandwidth, because the students installed the terminals on their terminals in order to improve the download speed of their own terminals, such as the software of P2P services such as Thunder, during this time. The network traffic is large and the network environment is very bad, which seriously affects the normal use of the campus network.

3 The New University Campus Network Security Issues Exist

3.1 Various Network Viruses Illegally Invaded

In view of the illegal intrusion of the computer management technology, the password in the system can be effectively cracked, and can exist normally in the Internet without being discovered, and then the appropriate timing is selected for destruction. Internet disruptors generally have the same computer processing rights as regular computer administrators, and can operate according to their own wishes, including the reading of computer information, etc. Of course, these are very small aspects; The use of its own authority to continually undermine the use of the Internet has led to the paralysis of computer systems.

3.2 Hardware Device Defects

From the perspective of the overall research of the Internet, there are some design loopholes in network hardware devices. Generally, the security problems caused by the Internet are not easily depended on, including some important local research management, and there are certain dangers, such as detailed parameter settings. Can not effectively verify the description, the role of the firewall can not be used well, whether there is a certain expected value, the specific operational management fluency, etc., and whether the parameter information has certain defects. Regardless of whether it is for the Windows desktop or the operation management system, there are loopholes in the use of the Internet. At present, the most used computer networks still have security problems. If the processing of the firewall does not meet the basic requirements well, the setup will not play a certain role, and even counterproductive. The reason is the connection between the firewall and the Internet, which is convenient for the outside world to attack the Internet directly.

3.3 Insecure Factors of Campus Network Users

Nowadays, students in school are the main body of the campus network. For students, the awareness of network security is generally not high. Even students with certain computer skills cannot control them. Other professional students will not It raises curiosity and thinking. They believe that maintaining campus network security is only the responsibility and obligation of colleges and universities and professional and technical personnel, and they cannot afford this responsibility. As a result, the existing campus network presents a larger stability.

In the actual operation process, the campus network is usually used as its own practice point. The unique curiosity of college students will encourage them to spend most of their time and energy to explore and analyze the campus network, and then on this basis, under what circumstances can each be carried out Real-time collection and acquisition of resource information.

From the perspective of the campus network itself, it is extremely easy to be broken. The campus administrators will apply various program strategies for the presentation of this situation. As a result, many crises are presented, and the existing offensive real-time control cannot be performed. The most serious problem is the irreparable damage to the existing Internet, which will lead to a vicious circle in the long run.

4 Campus Network Security Design Principles

When planning the university's network security system, we will follow the following principles and based on these principles. A complete and systematic overall network security solution:

4.1 System Design Principles

By analyzing the network hierarchy of information networks, security requirements elements and dynamic implementation process, a scientific security system and security model are proposed, and various security risks that may exist in the network are analyzed according to the security system and security model. Based on the implementation process, an overall network security solution is proposed to maximize the resolution of possible security issues.

4.2 Global, Balanced Principle

The design of the security solution starts from the overall situation, comprehensively considers the value of information assets and the security risks it faces, balances the relationship between the two, and adopts different strength security measures according to the value of the information assets and the magnitude of the risks. A security solution with the best performance-to-price ratio.

4.3 Feasibility and Reliability Principles

After adopting comprehensive network security measures, it should not have a great impact on the original network of the university and the application system running on the network, so as to effectively improve the network and the application system under normal conditions. The security strength of the network and applications ensures the security of the entire information assets.

4.4 Principles of Dynamic Evolution

The plan should formulate a unified technology and management plan for the university to realize the formulation of a unified security strategy, and realize the evolution of the entire network from the basic defense network to the defense network and the intelligent defense network to form a closed-loop dynamic evolution network security system.

5 Countermeasures and Implementation of Network Security Technology University

In view of the current situation of campus network security, combined with the development and changes of intelligent, systematic and service-oriented protection technologies, we believe that various protective measures are needed to build a safe equipment operating environment.

5.1 Firewall Deployment

The setting of the firewall system can realize the unified setting of the internal and external information connection, and comprehensively review and control the connection process, which greatly enhances the system's defense ability against Internet threats, and can uniformly reject the data information that has not been allowed yet. In a certain way, it can ensure that existing data information enters the campus network safely and reliably. The internal architecture of the firewall of the campus network security system is shown in Fig. 1.

This article will present the most widely used and best performing hardware platforms and internal architectures. Not only that, but even the government's work department and education industry have applied it; in the actual operation process, the existing factors can not have any impact on the firewall. In addition, for real-time intrusion detection of existing data information, it will be detected in the first time and the actual application of the best solution; for web filtering, traffic monitoring and state control, etc. Get one-to-one implementation to ensure network security and stability; routing function can be fully utilized here, which also allows the implementation of static routing strategy; in the actual operation process, the rendering of many malicious information will be fundamentally solved.

5.2 VLAN Technology Implementation Plan

Virtual LAN is actually a logical intranet that summarizes the information of each site in real time with reference to the actual needs. There is no restriction or limitation by geographical location and various factors. With the Virtual Local Area Network (VLAN) technology, the existing physical network can be optimally configured for logical subnets. In this form, all the data information summarized above will be forwarded to the VLAN site at the same time. management. The architecture diagram of the VLAN network topology of this college building is shown in Fig. 2.

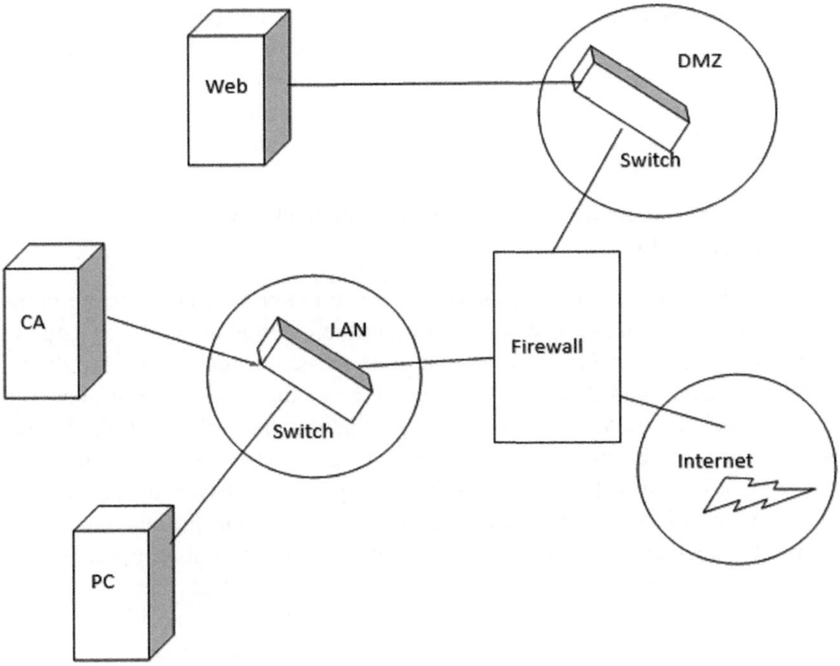

Fig. 1. Topology of the firewall in the network system

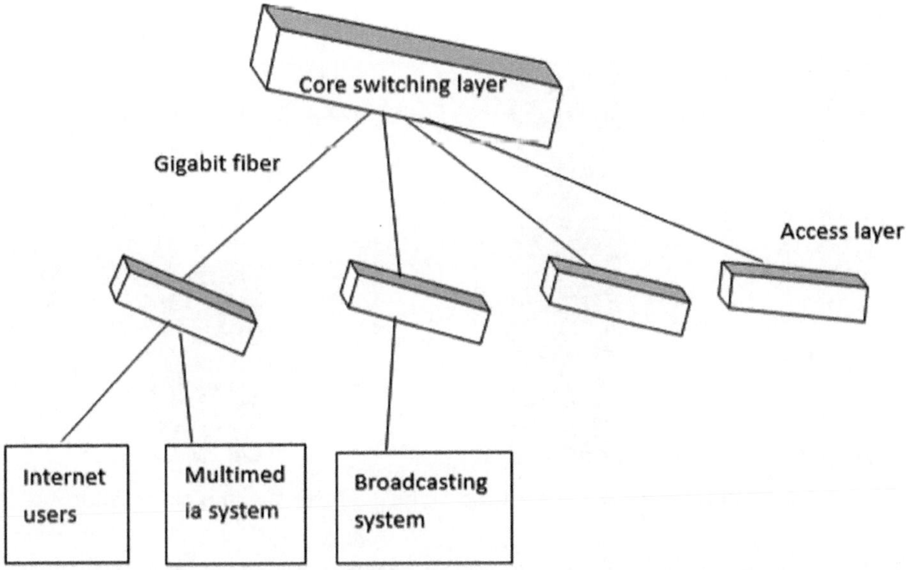

Fig. 2. VLAN structure topology of the teaching building

Colleges and universities will effectively integrate the current development status of the school, and divide them according to the differences of buildings, including broadcasting, campus cards and multimedia promotion rooms. This approach not only helps to promote the smooth flow of existing networks, but also ensures the security and stability of the internal information, laying the foundation for future operations.

5.3 Improve the Network Security Management of Colleges and Universities

Efficient, reliable and safe campus network construction also needs to be combined with the actual situation of the campus to build the campus network security system, and fully implement it in the actual operation process, in order to achieve the unified norms of the teachers and students in the school. From the perspective of Internet users, users should strictly abide by the laws and regulations and rules and regulations formulated by the state in the actual operation process. The following are the key documents summarized based on actual conditions, mainly reflected in the maintenance and management of computer network security in China. On the one hand, the phenomenon of implementing violations through the network is not allowed. If there is a situation contrary to the above description, it should be punished according to law, and resolutely put an end to the danger of campus network security accidents.

After the new scheme protection, the number of network problems in the existing scheme is lower than that in the previous scheme, as shown in Fig. 3.

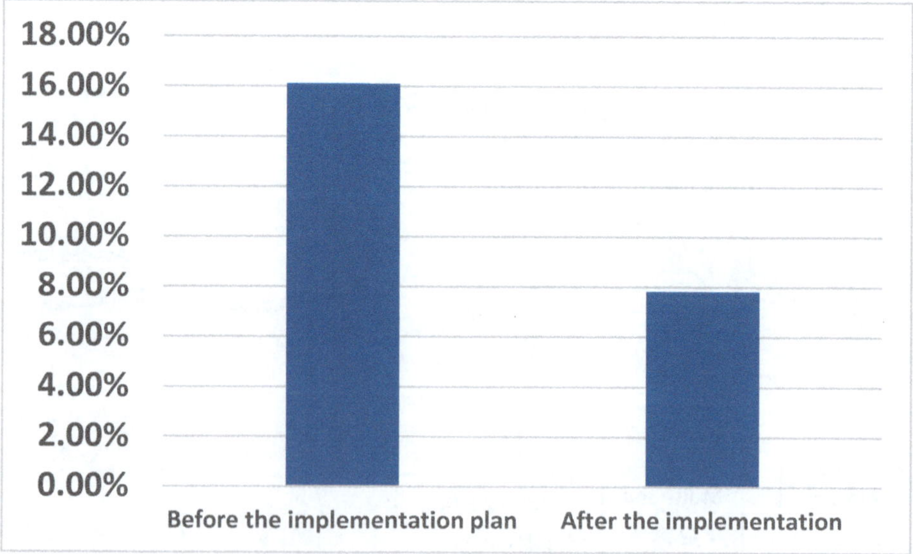

Fig. 3. Shows the frequency of monthly network problems before and after implementation measures

It can be seen from Fig. 3 that after the construction of the university campus network. The network topology of this university considers the division management of functions in different regions and different departments within the school, and realizes the effective management of the campus network. The problems appearing are obviously reduced. At the same time, in order to improve the security level of the campus network, in the actual operation process, the existing technologies are practically applied, mainly including firewall technology, etc., thereby promoting the increase of the network security factor. Network security construction is actually the effective combination of network security technology and management work. It not only requires the effective implementation of various technologies, but also comprehensively implements all aspects of management and abides by the various rules and regulations.

6 Conclusion

With the popularization and development of the Internet, the informationization and digital construction of colleges and universities are also accelerating. As an important infrastructure for informationization and digital campus, college campus network provides a good guarantee for the school's education, teaching and scientific management. The construction of campus network security is related to the future and development of the school. Building a safe and reliable campus network can not only rationally, safely and effectively utilize the various resources available in the school, but also improve the network working environment for easy management and maintenance and improve the reliability of the network.

References

1. Ali MNB, Rahman ML, Hossain SA (2014) Network architecture and security issues in campus networks. In: Fourth international conference on computing, communications and networking technologies. IEEE, pp 1–9
2. Dong YH, Na BI, Wang CX, et al (2016) 802.1 x-based design and implementation of campus network security certification. Jilin Norm Univ J
3. Li C (2015) Design and implementation of campus network security model based on VPN and PKI. Value Eng
4. Fu H-L (2015) Campus network security technology and prevention strategies. J Guangxi Univ Nationalities
5. Geng F, Ruan X (2017) Campus network information security risk assessment based on FAHP and matter element model, 298–306
6. Li Z (2014) Campus network security system design and optimization of high-performance computing environments under. Software
7. Xu X, Liu W, Cui W (2014) Design and implementation of campus network security system based on honeypot. Netw Secur Technol Appl
8. Zhao Z (2014) Based on VPN technology in constructing the campus network security system. Autom Instrum

9. Wang Y (2014) The construction of campus network security defense system analysis. Inf Secur Technol 164(4):1291–1303
10. Ma Y, Zhang Y (2014) Analysis of campus network security reform based on the core router. Netw Secur Technol Appl
11. Lan Y, Chen Z (2016) The research on the OSPF security optimizing of campus network. In: International conference on network and information systems for computers. IEEE, pp 592–594
12. Ye Z, University LT (2015) Problems and countermeasures on the campus wireless network security. Wirel Internet Technol
13. Lu H (2018) The construction of network security system of SRTVU campus network under the requirement of level protection. Chin Comput Commun
14. Gao W, Xu H, Ning YW, et al (2018) Research on campus network security operation center based on security situational awareness platform—taking the fourth military medical university as an example. Comput Technol Dev
15. Zhang B (2017) Research on campus network security evaluation in colleges and universities. Value Eng
16. Yang QC (2017) Construction and application of campus network security system based on dynamic security model. J Lanzhou Univ Arts Sci
17. Liu Z, Center IC, Polytechnic GI (2017) Research on campus network security operation and maintenance. Chin Comput Commun
18. Huang C (2018) Application of network firewall technology in campus network security. Heilongjiang Sci
19. Huang H, Zhu K (2014) Analysis and research campus network information security. Netw Secur Technol Appl
20. Lei X (2017) The application of network security technology in campus network. Chin Comput Commun

The Safety Calculation and Simulation Analysis of Guyed Travelling Carriage for Main Beam of Polonggou Bridge

Xiaojie Wu[1(✉)] and Yuebo Zhang[2]

[1] City College of Wuhan University of Science and Technology, Wuhan 430053, China
2987623713@qq.com
[2] Hubei Provincial Communications Planning and Design Institute CO., LTD., Wuhan 430051, China

Abstract. Polonggou bridge is the key project for "TongMai natural barrier" section of Sichuan-Tibet road. It's currently rare for the high altitude, terrible geological conditions and construction difficulty. The simulation of the main girder guyed travelling carriage of Polonggou Bridge by Midas civil is introduced in this paper. The tress and displacement of the basket are calculated and the results show that the construction safety is well met. In view of the research lack for the important parts of the curved head and C beam, simulations for them are conducted using nonlinear analysis module Midas FEA in this paper. The results show that all these parts meet the requirements well. It provides a guarantee for the safety of the project.

Keywords: Guyed travelling carriage · Modeling · Safety · Nonlinear analysis

1 Foreword

Sichuan-Tibet Highway is located between the snow-capped mountains and plateaus of Tibet. It is extremely difficult and dangerous all the way, especially for the "Tongmai natural risk" in the territory in Linzhi County. In this area, narrow mountain passes and deep Hollows are continuous. The mudslides, the landslides and other disasters occur frequently. It is the most dangerous section of the entire Sichuan-Tibet Highway. Polonggou bridge as the key project and Landmark project of this road can solve the traffic problems that have plagued the local area for many years, and truly realize the people there to travel freely.

The total length of the bridge is 743 m, and the bridge span is arranged as: 156 m + 430 m + 156 m, the Structural form is double-tower and double-cable mixed bridge. The tower form is a type A column above the bridge deck, and merged into a single column below the bridge deck. The group-pile foundation is adopted. The middle span is a steel-concrete mixed beam of the I-shaped section steel with a prefabricated 280 mm thick concrete plate on it, and the full width of the main beam is 14.3 m.

Z. Xu et al. (Eds.): CSIA 2019, AISC 928, pp. 559–568, 2020.
https://doi.org/10.1007/978-3-030-15235-2_80

The side-span cantilever main beam is cast-in-place concrete with double-sided rib section, and the height of the beam center is 2.62 m, the full width is 13.8 m.

2 Brief Introduction of Pull Rope Guyed Travelling Carriage Research

2.1 Introduction of the Guyed Travelling Carriage

The pull rope guyed travelling carriage is the main construction structure of the cantilever cast-in-place construction process of the long-span cable-stayed bridge. It usually consists of several main systems as follows: load-bearing system (truss system), cable system, template system, walking system and positioning anchorage system etc. The pull rope guyed travelling carriage is provided with good controllability, flexible adjustability, high accuracy and reliable quality, so that it's widely used in the construction of modern cable-stayed bridges. The long platform and the short platform are both available based on construction requirements. And according to the shape of the main truss of the load-bearing system, it can be divided into rhombus and triangles etc.

2.2 Research Status of the Guyed Travelling Carriage

Due to the prominent advantages of the pull rope guyed travelling carriage in cable-stayed bridge construction, the research on it has been very hot in recent years. However, most of the current research focuses on the following aspects [1–9]:

(1) Research on the design of guyed travelling carriage structure
According to the characteristics, dimensions and construction requirements of different bridges, the composition of the guyed travelling carriage system and the connection between the various components are adjusted to study the adaptability between the guyed travelling carriage form and the bridge structure.
(2) Research on the walking system and the anchoring control
The optimizations of the improvement of the walking system and the anchoring system are mainly researched to reduce the walking resistance, to accurately control the walking alignment and to reinforce the stability of the anchoring system. These studies are aimed to ensure the efficient and smooth operation of the guyed travelling carriage structure.
(3) Simulation and calculation of the whole guyed travelling carriage structure
The simulation and calculation of the whole guyed travelling carriage structure are carried out to ensure that the overall stress and the overall deformation of it meet the standard and construction accuracy requirements precisely.

The simulation and calculation of the whole guyed travelling carriage structure are carried out to ensure that the overall stress and the overall deformation of it meet the standard and construction accuracy requirements precisely.

The aspects of research above have been carried out much deeply, and it can basically meet the construction of cable-stayed bridge under normal conditions. In recent years, as the traffic network becomes more and more intensive, many new high-standard roads and railways are frequently built in the complex geological condition areas which are bypassed in the route selections before. Super span cable-stayed bridges are built in large numbers built in particularly harsh Geological conditions. Higher requirements on the cable-stayed bridge guyed travelling carriages are put forward. The damages of partial key parts such as type C beam and Shear keys are often occurring when the overall stress and the overall deformation of the basket are perfectly meet. This may now and then interrupt the construction process.

3 Structure and Load Analysis of Polonggou Bridge Guyed Travelling Carriage

3.1 Overall Composition of the Guyed Travelling Carriage Structure

According to the construction plan, Cantilever pouring construction for the side-span main beam (B2–B14 sections) of the bridge on the Chengdu side and B2–B14 sections on the Lhasa side are carried out by using guyed travelling carriage. The long-desk basket structure type [10] is employed, and the overall layout is shown in the Fig. 1.

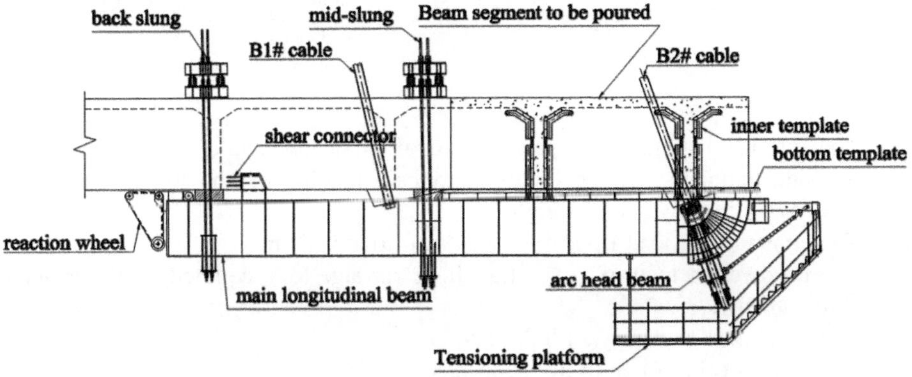

Fig. 1. Overall layout of the guyed travelling carriage

The load-bearing system consists of main longitudinal beam, secondary longitudinal beam and horizontal beam, they are connected by 10.9 level high-strength bolts and equal strength welding.

The cable drawing system consists of tensioning mechanism, cable length device and arc distribution beam, spherical bearing plate and spherical cushion plate are set up to adapt to the angle change of the cable in the horizontal direction of the bridge.

The traveling system consists of C-type beam, vertical jack, reaction wheel, slide, jacking jack and reaction seat. A sliding shoe pasting MGE slide plate and stainless steel plate is setting under the C-type beam.

The positioning anchorage system consists of shear connector head slung, middle slung and back slung. When the basket moves forward C-type beams at the both side must keep moving forward at the same time. Continuous observation must be carried out to correct the deviation of the guyed travelling carriage along the transverse cross of the bridge.

The template system consists of bottom template, side template and inner template. The bottom and the side template are installed with guyed travelling carriage, the inner template divided into blocks in the factory, is integrated with guyed travelling carriage on site.

3.2 Design Load

(1) The concrete load (q_1): the guyed travelling carriage is designed to construct the sections from B2–B16 of the main beam. So, the concrete load is calculated according to the main beam drawing taking into account 5% of the concrete excess load. The deadweight of each section is summarized in Table 1.

Table 1. Concrete deadweight of every beam section

Number	Length (m)	Concrete volume (m^3)	Weight (kN)
B2–B11	8.5	128.5	3341
B15	8.5	152.4	3962.4
B16	8.5	136.7	3554.2

(2) The template load (q_2): this part consists of the deadweight of the inner template, the side template and the bottom template, loading on the longitudinal beam and cross beam by actual weight.
(3) The construction load (q_3): it is calculated to 1.5 kN/m^2.
(4) The deadweight load of guyed travelling carriage (q_4): counted by the Program automatically.
(5) The first cable force (q_5): 1302.4 kN.
(6) The second cable force (q_6): 2214.4 kN.

4 Overall Simulation Analysis of the Guyed Travelling Carriage

4.1 Working Conditions Division

The load changes obviously and the main load member changes constantly during the different construction processes, so that the system and state of force change frequently and complex. Due to this condition, we need divide the whole process into several working conditions. The loading and calculation of per working condition is carried out separately. According to the main control stage, the calculation and analysis of the guyed travelling carriage is divided into 5 stages as follows:

Working condition 1: the installation in position of the guyed travelling carriage, the first time for cable force adjustment, considering loads: q_2, q_3, q_4, q_5;

Working condition 2: 50% concrete poured, considering loads: $0.5q_1$, q_2, q_3, q_4, q_5;
Working condition 3: the second time for cable force adjustment, considering loads: $0.5q_1$, q_2, q_3, q_4, q_6;
Working condition 4: 100% concrete poured, considering loads: q_1, q_2, q_3, q_4, q_6;
Working condition 5: the carriage putting down and travelling, considering loads: q_2, q_4;

It is shown that the working conditions from 1 to 4 are the pouring and forming process of the cantilever beam, the loads are very large and the stress states are dangerous. So, the most important thing is to control the general stress of the carriage in the working conditions 1–4. The working condition 5 is the laying down and travelling process of the guyed travelling carriage. The loads are obviously small, so it is in the safe stress state. Therefore, the key problem is the deformation control of the main longitudinal beam.

4.2 Simulation Calculation of the Whole Carriage Structure

The general calculation and simulation analysis model is established by using the professional structural analysis software Midas Civil according to the structure and force characteristics of guyed travelling carriage, it is shown in the Fig. 2.

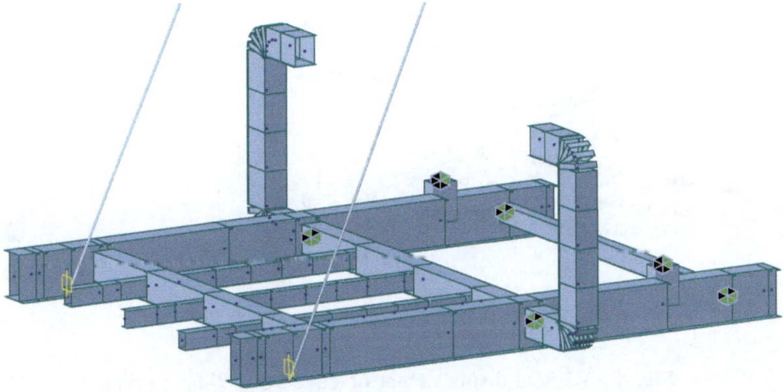

Fig. 2. General calculation model of the guyed travelling carriage

Taking the largest weight section B15 stages as an example, the calculation and analysis results of all the working conditions are Obtained when cantilever casting process is carried out. The worst results of the Structural Stress, the vertical displacement and the reaction forces are respectively shown as the Figs. 3, 4 and 5.

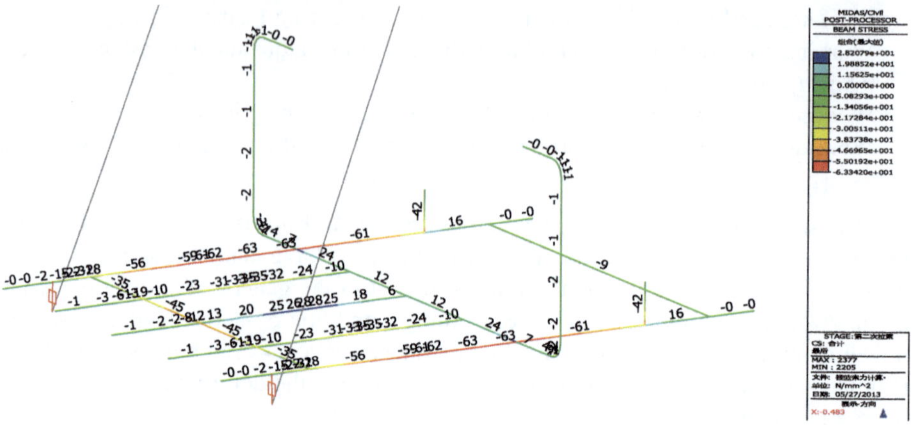

Fig. 3. Stress of guyed travelling carriage

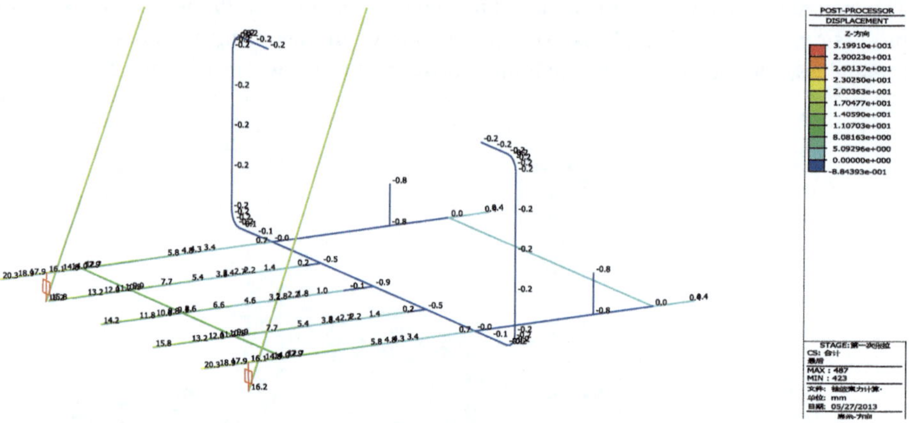

Fig. 4. Vertical displacement of guyed travelling carriage

The Maximum stress and displacement of the horizontal beam and the main longitudinal beam in the working conditions 1 to 4 are summarized in Table 2. The reaction forces of the mid-slung and the back slung and the shear connector in the main vertical beam are also summarized in Table 2.

It is clearly shown that the stress changes of the main members of hanging basket such as main longitudinal beam, horizontal beam and secondary longitudinal beam fit the objective law of stress change.

For the main longitudinal beam: in the first time for cable force adjustment, the cable force is loaded directly on it and produce stress correspondingly. And then, the stress produced by first concrete pouring adding up in the opposite direction, it results

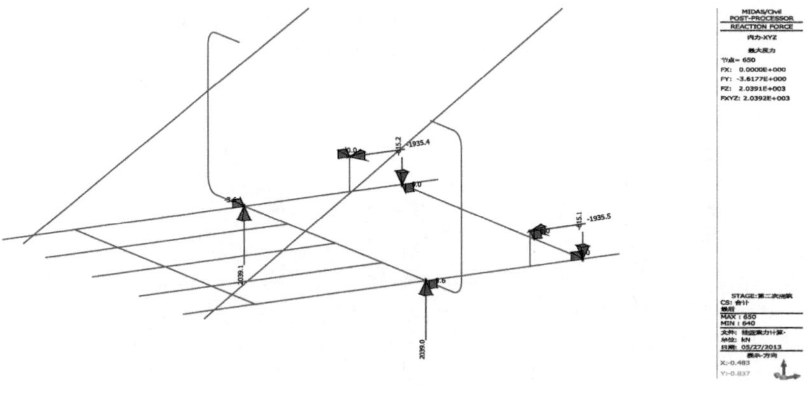

Fig. 5. Reaction force of guyed travelling carriage

Table 2. Results of the section B15

	Main longitudinal beam		Horizontal beam		Reaction force		
	Stress	Displacement	Stress	Displacement	Mid-slung	Back slung	Shear connector
First cable tension adjustment	101	32.2	27.9	15.7	−508	461	1090
First pouring	46.0	−1.2	68.8	−11.7	1224	−326	1174
Second cable tension adjustment	97.8	32.2	71.0	8.0	254	186	1841
Second pouring	77.9	−2.4	134	−23.5	2039	−615	1936

the total stress number decreases. In the same, the stress increases in the second time for cable force adjustment, it decreases with the second concrete pouring.

For the horizontal beam and the secondary longitudinal beam: the cable force is only loaded indirectly by the main longitudinal beam so that in the cable force adjustment for two times, the stresses of them are effected little. The main load for them is the weight of concrete pouring, it obviously shown in the Table 2 that the stresses increase up.

A line graph of the stress change of the longitudinal beam, horizontal beam and secondary longitudinal beam in the working conditions 1 to 4 is drawn as Fig. 6. It shows that all the stress changes of the members are consistent with the law and it is well proved the correction of the simulation calculation.

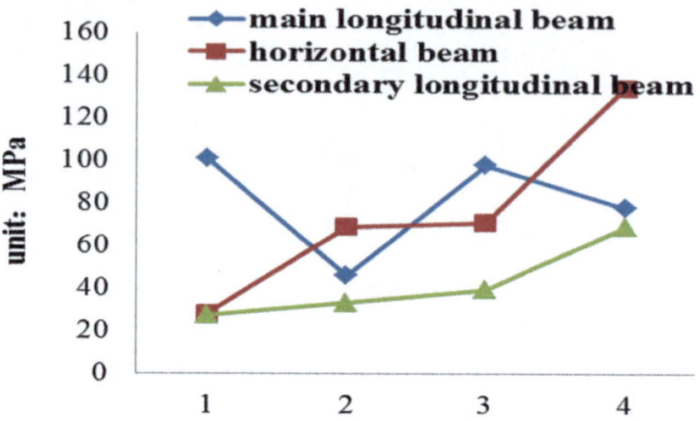

Fig. 6. Stresses of main longitudinal beam, horizontal beam and secondary longitudinal beam in all working conditions 1 to 4

4.3 Calculation and Analysis of the Putting Down and Travelling of Carriage Structure

According to the divisions for working conditions, displacement control is the main problem when the guyed travelling carriage is put down and travelling. The displacement results of each main member are calculated and drawn in the Fig. 7. It is shown that the maximum displacement in all the structure is 34 mm, this is within the permitted limits of deflection for steel members [11]. The whole structure is safe.

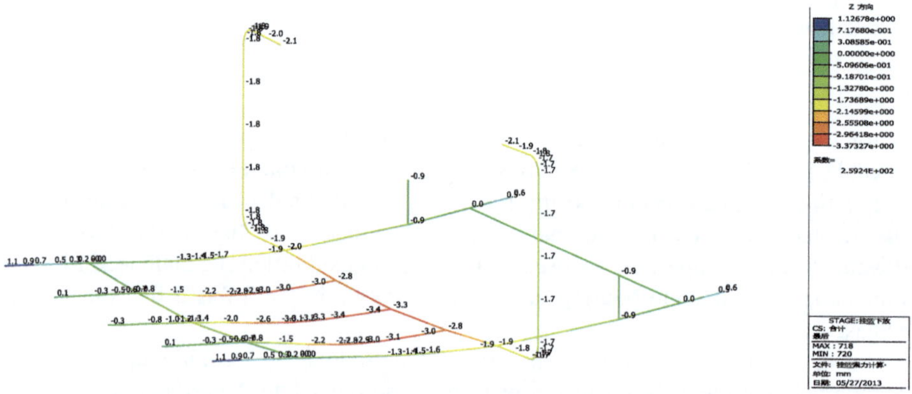

Fig. 7. Vertical displacement of basket in moving down condition

5 Nonlinear Simulation Analysis of Locally Important Sites

The calculation of the general analysis for the guyed travelling carriage during the whole cantilever casting construction shows that the maximum stress in the concrete pouring and cable force adjustment processes is at the arc head. The maximum

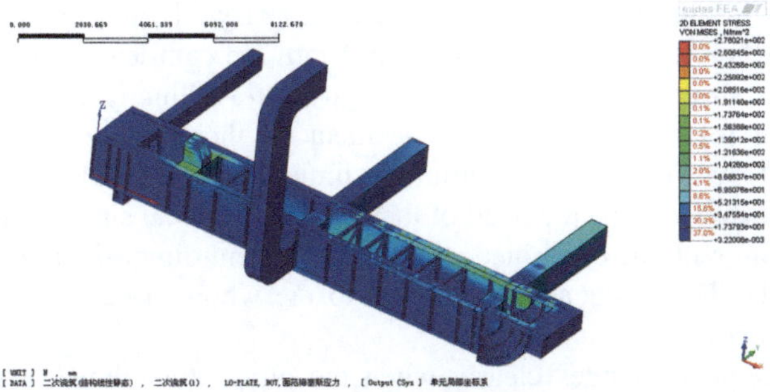

Fig. 8. Van-Messes stress

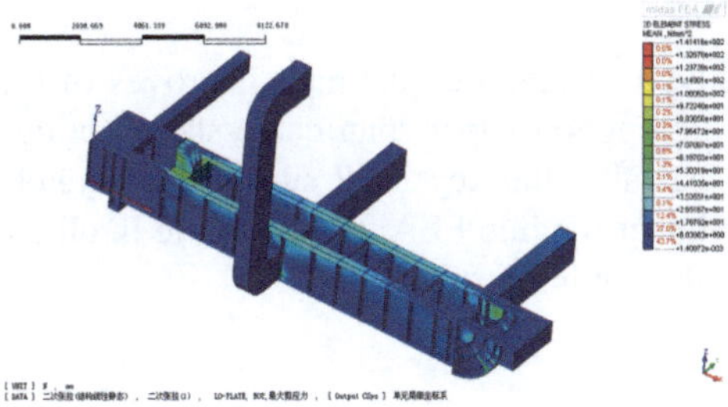

Fig. 9. Maximum shearing stress

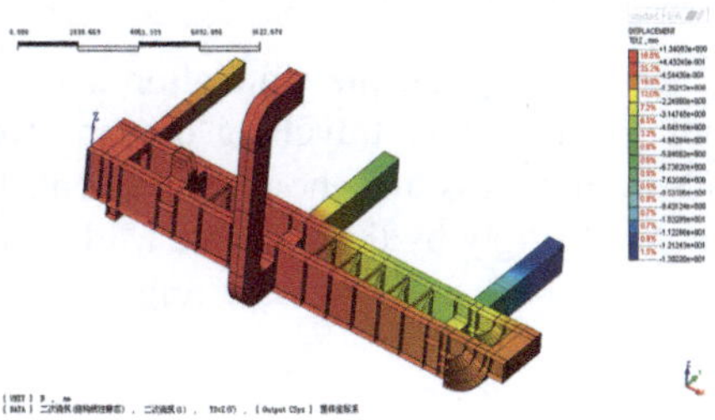

Fig. 10. Vertical displacement

displacement occurs at the C-type beam in the putting down and travelling working condition. It is obviously that the arc head and C-type beam are the key locally important sites. More targeted and painstaking analysis and calculation should be done for them by the special nonlinear analysis Module—Midas FEA. First, an accurate model should be made to simulate the detailed structure for the arc head and C-type beam. Second, nonlinear analysis calculation is carried out by the software Midas FEA module.

The results of local arc head by accurately simulation are shown in the Figs. 8, 9 and 10. It is clearly that the local maximum Van-Messes stress is 173.76 MPa, the maximum shearing stress is 106.06 MPa. They are both within the limits of the steel material. The maximum downward linear displacement is −9.43 mm, the maximum upward linear displacement is 1.34 mm. They are also in the limits.

6 Conclusion

According to the calculation results of the general analysis of the carriage structure and detail nonlinear simulation analysis of the local key site, we can get conclusions as follows:

1. The general limits of stress and displacement is well satisfied in every working condition, when the sections of the main beam are cantilever casting.
2. Stress of the structure reduces, when the guyed travelling carriage is putting down and moving forward, but the displacement of the longitudinal beam increases obviously. However, it is still within the limits and can satisfy the requirement.
3. The dangerous site is the arc head of the carriage. Its local single maximum stress is generally safe with the combined stress in only 2 cm dimensional range beyond the limits a little. This influence is minimal to the whole structure in the entire construction process.
4. The displacement changes clearly when the guyed travelling carriage is moving forward without the construction loads. The displacement downward decreases significantly, the reverse displacement upward even appears. But, the ones in both directions are not big enough to go beyond the limits, neither. And the construction needs are well satisfied.

In summary, all the simulation and analysis results are qualified. It proves of the safety of the guyed travelling carriage for Polonggou super-span cable-stayed bridge both in the stress and the displacement. Especially, the key local sites are accurately simulated singly by the software Midas nonlinear module FEA, they are safe in all the aspects. And it can provide a reference for the similar structure.

References

1. Zhang Y (2015) Key points analysis of the design and construction of bridge hanging basket. Constr Des Proj 10:104–106
2. Tang H, Liu C (2013) Cantilever casting technology of compound guy triangle hanging basket for variable wide cross-section bridge. Constr Technol 42(23):34–37
3. Hu W (2009) Construction techniques of guyed form traveler used for yangtze river bridge in Zhong county of Chongqing. Bridge Constr 1:52–55
4. Duan B, Yan J (2017) Design and installation construction technology of ultra-wide and ultra-weight guyed travelling carriage. Traffic Sci Technol 01:33–36
5. Liu C (2015) The design and simulation calculation for girder of cable-stayed bridge of front support point hanging basket for main girder of fifth bridge. J Zhejiang Inst Commun 16 (4):1–4
6. Lai M, Cai W (2015) Research on technology of travelling system of compound guyed travelling carriage. Sichuan Constr Sci Res 6(41):241–244
7. Xiaorong W (2016) Analysis of super-span cable-stayed bridge guyed travelling carriage. Eng Constr 5:101–103
8. Deng X (2016) Design of guyed travelling carriage using in the single-cable-plane concrete cable-stayed bridge. Transp World 6(16):35–38
9. Luo L (2016) Research on the general stress of guyed travelling carriage for double-cable-plane prestressed concrete cable-stayed bridge. Transp Technol 09:45–48
10. Zhou S, He Z (2001) Calculation manual of road and bridge construction. People's Transportation Press, Beijing, p 66
11. People's Republic of China Ministry of Construction (2017) Standard for design of steel structure (GB50017-2017). China Construction Industry Press, Beijing, p 43

Human Action Recognition Based on Fusion Features

Shiqiang Yang(✉), Jiangtao Yang, Fei Li, Guohao Fan, and Dexin Li

School of Mechanical and Precision Instrument Engineering,
Xi'an University of Technology, Xi'an 710048, Shaanxi, China
yangsq@126.com

Abstract. Human action recognition has a wide range of application prospects in areas such as artificial intelligence and human-computer interaction. Action feature models and action recognition models are the basis of human action recognition. Based on the simplification of human skeleton model, the complementary features information such as the main joint angle, speed and relative position of the human body joint are extracted and fused to describe the behavioral gestures. And the action is expressed with the gesture series. A behavioral action model is established. In order to facilitate calculating, Fourier interpolation is performed on each action sample in the action database which taking the most characteristic dimension of the action video as the standard to keep the action samples feature dimensions consistent and normalized. And the principal components are used to extracting the main components of the feature, reducing the feature dimensions and redundant information. A one-to-many multi-category action recognition model was established based on the theory of support vector machines. The action recognition experiment was carried out with the open human action video database. The results showed that the algorithm has good adaptability and practicality.

Keywords: Feature fusion · Action recognition · Support vector machine

1 Introduction

Human action recognition is a research hotspot in the field of machine vision, which involves feature extraction, feature representation and classification methods. It has a wide range of applications in artificial intelligence, robotics, human-robot interaction [1] and intelligent video surveillance [2, 3].

The construction of action feature model and action recognition model is an important basis for human action recognition. Human actions are complex in time and space. Each action needs to consider the differences in shape, angle, direction, etc. The action features should be able to fully express the structural information contained in the action [4]. Human body structure model features can represent human body movements. It uses a model to describe the human postures, three-dimensional state of the human body, and a series of changing postures [5–8]. The joint position can be used to calculate the distance between each joint, forming a distance matrix as a feature of each frame [9].

Z. Xu et al. (Eds.): CSIA 2019, AISC 928, pp. 569–579, 2020.
https://doi.org/10.1007/978-3-030-15235-2_81

The human action recognition model obtains a classifier by training the feature data to perform classification and recognition. Support vector machine (SVM) is a kind of probability and statistical learning algorithm, mainly used for classification and prediction. Li et al. [10] used the support vector machine to classify skeleton features based on the human skeleton structure. Neili et al. [11] used the skeletal data extracted by Kinect camera to perform attitude prediction based on joint features. It also uses convolutional neural networks for pose estimation, and performing pose classification based on multi-class support vector machines.

A human body description model is constructed by extracting and fusing the features of the joint's angle, displacement and velocity. Principal component analysis was used to extract the main components and reduce the dimension of the fusion feature. A multi-classification action recognition model is established based on the support vector machine theory. The human body movement is identified to verify the validity of the model with the open human action database.

2 Human Body Description Model

2.1 Action Feature Model Construction

The human body feature model is constructed based on the human body structure model. In the human body structure model, there are more joint points in the human body. In order to reduce the complexity of human body action representation, a human body structure model was constructed using 15 body joint points, as shown in Fig. 1.

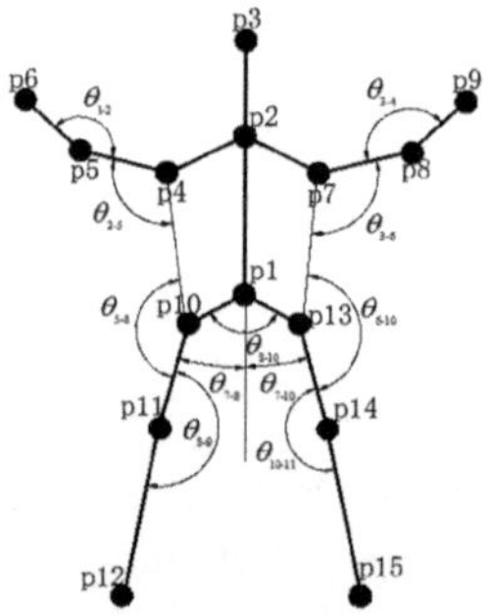

Fig. 1. Skeleton information diagram

For the convenient of calculation, the corresponding joint points are numbered as p1–p15 as shown in Fig. 1. Based on the position information of these joint points, features such as joint angle, displacement, and speed can be calculated.

2.2 Joint Angle Characteristics

The three-dimensional coordinates (x, y, z) of 15 adjacent joint points constitute 11 vectors, such as V_1: p5-p6, V_2: p4-p5, V_3: p8-p7, V_4: p9-p8, V_5: p10-p4, V_6: p13-p7,

$\boldsymbol{V_7}$: p3-p1, $\boldsymbol{V_8}$: p11-p10, $\boldsymbol{V_9}$: p12-p11, $\boldsymbol{V_{10}}$: p14-p13, $\boldsymbol{V_{11}}$: p15-p14. Calculate the angle between the vectors in the three-dimensional space, and obtain the angular features of the body pose in each frame of image. The angle θ_{i-j} between any two vectors V_i and V_j is as follows:

$$\theta_{i-j} = \frac{V_{i1} \times V_{j1} + V_{i2} \times V_{j2} + V_{i3} \times V_{j3}}{\sqrt{V_{i1}^2 + V_{i2}^2 + V_{i3}^2} \times \sqrt{V_{j1}^2 + V_{j2}^2 + V_{j3}^2}} \tag{1}$$

$$V_i = V_{i1}x + V_{i2}y + V_{i3}z \tag{2}$$

$$V_j = V_{j1}x + V_{j2}y + V_{j3}z \tag{3}$$

The extracted angles are $\theta_{1-2}, \theta_{3-4}, \theta_{2-5}, \theta_{3-6}, \theta_{5-8}, \theta_{6-10}, \theta_{8-10}, \theta_{7-8}, \theta_{7-10}, \theta_{8-9}, \theta_{10-11}$. Human action in video consists of multi-frame human gestures. By connecting a series of human postures, continuous changes in the angle of the human joint can effectively represent the human body's movements. Each angle in the image of the i-th frame in the video constitutes a vector, describing the posture of the human body Z_i,

$$Z_i = [\theta_1, \theta_2, \theta_3, \theta_4, \theta_5, \theta_6, \theta_7, \theta_8, \theta_9, \theta_{10}, \theta_{11}] \tag{4}$$

The sequence of poses constitutes the angle D of human body action:

$$D = [Z_1, Z_2, Z_3, Z_4, Z_5 \cdots Z_n]^T \tag{5}$$

Where n is the number of video frames.

2.3 Joint Displacement Characteristics

The six joint points with the most obvious position change in the anatomy model are p9, p8, p5, p6, p15 and p12, as shown in Fig. 1. Calculate the distance of these joint points to the joint points of the spine (p1) to describe the human body posture in each frame of the image. The position of different joint points in space is different in three dimensions and the Euclide distance from the reference point is similar. The 6 relative positions of the three joint points to the waist joint points is the reference. Calculate the distance from the six joint points to the three dimensions of the joint points of the spine. Then construct an 18-dimensional feature vector to represent the human body pose for each frame t_n in the image.

$$\begin{cases} d_i^x = p_i^x - p_1^x \\ d_i^y = p_i^y - p_1^y \\ d_i^z = p_i^z - p_1^z \end{cases} \qquad i = 5, 6, 8, 9, 12, 15 \tag{6}$$

$$f_n = [d_5^x, d_5^y, d_5^z, d_6^x, d_6^y, d_6^z, d_8^x, d_8^y, d_8^z, d_9^x, d_9^y, d_9^z, d_{12}^x, d_{12}^y, d_{12}^z, d_{15}^x, d_{15}^y, d_{15}^z, t_n] \tag{7}$$

Where d_i^x, d_i^y and d_i^z are the distances from the six joint points of each frame to the X, Y, and Z axes of the joint points, respectively. The feature vector f_n is the human pose feature of the n frame image in the video. Normalize displacement to reduce relative position error due to different heights of the human body. Use the 6 joint points in each frame of the image divided by the distance between the neck and the spine joint:

$$f_n' = f_n / \|d_n - d_s\|_2 \tag{8}$$

Where f_n' is the pose characteristics of the nth frame image in the normalized video. $\|d_2 - d_1\|_2$ is the distance from the neck joint point to the joint point of the spine. Human body action consists of multiple frames of human body posture. Connected attitude feature sequence describes human action displacement characteristics F:

$$F = [f_1', f_2', f_3', \ldots, f_n'] \tag{9}$$

2.4 Joint Speed Characteristics

Similarly, select 6 joint points with obvious speed changes: p9, p8, p5, p6, p15 and p12. The relative velocity v_{ij} of each joint point is the ratio of the distance between every two adjacent frame points coordinates of the six joint points and the time.

$$v_{ij} = \frac{\|d_{ij} - d_{ij-1}\|_2}{t} \quad i = 5, 6, 9, 12, 15; \;\; j = 1, 2, \ldots, p \tag{10}$$

Where i is the joint point number, p is the number of video frames. The human body gesture v_j in each frame of the video is:

$$v_j = [v_{5j}, v_{6j}, v_{8j}, v_{9j}, v_{12j}, v_{15j}] \qquad j = 1, 2, \ldots, p \tag{11}$$

The speed of human body action velocity V of joint velocity is:

$$V = [v_1, v_2, \ldots, v_n]^T \tag{12}$$

3 Feature Fusion

The three different complementary features built together can more fully represent the human body movements. Perform Fourier interpolation on the description features in each action sample to transform the input data into the frequency domain. Then use the Fourier inverse transform as the time domain with the sample with the most feature dimension in the same kind of sample as the standard. Use the normalization algorithm to preprocess the features after Fourier interpolation. The multi-feature combination model obtained by combining the angular features of the human action, the action position features and the action velocity characteristics are as follows:

$$X_i = [D_i, F_i, V_i] \quad i = 1, 2 \cdots m \tag{13}$$

Where the X_i representing the multi-feature combination model to the i-th sample, D_i, F_i, V_i respectively are angle, relative position and velocity characteristics of the i-th sample. After the combination of multiple features, the dimension of the action feature is 4300 dimensions. Principal component analysis is used to process the data to extract the main components. The calculation steps are as follows:

(1) Standardizing the raw data

$$X = [X_1, X_2, \cdots, X_m]^T = \begin{bmatrix} c_{11} & c_{12} & \cdots & c_{1n} \\ c_{21} & c_{22} & \cdots & c_{2n} \\ \vdots & \vdots & \vdots & \vdots \\ c_{m1} & c_{m2} & \cdots & c_{mn} \end{bmatrix} \tag{14}$$

The value of the c_{ij} is the j-th indicator of i-th comment object. m is the number of samples, n is the feature dimension, convert each indicator value c_{ij} into a standardized value $\tilde{c}_{ij}$ as follow

$$\tilde{c}_{ij} = \frac{c_{ij} - \mu_j}{s_j}, \ i = 1, 2, \ldots, m; j = 1, 2, \ldots, n \tag{15}$$

The sample mean μ_j of the feature dimension of the i-th column in the matrix is:

$$\mu_j = \frac{1}{m} \sum_{i=1}^{m} c_{ij} \tag{16}$$

The sample standard deviation s_j of the i-th column in the matrix is:

$$s_j = \sqrt{\frac{1}{m-1} \sum_{i=1}^{m} \left(c_{ij} - u_{ij}\right)^2} \quad j = 1, 2, \ldots, n \tag{17}$$

The corresponding standardized indicator variable is:

$$X_i = \frac{X_i - u_i}{s_i}, \ i = 1, 2, \ldots, n \tag{18}$$

(2) Calculate the correlation coefficient matrix $R = (r_{ij})_{n*n}$, Where

$$r_{ij} = \frac{\sum_{k=1}^{n} c_{ki} * c_{kj}}{m-1}, i, j = 1, 2, 3 \ldots n \tag{19}$$

r_{ij} is the correlation coefficient between the i indicator and the j indicator. Where $r_{ii} = 1$, $r_{ij} = r_{ji}$ in the formula.

(3) Calculate eigenvalues and eigenvectors
Eigenvalue $\lambda_1 \geq \lambda_2 \geq \ldots \geq \lambda_p \geq 0$ of matrix R, the corresponding feature vectors are $\mu_1, \mu_2, \ldots, \mu_n$, among them $\mu_i = [\mu_{1i}, \mu_{2i}, \ldots, \mu_{ni}]^T$, these feature vectors make up n new variable q_i, $i = 1, \cdots, n$,

$$\begin{aligned} q_1 &= \mu_{11}c_1 + \mu_{21}c_2 + \ldots + \mu_{n1}c_n \\ q_2 &= \mu_{12}c_1 + \mu_{22}c_2 + \ldots + \mu_{n2}c_n \\ &\ldots \\ q_n &= \mu_{1n}c_1 + \mu_{2n}c_2 + \ldots + \mu_{nn}c_n \end{aligned}$$

Where, q_1 is the first principal component, q_2 is the second principal component,…, q_n is the n-th principal component. Select $k(k \leq n)$ principal components and calculate principal component contribution rate and cumulative contribution. The information contribution rate b_k of the principal component q_k is:

$$b_k = \frac{\lambda_k}{\sum_{i=1}^{m} \lambda_i}, i = 1, 2, 3 \ldots n, k \leq n \tag{20}$$

The cumulative contribution rate ∂_k of the principal component $q_1, q_2, \ldots q_k$ is:

$$\partial_k = \frac{\sum_{i=1}^{k} \lambda_i}{\sum_{i=1}^{n} \lambda_i} \quad k \leq n \tag{21}$$

The principal component analysis is used to reduce dimensionality of multidimensional feature data. First, the characteristic matrix variable $q_1, q_2, \ldots q_k$ with a cumulative contribution rate ∂_k of 95% is used as the principal component. They are used to replace the original feature matrix variables. Finally, the feature matrix composed of principal components after dimension reduction is used as the action feature of training and testing during recognition. The Principal Component Analysis method reduces the dimension and obtains the i-th action description model X_{Pi} of the action sample is:

$$X_{Pi} = [q_1, q_2, \cdots, q_k]^T, i = 1, 2, \cdots m, k \leq n \tag{22}$$

4 Action Recognition Model

4.1 Support Vector Machine Principle

Support Vector Machine is a two-class algorithm. A kernel function is introduced to map the sample data into a high-dimensional nonlinearly space. The optimal classifier is obtained by finding the support vector $SV_1, SV_2 \ldots SV_n$ and constructing the optimal hyper plane.

For the linear indivisible problem of training set such as behavior recognition, first define a kernel function $k(X_{Pi}, X_{Pj})$ and penalty parameters C, then construct and solve the optimization problem:

$$\begin{cases} Q(\alpha) = \min\left(\frac{1}{2}\sum_{i=1}^{n} \alpha_i\alpha_j O_i O_j k(X_{Pi} \cdot X_{Pj}) - \sum_{i=1}^{n} \alpha_i\right) \\ s.t \quad \sum_{i=1}^{l} R_i\alpha_i = 0 \end{cases} \tag{23}$$

Where $O_i, O_j \in [1, -1]$ is labels for two different types of actions, α_i is the Lagrange multiplier, l is the number of Lagrange multiplier, $0 \leq \alpha_i \leq C$, $i = 1, 2, \ldots l$, X_{Pi} and X_{Pj} are two types of action samples respectively. So we can get the optimal solution $\alpha^* = \left(\alpha_1^*, \cdots, \alpha_l^*\right)^T$. Select a positive component $0 \leq \alpha_j^* \leq C$ in α^*, and calculate the threshold. The classification decision function is:

$$f(X_P) = \mathrm{sgn}\left((w^i)^T \varphi(X_p) + b^i\right) \tag{24}$$

Select different kernel functions in nonlinear classification to map multidimensional features into different high-dimensional spaces. Then solve different classification models. Training with different kernel functions will have different recognition and classification effects. Use different kernel functions to calculate the accuracy of human action recognition. Finally, the kernel function with the highest recognition accuracy is selected by the experimental test. The kernel function is selected as the Gaussian kernel function:

$$k(X_{Pi}, X_{Pj}) = \exp\left(-\frac{\|X_{Pi}, X_{Pj}\|^2}{2\sigma^2}\right) \tag{25}$$

Where, parameter σ is the core width. This value plays an important role in the Gaussian function. If it is too large, it will cause the kernel function to have no effect. If it is too small, the planning ability of the function will be weakened and the problem of over-fitting will occur.

4.2 Multi-class Support Vector Machine

Support Vector Machine is a two-class algorithm. It constructs multiple classifiers by combining multiple two classifiers. Multiple categories can be converted into a two-category problem classification using a one-to-many approach.

During training, for *n* categories of sample data, it need to train *n(1 + n)/2* SVM class two-classifiers. It constructs the sample data of the *i*-th SVM sub-category as a positive class, and other sample data which is not in the *i* category is marked as a negative class. During the test, the values of the discriminant functions are calculated separately for the test data. If only one classifier outputs a positive value, the result can be directly judged as the corresponding classifier number, otherwise, the category corresponding to the maximum value of the discriminant function is selected as the category of the test data. For known training sets *T*,

$$T = \{(X_1, O_1), \ldots, (X_n, O_n)\} \in (X \times O)^n$$

Where $X_i \in X = \mathbb{R}^n, O_i = \{1, \ldots, n\}, i = 1, \ldots, n * (n+1)/2$. For $j = 1, \ldots, n$, think of the j-th class as a positive class, the remaining n − 1 classes are negative. According to n types of actions, the $n * (n+1)/2$ class decision function is trained separately:

$$f_i(X_P) = (\mathrm{w}^i)^{\mathrm{T}} \varphi(X_p) + b_i, i = 1, 2, 3 \cdots n * (n+1)/2 \tag{26}$$

if

$$f_n(X_P) = (w^n)^T \varphi(X_P) + b_n \geq +1$$

then

$$(w^i)^T \varphi(X_P) + b_i \leq -1, i = 1,\ 2, \cdots n-1, n+1, \cdots n * (n+1)/2 \tag{27}$$

When the value of the decision function is a positive value, the predicted action category belongs to the category.

5 Experimental Simulation

In the Matlab2010b environment, based on the open database the Microsoft Research Cambridge [12], Select 6 joint points with obvious change rate among 15 joint points: left elbow (p9), left hand (p8), right elbow (p5), right hand (p6), left foot (p15) and right foot (p12) as a joint points of velocity feature and displacement features. There are six different human actions in the database: action 1 (start), action 2 (lower), action 3 (next), action 4 (bow), action 5 (throwing) and action 6 (kick). There are about 100 actions in each sample, and the total number of samples is about 600.

Perform Fourier interpolation and normalization preprocessing on the angular features of each action, i.e., the human pose sequence. The angular feature dimension of each action sample is 1300 dimensions, and the eigenvalue is ranges between −1 and 1. The various angular characteristics of the action are shown in Fig. 2.

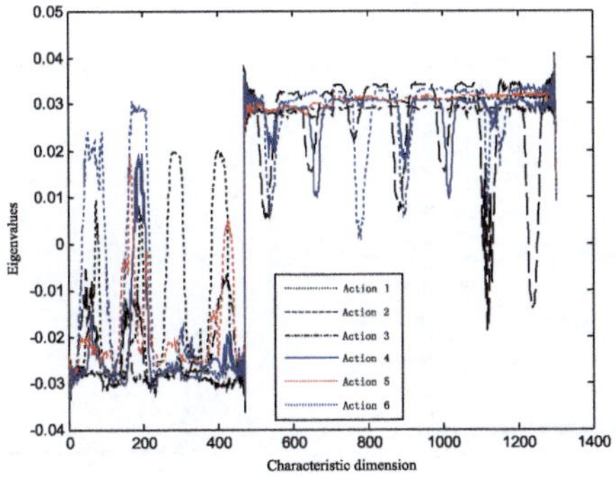

Fig. 2. Angle characteristics of the six movements

Extract the relative speed and position of the selected six key joint points. The speed characteristics and relative position characteristics of the six types of movements achieved. After Fourier interpolation and normalization preprocessing, the relative position and velocity characteristic dimensions of each action are unified to 2000 and 1000 dimensions respectively. The relative position and velocity characteristics of the six movements are shown in Figs. 3 and 4 respectively.

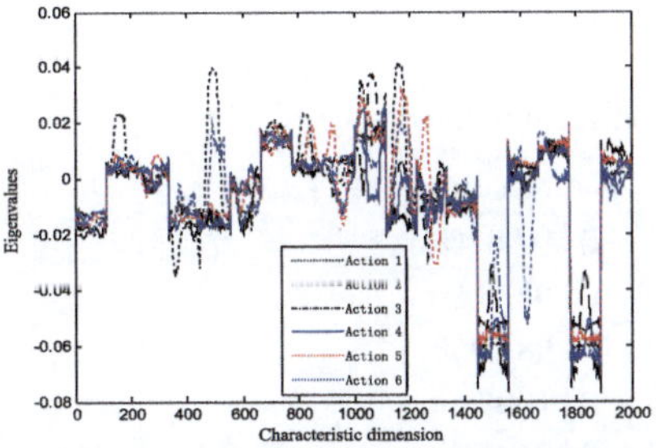

Fig. 3. Relative position characteristics

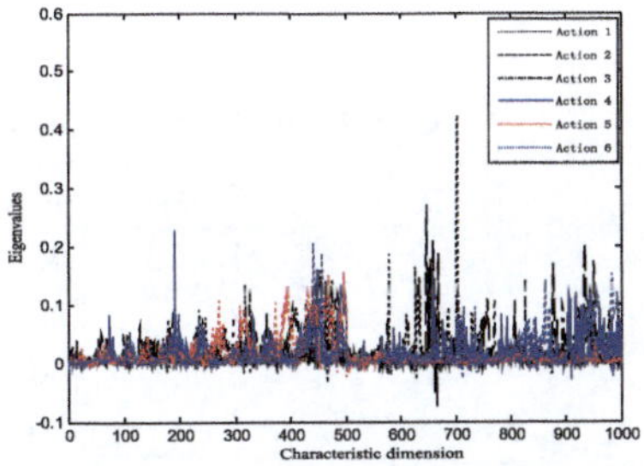

Fig. 4. Joint speed characteristics

By constructing three features, the angular feature matrix D, the relative position feature matrix F and the velocity feature matrix V of the human body action achieved. A multi-feature matrix X combining three feature matrices is a matrix of 600 * 4300. The number of samples is 600, and the feature dimension of each action sample is 4300. Key features are extracted with Principal Component Analysis algorithm, and the contribution rate of the top ten principal components in the mapping space as shown in Fig. 5. It can be seen that after the feature data is mapped into the new space, the cumulative contribution rate of the top ten main components with the highest contribution rate in the feature data is about 55%. Taking the feature that the cumulative contribution rate in the extraction matrix reaches 95% as the main component, as shown in Fig. 6. The body composition characteristic matrix of the main components is 600 * 200 dimensions. The original 4300 dimensions of each action sample are reduced to 200 dimensions. Redundant information is reduced.

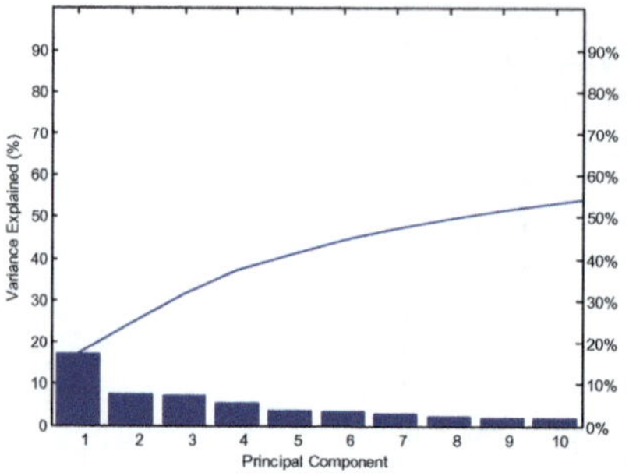

Fig. 5. Top ten principal component contribution

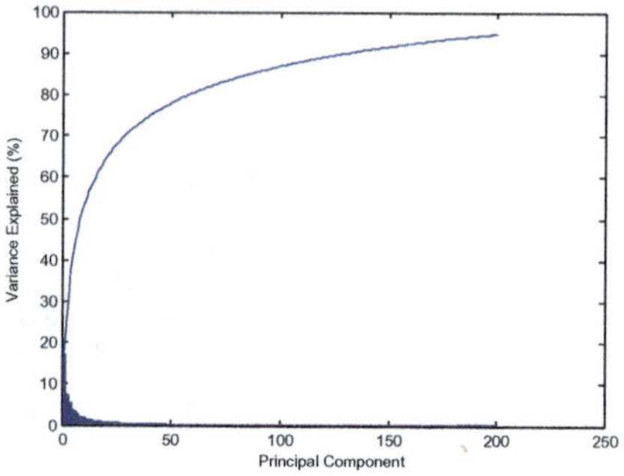

Fig. 6. The main component

Gaussian kernel function parameters is $g = 1$, $\sigma^2 = 1$, c = 0.2 in support vector machine classifier. The 300 samples were randomly selected from the database as training samples, and the rest were used as test samples. The average accuracy of the obtained action recognition and the recognition accuracy of each type of action are shown in Table 1.

Table 1. RBF kernel function human action recognition accuracy table

Category	1	2	3	4	5	6	Accuracy
1	46	0	0	1	0	0	97.87%
2	0	48	0	1	0	0	97.96%
3	0	0	50	0	0	1	98.04%
4	0	0	0	44	0	0	100%
5	0	0	1	1	49	0	96.08%
6	0	0	0	1	0	53	98.15%
Average							98.31%

The categories in Table 1 represent a total of 295 test samples with six actions, action class tags 1, 2, 3, 4, 5, and 6 are assigned respectively. The numerical value indicates the action which each row belongs to, and it is identified as the number of samples of the action class in the corresponding column. The accuracy rate indicates the correct recognition rate of the action class shown in each row. The value of the diagonal line identifies the correct number of samples for each action, the rest are the number of misidentified samples. The average action recognition accuracy rate is 98.31%.

6 Summary

Based on the simplified anatomical model, the angle, relative position and velocity characteristics of the joint are extracted. And Fourier interpolation and normalization preprocessing are performed, and a fusion feature model is constructed to describe the human body posture and action. The principal component analysis method is used to extract the main feature information, and designing a support vector machine action classification recognizer based on the fusion feature and one-to-many classification. Finally, model training and action recognition test is held with the public action database, and it obtains a better recognition effect. It shows that the method is effective and feasible.

Acknowledgments. This research was supported by National Natural Science Foundation of China (Grant No. 51475365), and Shaanxi Natural Science Basic Research Project (Grant No. 2017JM5088).

References

1. Cheng W, Xue Y, Liu C et al (2017) An efficient method of recognition human motion in human-robot interaction system. In: International conference on electrical engineering and automation, pp 785–793
2. Li R, Wang L, Wang W (2014) A review of research on human action recognition. Pattern Recognit Artif Intell 27(01):35–48 (in Chinese)
3. Lillo I, Niebles JC, Soto A (2016) Sparse composition of body poses and atomic actions for human activity recognition in RGB-D videos. Image Vis Comput 59(3):63–75
4. Li Y, Wang S, Tian Q et al (2015) A survey of recent advances in visual feature detection. Neurocomputing 149(PB):736–751
5. Chang X, Ma Z, Lin M et al (2017) Feature interaction augmented sparse learning for fast kinect motion detection. IEEE Trans Image Process 26(8):3911–3920
6. Raptis M, Kirovski D et al (2011) Real-time classification of dance gestures from skeleton animation, vol 8, pp 147–156
7. Shimada A, Taniguchi RI (2012) Gesture recognition using sparse code of hierarchical SOM. In: International conference on pattern recognition, pp 1–4
8. Zhao X, Li X, Pang C et al (2013) Human action recognition based on semi-supervised discriminant analysis with global constraint. Neurocomputing 105(105):45–50
9. Vieira AW, Lewiner T, Schwartz WR et al (2012) Distance matrices as invariant features for classifying MoCap data. In: International conference on pattern recognition, pp 2934–2937
10. Li X, Zhang Y, Liao D (2017) Mining key skeleton poses with latent SVM for action recognition. Comput Intell Soft Comput 12(1):1–11
11. Neili S, Gazzah S, Yacoubi MAE et al (2017) Human posture recognition approach based on ConvNets and SVM classifier. In: International conference on advanced technologies for signal and image processing, pp 85–91
12. https://www.microsoft.com/en-us/download/details.aspx?id=52283&from=http%3A%2F%2Fresearch.microsoft.com%2Fen-us%2Fum%2Fcambridge%2Fprojects%2Fmsrc12%2F

Effective Teaching Design Scheme for Network Troubleshooting

Xiaoqi Yin(✉)

Heilongjiang College of Business and Technology, Harbin 150025, China
3327591490@qq.com

Abstract. With the increasing demand for application-oriented talents in society, the specialized training of network engineering talents has become a growing concern in application-oriented colleges. This paper analyzes the principle of effective teaching, based on the network engineering skill course, deeply studies the network troubleshooting method and network fault classification technology. Through the integration of network troubleshooting courses and effective teaching, practical and effective teaching cases are formed. In the actual teaching, the program is actively applied to the practical teaching to improve the teaching effect, and it also provides a reference for the curriculum construction of other specialties.

Keywords: Network failure · Effectivity teaching · Applied undergraduate course

1 Introduction

Network is the most important invention in the history of human development, which improves the development of science and technology and human society [1–3]. In the 21st century, mankind has entered the information age in an all-round way. The important features of the information age are digitalization, networking and informatization. To realize informatization, we must rely on a perfect network, because the network can transmit information very quickly. Therefore, the network has now become the lifeblood of the information society and an important basis for developing the knowledge economy. The network has had an immeasurable impact on many aspects of social life and on the development of social economy [4, 5].

Today, more and more business applications are running on the network architecture, ensuring the continuous, efficient and safe operation of the network has become a huge challenge for network managers. Despite careful deployment and strict security policies, and despite more and more investment in network management, network problems still emerge one after another [6].

In the actual teaching process, the teaching work of the network troubleshooting course must be closely combined with the current actual network situation, teaching in view of the complex network environment, and at the same time [7], the professional skills training should run through the whole teaching process. However, teachers should not limit themselves to rigid teaching skills, but make full use of various

Z. Xu et al. (Eds.): CSIA 2019, AISC 928, pp. 580–589, 2020.
https://doi.org/10.1007/978-3-030-15235-2_82

teaching methods to actively mobilize students' learning enthusiasm and effectively carry out teaching work in the limited classroom time [8–10].

2 Related Notion

2.1 Network Failure

Network failure refers to the state that the network cannot provide normal service or reduce the quality of service due to hardware problems, software vulnerabilities, virus intrusion, etc.

2.2 Applied Undergraduate Course

Since the 1980s, a new trend has gradually taken shape in the field of higher education, which is to attach universal importance to practical teaching and to strengthen the cultivation of applied talents. Many colleges and universities in China have also paid attention to strengthening the practical environment in the exploration of educational and teaching reform in recent years, because people have realized more and more clearly that practical teaching is an important link in cultivating students' practical ability and innovation ability, and is also an important way to improve students' social professional quality and employment competitiveness. Application - oriented undergraduate refers to an undergraduate college with application-oriented orientation rather than scientific research-oriented orientation. It is an exploration of an educational model combining new undergraduate education with new-level higher vocational education. Application - oriented undergraduate education has played a positive role in meeting China's economic and social development, in meeting the needs of high-level application-oriented talents and in promoting the popularization of higher education in China.

2.3 Effective Teaching Theory

Effective teaching theory is an important branch of pedagogy. It is both a theoretical science and an applied science. And it should not only study the phenomena and problems of teaching, reveal the general rules of teaching, but also study the methods, strategies and techniques to solve practical problems of teaching by using and following the rules. It is not only a descriptive theory, but also a prescriptive and normative theory. "Effective" is the essential characteristic of teaching, the core idea of the current curriculum reform, and the inevitable requirement of realizing the connotation development of education.

2.4 Effective Teaching of Network Troubleshooting Course

Teachers are the guides in the classroom, playing a guiding role in the teaching process, fully mobilizing students' learning enthusiasm, helping them find problems, explore problems, solve problems, what to give students and what to gain from this lesson. The complexity of Internet technology, topology and application determines that network managers must: ensure the stable operation of the network; Master troubleshooting methods; Familiar with the possible failure points of various protocols and quickly locate and eliminate the failures. Network troubleshooting course is the core skill course of network engineering specialty. In the effective teaching, failure analysis is taken as the introduction point. Students independently analyze and explore the causes, supplemented by teacher guidance, emphasize cooperation among students, enhance participation awareness and solve problems together. Constantly cultivate students' overall analysis habit, thinking spirit and self-study ability.

3 Relevant Analysis of Troubleshooting

In recent years, the probability of network failure is increasing year by year. The network manager must collect relevant information according to the fault phenomenon, analyze the fault point by using the structured theoretical method, formulate the troubleshooting solution, and troubleshoot through the set steps.

3.1 Fault Classification

Due to the diversity and complexity of network faults, the classification methods of network faults are also different. Network failures generally fall into two categories: connectivity failures and performance problems. Connectivity failure is the most serious network failure because the network is interrupted and the service cannot be carried out. Performance failure means that the performance of the network decreases, the transmission rate slows down, and the service is affected to some extent, but not interrupted. The classification of network faults and common faults are shown in Table 1.

Table 1. Classification of network faults and common faults

Connectivity failure	Hardware, media, power failure
	Configuration error
	Device compatibility error
Performance issues	Network congestion
	The route to the destination is not optimal
	Insufficient power supply
	Routing loop
	Network instability

3.2 Classification of Exclusion Methods

It is a great challenge for network maintenance and management personnel to pay attention to the order and method of troubleshooting, to properly maintain the network and to ensure that problems can be quickly and accurately located and resolved after a failure occurs. This requires not only an in-depth understanding of the relevant network protocols and technologies, but also the establishment of a systematic troubleshooting idea and its rational application in practice to isolate, decompose or reduce the scope of troubleshooting a complex problem, so as to repair the network failure in time. The troubleshooting methods are shown in Table 2.

Table 2. Troubleshooting methods

Category	Description
Hierarchical troubleshooting	The OSI model is examined layer by layer
Block troubleshooting	Fault classification filing
Piecewise troubleshooting	The fault point is determined in sections
Replacement method	Replace suspected faulty equipment

3.3 Effective Teaching

Improving classroom effectiveness teaching is the key and fundamental requirement of deepening curriculum reform at present. In teaching practice, seize the opportunity and reflect in time to think about which teaching designs have achieved the expected results, and which links need to be improved in the future. Effective teaching is designed from the following five aspects:

Teaching Communication
Through effective communication to promote the effective achievement of classroom teaching objectives and enable students to acquire more effective knowledge. Teachers should use the teaching style reasonably according to different teaching environments and students' personality characteristics to improve the effectiveness of teaching.

Classroom Management
Good classroom discipline, high concentration of students' attention and active thinking. Both teachers and students have full enthusiasm and speak enthusiastically in class. In a warm classroom atmosphere, students keep a cool head, listen to their classmates' speeches, and think nervously and deeply.

Classroom Questioning
First of all, pay attention to the pertinence of questions, radiation and individual differences. Secondly, pay attention to providing students with time and space for thinking. Finally, pay attention to the positive situation of feedback and encourage students to think creatively.

Application of Teaching Media
The successful and effective use of teaching media in curriculum teaching should be carefully designed and follow certain application steps. There are mainly the following aspects: analyzing students' background; Formulate teaching objectives; Preparing teaching media; The use of teaching media; Stimulate students' reaction; Evaluate the teaching effect.

Time Arrangement
First, reasonably arrange the time density of classroom teaching and grasp the rhythm of classroom teaching. Secondly, optimize the teaching sequence.

4 Design Plan

4.1 Troubleshooting Design

Observe and Describe the Fault Phenomenon
In person, observe the user's demonstration of the failure, ask the user about the operation behavior before the failure, the first time, frequency, scope of influence of the failure and whether to change the node or network before the failure occurs, and sort out and record the failure phenomenon.

Collect Information About Possible Causes of Failure
Check the network maintenance log and ask other network administrators about the relevant change information in the recent fault area. Collect fault-related information through the network management system, network equipment diagnostic commands, operating system diagnostic commands, protocol analysis tools, and network testing instruments.

Develop Solutions
According to the cause of the problem, the solution is worked out according to the priority order, from the possibility to the possibility, so as to improve the efficiency of the investigation.

Implement Solutions One by One
The solution countermeasures are implemented one by one and the network state is observed until the fault returns to normal.

Record the Troubleshooting Process
The process of troubleshooting should carefully record the results of each parameter change. After handling the failure, we should find out the cause of the failure and formulate corresponding countermeasures to avoid the recurrence of similar failure as much as possible. Meanwhile, we should record the network management log for future reference and accumulate operation and maintenance experience.

The network troubleshooting procedure flow is shown in Fig. 1.

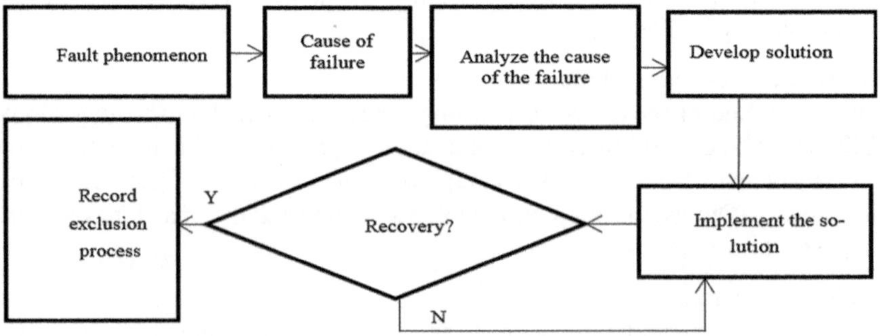

Fig. 1. Flow chart of network troubleshooting steps

4.2 Effective Teaching Model Design

Effective teaching refers to a personalized teaching activity in which teachers creatively and comprehensively utilize all teaching methods, methods and strategies conforming to the teaching rules, teaching principles and advanced teaching concepts in order to achieve teaching objectives and promote students' development to optimize teaching links, improve teaching process and strive to improve teaching effect. The first problem that the effective teaching exploration attempts to solve is the teaching effect, whether the teaching activities can achieve the predetermined teaching goal, that is, to promote the all-round development of students with high quality and efficiency. In fact, the ultimate goal of effective teaching is to improve the teaching effect and promote students' all-round physical and mental development. In other words, if there is no teaching effect and quality, there will be no problem of effective teaching. Effective teaching has no fixed pattern or program, but is a personalized, creative, highly flexible and non-fixed pattern of teaching activities. The reason why this kind of teaching activity can be called "effective teaching" is that its implementation helps to improve the effect and quality of teaching and to achieve the intended teaching goal. Although the realization of effective teaching requires the participation and cooperation of students and close attention to the physical, mental and behavioral changes of students, it is mainly a kind of teaching behavior, teaching concept, teaching creativity and innovation of teachers. Whether such teaching behavior can arouse students' enthusiasm for participation and learning motivation, and whether it can achieve the intended teaching goal depends on the flexible control and creative use of teaching rules, teaching principles and teaching concepts by teachers. That is to say, in effective teaching, the development and change of students is only a display screen of whether this teaching behavior is effective or not, and teachers are the implementers and responsible subjects of effective teaching activities. The effective teaching model is shown in Fig. 2.

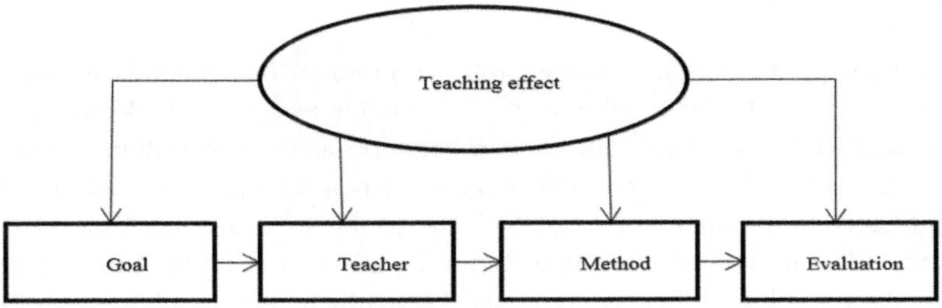

Fig. 2. Effective teaching model

Focusing on the core of "teaching effect" to regulate teachers' teaching behavior and teaching methods is the unique internal framework and operation mechanism of effective teaching, which constitutes a dynamic cycle of regulating and controlling teaching methods for the sake of improving teaching effects. From the perspective of teaching activities as a whole, the above five elements of effective teaching - effect, goal, teacher, method and evaluation are inseparable. The five links of the normal operation of effective teaching cannot be carried out normally without any of them.

With the emergence of various new technologies, the comprehensive application of data, audio and video integrated transmission, and the coexistence of modern network and traditional network. The effective teaching design of network troubleshooting is based on four troubleshooting methods and conducted under the guidance of effective teaching. A variety of teaching methods are introduced and the teaching results of each process are fed back in time to implement effective teaching in the whole teaching process of network troubleshooting course. The effective teaching design for network troubleshooting is shown in Fig. 3.

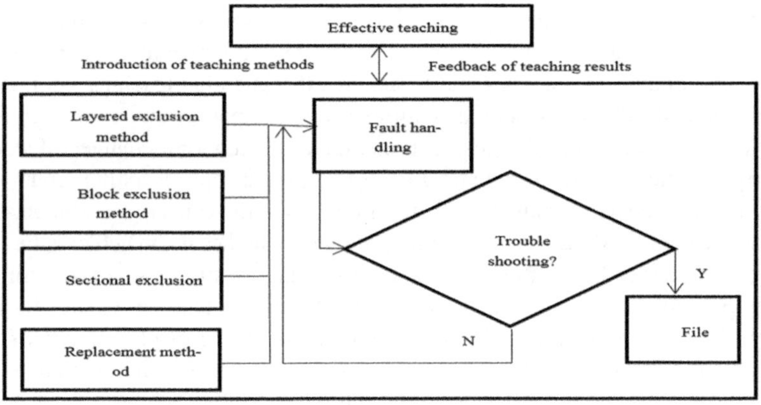

Fig. 3. Effective teaching design for network troubleshooting

5 Case Application

OSPF routing protocol is a link-state routing protocol that belongs to the internal gateway protocol and works within a single autonomous system. Dijkstra algorithm is used to calculate the shortest path to each network, and the algorithm is executed to quickly converge to the new loop-free topology when detecting link changes (such as link failure). The common failures of OSPF are: unable to form a neighbor relationship; Adjacency relations stagnate; Routing cannot be advertised; Routing cannot be added to routing table; SPF repeat calculation, etc. Taking the neighbor relationship failure as an example, the main reasons for the neighbor relationship failure are: whether the interface starts OSPF or not and whether it is a silent interface. Whether the ACL rejected the hello message; Timer settings, regional settings, verification of configuration matches, etc. The process of neighbor relationship troubleshooting is shown in Fig. 4.

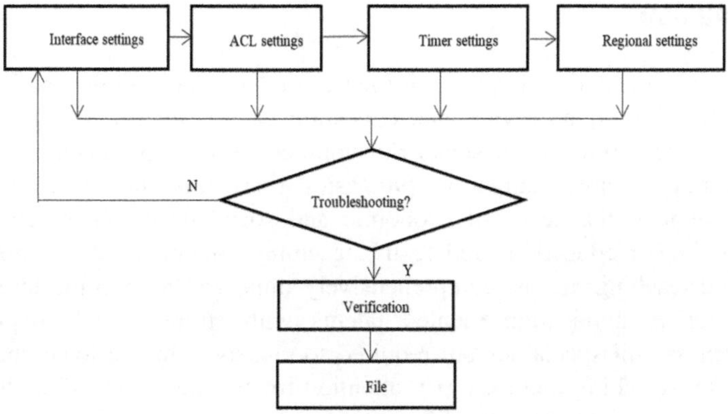

Fig. 4. Flow chart of neighbor network troubleshooting

There are many reasons for neighbor network failure, which are not listed here. With the continuous application of OSPF routing in various networks, failures occur more and more frequently and the reasons are more and more complicated. In the process of training applied undergraduate talents, the training of skills should not only increase the intensity of experiments, but also pay attention to teaching methods and fully integrate effective teaching into the teaching system. From this, we can get the design model of effective teaching of network troubleshooting course, as shown in Fig. 5.

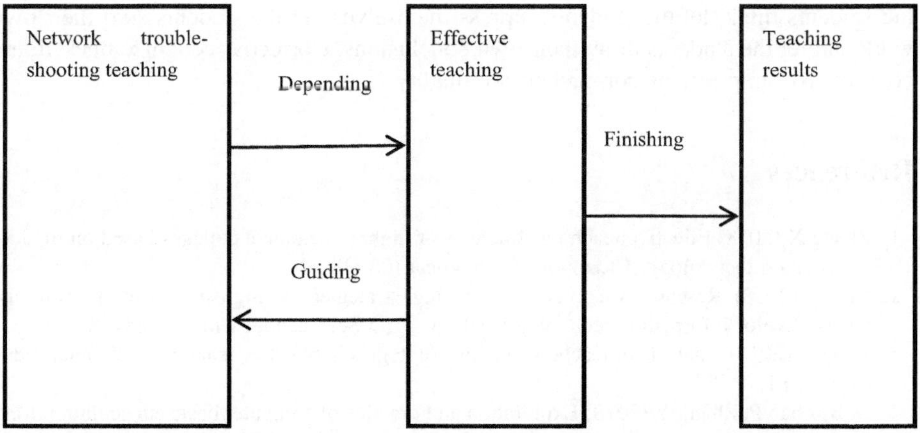

Fig. 5. Design model for effective teaching of network troubleshooting course

6 Conclusion

The application-oriented undergraduate focuses on the word "application" and requires that the talent view, quality view and education view, which reflect the spirit of the times and the requirements of social development, should be taken as the guide to construct a new discipline direction, professional structure and curriculum system to meet and adapt to the needs of economic and social development under the new situation of higher education, update the teaching content, teaching links, teaching methods and teaching means, comprehensively improve the teaching level and cultivate high-quality application-oriented talents with strong social adaptability and competitiveness. All specialties are required to closely combine local characteristics, reasonably use teaching methods, pay attention to students' practical ability, and cultivate application-oriented talents diligently.

In the teaching of the course of network troubleshooting, the emphasis has shifted from imparting knowledge to cultivating students' learning ability. Teachers should not only pay attention to the results of students' learning, but also pay more attention to students' learning process and promote students to learn autonomous learning and cooperative learning. Guide students to explore learning, let students experience, feel and understand the process of knowledge generation and development, cultivate students' professional quality and innovative thinking ability, attach importance to students' sustainable development, and cultivate students' lifelong learning ability. Taking the network troubleshooting course as an example, we must renew the educational concept under the guidance of the concept of application-oriented curriculum standards, truly change the injection teaching into heuristic, change students' passive listening to lectures into active participation, and change the simple knowledge teaching into equal emphasis on knowledge and ability. In teaching, let the students observe, let the students think, let the students express themselves, let the students start their own work and let the students draw their own conclusions. Correctly recognize oneself and continuously improve its comprehensive quality.

References

1. Zhang X (2018) Effective teaching strategies of higher vocational colleges based on modern information technology. J Liaoning High Vocat (08):38–40
2. Chu J (2018) Research on effective teaching strategies in english classroom from the perspective of information technology. J Jilin Agric Sci Technol Univ (01):84–86
3. Jiang X (2018) Analysis of teachers' leading strategies in effective teaching. J Yanbian Educ Coll (01):130–132
4. Li B, Zhao P, Zhang Y (2018) Exploration and practice of computer basic curriculum reform in applied undergraduate universities. Microcomput Appl (01):22–25
5. Wang Y, Shen B (2016) Analysis and investigation of network faults in university campus network. Electron Technol Softw Eng (15):15
6. Liu X (2018) Research on computer network fault and maintenance technology (05):40–41
7. Qi Y (2018) Practical research on the effectiveness of college english curriculum based on project teaching. J Shandong Agric Eng Univ (08):159–160

8. Yao Q (2017) Exploration and research on effective teaching of SPOC course, higher vocational education. J Tianjin Vocat Inst (05):83–86
9. Liu Q (2016) TPACK framework supported by the high school information technology course teaching design method. Henan Normal University
10. Chen W, Lu W (2010) Effective teaching begins with effective teaching design – taking geography teaching design as an example. Fujian Basic Educ Res (11):40–42

Research on Fatigue Test Based on Mobile Phone Gesture Design

Fujun Ma[1], Yan Liu[2(✉)], Fanghao Song[2], Junfeng Li[1], and Chhunli Liu[1]

[1] School of Mechanical-Electronic and Vehicle Engineering, Weifang College, Weifang 261061, China
[2] School of Mechanical Engineering, Shandong University, Jinan 250061, China
kagaomei@163.com

Abstract. At present, mobile phone as mobile Internet terminals are still on the rise in the world. The operation mode of mobile phone is very important for every user. In order to accurately locate the rationality of operation mode of mobile phone, it is necessary to use certain experimental data to find the basis for gestures design. However, it lacks the hand ergonomics research on the operation of the mobile product carriers, which needs fatigue testing and data support. Therefore, under the condition of bracing elbow with cellphone, this paper uses the surface electromyography signal measurement method to test the gesture fatigue, selects the muscle of thenar, hypothenar, the extensor digitorum and flexor digitorum superficialis as test objects, and tests 3 most commonly gestures in daily life, combines with Borg fatigue evaluation method, evaluates comprehensively the degree of user's hand fatigue. It provides data supports for gesture design in the development process of mobile products to optimize the evaluation method of mobile gesture design and provide assistances for gesture designers.

Keywords: Surface electromyography signal measurement method · Borg fatigue evaluation method · Gesture design

1 Introduction

In modern society, with the rapid development of the Internet, mobile phone has become one of indispensable information exchange tools for people's daily life. In addition to information interaction, mobile phones can assist people in various social life information services such as entertainment, shopping, payment, communication. The latest survey by German data statistics Internet company found that Brazilians spend most time on mobile phones, averaging nearly 5 hours per day; Chinese rank second with 3 hours a day; the third to eight places are Americans, Italians and other nationals whom spent time on mobile phones is more than 2 hours a day. A British mobile technology consultant Tommy Ahonen found that, people look at the phone every 6.5 min on average. If the waking time counts 16 hours a day, people need to watch about 150 times [1]. Inappropriate use of mobile phones causes many health problems such as myopia, muscle fatigue and so on.

Z. Xu et al. (Eds.): CSIA 2019, AISC 928, pp. 590–599, 2020.
https://doi.org/10.1007/978-3-030-15235-2_83

With the popularity of large-screen mobile phones, the research of ergonomics is relative lagged. For example, in Apple's iOS application, the return button is located in the upper right corner of the mobile phone screen, which is difficult to touch during operation form user's feedbacks. Hence, the setting of double-click to pull half of the screen interface is added to facilitate the user's operation. This raises the problem of different effects of different sizes of mobile phone screens on ergonomics. Large-screen mobile phone, "phablet", generally refers to screen sizes ranging from 5 to 6.9 in. These phablets have gone beyond the customary, controllable size of palms [2]. Therefore, it is necessary to start with scientific research methods, carry out ergonomics research on mobile phone comfort, provide reliable data support for fatigue testing based on human physiological function and bases for the development of mobile products.

2 Anatomic Characters of the Hand

Hand is the main limb of human activity, located at the distal part of the wrist, and is the end structure of the entire upper limb. It is divided into 3 parts: palm, back of hand and fingers. The palm of hand is slightly quadrilateral and concave, which is the transitional area between the wrist and fingers. There are 3 groups of hand muscles: lateral group (muscle of thenar) including abductor pollicis brevis, flexor pollicis brevis, opponens pollicis and adductor pollicis; intermediate group including lumbrical muscles, palmar interosseous muscles and interosseous dorsal muscles; medial muscles (muscle of hypothenar) including abductor pollicis of little finger, flexor digiti minimi brevis and the muscle opponens digiti minimi [3]. The contraction of each muscle controls the related position movement of the hand by tendon sheath connections. After consulting relevant anatomical data and the guidance of professional doctor, combining characteristics of EMG signals acquisition, flexor digitorum superficialis and the extensor digitorum of forearms and the muscle of thenar and hypothenar of hands are finally selected for EMG tests (Fig. 1).

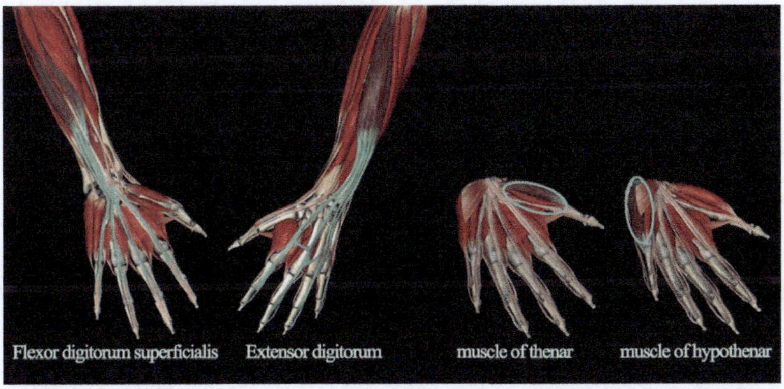

Fig. 1. 3D Body anatomy App shows flexor digitorum superficials and the extensor digitorum of forearm controlling four fingers, the muscle of thenar and hypothenar located in the palm of hand

3 Experimental Methods

This paper uses two fatigue testing methods: the sEMG signal measurement and the Borg fatigue evaluation method to assess comprehensively the fatigue degree of hands muscles.

3.1 The sEMG Signal Measurement Method

When the neuromuscular system is active, an electrical pulse sequence is triggered by α-Motoneurons in the spinal cord and conducted along the axon to superficial muscle fibers, which generates bioelectrical signals. These EMG signals are guided by the surface electrode, presented the original sEMG signal and decomposed into individual motor unit action potential trains. Through filtering, amplifying, visualizing and recording, the obtained one-dimensional voltage time sequence signals can reflect the activity of the neuromuscular system to some extent [4]. The voltage time series is called as sEMG signal, which has important practical value in clinical medicine, ergonomics [5], rehabilitation medicine [6] and so on. The sEMG has advantages of noninvasive measurement, simple operation, clear signal, easy to analyze, which is suitable for EMG signal test of superficial muscles.

The experiment uses the EMG wireless signal collector of Shandong University, applies the ErgoLAB software from Beijing Jinfa Technology Co., Ltd. to analyze the time domain and frequency domain.

Time domain analysis method, known as the waveform analysis or the time domain statistical analysis, utilizes a function or graph of time-varying signal amplitude to describe the signal. The statistical parameters, such as the peak value, the mean value, the variance and others of signal can be obtained [7]. Making use of its MAV, VAR, RMS as the characteristic quantities of EMG signal action pattern recognition [8, 9].

Frequency domain analysis takes frequency f as an independent variable to establish the relationship among the amplitude, phase and frequency of signal, which is mainly used to identify the periodic component of signal [7]. It utilizes the fast Fourier transform to convert EMG time domain signals into EMG frequency domain signals; analyzes the power spectral density of signal, determines the frequency band of EMG signal from power spectrum density. Research has shown that frequency domain characteristics of surface EMG signals are relatively stable [10], and widely used.

3.2 The Borg Fatigue Evaluation Method

The Swedish psychologist Gunnar Borg first proposed the concept of "subjective physical sensation" and a "rating of perceived exertion (RPE)" test method in 1972, as shown in Table 1. Since the subjective assessment of the body's physical status is derived from its material basis, the RPE score of the Borg table relates closely to the current physiological and biochemical indicators (such as metabolic rate, heart rate, hormone level, etc.) [7]. Due to the short test time, the experiment needs to combine with the user's daily experience. The experiment consults user's experience of using mobile phone, let users describe their feelings, combines with the fluctuation of EMG signals, analyzes and judges the degree of the user's muscle fatigue, improves the experimental completeness.

Table 1. Brog's RPE score

Descriptor	Rating
No exertion at all	6
Extremely light	7
	8
Very light	9
	10
Light	11
	12
Somewhat hard	13
	14
Hard	15
	16
Very hard	17
	18
Extremely hard	19
Maximal exertion	20

4 Fatigue Experiments of the Hand EMG Signal

The participants are 5 postgraduates aged 20–35, all are right-handed. Their hand width and palm thickness are in line with the ratio of human body, and also meet the standard human body average size defined by the national standard GB/T 26158. This paper focus on common hand gestures including single-handedness and bi-handed operation. During the test, users takes several high frequency behaviors, such as browsing micro-blog, WeChat, playing mobile games to ensure the experiment has more authenticity.

The EMG fatigue test uses a 5.5-in. phablet and a 4.7-in. small-screen phone as tools. Two different screen sizes are tested with one-handed and bi-handed, each test time takes 5–30 min (Fig. 2).

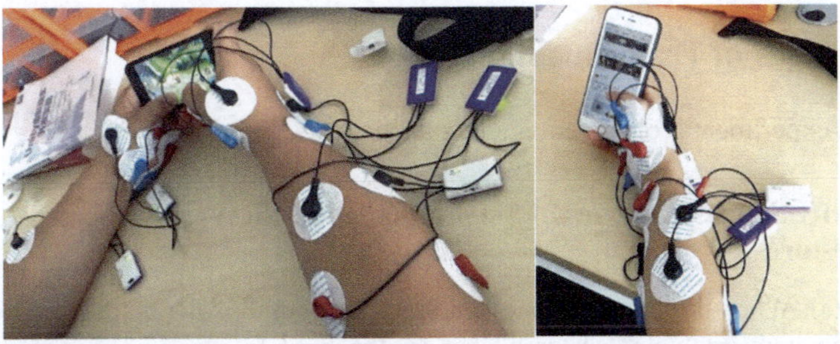

Fig. 2. Experimental scenes of using a 4.7-in. screen phone with bi-handed and single-handed

According to user's habits and gestures of using mobile phones, the test takes the posture of arms placed on table and handheld phones. There are 3 gestures are selected for analysis, as shown in Fig. 3. According to the term "thumb zone" proposed by Steven Hoober's book "Designing Mobile Interfaces" in 2011, the vertical screen is divided into three areas: comfortable reach, reach with a stretch and hard to reach. The one-handed concerned mainly with vertical control gestures of the comfortable reach zone and hard to reach zone, as well as the horizontal control gesture. Since the bi-handed gesture has better grip strength, the thumb can easily contact the most areas of screen, thus testes one gesture as picture (c) in Fig. 3.

(a) The sketch map of right-handedness operated vertical screen on comfortable reach area

(b) The sketch map of right-handedness operated horizontal screen on comfortable reach area

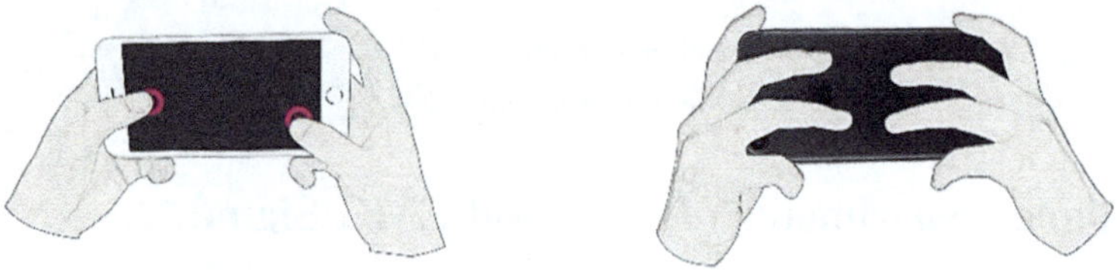

(c) The sketch map of bi-handed operated horizontal screen of mobile phone

Fig. 3. Gestures diagrams of single-handed and bi-handed operation

The EMG original and processed signals are obtained by sensors and software feedback. The EMG signal amplitude interval distribution histogram of time domain, the EMG signal power spectral density graph of frequency domain and related parameters are obtained by software analysis. Based on above, this paper analyzes data and summarizes EMG signals characteristics of these gestures.

5 Analyses of Experimental Results

Through experimental observation and statistic, it is found that:

5.1 EMG Signals Analyses of Single-Handed of the Vertical Screen Gesture

Gesture (a): When the user is in one-handheld, the thumb is the main operator. Outside the thumb comfortable reach area, that is, the hard to reach zone, there is a high frequency of EMG fluctuations. For example, when the return button in the upper left corner of the touch screen is pressed, the grip force is required by the muscle of thenar and hypothenar, and the influence of the muscle of thenar causes the thumb stretching, the contractions of flexor digitorum superficialis cause four fingers to bend. At this time, there are mutation EMG signals fluctuated from the muscle of thenar, hypothenar and flexor digitorum superficialis, as shown in Fig. 4.

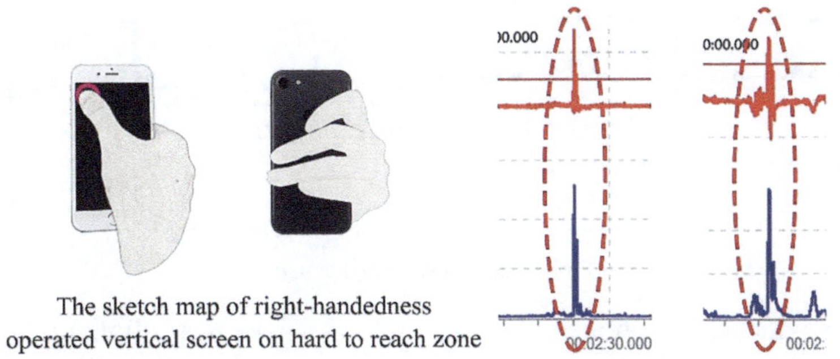

Fig. 4. The EGM signal of thenar and flexor digitorum superficialis of gesture (a) shows that the red frame marks the mutation signal when the thumb touches the upper-left corner area.

In the same behavior of browsing information, the maximum amplitude (Y Max) of the muscle of thenar EMG signal of the small-sized mobile phone is 145.81, and the value of the phablet is 324.35, which is about 2 times more than that of the smaller; The minimum amplitude (Y Min) is 0.10 and 0.12, which is about the same; the average amplitude, that is, the average level of the response physiological signal amplitude (Y Average) is 8.18 and 29.64, which is about 3.6 times; the amplitude variance (Y Variance) is an estimate of physiological signal amplitude deviating from the zero mean value is 77.65 and 4372.55, which is more than 56 times; the Y STD is 8.81 and 66.13, about 7.5 times; the Y RMS is 5208.46 and 1366.84.03, about 262 times; the Y MAV is 353937.46 and 378642855.71, which is about 1069 times. This shows that effects of different screen size on muscles fatigue degrees; the Y Min of the EMG signal is basically the same, but there are significant gaps among the values of Y Max, Y STD, Y RMS, etc. Therefore, when designing the UI of the phablet, the interface format of Internet products should be optimized by the area division of the gesture design, so that the user's interaction mode is more distributed rationally, the visual experience is improved, and the comfort of gesture operation is enhanced to achieve the optimized experience.

Figure 5 shows the EMG signals of user gestures at different sizes phones, there are the muscle of thenar, hypothenar, flexor digitorum superficials and the extensor digitorum from top to bottom. When holding a phablet, the force exerted by the muscle of hypothenar is obviously larger than that of the small-size mobile phone, because the phablet needs more force to keep the screen stable. The peak of mutation signals of the muscle of thenar of phablet is significantly higher than that of small-size, which is because the thumb needs more force to stretch in hard to reach zone.

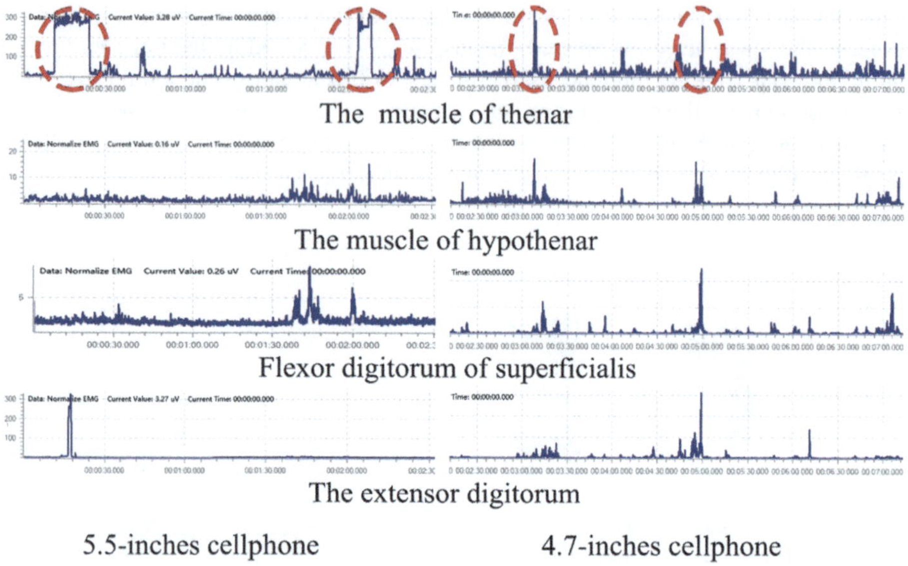

Fig. 5. Comparison of gesture EMG signals for different size phones

In Fig. 6, the experimental subject of the left picture takes the behavior of playing vertical screen single games. It has obvious traces of key operation after games over, such as mutation signal areas. The frequency of occurrence of mutation signal is closely related to the nature of games. The high frequency peak of the EMG signal of muscles of hypothenar can be seen after changing games. Therefore, when designing a single game with a vertical screen, designers should try to set control buttons in comfortable reach zone of the thumb to avoid too complicated operation, and make fingers have enough intervals to reduce the fatigue of gestures. The right picture of Fig. 6 takes the behavior of browsing micro-blog and WeChat Circle with relatively uniform mutation and continuous signals. It proves that the time for browsing each article is approximately the same, returning to the previous menu has clear mutation signal, browsing an article spends about 1 min. Therefore, in the design of such Apps, designers should set reasonably the length of information, distinguish the primary and secondary information, and highlight the title. Otherwise, it will reduce the user's reading patient.

5.2 EMG Signal Analysis of Single-Handed of Horizontal Screen Gesture

Gesture (b): The user prefers the gesture of holding the horizontal screen with one hand when watching the video or operating less activities. The widescreen display has better visual effect and less hand cover, which helps users to improve their experience. Compared with the vertical screen, there are fewer mutation signals, and the thumb touch operation is not used for a long time. Only EMG signals of grip strength of four fingers and the muscle of thenar and hypothenar is mostly continuous uniform.

Therefore, when designing the video application interface, most of control buttons should be lay out on the right side of the horizontal screen of mobile phones.

5.3 EMG Signal Analysis of Bi-handed Gesture

Gesture (c): The gesture of bi-handed operation is often used in activities such as playing games, typing, and editing documents, etc., which are more complicated and frequent. This experiment participators take behaviors of playing online games. The characteristic of gesture is bi-handheld mobile phone with sufficient grip and lifting force, using mainly the thumb to operate. Because the muscle of thenar is most frequently used, the EMG signal is the strongest and the mean square error is large. The muscle of hypothenar and the flexor digitorum superficials have mutation signals when the thumb touching hard to reach area. Therefore, when designing the software to match gesture (c), touch button should be designed in the comfortable reach area of the thumb. Such as the gesture design of game software.

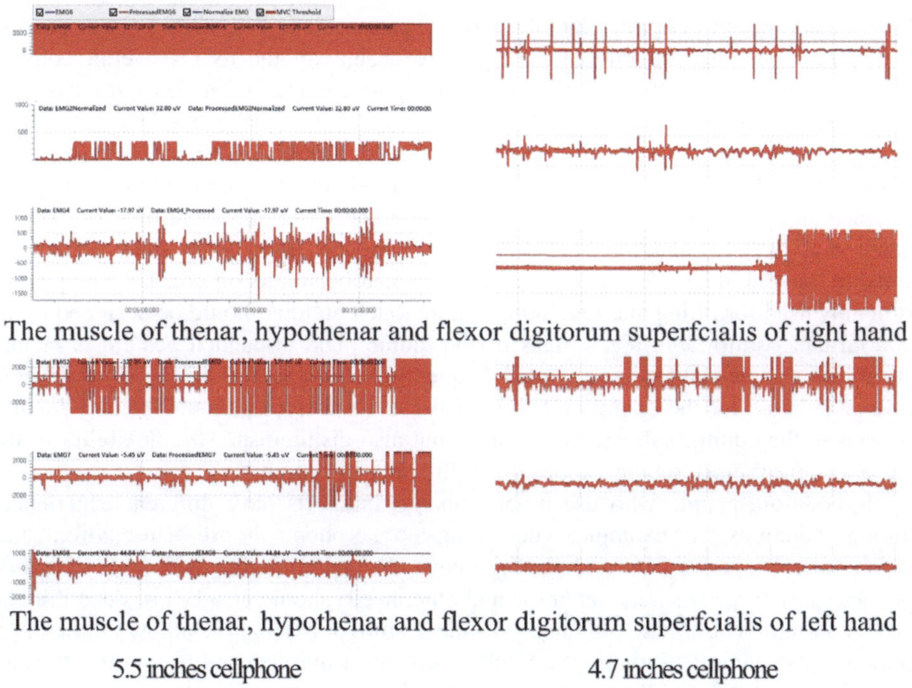

Fig. 6. Comparison of EMG signals for bi-handed gestures with different sizes mobile phones

As shown in Fig. 6, the signal amplitude of phablets on the left is significantly larger than that of the small-size phone on the right, indicating that the large-size mobile phone has better game scene effects, but it also increases the user's gesture fatigue. The game on the left is more complicated than the right, requiring frequent operations. Therefore, designers should pay attention to the click frequency setting of game gestures and find a balance between the sense of participation and fatigue. When designing game gestures for bi-handed, the operation should be simple. The convenient position of the touch button for the left thumb should be arranged in the lower left corner, such as the button that controls the forward direction of the game character; In order to cooperate with the flexible operation for the right thumb, the lower right corner can lay out buttons with higher complexity, Such as home button, attack button, set button, return button, etc. At the same time, designers should also add a handy choice in the settings, so that the buttons can be swapped to facilitate left-handed users to operate the game.

According to the Borg fatigue survey, some users said that: 30 to 40 min is between easy and tired, about 1 h, they will feel a little tired of hands. When playing games, the user's attention is more invested, and it will take about 2 h to feel obvious fatigue.

6 Conclusion

In summary, when designing mobile phone Apps, not only the user's visual process experience, but also the influence of gesture design on the user's overall comfort should be considered. Different sizes of mobile phone operating modes and areas have different divisions. The single-handed gestures are more suitable for browsing information, simple games, etc., especially for small-sized phone. When designing such mobile products, the arrangement of touch buttons should be formatted in comfortable to reach areas for thumbs, avoiding hard zone. Otherwise, excessive muscle force of hands will lead to muscles fatigue. The single-handed horizontal screen gesture is suitable for watching videos and other less operated behaviors on phablet. The gesture is mainly used for lifting and regulating. These touch buttons should be designed in the bottom and comfortable reach areas of the thumb. The bi-handed gesture is mainly suitable for games and other operated frequently activities. The button layout should also be suitable for the operating habits of the left and right thumb. Not only layouts buttons in the comfortable area of thumbs, but also distinguishes the flexibility of the left and right thumb, sets the appropriate difficulty of buttons.

In addition, people who use mobile phones in sports have different experiences during operations. For example, when using mobile phones in different environments such as kitchens, restrooms, and waiting rooms, the operation time and mode of mobile phones cause directly safety, attention and other needs should also be considered in the gesture design. This article only analyzes the gesture of placing the elbow on the table, does not involving the fatigue of the elbow and wrist muscles and testing the muscles on the arms that control these two parts. Considering the travel mode, many users will use the gesture of arm-suspension to operate the mobile phone, which will be supplemented in future experiments.

Acknowledgement. This work is supported by the project of Application Research on the Maker Platform for Weifang with 2015GX005, and the project of Research on the Integration of Interactive Information Construction Based on Ergonomics with ZR2015FL031.

References

1. Life Times. Globe Times, Beijing, 23 June 2018
2. https://designmodo.com/phablets-ux/
3. Shuwei L, Xirui L (2013) Regional anatomy. People's Medical Publishing Press, Beijing
4. Qingling L (2009) Study on sEMG based exoskeletal robot for upper limbs rehabilitation. Harbin University of Science and Technology, pp 1–2
5. Mihelj M, Nef T, Riener R (2007) ARMin II-7 DOF rehabilitation robot: mechanics and kinematics. In: Proceeding of the 2007 IEEE international conference on robotics and automation, pp 4120–4125
6. Lixun Z, Yong Y, Jinyu Z (2008) Experimental research on impedance control of arm rehabilitation robots. J Harbin Eng Univ 29(1):69–72
7. Hailin Z (2012) Measurement technology in mechanical engineering, vol 7. China Machine Press, Beijing
8. Han JD, Ding QC, Xiong AB et al (2015) A state-space EMG model for the estimation of continuous joint movements. IEEE Trans Industr Electron 62(7):4267–4275
9. Reddy NP, Gupta V (2007) Toward direct biocontrol using surface EMG signals: control of finger and wrist joint models. Med Eng Phys 29(3):398–403
10. Jihai Y, Luyao S (1998) The research of relations between median frequencies of needle electrode electromuogram (NEMG) from the agomist and antagonist muscle. Acta Biophysica Sinica 14(1):189
11. Lianshi F, Kaigang L (2002) Test method and application of common physiological and biochemical indicators for athletes' functional assessment, vol 11, no 2. People's Sports Publishing House of China, Beijing, p 66

Two-Echelon Location Routing Problem with Multi-fuzzy and Pick-Delivery Model and Algorithm

Hong Li(✉) and Sujian Li

School of Mechanical Engineering,
University of Science and Technology Beijing, Beijing 100083, China
Lihong7953@foxmail.com

Abstract. With the rapid development of B2C e-commerce, the logistics distribution system has a situation of high cost and poor service levels. In this paper, according to the characteristics of B2C, we proposed a multi-fuzzy two-echelon location routing problem with pick and delivery model. The model contains two fuzzy factors: the fuzzy customer behavior in the process of consumption and the fuzzy time of traffic in logistics distribution vehicle. In addition, model also consider the customer's time window constraint. The model we proposed can further optimize the logistics distribution system, and reduce the total cost of logistics system. For a variety of fuzziness and complexity, this paper designed the random fuzzy simulation algorithm to transform the fuzzy factors as the certainty factor, embedding tabu search of simulated annealing algorithm for problem solving. In the end, the experiment of algorithm and model design are verified.

Keywords: Multi-fuzzy factors · Pickup and delivery · Two-echelon location routing problem · Simulated annealing algorithm · Random fuzzy simulation algorithm

1 Introduction

With the rapid development of B2C e-commerce in China, there is phenomenon inconsistent with market development of online shopping in the B2C logistics distribution system, such as high logistics cost and poor service level and so on, which has attracted the attention of many researchers. The paper proposes 2E-LRP model with multiple fuzzy factors, introduces fuzzy demand and fuzzy running time in order to solve above problems, so as to select the site of new distribution facility node built by B2C e-commerce and plan and design integrated vehicle service routing on the basis of site selection. The paper reflects a variety of uncertainties by giving a definite range of fuzzy changes with fuzzy processing method against multiple fuzzy factors. In view of solving the algorithm, the paper proposes random fuzzy simulation algorithm and simulated annealing algorithm embedded in tabu search to solve the 2E-LRP with multiple fuzzy factors. In the end, the model and algorithm shall be checked by experimental instance.

Z. Xu et al. (Eds.): CSIA 2019, AISC 928, pp. 600–610, 2020.
https://doi.org/10.1007/978-3-030-15235-2_84

2 Mathematical Model

2E-LRP [1] is defined as follows: a secondary distribution network is composed of several alternative distribution centers, several alternative transfer stations and demand points for known geographical locations. The goods start from the distribution center, pass through the transfer station and reach the demand point in the process of distribution, of which the distribution center, transfer station and corresponding distribution routing constitute the primary distribution network, and transfer station, demand point and corresponding distribution routing constitute secondary distribution network, as shown in Fig. 1. Under the condition of knowing the capacity of the logistics facilities at all levels, the distribution vehicle capacity at all levels and the location of the demand point, the quantity and location as well as vehicle routing are required to be determined by 2E-LRP, thus minimizing the total logistics cost of the network. The total costs of logistics network include fixed cost of logistics facility node at all levels, fixed cost and driving costs of vehicles at all levels.

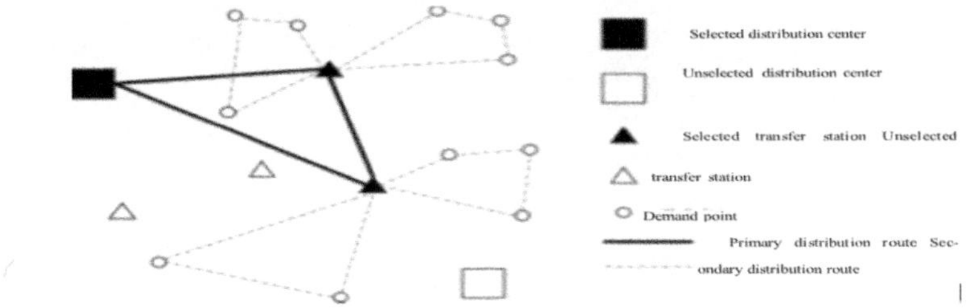

Fig. 1. Example of 2E-LRP network

3 Solving Algorithm

It is difficult to solve the optimal solution through accurate algorithm within the acceptable time limits as LRP2 problem belongs to NP-hard problem, so some heuristic algorithms have gradually become the research hotspots of scholars at home and abroad. The paper designs the random fuzzy simulation algorithm and simulated annealing algorithm embedded in tabu search (SA-TS) to solve the 2E-LRP with multiple fuzzy factors [2]. Random fuzzy simulation algorithm shall be used to carry out deterministic processing for customer fuzzy distribution, return quantity demand and fuzzy vehicle traveling time and convert fuzzy problems to the identified problems. With good convergence effect and insensitivity to initial parameter settings and other advantages, the simulated annealing algorithm has been applied to the research on LRP problem repeatedly. The optimization solution shall be implemented on the primary distribution network by tabu search that is able to quickly and efficiently solve the optimal solution or satisfactory solution of the primary distribution network due to its small relative scale. SA-TS adopts a bottom-up solution process, which first construct the neighborhood solution of secondary distribution network and then solve the primary distribution network solution by virtue of tabu search based on this in the process of iteration.

4 Algorithm Steps

In this algorithm, the random fuzzy simulation algorithm shall be first used to convert the fuzzy problems to identified problems and then generate initialization parameter and initial solution. The neighborhood solution of secondary distribution network is generated by neighborhood construction method during the process of iteration, then the primary distribution network solution is solved by tabu search algorithm in accordance with the neighborhood solution and the objective function value of the whole neighborhood solution is calculated in the end. The acceptance is executed and the optimal solution is updated in accordance with Monte Carlo acceptance criteria based on current solution and objective function value of neighborhood solution. Finally, update the temperature until meeting the algorithm termination condition, and return to the optimal solution.

The concrete steps of the algorithm are as follows and the algorithm process is shown in Fig. 2.

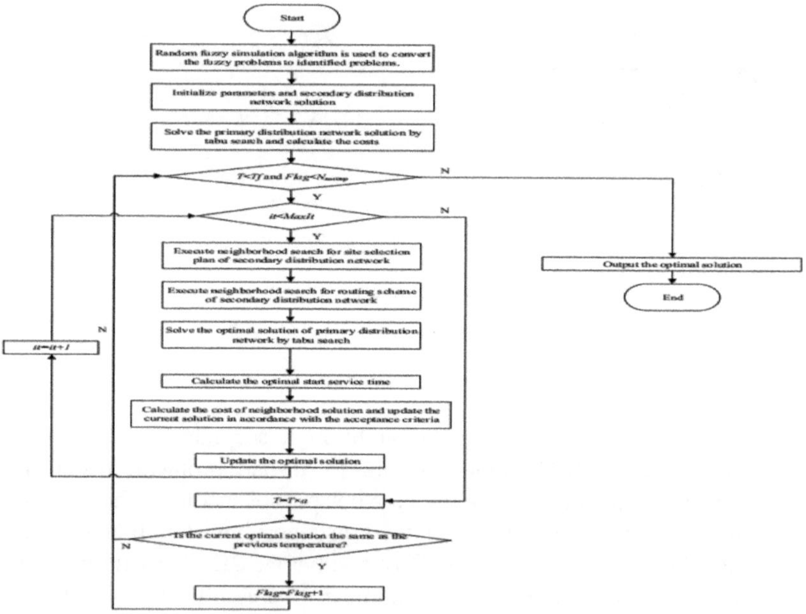

Fig. 2. Algorithm flow chart

5 Simulated Annealing Algorithm Embedded in Tabu Search (SA-TS)

(1) Coding

This model contains two levels distribution networks. The solution is solved by the same real number coding because the solution of two levels distribution network is similar. Taking secondary distribution network as an example, the coding schematic diagram of secondary distribution network of 10 demand points and 5 alternative transfer stations is shown in the Fig. 3 and can be converted into 3 paths:

1:11 -> 1 -> 2 -> 3 -> 11 Path 1: 11 -> 1 -> 2 -> 3 -> 11
2:11 -> 4 -> 5 -> 6 -> 11 Path 2: 11 -> 4 -> 5 -> 6 -> 11
3:12 -> 7 -> 8 -> 9 -> 10 -> 12 Path 3: 12 -> 7 -> 8 -> 9 -> 10 -> 12

11	1	2	3	0	4	5	6	12	7	8	9	10

Fig. 3. Example of coding with 10 customers and 5 satellites

(2) Initialization

The algorithm adopts the bottom-up solution process to initialize the secondary and primary distribution network respectively. On account of structural similarity of two levels distribution network, the initial solution is generated by the same greedy random initialization method in this paper. Taking the secondary distribution network as an example, the concrete steps of greedy random initialization are as follows:

(3) Neighborhood construction [3]

Because the problem contains the two levels distribution network with similarity, the same neighborhood construction method shall be used in the process of solving and an example of secondary distribution networks is given as follows. The neighborhood construction process of secondary distribution network contains two parts: status updates of transfer station and route update.

(1) Status updates of transfer station

The status of transfer station can be updated by closing, opening and exchanging the transfer station. The status of transfer station is not allowed to be updated again in a period after executing any method to update the status of transfer station because the status updates of transfer station will change the solution greatly, so only the routing can be updated to ensure that the optimal routing scheme can be found under the open state of the current transfer station.

(2) Route update

The routing can be updated by 2-opt*, exchange and insert. Select one route update method at random when the neighborhood of secondary distribution network is constructed.

2-opt*: two nodes are selected as 2-opt* start and stop nodes in the remaining nodes from the first node except the coding at random in the decoding of secondary distribution network, and the sequence of each node in the start and stop node shall be inverted.

Exchange: two nodes are selected as exchange node in the remaining nodes from the first node except the coding at random, and the location of two nodes shall be exchanged.

Insert: two nodes are selected as insert node in the remaining nodes from the first node except the coding at random and the first selected insert node shall be inserted behind the second selected insert node.

(4) Optimal start service time

Influenced by new generated neighborhood solution, the start service time of each customer is the vehicle arrival time, that is, the service time can be started at the earliest. However, sometimes the arrival time of the vehicle may be earlier than the upper limit of the earliest start service time window required by the customer. At this point, the vehicle can be properly waited to improve the customer satisfaction of the current customer point. The relationship between waiting time and customer satisfaction is shown in Fig. 4.

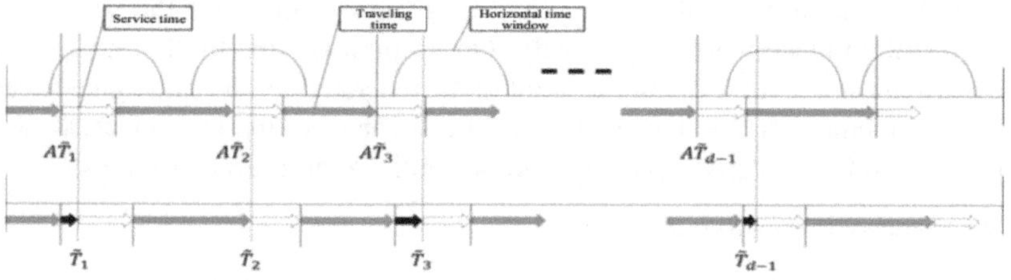

Fig. 4. Relationship between waiting time and customer satisfaction

6 Experimental Results and Analysis

6.1 Algorithm Validity Check

(1) Algorithm parameters

The parameters of the simulated annealing include: initial temperature (T0), final temperature (Tf), coefficient of temperature drop (α), internal circulation times (MaxIt), and maximum algebra not updated (Nnotimp). The parameters of tabu search include: maximum iterative algebra (TSMaxIt), maximum algebra not updated (TSNnotimp), maximum number of candidate solution set (MaxCandList) and length of tabu list (L) (Tables 1 and 2).

Table 1. Parameter values of SA

Algorithm	Parameters				
	TO	Tf	α	MaxIt	N_{notimp}
SA	50000	0.1	0.96	500	50

Table 2. Parameter values of TS Parame

Algorithm	Parameters			
	TSMaxIt	TSN_{notimp}	MaxCandList	L
TS	5	3	10	2

(2) Comparative analysis of experimental results

The validity of SA-TS algorithm is verified by instances set Nguyen and compared with the GRASP algorithm in literature GRASP WITH LEARNING PROCESS FOR A TWO-ECHELON LOCATION ROUTING PROBLEM and the MS-ILS algorithm in literature A multi-start iterated local search with tabu list and path relinking for the two-echelon location-routing problem, of which the specific results are shown in Table 3. The paper only provides solution results of instance 50-5-MN and convergence curve of objective function value for more experimental results, as shown in Fig. 5.

Table 3. Results for 2E-LRP instances set nguyen

Instances	BKS*	GRASP	Gap	MS-ILS	Gap	SA-TS	Gap
25-5 N	**80,370**	81,152	0.97%	**80,370**	0.00%	**80,370**	0.00%
25-5 Nb	**64,562**	64,572	0.02%	**64,562**	0.00%	**64,562**	0.00%
25-5MN	**78,947**	80,412	1.86%	79,593	0.82%	**78,947**	0.00%
25-5MNb	**64,438**	**64,438**	0.00%	**64,438**	0.00%	**64,438**	0.00%
50-5N	**137,815**	145,942	5.90%	138,126	0.23%	138,646	0.60%
50-5Nb	**110,094**	113,234	2.85%	111,290	1.09%	111,062	0.88%
50-5MN	**123,484**	126,313	2.29%	**123,484**	0.00%	**123,484**	0.00%
50-5MNb	**105,401**	106,033	0.60%	**105,401**	0.00%	105,846	0.42%
50-10 N	**115,725**	116,709	0.85%	116,132	0.35%	116,132	0.35%
50-10Nb	**87,315**	90,559	3.72%	**87,315**	0.00%	**87,315**	0.00%
50-10MN	**135,519**	137,321	1.33%	136,123	0.45%	136,337	0.60%
50-10MNb	**110,613**	110,703	0.08%	**110,613**	0.00%	**110,613**	0.00%

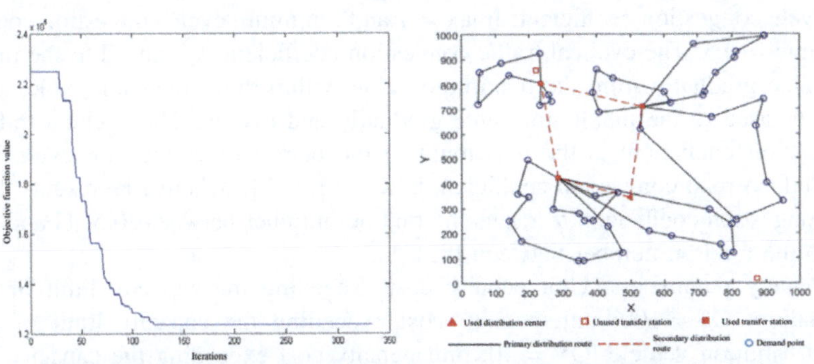

Note : BKS represents the present international known optimal solution.

Fig. 5. Example of solution with instance of 50-5MN and convergence trend graph

The experimental results show that the solving effect of SA-TS is the best when solving the scale problems of 25 and 50 customer points and the optimal solution can be found. Compared with the GRASP algorithm, the solving effect of SA-TS is better. The optimal results are obtained in all test instances. Compared to MS-ILS algorithm, the better results are obtained through such algorithm when solving the scale problem of 25 customer points, but the solving effect is similar when solving the scale problem of 50 customer points.

6.2 Model Validity Test

The paper adds the related data as experimental instance data in the instance 25-5N of instance set Nguyen to verify the validity of fuzzy 2E-LRP mode proposed in this paper and to solve through the algorithm proposed in this paper.

(1) Return amount: the paper adopts the return amount generating method proposed by Angelelli and Mansini in literature [9]. The return amount is generated in accordance with the customer ID one by one, and the return amount p*i* of the ith customer is as follows:

$$p_i = \begin{cases} (1+\alpha)\,d_i, & ii \text{ means odd number} \\ (1-\alpha)\,d_i, & ii \text{ means odd number} \end{cases}$$

Where, d*i* represents the delivery amount of the ith customer and α is the random number between [0.2, 0.8].

(2) Fuzzy distribution and return demand: trigonometric fuzzy number $\tilde{d}_i = (d_{1,i}, d_{2,i}, d_{3,i})$ and $\tilde{p}_i = (p_{1,i}, p_{2,i}, p_{3,i})$ of fuzzy distribution and return demand is generated in accordance with customer ID one by one.

$$\begin{array}{lll} d_{1,i} = (1-a)d_i & d_{2,i} = d_i & d_{3,i} = (1+a)d_i \\ p_{1,i} = (1-a)p_i & p_{2,i} = p_i & p_{3,i} = (1+a)p_i \end{array}$$

Fuzzy demand $\theta_d = 0.8$ and preference degree of fuzzy return amount $\theta_p = 0.8$.

(3) Fuzzy vehicle traveling time: the periodic road congestion parameters include: cycle time of congestion: Jc = 50, congestion cycle interval: Jg = 50, maximum cycle congestion coefficient: Jmax = 1 and minimum cycle congestion coefficient: Jmin = 0.5. The cyclical traffic congestion coefficient is reduced to the minimum value gradually from the maximum value within the congestion cycle, and then increased to the maximum value gradually and evenly. The cyclical traffic congestion coefficient is the maximum in the period of congestion cycle interval. Sudden road congestion coefficient is R = α1 − J β, where J represents the cycle congestion coefficient, α means the random number between [0.5, 1] and β refers to the random number between [0, 0.1].

(4) Penalty parameters: Unit penalty cost exceeding the capacity limit of transfer station: OS = 100; unit penalty cost exceeding the capacity limit of primary distribution vehicle: OV = 40; unit penalty cost exceeding the capacity limit of secondary distribution vehicle: OK = 25; unit penalty cost beyond the user's time window: OT = 5.

(5) Other parameters: vehicle running speed: 10; unit delivery service time: 1; unit return service time: 1

Where, the geographic details of transfer station node and primary fixed cost information of capacity constraint are shown in Table 4, and the geographic details of customer point and information about soft time window are shown in Table 5.

Table 4. Transfer station information

Transfer station	X-coordinate	Y-coordinate	Capacity constraint	Fixed cost
1	600.656	503.332	332	5527
2	463.966	356.072	359	6894
3	468.705	627.231	331	4405
4	294.486	905.849	317	5570
5	652.129	723.145	365	3650

Table 5. Customer point information

Customer point	X-coordinate	Y-coordinate	Soft time window
1	918.283	709.536	[0, 0, 95, 145]
2	577.06	134.427	[0, 5, 92, 142]
3	575.991	271.526	[0, 0, 96, 146]
4	579.94	694.336	[0, 18, 84, 134]
5	660.582	356.212	[0, 39, 124, 174]
6	236.192	174.807	[0, 25, 76, 126]
7	484.405	572.263	[0, 0, 96, 146]
8	600.786	740.53	[0, 0, 71, 121]
9	892.965	592.959	[0, 43, 121, 171]
10	287.499	391.33	[0, 13, 108, 158]
11	437.226	699.173	[0, 0, 84, 134]
12	499.022	721.307	[0, 0, 90, 140]
13	279.812	807.543	[0, 44, 144, 194]
14	328.354	457.081	[0, 17, 100, 150]
15	365.208	344.893	[0, 0, 50, 100]
16	737.698	785.66	[0, 43, 109, 159]
17	357.351	501.733	[0, 27, 113, 163]
18	438.39	414.691	[0, 29, 120, 170]
19	969.968	572.002	[0, 0, 65, 115]
20	631.052	133.572	[0, 0, 82, 132]
21	423.349	740.151	[0, 8, 64, 114]
22	873.183	145.156	[0, 45, 105, 155]
23	502.492	407.532	[0, 0, 86, 136]
24	377.464	254.122	[0, 0, 69, 119]
25	533.624	291.449	[0, 0, 92, 142]

The parameter setting of SA-TS algorithm is the same as the LRP-2E problem solving above-mentioned. The paper provides experimental result at a time for too much experimental data, as shown in Table 3. The solution result of instance 25-5 N after expansion is shown in Fig. 6.

The Table 3 shows that the SA-TS algorithm can effectively solve the expanded instance and optimize it through MATLAB software to obtain the approximate optimal solution in the end. The total cost of the optimization scheme is 90224.84 and No. 26 and 27 transfer stations are selected. 1 primary distribution vehicle and 6 secondary distribution vehicles are used, of which the routing is shown in Table 6. The result satisfies the minimum total cost principle of logistics distribution system in the whole process of logistics distribution, and the reasonable logistics network planning is carried out to make the resource distribute reasonably, showing that feasibility and practicability of the model.

Table 6. Operation result

Total cost	90224.84
Site selection of transfer station	No. 26 and No. 27 transfer station
Number of used primary vehicles	1
Number of used secondary vehicles	6
Primary distribution route	0 -> 26 -> 27 -> 0
1 Secondary distribution route 1	27 -> 23 -> 3 -> 25 -> 27
2 Secondary distribution route 2	27 -> 14 -> 17 -> 11 -> 21 -> 13 -> 18 -> 27
3 Secondary distribution route 3	27 -> 5 -> 22 -> 2 0-> 2 -> 27
4 Secondary distribution route 4	27 -> 15 -> 10 -> 6 -> 24 -> 27
5 Secondary distribution route 5	26 -> 4 -> 8 -> 12 -> 7 -> 26
6 Secondary distribution route 6	26 -> 16 -> 1 -> 19 -> 9 -> 26

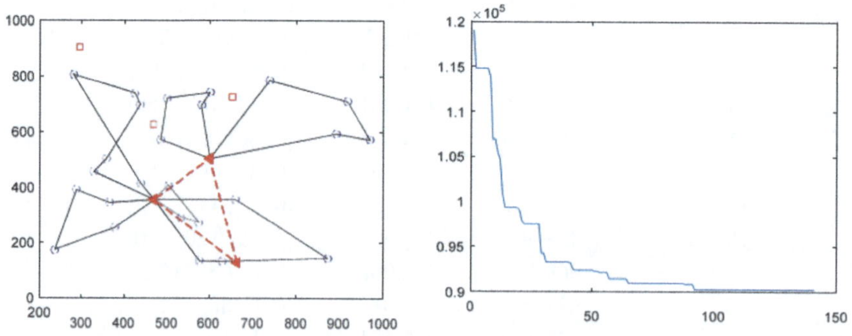

Fig. 6. Operation result

The experiment shows that the fuzzy 2E-LRP model proposed in this paper is right and the random simulation algorithm and SA-TS algorithm proposed in this paper are reasonable and effective for solving the above problem model, which is a good method to solve this kinds of complex problem.

7 Conclusion

Based on characteristics of B2C distribution network, the paper establishes multi- fuzzy 2E-LRP model for optimization. Targeted at minimum total cost of logistics distribution for solving, the model integrates site selection of secondary distribution network node and distribution vehicle routing planning, which reduces the total cost of logistics operation in general and considers the fuzzy customer demand, fuzzy vehicle traveling time and soft time window constraint. Combined with the B2C e-commerce return problem, the model comprehensively considers the touring access characteristics of the distribution vehicles, forward and reverse logistics, and introduces the vehicle dynamic load constraint to ensure the effectiveness of forward and reverse logistics operation. The random fuzzy simulation algorithm and simulated annealing algorithm embedded tabu search proposed in this paper are reasonable and effective. The algorithm can find the international optimal solution in standard example of solving 2E-LRP with multi-fuzzy factors, which proves the effectiveness and generality of the algorithm, and can provide a reasonable and effective approximate optimal solution of 2E-LRP with multi-fuzzy factors. The 2E-LRP model with multi-fuzzy factors and solving algorithm proposed in this paper can provide effective decision-making support for the decision-makers of B2C logistics distribution enterprises.

References

1. Deng XP, Zhou XM, Tian SH (2016) Research on integrated optimization of location and routing for B2C E-commerce logistics center. J Chongqing Univ Posts Telecommun (Nat Sci Ed) 28(4):593
2. Zhang XN, Fan HM, Li JF (2015) Bi-objective fuzzy location model and algorithm for the design of logistics distribution network in B2C e-commerce. Syst Eng - Theory Pract 35 (5):1202
3. Chen JM, Zeng B (2014) Artificial bee colony algorithm with variable neighborhood search and path relinking for two-echelon location-routing problem. Comput Integr Manuf Syst 20 (5):1228
4. Luo YB, Sun YM (2014) Capacitated location routing problem based on fuzzy time windows. Syst Eng (1):19
5. Angelelli E, Mansini R (2002) The vehicle routing problem with time windows and simultaneous pick-up and delivery quantitative approaches to distribution logistics and supply chain management, p 249
6. Arunkumar N, Ramkumar K, Hema S, Nithya A, Prakash P, Kirthika V (2013) Fuzzy Lyapunov exponent based onset detection of the epileptic seizures. In: 2013 IEEE conference on information and communication technologies, ICT 2013, art no 6558185, pp 701–706

7. Faig JJ, Moretti A, Joseph LB, Zhang YY, Nova MJ, Smith K, Uhrich KE (2017) Biodegradable kojic acid-based polymers: controlled delivery of bioactives for melanogenesis inhibition. Biomacromolecules 18(2):363–373
8. Lv Z, Halawani A, Feng S, Li H, Réhman SU (2014) Multimodal hand and foot gesture interaction for handheld devices. ACM Trans Multimedia Comput Commun Appl (TOMM) 11(1):10
9. Chen YZ, Tang FJ, Bao Y, Tang Y, Chen GD (2016) A Fe-C coated long period fiber grating sensor for corrosion induced mass loss measurement. Opt Lett 41:2306–2309

Analysis of FY-4 Satellite Lightning Data Processing Based on MATLAB

Xiangke Liu(✉), Leqi Chen, Meng Deng, and Peng Shang

Shandong Meteorological Disaster Prevention Technical Center, Jinan 250031, China
lxk81216@sina.com

Abstract. MATLAB is a powerful tool for data analysis. After the successful launch of Fengyun-4 satellite, how to use MATLAB to read the data of Fengyun-4 satellite has become a difficult problem for researchers. This paper establishes a simple MATLAB interface to realize the lightning processing and analysis system based on MATLAB, which can realize the reading and writing of lightning data of Fengyun No. 4 by MATLAB.

Keywords: NC data processing · Fengyun four · Lightning data · MATLAB

1 Introduction

With the rapid development of human society and economy and the gradual improvement of modernization level, especially the rapid development of high-tech industries such as electronics, computers and communications, the losses caused by lightning disasters are becoming more and more serious. The United Nations ranks lightning disasters as "one of the ten most serious natural disasters" and the International Electrotechnical Commission calls them "a major public hazard in the age of electronics". But lightning activity plays an important role in maintaining the circuit balance between the atmospheric ionosphere and the earth, and in the production of nitrogen oxides in the global nitrogen cycle. Due to the short-term and local characteristics of thunder and lightning weather and the limited convection information in the large-scale reanalysis field of driving numerical model, the prediction of convective weather is uncertain at present.

With the continuous development of lightning detection technology and the continuous deployment of lightning location network, people have been able to detect and locate all kinds of lightning discharge events, and obtained a wealth of lightning activity information. Lightning detection has the advantages of wide detection range (especially the ability to detect the ocean surface), less affected by terrain, continuous monitoring, and lightning detection equipment is cheaper than weather radar, but also easier to maintain, easy to deploy in a wide range. Therefore, lightning data have great potential in the monitoring, early warning and forecasting of severe convective weather systems. For example, a simple parameterization for simulating the global lightning distribution has been published in Price C. The convective cloud top height is used as a variable in the parameterization, and two lightning data sets are used to verify the

Z. Xu et al. (Eds.): CSIA 2019, AISC 928, pp. 611–618, 2020.
https://doi.org/10.1007/978-3-030-15235-2_85

parameterization for the difference between continental and oceanic thunderstorms: one global and one regional. In both cases, the lightning distribution and frequency simulated are very consistent with the observed lightning data. The parameterization can be used in global research of lightning climatology, the circuit of the earth, general cycle models for simulating global lightning activity, atmospheric NOx concentration and forest fire distribution in current and future climate. Moreover, it may even serve as a short-term forecast aid; in Estberg et al. [1, 2] multivariate linear regression models were used to predict peak lightning activity a year ago. Based on the cloud-to-mine strike data of the National Lightning Detection Network in 1985, a prediction model was proposed to determine the total monthly peak value in the Northwest Pacific. The six climate index is assessed as a potential predictor of lightning. A multivariate linear model is developed based on these indices. In the research results of Christian et al. [3] has been observing the continuous lightning in near space for more than six years. Statistical tests of lightning data show that nearly 1 billion 400 million lightning strikes occur throughout the earth every year. This year's flash counts translate into an average of 44 $\pm$ 5 lightning flashes per second worldwide, far below the traditional estimate of 100 flickers per second. The analysis of annual lightning distribution shows that lightning mainly occurs on land (88% on the continent and 12% on the ocean). An average of 78% of the global lightning activity occurs between 30° south latitude and 30° north latitude. A dominant summer peak in the Northern Hemisphere occurs in the annual cycle and evidence of a tropical-driven half-year cycle is found. With the deepening of lightning research, more and more [4–7] related lightning predictions are available.

Lightning frequency is not only the most basic characteristic parameter to measure the intensity of thunderstorm, but also the important input parameter of numerical model to simulate the generation and evolution of nitrogen oxides. There are obvious regional differences in lightning activities in different regions of the world, and many scholars have done a lot of research in this area [8–12]. The enrichment of lightning research results has directly promoted the improvement of lightning defense technology. By studying the parameters of potential stability index [13, 14], epidemic stability index [15, 16], lifting index [17, 18], a good lightning prediction model has been established.

Lightning research can not be separated from the collection of lightning data in the 1990s, the United States successfully developed and launched a high spatial resolution, accurate positioning, high detection efficiency of lightning detection instrument optical transient detector, so far has accumulated nearly 20 years of satellite lightning observation data. The flourishing development of lightning observation technology on ground and space and the continuous accumulation of lightning observation data make it possible for us to study lightning activities more comprehensively and thoroughly. At present, the ground-based low-frequency lightning location network has basically covered the populated areas in China, and can realize real-time monitoring of ground lightning activities in most parts of the country. In order to collect lightning data better, in 20116, China launched a new generation of geostationary orbit Quantitative Remote Sensing Meteorological Satellite Fengyun 4, which is equipped with a high-speed lightning imager (500 frames per second), which can continuously and real-time monitor lightning activities in China and its surrounding areas. On the one hand, it will broaden the coverage of lightning data, ensure the temporal and spatial continuity of lightning data, facilitate us to monitor lightning activities in China and its surrounding areas more comprehensively, and also put forward higher and more urgent requirements for how to use lightning data more effectively.

Because the satellite transmission signal is a disordered data stream, a simple and practical satellite data pre-processing system is designed to facilitate lightning prediction researchers. The system can separate the lightning data well and save the lightning data in a simple TXT format. At the same time, in order to display the lightning distribution intuitively. This paper also designs a visual interface to display the lightning distribution at different times.

2 System Processing Flow

China's new generation of geostationary satellite Fengyun-4 (FY-4) lightning imager (LMI) [1, 2] is the first spaceborne lightning imager developed independently in China. It is facing tremendous challenges in instrument development and product generation algorithm. Launched in 2016, from 0:00 on May 8, 2018, users in China and the Asia-Pacific region can officially receive data from the "Fengyun IV" A satellite. FY-4 satellite lightning imager is a real-time observation of severe convective weather phenomena in geostationary orbit, which realizes lightning imaging observation and obtains lightning distribution map in the observation coverage area. Through real-time and continuous observation of lightning, it realizes the monitoring and tracking of severe convective weather and provides lightning disaster warning. It can provide the basis for lightning forecast and global climate change research, provide the necessary scientific data for the observation, prediction and research of severe convective weather phenomena in the mainland of China and its surrounding areas, and provide the scientific basis for global atmospheric circulation research [19–21].

Fengyun-4 satellite transmission signal is a new type of data. In order to convert satellite data into data that researchers can read and study, it is necessary to convert this data stream. This paper is based on MATLAB to achieve data conversion. The whole system processing flow is shown in Fig. 1.

As shown in Fig. 1, the satellite signals are first received and archived, and the archived data is stored in the database as NC files. In this paper, the design system is based on this format of data design, first of all, the system is based on MATLAB NC data file reading module. The input data is processed by the read-in module to separate the parameters, and then the TXT document is saved. According to this document, the interface module designed by this system can read the data and load the map. According to the map design, the coastline and provincial boundary are loaded, then the longitude and latitude of lightning density are determined. Then the lightning density is loaded according to TXT document, and the lightning density is displayed in the form of image.

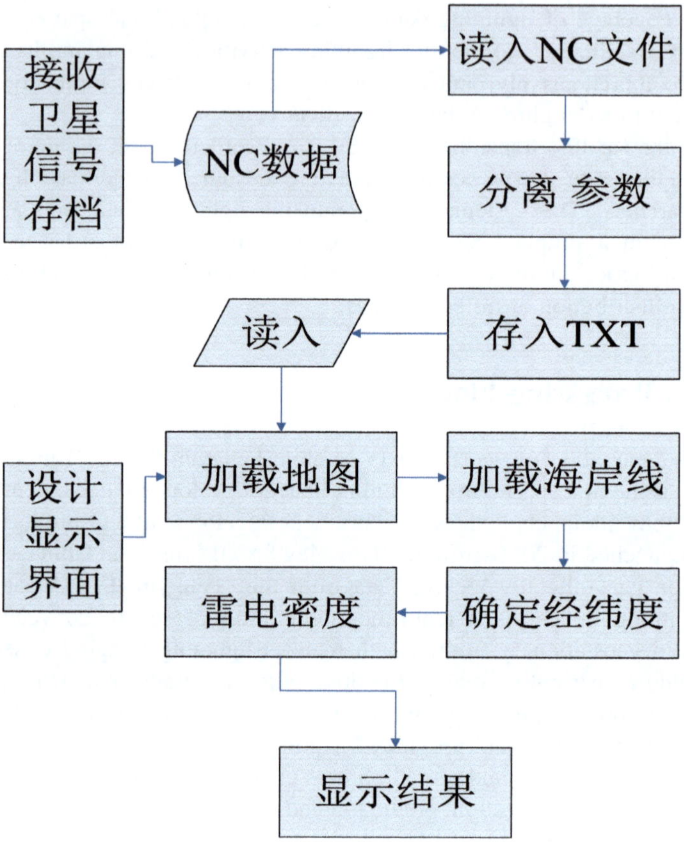

Fig. 1. System processing flow

3 NC Data Parameters and Preprocessing

This system processes NC format satellite signal documents, and the main NC file variable name variable description:

LON Event Longitude; LAT Event Latitude; ETT Event Time; EOT Event Observe Time;
ER Event Radiance; EFP Event FootPrint; EA Event Address; EGA Event Group Address;
EFA Event Flash Address; EXP Event X Pixel; EYP Event Y Pixel; DQF Lightning Event Data Quality Flag.

The data preprocessing steps are as follows:

Step 1: Determine the longitude and latitude range, if the user does not set, then the default longitude range is [70, 140], latitude range is [10, 60];

The second step is to generate TXT documents that need to be saved, and the default document name is sd8_txt + storage time string.

The third step: read the parameters in the NC file;

The fourth step is to write the read parameters into TXT documents, such as the time to read the document. MATLAB script is:

```
TT1=ncfilelist(ss).name(45:58);
                    TT2=ncfilelist(ss).name(82:83);
originalTime=[TT1(1:4),'-',TT1(5:6),'-',TT1(7:8),' ,TT1(9:10),':',TT1(11:12),':',TT1(13
:14)];
                    dt_Temp = datevec(originalTime);
                    dt_Temp(4) = dt_Temp(4) + 8; % plus 8 hours
                    dt_Temp(5) = dt_Temp(5) + str2num(TT2) - 1;
                    dt_Temp(6) = dt_Temp(6) + floor(ett(ii)/1000);
```

The final data saved to TXT is shown in Fig. 2.

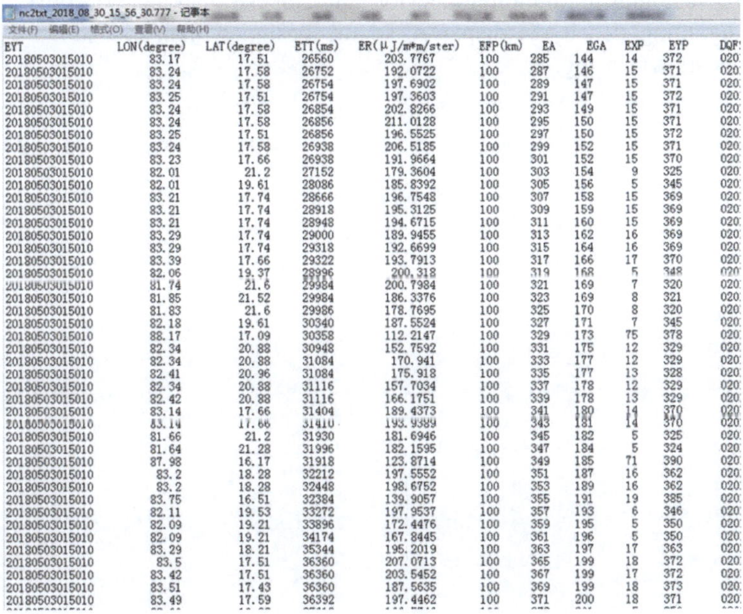

EYT	LON(degree)	LAT(degree)	ETT(ms)	ER(μJ/m*m/ster)	EFP(km)	EA	EGA	EXP	EYP	DQF
20180503015010	83.17	17.51	26560	203.7767	100	285	144	14	372	020
20180503015010	83.24	17.58	26752	192.0722	100	287	146	15	371	020
20180503015010	83.24	17.58	26754	197.6902	100	289	147	15	371	020
20180503015010	83.25	17.51	26754	197.3603	100	291	147	15	372	020
20180503015010	83.24	17.58	26854	202.8266	100	293	149	15	371	020
20180503015010	83.24	17.58	26856	211.0128	100	295	150	15	371	020
20180503015010	83.25	17.51	26856	196.5525	100	297	150	15	372	020
20180503015010	83.24	17.58	26938	206.5185	100	299	152	15	371	020
20180503015010	83.23	17.66	26938	191.9664	100	301	152	15	370	020
20180503015010	82.01	21.2	27152	179.3604	100	303	154	9	325	020
20180503015010	82.01	19.61	28086	185.8392	100	305	156	5	345	020
20180503015010	83.21	17.74	28666	196.7548	100	307	158	15	369	020
20180503015010	83.21	17.74	28918	195.3125	100	309	159	15	369	020
20180503015010	83.21	17.74	28948	194.6715	100	311	160	15	369	020
20180503015010	83.29	17.74	29000	189.9455	100	313	162	16	369	020
20180503015010	83.29	17.74	29318	192.6699	100	315	164	16	369	020
20180503015010	83.39	17.66	29322	193.7913	100	317	166	17	370	020
20180503015010	82.06	19.37	28996	200.318	100	319	168	5	348	020
20180503015010	81.74	21.6	29984	200.7984	100	321	169	7	320	020
20180503015010	81.85	21.52	29984	186.3376	100	323	169	8	321	020
20180503015010	81.83	21.6	29986	178.7695	100	325	170	8	320	020
20180503015010	82.18	19.61	30340	187.5524	100	327	171	7	345	020
20180503015010	88.17	17.09	30358	112.2147	100	329	173	75	378	020
20180503015010	82.34	20.88	30948	152.7592	100	331	175	12	329	020
20180503015010	82.34	20.88	31084	170.941	100	333	177	12	329	020
20180503015010	82.41	20.96	31084	175.918	100	335	177	13	328	020
20180503015010	82.34	20.88	31116	157.7034	100	337	178	12	329	020
20180503015010	82.42	20.88	31116	166.1751	100	339	178	13	329	020
20180503015010	83.14	17.66	31404	189.4373	100	341	180	14	370	020
20180503015010	83.14	17.66	31410	193.9389	100	343	181	14	370	020
20180503015010	81.66	21.2	31930	181.6946	100	345	182	5	325	020
20180503015010	81.64	21.28	31996	182.1595	100	347	184	5	324	020
20180503015010	87.98	16.17	31918	123.8714	100	349	185	71	390	020
20180503015010	83.2	18.28	32212	197.5552	100	351	187	16	362	020
20180503015010	83.2	18.28	32448	198.6752	100	353	189	16	362	020
20180503015010	83.75	16.51	32384	139.9057	100	355	191	19	385	020
20180503015010	82.11	19.53	33272	197.9537	100	357	193	6	346	020
20180503015010	82.09	19.21	33896	172.4476	100	359	195	5	350	020
20180503015010	82.09	19.21	34174	167.8445	100	361	196	5	350	020
20180503015010	83.29	18.21	35344	195.2019	100	363	197	17	363	020
20180503015010	83.5	17.51	36360	207.0713	100	365	199	18	372	020
20180503015010	83.42	17.51	36360	203.5452	100	367	199	17	372	020
20180503015010	83.51	17.43	36360	187.5635	100	369	199	18	373	020
20180503015010	83.49	17.59	36392	197.4462	100	371	200	18	371	020

Fig. 2. Preprocessing result TXT document data format

4 Data Read and Display

After pretreatment, the data is transformed into TXT format. This document contains the lightning data needed. From this data, the lightning data can be read and displayed. First, a rectangular frame is drawn according to the input rectangular frame, and the lightning density is displayed in the rectangular frame. Then the lightning density is calculated. The calculation method of lightning density of the system is as follows:

```
disp(' Please wait for the lightning density according to the latitude and longitude of your input.......');
midu=0;
for ii=1:length(rlon)
    if rlon(ii)>=sslon && rlon(ii)<=eelon && rlat(ii)>=sslat && rlat(ii)<=eelat
        midu=midu+1;
    end
end

R=6371;SS=(eelon-sslon)*2*pi*R/360 * (eelat-sslat)*2*pi*R/360;
set(handles.edit8,'String',midu/SS);
disp([' The time the program runs from the beginning to the present. (s):',num2str(toc)]);
disp([' ']);
```

After calculating the lightning density, the Chinese map and coastline are loaded and the lightning density is displayed in the designed image window. The initial interface is shown in Fig. 3.

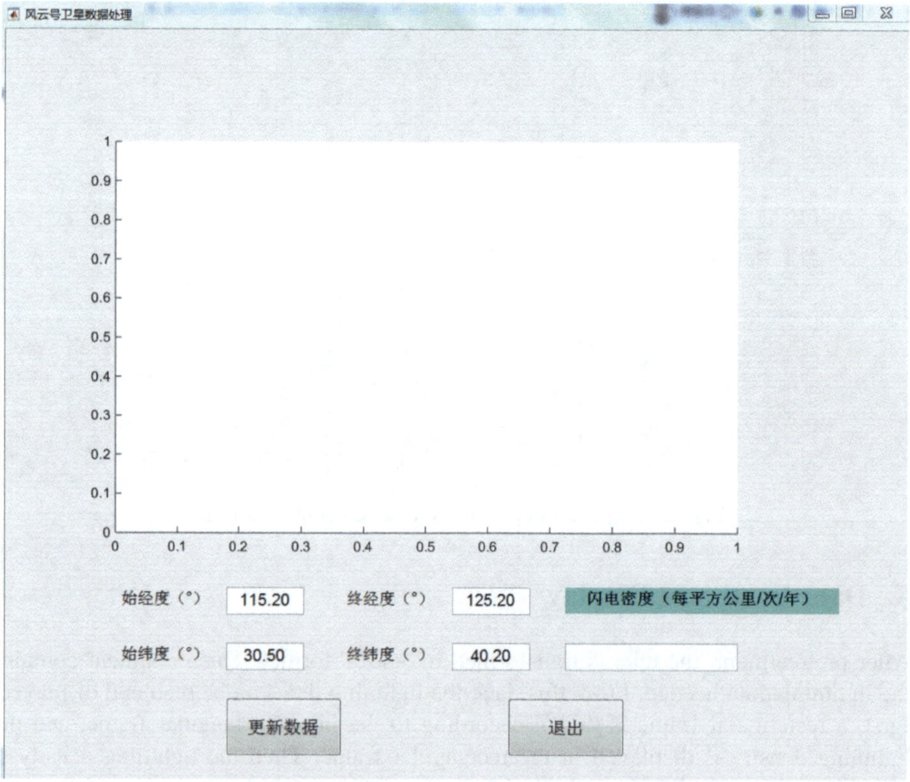

Fig. 3. Initial interface

As shown in Fig. 3, the lightning density data is displayed above the interface, and the lightning density related attributes are displayed in the lower part. The update data button is the loading TXT data section. When the button is clicked, the lightning density will appear, as shown in Fig. 4.

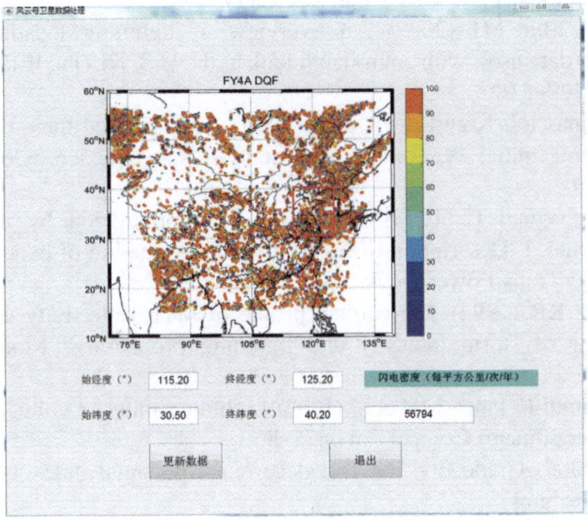

Fig. 4. After loading thunderbolt TXT data

5 Conclusion

The accuracy of lightning prediction needs to improve the accuracy of lightning detection satellite. However, with the updating of meteorological satellite, although it brings more accurate detection accuracy, it also brings difficulties for researchers to read the data. This paper mainly reads the data of Fengyun 4 satellite and displays it in the form of images. It is convenient for researchers to use.

Acknowledgments. This research was supported by Shandong Province Weather Bureau project number: (project number: 2017sdqxm09).

References

1. Price C, Rind D (1992) A simple lightning parameterization for calculating global lightning distributions. J Geophys Res Atmos 97(D9):9919–9933
2. Estberg GN, Norris S, Hutten KM (2009) Year-in-advance prediction of lightning activity for the pacific northwest. Northwest Sci 82(2):151–157
3. Christian HJ (2001) Global frequency and distribution of lightning as observed from space. In: AGU fall meeting. AGU Fall Meeting Abstracts, pp ACL 4-1–ACL 4-15
4. Christian HJ (2002) Global frequency and distribution of lightning as observed from space by the optical transient detector. J Geophys Res Atmos 108(D1):ACL 4-1–ACL 4-15

5. Cummins KL, Murphy MJ, Bardo EA et al (1998) A combined TOA/MDF technology upgrade of the U.S. National Lightning Detection Network. J Geophys Res Atmos 103 (D8):9035–9044
6. Sponsor (1997) IEEE Guide for Improving the Lightning Performance of Transmission Lines. IEEE, pp 1–44
7. Renkin RA, Despain DG (1992) Fuel moisture, forest type, and lightning-caused fire in Yellowstone National Park. Can J For Res 22(1):37–45
8. Cummins KL, Mur MJ (2009) An overview of lightning locating systems: history, techniques, and data uses, with an in-depth look at the U.S. NLDN. IEEE Trans Electromagn Compat 51(3):499–518
9. Rison W, Thomas RJ, Krehbiel PR et al (1999) A GPS-based three-dimensional lightning mapping system: initial observations in central New Mexico. Geophys Res Lett 26 (23):3573–3576
10. Dellera L, Garbagnati E (1990) Lightning stroke simulation by means of the leader progression model. I. Description of the model and evaluation of exposure of free-standing structures. IEEE Trans Power Deliv 5(4):2009–2022
11. Zipser EJ, Lutz KR (1994) The vertical profile of radar reflectivity of convective cells: a strong indicator of storm intensity and lightning probability? Mon Weather Rev 122 (8):1751–1759
12. Nucci CA, Rachidi F, Ianoz MV et al (1993) Lightning-induced voltages on overhead lines. IEEE Trans Electromagn Compat 35(1):75–86
13. Fengjiao W (2009) Grade division and occurrence potential index of lightning disasters. Meteorol Sci Technol
14. Sá JAS, Almeida AC, Rocha B Lightning Forecast Based on the Hourly Evolution of the Convective Available Potential Energy (CAPE). Icae.net.br
15. Acharya NV, Rao PSN (2013) A new voltage stability index based on the tangent vector of the power flow jacobian. In: Innovative smart grid technologies - Asia. IEEE, pp 1–6
16. Vallabhan SC (1995) Application of artificial neural networks for voltage stability evaluation of power systems. Memorial University of Newfoundland
17. Mazany RA, Businger S, Gutman SI et al (2002) A lightning prediction index that utilizes GPS integrated precipitable water vapor. Weather Forecast 17(5):1034–1047
18. Dementyeva SO, Ilin NV, Mareev EA (2015) Calculation of the lightning potential index and electric field in numerical weather prediction models. Izv Atmos Oceanic Phys 51 (2):186–192
19. Huang F, Guo J, Feng X (2009) Simulated calculation of false alarm rate and detection rate of lightning imager. J Photonics 38(12):74–78
20. Shulong B, Shaofan T, Li Y et al (2012) Real-time detection of transient point source multi-target lightning signals in geostationary orbit. Infrared Laser Eng 41(9):2390–2395
21. Cao D (2016) Research progress on monitoring principle and product algorithm of lightning imager for geostationary satellite FY-4. Progress in meteorological science and technology

Design of Wireless Sensor Network Node Based on CC2430

Jintao Yu[1(✉)], Liangliang Yu[2], and Cheng Wang[3]

[1] School of Computer and Information Engineering, Harbin University of Commerce, Harbin 150086, China
hityjt@sina.com

[2] Beijing Aerospace Automatic Control Institute, Beijing 100854, China

[3] Beijing Satellite Manufactory Co., Ltd., Beijing 100094, China

Abstract. Wireless sensor network is a wireless network composed of a group of sensor nodes in the form of self-organization. This paper's main control unit is CC2430. It's designed for wireless sensor network system, whose basic structure is main node, beacon node of Wireless Sensor Network and mobile robots. The serial communication between the master node and the PC and the wireless communication method between the nodes are emphatically designed. The experimental results prove the methods described in this paper are correct and valid.

Keywords: Wireless sensor network · Node · CC2430

1 Introduction

The goal of the Wireless Sensor Network is to collaboratively sense, collect, process, and transmit information that the network covers the perceived object within the geographic area, and ultimately send this information to the owner of the network [1–3]. The sensor network can collaboratively monitor, sense and collect information of environments in the network distribution area in real time, and process the data to obtain detailed and accurate information. Through the collaborative information processing of these information, the sensor network accurate information, and then sent for the users who need this information.

The architecture of a typical sensor network consists of a distributed sensor node (cluster), a receiver transmitter, Internet and a user interface [4], etc. Such as Fig. 1. The composition and functions of the sensor network node include the following four basic units: a power supply portion, a communication unit, a processing unit and a sensing unit. The nodes form a wireless network in a self-organized way to collect, detect and process the information and transmit the data through the Sink node link through the multi-hop network. This way can ensure real-time data processing and coordination with the node. As the same time, it's transferred to the Remote Control Manager Center [5–7]. On the contrary, management center can also be real-time control and operation of the network node.

Z. Xu et al. (Eds.): CSIA 2019, AISC 928, pp. 619–625, 2020.
https://doi.org/10.1007/978-3-030-15235-2_86

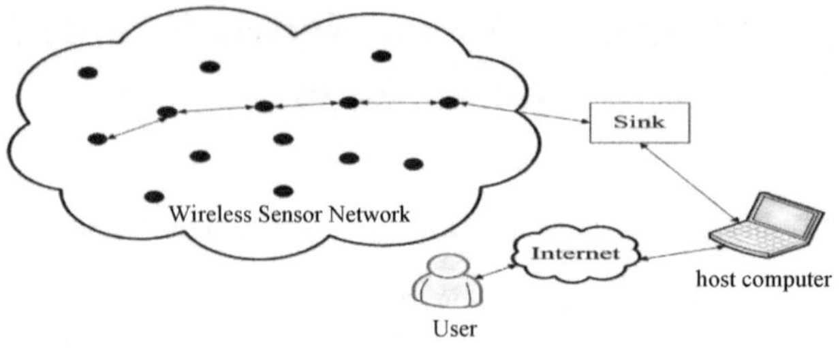

Fig. 1. The basic composition of wireless sensor network

2 CC2430 Chip Introduction

In this paper, CC2430 chip integrates wireless communication, data processing and signal acquisition, and is used as the main control unit of beacon nodes and mobile nodes, and is used to develop beacon nodes and mobile nodes with ultra-low power consumption and strong computing power and reduce the cost of single-node hardware. It's useful to prepare for the final practical.

This paper is use a 'ZIGBEE' wireless microcontroller CC2430, which is based on the 8051 microprocessor as the core of the wireless microcontroller, also known as "RF-on-chip system". It integrates IEEE802.15.4 compliant 2.4 GHz RF transceiver with excellent wireless reception sensitivity and strong anti-jamming capability. It also integrates 14-bit analog-to-digital Converted ADCs with two powerful USARTs supporting several sets of protocols. With low power consumption and high performance, it can well meet design requirements.

3 Node Structure Design

In this design, the nodes of the wireless sensor network system are mainly composed of the main node, the mobile node and the beacon node [8], such as Fig. 2.

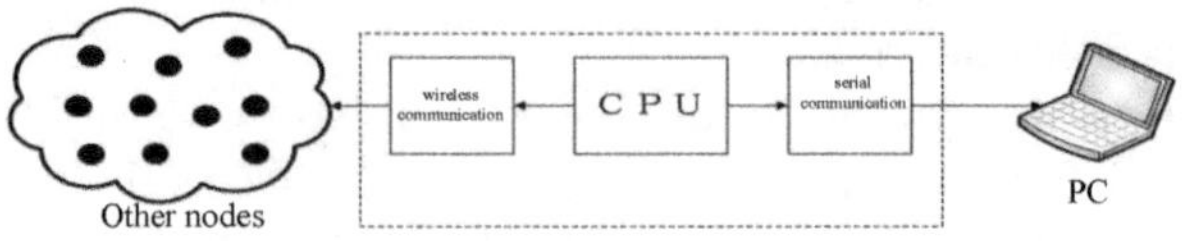

Fig. 2. Master node function diagram

A. Design of main node structure

The main function of the master node is to complete the beacon node data integration, as well as the beacon node, mobile node and PC to transfer information between each other. When the area of the wireless sensor is expanded, it will be divided into several small areas, and the main node in each small area will serve as a symbol of identifying the area and communicating with the master node of the next area [9, 10].

In this paper, the master node obtains the current direction and position of the robot in the wireless way and sends it to the PC through the serial port. It receives the instructions of the PC and sends it to other nodes in the wireless way.

The serial port can be used to realize the communication between the master node and the PC.

PC has a standard RS-232C serial port, the signal level is RS-232C level. RS-232C uses negative logic, logic "0" for the +5 V ~ +15 V and logic "1" for the −5 V ~ −15 V.

P0.2 (11) and P0.3 (9) of the single-chip microcomputer CC2430 are the serial data receiving end RxD and transmitting end TxD, the receiving and sending signals are TTL level. The logic "1" and the logic "0" of the TTL level are 3 V and 0 V, respectively.

So the microcontroller and PC serial communication, the need to use the chip MAX3232 RS-232C level and TTL level conversion, such as Fig. 3.

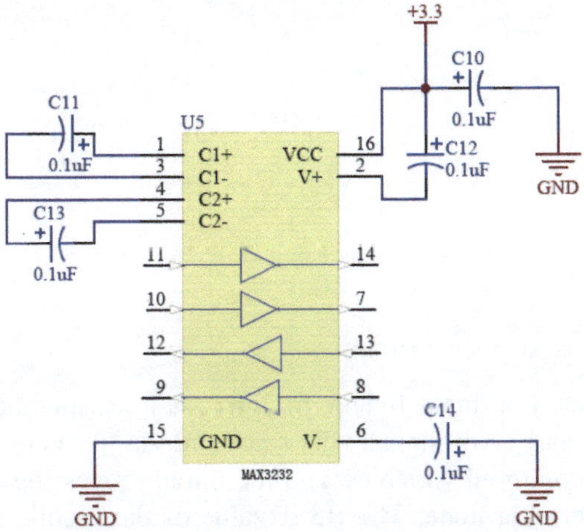

Fig. 3. MAX3232 application circuit

B. Mobile node structure design

The mobile node is a two-wheeled robot with two drive wheels and a support wheel, such as Fig. 4. Driving wheel through the DC motor drive can be implemented forward, backward turn and other posture movement. Support wheel in the rear, it can make 360° within the free rotation and can maintain the level of the body, without affecting the flexibility of movement. In addition to the modules required by other nodes, the mobile robot controls various movements of the robot. The course control module detects and controls the course of the robot. The navigation control module guides the robot to move to the target node, structure diagram such as Fig. 5.

Fig. 4. Mobile robot physical picture

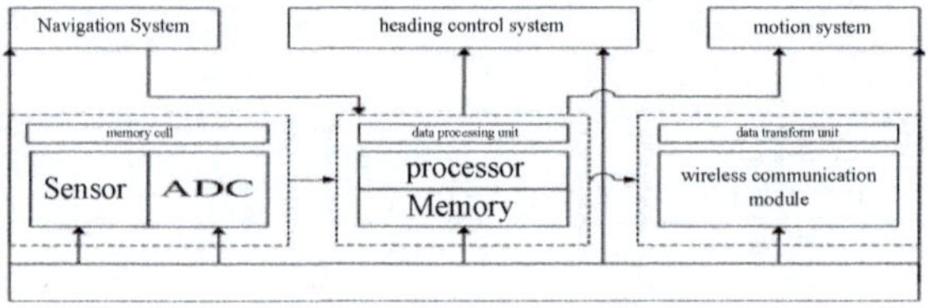

Fig. 5. Mobile robot structure diagram

C. Design of beacon node structure

The beacon node has two main functions. First, as a sensor node, connected to the sensor measuring a variety of signals, the sensor needs to monitor the required monitoring area of the monitored variables, and the initial processing, and then sent to the wireless node to the main node; The RSSI value of the mobile node is measured to determine the current location of the mobile node and to navigate the mobile node.

Since the beacon nodes are basically the same in structure, each node is identified by a different identifier, that is, an address, called a node address or a node ID. When the beacon node communicates with other nodes, the other nodes determine from which node the data comes from.

CC2430 wireless microcontroller integrates all the 802.15.4 short-range wireless communication standard required high-frequency circuit part, CC2430 receiver is based on low-IF structure on the RF signal received from the antenna by the low-noise amplifier and amplified by the Down conversion to 2 MHz of the IF signal. IF signal by filtering, amplification, through the A/D converter into a digital signal. Demodulation, channel filtering, automatic gain control in the digital domain to complete in order to obtain high accuracy and space utilization, such as Fig. 6. The integrated analog channel filter allows good coexistence of different systems operating in the 2.4 GHz

ISM band. In transmit mode, bitmapping and modulation are performed according to the IEEE 802.15.4 specification. Modulation is accomplished digitally.

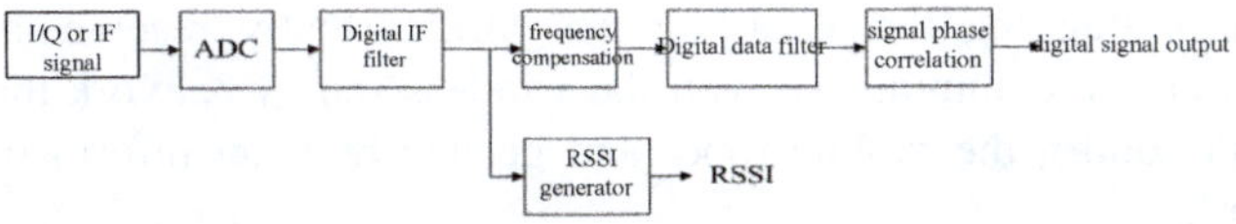

Fig. 6. CC240 wireless receiver

4 Experiment and Result Analysis

Serial communication is very important, because serial port can be wireless sensor networks and man-machine interface linked together, the user can observe the PC through the wireless sensor network is working properly or not. Serial test is divided into three stages,

1. By CC2430 through the serial port to send a string of fixed data, through the serial debugging assistant to see if there is data, and data is normal.
2. By the serial debugging assistant to send a string of data received by the CC2430 and then issued by the serial port, and then by the serial debugging assistant to observe whether the data is just the same with the data issued, such as Fig. 7.

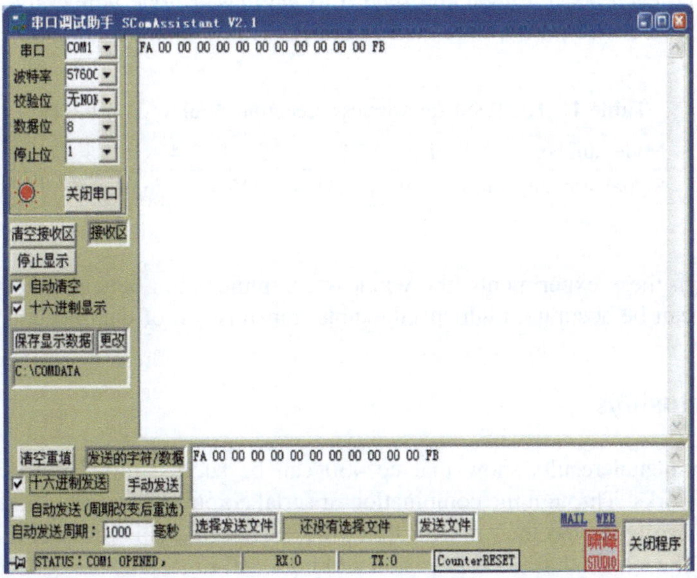

Fig. 7. Serial assistant serial debugging diagram

3. Through the PC data management system and the main node serial port communication, for sending control instructions and receive RSSI information of each node.

After these debugging, the serial port can work normally, receive and send data all normal, the user can communicate with the wireless sensor network through PC. The users also can control the mobile robot and get the relevant information of wireless sensor network.

The results of wireless communication is not easy to direct observation, so need to communicate through the above-mentioned serial port to help observe the results, the specific steps are as follows:

1. The wireless communication module has one red-green LED, one for sending indication (red) and the other for receiving indication (green). When the wireless module does not work, the two LEDs are in the off state; when the wireless module sends out a string of data, the red light flashes once; when the wireless module receives a string of data, the green light flashes once.
2. Using two modules, one (node 1) connected with the PC through the serial port, another (node 2) with its wireless communications. PC sends a serial data to Node 1 using the serial assistant. Node 1 sends the data to Node 2 wirelessly. After Node 2 receives the data, it sends the data back to Node 1 wirelessly. Sent to PC, the PC through the serial port assistant can observe whether the data is just the same with the data sent, if the same is to prove that the success of wireless transceiver.
3. Each beacon node and the robot wirelessly connect to obtain the RSSI of the mobile node and each beacon node and send it to the master node wirelessly. And finally through the serial port will be sent to the PC RSSI, such as Table 1.

Table 1. The RSSI for wireless reception displayed on the PC

Node number	1	2	3	4	5
Signal strength (dbm)	−49.23	−31.04	−35.97	−76.03	−43.99

Through these experiments, the wireless communication between the nodes work normally, can be accurate, high-speed, stable transmission of data needed to send.

5 Conclusions

The experimental results show that cc2430 can be successfully applied to wireless sensor networks. Through the combination of serial communication and indicator light, it is successfully solved the problem t that the wireless node is difficult to measure in real time. The system has good precision and high real-time, and can collect the data of every child node steadily. The robot can rely on the system designed in this paper for accurate displacement and pose adjustment.

Acknowledgment. This work was supported in part by the Program of Research and Development of Applied Technology of Heilongjiang Province (China) (Grants no. GC13A310) and by the Program of Research and development of Applied Technology of Harbin (China) (Grants no. 2013RFLXJ020).

References

1. Sibley GT, Rahimi MH, Sukhatme Robomote GS (2002) A tiny mobile robot platform for large-scale ad-hoc sensor networks. In: Proceedings of the 2002 IEEE international conference on robotics and automation, Washington D.C., USA, May 2002, pp 1143–1148
2. McMickell MB, Goodwine B, Montestruque LA (2003) MICAbot. A robotic platform for large-scale distributed robotics. In: Proceedings of the IEEE international conference on robotics and automation, Taipei, Taiwan, September 2003, pp 1600–1605
3. Bergbreiter S, Pister KSJ (2003) CotsBots. An off-the-shelf platform for distributed robotics. In: Proceedings of the IEEE/RSJ, international conference on intelligent robots and systems, Las Vegas, Nevada, USA, October 2003, pp 1632–1637
4. Akyildiz I, Su W, Sanakarasubramaniam Y et al (2002) Wireless sensor networks: a survey. Comput Netw 38(4):393–422
5. Zhao H, Zhao J, Liu Z, Ji S (2009) Design and implementation of a wireless sensor network node. J Northeast Univ 30(6):809–812
6. Qin L, Hu R, Li Z (2009) Research on node energy saving in wireless sensor networks. J Wuhan Univ Technol 31(3):369–371
7. Chen J, Li CG, Ning X (2009) Node location of wireless sensor networks based on mobile. Chin J Sens Actuators 22(1):121–125
8. Yick J, Mukherjee B, Ghosal D (2008) Wireless sensor network survey. Comput Netw 52 (12):349–350
9. Fakcharoenphol J, Laekhanukit B, Nanongkai D (2014) Faster algorithms for semi-matching problems. ACM Trans Algorithms (TALG) 10(3):1450–1455
10. Levin L, Segal M, Shpungin H (2013) Cooperative data collection in ad hoc networks. Wirel Netw 19(2):1043–1048

A Preliminary Study on the Introduction and Disposal of Pollutants from Hazardous Waste Transfer and Single Billing System

Dongfang Zhang(✉) and Songming Wang

Jiangsu Maritime Institute, Nanjing, China
zhangdf8103@163.com

Abstract. The construction, repair and dismantling of ships is a complete industrial chain. With the gradual increase of world trade volume, the demand for ships in the world is gradually growing. The increase in the demand for the number of ships means that the number of ships that need to be repaired and the number of ships that need to be dismantled are also increased. In the process of construction, repair and dismantling of ships, inevitably, pollutants are generated. Behind the development of the shipbuilding industry, hidden dangers of huge pollutants prevention and control. At the same time, with the continuous development of the shipping industry, the repair and dismantling of ships is becoming more and more frequent. In the process of ship repair or scrapping, it is likely that a fuel leakage accident will occur near the shipyard. Therefore, the provision of oil spill emergency equipment is also an effective way to prevent the pollution caused by shipyard pollutants. Based on the practical experience of China's hazardous waste transfer and single-stock system, China can refer to the establishment and improvement of the ship pollutants receiving and disposal joint supervision system to fill the regulatory loopholes, realize the closed-loop management of ship pollutants, prevent and mitigate environmental pollution.

Keywords: Ship pollutants · Joint order system · Pollution prevention

1 Introduction

As humans have become more aware of environmental protection, many environmental pollution problems such as smog and water pollution have caused widespread social concern. There are also many reasons for environmental pollution, mainly industrial production pollution, agricultural production pollution, and transportation pollution. How to alleviate and control environmental pollution has become a common concern of the government, non-governmental organizations, the media and the public [1]. The situation of the ship's polluted water environment is very serious. How to effectively fulfill the duties of pollution control of the competent authorities has become a major issue for the sustainable development of the port economy. This article describes the background, purpose and significance of the study, summarized the study of relevant scholars at home and abroad, then ship pollution prevention management from the

Z. Xu et al. (Eds.): CSIA 2019, AISC 928, pp. 626–632, 2020.
https://doi.org/10.1007/978-3-030-15235-2_87

status quo in port, ship pollution introduce the concepts and types of hazards, including oil pollution, noxious liquid The effects of material pollution, domestic sewage and domestic garbage pollution, tail gas pollution, solid waste pollution, etc. on water, atmosphere, personal health and social and economic development [2, 3]. Summarizes the effective measures and experience in the Marine Pollution Emergency management in taken further found port ship pollution prevention and control issues [4, 5]. According to the problem, the solution strategy is proposed to realize the closed-loop management of the ship's pollutants, and prevent and mitigate environmental pollution.

2 The Core Concept of the Port of Pollution from Ships

Ship pollution mainly refers to the irreversible pollution of the marine environment caused by the leakage of fuel and dangerous goods carried by the ship due to objective or human factors in the whole process of construction, operation and scrapping [6]. From the point of view of pollution categories, ship pollution is very diverse, such as domestic garbage and domestic sewage pollution, oil pollution, ship tail gas pollution, ship-borne solid waste pollution, and hazardous substances carried by ships.

2.1 Ship Oil Pollution

For the oil pollution of ships, there are two major categories, one is the oil spill pollution of the ship; the other is the oily sewage from the ship. From the point of view of the hazard of oil spills from ships, it is mostly in the port economic development and marine ecological environment hazards [6, 7]. After carrying ship petroleum and its derivatives by a fuel leak into the sea, sea oil film is formed, separated in the oxygen exchange between the air and the water, at the same time weaken the sun, hindered photosynthesis aquatic plants, resulting in The water body is severely deprived of oxygen, causing fish and other organisms to die due to lack of oxygen, which in turn causes a series of bio-chain reactions. Relevant reports indicate that since the 1970s, nearly 1,000 species of extinct marine species have been extinct due to oil pollution, and marine life has fallen by nearly 40%. Marine products cultured in coastal areas will be greatly reduced due to oil pollution, and contaminated seafood may cause a public food crisis after entering the market [8]. The impact of oil spills on ships is also very large. Due to the need to clean up polluted waters, it will inevitably affect the frequency of entry and exit of ships and bring greater economic losses to port production.

2.2 Ship Domestic Garbage and Domestic Sewage Pollution

The pollution of domestic garbage of ships is mainly the domestic sewage and domestic garbage generated by the life of the crew in the daily operation of the ship [9, 10]. Ship domestic sewage refers to all kinds of waste and sewage, which are mostly composed of sewage discharged from toilets, urinals, medical offices and ward baths [10]. The domestic sewage of the ship is rich in oxygen-consuming organic matter, and also carries various pathogenic microorganisms and parasites. When the dissolved oxygen in the seawater is sufficient, these organic substances are oxidized to form CO, NO, etc., so that the water body is deprived of oxygen. When the seawater is deprived of oxygen, these organic substances undergo anaerobic reaction to produce organic acids and gases such as CH4, H2S, NH3, etc., resulting in a decrease in water quality, which in turn leads to serious death of marine organisms [11]. Ship garbage refers to work supplies, daily necessities, food wastes that are handled during the operation of the ship during operation, mostly food waste and plastic waste and other garbage, mainly padded materials, packaging materials, etc., crew, passengers Food residues formed in daily life, waste from daily consumer goods, etc.

2.3 Ship Exhaust Pollution and Its Hazards

Ship exhaust emissions mainly include NOx, SOx and other harmful substances. The relevant theoretical estimation results show that if the power of a low-speed diesel engine is equivalent to 5000 kw and it runs for 5000 hours per year, the atmospheric pollution products emitted will be about 500 tons, and there will be more than 1,000 tons of remaining combustion products [12, 13]. At present, the total diesel power of the world reaches hundreds of millions of kilowatts, and the emissions correspond to hundreds of millions of tons. In addition, the refrigerant used in air-conditioning equipment on board is mainly based on Freon, and the destruction of the ozone layer by Freon, which is discharged into the atmosphere every year due to equipment, is also worrying. The photosynthesis of NOx and HC exists in the exhaust emissions of the main engine of the ship. After the light, the photochemical smog will appear in the lubricating oil, and the stimulating effect is strong [14]. It is prone to crop death, sore throat, red eyes and other phenomena, and is a carcinogen [15]. Nitrogen oxides (NOx) and sulfur oxides (SOx) emitted by ships exceeding the standard will form acid rain in the atmosphere, endangering sea surface or ground organisms and crops, and endangering marine life.

3 Transfer of Hazardous Waste Manifest System's Limitations

3.1 Laws and Regulations Are not Perfect, Law Enforcement Lacks Legal Grasp

The joint order system is an innovative measure, and there are no supporting laws and regulations to be revised [16]. When the maritime and environmental protection authorities promote the system, the law is insufficient. In the implementation, it mainly relies on visiting and exposing the means of publicity to enhance the sense of responsibility of the terminals and pollutant receiving units. The units that do not cooperate with the implementation lack legal basis in law enforcement.

3.2 It Is Difficult to Fundamentally Solve the Problem of Ship Pollutant Receiving and Disposal

The joint order system has effectively supplemented the "absence" of the past supervision, but it cannot fundamentally solve the problem of insufficient reception capacity of pollutants in the national port [16]. In order to meet the pollutant treatment needs of ships arriving in Hong Kong, it is necessary to supervise the port terminal to implement the main responsibility of the anti-pollution and equip the corresponding pollutant receiving facilities and equipment. According to the "Regulations on the Prevention and Control of Marine Pollution in Inland Waters" and the "Regulations on the Prevention and Control of Marine Environment Prevention and Control of Ships and Their Working Activities in the People's Republic of China", the port shall be equipped with equipment for receiving pollutants from ships [17]. As most of the ports are located in more remote urban areas, the investment in pipeline network transformation is large [18]. Some local governments and port departments have difficulties in promoting the construction of port pollutants receiving capacity, and the construction of terminal pollutants receiving and disposing capacity is slow.

3.3 The Management Object Is Limited to Hazardous Waste, Washing Water, Still Need to Be Further Improved

The implementation of the joint order system is only for residual oil, oily sewage and hydrocarbon water mixture. Other ship pollutants such as ship garbage and domestic sewage are not included in the management of the joint order, and the linkage with the relevant competent authorities needs to be further strengthened. The maritime, environmental protection, port, urban management, housing construction and other competent authorities have not unified the responsibility for the supervision of ship pollutants. The law enforcement borders of various departments are vague, and environmental protection, maritime affairs, ports, housing construction, and urban management are unified nationwide [19]. It is difficult to coordinate the linkage supervision mechanism.

3.4 The Pressure of Supervision Is Huge

According to the statistics from the US Environmental Protection Agency, the annual working time for running the system in the United States is about 4.3 million hours, the working cost is about 6 million US dollars, and the preparation of the paper joint order takes about 40% of the working time, processing and statistics. 70% of the work is required. According to the "Annual Report on the Prevention and Control of Environmental Pollution by Solid Wastes in Large and Medium Cities in 2017", the national industrial hazardous waste production amount is 33.446 million tons. The huge amount of production of hazardous waste relies only on traditional paper-linked supervision, and the work efficiency is low. The pressure to prevent environmental risks of hazardous waste is enormous.

4 The Receiving Ship Pollutants Recommended Disposition of the Single Management and Reference

4.1 US New York - New Jersey Green Port Management Experience

The US-New York-New Jersey port is located in the eastern part of the United States and is connected to domestic and global consumption. The port provides a large number of jobs for the local area, about 200,000. For residents of New Jersey and New York, the Hong Kong side has continuously invested capital, mainly in the improvement of facilities and port expansion, combined with a large number of environmental protection. The implementation of the measures has realized the burden and pressure of residents and environmentalists on green port policies. At this stage, the port has achieved good results.

For the US management mechanism, whether it is the operation or management of the port, the jurisdiction is related to the state government, not the federal government. The port authority created by the state government marks its regulation of the federal government's use of port assets, as well as the construction of port coastline resources. As the leading department of the Port Authority, the Management Committee is mainly responsible for port operations, employment promotion, economic development and taxation. The New York-New Jersey Port Authority used the port environmental management system to achieve the construction of a green port. From the perspective of the port operator, the treatment of port sewage through the sewage treatment system has strengthened the utilization rate of renewable equipment to a certain extent, and the equipment that will bring serious environmental pollution is gradually replaced by modern equipment, and the harmful gas emissions are eliminated. It has been reduced and the results have been obvious. From the point of view of ship ballast water discharge, the Port Authority and the New Jersey Ocean Fund have jointly handed out books on ballast water knowledge and proposed that the ship should be replaced in the open sea before arriving at the port. Species prevent the entry of freshwater waters within the port.

4.2 Improve Information Technology

From the perspective of the regulatory authorities, the hazardous waste transfer joints only involve environmental protection. The ship pollutants receiving and disposing bills cover environmental protection, port and shipping, maritime, sanitation, water, etc.; from the perspective of pollutant types, hazardous waste transfer Only the hazardous wastes in the National Hazardous Waste List are involved. The ship pollutants receiving and disposing bills include hazardous waste, waste water, domestic waste and industrial waste. The regulatory process for the receipt and disposal of ship pollutants is more complicated than the transfer of hazardous waste. The operation process of the paper transfer single system is very complicated, and the manual information required for the joint order is huge, which cannot meet the needs of current hazardous waste management. Considering the complexity of ship pollutants receiving and disposing of joint orders, it is recommended that the ship pollutants receiving and disposing joint order should focus on establishing a national ship pollutant receiving and disposing joint system at the beginning of the system design, and the relevant regulatory authorities can grasp the ship pollutants. The specific conditions of receiving, transshipment and disposal will comprehensively improve the supervision capability of ship pollutants and improve the efficiency of supervision.

4.3 Strengthen the Anti-pollution Infrastructure

First, actively strive for financial security. It is suggested that the relevant government departments actively strive for the central government's capital project budget in the annual budget, establish a national and local pollution emergency special fund, and at the same time clarify the local government's main position in the sudden oil spill emergency, and actively seek local government environmental protection fund investment. In addition, it is necessary to strengthen the social responsibility of water pollution prevention of water terminals and water transport units, and promote the collective purchase of shared reserve materials.

The second is to standardize the market-oriented operation of ship anti-pollution units. It is recommended to establish one or more comprehensive and professional anti-pollution companies, adopt market-oriented operation methods, actively strive for national policy support and financial subsidies, and extensively absorb the participation of port and shipping units, shipping companies and the public, and maximize the use of the market. Economic means promote the construction of pollution prevention infrastructure and standardize management. Give full play to its important role.

5 Conclusion

Through the data and management status of ship anti-pollution supervision in recent years, the problems and causes of port ship pollution prevention and control are analyzed. Focusing on these issues, this paper proposes targeted countermeasures through in-depth research, applying sustainable development theory, externality theory, government intervention theory, safety management theory, and combining domestic and international advanced port pollution prevention and control experience. In order to better protect the port environment and prevent ships from polluting waters and air environment, relevant government departments have increased the importance attached to ship pollution prevention management, extensively carried out legal education, and enhanced environmental awareness and legal awareness among crew members and water workers. Give full play to the supervision role of the media and the masses in the society, and strive for support and investment from all walks of life in the environmental protection of port waters in various aspects, actively guide relevant departments and enterprises to strengthen environmental protection work, and continuously enrich the anti-pollution infrastructure of the jurisdiction from the pollution emergency response system. Professional emergency response team building.

References

1. Chen DH, He J-J, Zhang GH et al (2014) A preliminary study on the characteristics of fine particle pollution characteristics for different weather type in the Guangdong atmospheric supersite. Geochimica 43(3):217–223
2. Zhao DJ, Tan C, Peng XC et al (2015) Preliminary study on the pollution of rainwater in the initial stage of Shijing river Basin in Guangzhou. J Guangdong Techn Coll Water Resour Electric Eng
3. Xia G, Liao R (2017) A preliminary study on the course teaching reform of the solid waste treatment and disposal from engineering education professional certification. Sci Technol Inf
4. Greco E, Ciliberto E, Cirino AME et al (2016) A new preparation of doped photocatalytic TiO 2, anatase nanoparticles: a preliminary study for the removal of pollutants in confined museum areas. Appl Phys A 122(5):1–6
5. Peng X, Zheng H, Zhao D et al (2015) Preliminary study on the source pollution of rainwater at the initial stage of Guangzhou. Guangdong Water Resour Hydropower
6. Hajizadeh Y, Mokhtari M, Faraji M et al (2017) Trends of BTEX in the central urban area of Iran: a preliminary study of photochemical ozone pollution and health risk assessment. Atmos Pollut Res 9(2):220–229
7. Zhang L (2014) A preliminary study on mercury pollution in fish and shellfish from Jiaozhou Bay in North of China. Adv Mater Res 864–867:234–238
8. Shi Y (2017) A preliminary study on the community intervention methods for the victims assistance of water pollution in China rural areas from the perspective of system theory. Innov Sci Technol
9. Liang C, Liang X, University F (2016) A preliminary study on the legal issues concerning the criminalization of seafarers in marine pollution accidents—from the perspective of the laws of EU, some of its members, and that of the US as well as some related international conventions. Chin J Marit Law
10. Meng J, Xiang Y, Ren Y (2018) Preliminary study on the heavy metal pollution sources and its effects on health in China. Energy Energy Conserv
11. Honghui LI, Zhao S, Jia M, et al (2016) Preliminary study on FEPs selection and scenario development of HLW geological disposal. World Nuclear Geosci
12. Shi J, Deng H, Wang M (2016) Preliminary studies on the pollution levels of antibiotic resistance genes in lower reaches of the Yangtze river. J Comput Theor Nanosci 13(9):5971–5974
13. Jiang Y, Liu J, Bi P (2016) Preliminary study on pollution determination of aluminum in food and food sources of aluminum. Modern Food
14. Cong B (2016) A preliminary study on the forensic appraisal of injury caused by environmental pollution. Chin J Forensic Sci
15. Zhao Q (2014) Preliminary study on the control effect of four kinds of pollution-free agricultural chemicals on powdery mildew of pumpkin. Heilongjiang Agric Sci
16. Cao YQ (2015) A preliminary study on drug self disposal mode training combined with telephone follow-up regularly in preventing the relapse of schizophrenia. J Qiqihar Univ Med
17. Shi-Yan HE, Zhang XS, Gao SB et al (2017) Preliminary study on the biological characteristics and non-pollution control of Sacada SP. Guizhou Forestry Sci Technol
18. Jiang G, Yuan M, Li M et al (2017) A preliminary study on rapid source appropriate of fine particles pollution in dust weather. J Green Sci Technol
19. Jun WU, Zheng J, Tao F et al (2018) Preliminary study on the theory in the treatment of diseases related to environmental pollution from turbid toxic material. J Gansu Univ Chin Med

Research on the Integration of Marketing Management and Big Data Technology

Dian Jia(✉)

School of Business Administration, Qinghai Nationalities University, Xining 810007, Qinghai, People's Republic of China
jiatianabc@126.com

Abstract. At present, with the development of big data, the reform and development of enterprises are facing new situations and tasks. The mode of transaction will be upgraded gradually, and there will be many flexible and independent trading modes including product + Internet + credit + futures + retail + wholesale. As a result, enterprises are facing increasing pressures in exploiting product markets and preventing business risks. At the same time, they are facing severe challenges such as the reduction of high-quality customers, the decline of market share and the loss of high-quality talents. Especially, they have put forward higher requirements and new challenges for marketing services. For enterprises, it is necessary to use big data technology to build a technical support platform for marketing services, and to analyze the characteristics and needs of marketing customers' products, it can provide data support for the formulation of product marketing programs, and provide important data support for enterprises to seize the market and users, so as to improve the economic efficiency of enterprises, and constantly improve the level of customer service and service quality. Based on this, the research of this topic has very important practical significance.

Keywords: Marketing management · Data technology · Integration innovation

1 Brief Introduction

In the era of big data, it is a revolution of science and technology, and also a social revolution and cognitive revolution [1]. At present, big data is increasingly exerting an important and far-reaching impact on global production, circulation, distribution, consumption activities, economic operation mechanism, social life style and national governance capacity. How to break through the key technology of big data, how to use big data to promote economic development and improve social governance, and how to promote the development of big data while ensuring information security, have become a hot issue of universal concern in all countries and all walks of life [2]. Enterprise marketing management includes project management, material management, report management, contract management, warehouse management, plan management, market information management, marketing management, system management, settlement management, account management, etc. The development and application of big data technology also brings certain challenges to enterprise marketing management mode.

Z. Xu et al. (Eds.): CSIA 2019, AISC 928, pp. 633–639, 2020.
https://doi.org/10.1007/978-3-030-15235-2_88

With the advent of the era of big data, competition among enterprises has begun to turn to the Internet platform, and commodity data is growing with the continuous expansion of the sales scope, which puts forward higher requirements for the comprehensive management of enterprise business [3, 4]. This paper takes warehouse management and marketing management as examples to analyze the application of big data technology in enterprise marketing management.

2 Big Data Technology Applications

In the past few years, information technology has been infiltrating into different fields such as society, economy and life, bringing forth new ideas, and producing many new technologies. For example, mobile information technology, Internet of Things technology, cloud computing and so on, with the support of a series of emerging technologies, social media began to play a role, resulting in a variety of virtual services, making the use of information more and more profound, and expanding the scope of the use of information [5, 6]. The emergence of new technologies and application modes has led to a sharp increase in global data volume. Computing power is the core of the computer age, which is mainly reflected in the processing of large amounts of data. In addition, in the process of processing, it is necessary to dig and analyze the information data in depth, so as to obtain more valuable information, and apply this information to business activities, resulting in a new business model. In the era of big data, great changes have taken place in the marketing mode of enterprises. Simply speaking, big data refers to the sum of all kinds of information data. It refers to those data that exceed the processing capacity of traditional databases. Usually, the amount of data is more than 10 TB. In the process of Internet popularization and rapid development, the social application of cloud computing technology and big data technology is more and more widely. In the process of increasing, decreasing and mining data, data is changeable. Enterprises or third-party service organizations can use these data information to provide corresponding services for the marketing of enterprises. For example, consulting, strategy and other services, also known as big data marketing. Data mining is the most important part in the process of big data marketing. There is a close relationship between data mining and enterprise marketing. And marketing is developed on the basis of data mining and database marketing [7]. The key of marketing is to find out the needs and preferences of customers and collect a large amount of customer information data. And build a customer-centered data warehouse, find valuable information from a large number of information data, and analyze customers' consumption needs, consumption behavior, psychological characteristics, product preferences, service requirements, so as to provide targeted services and products to different customers, which is a marketing skill with high investment and high returns.

3 The Background of Big Data and the Characteristics of Marketing Management

Open data capability is a virtual operation mode of large data capability platform [8]. Third parties rent platform space, computing and data resources, and carry out data analysis and mining applications to meet their business needs or provide business information services to their customers. To achieve the capability of big data, the following are analyzed.

3.1 Insight into Customer Needs and Enhance Customer Perception

Insight into customer characteristics, accurately grasp customer needs. Based on big data, we construct a multi-dimensional classification of high-definition customer portraits, insight into customer consumption behavior and demand preferences, carry out customer-centered precision, efficient package promotion, service maintenance, flow promotion, targeted marketing and other activities to meet customer needs and enhance customer perception [9, 10]. And reserve foreign cooperation data value diversification, diversification of service delivery capabilities, such as large data platform based on the complete portrait of customers, one-to-one personalized advertising for customers, through text messaging and other means to send catering, entertainment, shopping and other life information recommended to customers.

3.2 Focus on Big Data and Carry Out Precision Marketing

Change the past extensive marketing and maintenance policy formulation and implementation, accurately identify, precise delivery, real-time tracking, circular optimization, to achieve the wisdom of marketing maintenance, precise resource delivery and experience case sharing [11, 12]. Strict pre precision planning, precise control in the matter, optimization after the cycle.

3.3 To Meet the Integrated Operation, Help Service Sink

To build a leading, standardized, top-down integrated precision marketing maintenance system, while supporting provincial and local network innovative marketing and maintenance needs, to help service sink. Maintaining integrated operation capability of marketing includes standardizing customer label system and accurately discerning customer needs; strengthening the landing of operation and maintenance system to ensure integrative and intelligent operation; unifying customer contact standards and establishing enterprise service image; sharing whole-network marketing cases, replicating and promoting successful experience. Intensive farming and meticulous farming services sink, including supporting provincial, prefectural and multi-professional personalized labels; supporting provincial, prefectural and multi-professional personalized templates; supporting provincial, prefectural and multi-professional personalized process configuration; supporting provincial and municipal implementation channels docking, such as call center, short message, business hall, grid manager, etc.

4 Current Problems in Marketing Management of Chinese Enterprises

The restrictive factors in the traditional marketing process of enterprises have always been an important part of the development process of enterprises. Efficient marketing can improve the economic efficiency of enterprises and promote the sustainable development of enterprises. In the traditional mode of enterprise marketing, the restrictive factors are mainly reflected in the marketing environment, such as the marketing environment expanding through the market content, and the changing factors of the enterprise itself, which has a great impact on the marketing process of the enterprise. Enterprises must also rely on the market environment when developing marketing, so they are limited. Specifically, the main part of the traditional marketing problems of enterprises is illustrated in Fig. 1.

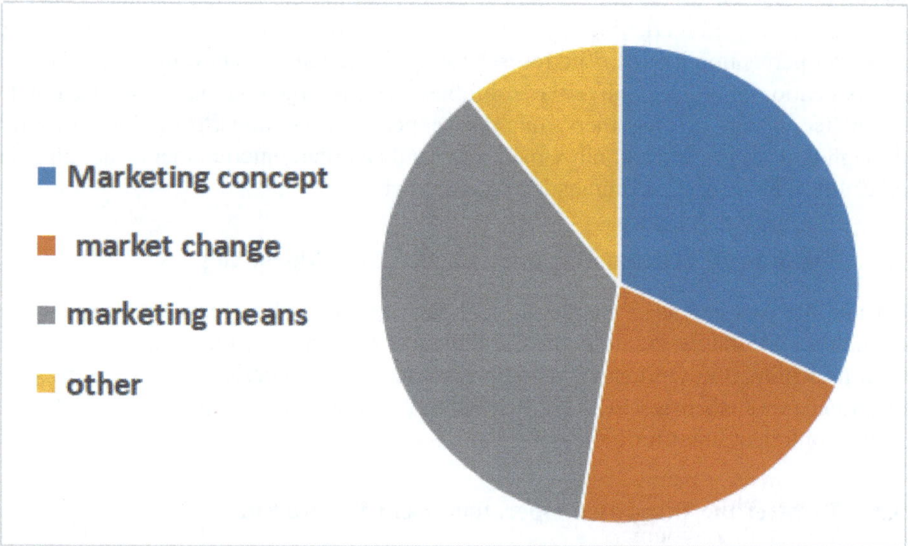

Fig. 1. The proportion of traditional marketing problems in enterprises

4.1 Marketing Concept Is Obsolete

In the traditional marketing process, many enterprises focus on advertising. With the advent of the era of big data, the disadvantages of traditional promotional activities are gradually highlighted [13]. For example, the lack of instant communication among users in various promotional activities leads to the lack of thorough understanding of telecommunications business products and services, which makes users reluctant to easily try new products and services [14]. In addition, because people's lifestyle has changed, so the traditional marketing methods are ineffective, people spend more and more time using the network, and the traditional marketing methods will inevitably reduce the efficiency of communication between enterprises and customers.

4.2 The Market Is Changing Rapidly

Research is an important part of enterprise marketing [15, 16]. Through research, we can grasp the basic market information and customer needs, so as to provide customers with more products and services that meet their psychological expectations [17, 18]. In the process of market research, the traditional marketing method needs a lot of manpower, financial and material resources, and the results are not satisfactory. Moreover, the current market environment is becoming more and more complex, and it is more difficult to carry out market research and analysis. Therefore, we must strengthen the innovation of market research model in the new era.

4.3 Single Marketing Means

Marketing means refer to the process of promoting products or services of enterprises to users. In the process of marketing, marketing means are relatively single. For example, marketing advertisements mainly stimulate audiences' hearing or vision through television, radio, newspapers, magazines and other forms, thus imposing various kinds of information on audiences.

This marketing method is often difficult to achieve good results, and many users do not have a deep impression when they receive information.

5 Enterprise Marketing Management and Big Data Technology Integration Innovation

5.1 Deep Integration of Big Data Within Enterprises

The internal big data of enterprises mainly include some business process information of enterprises, such as production plan information, inventory information, raw material procurement information, and information applied and created by daily activities such as activities news, work experience, exchange records and experience sharing of employees [19]. This part of information can be collectively referred to as knowledge and communication information of enterprises, which may contain significant potential value of enterprises. The application of big data technology in the two internal information data of an enterprise can integrate the business process of the enterprise, develop from inventory management system, material requirement planning (MRP) to enterprise resource planning (ERP), and integrate the dispersed business information into the enterprise system so as to realize information sharing and communication oriented to business process. On the basis of this, cross sectoral collaboration is used to achieve the optimization of business processes and the improvement of decision-making ability. After optimizing the process and decision-making ability, enterprises can improve their management efficiency, innovate their management mode, ultimately improve their competitiveness, and gain a place to survive in the new normal economy.

5.2 Deep Integration of External Big Data of Enterprises

The external data of enterprises also includes two dimensions. Firstly, the production capacity information, supply chain information, transaction quotation information, order information and other information directly related to upstream and downstream transactions of enterprises; secondly, the market and environmental information faced by enterprises, such as customer service and satisfaction information, market demand information, raw material price trend information, national policy information and consumer preference information. In the traditional enterprise management, each part of the enterprise information is separated from each other, the business between upstream and downstream enterprises is independent, and the degree of integration is not high. The development of inter-organizational information system has brought about the integration of supply chain information, which can be gathered in the same platform through big data technology, and the information systems of partners can be docked, or the interconnection of enterprise information systems can be realized by means of electronic market platform. Business information of upstream and downstream enterprises can be exchanged automatically through standardized interfaces, so that enterprises can make business decisions on the basis of fully grasping upstream and downstream information. To achieve collaborative optimization among trading partners, deepen trust among partners, and timely change enterprise's R&D, production and development strategy according to changes in external policy environment and consumer demand, develop products to meet consumer demand, and identify the best partners to improve enterprise's competitive advantage. The attributes of mobile Internet based on users should firstly make statistics on users' network consumption affordability, which can be obtained by users' traffic usage in a certain period of time. Secondly, the software on the user's mobile phone is identified and recorded, and the proportion of traffic users put into each software every day is counted, as well as the download rate and usage rate of the software. Then, we can effectively identify the online content of different users, such as websites that users have visited, keywords used in search engines, products and keywords that have been browsed in electronic shops, videos that users often watch and time spent online. Finally, the actual location information of different users is positioned and identified, including ordinary residential areas, high-end residential areas, office buildings, administrative areas and so on.

6 Summary

The evolution of big data technology has never stopped, and the scope of big data applications is also expanding. At present, the huge amount of data in the telecommunications industry has not yet formed an effective information asset, which contains the value and opportunity to be explored. Basic operators can focus on their own real and special data resources, and enhance the skills of big data analysis and mining. So as to enhance operational efficiency, enhance network quality, revitalize strategic assets and maximize data value. The development of big data requires the opening of the industrial chain and the wider data aggregation. For this reason, basic operators will cooperate with all parties in the industry to create a big data ecosphere for enterprise marketing and achieve win-win cooperation. Big data will become a new engine for enterprise development.

Acknowledgments. This research was supported by Scientific Research Project of Hubei Provincial Department of Education (Grant No. B2013209), and 2013 Wuhan Textile University Students' Innovative Training Program (project number: 2013CXXL070).

References

1. Lorenzini M (2014) Independent hotels leverage maestro PMS yield management and big data BI to increase revenue
2. Sheng J, Amankwah-Amoah J, Wang X (2017) A multidisciplinary perspective of big data in management research. Int J Prod Econ 191:97–112
3. Martensen A, Mouritsen J (2014) Prioritising investments in marketing activities to improve business performance. Total Qual Manag Bus Excell 25(5–6):582–601
4. Bui DM, Huynhthe T, Lee S et al (2016) Replication management framework for HDFS based on prediction technique. In: International conference on advanced cloud & big data. IEEE, pp 58–63
5. Yang YY (2018) Xian International University. Design and implementation of integrated marketing system based on cloud computing and big data. Electron Des Eng
6. Choi T, Wallace SW, Wang Y (2017) Big data analytics in operations management. Prod Oper Manag
7. Fuchs M, Höpken W, Lexhagen M (2014) Big data analytics for knowledge generation in tourism destinations – a case from Sweden. J Destination Mark Manag 3(4):198–209
8. He J-S (2017) The research on marketing management of electric power enterprises in the background of large data. Telecom Power Technol
9. Ren JJ, Polytechnic S (2017) An analysis of application of big data in marketing management. J Jiangxi Vocat Tech Coll Electricity
10. Yuan YL, Ho CI (2015) Rethinking the destination marketing organization management in the big data era. In: ASE Bigdata Socialinformatics. ACM, p 60
11. Guan W (2016) Research on marketing management innovation of electric power enterprise based on the background of large data. Electron Test
12. Lichy J, Kachour M, Khvatova T (2017) Big data is watching YOU: opportunities and challenges from the perspective of young adult consumers in Russia. J Mark Manag 33(9–10):719–741
13. Zhou J, Lin HM, Shen HS et al (2016) Marketing management of enterprises under the big data background. Times Agric Mach
14. Wang C (2016) Research on the innovation of marketing management of electric power enterprises under the background of large data. Electron Test
15. Chen X, Lv Z, Fu Y et al (2018) Discussion on marketing management innovation of power supply enterprises in big data era. Mod Ind Econ Informationization
16. Huang J (2016) Cloud services platform based on big data analytics and its application in livestock management and marketing. In: International conference on information science and cloud computing, p 063
17. Yin H (2015) The impact of big data on chinese insurance industry. J Insur Prof Coll
18. Galletta A, Carnevale L, Celesti A et al (2018) A cloud-based system for improving retention marketing loyalty programs in industry 4.0: a study on big data storage implications. IEEE Access
19. Chang CL, Mcaleer M, Wong WK (2018) Big data, computational science, economics, finance, marketing, management, and psychology: connections. Social Science Electronic Publishing

Analysis of Printing Talents Training Mode Under the Background of Internet+

Yongxin Liu(✉) and Guohua Song

Media Academy, Jilin Engineering Normal University, Changchun, Jilin, China
35249827@qq.com

Abstract. China is a big printing country in the world. In the new historical period, a large number of highly skilled personnel are needed to promote the transformation and upgrading of printing industry and realize the leap from a big printing country to a powerful printing country. Limited to the training conditions, teachers' skills and students' participation in enterprise training, there is a big gap between the printing professionals and the talents needed by enterprises. By analyzing the requirements of the transformation and upgrading of printing industry for professionals, this paper probes deeply into the shortcomings of current colleges and universities in the cultivation of printing professionals. From the aspects of adjusting the training specifications, curriculum system, quality education, innovative education and the construction of teaching staff, this paper considers the cultivation of high-quality talents adapted to the requirements of industrial transformation and upgrading.

Keywords: Innovation · Packaging and printing industry · Talent training

1 Introduction

Since the reform and opening up, China's printing industry has made rapid development, the printing industry structure has been gradually optimized, green printing has achieved remarkable results, but there are also shortcomings such as the green development mechanism needs to be improved [1, 2], the comprehensive management level is relatively extensive [3, 4], and the quality of practitioners needs to be improved [5–8]. With the transformation of the Internet+ era [9], the new generation of information technology is widely applied, and the "made in China 2025" will be implemented in depth [10, 11], and the printing industry will further develop towards digitalization, intellectualization and greening [12–15]. In the context of Internet+, we must also soberly see the reasons for the misplacement of the school's professional setup and the rapidly changing market demand. On the one hand, the market changes rapidly. On the other hand, the adjustment of professional settings is slow, which makes the supply and demand disunity, and to a certain extent, it makes it difficult for students to adapt to the market demand when they are employed.

Z. Xu et al. (Eds.): CSIA 2019, AISC 928, pp. 640–647, 2020.
https://doi.org/10.1007/978-3-030-15235-2_89

2 Internet+ Era Background Industry Development

The characteristics of the Internet era are based on the modern information technology including computer technology, which makes the content of information spread to all parts of the world in the shortest time [16]. The development of the Internet cannot be separated from information. With the help of information technology, people can easily obtain more information content and use it in all walks of life and in all fields. However, the emergence of the Internet and information technology also puts forward new requirements for the connotation and extension of enterprise management. In the Internet era, through the use of various modern and specialized network software, enterprises can be more efficient in helping them to complete many complex manual operations. In modern enterprise human resource management, cloud computing, big data and other technologies can help them complete the related work with the shortest time and the least cost. Especially by using the APP client to push the product information and related activities information to customers in time, enterprises can help them grasp the fresh information in the first time and complete the purchase of products and other activities. Printing industry has also produced a new way of playing.

2.1 Personal User Perspective Analysis

The biggest gathering place for personalized trading is Taobao and Tmall. Several well-known personalized printing and photo printing companies in China have invested a lot of money and energy in the printing e-commerce website, but the biggest source of business is Taobao and Tianmao [17–19]. In fact, as an incremental printing market, individualized printing itself has little influence on the existing printing market pattern.

2.2 Analysis of Internet Disk Providers

One thing that should be paid attention to is that although it has no possibility to compete for business cards, leaflets and picture albums in joint printing [20], if more product lines are expanded, they may break into the disturbance in joint printing by circuitous means. There are four words "how fast, how good and how economical" in the field of commercial circulation. Those who are familiar with joint printing should have a strong feeling about "fast and economical", and only those who can achieve fast delivery and have cost advantages can gain the upper hand in the competition. If we can achieve "more and faster" or "more and better" in this market, these enterprises will be able to gain a firm foothold through a large number of non-print products under the circumstances of controlling costs and profit margins.

2.3 Analysis of Cooperative Empowerment Service Objects

The concept of software as a service (SaaS) may be unfamiliar to many printing colleagues in graphic stores and printing factories. In fact, if ERP is standardized and cloud-based, it is a kind of SaaS. Using software as a service, not a product, is rarely seen in print companies. There are two SaaS in the transition from map printing to e-commerce in the whole industry chain. The first one is to provide Tujie software from order flow, document flow, capital flow to logistics management for graphic and text fast printing shop. The second is to provide printing chain software for printing plants from order flow, document flow, capital flow, production flow to logistics management. As an output center of design capability, a picture is embedded into the design module of Tujie and Printing Chain to help the graphic shop and printing factory improve their design capability. A picture, graph, and printing chain are mutually opened, and eventually form an ecological closed loop. If e-commerce platform for end-users forms a competitive relationship with the industry in the period of map-changing network, then the whole industry chain e-commerce ecology and industry of a map is a cooperative enabling relationship. Cooperative empowerment is the essence of SaaS model in printing industry. Through Internet technology and artificial intelligence, it can help the industry to improve efficiency and enhance capacity, and help the transformation and upgrading of printing industry.

3 Requirements for Talents in Printing Industry Under the Internet Background

In the Internet era, industrial transformation and upgrading involve two meanings, that is, industrial transformation and industrial upgrading. Industrial transformation is to adjust industrial structure through policy measures; industrial upgrading is to improve industrial structure, industrial quality and industrial efficiency. The essence of the transformation and upgrading of printing industry is to enrich the connotation of printing, to realize the transformation from simple product processing with low added value to modern service-oriented industry with high added value, to change the business model based on cheap labor resources to the business model with capital and technology as the core, and to rely on science and technology to advance. Step by step, realize the transformation from low productivity and high energy consumption production mode to high productivity and environment-friendly production mode; utilize multi-resources to realize the transformation from single printing production to multi-development. The transformation and upgrading of printing industry puts forward new requirements for vocational education, especially higher vocational education.

3.1 The Demand for Printing Technology Professionals Is Diversified

With the promotion of industrial transformation and upgrading, the extension and connotation of printing have changed. The printing industry chain is constantly expanding and the value chain is constantly upgrading. The demand for professional and technical personnel is diversified. It is not only necessary to train high-quality front-line operators. More importantly, it is necessary to train compound talents such as product creative talents, market creative talents, technological innovative talents and management personnel.

3.2 To Put Forward Higher Requirements for Practitioners' Ability to Apply New Technologies, Scientific and Technological Innovation Is the Driving Force to Promote Industrial Transformation and Upgrading

For the printing industry, technological research and development and application of new technologies are the only way out. Therefore, cultivating talents who are good at technology research and development, product innovation, management innovation and technology application is the real support force to promote industrial transformation and upgrading. With the application of digitization and low-carbon technology, printing enterprises will surely step out of the low-end "labor-intensive and extensive" industry chain, towards the middle and high-end "capital and technology intensive" of the industrial chain, the demand of general printing workers will shrink. High-skilled talents, high-tech talents and compound talents who know technology, can operate and manage well will become the goal pursued by the printing industry and the main body of human resources competition in the industry.

3.3 To Put Forward Higher Requirements for the Professional Quality of the Practitioners Is the Comprehensive Expression of Their Professional Awareness and Adaptability, and Is the Inherent Quality and Ability of Their Professional Activities

The transformation and upgrading of printing industry not only put forward new and higher requirements for the knowledge ability and skill level of printing practitioners, but also put forward new expectations for the professional quality of printing practitioners. As the carrier of inheriting civilization and spreading culture, printing requires practitioners to have strong sense of social responsibility, correct world outlook and outlook on life, and good media literacy. The extension of printing industry chain, the development of new fields, the adjustment and maintenance of digital equipment, the solution of technological problems and the fine management of production all require practitioners to have the foundation of sustainable development, the ability of re-learning, and noble professional ethics and professionalism.

4 The Path of Talent Training in Printing Universities Under the Background of Internet+

4.1 Carry Out Various Forms of School Enterprise Cooperation

In order to improve the quality of printing personnel training, the school actively develops personnel training programs, professional curriculum standards, construction of training rooms, compilation of training textbooks and so on. Strengthen in-depth exchanges and cooperation with enterprises, regard enterprises as training bases for teaching, and let students have a buffer period before graduation, which is conducive to meeting the needs of enterprises' posts more quickly, and gradually form a rapid response mechanism for professional construction adapted to the market.

(1) The participation of enterprises at the source, integration of production and education, and the establishment of a mechanism for professional construction to respond quickly to the market. Deep cooperation between schools and enterprises, enterprises at the source to participate in the development of new professional skills. Both sides regularly investigate and research, enterprises will bring information, information and technology in the adjustment of market dynamic changes into schools, and help schools train teachers, guide students' practical training and practical activities. (2) Introducing enterprises into schools, transmitting market intelligence and shortening the development cycle. Enterprise personnel dominate curriculum development, and schools are responsible for transforming into standard teaching texts to ensure that professional training objectives meet enterprise employment standards. Enterprises help schools plan training positions, locate functional areas, provide information on production equipment, rationally add facilities and equipment, and ensure the smooth implementation of new professional training programs. Engineers go deep into the training class. The impact of diversified school enterprise cooperation on students' enthusiasm is shown in Fig. 1.

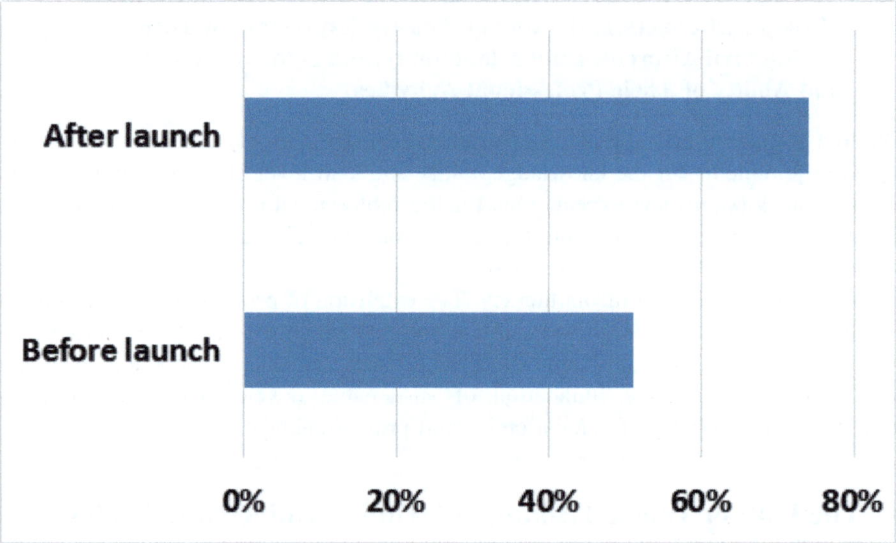

Fig. 1. Comparison of students' enthusiasm before and after launching

4.2 Keep up with the Development of the Industry and Innovate the Mode of Personnel Training

Under the background of digitalization, intellectualization and integration of printing industry, Shanghai Press and Publishing Vocational and Technical School continuously optimizes its talent training mode. In terms of professional (skill orientation) setting, appropriate combination of different categories such as pre-press production, printing operation and post-press processing will provide potential for the multi-purpose development of high-skilled talents. In terms of personnel training, we pay special attention to the innovation of the following three aspects: first, we should pay attention to the training of innovative talents, second, we should pay attention to the construction of cross-integration knowledge system, and third, we should pay attention to the training of compound talents.

4.3 The Teaching Staff Is Implemented

The key to the teaching plan is to ensure that the training of talents in Colleges and universities meets the needs of industrial transformation and upgrading. To cultivate high-quality professional and technical talents to meet the needs of industrial transformation and upgrading, we must have a high-level teaching staff to adapt to them. Suggestions are made in the following aspects: Firstly, strengthen the construction of teachers' morality and style, cultivate teachers' sense of responsibility of loving their posts and devoting themselves to their work. Teachers' moral accomplishment directly affects the cultivation of students' professional qualities and their future growth; secondly, strengthen the cultivation of teachers' key abilities, especially the cultivation of innovative and sustainable development abilities; Thirdly, we should strengthen the cultivation of "double teachers' quality" so that teachers have solid professional foundation and strong practical skills. Fourthly, we should strengthen the cultivation of teachers' teaching methods, requiring teachers to have insight into students' psychology and "teach students in accordance with their aptitude" according to the laws of higher education.

4.4 Establish Teaching Quality Monitoring and Evaluation System

The main responsibility of schools and enterprises should be implemented, and the quality monitoring and evaluation system of talent training should be constructed. Arrange full-time teachers to be "academic tutors", employ brand graduates to be "alumni tutors", recognize enterprise technicians as "industry tutors". Three tutors follow up and guide the apprentices in an all-round way. Innovation of assessment methods can realize the transformation of evaluation subject from single teacher subject to multiple subject, from single school standard to comprehensive professional ability, and from result assessment to combination of process and result assessment.

5 Summary

Printing industry can continue to develop from this point of view. Professional colleges and universities should pay more attention to the education of world outlook and values while teaching students to learn professional knowledge. Let our successors first love this industry and be willing to plan to provide security. Such enterprises will surely be invincible on the way forward. When the country attaches great importance to the training of highly skilled talents, schools must continue to reform and innovate, integrate training, and always meet the needs of market development, so as to meet the transformation and development of industries under the background of Internet plus.

References

1. Wang X (2017) Analysis of the mode of editing and publishing talents in the background of digital publishing. J Heilongjiang Inst Technol
2. Hu Y-D (2018) The innovative reform to talent training mode for application-oriented technology environmental design specialty by "combination of production and education". J Changchun Inst Technol
3. Hu W (2014) On cultivation of printing high- quality talents under the background of industrial transformation and upgrading. J Shaoxing Univ
4. Fang M (2015) An empirical study on the talent training mode of international hotel management in higher vocational education—a case study of tourism college of Zhejiang China. J Jilin Teach Inst Eng Technol
5. Xiong L, Wang R, University JV (2016) Analysis on the reform of training mode of automobile specialty. Jiangsu Sci Technol Inf
6. Zheng Y (2017) Research on the cultivation of internationalization foreign language talents in higher education under the "Internet plus" background. J Jiamusi Vocat Inst
7. Fan J (2015) The construction and exploration of the training mode of printing talent team in basic level with target leading. China Packag Ind
8. Feng Y (2018) Exploration and analysis on international talents training mode and path under the background of "the Belt and Road initiative". J Tianjin Vocat Inst
9. Zhang Y (2015) Rule of law of the restricting of self-determination right of recruiting students of public universities in the reform background. J Nat Acad Educ Adm
10. Jiang HX, Polytechnic HI (2017) Research on the training pattern of marketing talents in higher vocational education under the background of internet. Educ Teach Forum
11. Zhu J, Wang C (2016) Analysis on the training mode of nursing talents under the background of comprehensive health. China Modern Doct
12. Yan SX (2018) Analysis on the talent cultivation mode of electric automation technology program in the background of industry 4.0. J Guangdong Commun Polytech
13. Wang L, School FL, University SN (2016) Study on the talent training mode of the Russian majors in the ordinary universities under the background of The Belt and Road. J Anshan Normal Univ
14. Wang Z, University J (2015) On the thoughts of training undergraduates' research and innovation capacity in universities of ethnic areas under the background of regional development. Educ Teach Res
15. Huang XY, Qin FM (2017) The foundations and mode of the cooperation on production capacity between China and Western Asian countries under the background of "the Belt and Road". J Xinjiang Univ
16. Wang S, Song L (2017) Promoting the revitalization of Liaoning old industrial base with institutional innovation—based on the theoretical background of supply-side reform. J Shenyang Normal Univ
17. Zhao CS, Ren JL, Chen YB et al (2018) Spatial-temporal coupling and driving forces of China's provincial tourism industry and regional development under the background of comprehensive tourism. China Popul Resour Environ
18. Chen XL (2018) Research on school enterprise collaborative education model based on integration of production and education under the background of Guangdong-Hong Kong-Macao Greater Bay Area—taking hotel property management service as an example. Educ Teach Forum

19. Zhu L, Xue LY, Chen QG et al (2018) Reform and exploration of the project-driven measurement and control technology and the personnel training mode in the background of engineering education. Educ Teach Forum
20. Dong H, Liu D, Guanxiong MA et al (2018) The adaptability strategy and business model of power grid enterprises participating in competitive business under the background of power system reform. Power Syst Big Data

Development and Future Trends of Internet of Things

Wencheng Liu, Fei Li(✉), and Yan Li

College of Mechanical Engineering, Jilin Engineering Normal University, No. 3050 Kaixuan Road, Changchun, China
383917380@qq.com

Abstract. The rise of the Internet of Things (IoT) is a product of the emergence of the mobile Internet and the new generation of information technology. It is of great significance to the upgrading and transformation of industries, agriculture, and cities. As one of the cutting-edge technologies with great application value, the IoT is characterized by its green, intelligent, and sustainable development, which makes it a promising foundation and an important engine for future technological progress and economic development. This paper starts with the status quo of the development of the IoTs, takes the national development policy as a basis, and focuses on the technical difficulties and application bottlenecks of the IoTs. The key technologies and application areas of the IoTs are also introduced to give a reasonable prediction of its future trend. The development of the IoTs will play an active role in the promotion of industries, agriculture, and cities.

Keywords: IoT · NB-IoT · Industrial information · Smart city

1 Introduction

The Internet of Thing is entering a new stage with everything connected. Smart devices such as smart wearable devices, smart home appliances, intelligent networked vehicles, and intelligent robots will be connected to the network to form massive amounts of data, and related applications will also show explosive growth. The production, life, and social management has been further transformed toward intelligence, refinement, and networking [1]. New technologies such as the 5th generation mobile communications technology (5G) and narrow-band Internet of Things (NB-IoT) provide a powerful infrastructure support capability for the connection of everything [2]. The ubiquitous access, efficient transmission, massive heterogeneous information processing and equipment intelligent control of things, as well as the technical and security issues arising therefrom, all impose higher requirements on the development of IoT technologies and applications.

The vertical industry market for the IoTs is also emerging. Manufacturing has become an important application area of the IoTs. Related countries have proposed the concept of "Industrial Internet" and "Industry 4.0" [3]. Our country proposes to build a strong manufacturing country and a strong network, and promote structural reforms on the supply side. The IoT intelligence information technology represented by the Cyber

Z. Xu et al. (Eds.): CSIA 2019, AISC 928, pp. 648–656, 2020.
https://doi.org/10.1007/978-3-030-15235-2_90

Physical System (CPS) will play an important role in the transformation and upgrading of the manufacturing industry in terms of intelligence, networking, and service [4]. The demand for car networking, health, home, smart hardware, and wearable devices is more active, driving the convergence of IoT and other leading technologies. New breakthroughs have been made in technologies such as artificial intelligence, virtual reality, autonomous driving, and intelligent robots. The construction of smart cities has become a global hotspot and the IoT is the basic element and modular unit in the smart city architecture. IoT has become the key infrastructure and important support for realizing the "automatic perception, rapid response, and scientific decision-making" of smart cities [5].

With the vigorous development of the IoTs, the IoT technology and applications still face many problems. NB-IoT is a mobile IoT implementation technology based on licensed spectrum and is currently the only international standard with unparalleled advantages, but it also faces competition from other technologies such as LoRa [6, 7]. NB-IoT mainly relies on the operator's infrastructure for protocol interconnection. LoRa is a more flexible and autonomous network that can be deployed wherever it is needed. It also has a faster pace of maturity in the industry chain and commercial use. Currently, many countries and regions have a certain scale of deployment in the world, and the technical operation and business model are also relatively mature and the cost is relatively controllable. In addition, at present, China's mobile IoT-related applications are still in the experimental and exemplary exploration stage, the scale of application is still relatively small, innovation is still insufficient, the technical difficulties will gradually increase. At this stage, equipment manufacturers such as Huawei, Ericsson and chip suppliers such as Qualcomm are more active [8]. However, demand-side applications, such as the water and electricity public service provision industry, have not yet matured and need further improvement. In addition, the NB-IoT project involves multiple links in the industry chain, including chip vendors, module manufacturers, terminal manufacturers, communication equipment suppliers, platform parties, and operating companies. The collaboration between each other needs to be further improved. Therefore, the NB-IoT industry chain is limited due to factors such as frequency bands and operators, and the promotion of large-scale commercialization needs a time course. In summary, breaking through the key core technologies and implementing the key areas as soon as possible are the primary tasks for the development of the IoTs. The following sections will describe the key technologies of the IoTs, the application of key areas, and the development trend.

2 Key Technologies of IoTs

2.1 Sensor Technology

Sensors are the components to realize information perception in the IoTs, and their role is equivalent to the five senses of human. With the introduction of similar concepts such as Industry 4.0 and smart manufacturing, the global sensor market has once again been widened. It is predicted that from 2016 to 2021, the compound annual growth rate of the sensor is expected to be 11%, and the market size will reach 190.6 billion U.S.

dollars by 2021. At present, the global sensor market is dominated by related companies in the United States, Japan, and Germany, such as MEAS Sensors, Honeywell, Keller, Emerson Electric, General Motors, etc. However, the sensor market in the Asia-Pacific region also appears a rapid growing trend [9]. In recent years, China has invested in sensor-related industries to increase the share of state-owned sensors and instrumentation, especially the market share of high-end sensor types. Currently, domestic sensor and instrumentation companies are researched by Shenyang Instrument Research Institute, Shenzhen Tsinghua University, Henan Hanwei Electronics, etc. and achieved a certain degree of performance currently as representatives.

The development of core elements of sensors requires the testing of sensitive materials such as biomaterials, graphene, and specialty functional ceramics to capture the first-mover advantage in leading-edge sensitive materials. Strengthen the research of sensitive mechanisms, structures, and packaging processes for silicon-based sensors, and accelerate the research and development of all types of sensitive components. At the same time, in order to realize sensor integration, miniaturization, and low power consumption, it is necessary to develop technologies and processes for the integration of sensors of the same and different types. Support the formation of different types of sensitive chips based on MEMS technology and thin film process technology. Carried out a variety of different forms of packaging and packaging process innovation. Support the development of a modular device with external energy self-collection, power-down sleep self-start and other energy storage and power control. In addition, high-performance sensor products and application technologies such as inertial, pressure, magnetic force, acceleration, light, image, temperature, humidity, and distance must be developed for key application areas.

2.2 Operating System Technology

The IoT operating system is a file system for industrial control, aerospace, smart home, wearable devices, and other key areas, and is adapted to the characteristics of the IoTs. It includes network protocol stacks and other peripheral modules as well as various types of development interfaces and tools. Enterprises are thus supported to launch open-source operating systems and open kernel development documents. Users are encouraged to carry out a secondary development of operating systems, and thus the migration of mobile terminal operating systems to IoT terminals will be promoted.

Take industrial control as an example, industrial control is the purpose of industrial process detection, control, optimization, scheduling, management, and decision-making. It is mainly embodied in PLC, DCS, and other product forms. At present, the PLC market can be divided into three schools in the United States, Europe, and Japan, among which the United States has AB, General Electric, Modicom, etc.; Europe has German Siemens, AEG, French TE; Japanese PLC Manufacturers include Mitsubishi, Omron, Panasonic, Fujitsu, South Korea's Samsung, LG, etc., these manufacturers' products occupy the main PLC market share. China's PLC market is still dominated by foreign products, the number of products and production scale of domestic PLC manufacturers have yet to be improved. DCS manufacturers are mainly concentrated in the United States, Japan, and Germany. China's DCS market has a certain share of foreign DCS products, such as Honeywell and Yokogawa's DCS products. However, in recent years, the market share of domestic DCS has also been increasing rapidly. A number of excellent DCS products have emerged in China, such as Beijing and Lishi, Shanghai Xinhua, Zheda Zhongguan, Zhejiang Weisheng, Aerospace Measurement and Control, and Electronics. Hospital and Beijing Kangtuo Group. The DCS systems developed and produced by these enterprises not only increase the number of varieties, but also reached the international advanced level [4].

2.3 Cloud and Edge Computing Technology

Current forecasts indicate that by 2020, smart sensors and other IoT devices will generate at least 507.5 Zettabytes (Zb) of data [10]. Trying to complete such a heavy calculation in a different place is bound to be difficult to achieve. To fully unlock the infinite potential of real-time intelligence, event-driven analysis and decision-making processes must be brought closer to interaction and data generation—in other words, to the edge of the network. Therefore, there is a need for mobile terminals to focus on the support for human-machine interaction, micro-smart sensors, MEMS sensor integration, ultra-high frequency or microwave RFID, and fusion communication modules. Combining typical application scenarios such as industrial, intelligent transportation, and smart cities, we will break through the key technologies of data mining and visualization in the IoTs to form specialized application software and services.

For the problem of high data exchange cost for data sharing services and data collection, cloud computing technology can be used to achieve compression, indexing, storage, and multidimensional query of massive high-frequency data. Cloud computing technology can optimize three key attributes in the IoTs: computing resources, storage space, and communication bandwidth [11]. Computing resources refer to virtualized cloud servers in physical servers, which can provide enterprises with flexible computing resources. When the number of IoT devices increases, enterprises can expand the number of cloud hosts as needed. The structured data returned by each networked sensor will be stored in the database. For the unstructured data such as pictures, audio and reports returned, the cloud computing vendor also provides object storage services. At present, deep learning has made breakthroughs in the fields of pictures and natural language processing. Besides, tons of data can also be measured and analyzed by the IoTs, providing significant information with the help of artificial intelligence in the future. Communication, as the name suggests, is to ensure that the information is returned to the channel, in addition to ensuring cross-region server communication needs. In summary, due to the on-demand payment of resources such as elastic computing and storage, cloud computing not only reduces the operating costs of the IoT cloud platform, but also provides guarantees for business growth.

3 Applications in Key Areas

3.1 Intelligent Manufacturing

Industrial IoTs realize the flexible allocation of manufacturing materials, on-demand execution of manufacturing processes, rational optimization of manufacturing processes, and rapid adaptation of manufacturing environments through the interconnection of industrial resources. In order to build a service-driven new industrial ecosystem. Industrial IoT exhibits six typical features: intelligent sensing, ubiquitous connectivity, precision control, digital modeling, real-time analysis, and iterative optimization [4].

China's manufacturing industry is faced with the strategic task of improving production efficiency, achieving energy-saving emission reduction, and completing industrial restructuring. The Industrial IoTs will bring about profound changes in the production, operation and management modes of enterprises. Smart manufacturing is based on the deep integration of a new generation of information and communication technologies with advanced manufacturing technologies. It penetrates through all aspects of manufacturing activities such as design, production, management, and service, and has functions such as self-awareness, self-learning, self-decision-making, self-enforcement, and self-adaptation. New production methods and implementation of industrial IoT provide the cornerstone for smart manufacturing. With the help of industrial IoTs, smart manufacturing will allocate supply chain resources rationally to improve production and service efficiency, and realize the innovation of intelligent manufacturing model.

Today's industrial IoTs needs to face the needs of supply-side structural reforms and manufacturing transformation and upgrades, develop cyber physical systems and the industrial Internet, and promote the transition from manufacturing and operation management to intelligence, refinement, and networking. RFID technology is used to electronically identify relevant production materials, realize intelligent management of production processes and supply chains, and use sensors and other technologies to enhance real-time acquisition and data analysis of production status information, improve efficiency and quality, and promote safe production and energy saving. row. Through the ability to preset sensing, positioning, and identification in the product, remote maintenance of products can be realized and the transformation of manufacturing services can be promoted.

3.2 Smart City

The smart city is a revolution in urban management and development model innovation. It is a higher form of modern city integration and development. Its core lies in the use of modern information and communication technologies to build a ubiquitous high-speed converged network, intelligent sensing environment, and super-power computing capabilities. As a result, the management mechanisms of urban information systems will be reformed, urban service and quality of life will be improved, the development of high-end industries and the transformation of economic will also be promoted [5].

According to the ranking of China's top 20 super smart cities released by Deloitte, Shenzhen City won the championship of smart cities with 76.48 points. In terms of technological capabilities, intelligence, and ability to innovate, Shenzhen has achieved high scores to lead the country [12]. Especially in the field of technology, Shenzhen is one of the demonstration regions for the IoTs in China, and it is in the leading position in the country in the construction of the IoT and cloud computing platform. In 2016, the penetration rate of smart mobile users in Shenzhen was 73.8%, and the penetration rate of internet broadband users was 89%. It was the first batch of "Broadband China" model cities in China. In the environment in which the municipal government and the people of the city are actively creating smart cities, a large number of enterprises have been attracted to participate in the construction of intelligent infrastructure and the rapid development of high-tech industries has been promoted. The IoT industry, as a newly emerging strategic industry supported by Shenzhen, is hailed as the third information industry revolution following the computer, Internet, and mobile communications networks. The Guangdong, Hong Kong, and Macao Dawan District will serve as a pilot field for the "One Belt and One Road" construction and will lead the new pattern of China's open economy. Currently, the IoTs has become a new engine for the transition from manufacturing to smart services and economic growth.

Therefore, in order to build a smart city and improve the quality of people's lives, smart elder and child care, telemedicine and health management, intelligent control of home appliances, home security, smart metering for water, electricity, gas, and home air purification should be carried out for the public's needs of safety, comfort, and functional diversity. Through specifying the underlying communication technology, device interconnection, and application interaction, the interoperability of products from different manufacturers can be promoted, and thus a breakthrough of the smart home technology and products can be achieved.

3.3 Sharing Economy

As a new economic form, a new resource allocation method and a new development concept, the sharing economy embodies the inherent requirements of concept innovation, technological innovation, model innovation, and institutional innovation [13]. In recent years, the sharing economy in China has developed rapidly, new patterns of new industries have emerged vigorously, and innovative companies have continued to emerge as a bright business card for Chinese innovation. The continuous maturation and innovative application of a series of information technologies such as mobile Internet, IoTs, big data, cloud computing, location-based services (LBS), mobile payments, and near-field communications are the key supporting conditions for the explosive growth of the sharing economy. As an original product of China, sharing bicycle is another major application innovation of information technology in the field of transportation, and has become the brightest star of the sharing economy in 2017.

The development of sharing bicycles has strongly promoted the upgrading of bicycle industry technology and IoT technologies, including lightweight integrated body design, vacuum saddles, aluminum alloy hubs, NB-IoT IoT smart locks, and unique solar panels. Double density structure of special tires and so on. Incomplete statistics [14], the cumulative number of patent applications sharing bicycle companies more than 300, such as OFO Xiao Huang bicycle's products accumulated 25 iterations, the latest "OFO Little Hornet" carried out 101 user experience research. Ofo In the iterative upgrade of vehicle quality, integrated vacuum saddles, silent chain covers, integrated 37-style cranks, aluminum alloy palindrums and other industry-leading technologies have been adopted. At the same time, ultra-light running shoes tires have been jointly developed with global chemical giants. The concept and material of running shoes applied to bicycle tires. NB-IoT based "smart locks" jointly developed by Ofo, China Telecom and Huawei are also actively put into use, users can "Open within seconds" the bicycles with the support of intelligent hardware. Xiao Huang bicycle has built a leading platform with mobile IoT smart locks and smart analysis of traffic, within which the IoT is as the carrier and the artificial intelligence is as the core. The "Singularity" system that shares the big data platform for travel uses "people, vehicles, and locations" as the core, and can use artificial intelligence algorithms to analyze the accuracy of user repairs, send information for repairs to the system within 1 s.

4 Future Trends

At present, the industry of mobile IoTs in China already has a relatively large scale, forming a relatively complete industrial chain including hardware, software, system integration, network operations, and application services. In the coming years, the accelerated development of NB-IoT chips will drive the development of other links in the mobile IoT industry chain. The chip is the core of NB-IoT and the world's major chip design companies have already invested in its research and development. There are currently three types of chip companies involved in the NB-IoT field. One is a communications chip company, such as Qualcomm, Huawei, ZTE, etc.; the other is the computing chip company such as Intel; and the third is the wireless chip company such as Nordic, Qorvo, etc. Among them, Qorvo's PA module for NB-IoT will be available soon. Ericsson and Qualcomm are also actively developing related products. In terms of downstream terminals, at present, dozens of terminal vendors, including instrument companies and wearable device companies, are involved in the cooperation and testing of Huawei's NB-IoT chips. The mass production of Huawei's NB-IoT chip will accelerate the research and development of other companies in this field, and the industrial chain will be further improved [15].

The mobile IoTs in China have always been the world's first echelon in terms of technology research and development, standard setting, and experimental promotion. In 2018, a number of more mature applications will be formed in the fields of intelligent transportation, smart municipal administration, smart logistics, and smart homes. A number of mature operating service platforms and business models will emerge. For example, intelligent parking solution based on NB-IoT can easily realize real-time monitoring of parking behavior, coordinate the supply and demand of parking spaces in real time, and remotely book and sublet parking spaces to achieve scientific guidance for traffic and efficient use of parking resources. It can even be applied to intelligent urban areas such as smart street lamps, smart bins, smart covers, etc.; smart meter reading based on NB-IoT can remotely and automatically collect water, electricity, gas, heat, and other data through a cellular network. The smart metering terminals can be continuous because of low power consumption. Working for many years, can save a lot of existing labor costs, and because the water, electricity, gas, heat and other public utilities involve hundreds of millions of users, the market prospect is very broad; for the transport of some large or valuable goods such as container transport, cold chain transport, etc. In other words, GPS, Beidou and other positioning information can be transmitted through the NB-IoT system to track the location and status of people, cargo, and even animals in real time across the country, helping to respond to emergencies in a timely manner. The combination of NB-IoT network, consumer electronic products and wearable devices can help locate the elder, children, disabled people, and pets. Intelligent sensors based on NB-IoT technology can work in various risky and harsh environments for a long time, providing timely monitoring data of water quality, soil, air and various pollutions to discover and prevent environmental problems. Smart agriculture needs to collect a large number of environmental values such as atmospheric pressure, humidity, and temperature, as well as light, soil, and water quality. Livestock husbandry needs to collect data on pasture growth and climate conditions, and NB-IoT technology can meet these needs effectively [16].

5 Conclusion

This paper starts with the concept, history and current situation of the Internet of Things technology, and analyzes the key technologies in the Internet of Things research, including sensor technology, operating system technology and cloud computing, edge computing technology. Then, taking smart manufacturing, smart cities, and sharing economy as examples, the applications of IoT technology in key areas are introduced. Finally, by analyzing the policy and technology development trends of the Internet of Things, this paper looks forward to the development prospects of the Internet of Things technology and its possible role in the promotion of human life, industrial development, and technological progress.

References

1. Tao F (2017) Key technologies and trends of the Internet of Things. Electron Technol Softw Eng 19:14
2. Hequan W (2017) Technology and application progress on Internet of Things. Chin J Internet Things 1(1):1–6
3. An Z, Io T (2015) The accelerating booster of Industry 40. Commun World Wkly (8):21–22
4. Xifan Y, Miao Y, Yong C (2014) Connotation, architecture and key technologies of Internet of manufacturing things. Comput Integr Manuf Syst 20(01):1–10
5. Hui F, Sheng C (2016) Application and thinking of internet of things in smart city. China CIO News 1:85
6. Yulong Z, Xiaojin D, Quanquan W (2017) Key technologies and application prospect for NB-IoT. ZTE Technol J 23(1):43–46
7. Yang W, Xiangming W, Zhaoming L (2017) Emerging technology for the Internet of Things - LoRa. Inf Commun Technol 1:0–10
8. Shou G (2017) Huawei promotes digital transformation of the industry through edge computing IoT solutions. Autom Panorama 34(11):54–55
9. Wenjie X (2017) Brief discussion on the characteristics and classification of intelligent sensors in IoT. Process Autom Instrum 38(3):44–47
10. Rong H (2017) Applications and future challenges of edge computing. Autom Panorama 34 (2):52–53

11. Hongen D (2016) The application of cloud computing and internet of things in the construction of smart cities. China CIO News 11:33
12. Shenzhen High-Tech Growth Top 20 Report (2017) Deloitte Research
13. Jeremy R (2015) Towards the Internet of Things and sharing economy. Bus Res 2:14–21
14. Yaqing Y, Tao L (2017) Talking about the development of sharing economy from the shared bicycle. China Telecommun Trade 7:22–25
15. Wenchao H (2017) Current situation and development trend of NB-IoT low rate narrowband communication technology. Electron Test 6:0–28
16. Daoliang L, Hao Y (2018) State-of-the-art review for internet of things in agriculture. Trans Chin Soc Agr Mach 49(1):1–20

The Impact of Data Quality on Neural Network Models

Chunmei Li[1(✉)], Zhao Li[1], Xu Jun[2], and Wei Pi[1]

[1] Department of Computer Technology and Application, Qinghai University, Xining, China
Li_chm0422@sina.com
[2] China Mobile Communications Group Guangxi Co. Ltd., Guilin, China

Abstracts. This paper introduces the neural network evaluates Sanjiangyuan grassland degradation degree evaluation model from three aspects: data source, the neural network design and experimental process analysis. It focuses on analyzing the impact of training data quality on neural network quality and points out that the importance of data quality in big data era.

Keywords: Grassland degradation · Evaluation · Neural network · Big data

1 The Introduction of Data Sources

In the era of big data, every filed produces a lot of data every day. If the data cannot serve humans, it may become useless data or even garbage. However, when we make good use of these big data, we may turn waste into treasure. The Qinghai area on the Qinghai-Tibet Plateau, the roof of the world, is the hinterland of the Three Rivers. It is the birthplace of the Yangtze River, the Yellow river and the Lancang River [7–9]. The climate here will affect the climate of the middle and lower reaches of the Yellow River and Yangtze River and even central and south Asia. Therefore, Sanjiangyuan eco logical protection is one of the important strategies in the country. It has invested a lot of money over years to do this great and arduous task. As a unique type of grassland, Sanjiangyuan alpine meadow has become the research object of many researchers. The local grassland station, animal husbandry office and ecological protection workers and researchers collect a large amount of work data every year, recording their efforts over the years [10].

2 Neural Network Design

The study of grassland [1, 2] is a hot issue for many grassland experts. In the literature 1 and 2, remote sensing techniques were used to identify grassland categories and grassland productivity. Aiming at the causes and influencing factors of alpine meadow grassland degradation in Sanjiangyuan area [3], many experts have done a lot of research. The research results will be different because of different research methods, different research areas and different points of view. In the literature 3, the degradation

Z. Xu et al. (Eds.): CSIA 2019, AISC 928, pp. 657–665, 2020.
https://doi.org/10.1007/978-3-030-15235-2_91

of grassland was studied from three aspects: healthy meadow(hm), degraded plaque (dp), and 2-year Zokor mound(zm2). This paper takes the viewpoint of another class of experts as an example to study the decision-making method of grassland degradation degree. According to this view, the classification of alpine grassland in the Sanjiangyuan area is divided into five levels, namely, non-degraded, mildly degraded, moderately degraded, severely degraded, and extremely degraded, and there are five main factors influencing these five levels. It is the coverage of convex, the proportion of edible forages, the proportion of degraded indicator species, the soil organic matter content of 0-10 cm, and the situation of rodents [4].

The neural network structure [5–7] is shown in the following figure, tbdgd (the coverage of convex), ksmcbl (the proportion of edible forages), thzsczb (the proportion of degraded indicator species), tryjzhl (the soil organic matter content), shqk (the situation of rodents) five important factors affecting grassland degradation are used as the five input nodes of the neural network, and the five degradation categories of class1 (non-degraded), class2 (mild degradation), class3 (moderate degradation), class4 (severe degradation) and class5 (extreme degradation) are the five output nodes of the neural network. The determination of the hidden nodes are determined by experiments with a large amount of training data. In this study, more than 20000 pieces of data were collected, analyzed and pre-processed, and part of them were used as neural network design, training and testing for the first grassland evaluation, some were used for supplementary evaluation, Fig. 1 is the neural network structure of two structures with implicit nodes 18 and 4 in the random screenshot in the experimental process. The green part is the input of the neural network, the yellow part is the output of the neural network, and the middle red part is the hidden node.

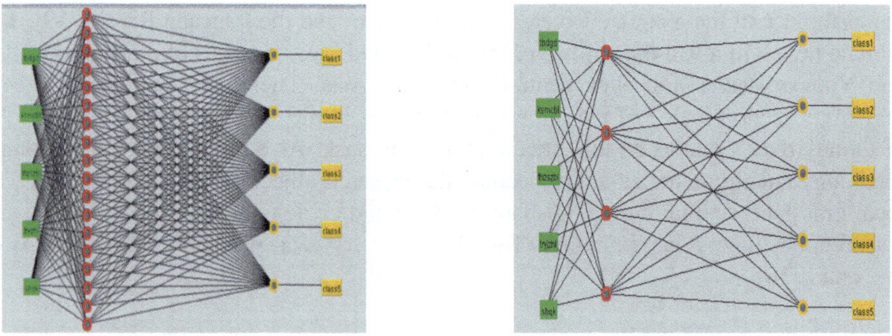

Fig. 1. Neural network structure with hidden nodes 18 and 4

3 The Impact of Training Data Quality on the Quality of Neural Network Models

The neural network structure in the above figure is not unique, because the quality of the neural network model has a great relationship with the quality of the training data. When the data quality is not good enough, adjusting the various parameters of the neural network can make the neural network model improve the accuracy of training data; But the neural network model is not too accurate for this test data; when the data quality is good enough, the neural network model will have good predictive power. Data quality includes many factors, such as the amount of data, the format of the data, the distribution of the data and so on. The data here includes training data and test data.

3.1 The Impact of the Scale of Training Data on the Quality of the Network Model

In experiment 1, we used a dataset of size 40 as the training set for training. By adjusting the neural network parameters, the correct rate of the network model can reach 100%, and 5 data are randomly selected for testing. The error rate is 90%, that is, the correct rate is 10%.

When we continue to increase the data size, the test accuracy rate continues to increase. Until the data model scale reaches more than 10000, the correct rate of the training network model is still 100%, and the correct rate of the test set continues to grow, and finally reaches 98%.

Table 1 below. (Testing the correct rate of growth of the network model for each data size.)

Table 1. Data size increases the correct rate of the test set

Training data size	Test set correct rate
40	10%
100	12%
1000	52%
5000	61%
80000	82%
15000	98%

The distribution maps of the training and test sets in the above table are normal distribution.

3.2 The Impact of Training Data Format on the Quality of the Network Model

When we chose the original data for experiment, the bp neural network we wrote can recognize this format, but the correct rate of the model is not very high. However, this format is not recognize in weaka. The original data in the experiment shown in Table 2 below.

Table 2. Original data

tbdgd	ksmcbl	thzszbl	tryjzhl	shqk	class
7	75	10	11.3	0.08	class1
8	76	9	12.3	0.09	class1
9	78	9.8	13.4	0.089	class1
9.3	81	9.1	11.3	0.015	class1
...	...	...	...	...	...
17	70	16	9.4	0.18	class2
18	61	22	9.2	0.11	class2
21	60	13	9.3	0.12	class2
25	54	25	10.8	0.2	class2
...	...	...	...	...	...
36.1	49	43.1	8	0.26	class3
39.4	54.9	48.4	8.9	0.25	class3
40.1	53.4	46.1	9	0.27	class3
...	...	...	...	...	...
58.1	24	59.4	7.28	0.44	class4
59.3	26	58.4	6.03	0.45	class4
60	25	56.1	6.37	0.46	class4
61	22	57.3	6.34	0.47	class4
61.63	22.3	58.2	6.24	0.48	class4
...	...	...	...	...	...
73	3	82	3.9	0.89	class5
72	13	76	3.4	1	class5
90	5	75	3.9	0.97	class5
...	...	...	...	...	...

We format the original data using the formula minimum and maximum normalization, and the minimum and maximum normalization is shown in the formula (1):

$$v' = \frac{v - \min_a}{\max_a - \min_a} \tag{1}$$

Where v is the original data, $\min_a$ is the minimum value of the column data, and $\min_a$ is the maximum value of the column data.

After the minimum and maximum normalization, all data is defined in the range of [0, 1]. The advantage of this is that all data is collectively enlarged or collectively reduced, so that the data falls within a uniform range, but the relationship between the data is not going to change. The data after the minimum and maximum normalization is shown in Table 3 below.

Table 3. Formatted data

tbdgd	ksmcbl	thzszbl	tryjzhl	shqk	class
0.816327	0.080808	0.785714	0.048148	0.899194	class5
0.826531	0.080808	0.959184	0.12963	0.989919	class5
…	…	…	…	…	…
0.642857	0.313131	0.734694	0.211111	0.788306	class4
0.642857	0.313131	0.734694	0.211111	0.465726	class4
…	…	…	…	…	…
0.276531	0.444444	0.358163	0.296296	0.304435	class3
0.285714	0.444444	0.387755	0.237037	0.385081	class3
…	…	…	…	…	…
0.119388	0.636364	0.244898	0.311111	0.102823	class2
0.119388	0.636364	0.142857	0.311111	0.102823	class2
…	…	…	…	…	…
0.039796	0.787879	0.041837	0.459259	0.056452	class1
0.039796	0.787879	0.041837	0.648148	0.079637	class1
…	…	…	…	…	…

3.3 The Impact of the Distribution of Training Data Sets on the Network Model

When we choose data that is randomly distributed in different types, that is, the data in the training set is unevenly distributed among the five categories, and most of them are concentrated in one class, such as class 1, or most of them are concentrated in an area of one class. The data in other classes and other areas are not selected. For example, if the training set is as shown in Table 4, the training result is that the correct rate of the model is 100%, and the test data does not fall within the test range, the error rate is 100%, that is, the correct rate is 0%. Only when the test data are distributed in a densely distributed area of the training data, the test results are correct, otherwise it is wrong. Some data distributions in the experiment are shown in Figs. 2 and 3. Figure 2 shows that 15000 pieces of data are distributed in each class. Figure 3 shows that 10000 pieces of data are mainly distributed in categories 3, 4 and 5.

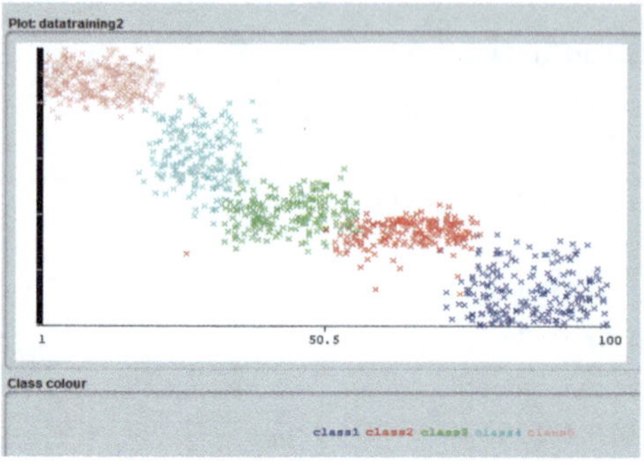

Fig. 2. The five categories of data are basically evenly

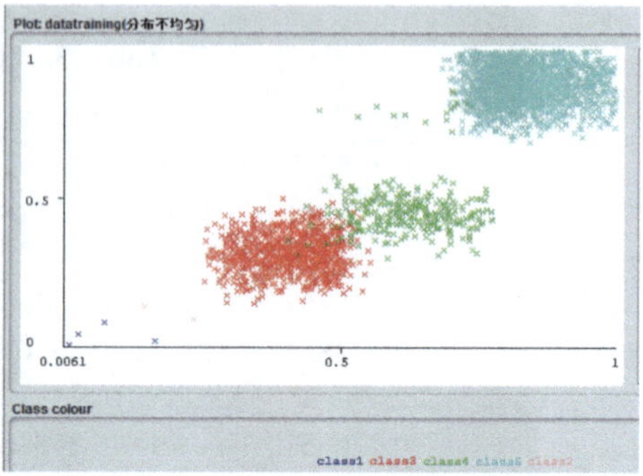

Fig. 3. Shows the data in the 3, 4 and 5 categories

The above two distributed data sets are modeled separately as training sets, and we using the same test sets for testing. The test set data size is 2000, and the number of distribution is same in each category. The experimental results are shown in Table 4 below.

Table 4. Test results of the same test set for different network models.

Neural network model	The set	Test results
Model with 15000 data training	2000 tests (evenly distributed data)	The correct rate is 98%
Model with 15000 data training	2000 tests (basically distributed in category 1)	The correct rate is over 98%
Model with 15000 data training	2000 tests (basically distributed in category 3)	The correct rate is over 90%
Model with 10000 data training	2000 tests (evenly distributed data)	The correct rate is 58%

When using the data set with sufficient data and basically evenly distributed data of each kind as the training set, the accuracy rate of the test data distribution is more than 90%.

4 Test Data Set Impact in the Quality of the Network Model

The data sets under the above two distributions are separately modeled as training sets, and tests with different test sets. The test set data size is 2000, and the number of distributions is the same in each category. The experimental results are shown in Table 5 below.

Table 5. Neural network scale and test accuracy

Neural network model	Test set	Test results
Model with 15000 data training	2000 tests (evenly distributed)	The correct rate is 98%
Model with 10000 data training	2000 tests (concentrated in categories 3, 4 5)	The correct rate is 97%

The above two experiments show that one-side or partial data can only come to a one-side conclusion. To draw a reasonable and generalizable conclusion, it is necessary to collect a comprehensive data set and use a comprehensive test set to test the network model after successful. It is a good model.

A good model requires a lot of test data for testing and checking. Model that cannot withstand testing are meaningless. The quality of the test data set also affects the quality of the model. The test result of a one-side data set, no matter how high the accuracy rate is, it is useless.

5 Other Parameters Affects the Network Model

5.1 Impact of Training Time on Training Model

Training time has a great impact on the training model. When the network training time is not enough, the network model that tends to end in advance is incomplete. The trained network model is not universal, and there is no generalization ability, so the purpose of training is lost. When the training time is long enough, a more stable network model can be trained.

5.2 The Impact of the Number of Hidden Nodes on the Training Model

The number of implicit nodes also has great influence on the network model. It is not that the more hidden nodes are, the more stable the network is. The determination of hidden nodes is different because of different research objects. The calculation formula of implicit nodes is given in many literatures, which is done in Document 4. These formulas are not universal and versatile. In any research filed, a large number of experiments are needed to determine the number of hidden nodes. In this study the number of hidden nodes started from 2 and gradually increased. When the number is small, the correct rate of the network model test is low. When the number of hidden nodes increase gradually, the correct rate gradually increase. When is increases to a certain extent, during the process of increasing, the result is that the test accuracy is gradually decline. It's not that the more hidden nodes, the better. Taking a set of training sets of size 1000 as an example, the process determining hidden nodes is as shown in Fig. 4. As the number of hidden nodes increase, the error gradually decreases to a minimum and the gradually increases.

Number of hidden nodes	Number of input nodes	Number of output nodes	error
2	5	5	big
3	5	5	smaller
4	5	5	smaller
5	5	5	smaller
6	5	5	minimum
7	5	5	lager
8	5	5	lager
9	5	5	lager
10	5	5	lager
11	5	5	lager
12	5	5	big

Fig. 4. Implicit number of nodes and error

5.3 Impact of the Number of Iterations on the Network Model

The number of iterations is too small or too much, the number of iterations is too small, and the result is too short. To end the training too early, the resulting network is incomplete, not universal, iteratively too long, sometimes over-fitting, but the result is that the test accuracy is not high.

6 Conclusion

No matter what the research object is and what the research field is, for the research field where there is no formula to calculate and derive, it is necessary to use experience values to make certain predictions. The high quality is reflected in the following aspects: firstly, the data is the correct empirical value, and the wrong data training is also wrong; secondly the data must be comprehensive in the entire research field, not one-sided or partial, partial data can only lead to partial conclusions, which does not represent universality. thirdly, the data format should meet the requirements of the algorithm. It is necessary to format the data. Collective enlargement or collective reduction will not affect the connection and relationship between the data; fourthly, the choice of test data is also very important. The model trained by comprehensive test has universality and generalization ability.

References

1. Liu B, You G, Li R, Shen W, Yue Y, Lin N (2015) Spectral characteristics of alpine grassland and their changes responding to grassland degradation on the Tibetan Plateau. Environ Earth Sci 74(3):2115–2123
2. Chang J, Ciais P, Viovy N, Soussana J-F, Klumpp K, Sultan B (2017) Future productivity and phenology changes in European grasslands for different warming levels: implications for grassland management and carbon balance. Carbon Balance and Manage. 12(1):11. First Online: 04 May 2017

3. Ma L, Yao Z, Zheng X, Zhang H, Wang K, Zhu B, Wang R, Zhang W, Liu C (2018) Increasing grassland degradation stimulates the non-growing season CO2 emissions from an alpine meadow on the Qinghai-Tibetan Plateau. Environ Sci Pollut Res 25(26):26576–26591
4. Li C, Wang Y, Tian F, Zhou X, Cui P Cluster Comput. https://doi.org/10.1007/s10586-018-1698-x
5. Li C, Wang G, Yang G et al (2015) Application of BP neural network in the study of natural dyesensitizing agent for solar cells. Exp Technol Manage 32(3):50–56
6. Qing Z, Xueli W, Ting Z et al (2016) Study on prediction of water quality index of honghu lake based on BP neural network. Wetland Sci 14(2):212–217
7. Li B, Wang J, Li J (2017) A prediction model based on neural network. Stat Decis-Making 478:43–46
8. Sun H, Lin C, Li X et al (2013) Analysis of vegetation community structure and productivity of different degraded grassland in alpine meadow in sanjiangyuan area. Anim Husbandry Veterinarian Heilongjiang Province 10:1–3
9. Bai X, Wei J, Xie H (2017) Characteristics and effects of dry-wet change in sanjiangyuan area. J Ecol 37(24):1–14
10. Lei X (2017) The application of network security technology in campus network. China Comput Commun

Research on Improvement Method of RF Phase of Antenna Tracking System

Haidong Zou(✉) and Hongmin Li

China Satellite Maritime Tracking and Control Department, Jiangyin, Jiangsu Province, China
605571975@qq.com

Abstract. The sum and difference branch phases of satellite monopulse tracking system are easy to change greatly with the change of geographical location and ambient temperature and humidity, which leads to the performance degradation of servo tracking system. Through the analysis of the principle of monopulse tracking and the realization principle of RF phase calibration system, the basic factors affecting its key performance are found out. Through practical application, the key performance indicators are improved to meet the demand of dynamic conditions.

Keywords: Radio frequency · Phase calibration · Antenna · Tracking system

1 Introduction

As an effective auxiliary means, RF phase calibration technology can improve the tracking performance of monopulse satellite communication system [1–3]. According to its design principle, its realization method is to install a small radio frequency phase correction auxiliary antenna in the fixed position of the antenna to simulate the satellite star. Because its position is fixed and the whole radio frequency loop is fixed, a phase correction loop is constructed artificially. On the basis of manual calibration, the system needs to store its reference value in time. When the following changes in geographic location and ambient temperature and humidity lead to the decrease of cross-coupling degree of the system, these dynamic variables can be quickly eliminated by using the system to get the initial value of RF phase correction, which makes the servo system achieve a better tracking effect [4].

Therefore, further research on this system, taking full account of the various factors involved in the actual use environment, and solving the problems in the current use environment, can promote the follow-up work.

2 RF Alignment Technology

2.1 Principle of Monopulse Tracking System

When the monopulse tracking system is adopted, the sum and difference signals are used to obtain the error voltage of the antenna in the direction of azimuth axis and pitch

Z. Xu et al. (Eds.): CSIA 2019, AISC 928, pp. 666–672, 2020.
https://doi.org/10.1007/978-3-030-15235-2_92

axis. In order to demodulate the error voltage accurately, it is necessary to keep the phase of the difference signal consistent, otherwise cross-coupling will occur, which directly affects the tracking accuracy of the antenna [5]. During the course of navigation, the phase of sum and difference channels will change with the changes of ambient temperature and geographic location and the polarization adjustment when working on-line polarization satellite.

For the differential mode feed of TE21 mode, the pitch error signal EL and azimuth error signal AZ are added orthogonally into the differential signal in the differential branch [6].

The expressions for signal and differential signal are:

$$\Sigma = b \cdot \cos(\omega t + \xi_1), \Delta = b \cdot \mu \cdot \theta \cdot \cos(\omega t + \varphi + \xi_2) \tag{1}$$

In the formula above, b is the amplitude of the signal and μ is the normalized error,

$$\mu = \Delta \Big/ \sum = \sqrt{\mu_1^2 + \mu_2^2}, \varphi = \tan^{-1}(\mu_2/\mu_1) \tag{2}$$

The frequency of the beacon is received by the omega; zeta 1 and zeta 2 are the initial phase when the single channel is combined with the difference signal, theta is the angle of the antenna deviation from the satellite.

The synthesized signal is transmitted in one channel, and the signal is equally amplified, the same frequency conversion and the same time delay when the signal is transmitted to the differential signal. The channel pair has the same effect on the signal and the differential signal [7]. After the AGC amplifier completes the normalization of the difference signal pair and the signal, in the phase locked and amplitude demodulation, the narrow band filter tracking characteristic of the superheterodyne phase locked loop is used to lock the loop in the reference source frequency, so that the frequency, phase and reference source of the signal are kept fixed. Then, amplitude synchronization detection and error signal demodulation are accomplished by reference source [8]. The intermediate frequency output of the phase locked loop is as follows:

$$\Sigma + \Delta' = k \cdot \cos \omega_1 t + b \cdot \mu \cdot \theta \cdot \cos(\omega_1 t + \varphi + \gamma) \cdot c(t) \tag{3}$$

Formula: K is fixed value, determined by AGC reference voltage, and Omega 1 is the frequency and signal angle frequency after channel down conversion.

When the reference source signal is u (T) = sin (ω_1t), the phase shifter is adjusted to make $\gamma = \gamma'$, and the phase detection output is:

$$r_1(t) \cdot \left(\sum + \Delta'\right) = -\frac{\mu}{2} \cdot k \cdot \theta \cdot \sin\varphi \cdot c(t) = -\Delta_{EL} \cdot \mathrm{c}(t) \tag{4}$$

$$r_2(t) \cdot \left(\sum + \Delta'\right) = \frac{\mu}{2} \cdot k \cdot \theta \cdot \cos\varphi \cdot c(t) = \Delta_{AZ} \cdot \mathrm{c}(t) \tag{5}$$

It can be seen that after the phase demodulation, the error signal has been demodulated, but only in the form of communication [9]. The signals expressed by the $\Delta_{\mathrm{EL}}c(t)$ are amplified by inverse phase. The signals expressed by the $\Delta_{\mathrm{AZ}}c(t)$ are amplified in the same phase. A large number of signals are set to K1. The amplified signal enters the synchronous detector. The essence of the synchronous geophone is the multiplier, and the multiplier gain is 1, then the multiplier is added to the multiplier.

$$\Delta_{EL} = \frac{\mu}{2} \cdot k \cdot \theta \cdot \sin(\gamma - \gamma') \cdot \cos\varphi \tag{6}$$

$$\Delta_{AZ} = \frac{\mu}{2} \cdot k \cdot \theta \cdot \cos(\gamma - \gamma') \cdot \cos\varphi \tag{7}$$

It can be seen from the above two expressions that the signal loss is 0 EL and the error voltage drop is cos times the original $\cos(\gamma')$, resulting in a cross coupling. The cross coupling is

$$H_{A-Z} = \Delta_{AZ}/\Delta_{EL} = \cos(\gamma - \gamma')/\sin(\gamma - \gamma') = \cot(\gamma - \gamma') \tag{8}$$

Similarly, if gamma gamma′, gamma gamma′ + n PI (n is integers), the cross coupling is

$$H_{E-A} = \Delta_{EL}/\Delta_{AZ} = \cot(\gamma - \gamma') \tag{9}$$

From the above analysis, it can be seen that if the phase shift adjustment of the phase shifter is inaccurate [10], or the phase shift of the differential network is added before and the relative phase shift of the differential signal causes the gamma ray gamma, the cross coupling is produced, the cross coupling equals the difference $(\gamma - \gamma')$, and the simultaneous error of the cross coupling is reduced to the original $\cos(\gamma - \gamma')$ times.

2.2 Principle of RF Alignment Technology

A small antenna is installed at a suitable position on the main surface of the satellite communication antenna. The beam center of the small antenna is aligned with the satellite communication antenna's sub-surface. Because the small antenna deviates from the satellite communication antenna's center, both sum and difference signals are generated in the feed of the satellite communication antenna, and the relative phase shift of the difference channel can be detected. The small antenna can adopt pyramidal horn antenna, dipole antenna, conical helical antenna and planar helical antenna. The microstrip planar helical antenna has the advantages of small size, easy installation and low cost, and is easy to realize.

It is assumed that the corresponding sum difference channel gain is $G = (k \cdot \mu \cdot \theta)/2$, for the star calibration phase. The output error can be obtained from the analysis of the preceding section.

$$\Delta_{AZ} = \frac{\psi}{2} \cdot k \cdot \theta \cdot \cos(\gamma + \lambda), \Delta_{EL} = \frac{\psi}{2} \cdot k \cdot \theta \cdot \sin(\gamma + \lambda). \tag{10}$$

Formula: the small antenna deviates from the side angle; the gamma phase shifts relative to the difference channel; the lambda is the synthetic error angle of the small antenna relative to the guard antenna.

The azimuth error voltage delta AZ and the pitching error voltage delta El can be obtained by the receiver. Thus, the sum of the difference channel gain G and the differential channel synthesis phase shift beta can be obtained as follows:

$$G = \frac{\psi}{2} \cdot k \cdot \theta = \sqrt{\Delta_{AZ}^2 + \Delta_{EL}^2}, \beta = \gamma + \lambda = \tan^{-1}(\Delta_{EL}/\Delta_{AZ}). \tag{11}$$

It is assumed that the channel gain becomes G' after a period of time, and the phase shift of the channel changes to gamma. If the phase value does not change, the output error of the receiver will be cross coupled, the demodulation sensitivity will change, and the phase is needed again.

The small antenna is connected with the signal source, and the signal source is set to the beacon frequency to be used. The error voltage output by the receiver is used to calculate the new sum and difference channel gain G′ and sum and difference channel synthesize phase shift beta′. Combining the former sum and difference channel gain G and sum and difference channel synthesize phase shift beta, the tracking receiver's relative phase value changes as follows: the new phase correction value is $\gamma' = \gamma + (\beta' - \beta)$, the new channel gain is $K' = K \cdot G/G'$. In this way, using γ' and K' to reset the receiver parameters, the phase correction is completed.

3 Factors Affecting Accuracy of Antenna Calibration

According to the theoretical deduction process mentioned above, the digital tracking receiver can calculate the deviation error voltage of small antenna relative to the center of the sub-surface according to the relative phase shift value of existing calibration in the RF phase calibration mode, and then calculate the relative phase shift lambda of small antenna relative to the satellite antenna. In the process of correcting the value of RF phase calibration, the digital tracking receiver can calculate the relative phase shift lambda of small antenna relative to The digital tracking receiver defaults that the relative phase shift value of the small antenna relative to the satellite antenna remains unchanged. Based on the obtained error voltage values AZ and el, the new relative phase shift value of the sum and difference channels is calculated. The possible factors affecting the phase change of the digital tracking receiver are the phase shift change of the phase shifter and the sum and difference network. The relative phase shift value of the sum differential channel before and after the network is changed.

Based on this, we can see that if the small antenna has an invisible displacement between the initial value and the correction value, and the system still considers that its relative phase shift remains unchanged, it will inevitably bring relative error phase shift to the system; the radio frequency signal of the small antenna is transmitted by line by default and reflected by the side of the main antenna. After entering the feed network, if there is a certain occlusion near the main plane of the antenna, it may appear that the reflected signal is much larger than the direct signal into the feed network, and this state is extremely unstable, which affects the calculation of the actual phase shift value of the digital tracking receiver. In addition, the relative variation of small antenna gain should also be regarded as a possible variable as a factor affecting the system.

3.1 Antenna Angle Analysis

In order to verify this problem, by changing the azimuth angle of the antenna, the changes of A and E before and after the initial value of the RF phase calibration of a satellite antenna are analyzed to understand their influence on the actual phase calibration results.

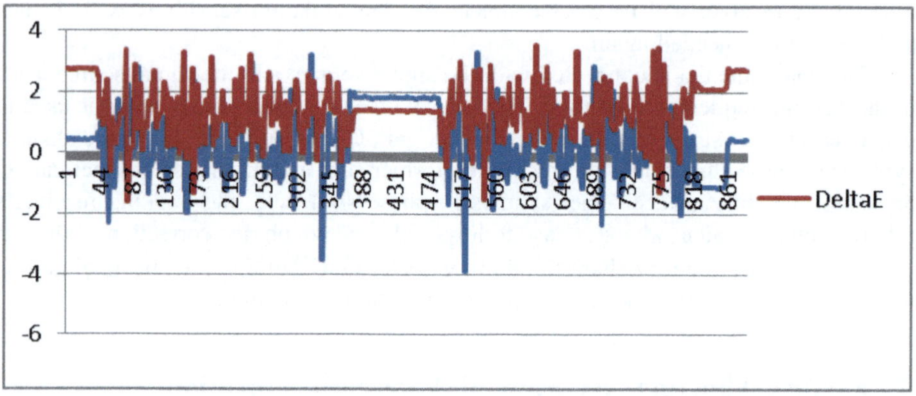

Fig. 1. Effects of attenna angle variation on Delta A and Delta E before.

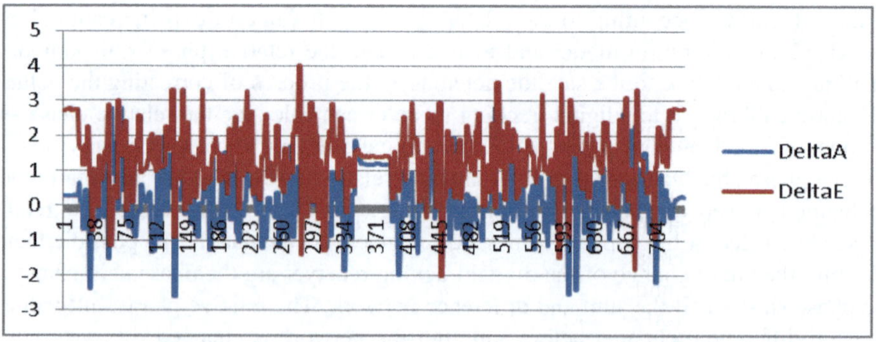

Fig. 2. Effects of attenna angle variation on Delta A and Delta E after.

Figures 1 and 2 are the results of the test. During the test, the pitch angle of the antenna is maintained at 90°, and the azimuth angle of the driving servo rotating antenna is from 0° to 310°. After a short stay, the antenna returns to 0° at the original speed.

It can be seen from the figure that the effect of the satellite radome on the phase correction results of the small RF phase correction antenna is very obvious, but it has nothing to do with whether or not to carry out the phase correction. Therefore, when using RF calibration, the antenna azimuth and pitch angle must be strictly fixed.

3.2 Analysis of the Influence of RF Antenna Output Signal Level

For the verification of this method, a relative method is adopted, that is to verify the effect of the change of input level of signal source on the change of error voltage demodulated by digital tracking receiver. For a satellite antenna, the attenuator connected to the back end of the signal source is removed and the signal source is directly input into the RF phase correction antenna. Then, the attenuator is connected to get the error voltage of the tracking receiver.

The test results are shown in Table 1:

Table 1. System test results.

Items	DTR	ΔA	ΔE	AGC	SNR
Without attenuator	A	0.24	0.52	−56.67	75.67
Without attenuator	B	0.23	0.52	−57.47	75.77
With attenuator	A	0.24	0.56	−76.80	55.45
With attenuator	B	0.23	0.56	−78.14	55.48

From the results of the table, we can know that the change of output signal level of small antenna will only affect the level and SNR of the final demodulated signal of tracking receiver, but the effect on the final error voltage is very small (almost no effect on the value of A, about 7%). From the point of view of system usage, even the error term can be fine-tuned by changing the input signal level of the signal source when the initial value and the final correction value of the RF phase correction are taken. Finally, the influence of the phase change caused by the change of the gain of the test link can be eliminated.

Based on the above analysis of the influence on the angle of satellite antenna, the position of radio frequency antenna and the gain of radio frequency link, we can fix the angle of satellite antenna after opening the radio frequency phase calibration mode, and take technical fixing and protection measures for the radio frequency phase calibration antenna, and observe the digital tracking. The AGC (electric field strength) and SNR (signal-to-noise ratio) values of demodulated signals in the tracking receiver are used to adjust the signal strength in the link, so as to improve the overall stability of the RF phase correction loop.

4 Conclusion

This paper focuses on the technical improvement of the RF calibration system of the Wei Tong antenna. By analyzing the technical principle of monopulse tracking and RF phase calibration, the key factors affecting the performance of servo tracking system are analyzed. Through the experimental verification, it can be seen that in the RF phase correction mode, the performance of the RF phase correction system can be effectively improved by fixing the angle of the satellite antenna, protecting the position of the RF phase correction antenna, and adjusting the signal strength in the tracking receiver to maintain consistency.

References

1. Hao C, Luo C, Liu H (2016) Summary of recent developments in satellite communications. Appl Electron Technol 42(8):11–15
2. Wei Y (2017) Application status and development trend of satellite communication technology at home and abroad. Commun World 10:34–35
3. Cheng S, Wei W (2016) Application status and development trend of satellite communication technology at home and abroad. Electron Design Eng 24(5):110–112
4. Zuo Y, Liu C, Jie Z et al (2016) A new design method of PI controller for speed servo system of permanent magnet synchronous motor. J Electr Technol 31(13):180–188
5. Zhai S, Wang C, Li T et al (2016) Driving control of space high precision scanning servo system. Opt Precis Eng 24(5):1104–1111
6. Yu M-Y, Lu A-N (2016) An on-board automatic phase correction monopulse direction finding and positioning method. Commun countermeasure (3):6–9
7. Li Z, Liu S (2017) Dynamic polarization calibration of dual-channel tracking receiver. Telecommun Technol 57(4):469–473
8. Xie L, Li F, Li B (2018) Optimizing the algorithm of automatic phase calibration for tracking receiver based on ACU. China New Commun (9)
9. Zhang Z, Geng D (2015) Research and realization of dynamic phase calibration technique for shipborne antennas. Radio Commun Technol (01)
10. Yang S, Lin X (2011) Design and Implementation of automatic phase calibration in monopulse tracking system. Inf Commun (03)

Scientific Researcher Credit Evaluation Model Design

Yan Zhao[1], Li Zhou[1], Bisong Liu[1], Zhou Jiang[1], and Jing Wang[2](✉)

[1] Social Credit Branch, China National Institute of Standardization, Beijing, China
[2] Business School, Nankai University, Tianjin, China
wangjingteda@nankai.edu.cn

Abstract. Scientific research credit is an important part of social credit. From the view of the current scientific research credit system framework, the foundation and most important parts of the construction of scientific research credit system are establishing evaluation index system and evaluation method. In this paper, based on the definition of scientific research credit and the reference of western classical credit theory, a 4C model for scientific researcher credit evaluation, namely, Character (The willingness of scientific researchers to be honest and trustworthy), Capacity (The ability of scientific researchers to fulfill their scientific research commitments), Conduct (The behavior of scientific researchers to meet scientific research commitments) and Condition (Scientific research environment) has been proposed. Also, scientific research credit evaluation indices have been designed and the Delphi method and AHP were suggested to determine the weights of scientific researcher credit evaluation index. Furthermore, scientific research credit grades are given in this paper.

Keywords: Credit · Scientific research credit · Credit evaluation · Index design · Delphi method · AHP

1 Introduction

The construction of scientific research credit is an important part in the construction of the whole social credit system. The policy document issued by the State Council in 2014, named "Outline of Social Credit System Construction Plan (2014–2020)" [1], clearly points out that it is necessary to strengthen the credit construction in the fields of education and scientific research, explore to establish credit evaluation system for scientific research institutions, science and technology associations and researchers. Also, incorporating credit evaluation results into the process of establishing scientific research projects, evaluating and appointing technical posts and commending model researchers will help to solve such problems as academic fraud, paper plagiarism and academic misconduct.

During the transitional period from the 12th Five-Year Plan to the 13th Five-Year Plan, the State Council issued a series of policies (upper-level law) such as Guofa (2014) 11 [2] and Guofa (2014) 64 [3], and put forward to optimize the layout of the central financial science and technology plan (special projects, funds, etc.), integrate

Z. Xu et al. (Eds.): CSIA 2019, AISC 928, pp. 673–681, 2020.
https://doi.org/10.1007/978-3-030-15235-2_93

and form five types of science and technology plan (special projects, funds, etc.). Correspondingly, major changes have taken place in the allocation of scientific research funds and mechanisms. At the same time, the above-mentioned documents clearly state that we should strengthen the supervision of scientific research projects and funds, improve the management of scientific research credit, establish a scientific research credit record system covering the whole process of compiling guidelines, project application, project establishment, execution, evaluation and acceptance, carry out credit rating for all subjects, implement classified management and share credit evaluation information. All in all, the evaluation information and evaluation results should be fully employed to implement the blacklist system and the responsibility investigation mechanism.

Recently, General Office of the CPC Central Committee and General Office of the State Council jointly issued a policy document about scientific research integrity named "Opinions on Further Strengthening the construction of scientific research integrity" [4], which clearly points out that scientific research integrity information system shall be established and improved. The Ministry of Science and Technology, in conjunction with the Chinese Academy of Social Sciences, will establish and improve an scientific research integrity information system in natural sciences, philosophy and social sciences covering the whole country, and record the integrity of scientific researchers, relevant institutions and organizations. The key points are to study and establish scientific and reasonable evaluation index and method model of scientific research integrity, which are suitable for different types of scientific research activities and objects, and to clarify evaluation methods, periods and procedures. Integrity evaluation of the project leaders, consultants and appraisal experts involved in scientific and technological plans (projects), as well as the relevant responsible subjects such as project management professional institutions, scientific research institutions and intermediary service agencies is of vital importance.

In all, making full use of credit evaluation results on scientific research to encourage trustworthiness and punish dishonest, can enhance credit awareness of the scientific research subject, thus promote the overall construction and development of the whole social credit system.

2 Scientific Research Credit Evaluation Index Design

Scientific research credit refers to the willingness and capacity of the subjects of scientific research. In particular, it refers to the extent to which scientific research subjects seek truth from facts, do not deceive, do not falsify, abide by relevant laws and regulations, abide by scientific values, scientific spirit and code of conduct of scientific activities, and the willingness and capacity to meet scientific research commitments [5]. Among them, the scientific research subjects include not only scientific researchers, scientific research organizations, but also consultants, appraisal experts, project management professional institutions and intermediary service agencies. All kinds of scientific research subjects differentiate in designing scientific research credit evaluation index. Therefore, this paper only focuses on credit evaluation scientific researchers, including index design and determining weight.

2.1 Scientific Research Credit Evaluation Theory

For a long time, western economists have made great achievements [6] on the measurement of customer credit level and quality, such as C factor theory, 3F factor theory, 5P factor method, 6A factor method, 10 M factor method. These theories analyze the characteristics of credit elements from different angles, and are widely used in enterprise credit evaluation, as well as the management of individual customers' credit risk level. 3C factor theory includes character, capacity and capital, and 4C factor theory includes 3C factor theory plus collateral. Similarly, 5C factor theory includes 4C factor theory plus condition and 6C factor theory includes 4C factor theory plus coverage insurance. 3F factor theory is evolved from 6C theory. Specifically, character, capacity are merged into management factor; capital, collateral are merged into financial factor, whereas condition and coverage insurance are merged into economic factor. 5P factor method indicates personal factor, purpose factor, payment factor, protection factor and perspective factor. 6A factor method means economic aspects, technical aspects, management aspects, organizational aspects, commercial aspects and financial aspects. 10 M factor method denotes man, money, machinery, market, management, material, manufacturing, method, making plan and margin. The above western credit theories are shown in Fig. 1.

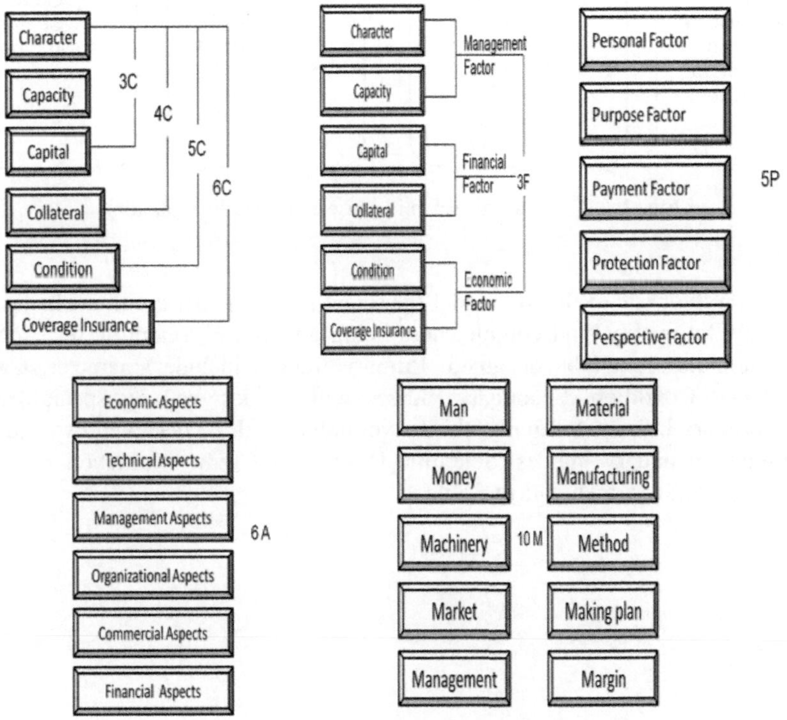

Fig. 1. Western credit theories

2.2 Scientific Researcher Credit Evaluation Model and Index Design

Based on the definition of scientific research credit and the reference of western classical credit theory, this paper proposes a 4C model for scientific researcher credit evaluation, namely, Character (The willingness of scientific researchers to be honest and trustworthy), Capacity (The ability of scientific researchers to fulfill their scientific research commitments), Conduct (The behavior of scientific researchers to meet scientific research commitments) and Condition (Scientific research environment). 4C model for scientific researcher credit evaluation is shown in Fig. 2.

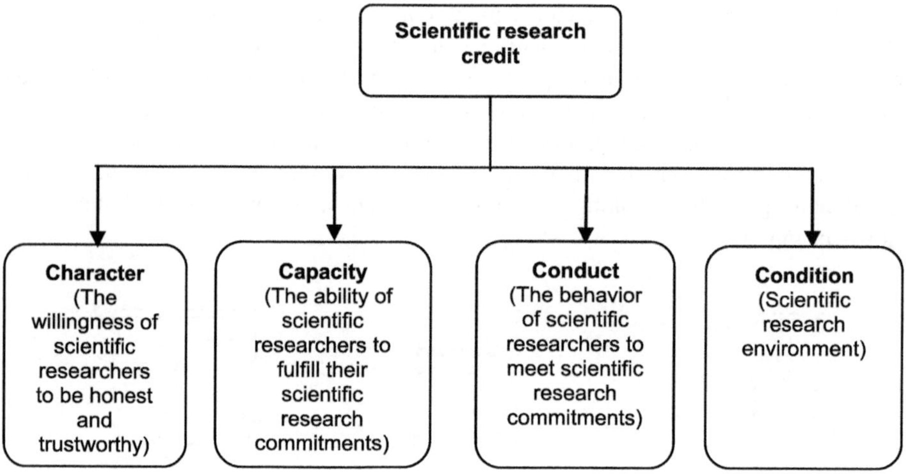

Fig. 2. 4C model for scientific researcher credit evaluation

Then, on the basis of the 4C model for scientific researcher credit evaluation, with scientificity, rationality, and completeness being guaranteed, scientific research credit evaluation indices shall be designed. Primary indices include Character, Capacity, Conduct and Condition. Secondary indices will be designed by specializing and refining primary indices. Similarly, third level indices will be designed by specializing and refining secondary indices. Scientific Researcher Credit Evaluation Indices and index interpretation are shown in Table 1.

Table 1. Scientific researcher credit evaluation index design

Name of primary indices	Name of secondary indices	Name of third level indices	Index interpretation
Character (The willingness of scientific researchers to be honest and trustworthy)	Scientific research awareness	NA	Is there a sense of honesty and trustworthiness, such as the initiative to learn the "Guide to Integrity in Scientific Research Activities", the norms and requirements that should be followed in scientific research activities?
	Scientific research interests	NA	Interest in scientific research work
	Scientific research spirit	Matter-of-fact attitude	Seeking truth from facts, persistently pursuing truth and steadfastly adhering to the truth
		Spirit of innovation	Actively explore, strive to develop, and persist in innovative thinking
		Team spirit	Humility, tolerance, solidarity and cooperation to create a career
		Spirit of self-sacrifice	Willing to be lonely, courageous, dedicated and perseverant, dare to overcome difficulties and not to achieve their goals
		Self-discipline spirit	Abide by professional ethics, abide by discipline, abide by the law, be honest and trustworthy, never plagiarize, tamper with, and falsification
	Pressure	Life pressure	Economic pressure and so on
		Working pressure	Scientific research project pressure, thesis pressure, professional title promotion pressure, surrounding environmental pressure
Capacity (The ability of scientific researchers to fulfill their scientific research commitments)	Existing knowledge reserve	Education	Education of scientific researchers
		Qualifications	Qualification certificates obtained by scientific researchers
		Train	Training experience of scientific researchers
		Project experience	Projects previously hosted or participated by scientific researchers

(*continued*)

Table 1. (*continued*)

Name of primary indices	Name of secondary indices	Name of third level indices	Index interpretation
	Innovation ability	Creative thinking ability	Creative thinking ability of scientific researchers
		Ability to identify and solve problems	Ability to identify and solve problems of scientific researchers
		Ability of analysis and argumentation	Ability of analysis and argumentation of scientific researchers
		Logical thinking ability	Logical thinking ability of scientific researchers
		inductive ability	inductive ability of scientific researchers
		Ability to practice and operate	Ability to practice and operate of scientific researchers
	Expression and communication skills	Ability to communicate with others and teamwork	Ability to communicate with others and teamwork of scientific researchers
		Self expression ability	Self expression ability of scientific researchers
Conduct (The behavior of scientific researchers to meet scientific research commitments)	Is there any misconduct in scientific research in the past?	NA	If there is any misconduct (such as plagiarism and fraud, etc.) in scientific research in the past, please list time and details
	Past awards	NA	Past awards achieved by scientific researchers
	Published articles	NA	Published articles written by scientific researchers
	Published books	NA	Published books written by scientific researchers
	Patents	NA	Patents written by scientific researchers
	Experience of presiding over and participating in projects	Fund management of scientific research projects	Fund management of projects hosted or participated by scientific researchers
		Opening of scientific research projects	Opening of projects hosted or participated by scientific researchers
		Mid term assessment of scientific research projects	Mid term assessment of scientific research projects hosted or participated by scientific researchers
		Acceptance of scientific research projects	Acceptance of scientific research projects hosted or participated by scientific researchers

(*continued*)

Table 1. (*continued*)

Name of primary indices	Name of secondary indices	Name of third level indices	Index interpretation
		Conversion of scientific research projects achievements	Conversion achievements of scientific research projects hosted or participated by scientific researchers
	Participation in academic conferences	NA	Academic conferences attended by scientific researchers
Condition (Scientific research environment)	Experimental condition	Hardware and software devices	Hardware and software devices of laboratory
		Project fund	Project fund
		Talent team	Talent team
	Policy environment	NA	Scientific research policy environment
	Team academic atmosphere	NA	Academic atmosphere of the scientific research team
	Team incentive mechanism	NA	Team incentive mechanism of the scientific research team

3 Scientific Research Credit Evaluation Method

3.1 The Determination of Index Weights

The Delphi method and AHP (The analytic hierarchy process) can be employed to determine the weights of scientific researcher credit evaluation index.

The Delphi method is a structured communication technique or method, originally developed as a systematic, interactive forecasting method which relies on a panel of experts [7–9].

AHP (The analytic hierarchy process) is one of the multiple criteria decision-making methods that were originally developed by Saaty [10]. AHP can be employed to determine the weights of scientific researcher credit evaluation index.

3.2 Scientific Research Credit Grade

Credit ratings can address a corporation's financial instruments, i.e. debt security such as a bond, but also the corporations itself. Ratings are assigned by credit rating agencies, the largest of which are Standard & Poor's, Moody's and Fitch Ratings. They use letter designations such as A, B, C. Higher grades are intended to represent a lower probability of default.

Credit ratings also provide good reference for scientific research credit evaluation. Therefore, scientific research credit grades are summarized in Table 2.

Table 2. Credit evaluation scale for scientific research

Name of the grades	Range	Remarks
A	90 ~ 100	The scientific research credit (character, capacity, conduct and condition) of scientific researcher is exceptional. The probability of default in scientific research in the future is very low
B	75 ~ 89	The scientific research credit (character, capacity, conduct and condition) of scientific researcher is good. The probability of default in scientific research in the future is relatively low
C	60 ~ 74	The scientific research credit (character, capacity, conduct and condition) of scientific researcher is fair. The probability of default in scientific research in the future is relatively high
D	Below 60	The scientific research credit (character, capacity, conduct and condition) of scientific researcher is poor. The probability of default in scientific research in the future is extremely high

4 Summary

Scientific research credit is an important part of social credit. From the view of the current scientific research credit system framework, the foundation and most important parts of the construction of scientific research credit system are establishing evaluation index system and evaluation method. Therefore, in this paper, based on the definition of scientific research credit and the reference of western classical credit theory, this paper proposes a 4C model for scientific researcher credit evaluation, namely, Character (The willingness of scientific researchers to be honest and trustworthy), Capacity (The ability of scientific researchers to fulfill their scientific research commitments), Conduct (The behavior of scientific researchers to meet scientific research commitments) and Condition (Scientific research environment). Then, on the basis of the 4C model for scientific researcher credit evaluation, with scientificity, rationality, and completeness being guaranteed, scientific research credit evaluation indices have been designed. Secondary indices have been designed by specializing and refining primary indices. Also, the Delphi method and AHP were suggested to determine the weights of scientific researcher credit evaluation index. Furthermore, scientific research credit grades are summarized.

Acknowledgments. This research was supported by National Key R&D Program of China (Grant No. 2017YFF0207600) and National Natural Science Foundation of China (71302016).

References

1. Outline of Social Credit System Construction Plan (2014–2020), Guofa (2014) 21
2. Opinions on improving project management and fund management of central financial science research program me issued by The State Council, Guofa (2014) 11
3. Notice of deepening the management reform of the Central Financial Science and Technology Plan (Special Programs, Funds, etc.) issued by the State Council, Guofa (2014) 64
4. Opinions on Further Strengthening the construction of scientific research integrity. http://www.people.com. Accessed 31 May 2008
5. Zhou L, Zhao Y, Liu B-S, Jiang Z (2017) Research on scientific research credit standardization and its standards system frame. Stan Sci 3:11–14
6. Peprah WK, Agyei A, Oteng E (2017) Ranking the 5C's Of credit analysis: evidence from ghana banking industry. Int J Innovative Res Adv Stud (IJIRAS) 4(9):78–80
7. Dalkey N, Helmer O (1963) An experimental application of the delphi method to the use of experts. Manage Sci 9(3):458–467
8. Bernice BB (1968) Delphi Process: A Methodology Used for the Elicitation of Opinions of Experts. An earlier paper published by RAND (Document no: P-3925, 1968, 15 p)
9. Rowe G, Wright G (1999) The delphi technique as a forecasting tool: issues and analysis. Int J Forecast 15(4):353–375
10. Saaty TL (2008) Decision making with the analytic hierarchy process. Int J Serv Sci 1 (1):83–98

Practice on the Sustainable Development of Talent Cultivation Mode in the Context of Big Data

Limei Chen(✉)

Department of Information Engineering, Heilongjiang International University, Harbin 150080, China
3140025490@qq.com

Abstract. Nowadays, emerging majors including big data science and big data technology, artificial intelligence, robotics engineering, new engineering, and new media technologies are all full of digital media majors. Meanwhile, there are interdisciplinary intersections between different majors, so how to enable digital media professional to continue to balance development in new and old majors? This paper mainly analyzes and summarizes the multi-dimensional and multi-angle of the digital media technology professional talent training mode, enterprise talent demand, teachers and students in domestic universities. This laid a good foundation for the reform of the talent training of Heilongjiang International University's digital media technology and the practical exploration of the UI direction in the digital media profession. It is necessary to implement the new policy and actively explore a high-quality, professional, practical, and applied training model that focuses on students and adapts to industries and industries.

Keywords: Digital media technology major · Talent training mode · Sustainable development

1 Introduction

At the National Education Conference held in Beijing on September 10, 2018, General Secretary of the CPC Central Committee, President of the State and Chairman of the Central Military Commission Xi Jinping attended the meeting and delivered an important speech [1, 2]. Xi Jinping emphasized that since the 18th National Party Congress [3–6], we have comprehensively strengthened the party's leadership over education work around the fundamental issue of what kind of people to be cultivated, how to cultivate people and for whom to cultivate people. We must educate people with morality, strengthen ideological and political work at school, promote educational reform and accelerate to make up for the shortcomings in education. In January 2018, the Ministry of Education issued the "National Standards for Undergraduate Professional Teaching Quality in Ordinary Colleges and Universities", and the preparation of the "National Standards" to grasp three basic principles: first, highlight the student center [7, 8]. The second is to highlight output orientation. The third is to highlight continuous improvement. At the same time, all colleges and universities are required to

Z. Xu et al. (Eds.): CSIA 2019, AISC 928, pp. 682–691, 2020.
https://doi.org/10.1007/978-3-030-15235-2_94

revise the talent training program according to the "national standard" and cultivate diverse and high-quality talents. Earlier, there were a lot of research and examples of the models of talent training for different disciplines and majors [9, 10]. Based on the digital media technology major of Heilongjiang Foreign Languages Institute, this paper finds some problems in the existing talent training mode. At the same time, according to the student-centered education concept, the countermeasures and constructive measures with problems are given, and the sustainable development of the talent training mode of digital media technology is truly realized.

2 Data Collection and Question Analysis

2.1 Collection and Analysis of Students' Questionnaire Data

A total of 150 questionnaires were distributed to students of digital media technology in the first, second and third grades, and 126 valid questionnaires were returned. The questionnaire recovery rate was 84%. The following results were obtained.

(1) Understanding of the Major
By analyzing the students' questionnaires, it can be seen that the students themselves are not clear enough about their majors and their professional competence and talent development. As shown in Table 1. It can be seen from the results that only a few students understand their majors.

Table 1. Understanding of the major

No.	Main contents of the questionnaire (Professional direction)	The no. of students who are unfamiliar	Proportion
1	Whether understand the talent training direction of the major	96	76%
2	Whether understand the professional demands of the major	112	88%
3	Whether understand the core courses of the major	103	82%

(2) Professional Ability Requirements Need Urgent Improvement
It can be seen from the results in Table 2 that students pay more attention to the improvement of the professional ability of the major, but they do not pay enough attention to the humanistic quality.

Table 2. Understanding of the ability cultivated by the major

No.	Professional abilities needed	Proportion of population
1	Basic operation of computer	96%
2	Software for the major	100%
3	Independent innovation	60%
4	Teamwork	50%
5	Interpersonal exchange and communication	40%
6	Writing and reading abilities	20%
7	Art qualities	35%
8	Professional design and technologies	90%

(3) Feedback on Classroom Learning
Through feedback on the evaluation of classroom teaching, most of the students like the practical curriculum and have a lot to do with the teaching method of the teacher, as shown in Table 3.

Table 3. Students' feedback on classroom learning

No.	Problems	Proportion of population
1	Teaching is not vivid but is boring and hardly understood	78%
2	Teaching dominated by lecturing with little practice	80%
3	Some courses contain many theories that can be hardly understood	60%
4	Assessment of students is not timely	84%
5	The contents learned are irrelevant to life	40%
6	No interest in the course	30%
7	No interest in programming and cannot understand it	28%

2.2 Collection and Analysis of Teacher Questionnaire Data

A total of 18 full-time teachers and part-time teachers of digital media technology at Heilongjiang International University issued a questionnaire on the talent training model. The recovery rate of the questionnaire was 100%. Data collection and analysis were carried out mainly from the curriculum setting, teacher teaching methods, teaching evaluation and teachers' continuing education. (1) The professional instructors are all in line with the digital media majors. At the same time, the teachers' titles are all above the lecturers. The teachers are very experienced in teaching, and they also lead the students to participate in various competitions and have achieved good results. (2) Some teachers use the same lesson plan, and there is no difference in the learning ability of different students in different classes. (3) In the teaching methods and methods taught by teachers, most of the teachers mainly adopt the traditional teaching methods, and the student-centered teaching mode has not yet been reached. (4) Most teachers have strong desires for further study, and they hope to be able to train and study in enterprises and industries in order to improve teachers' teaching ability and enterprise volume data analysis to enhance students' practical ability.

2.3 Collection and Analysis of the Data of Business Questionnaire

The questionnaires for enterprises mainly collected data on graduates' employability, graduates' requirements for digital media technology industry, satisfaction with graduates, and cooperative enterprises. The results of the data show that the company is more recognized for the professional knowledge of graduates, but the lack of basic literacy, such as student teamwork, innovative professionalism and the positive expression of students' creative works. Take the employment ability of the UI direction in the digital technical specialty as an example, such as the list of employment ability and quality requirements in Table 4 UI.

Table 4. Employment ability and quality requirements for the UI direction

No.	Quality requirements	Proportion (100%)
1	Basic theoretical knowledge related to digital media and UI	12%
2	UI design, creative thinking, innovative consciousness	15.2%
3	Writing ability	5.6%
4	Grasping the core technology of UI design production	30.1%
5	Have a fine art foundation for painting, color, composition, etc.	8.6%
6	Have the concept of updating knowledge, lifelong learning and strong independent innovation ability	13.4%
7	High social responsibility and collaborative spirit	14.1%
9	Others	1%

3 Analysis of the Cultivation of Professional Talents of Digital Media Technology in Domestic Universities

3.1 Through the Network and Field Visits and Surveys, 20 Digital Media Technology Majors in Major Universities in China Have Carried Out a Detailed Analysis of the Talent Training Mode of Digital Media Technology, and the Graduation of Students in the Curriculum

It is worth noting that different colleges and universities will have different levels of talent training for digital media technology. This paper gives in-depth research and analysis on the quality of digital media professional training, professional ability, talent training orientation, major core courses, and graduate employment. (1) Graduation and career orientation, there are clear talents in all colleges and universities to train graduates, and the proportion of curriculum structure and practice curriculum will be different in the process of talent cultivation. This paper summarizes the employment direction of colleges and universities in the employment orientation of digital media technology talents, including animation direction, film and television direction, game development direction, UI direction, 3D modeling direction, visual design direction, WEB front-end development and so on. (2) In terms of quality training, all colleges and universities have set up and mainly focus on the comprehensive development of morality, intelligence, body and beauty. Students are innovative and innovative, full of artistic accomplishment and professionalism, capable of teamwork and an international vision. (3) The analysis and summary of the professional skills in the training objectives include solid professional theoretical knowledge, strong design ability, production ability, mastery of professional design software and specific professional competence in different directions as shown in Fig. 1.

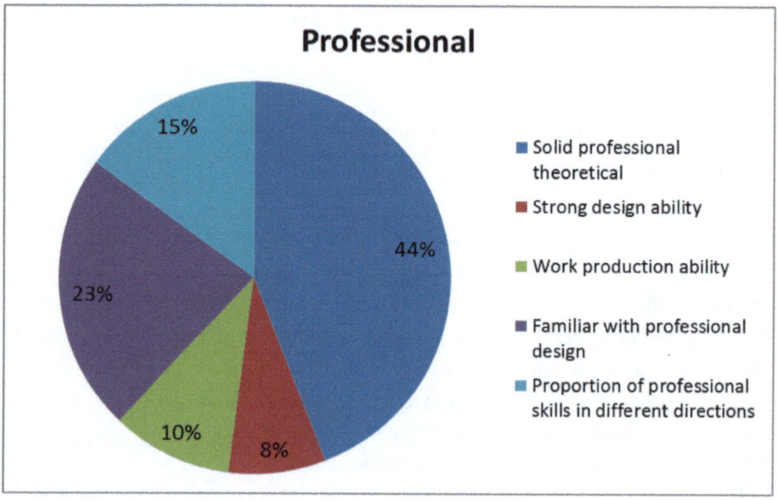

Fig. 1. Figure of professional proportion

3.2 A Comparative Analysis of Curriculum and Industry Needs

Through summarizing and summarizing the curriculum settings of digital media technology majors in colleges and universities, it can be seen that different colleges and universities have established professional basic courses, and most colleges meet the requirements of enterprises for digital media majors. Due to the rapid development of this major, there are great differences in the industry from north to south, and there are also differences in the professional curriculum. At the same time, from the analysis results, some colleges and universities are not able to adapt to the training of professional skills in the enterprise industry as soon as possible.

4 Optimization and Practice Exploration of Talent Cultivation of Digital Media Technology at Our College

On the basis of the above investigation, analysis and summary, the talent training mode of the digital media technology major and the core curriculum of the professional are optimized. The following is a detailed explanation of the UI direction in the digital media technology major.

4.1 Optimization of Talent Training Goals

The optimization of talent training objectives is mainly based on the establishment of talent training mechanisms and the construction of professional core courses, such as the professional knowledge, skill proficiency and comprehensive literacy of graduates from enterprises and industries. It is mainly from the cultivation of students' analytical ability and practical problem-solving ability, the cultivation of student teamwork ability and professionalism, the cultivation of innovative entrepreneurial consciousness, the cultivation of international vision, and the cultivation of culture and art. It pays more attention to the cultivation of students' social practice, strengthens the cultivation of students' artistic accomplishment, and cultivates the production, organization and promotion ability of participating projects, and provides a good platform for students to find jobs. It is necessary to pay more attention to the students' participation in the real-life learning of relevant companies, and to conduct practical internships of relevant companies every semester.

4.2 Optimization of Course Structure

The basic curriculum has been continuously strengthened. The basic curriculum incorporates art courses to meet the needs of the enterprise industry, and the professional core curriculum has been reset, which further highlights the characteristics of the professional core curriculum. For the teaching methods and methods, the students are the most important, and the number of courses and hours of practice courses are increased. The demand for employment in a certain direction of the digital media technology profession is tailored to the professional learning route. As shown in Fig. 2, the digital media technology professional UI direction learning roadmap.

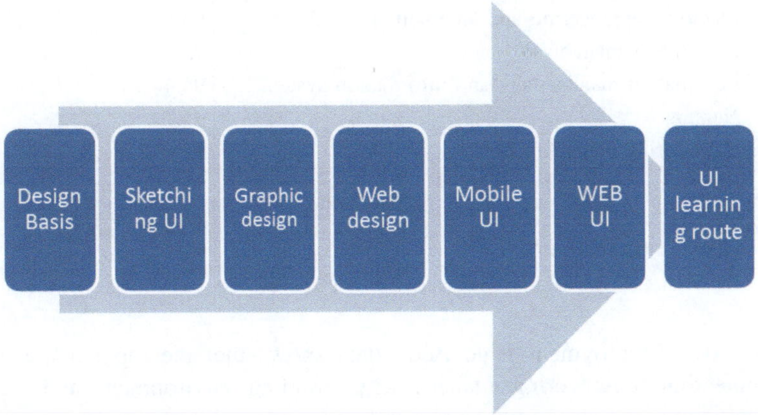

Fig. 2. Learning route for the UI direction of digital media technology major

With the direction of the digital media technology profession, students can have a clearer grasp of professional knowledge, and also have the characteristics of learning direction and purpose. At the same time, students have a good coordination of the teacher's curriculum and the convergence and cross-cutting of the curriculum. In addition to some basic courses, it also adds design expertise. For example, in the first phase of the design foundation, not only the course of the software required for the design, but also the skilled use of the software, the color matching, the commonly used software operation skills, such as image effects, mixed mode and reconstruction and deepening of key knowledge such as masks and path and channel maps. In the second stage, in addition to the sketching basic course, the hand-painting stage also adds the shading relationship and the UI drawing knowledge of the attention in the hand-drawn.

4.3 Graduates' Employment Rate and Employment Destination

(1) From the data of 2018 China University Student Employment Blue Book released by 2018, the top 10 statistics of graduates in the 2017 world are in the statistics of software engineering, energy and power engineering, electrical engineering and automation, logistics management, information management and information systems, nursing, engineering management, preventive medicine, gardening, digital media technology, employment rates are close to or exceed 95% can be seen in digital media technology, as shown in Table 5. (Source: Michaels - China 2017 graduates quality evaluation.)

Table 5. Employment rate of grade 2017 graduates after half a year

No.	Name of undergraduate major	Employment rate after half a year of graduation (%)
1	Software engineering	96.7
2	Energy and power engineering	95.8
3	Electrical engineering and automation	95.6
4	Logistics management	95.4
5	Information management and information system	95.4
6	Nursing	95.1
7	Engineering management	95.1
8	Preventive medicine	95.1
9	Garden	94.9
10	Digital media technology	94.9

(2) The 2017 Employment Blue Book data shows that the top ten are: software engineering, engineering management, building environment and equipment engineering, electrical engineering and automation, information management and information systems, nursing, thermal and power engineering, mechanical and electrical engineering, Logistics management, digital media technology, as shown in Table 6. (Source: Michaels - China 2016 graduates quality evaluation.)

Table 6. Employment rate of grade 2016 graduates after half a year of graduation

Name of undergraduate major	Employment rate after half a year of graduation (%)	Name of undergraduate major	Employment rate after half a year of graduation (%)
Software engineering	96.5	Rehabilitation therapy	94.5
Engineering management	95.9	Vehicle engineering	94.4
Building environment and equipment engineering	95.8	Prepaid medicine	94.4
Electrical engineering and automation	95.5	Safety engineering	94.3
Information management and information system	95.4	Geographic science	94.0
Nursing	95.4	Human resources management	93.9
Thermal energy and power engineering	95.4	Surveying engineering	93.9
Mechatronic engineering	95.3	Computer science and technology	93.9
Logistics management	95.3	Medical imaging	93.9
Digital media technology	94.9	Geographic information system	93.8

(3) The 2016 Employment Blue Book data shows that there are no digital media technology majors in the top ten rankings. As shown in Table 7. (Source: Michaels - China 2015 graduate quality training evaluation.)

Table 7. Shows the employment rate of graduates in 2015

Name of undergraduate major	Employment rate after half a year of graduation (%)	Name of undergraduate major	Employment rate after half a year of graduation (%)
Logistics management	96.6	Information engineering	94.6
Electrical engineering and automation	96.4	Computer science and technology	94.5
Software engineering	96.2	Mechatronic engineering	94.5

(continued)

Table 7. (*continued*)

Name of undergraduate major	Employment rate after half a year of graduation (%)	Name of undergraduate major	Employment rate after half a year of graduation (%)
Building environment and equipment engineering	95.7	Textile engineering	94.5
Geographic information system	95.6	Japanese	94.3
Human resources management	95.6	Surveying engineering	94.2
Nursing	95.5	Automation	94.2
Preventive medicine	95.4	Water supply and drainage	94.0
Geographic science	95.4	Car service engineering	94.0
Information management and information system	95.4	Network engineering	93.9

(4) The employment rate of digital media technology major at our school in 2015 was 95.5%, the employment rate in 2016 was 95.9%, and the employment rate in 2017 was 97.2%. From the above data analysis, it can be seen that there are many interdisciplinary subjects in digital media professional operations. However, from a development perspective, the demand for digital media professionals is still very high. Meanwhile, from the perspective of the employment rate and employment direction of our students, the training of digital media majors has certain effectiveness.

5 Conclusion

From the perspective of sustainable development of digital media technology professionals. A lot of research and data analysis and summary have played a good role in guiding and improving the digital media technology specialty in our university. At the same time, in the process of research, the digital media technology specialty of our university pays more attention to the concept of discipline integration, the importance of curriculum structuralization, and strengthens the thought of systematization and specialization. Thus formed the UI direction professional characteristics of digital media technology specialty, and guaranteed the application-oriented personnel training of student-centered digital media technology specialty. More importantly, from the survey of colleges and universities and the employment needs of enterprises, the warning should have the foresight of disciplinary development. At the same time, according to the needs of professions and enterprises, timely adjustment of talents makes the digital media technology specialty have sustainable development.

Acknowledgments. This work was supported by the youth special topic of Heilongjiang provincial education department (GBD1317104).

References

1. Li W (2017) Project driven innovation of education mode of career technical college industry university research institute. Teach For Reg 01:20–21
2. Tong H (2018) Research on the training mode of e-business compound talents based on post ecological chain. Vocat Tech Educ China 02:51–54
3. Lockspeiser TM, Kaul P (2016) Using individualized learning plans to facilitate learner-centered teaching. J. Pediatr. Adolesc. Gynecol. 29(3):34–40
4. Xu X, Wang A (2017) Deep integration of big data and multiple intelligences in education and teaching. Educ Theory Pract 37(25):32–35
5. Xu S, Jin Y (2018) How to implement the main position of enterprises in school enterprise cooperation in Vocational Education – an empirical analysis based on incentive effect. Vocat Tech Edu China 06:31–38
6. Xu Q (2011) Exploration of the core curriculum of digital media technology specialty. Nanjing Normal University
7. Zhou Y (2018) Research on learner centered training mode of digital media technology application professionals. Liaoning Normal University
8. Mike (Beijing) Co., Ltd. https://baike.baidu.com/item/2018
9. Hong G, Xi Z (2017) Research report on "learner centered talent training mode". Vocat Tech Educ China 31:50–58
10. Zhang Y, Zhao Y, He L, Bai Q (2017) Interdisciplinary teaching design and practice based on STEM. Mod Distance Educ Res 1:10–11

Research on the Automatic Test System of Motor Drive

Chen Tang and Maoyun Zhang(✉)

Changchun University of Science and Technology, Changchun, China
tangchen@cust.edu, 114816943@qq.com,
zhmaoyun@163.com

Abstract. The motor driver test system is designed to monitor the performance and working status of the motor controller, and discover the abnormal motor driver. It is used to assist product developers and testers in system testing. The system feedback the test results in the way of dynamic display. When there are abnormal conditions occur will be issued a warning beeping. Test content mainly includes the motor controller voltage value, current value, drive signal delay and the protection of delay and so on. After the test, the completion of the test platform will be collected, the data will be displayed in the LCD screen.

Keywords: Motor driver · Test · PFC · IGBT

1 Introduction

With the continuous development of computer application systems and the continuous improvement of industrial production requirements, motor drivers have been used in many fields. Motor drives have played an important role in various systems, once an error occurs, the work of the whole system will be hugely impacted [1–8]. The motor drive is composed of complex circuit, it is particularly prone to all kinds of human error. So the testing work of motor drives will play a vital role. Traditional testing is done manually by human, using instrumentation and observing the measurement data. Such testing is a heavy workload, cumbersome data statistics, time consuming and laborious, high cost and difficult to guarantee accuracy. It is unable to meet the requirements of rapid and accurate modern testing.

The motor driver test system is designed to monitor the performance and working status of the motor controller, and discover the abnormal motor driver. It is used to assist product developers and testers in system testing. The system feedback the test results in the way of dynamic display. When there are abnormal conditions occur will be issued a warning beeping. Test content mainly includes the motor controller voltage value, current value, drive signal delay and the protection of delay and so on. Through the keyboard keys to choose test content system test, the LCD liquid crystal display and a buzzer shows the results of the test. After the test, the completion of the test platform will be collected, the data will be displayed in the LCD screen.

Z. Xu et al. (Eds.): CSIA 2019, AISC 928, pp. 692–699, 2020.
https://doi.org/10.1007/978-3-030-15235-2_95

2 The System Implementation Scheme

The motor driver used at present generally determines the motor's running shape by sending the given value of the control motor and receiving the feedback signal, forming a closed loop control. According to the test requirements and various practical possibilities, the whole system is considered to be divided into six circuit boards. The overall framework is as follows (Fig. 1):

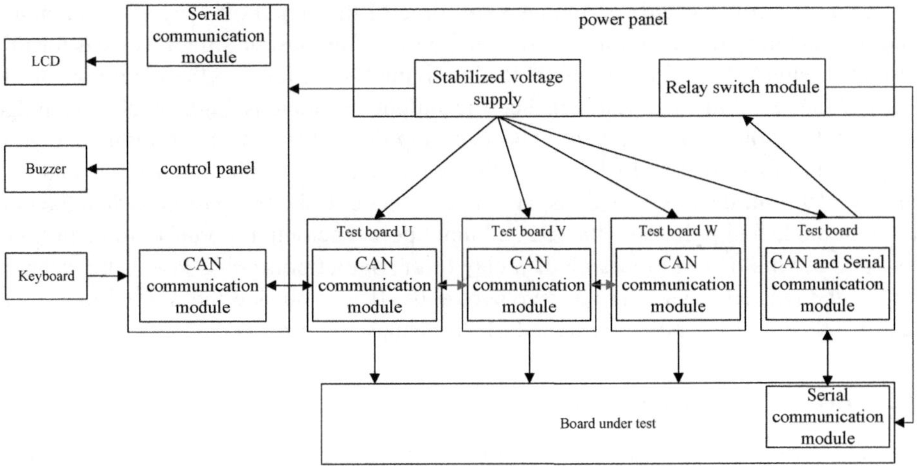

Fig. 1. System framework diagram

(1) Power supply panel: Convert the input voltage into the voltage required by the driver test system. There are several voltage stabilizing power supply circuits, voltage regulating circuits and relay switch module. The voltage input to the control panel and test panel is switched through the master single chip computer.
(2) Control panel: The process used to control the entire test, send test signals, and receive test values that process the feedback from the test panel. With the core of the main control chip, the serial communication and CAN communication circuit are designed at the same time to meet the communication between the control panel and the PC, the control panel and the test panel. The AD acquisition unit, the button and the LCD circuit are also important components of the control panel.
(3) Test panel: The content used to implement all tests. With the main control chip as the core, it also has serial communication and CAN communication circuit, AD acquisition unit, and also designed a protection unit for the IGBT signal. Since the lower arm test requires three boards to measure separately to prevent interference with each other, plus the upper arm test, four identical test boards are required.

3 System Test Process

There are several tests performed by the system with more complicated processes: PFC drive test, PFC error protection test, IPM test, current test, communication test, fan VSP voltage test and fan FG test.

3.1 The Drive Test of PFC

The PFC driver test mainly tests whether the PFC driver chip can output the IGBT drive signal normally and test the response time of the signal off. The system counter and test counter are cleared first at the beginning of the test and the bits used for this test are cleared. By pulling the IO port high, input a high level to the input port of the IR2127 driver chip, and test whether the output terminal is high level, and judge whether the voltage is sufficient. The output port at this time is configured in AD acquisition mode. If the test fails, switch the port back to analog input mode and exit the test. Turn off the power and display the error code. If the test passes, switch the port to input mode and pull down the IR2127 input port. The output port should be fast. Go low and use the timer to measure the delay of the output port going low. If the delay is within the normal range, the measurement ends. If it is outside the normal range, the test is exited and the power is turned off to display the error code.

3.2 Error Protection Test of PFC

The PFC error protection test is used to test whether the PFC can react in time after encountering an abnormal condition, pull the drive signal low, and protect the circuit. The system counter and test counter are cleared first at the beginning of the test and the bits used for this test are cleared. Pull the IR2127 input port high. At this point, the output port should output a high level, then pull the drive transistor on the front end of the IGBT low to create an error, and the timer starts counting until the IR2127 output port goes low, if the timer If the recorded delay is within the normal range, the measurement ends [9, 10]. If it is outside the normal range, the test is exited and the power is turned off to display the error code.

3.3 Test of IPM

The IPM is a module that integrates a power switching device and a driving circuit, and has a fault detecting circuit such as overvoltage, overcurrent, and overheat. The IPM can send a detection signal to the CPU. It consists of a high-speed, low-power dissipation and an optimized gate drive circuit as well as a fast-protection circuit that protects the IPM from damage even in the event of a load accident. The system counter and test counter are cleared first at the beginning of the test and the bits used for this test are cleared. Set the port after the current amplifier to the AD conversion mode. After a period of delay, inject the protection current to determine whether the value on the AD side is correct. If it is not correct, exit the test and display the error code. If the three-phase test is continued correctly, Use the timer to record the test time. If the time is reasonable, the test is completed. If the time is too long, the test will be exited and the error code will be displayed.

3.4 The Test of Temperature Sensor

The temperature sensor is mainly used to test the outdoor ambient temperature, coil temperature and exhaust temperature when the system is working. The voltage value of the temperature sensor is obtained by AD acquisition, and the real-time temperature is obtained by conversion, and compared with the standard temperature to see if it is at a normal value. If it is out of range, turn off the power and an alarm is given.

3.5 Current Measurement

The current test is used to test the system's protection against large current. When the current is tested, it provides a large current to the IPM protection module through capacitor charging and discharging. If the module works normally, it should be cut into self-protection mode, and IDC should output low power. The system counter and test counter are cleared first at the beginning of the test and the bits used for this test are cleared. Then control the IO port, turn on the relay to open, and inject a large current into the IGBT detection resistor, the timer starts counting until the IDC port level is pulled low [11, 12]. If the waiting time is too long, then the test is exited and the power is turned off to display the error code (Fig. 2).

3.6 Communication Measurement Test

The current test is used to test the system's protection against large current. When the current is tested, it provides a large current to the IPM protection module through capacitor charging and discharging. If the module works normally, it should be cut into self-protection mode, and IDC should output low power level. The system counter and test counter are cleared first at the beginning of the test and the bits used for this test are cleared. Then control the IO port, turn on the relay to open, and inject a large current into the IGBT detection resistor, and the timer starts counting until the IDC port level is pulled low. If the waiting time is too long, then the test is exited and the power is turned off to display the error code.

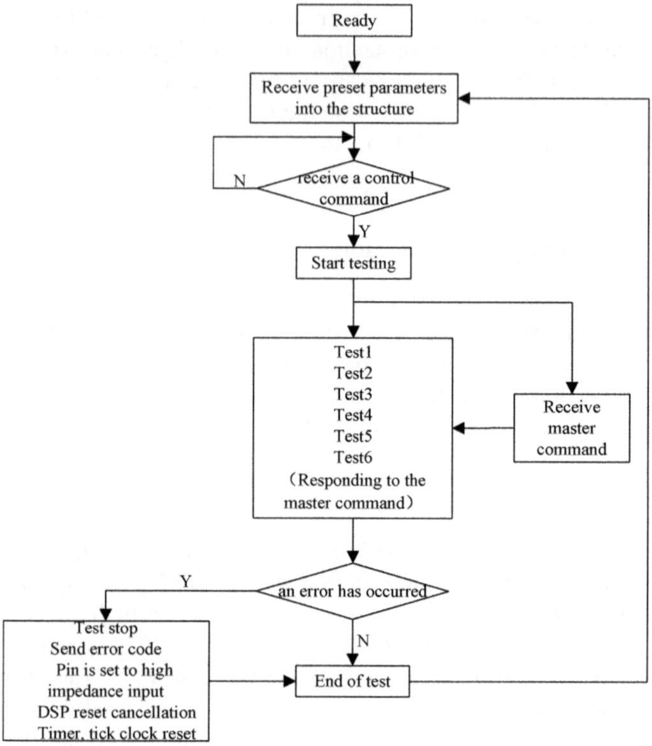

Fig. 2. Test flow chart

4 Test Results and Analysis

4.1 The Drive Test of PFC

There are two test points for the PFC driver test. One is the input port IN of the IR2127 driver chip. The other point is the output port HO of the R2127 driver chip. The yellow curve on the oscilloscope measures the signal of the input port IN. The blue curve test is The curve of the output port OUT, we can see that when the level of the input port IN is pulled high, the output port HO also outputs a high level, but the high level of the output is not an instantaneous rise, but a slow gradient High, this is to protect the circuit from reducing the impact. The system will wait for a period of time after inputting the high level. It is detected by the AD whether the output port is also high. When the level of the input port IN is pulled low, the output port HO is also. After outputting a low level, the system waits for a while after inputting a low level, and detects whether the output port is also low by AD. The results of the PFC drive test are standards compliant (Fig. 3).

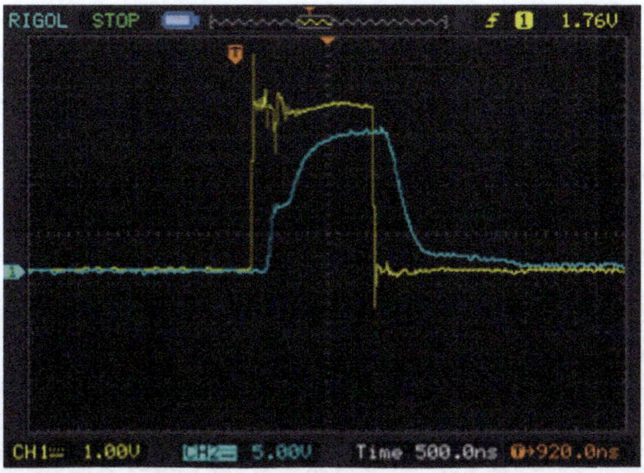

Fig. 3. The drive test of PFC

4.2 Error Protection Test of PFC

There are two test points for the PFC error protection test. One is the base of the IR2127 front-end transistor. This port is set to the IO port with drive capability, and the other is the output port of the IR2127. The IR2127 input port is set high at the beginning of the test. The signal outputted in this way is a high level, and then the driving transistor of the front end of the IGBT is pulled low to create an error, and the test output becomes a low-level protection delay. The oscilloscope picture can roughly estimate the delay to be 1. 2 us. The delay value obtained by the system after the timer count value is converted is 0. 1 us, which is not much different from the actual value. The delay value is within the standard value range (Fig. 4).

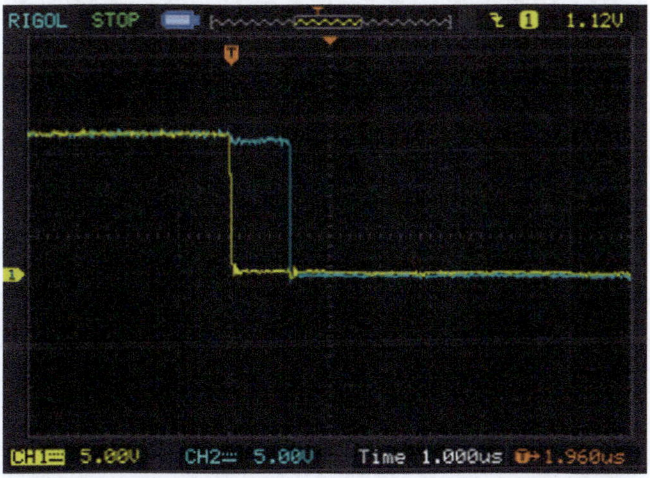

Fig. 4. Error protection test of PFC

4.3 Test of IPM

The IPM test will inject large currents from NU, NV, and NW separately to detect the voltage change of IDC. The test points are NU input and IDC output respectively. NV input and IDC output have NW input and IDC output. Since NU, NV, and NW change as long as one end changes, IDC changes are affected, so only a NU and IDC oscilloscope screenshot is shown as a representative (Fig. 5).

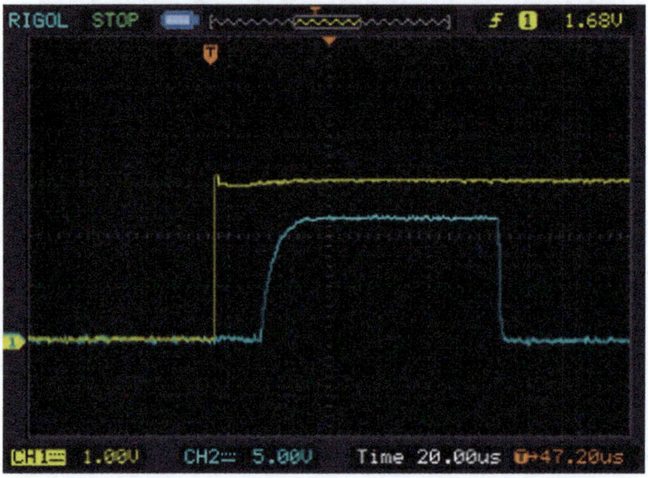

Fig. 5. Test of IPM

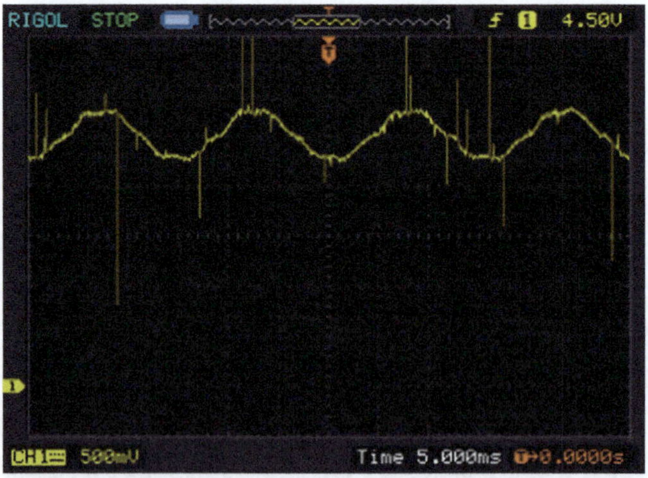

Fig. 6. Current measurement

4.4 Current Measurement

The waveform of the voltage value of the current sensor outlet collected by the motor during the startup process is as shown in the figure below. It can be seen that the current is sinusoidal during the startup process, and the current value is large, the startup is successful, and the test result of this startup is also normal startup (Fig. 6).

5 Conclusion

This paper designs and implements the motor driver test system, which is used to test various functions of the motor driver. It tests various working conditions of the motor driver, simulates the voltage and current environment required by the motor driver, and tests its function and actual performance. A series of analysis and judgment have a more objective understanding of the motor driver under test.

References

1. Hemmer R (2009) Intelligent IGBT drivers with exceptional driving and protection features. In: European conference on power electronics and applications
2. Wang Z, Shi X, Tolbert LM et al (2013) A fast overcurrent protection scheme for IGBT modules through dynamic fault current evaluation. In: IEEE applied power electronics conference and exposition (APEC)
3. Tian S, Zhang Y, Yang X, Jiang J (2010) Research on mutual-feeding mutual-driving test-bed for permanent magnetic synchronous motors. In: Proceedings of the 9th international PCIM China conference. Mesago PCIM GmbH Stuttgart, p 6
4. Shimizu M (2017) Implementation of impedance control system using joint velocity servo driver. In: The proceedings of JSME annual conference on robotics and mechatronics (Robomec), vol 2017
5. Yan J, Yan Y, Zhao X, Zhao Z (2012) Simulation study on automatic testing method of one type of cannon servo driver. In: 2012 7th international power electronics and motion control conference (IPEMC)
6. Yasa Y, Sahin E, Acar C, Gozutok A, Firat E, Mese E (2013) Servo motor driver design for high performance applications
7. Yuan X, Fang Z, Nie R, Li H (2011) Hardware design of a full-digital AC-servo driver based on TMS320F2812. In: 2011 second international conference on mechanic automation and control engineering (MACE)
8. Qin H, Shen A, Zhang Q, Wang W (2009) Analysis and compensation of dead time effects for the AC servo driver. In: International conference on electrical machines and systems, ICEMS 2009
9. Choi U-M, Blaabjerg F, Lee K-B (2015) Study and handling methods of power IGBT module failures in power electronic converter systems. IEEE Trans Power Electron 30:2517–2533
10. Lutz J, Basler T (2012) Short-circuit ruggedness of high-voltage IGBTs. In: International conference on microelectronics
11. Hu L, Sun C, Chen Y, Xiao M, Ai S (2018) Short-circuit fault detection for high-power IGBT. Trans China Electrotech Soc 33(11):2592–2603
12. Masi A, Butcher M, Losito R, Picatoste R (2013) DSP based smart sensorless stepping motor driver for LHC collimators. IEEE Trans Nuclear Sci 60(5):3514–3520

Remote Teaching System Based on Cloud Platform

Shuyan Wang[1(✉)], Xiaohui Li[2], and Lingyun Meng[1]

[1] Jilin Agricultural Science and Technology College, Jilin 132000, Jilin, China
wsy_jl@126.com
[2] Changchun University, Changchun 130022, Jilin, China
xiaohui.li_77@hotmail.com

Abstract. With the development of computer information technology, the concept of cloud computing has been put forward, and the use of cloud computing service platform has achieved good results. The deployment of the remote teaching system in the cloud platform solves the resource sharing problem, which is conducive to the use of self-learning and in-class teaching.

Keywords: Cloud computing · Internet · Java · Cloud platform · Network architecture · The cluster storage

1 Introduction

The Internet+ model has become a new mode of teaching development. For schools, the development of a new type of teaching assistant system is the need of The Times as well as the need of teaching development itself. Under the mode of Internet+, there are many digital management systems for schools, and each system is relatively independent, and each system must pass relevant processes and links. Therefore, multiple systems are mixed together, which can easily lead to confusion and disordered system management. And some systems require a server to be configured, and only one operating system can be installed on a server. Servers are very powerful, but often have low traffic, which results in a huge waste of resources and low utilization of computer processors and hardware [1]. With the development of computer information technology, the concept of cloud computing has been put forward, and the use of cloud computing service platform has achieved good results. The advantage of cloud platform lies in the better utilization of the computing capacity of the cloud platform of machine cluster. The computing work that could not be completed by a single server can be completed by the cooperation of various machines in the cloud platform, which greatly improves the ability of cluster to process data. Combined with the characteristics of the school's multi-system management, such as high efficiency and coordination between various departments, the concept of cloud computing is introduced into the multi-system management, and the management project in the cloud platform is in line with the reality [2].

Remote teaching system is a teaching system based on cloud platform. The system is deployed in the cloud platform and Shared cloud resources with other systems under

Z. Xu et al. (Eds.): CSIA 2019, AISC 928, pp. 700–705, 2020.
https://doi.org/10.1007/978-3-030-15235-2_96

the management and scheduling of the cloud platform [3]. The system can be used as a platform for students to take elective courses by themselves or as an after-class auxiliary teaching system for in-class teaching.

2 System Structure

The design of this system architecture diagram is shown in Fig. 1. The preliminary work of the system is to build a cloud computing platform, assign data storage and data processing tasks to each node in the cluster for processing, and realize the submission, processing and query of tasks through the command line and Java interface. When the task is finished, the results from each node are aggregated and then stored in the distributed file system HDFS [4]. The system architecture can be divided into four layers:

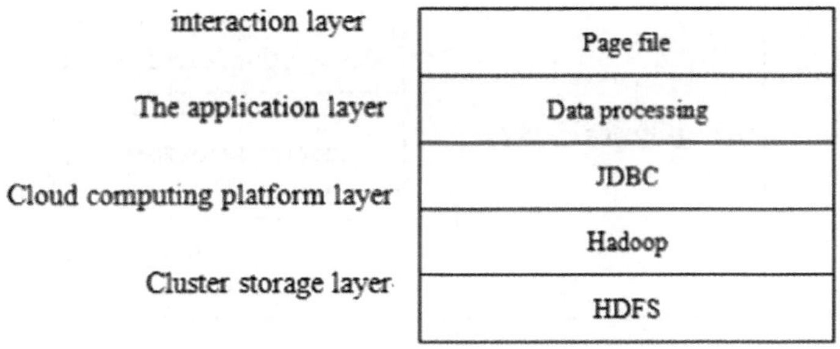

Fig. 1. System architecture diagram

3 Cloud Computing Platform Structure

3.1 Internal Hierarchical Structure of Cloud Platform

The interior of cloud platform is divided into different layers according to the functions provided. The same layer deals with the respective problems. The functions of each layer are independent from each other, and corresponding interfaces are defined for data communication. From the bottom to the top, they are: physical layer responsible for system communication, unified resource layer responsible for resource integration, platform layer obtained after cluster virtual and relevant application layer on platform layer [5]. See Fig. 2.

(1) At the bottom is the physical layer, which is composed of the basic hardware facilities and communication facilities of the cloud platform. The physical layer includes computers, storage devices, network devices and other facilities to provide hardware service support for the cloud platform and guarantee the communication function of the system [6].

The school provides the physical layer resources uniformly, which can be used by all departments of the school. These resources are managed uniformly. After the cluster

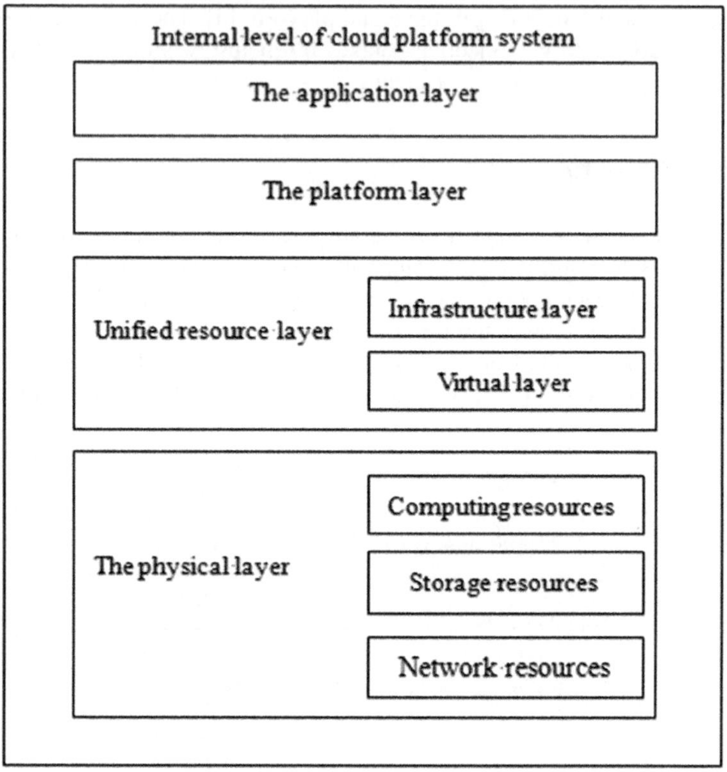

Fig. 2. Internal hierarchical structure of cloud platform

virtualization, they are divided according to the needs of different departments. Unified resource maintenance and scheduling mechanism reduces resource maintenance cost. At the same time, when a system is completed and the historical data of the system is saved, the virtual server occupied by the system can be recycled, which can be unified management for allocation to other systems. At the same time, all departments can share cloud data to improve resource sharing capacity and the department's ability to coordinate and handle affairs.

(2) Virtual layer services (dSaaS) mainly virtualize the infrastructure as required into several virtual facilities (virtual machines), whose storage and computing power can be adjusted again according to different requirements [7].
(3) The platform layer provides services equivalent to the operating system level. It can usually configure the specified platform for users according to their different platform requirements, such as Windows platform, Linux platform and so on. Users can process their own transactions and compute on the platform [8].
(4) Application layer is the service software above the platform layer, which is centrally deployed on the cloud platform and used by users [9]. The advantage of unified software deployment is to avoid problems caused by different versions. In addition, from the perspective of security, as the data of service software is stored

in the cloud, end users can only log in to the cloud system or use the service software through the browser, and cannot download the data privately, which is more conducive to data protection and the security of relevant confidential information.

3.2 Cloud Platform Network Architecture

The remote teaching system based on cloud platform is for the unified operation of the school. The existing hardware resources of the school are unified and virtualized out of the database server resources for the installation of projects, database operating system and storage of project data. Application server resources are responsible for running web applications, interface server resources, web server resources, and so on. When a user accesses the cloud platform, it is first verified through the firewall. If it is a campus user, it is allowed to access; if it is an external user, it is denied. Access to the web server, which communicates with the data storage server and returns the communication results to the web server, and finally to the user. The network architecture of cloud platform is shown in Fig. 3.

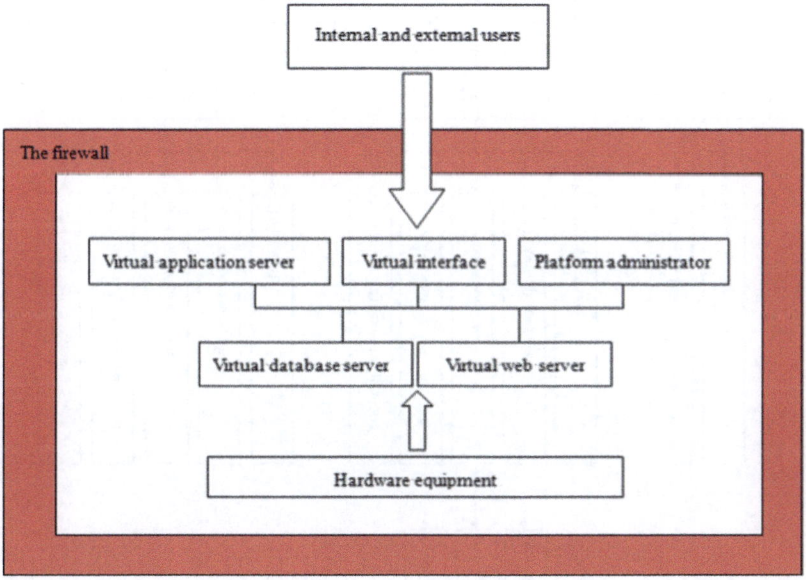

Fig. 3. Cloud platform network architecture diagram

4 Cluster Storage

HDFS, a distributed file storage system, can improve fault tolerance and writing performance by partitioning data storage. Each block is divided into 128M for sequential storage of data, recording its offset. Therefore, even if a certain node of the stored procedure fails, the data after recovery can be stored according to its offset and block number to ensure the normal operation of the whole system and prevent data loss. These features can well solve the storage and security problems of a large number of course-related data. Therefore, this module mainly implements the storage and management of course data and the upload and download of the data through HDFS [10]. On the other hand, in a Hadoop cluster, both the Mapreduce computational model and the RDD abstract model need to read the data to be processed from the distributed file system HDFS while performing tasks. When the task is finished, you also need to store the results in HDFS [11].

5 Functional Modules

The system is divided into three user rights: administrator, teacher and student. The main permissions of each module are shown in Fig. 4. This part is designed in the application layer and interaction layer of the system structure.

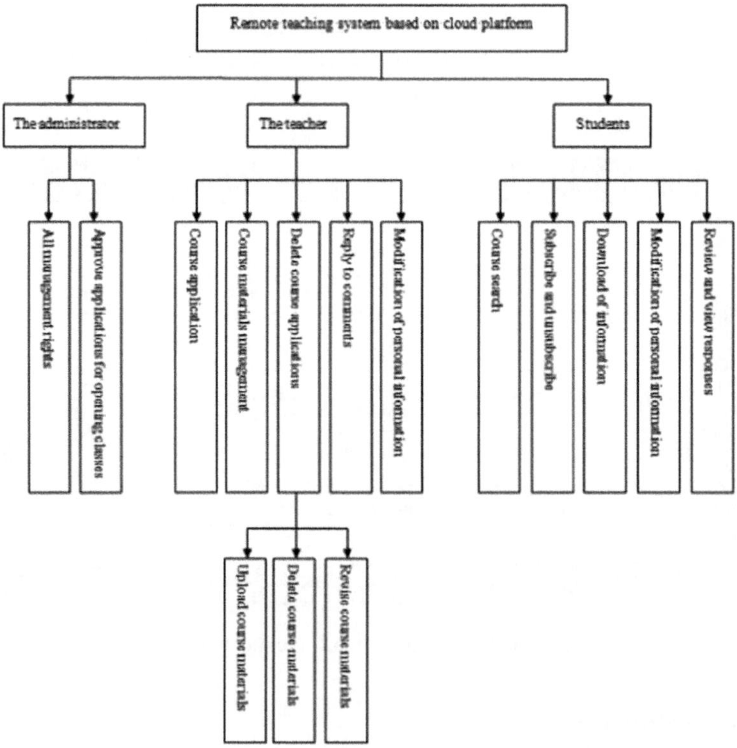

Fig. 4. Function module diagram of remote teaching system based on cloud platform

6 Conclusion

Cloud is a product of the new era. It integrates information resources to the greatest extent and provides the possibility for multiple systems in the university to share resources. Remote teaching system is one of the systems deployed in cloud platform. This system follows the design idea of MVC and adopts the SSH box multi-frame integration scheme and MySql database. Cloud platform based remote teaching system makes teachers' teaching and students' learning become efficient and convenient.

References

1. Tong B, Chen J, Zhang S (2017) LTR control based on GA-BP neural network for Hammerstein model of pH neutralization process. In: 2017 29th Chinese control and decision conference (CCDC). IEEE, pp 5494–5500
2. Wang Q, Shi Y, Shi C, Zhu X Design and implementation of TCM data mining system based on cloud platform hadoop. Comput Appl Softw 35(10)
3. Tang C, Lu P (2018) Design of remote energy saving control platform system based of cloud computing technology. Comput Meas Control 26(10)
4. Yu, L Development and design of hospital information management system based on client/server structure. Electron Design Eng 26(21)
5. Feng, X (2018) Construction and realization of the university library information sharing system based on cloud computing. Electron Design Eng 26(18)
6. Dou J, Yu W, Cao J, Liu Y (2018) Engineering material inquiry system based on cloud computing platform. J Comput Appl 38(S1)
7. Zhou J, Li W (2018) Research on logistics block chain consensus algorithm based on cloud computing. Comput Eng Appl 54(19)
8. Zhang L Design and research of network security attack and defense experiment platform based on cloud computing technology. Electron Design Eng 26(17)
9. Yang G, Gao, Xue T (2018) Design and implementation of community elderly monitoring system based on cloud platform. Mod Electron Tech 41(20)
10. Ma M (2017) The research on the remote collaboration of the Chinese medical science project based on the cloud platform. Mod Electron Tech 1(4)
11. Xiong L, Chen Z, He C (2017) Design and implementation of the personnel management. System based on cloud computing. Softw Eng 20(12)
12. Youth Fund Project of Jilin Agricultural Science and Technology College Project Number: 2017 No 113

The Review of Risk Identification of E-Commerce Supply Chain Under the Network Economy

Ping Kang[1(✉)] and Tianyu Zhang[2]

[1] School of Management, Wuhan University of Technology, Wuhan 430070, People's Republic of China
kpkangping@126.com
[2] Bank of China, Kaifeng 475000, China

Abstract. For the past few years, with the supply chain more and more important to the enterprise development, the supply chain has gradually become a undeveloped important research object. However, many risks still exist in the development of the supply chain, especially in the context of the prevailing network economy. As more and more companies use e-commerce platforms, new development environments will create new challenges for the growth of these e-commerce supply chains. This article will combine the characteristics of the network economy, based on previous scholars' research, and review the supply chain risk and prevention under the network economy, then lay a solid theoretical background for the following research.

Keywords: Internet economy · E-commerce supply chain · Supply chain risk · Risk identification

1 Introduction

With the progress of technology and the maturity of society, the competition between enterprises now no longer stays at the price competition, but between supply chains. A supply chain is a network of companies or departments that are obtained from raw materials, processed into semi-finished products, and delivered to customers. Supply chain is to grasp the flow from the source of suppliers to the end of users through value-added processes and distribution channels. It starts with supply and ends with consumption. With the enthusiasm of the supply chain, more and more enterprises constitute the enterprise "ecosystem". The supply chain is the "food chain" of this system. The reason why the supply chain is deeply valued by enterprises and academia is because of the economic effects that supply chain management can bring to enterprises: Linking economic effects. It means connecting a large number of market entities through an information network, establishing a new type of competitive synergy, and creating a new economic effect that is different from economies of scale and economies of scope [1]. The integration effect of the value chain: the supply chain itself is a value chain, but beyond the value chain. It optimizes the value chain by planning, coordinating and controlling logistics, information flow and capital flow throughout the

Z. Xu et al. (Eds.): CSIA 2019, AISC 928, pp. 706–713, 2020.
https://doi.org/10.1007/978-3-030-15235-2_97

supply chain, creating maximum customer value and achieving value chain integration. The comprehensive economic effect of the process: The management of the supply chain is essentially a business process reengineering that relies on the value chain. It organizes the node enterprises around the business process, and obtains the comprehensive economic effects of the process through the control and coordination of the process.

On the other hand, the development of the network economy is unstoppable. The rising proportion of China's network economy in GDP shows the status and role of the network economy in the national economy. The data shows that as of June 2017, the number of Chinese netizens has reached 750 million, which is close to the total population of Europe. And the huge population base provides a broad space and market for the development of the network economy [2]. Moreover, China's network infrastructure has gradually improved, and the conditions for information soft environment such as database, information transmission, information application system, information technology industry, and information human resources have been greatly improved. With the continuous improvement of network infrastructure, the continuous upgrading of mobile terminals and the increasing number of Internet users, more and more enterprises are beginning to use the Internet. Enterprises actively participate in the system construction of e-commerce and conduct trade activities through e-commerce. Internet companies continue to grow and develop, and a large number of well-known Internet companies have emerged, such as Alibaba, Tencent, Baidu, Sohu, Netease, and Ctrip. China's network economy has shown a good momentum of development.

It is this change in the background that makes the environment for supply chain development changeable. Whether it is from the external overall development environment or the internal technology development environment and information sharing environment, it has brought more uncertainty to the growth of the e-commerce supply chain. It also challenges the progress of the supply chain and is one of the reasons for the high risk of supply chain. Therefore, when we are in this big background, we will find that the supply chain is a double edged sword. On the one hand, it can bring different economic effects and efficient profits to enterprises. On the other hand, competition between supply chains is becoming increasingly fierce. Moreover, driven by various factors, including information asymmetry, information untruth, market uncertainty and so on, the cooperation between some relevant organizations will also have different types of risks [3]. These risks will also become potential threats to its development. For example, the vulnerability of the chain system is utilized, which not only damages the system itself, but also causes losses of upstream and downstream enterprises and the entire supply chain [4]. Therefore, supply chain risk management has become the focus of academic research. This paper will also combine the meaning and characteristics of the network economy from the perspective of literature review. And in the context of the development of network economy, we identify risks faced by e-commerce supply chain. It lays the foundation for the subsequent risk assessment, risk countermeasures and risk assessment measures in the risk management process.

2 The Development Trend of the Network Economy

The Internet economy is also the Internet economy, including the Internet-based economic form, the economic form of applying the Internet, and the economic form of providing Internet applications [5]. In the final analysis, it is the new integration of the Internet in the economic form, which is different from the traditional nomadic economy, agricultural economy, and industrial economy. The basic industries of the network economy include the information technology industry and the information service industry. Among them, the information technology industry can be subdivided into computer hardware industry, software industry and information media. The information service industry includes news, consulting, agency, telecommunications, and networking. From the classification of basic industries in the network economy, it can also be clearly seen that the biggest advantage of the network economy is the dissemination, collection and acquisition of information. The network economy uses two-way information communication to obtain timely and effective information in the economy. This is crucial for timely business information, and timely access to effective information is convenient for making correct decisions and plans.

There is no uniform definition of the meaning of the network economy. However, in terms of form, the narrow sense of the network economy is a new type of economy based on computer networks and an economic relationship network [6, 7]. It is an economic activity based on the Internet, such as e-commerce, network enterprises, network consumption, etc. The broad-based network economy refers to the economic activities that use information networks (including the Internet, extranets, intranets, etc.) as the platform, information technology and information resources as the characteristics, and information and knowledge play an irreplaceable role [8]. In short, the network economy is an organization that uses the Internet as a platform, using information as a carrier, and using new methods of information dissemination, collection, and acquisition to conduct business transactions. On the other hand, the network economy is not only related to the information economy, but also has a close relationship with the network economy. However, it is significantly different from the agricultural economy and the industrial economy. It uses the Internet as a platform and takes information as the center. Through the transmission and collection of information, it makes timely positioning and decision-making on economic related information [9]. In today's era of rapid Internet development, the network economy is gradually gaining popularity among the people and becoming an important economic way to promote enterprise development and even GDP growth throughout the country. At the same time, the Internet has become an indispensable element in network economic activities. In the process of Internet economic development, we compare and summarize the characteristics of the network economy. The development characteristics of the network economy limit the application of traditional risk identification in the e-commerce supply chain.

The network economy is knowledgeable, time-sensitive, global and permeable. First of all, the network economy is different from the agricultural economy and the industrial economy. Most of the agricultural economy relies on manpower. People grow crops and grain crops, plough and weave, and they can be self-sufficient in food, clothing and housing. At the same time, weather and climate become important factors. The prosperity of agriculture determines whether the country prosperous or not, and determines the lifestyle and quality of life of the people. The second industrial revolution opened the door to the era of industrial economy. Under the industrial economy era, various steam looms and other machines became important production subjects, but they also required human operation and cooperation. However, in the era of network economy, knowledge has become an important element. The most important thing that the network economy relies on is intelligence rather than physical strength. Knowledge management has become especially important in the management activities of a company. This is the knowledge of the network economy. Secondly, the network economy is based on the Internet network, based on the social network, and the information-centered economic activities. Time-sensitive information plays an irreplaceable role in the Internet economy. The Internet economy is a real-time economy with an unprecedented pace of acceleration. Coupled with the timeliness of economic information itself, the timeliness of the network economy is particularly prominent. Furthermore, the scope of the network economy has expanded to a global scale, and informationization has made the space for economic activities smaller. In this era of economic globalization and world integration, the development of the network economy has gradually broken through national boundaries and truly spread economic information on a global scale. Finally, the network economy is different from the nomadic economy, the agricultural economy, and the industrial economy. However, in the era of rapid Internet development, information technology has gradually penetrated into the traditional economy as an important part of the network economy [10]. The integration of information technology and traditional economy, the integration of industries, the combination of network information and artificial intelligence, and power machine technology have led to the rapid development of emerging industries under the network economy.

3 E-Commerce Supply Chain Research

Recently, with the rapid growth of the network economy, the extensive blending of e-commerce models and supply chain management has become a trend in the field of scientific research related to operational management. E-commerce supply chains are more fully applied to ICT than traditional supply chains. It also greatly improves the efficiency of the entire supply-chain management, making the management of e-commerce supply chain and traditional supply-chain management different. Hausman [11] believes that electronic supply-chain management is to further expand the scope of actions and capabilities contained in the supply chain by improving products and services to meet the end consumers' needs in the supply chain. Giménez and Lourenco [12] believe that the electronic supply chain is the automation of the main business processes from the upstream of the chain to the user through the use of Internet technology. It also improves the quality of products and services, accelerates the transfer of information in the supply chain, and creates additional value for companies in the supply chain. Lancaster et al. [13] believe that electronic supply chain management is to integrate information technology and, more importantly, Internet technology to promote the integration of business processes and promote faster information exchange between relevant nodes in the supply chain. Frohlich and Westbrook [14] believe that the application of Internet information technology in supply chain management can be extended from the direction of demand and supply strategy. This view gives two directions for e-commerce supply-chain management under the background of network economy.

4 E-Commerce Supply Chain Risk Identification

As we all know, in today's competitive global business environment, supply-chain management is the essential point to the success of any business alliance. Aiming at maximizing profit and optimizing operations, many supply chain alliances continuously improve the real-time response speed and operational efficiency of customer demand in the supply chain. However, while the lean and agile of the supply chain bring unparalleled competitive advantages to members of the supply chain, it also brings many risks that are uncertain from various dimensions. It will lead to a more fragile supply chain, and only timely identification and assessment of supply chain risks can effectively prevent losses caused by supply chain breaks. Supply chain risk management begins with the identification of potential sources of risk that affect supply chain performance. Understanding and understanding the sources of risk that disrupt supply chains and assessing their impact on the supply chain can help companies build efficient supply chain networks and make appropriate decisions that mitigate their adverse impact on the supply chain. The risk factors that cause supply chain failure or breakage are diverse and, in general, can be divided into endogenous and exogenous risks, depending on the source of the risk. Dalmas [15] and others believe that the e-commerce supply chain relies on e-commerce technology, and its risk is more virtual and uncertain than the traditional supply chain. Although the traditional supply chain risks are also complex and diverse, the e-commerce supply chain risk sources are more complex and diverse. Wang [16] believes that the risks of the e-commerce supply chain mainly come from the risks of information flow, capital flow and logistics. Risks in information flow include the risks brought by the inaccuracy of the information itself and the uncertainty of the information delivery medium and the information risks caused by the moral hazard of related companies in the supply chain. The risk of capital flow includes the risks of the company's own financial risks and the deviation of the flow of funds. The logistics risks are mainly from the risks caused by rapid response [17].

Supply chain risk under the network economy is manifested in many aspects, and the perspective of identifying supply chain risks is also important. Nie [18] summarized the risks brought by the 020 model to the supply chain under the network economy from four aspects: supply chain integration, supply chain informationization, supply chain operation process and organizational risk. Zhong [19] proposed that the challenges faced by supply chain risks include the difficulty of controlling risks, hidden dangers in e-commerce platforms, user management needs to be improved, and funding problems have not yet been resolved. Liu [20] analyzes the risk of rural e-commerce supply chain in Hebei Province from four aspects: demand risk, environmental risk, information risk and structural risk. Similar to the research perspective, Luo [21]'s supply chain risk exploration summarizes the supply chain risks under the network economy into four categories: credit risk, operational risk, environmental risk and security risk. It also studies the prevention of cultural credit risk and operational risk. In addition, some scholars discuss the risks of a certain feature of the network economy or a module in the supply chain process, or analyze the specific risks of a certain product supply chain by case. For example, Huang [22] explores supply chain risks from two major aspects: the inherent supply chain risk of the enterprise and the supply chain risk under the characteristics of modern information technology caused by Internet development. Among them, the risks faced by the supply chain under the network economy (mainly discussing the information technology characteristics of the network economy) include the risk of system failure, moral hazard of internal personnel, moral hazard of each partner in the supply chain, and social moral hazard. Li [23] explored supply chain risks from the perspective of e-commerce supply chain financing model under the network economy, mainly including supply chain risks under the two financing modes of network orders and network warehouse receipt financing. Wang [24] took a case analysis and analyzed the risks of the XP smart TV supply chain in the context of Internet development. In addition to market uncertainty, changes in economic cycles, macroeconomic policies, laws and natural disasters, there are inherent risks such as lack of supply chain leadership, frequent flow of people and supply chain management model, limited information sharing channels and insufficient professionalization of supply chain management.

5 Conclusion and Prospects

In general, whether it is to analyze the various risks faced by the supply chain in the network economy from a holistic perspective, or to intercept a module of the supply chain process, or to explore the supply chain risk based on a certain feature of the network economy. The risks mentioned in these studies will become direct or potential risks in the development of enterprise supply chain under the network economy. All need the risk identification work of the company's supply chain, analyze the risk source and identify the risk, and actively face these risks with confidence. These studies will have certain reference value and reference significance for enterprise supply chain risk management in the current rapid development of network economy.

However, the speed of the development of the network economy is staggering and unpredictable. Moreover, through literature research, it is found that most of the literature tends to be limited to a single problem from a single perspective, resulting in a chaotic research perspective on e-commerce supply chain management, and does not form a complete system as a whole. Research on traditional supply chain risk management has been relatively mature. For the e-commerce supply chain under the e-commerce model, most of the existing researches focus on the expansion and application of e-commerce in various industries, and there are few in-depth studies on e-commerce supply-chain risks. E-commerce supply chain and traditional supply chain have both commonality and uniqueness. Therefore, based on the previous scholars' research on traditional supply chain risk, combined with the characteristics of network economy and the development status and research status of e-commerce supply chain risk, it has certain practical significance to identify and evaluate the supply chain risk of e-commerce enterprises.

References

1. Ji Y (1998) The externality of network economy and the economic efficiency of connection. Quant Econ Econ Technol (8):61–64 (in Chinese)
2. The 40th statistics on the development of China's Internet, [EB/OL] [2017]. http://www.cnnic.net.cn/
3. Ma S (2003) How to prevent supply chain risks. Chin Comput Users (1):21 (in Chinese)
4. Ding W, Liu K, He G (2003) Supply chain risk research. Chin J Saf Sci (4):64–66 (in Chinese)
5. Zhang Y (2016) Analysis on the transformation of marketing strategy and the countermeasures under the network economic times (9):16–18
6. Hu Z (2014) Study on the marketing strategy of the small and medium-sized enterprises in the network economy 《Shandong Trade Unions' Tribune》
7. Wang Y (2012) Corporate marketing strategy choice and innovation under the network economy. Reform Strateg
8. Sun J (2012) Studies on Chinese enterprise management mode and marketing channel reform under the network economy circumstance. Int Bus Manag 4(2)
9. Barua A, Pinnell J, Shutter J et al (1999) Measuring the Internet economy: an exploratory study
10. Barua A, Whinston AB, Yin F (2000) Value and productivity in the Internet economy. Computer 33(5):102–105
11. Hausman WH (2012) Supply chain performance metrics. In: The practice of supply chain management: where theory and application converge. Springer, Boston, pp 61–73
12. Giménez C, Lourenco HR (2013) E-supply chain management: review, implication and directions for future research. In: Working papers
13. Lancaster S, Yen DC, Ku CY (2014) E-supply chain management: an evaluation of current web initiative. Inf Manag Comput Secur 14(2):167–184
14. Frohlich MT, Westbrook R (2011) Demand chain management in manufacturing and service: web-based integration, drivers and performance. J Oper Manag 20(6):729–745
15. Dalmas M, Giaroa S, Zamboni A, Bezzo F (2011) Strategic design and investment capacity planning of the ethanol supply chain under price uncertanity. Biomass Bioenergy 35(5)
16. Wang L (2010) Characteristics and classification of supply chain risk in e-commerce environment. Sci Technol Entrepreneurship Mon 23(4) (in Chinese)
17. Su Y (2015) Research on risk identification and evaluation of supply chain based on O2O e-Commerce enterprises. Wuhan University of Technology (in Chinese)
18. Nie R (2016) Supply chain risk and prevention in O2O mode. J Bus Econ (4):72–74 (in Chinese)
19. Zhong X (2016) Supply chain risk management in the age of "Internet+"—taking coal traders as an example. China Bus Rev (16):137–138 (in Chinese)
20. Liu Z, Sun Q, Guo D (2017) Research on supply chain risk prevention mechanism in rural e-Commerce in Hebei Province. Modern Mark (Mag) (8):222–223 (in Chinese)
21. Luo Y, Wang Y (2018) Discussion on financial risks and prevention of internet supply chain. Logist Eng Manag (1):125–127 (in Chinese)

22. Huang W (2005) Internet-based enterprise supply chain risk and its prevention. China Manag Inf (a09):60–64 (in Chinese)
23. Li P (2018) E-commerce supply chain financing model and risk prevention. Accountant (1) (in Chinese)
24. Wang H (2016) Internet+ start-up intelligent hardware company supply chain risk management research. University of International Business and Economics (in Chinese)

Research on Multiple-Degree-of-Freedom Manipulator

Zezhong Chen(✉)

Southeast University, Nanjing, Jiangsu, China
1285373249@qq.com

Abstract. In industry, the work of a manipulator is the same as that of a human arm, which can be grasped and transported according to a specific structure. The manipulator system with multiple degrees of freedom is also called robot, which simulates the movement of human arms. Robot manipulator is composed of several links and joints. It is a complex dynamic system with multi-input and multi-output characteristics, and it has a complex coupling relationship with non-linear characteristics. In order to realize the dynamic simulation of the manipulator and the controller design of the manipulator, the dynamic analysis of the manipulator is an effective means. For this reason, a design of multiple-degree-of-freedom manipulator control system is proposed based on PLC. Firstly, the hardware of the system is designed, and the block diagram of the hardware is obtained. Then, the software of the system is designed. Finally, the experiment is carried out. The experimental results show that the design of the system has the advantages of stable operation and accurate positioning, and it will promote the future social and economic development of our country.

Keywords: Multiple-degree-of-freedom · Industrial robot · Manipulator

1 Introduction

On May 8, 2015, the State Council promulgated the first ten-year action plan for China to implement the strategy of manufacturing power, Made in China 2025, which clearly defines intelligent manufacturing as the core of strategic emerging industries that the state focuses on cultivating and developing, and as the guiding principle for the transformation from "Made in China" to "Made in China" [1]. As a branch of the core equipment of intelligent manufacturing, industrial robots have become the automation tools of flexible manufacturing systems, automated factories and computer integrated manufacturing systems, and an important symbol to measure the level and core competitiveness of a country's manufacturing industry [1–4]. Industrial robot is an automatic, reprogrammable, multi-purpose three-axis or multi-axis programmable manipulator which is fixed or mobile in industrial automation [2–9]. According to ISO definition, industrial robot is a multi-joint manipulator or multiple-degree-of- freedom machine facing the industrial field [5, 6]. It can automatically perform work, and it is a machine that realizes various functions by its own power and control ability [7, 10–16]. It can be commanded by human beings or run according to pre-programmed programs [17]. Modern industrial robots can also act according to principles and programs

Z. Xu et al. (Eds.): CSIA 2019, AISC 928, pp. 714–721, 2020.
https://doi.org/10.1007/978-3-030-15235-2_98

formulated by artificial intelligence technology. Industrial robots are a new and high technology which combines computer, cybernetics, mechanics, information and sensing technology, artificial intelligence, bionics and other disciplines. Industrial robots are highly accurate, reliable and consistent [8]. It is suitable for working in monotonous, repetitive, harsh and dangerous environment [18].

In this regard, this paper establishes a three level industrial automation multiple degree of freedom manipulator system. The field acquisition layer completes the electric parameter acquisition of each electrical equipment; the central control layer completes the functions of data processing, communication and control; and the data presentation layer completes the functions of picture presentation, management, alarm and publication[19]. In addition to the automation of conventional manipulator system (telemetry, remote control, remote communication, remote adjustment), the system can also replace manual operation in high temperature and dangerous operation area, and can change relevant parameters at any time according to the changes of work-piece and the requirements of movement process. It can reduce workers' labor intensity and increase labor productivity [20].

2 The Basic Structure of Industrial Robots

Generally speaking, industrial robots consist of 3 main parts and 6 subsystems. The 3 parts are the mechanical part, sensing part and control part. The six subsystems can be divided into mechanical structure system, driving system, perception system, Robot-Environment interaction system, human-computer interaction system and control system [9–13]. The block diagram of the industrial robot is shown in Fig. 1.

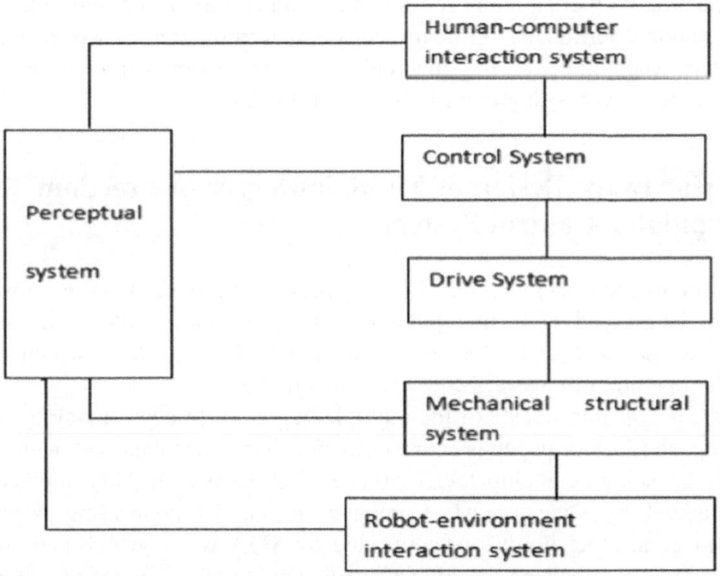

Fig. 1. Block diagram of industrial robot

The mechanical part is the body part of the robot, which is used to realize various actions. It is divided into mechanical structure and driving system. The mechanical structure is composed of many mechanical connecting rods connected by joints [14, 15]. It consists of four parts: fuselage, arm, wrist and hand. Each part has several degrees of freedom. The end-effector is an important component directly mounted on the wrist. It can be a multi-fingered gripper or a working tool. The driving system is a device to make all kinds of mechanical parts move. Its function is to provide the driving force for each part of the robot and its joints.

Sensing part is used to perceive internal and external information. It is divided into two systems: perception and Robot-Environment interaction. The perception system is composed of internal sensor module and external sensor module. It is used to transform all kinds of internal state information and environmental information of the robot from signals to data and information that can be understood and applied by the robot itself or between the robots. In addition to the need to perceive mechanical quantities related to its working state, such as displacement, speed, acceleration, force and moment, visual perception technology is an important aspect of industrial robot perception. Robot-environment interaction system is a system to realize the interconnection and coordination between industrial robots and equipment in the external environment [16, 17]. Industrial robots and external equipment are integrated into a functional unit, such as manufacturing unit, welding unit, assembly unit and so on. It can also be the integration of multiple robots, machine tools or parts storage devices into a functional unit capable of performing complex tasks.

The control part is the brain of the robot which is the main factor to determine the function and performance of the robot and make the controlled object produce the desired behavior of the controller [18–20]. It is divided into two systems: human-computer interaction and control. Human-computer interaction system is a device that enables operators to participate in robot control and communicate with robots, such as computer standard terminals, command consoles, information display boards, danger signal alarms, teaching boxes, etc. In simple terms, the system can be divided into two parts: instruction given system and information display device.

3 The Hardware Design of Multiple-Degree-of-Freedom Manipulator Control System

The hardware of the system designed in this paper collects the feature information of the object to be grasped by the manipulator through the data collector, and transmits the collected information data to the next unit, that is, the controller, which effectively controls the operation process of the whole manipulator.

The data acquisition device of the multiple-degree-of-freedom manipulator control system proposed in this paper is more powerful than other data acquisition devices, because the latest high-end chip TAL7250 is used in the system. This chip is developed and popularized by Danish TAL Company. It has 14 connecting points. When working, 14 connecting points are connected by SMA wires, which can collect data together. It greatly enhances the efficiency of data acquisition, reduces the working time and improves the accuracy of work. The data acquisition accuracy of TAL7250 can be as high as 15 bits, and the acquisition process delay is very short, only 3 μs, which can be neglected. This chip has very low requirement for working environment. Usually when the voltage is over 15 V and the current is over 50 A, it can be implemented continuously and steadily.

The transmitter is to transmit the collected feature information to the controller. This process requires very high transmission efficiency. Therefore, the hardware transmitter designed in this paper chooses D6W52 transmission chip which Thced Company of France promoted and listed in 2017. Although the chip has not been put on the market for a long time and its performance and effect need to be studied, it uses the highest-end transmission technology, and has eight networks working together to complete the transmission work. The transmission time is very short. Generally, 15-bit data can be transmitted to the next control unit only after 5 μs of transmission time.

Data controller is the final unit of the whole hardware system, and it is also the core part. The main control system of this controller uses PLC crystal output tube FX2 N. The external connection controller has one serial port and three interfaces (PWM interface, SPI interface, I/0 interface). It receives the information and data transmitted by the transmitter through the wireless transceiver.

The main control chip is used to complete the control, PWM interface, SPI interface and I/0 interface. The display module, input module and control module are connected at the same time. The 45 control terminals use PLC algorithm to control the work of the manipulator at the same time. The comparison results of the two systems are shown in Fig. 2.

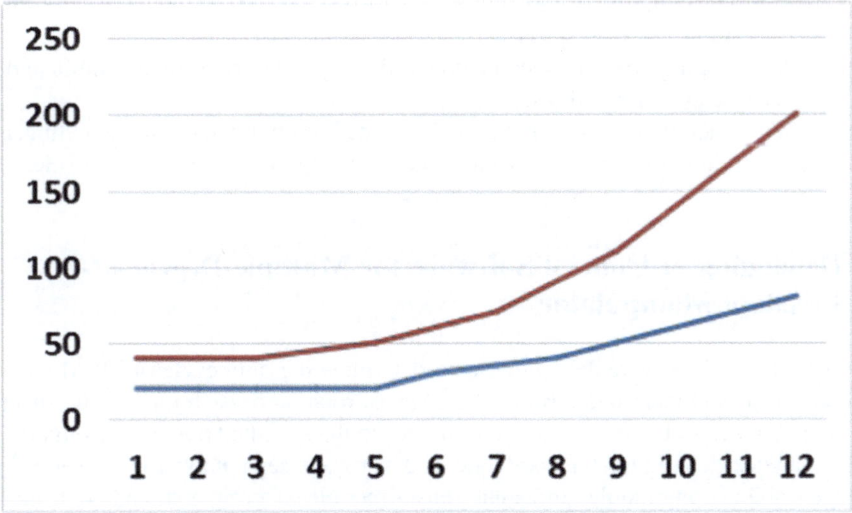

Fig. 2. Comparison results of two systems

The vertical motion is divided into two stages: the reset stage and the positioning stage. In the reset stage, when the manipulator moves through the motor, the whole arm is in a reversed state until the arm touches the whole switch formed by the limit. The positioning stage refers to the movement of the arm from the original position when the manipulator moves through the motor, and the position of the arm is measured by the technical switch, and the behavior ends when the arm reaches the designated position. When joint drive is set as sinusoidal function, its displacement, velocity and acceleration are similar to sinusoidal wave. For heavy-duty, high-precision or fast-moving manipulators, this trajectory or trend can be used to optimize the design of manipulator structure.

4 Requirements for Control Process of Manipulator

(1) Press the start button to detect the initial position and initial state of the manipulator.
(2) When the initial detection is completed, the rotary cylinder moves to the left station, and the lifting cylinder of the manipulator descends.
(3) When the lifting cylinder descends to the lower limit, the pneumatic air claw is opened, and the work-piece is grabbed from the left station.
(4) After lifting for one second, the lifting cylinder begins to rise.
(5) After lifting limit, the lifting cylinder is stopped and the rotary cylinder is turned.
(6) After the rotary cylinder is rotated to the right limit, the telescopic cylinder extends to the right.
(7) Right shift to right limit position (right working position), telescopic cylinder stop working.
(8) The lifting cylinder moves downward to the lower limit position, and the air claw loosens the work-piece.
(9) After stopping for one second, the lifting cylinder rises to the limit and the telescopic cylinder shrinks.
(10) After retracting the telescopic cylinder to the left limit, the revolving cylinder gas reversely turns to the left position and starts the next wheel action cycle.

5 Debugging of Control Software for Multiple-Degree-of-Freedom Manipulator

The main task of software debugging is to determine the duty cycle of PWM wave by trial and error, and then to determine the angle of rotation of each joint to the specified position. After repeated trials, each joint moves to the specified position throughout the process. After debugging the hardware and software separately, the power supply, voltage stabilization module, ultrasonic sensor module, steering gear and controller are connected as a whole according to the system overall connection schematic and device specific interface and function. Keil, the software of STM32, writes the control program to the controller through J-Link simulator, observes the power indicator lamp on the development board and the prompts on the software, and after downloading the program, the reset key can be used to carry out the operation of moving.

Turn on the switch of the power module and the power switch on the development board, and you can hear the rudder of the control claw making a slight noise, while the claw has a small opening and closing action, which is the result of the circular scanning of the ultrasonic sensor. The connection is correct and the steering gear is also supplied with electricity. The moving work-piece keeps approaching the specified position, and the manipulator does not move. When the work-piece reaches the designated position, the ultrasonic sensor sends control signals to the controller, the manipulator moves, completes a series of actions such as grasping, handling and placing, and then returns to the initial position and enters the waiting state. During the whole debugging process, due to various disturbances from inside and outside of the system, there are slight errors in the rotation angle of each joint. The most serious factor causing position errors is the unstable voltage of the whole system during operation due to the use of adjustable buck. The solution is to measure the input voltage of the joint drive system with a multimeter before debugging, correct it, and then put it into use after debugging several times to determine that the system is more accurate.

During the whole machine debugging process, it has not been very smooth, and there have been many problems. When debugging problems occur, the general step is to debug the hardware separately and check the hardware indirect line first, and then check whether the program is correct after confirming that there is no problem with the hardware. The problems and solutions are as follows:

(1) When the system and MCU are powered, the program is downloaded to the development board through J-Link simulator. After pressing the reset key, the arm that should be upright is not upright. There is obvious bending at the shoulder joint. At the same time, the steering gear buzzes. When the arm is straightened by hand, the sound is gone. The above phenomenon shows that the shoulder joint actuator's torque is not large enough to make it stand upright, so it needs to be replaced by a larger torque actuator. Torque is an important data in the selection of steering gear. The weight of the workpiece and the weight of the manipulator should be fully taken into account, and the steering gear should be selected with a certain margin.

(2) When the power supply of the system and the development board is supplied, the reset key is pressed after preliminary judgment of no abnormality by the indicator lamp. The whole system does not work according to the normal procedure. After careful investigation, it is found that there is no common ground between the controller and the manipulator arm. If the hardware is changed a little bit, the problem can be solved after the controller and the manipulator are shared. Because if there is no common ground between the controller and the manipulator, the servo cannot correctly identify the high and low levels of the PWM wave output by the controller, which leads to its failure to work properly.

(3) After many debugging, the whole system does not work normally. No problem was found in power failure inspection. However, when this problem occurred again, the power supply voltage was measured with a multimeter. It was found that the on-load voltage was too low, and the no-load voltage was normal after power failure, which indicated that the whole manipulator system was working with a large load. And when the power supply voltage is low, the steering gear

cannot work normally. Therefore, during the application of the system, the battery should be charged on time and the output voltage of the step-down module should be increased appropriately.

6 Summary

The control system of multiple-degree-of-freedom manipulator integrates PLC technology as the core, pneumatic drive and touch screen control. It can not only realize simple handling between two points, but also satisfy the arbitrary selection of material rotation direction in the process of transmission. It has the characteristics of simple structure, convenient operation, low cost and high efficiency. It has a wide range of applications in modern machinery, metallurgy, chemical engineering and other engineering fields. At the same time, the touch screen controller technology will be integrated, and it will have broad application prospects.

References

1. Jing NI, Cheng LP, Liu XQ et al (2014) Design of hydraulic servo manipulator with five degree of freedom. J Mech Electr Eng
2. Shen H, Yang L Zhu X et al (2015) A method for structure coupling-reducing of parallel mechanisms. 9246:199–210
3. Ullah MI, Ajwad SA, Islam RU et al (2014) Modeling and computed torque control of a 6-degree-of-freedom robotic arm. In: Conference on robotics and emerging allied technologies in engineering. IEEE, pp 133–138
4. Luo Q, Li X (2017) Research of the active vibration suppression of flexible manipulator with one degree-of-freedom
5. Sun J, Liu X, You L et al (2016) Design of multiple-degree-of-freedom robotic manipulators based on ARM. Microcomput Appl
6. Zhang D, Gao Z (2015) Mechanism and actuation hybridization for a four degrees-of-freedom parallel manipulator. Int J Mech Mater Des 11(3):301–308
7. Wang X (2015) Design and research of robot manipulator with four degrees of freedom. Mod Manuf Technol Equip
8. Luo Q, Li X (2017) Research of the active vibration suppression of flexible manipulator with one degree-of-freedom. In: International conference on intelligent robotics and applications. Springer, Cham, pp 624–636
9. Wang W, Peng X (2017) Realization of control system for flexible manipulator with multiple degree of freedom. Electron Test
10. Montazeri A, West C, Monk SD et al (2017) Dynamic modelling and parameter estimation of a hydraulic robot manipulator using a multi-objective genetic algorithm. Int J Control 90 (4):661–683
11. Xiang'An BU, Sun QK (2017) Design and research on wall painting manipulator with multi DOF. Dev Innov Mach Electr Prod
12. Shen Y, Yuan Z (2017) Design for multi-fingered picking manipulator based on FPA control system. J Agric Mech Res
13. Cong F, Zhang Q (2017) Research of the collision problem for dual robot coordination system. J Mech Transm

14. Li DQ, Hong HJ, Jiang XL (2017) Dynamics modeling, control system design and simulation of manipulator based on lagrange equation. In: Mechanism and machine science. Springer, Singapore
15. Wang G, Guo F, Sun H (2016) Analyses and linkage control methods for fast implementation of multiple degree of freedom manipulators. In: International conference on computer science & education. IEEE
16. Xin G, Deng H, Zhong G (2016) Closed-form dynamics of a 3-DOF spatial parallel manipulator by combining the Lagrangian formulation with the virtual work principle. Nonlinear Dyn 86(2):1329–1347
17. Wu S, Wu T (2017) Modeling and simulation of kinematic system for multiple-degree-of-freedom manipulator. Autom Instrum
18. Liu Z, Xiong L, Zhai T et al (2015) Design of six degree of freedom picking manipulator. J Agric Mech Res
19. Zhang Z, Zou ST, Jiao LI et al (2015) Three-dimensional visual simulation research on six degrees of freedom manipulator. Comput Simul
20. Liu X, Liu Y, Zhu L (2018) Research on intelligent control system of manipulator based on multi degree of freedom. Wirel Pers Commun 5:1–17

Research on Green Exhibition Technology and Construction of the Application Evaluation System

Aimin Wei(✉)

Guangzhou City Construction College, Conghua Road, No. 166, Guangzhou 510925, Guangdong Province, China
524496523@qq.com

Abstract. With the vigorous development of the country, green exhibition has become one of the focuses of the exhibition industry. However, there are few empirical studies on green exhibition in domestic and foreign academia. Especially from the perspective of participants' systematic perception of green exhibition, discussing the impact of green exhibition operation mode on participants and exhibition brand value is a gap to be filled urgently in this research field. Through empirical analysis, this paper explores the perception of green exhibition operation mode on exhibition organizers' environmental responsibility performance and its impact on exhibition brand value. The results show that the green exhibition operation mode can enhance the recognition of exhibition organizers to fulfill their environmental responsibilities, and enhance the brand image perception and brand loyalty of the exhibition. Finally, the development strategy of green exhibition is also put forward.

Keywords: Green exhibition · Technology research · Application evaluation

1 Brief Introduction

Comparing with foreign countries' attention to green exhibitions since 1990s, green exhibitions can take into account the interests and demands of participants as much as possible. On the other hand, green exhibitions can continuously reduce energy consumption and carbon emissions of exhibition activities by using various technologies and management means to achieve environmentally friendly development [1, 2]. From the function or influence of green exhibition, green exhibition can produce important economic value and environmental benefits. For example, many scholars elaborated the environmental benefits of green exhibition from the perspectives of sustainable development, low-carbon economy and circular economy [3–7]. In terms of economic value, green exhibition can bring economic benefits, efficiency improvement, cost savings, but also can create a positive public image and enhance brand value [8–10]. However, most of the above judgments remain at the viewpoint level, and there are still few empirical studies related to the economic benefits of green exhibitions.

When discussing the development strategy of green exhibition related to organization and management, scholars at home and abroad mentioned that the most

Z. Xu et al. (Eds.): CSIA 2019, AISC 928, pp. 722–729, 2020.
https://doi.org/10.1007/978-3-030-15235-2_99

frequently used technology and materials of environmental protection and energy saving were used for exhibition [10–13]. Secondly, many people mentioned the paperless processing in marketing and customer service. Some scholars emphasized that low-carbon transportation organization and management is one of the important contents of green exhibitions, including avoiding traffic congestion and encouraging participants to take special cars. Some scholars believe that the green accommodation facilities designated by the exhibition are related to the low carbonation of exhibition [14, 15]. Some scholars and industry organizations have suggested that we should try to accumulate and publish the carbon emissions of the exhibition, and adopt incentives to select the best energy-saving and low-carbon exhibitors and individuals, so as to enhance participants' awareness of environmental protection. In addition, some scholars put forward the green theme strategy of the exhibition, that is, the theme design of the exhibition should have the local industrial foundation, social impact and sustainable development prospects. Green human resource management is also one of the green exhibition strategies mentioned, which mainly refers to green recruitment, green staff training and development of exhibition enterprises.

2 The Definition of Green Mice

Green Convention and Exhibition refers to the comprehensive consideration of all aspects involved in Convention and Exhibition activities, including conference site, service, catering, accommodation, transportation, activities and material supply, based on the theory of circular economy. We should regard the conservation of resources and environmental friendliness of exhibition activities as a systematic project, so as to rationally develop and utilize various resources and protect the environment. By reducing pre-exhibition resources investment and effective utilization of renewable resources, strengthening the recycling of resources in exhibition, reducing the discharge of waste after exhibition, the harmonious development of exhibition economy and environment can be realized.

2.1 Green Exhibition and Exhibition

Since the 115th Canton Fair began to implement the green development plan, the green exhibition trend of Canton Fair has realized the transformation from the bad environment, serious pollution waste and high safety risk of the pavilions in the past, to the safe, efficient and greatly reduced pollution waste. In the first and second phases of the 117th Canton Fair, the green booth penetration rate reached 75.6% and 71.5% respectively; in the 118th Canton Fair, the green booth penetration rate reached 51 758, and the green booth penetration rate reached 85.6%. Compared with the same period in the previous period, the waste materials of the first and second special booth construction decreased by 27.86% and 31.13%, respectively. The air quality of the exhibition hall was significantly improved, and the green booth exhibition effect was further improved.

2.2 Green Exhibits

In order to promote the participation of green exhibits, the Canton Fair actively cultivates green exhibition themes, creates green exhibition areas, and also sets up new energy exhibition areas. The following rules are formulated for arranging the location of ordinary exhibitions: Under the same conditions, exhibits belonging to environmental and energy-saving products and enterprises with innovative technology or independent intellectual property rights should give priority to booth location. According to statistics, as early as the 109th Canton Fair, exhibits brought by participating enterprises mainly focused on energy conservation, environmental protection, low carbon, green theme, energy conservation, environmental protection, green and other products, the proportion of more than 40%. On the one hand, the green environmental protection characteristics of exhibits can give consumers high-tech, tasteful and trendy customer experience. On the other hand, the recyclability of environmental protection exhibits is conducive to reducing production costs and improving added value of products.

2.3 Green Services

The green service provided by the Canton Fair aims at creating a fast, efficient and waste-reducing service system, maximizing the utilization of resources, strengthening the construction of information technology in organizing exhibitions, on-site management and service links, and improving the paperless office level. The efficient service system of the Canton Fair mainly includes three aspects: One is to upgrade mobile application services, launch mobile terminal applications on official websites, upgrade the functions of Wechat service and add "exhibitor dynamic" information services; the other is to integrate on-site information services and set up a new Canton Fair network center. Strengthen customer data acquisition and behavior analysis, launch the Canton Fair data dynamic display system, promote the basic work of the Canton Fair index construction; Third, upgrade the software and hardware security measures in an all-round way. These transformation can enhance service and reduce the consumption of manpower and material resources. For example, the "Brand Shop Window" section of the 118th Canton Fair on its official website is a newly established online exhibition project for brand enterprises. All exhibits in the "Brand Shop Window" of the Canton Fair will be the official target of publicity. The Canton Fair will select high-quality enterprises and products from time to time, and promote them to customers at home and abroad through official online social media accounts such as Wechat and Facebook. No longer use paper publicity materials, using the way of Internet push, advocate paperless, efficient, environmentally friendly and green.

2.4 Green Venues

The Canton Fair promotes the construction of green exhibition and green exhibition hall by speeding up the energy-saving transformation of exhibition hall equipment and facilities, reducing the consumption of water, electricity, gas and other resources and energy. At the same time, the integrated management system including quality management system (ISO 9001), environmental management system (ISO 14001) and occupational health and safety management system 60 (OHSAS 18001) was introduced to comprehensively identify environmental and hazard factors and take the lead in environmental quality testing and ecological management. We should introduce environmental and occupational health and safety management system and pay attention to individual environmental health. The Canton Fair not only pays attention to the "green pavilion" in the material of the pavilion, but also creates the "green pavilion" in the visual sense. The main landscapes around the pavilion are optimized for environmental landscaping. A total of 5100 m^2 of greening area and 952 vegetation plants have been planted, which further improves the level of environmental greening landscape of the exhibition hall, and creates an evergreen exhibition hall in four seasons, so that exhibitors and businessmen can happily negotiate in a beautiful and quiet environment.

3 The Status of Green Exhibition Research Abroad

The research on green tourism abroad mainly focuses on two aspects. One is the Construction Guided by green exhibition. Many foreign scholars have carried out detailed research. A series of practical guiding principles have been worked out: crisci of cat yard has analyzed a five-day meeting attended by 2,500 people, which may result in a great deal of waste of plastics, paper waste and resources, and pointed out some concrete implementation steps of Green Convention and exhibition, such as the use of environmental protection paper publicity materials, and the selection of conference venues, accommodation hotels and restaurants should follow the principle of proximity, and it is best to arrive on foot, recycle recyclable waste paper at the end of the meeting and so on. The second is the construction of green ecological venues. When developing green exhibition activities, some foreign countries introduce ecosystem into the design of venues. It integrates venues, commerce, hotels, entertainment, transportation and other exhibition facilities to form a system. Germany emphasizes the distance between the exhibition hall and service facilities such as hotels, restaurants, commerce and so on in the overall planning of exhibition halls.

When the exhibition hall is constructed, its function area and form layout are optimized, that is to say, the traffic is normal during the exhibition period and the traffic volume is reduced. When Britain develops the exhibition center, it keeps an open space in front of the exhibition hall, with paving, water surface, sculpture and greening, which beautifies and purifies the environment. At the same time, the bottom of the building can be used as a shared space, which can make the exhibition space run through and stretch, increase the capacity area of the venue and reduce the use of land. In France, when building large-scale exhibition venues, agricultural landscape is introduced, and farmland is used as a natural isolation zone for each exhibition venue. PetcDayl also put forward the idea of all green grass buildings as exhibition venues. In addition, the development of green exhibitions abroad is also related to circular economy, using 3R0 principle to guide the construction and operation of venues, reduce resource consumption, improve the utilization efficiency of various resources and reduce pollution.

Economic development research, strengthen exhibition publicity and marketing, improve the propaganda of Zhengzhou exhibition city, innovate service mode, and strengthen the combination of social publicity and marketing.

3.1 Develop New Ideas for Exhibition

China's exhibition industry has been developing for a short time. In the process of exhibition, many enterprises regard meeting the needs of exhibitors as the primary task of their work, thus neglecting the real protagonists in exhibition activities. We need to change the concept of exhibition, from serving exhibitors to serving exhibitors, because the quality and quantity of the audience directly affect the satisfaction of exhibitors to the exhibition, and ultimately determine whether the success of the exhibition is to promote the development of Zhengzhou's exhibition industry. We should start with new ideas, systems and concepts to create an environment conducive to the development of the exhibition industry step by step. The development of the exhibition industry cannot be separated from the support of relevant industries. Therefore, vigorously developing the overall marketing of destination is the only way to develop Zhengzhou's Convention and Exhibition economy, taking Zhengzhou as a whole, effectively integrating and utilizing resources, making use of various channels to publicize Zhengzhou's software and hardware environment, strengthening infrastructure construction, developing characteristic industries, and becoming the foundation and guarantee of the development of convention and Exhibition industry. We should combine tourism with exhibition industry and vigorously develop exhibition tourism. As a tertiary industry, exhibition industry can promote the development of catering, accommodation, shopping, entertainment, transportation and other industries, so as to promote each other and develop together.

3.2 Exhibition Talents

Attaching importance to the training of exhibition talents, realizing the long-term development of exhibition industry, the construction of exhibition specialty and the training of talents, promoting the development of exhibition in Zhengzhou, and developing the direction of exhibition specialty in relevant undergraduate and vocational colleges. To cultivate students' knowledge of conventions and exhibitions in foreign languages, computers and management, and to speed up the cultivation of compound talents. In addition, schools should cooperate with exhibition enterprises to establish training bases so as to combine theory with practice. The concept of exhibition talents is a general concept, involving a variety of majors. It is necessary to strengthen the practical ability of exhibition students and expand the training of high-quality professional planning and management talents.

4 Research and Application Assessment

The perception of the operation mode of green exhibition can be summarized into three aspects. The operation mode of green exhibition can be divided into three modules by factor analysis: green promotion, green building and green travel. Among them, the top three items with the highest degree of identification are: The proportions of LED energy-saving lamps, exhibitions promoting the use of mass transit and exhibitions using recycled materials to produce propaganda materials are shown in Fig. 1.

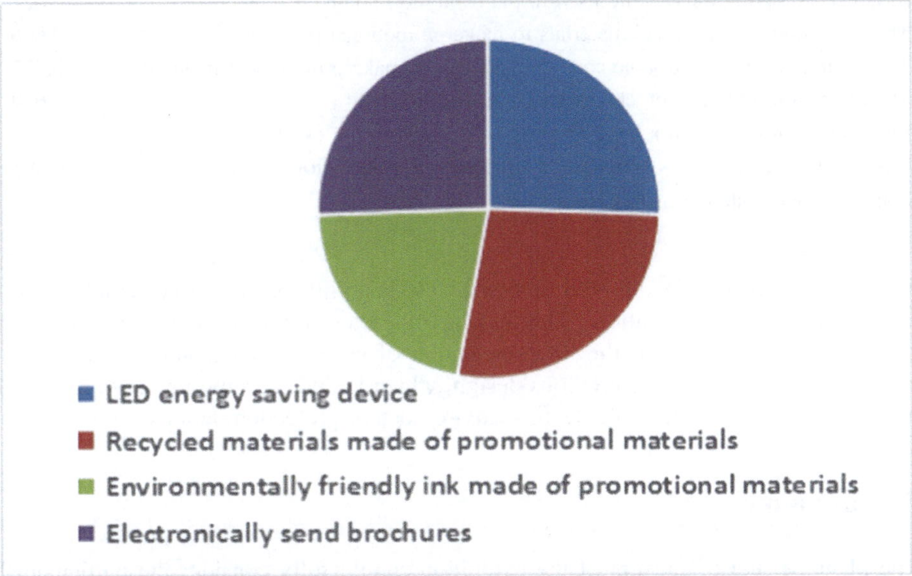

Fig. 1. Proportion of promotional materials made of recycled materials

The factor load factor table of Green Convention and Exhibition operation mode is shown in Table 1.

The above research finds that the composition of green exhibition is slightly different from that in the previous literature and industry guide, which is embodied in the combination of accommodation and transportation travel into green travel, and the induction of carbon emission measurement and award evaluation into green promotion. It can be seen that the green exhibition operation from the perspective of participants' perception mainly involves three aspects: travel, construction and promotion.

Respondents more agree with the environmental responsibility of exhibition organizers, the overall average of which is 3.84. Among them, the items with higher recognition are: the exhibition pays attention to promoting waste recycling; the exhibition implements green procurement to provide environmental protection and energy-saving products; there are facilities for recycling items, etc., the average value of facilities for recycling is 4.04, 4 and 4.01 respectively. Visible, respondents for

Table 1. Factor load factor table of Green Convention and Exhibition operation mode project Green promotion Green building Green Travel

The exhibition will arrange public transport for the audience and the public to the venue	0.686
The exhibition arranged a public transport to the hotel for exhibitors	0.667
Exhibitors will be invited to the hotel	0.625
Most of the materials used for exhibition are reusable materials	0.664
The exhibition used the product of environmental label	0.781
No carpet or second-hand carpet is used for the exhibition booth	0.515
LED energy saving devices are used in the exhibition booth	0.622
The exhibition uses recycled materials to make promotional materials	0.666
The exhibition uses environmentally-friendly ink to make promotional materials	0.523
The exhibition will send brochures and data electronically	0.620
Environmental protection related awards will be held in the exhibition	0.631
Carbon dioxide generated from the calculation of the exhibition	0.791

Source of data: collation of this study

exhibition organizers in the environmental responsibility of the implementation of a higher degree of recognition. The results of factor analysis also show that the respondents' perception of the implementation of corporate environmental responsibility is consistent with the previous design, which includes promoting environmental protection concepts and implementing environmental protection actions.

5 Summary

The planning and organizers of the exhibition should fully consider the participants' perception characteristics of the specific measures of the green exhibition when developing the green exhibition. Strengthen the planning and design of the respondents in a strong sense of the field, such as transportation, construction, promotion and so on. The main points of its planning are to reduce the pressure on the environment of exhibition activities, and to make these green initiatives explicit by means of green certification marks, etc. For example, introducing pure electric bus into convention and Exhibition traffic service, highlighting environmental protection labels on materials such as building materials, and considering the need of developing attractive marketing or app for online exhibition to realize green marketing in the promotion of Green Convention and Exhibition operation mode. Through the exhibition organizers' environmental responsibility, we finally achieve the effect of enhancing the brand value of mice. To this end, exhibition organizers should also take measures to let participants and society know that exhibition organizers take the initiative to assume environmental responsibility inside and outside the exhibition activities, and incorporate environmental responsibility into corporate culture propaganda messages. Only in this way can the exhibition participants' perception and loyalty to the exhibition brand be enhanced more effectively. The research also proves that the operation mode of green exhibition can have a significant impact on the brand awareness and perceived quality of the exhibition. The results also show that green travel is more conducive to enhance the brand awareness of the respondents, while green promotion has a slightly higher impact on the quality perception of the exhibition.

Acknowledgments. This article is supported by the National Study Abroad Fund "2016 Higher Education Teaching Law to study abroad project" (project number: stay blonde [2016] 5054).

Guangdong Brand Specialty Construction Subsidy Project (Grant No. 2016gzpp016), Guangdong Provincial Higher Vocational Education Construction and Real Estate Specialty Teaching Steering Committee and Guangzhou City Construction College Key Project in 2017 "Research on Green Exhibition Technology and Construction of the Application Evaluation System" (Grant No. Z201703).

References

1. Wei Y, Zhu Y (2013) A probe into the sustainable development of low carbon economy, green design and exhibition industry discussion. Bus Times
2. Li XD, Su L et al (2015) The props of the display props under the green environmental protection concept continued design strategy research—taking the canton fair as an example. Art Sci Technol
3. Song B (2010) Green exhibition evaluation index system and comprehensive based on circular economy theory research on evaluation methods. Dongbei Univ Financ Econ
4. Sun M, Zhang H (2006) The construction of green exhibition system and its development. Future Dev
5. Chen Y (2011) Some problems and countermeasures for developing China's green convention and exhibition industry. Int Bus Res
6. Zeng X (2014) Analysis of green exhibition design. Art Lit 3:244
7. Li X (2013) Feasibility study on practicing green exhibitions in Inner Mongolia. Mod Market
8. Wu T (2013) Low carbon environmental protection concept and its practice in exhibition design. Mod Bus
9. Weng S (2014) Economic analysis of corporate environmental responsibility. Ecol Econ (Chinese version)
10. Zhu C (2016) Discussion on the shaping of exhibition brand and sustainable development strategy. Chin Market
11. Zhang H, Sun M, Wang C (2007) Analysis of exhibition development model based on circular economy. J Shanxi Univ Financ Econ
12. Cai M (2014) Research on the mechanism of Hunan green exhibition industry based on circular economy theory. J Hunan Univ Bus
13. Wei Y, Zhu Y (2013) Low carbon economy, green design and sustainable development of exhibition industry. Bus Era
14. Cai M, Greens MK, Tang J (2015) Review of the concept and practice of international green exhibitions. J Shanghai Univ Int Bus Econ
15. Lul L (2013) Research on the development of convention and exhibition industry under the guidance of low carbon economy. SME Manag Technol J

